High Performance Computing in Science and Engineering '10

Wolfgang E. Nagel · Dietmar B. Kröner ·
Michael M. Resch
Editors

High Performance Computing in Science and Engineering '10

Transactions of the High Performance Computing Center, Stuttgart (HLRS) 2010

Springer

Editors
Wolfgang E. Nagel
Zentrum für Informationsdienste
und Hochleistungsrechnen (ZIH)
Technische Universität Dresden
Zellescher Weg 12-14
01069 Dresden
Germany
wolfgang.nagel@tu-dresden.de

Dietmar B. Kröner
Abteilung für Angewandte Mathematik
Universität Freiburg
Hermann-Herder-Str. 10
79104 Freiburg
Germany
dietmar@mathematik.uni-freiburg.de

Michael M. Resch
Höchstleistungsrechenzentrum
Stuttgart (HLRS)
Universität Stuttgart
Nobelstraße 19
70569 Stuttgart
Germany
resch@hlrs.de

Front cover figure: Temperature isosurface inside a lignite-fired utility boiler with a cross-sectional area of 23 m × 23 m and a height of 150 m. Institut für Feuerungs- und Kraftwerkstechnik, Universität Stuttgart

ISBN 978-3-642-15747-9 e-ISBN 978-3-642-15748-6
DOI 10.1007/978-3-642-15748-6
Springer Heidelberg Dordrecht London New York

Library of Congress Control Number: 2010938287

Mathematics Subject Classification (2010): 65Cxx, 65C99, 68U20

Cover design: WMXDesign GmbH

Printed on acid-free paper

Springer is part of Springer Science+Business Media (www.springer.com)

Preface

The Gauss Centre for Supercomputing (GCS) – linking the three national supercomputing centres HLRS (Stuttgart), NIC/JSC (Jülich), and LRZ (Garching) – has established itself as the driving force for high-performance computing (HPC) in Germany, and, even beyond, in Europe. Based on the agreement (Verwaltungsabkommen) between the Federal Ministry of Education and Research (BMBF) and the state ministries for research of Baden-Wuerttemberg, Bavaria, and North Rhine-Westphalia, the overall budget of 400 Million Euros has been allocated – shared equally between federal and state authorities in a five-year time period – to establish the next generation of HPC systems at the GCS. Since the installation of the IBM Blue Gene/P at NIC/JSC in May 2009 – the first phase of the HPC Tier-0 resources – the GCS provides the most powerful high-performance computing infrastructure in Europe today.

It is a pleasure to announce that the next major steps of this agreement will be taken at the HLRS in 2011 and 2013. In 2009/2010, due to the support of the state ministry for research of Baden-Württemberg, the available CPU resources were increased tenfold in an intermediate step, by replacing many of the NEC SX-8 nodes with NEC SX-9/12M192 and installing a remarkably large Intel Nehalem cluster. This has brought a peak performance of 80+ TFLOPs to the users of HLRS. Now, in October 2010 the next round of the GCS infrastructure has been contracted, and the system at the HLRS will be delivered by Cray Inc. in 2011 and 2013. The first delivery phase will be a large Cray XT-6 with a peak speed of roughly 1 PFLOPS. In 2013, in the second delivery phase the final system will be installed, and its peak performance will be roughly 5 PFLOPS, more than twice as fast as the current number-one TOP500-system, and more than 50 times faster than the current systems at the HLRS. After the first upgrade of the HLRS system, the LRZ will upgrade its systems accordingly. The plan is to have a Tier-0 HPC system within the GCS operating at any time within the five year period.

As part of the GCS, HLRS also participates in the European project PRACE (Partnership for Advances Computing in Europe), extending its reach to all European member countries. Within the PRACE project, the GCS will donate access to high performance computing resources valued at 100 million Euros. These PRACE-activities align well with the activities of the HLRS in the European HPC infrastructure project DEISA (Distributed European Infrastructure for Supercomputing Applications) and in the European HPC support project HPC-Europa2.

While the Gauss Centre for Supercomputing successfully addressed the needs on the high end, it was clear from the beginning that an additional layer of support was required to maintain the longevity of the Centre with a network of competence centers across Germany. This gap is addressed by the Gauß–Allianz, where regional and topical centers team up to create the necessary infrastructure, knowledge and the required methods and tools. The mission of the Allianz is to coordinate the HPC-related activities of its members. By providing versatile computing architectures and by combining the expertise of the participating centers, the ecosystem necessary for computational science is created. Strengthening the research and increasing the visibility to compete at the international level are further goals of the Gauß–Allianz.

At the moment, the entire HPC community in Germany is awaiting the funding decisions of the second BMBF HPC-call of May 2010. This call is directed towards proposals to enable and support petascale applications on more than 100,000 processors. While the projects of the first funding round started in early 2009, the follow-up call has been delayed by more than 18 months. Nevertheless, all experts and administration authorities continue to see the strong need for such a program, given that the main issue seen in nearly all applications is one of scalability. The strategic plan involves spending another 20 Million Euros each year over the next three years for projects to develop scalable algorithms, methods, and tools to support massively parallel systems. This may seem like a very large investment. Nevertheless, in relation to the investment in hardware in Germany over this five years period, it is still comparatively small. And furthermore, it will produce brains – brains we will need in order to use the newly developed innovative methods and tools to accomplish technological breakthroughs in scientific as well as industrial fields of application. Even more, the target will not be the Petascale only but also Exascale systems. As we all know, we do not only need competitive hardware but also excellent software and methods to address – and solve – the most demanding problems in science and engineering. The success of this approach is of significant importance for our community and will also greatly influence the development of new technologies and industrial products; beyond that, this will finally determine whether Germany will be an accepted partner alongside the leading technology and research nations.

Since 1996, the HLRS has supported the scientific community as part of its official mission. Just as in the past years, the major results of the last 12 months were presented at the 14th annual Results and Review Workshop on High Performance Computing in Science and Engineering, which was held on October 4–5, 2010 at the Stuttgart University. The workshop proceedings contain the written versions of the research work presented. The papers were selected from all projects running at the HLRS and the SSC Karlsruhe during the one-year period beginning October 2009. Overall, 44 papers were chosen from Physics, Solid State Physics, Reactive Flow, Computational Fluid Dynamics (CFD), Transport and Climate, and numerous other fields. The largest number of contributions – as in many previous years – came from CFD with 16 papers. Even though such a small collection cannot entirely represent such a vast area, the selected papers demonstrate the state-of-the-art in high performance computing in Germany. The authors were encouraged to emphasize computational techniques used in solving the problems examined. This often forgotten aspect was the major focus of these proceedings. Nevertheless, the importance of the newly computed scientific results for the specific disciplines should not be disregarded.

We gratefully acknowledge the continuing support of the federal state of Baden-Württemberg in promoting and supporting high performance computing. Grateful acknowledgement is also due to the German Research Foundation (Deutsche Forschungsgemeinschaft (DFG)), as many projects pursued on the HLRS and SSC computing machines could not have been carried out without its support. Also, we thank Springer Verlag for publishing this volume and, thus, helping to position the local activities in an international framework. We hope that this series of publications contributes to the global promotion of high performance scientific computing.

Stuttgart, October 2010

Wolfgang E. Nagel
Dietmar Kröner
Michael Resch

Contents

Physics
P. Nielaba 1

Spin-Liquid Phase in the Hubbard Model on the Honeycomb Lattice
Z.Y. Meng, T.C. Lang, S. Wessel, F.F. Assaad, and A. Muramatsu 5

Massive and Massless Four-Loop Integrals
J.H. Kühn, P. Marquard, M. Steinhauser, and M. Tentyukov 19

Ligand Protected Gold Alloy Clusters as Superatoms
M. Walter 29

The Chiral Critical Surface of QCD
Ph. de Forcrand and O. Philipsen 43

Mesoscopic Simulations of Polyelectrolyte Electrophoresis in Nanochannels
J. Smiatek and F. Schmid 53

The SuperN-Project: An Update on Core-Collapse Supernova Simulations
B. Müller, L. Hüdepohl, A. Marek, F. Hanke, and H.-Th. Janka 69

Higgs Boson Mass Bounds from a Chirally Invariant Lattice Higgs-Yukawa Model
P. Gerhold, K. Jansen, and J. Kallarackal 85

Dust, Chemistry & Radiation Transport in MRI-Turbulent Protoplanetary Discs
M. Flaig, W. Kley, R. Kissman, and P. Ruoff 103

Solid State Physics
W. Hanke 117

Organic-Metal Interface: Adsorption of Cysteine on Au(110) from First Principles
B. Höffling, F. Ortmann, K. Hannewald, and F. Bechstedt 119

Ab-initio Characterization of Electronic Properties of PbTe Quantum Dots Embedded in a CdTe Matrix
R. Leitsmann and F. Bechstedt 135

Si(111)-In Nanowire Optical Response from Large-scale Ab Initio Calculations
W.G. Schmidt, S. Wippermann, E. Rauls, U. Gerstmann, S. Sanna, C. Thierfelder, M. Landmann, and L.S. dos Santos 149

Laser Ablation of Metals
J. Roth, C. Trichet, H.-R. Trebin, and S. Sonntag 159

Conductance and Noise Correlations of Correlated Nanostructures
A. Branschädel and P. Schmitteckert 169

Cu Substitutionals and Defect Complexes in the Lead-Free Ferroelectric KNN
S. Körbel and C. Elsässer 181

Reacting Flows
D. Kröner 189

Scalar Mixing in Droplet Arrays in Stagnant and Convective Environments
A. Kronenburg, M.R.G. Zoby, S. Navarro-Martinez, and A.J. Marquis . . 191

Euler-Lagrange Simulation of a LOX/H2 Model Combustor with Single Shear Coaxial Injector
M. Lempke, P. Gerlinger, M. Rachner, and M. Aigner 203

Simulation of Triflux Heat Exchangers in Utility Boilers
A. Matschke, M. Müller, U. Schnell, and G. Scheffknecht 217

Computational Fluid Dynamics
S. Wagner 229

Direct Numerical Simulation of Swept-Wing Laminar Flow Control Using Pinpoint Suction
T.A. Friederich and M.J. Kloker 231

A Numerical Study of Turbulent Stably-Stratified Plane Couette Flow
M. García-Villalba, E. Azagra, and M. Uhlmann 251

DNS of Unsteady Heat Transfer Increase on a Curved Surface Due to Wake-Induced Turbulence
D. von Terzi, L. Venema, H.-J. Bauer, and W. Rodi 263

Application of a Novel Turbulence Generator to Multiphase Flow Computations
C. Huber, H. Gomaa, and B. Weigand 273

Numerical Investigation on the Deformation of Droplets in High-Pressure Homogenizers
K. Kissling, S. Schütz, and M. Piesche 287

Direct Numerical Simulation of Sediment Transport in Turbulent Open Channel Flow
C. Chan-Braun, M. García-Villalba, and M. Uhlmann 295

Grid Sensitivity of LES Heat Transfer Results of a Turbulent Round Impinging Jet
S.O. Neumann, N. Uddin, and B. Weigand 307

Large Eddy Simulations of a Jet in Crossflow
F.C.C. Galeazzo, P. Habisreuther, and N. Zarzalis 327

The Impact of Secondary Mean Vortices on Turbulent Separation in 3D Diffusers
D. von Terzi, H. Schneider, and H.-J. Bauer 339

Time-Dependent Three-Dimensional Simulation of the Turbulent Flow and Heat Transfer in Czochralski Crystal Growth Including the Three-Phase Boundary Movement
A. Raufeisen, M. Breuer, T. Botsch, and A. Delgado 353

Numerical Investigation of Shock Wave Boundary-Layer Interaction Using a Zonal RANS-LES Ansatz
B. Roidl, M. Meinke, and W. Schröder 369

Large Eddy Simulation of the Cyclic Variations in an Internal Combustion Engine
F. Magagnato, A. Walcker, and M. Gabi 385

CFD-CSD-Coupled Simulations of Helicopter Rotors Using an Unstructured Flow Solver
F. Bensing, M. Keßler, and E. Krämer 393

Wake Signature of Finite-Span Flapping Rigid Wings
J.E. Guerrero 407

Computational Design Study of a 3D Hypersonic Intake for Scramjet Demonstrator Testing
B. Reinartz and M. Behr 429

Characterization of Mixing in Food Extrusion and Emulsification Processes by Using CFD
M.A. Emin, K. Köhler, M. Schlender, and H.P. Schuchmann 443

Transport and Climate
C. Kottmeier 463

Modelling Regional Climate Change in Germany
P. Berg, H.-J. Panitz, G. Schädler, H. Feldmann, and C. Kottmeier 467

Modelling the Extratropical Transition of Tropical Cyclones and Its Downstream Impact
C.M. Grams and S.C. Jones 479

Global Long-Term MIPAS Data Processing: Some Aspects of the Dynamics of the Atmosphere from Lower Stratosphere to Lower Thermosphere
M. Kiefer, B. Funke, U. Grabowski, and A. Linden 501

Miscellaneous Topics
W. Schröder 515

Computer Simulation for Building Implosion Using LS-DYNA
G. Michaloudis, S. Mattern, and K. Schweizerhof 519

Quaero Speech-to-Text and Text Translation Evaluation Systems
S. Stüker, K. Kilgour, and J. Niehues 529

Molecular Modeling of Hydrogen Bonding Fluids: Transport Properties and Vapor-Liquid Coexistence
J. Vrabec, G. Guevara-Carrion, T. Merker, and H. Hasse 543

Software Framework UG: Parallel Simulation of a Three-Dimensional Benchmark Problem for Thermohaline-Driven Flow
M. Lampe, A. Grillo, and G. Wittum 553

Tailored Usage of the NEC SX-8 and SX-9 Systems in Satellite Geodesy
M. Roth, O. Baur, and W. Keller 561

A Geodynamic Model of the Evolution of the Earth's Chemical Mantle Reservoirs
U. Walzer and R. Hendel 573

Three-Dimensional Simulation of Rarefied Plasma Flows Using a High Order Particle in Cell Method
J. Neudorfer, T. Stindl, A. Stock, R. Schneider, D. Petkow, S. Roller, C.-D. Munz, and M. Auweter-Kurtz 593

Physics

Prof. Dr. Peter Nielaba

Fachbereich Physik, Universität Konstanz, 78457 Konstanz, Germany
peter.nielaba@uni-konstanz.de

The contributions to the HLRS proceedings present the results of large scale simulations for elementary particle models, nano-systems, soft matter systems and astrophysics phenomena. Several important results have been achieved by the computer time granted at the HLRS, in several cases resulting in publications in prestigious journals like Nature and Physical Review Letters.

Z.Y. Meng, T.C. Lang, S. Wessel, F.F. Assaad, and A. Muramatsu (Stuttgart University and Würzburg University) have analyzed the Hubbard model of spin-1/2 fermions on the honeycomb lattice at half-filling using large-scale quantum Monte Carlo simulations. The authors find that the weak coupling semimetal and the antiferromagnetic Mott insulator at strong interaction are separated by an extended gapped phase in an intermediate coupling regime. Exploring excitation gaps, various correlation functions as well as probing for flux quantization, they conclude that a quantum spin liquid, lacking any conventional order, emerges with local charge and spin correlations, best described by a resonating valence bonds state.

J.H. Kühn, P. Marquard, M. Steinhauser and M. Tentyukov from the KIT Karlsruhe have investigated massive and massless four-loop integrals, the computations were mainly performed on the Landeshöchstleistungsrechner XC4000. The problems treated within their project aim for the evaluation of so-called Feynman diagrams which in turn lead to quantum corrections within a given quantum field theory like Quantum Electrodynamics or Quantum Chromodynamics but also supersymmetric theories. The typical CPU time reaches from several hours to several months depending on the concrete problem under consideration. In order to be able to manipulate huge expressions a special tool is necessary. The workhorse of the authors for such calculations is the computer algebra program FORM and its parallel versions ParFORM and TFORM. The parallelization concept for FORM is quite simple: The original expression is divided into several pieces which are then distributed to the individual processors or cores (workers). Once the workers have finished their job the resulting expressions have to be collected by one processor which combines the results. A computer architecture running ParFORM or TFORM requires a fast connection to the (in general) local hard disks of the order of one terabyte per core.

Michael Walter from the University of Freiburg studied the properties of clusters by density functional theory (DFT). The DFT calculations used in his studies were performed at the RZ Karlsruhe with the real-space grid code GPAW using a generalized gradient approximation, and the Kohn-Sham states were represented via the projector-augmented wave method. The author shows in his contribution for two experimentally characterized clusters that the super-atom-picture found for pure Au clusters applies equally well to protected Au alloy clusters. The stability of these clusters is a consequence of the 8-electron shell closing, where the elements Ag and Au donate one and the elements Pd and Pt donate no electron to the set of delocalized electrons. The author proposes the stability of clusters similar to the stable thiol protected $Au_{25}(SR_3)_{18}$ via the replacement of one of the Au atoms by X=Pd, Ag and Cd, and he shows, that the stability of the well known carbonyl protected nickel-silver/gold clusters is governed by delocalized electronic shell closings. The clusters not only separate sterically, but also electronically into nickel-carbonyl and silver/gold subsystems.

Ph. de Forcrand from the ETH Zürich and O. Philipsen from the University of Münster have calculated the critical surface bounding the region featuring chiral phase transitions in the quark mass and chemical potential parameter space of quantum chromo dynamics (QCD) with three flavours of quarks. Their calculations are valid for small to moderate quark chemical potentials, $\mu \leq T$. For their Monte Carlo simulations the authors used the standard Wilson gauge and Kogut-Susskind fermion actions. Configurations are generated using the Rational Hybrid Monte Carlo (RHMC) algorithm. The simulations have been performed on the NEC SX-8 at the HLRS in Stuttgart and the EEGE Grid at CERN. An estimate of the Binder cumulant for one set of mass values consisted of at least 200k trajectories, and the estimate of a critical point required at least 500k trajectories.

Jens Smiatek from the University of Münster and Friederike Schmid from the University of Mainz in their project focused on the DPD simulation of coupled electrohydrodynamic phenomena on the microscale like polyelectrolyte dynamics in microchannels in external electric fields. The effects of electroosmotic flow and slippage combined with polyelectrolyte electrophoresis have been investigated in detail by taking full account of hydrodynamic and electrostatic interactions. All simulations in this work have been carried out by extensions of the software package ESPResSo (An Extensible Simulation Package for Research on Soft matter). One of the programs advantages is its high performance MPI-parallelisation implemented for simulations on supercomputers. The simulations have been run on the NEC SX-8 Cluster at the HLRS. The authors show that the product of the inverse screening length and the slip length massively influences the electroosmotic flow and therefore the total mobility of the polyelectrolyte. An important result of their study is that the characteristics of the boundaries have to be taken into account for a proper description of the polyelectrolyte migration dynamics. Even a negative mobility for certain parameter sets can be achieved, which has been observed in recent experiments. The characteristics of the channel walls could be used to significantly enhance flow profiles, which offers the possibility to reduce the time which is needed for polymer migration or separation techniques. This could be an important aspect for future applications in microchannels or micropumps to accelerate the measuring time in experiments.

B. Müller, L. Hüdepohl, A. Marek, F. Hanke, and H.-Th. Janka from the MPI for Astrophysics in Garching have investigated two-dimensional (core collapse) supernova by simulations and give an overview on the relevant equations and the algorithm for its solution that are employed in their code, and report on their efforts to improve the physics in their supernova code VERTEX as well as its the computational efficiency. Recent results of simulations performed on the NEC SX-8 at the HLRS include the first multi-dimensional general-relativistic neutrino transport simulations conducted with a new extension of the VERTEX code as well as simulations of neutron star cooling over several seconds for different nuclear equations of state.

Philipp Gerhold, Karl Jansen, and Jim Kallarackal from the Humboldt University and DESY Zeuthen considered a chirally invariant lattice Higgs-Yukawa model based on the Neuberger overlap operator. The model is evaluated using PHMC-simulations and the authors present final results on the upper and lower Higgs boson mass bound. The question of a fourth generation of heavy quarks has recently gained attention and the authors illustrate preliminary results of the Higgs boson mass bounds within this framework. The authors as well discuss their progress on properties of the Higgs boson with respect to its unstable nature, such as the decay width and the resonance mass of the Higgs boson.

Markus Flaig and Patrick Ruoff from the University of Tübingen studied dust, chemistry and radiation transport in magneto-rotational instability (MRI)-turbulent protoplanetary discs. The authors aim at setting up 3D protoplanetary disc models that include all the physically relevant factors, namely magnetic fields, radiation transport, chemistry and dust, in a self-consistent manner. They present results from radiative models (neglecting dust and chemistry), where for the first time radiation transport has been included into a 3D turbulent protoplanetary disc model. Their models achieve a quasi-steady state of saturated turbulence, where the turbulent heating is balanced by cooling due to radiation transport. For sufficiently high resolution, the turbulent saturation level shows a trend to converge towards a value of $\alpha \sim 2$.

Spin-Liquid Phase in the Hubbard Model on the Honeycomb Lattice

Z.Y. Meng[1], T.C. Lang[2], S. Wessel[1], F.F. Assaad[2], and A. Muramatsu[1]

[1] Institut für Theoretische Physik III, Universität Stuttgart, Pfaffenwaldring 57, D-70550 Stuttgart, Germany
[2] Institut für Theoretische Physik und Astrophysik, Universität Würzburg, Am Hubland, 97074 Würzburg, Germany

Summary. The Hubbard model encapsulates the physics of strongly correlated quantum systems in its most basic form. It has been studied intensively in the context of the high-temperature superconductivity. A number of novel phases were recently proposed for Hubbard-like models on the honeycomb lattice, the structure of graphene. We analyzed the Hubbard model of spin-$\frac{1}{2}$ fermions on the honeycomb lattice at half-filling using large-scale quantum Monte Carlo simulations. We find that the weak coupling semimetal and the antiferromagnetic Mott insulator at strong interaction are separated by an extended gapped phase in an intermediate coupling regime. Exploring excitation gaps, various correlation functions as well as probing for flux quantization, we conclude that a quantum spin liquid, lacking any conventional order, emerges with local charge and spin correlations, best described by a resonating valence bonds state.

1 Overview

The work that we present in the remainder of this report was published recently in Nature [1]. Apart from the work detailed below, we considered in the last grant period the following topics: We analyzed spin textures induced by magnetic impurities in two-dimensional quantum antiferromagnets by using quantum Monte Carlo simulations [2]. We showed that for weak coupling of the impurity spin to the host magnet, the antiferromagnetic order is enhanced throughout the host system, whereas a strong impurity coupling leads to an overall reduction of the antiferromagnetism apart from a local enhancement of the order parameter in the closest vicinity of the impurity spin. Furthermore, we studied the finite-temperature ordering of defect-induced magnetic moments in graphene [3], based on an effective spin-model that accounts for the long-ranged RKKY interactions mediated among the moments by the conduction electrons. We verified the mean-field character of the finite-temperature ordering transition, and analyzed the dependence of the Néel temperature on the defect concentration. This exhibited a crossover in the system's response

W.E. Nagel et al. (eds.), *High Performance Computing in Science and Engineering '10*, DOI 10.1007/978-3-642-15748-6_1,

between a low-doping shortest-distance behavior and a high-doping mean-field regime. Furthermore, we analyzed the magnetocaloric effect in spin-S chains by using a combination of quantum Monte Carlo simulations and exact numerical diagonalization [4], extending previous investigations towards the regime $S \geq 1$. Moreover, within the context of ultra-cold atoms on optical lattices, a major current effort aims at realizing within the Mott-insulating regime an antiferromagnetically ordered state. With respect to optimizing the cooling efficiency, it is important to obtain accurate estimates of the critical entropy required for the antiferromagnetic Néel state to resist thermal fluctuations. We obtained accurate quantum Monte Carlo estimates for this critical entropy for quantum magnets on simple-cubic lattices, and compared our numerical findings to recent estimates on this quantity [5]. We also devised a method for reaching low entropy states for fermions on optical lattices. That method, which we termed quantum distillation, consists on simply letting the fermionic system expand while the strength of the contact interaction U is maintained at a value much larger than the bandwidth of the fermions. Our simulations showed that essentially a Fock-state with 2 fermions per site is left behind the expanding cloud, and hence a very low entropy state [6].

2 Introduction

The current interest in spin liquids (SL) goes back to seminal work by P. W. Anderson on resonating valence bonds states (RVB), their relevance to the antiferromagnetic (AF) quantum Heisenberg model on low-dimensional lattices [7, 8], and implications on a possible mechanism for superconductivity in the cuprates [9, 10]. There is however compelling evidence, that on the square [11], triangular [12] and honeycomb lattice [13], the nearest-neighbor Heisenberg model does not realize such SL states, but instead long-range magnetic order survives quantum fluctuations. The situation is less clear for the Hubbard model, to which the Heisenberg model provides the low-energy effective theory at half-filling and strong coupling. On the square lattice, a nested Fermi-surface induces an AF state at half-filling for any finite value of the onsite repulsion [14, 15]. For the triangular lattice, there are indications for a SL phase within an intermediate coupling regime [16, 17]. Concerning the bipartite honeycomb lattice, previous studies [18–22] suggested that in the absence of nesting effects, a single quantum phase transition separates the paramagnetic weak-coupling semimetal (SM) phase from a strong-coupling AF Mott insulator (MI), without the details of this transition having been explored.

More recently, field-theoretical studies and lattice gauge theory simulations addressed the nature of an interaction-driven quantum phase transition between a SM and a MI on the honeycomb lattice [23–28]. Such studies focus on the charge sector, and proceed within a relativistic low-energy theory derived from the linear dispersion observed, e.g., in graphene around the Dirac points at the Fermi energy. On the other hand, various proposals have been put forward that additional phases, e.g. (algebraic) SL [25], charge density wave order [29, 30], quantum (spin) Hall states [30–32], or superconductivity [29, 33–36] emerge in Hubbard-like models on the honeycomb lattice at, or near half-filling. Given these developments, it is thus important to explore the ground state properties in the intermediate coupling regime of the original lattice model, in particular, given the absence of a sign-problem when applying unbiased large-scale quantum Monte Carlo (QMC) simulations in the half-filled case.

3 Model and Method

Before presenting our results from such an unbiased approach to the physics of interacting electrons on the honeycomb lattice, in this section, we introduce the model and the numerical method. The Hamiltonian of the spin-$\frac{1}{2}$ Hubbard model on the honeycomb lattice reads

$$H = -t \sum_{\langle i,j \rangle,\alpha} (c^\dagger_{i\alpha} c_{j\alpha} + c^\dagger_{j\alpha} c_{i\alpha}) + U \sum_i n_{i\uparrow} n_{i\downarrow}, \tag{1}$$

where $c_{i\alpha}$ ($c^\dagger_{i\alpha}$) denotes the annihilation (creation) operator for fermions of spin $\alpha = \uparrow, \downarrow$ on lattice site i; $n_{i\alpha} = c^\dagger_{i\alpha} c_{i\alpha}$, t sets the nearest-neighbor hopping amplitude, and $U \geq 0$ the strength of the onsite repulsion.

Our notations on the bipartite honeycomb lattice with the two sublattices A and B and a two-site unit cell in real space, as well as the momentum space are shown in Fig. 1a and b. We also introduce $c^\dagger_{\mathbf{x}A\alpha}$ and $c^\dagger_{\mathbf{x}B\alpha}$ ($c_{\mathbf{x}A\alpha}$ and $c_{\mathbf{x}B\alpha}$) to denote creation (annihilation) operators for fermions of spin $\alpha = \uparrow$ or $\downarrow$, on the lattice site that belongs to the sublattice A and B respectively, within the unit cell at position $\mathbf{x}$. Likewise, $n_{\mathbf{x}a\alpha} = c^\dagger_{\mathbf{x}a\alpha} c_{\mathbf{x}a\alpha}$ and $n_{\mathbf{x}a} = \sum_\alpha n_{\mathbf{x}a\alpha}$ denote the local density operators, and $\mathbf{S}_{\mathbf{x}a} = \frac{1}{2} c^\dagger_{\mathbf{x}a\alpha} \boldsymbol{\sigma}_{\alpha,\beta} c_{\mathbf{x}a\beta}$ the local spin operators, where $\boldsymbol{\sigma} = (\sigma_x, \sigma_y, \sigma_z)$ is the vector of Pauli matrices with $a \in \{A, B\}$. The corresponding operators in k-space are obtained from the Fourier transformation

$$c_{\mathbf{k}a\alpha} = \frac{1}{L^2} \sum_{\mathbf{x}} e^{-i\mathbf{k}(\mathbf{x}+\mathbf{x}_a)} c_{\mathbf{x}a\alpha}, \tag{2}$$

where $\mathbf{x}_A = (0,0)$ and $\mathbf{x}_B = (0,a)$, with a the distance between neighboring lattice sites. Similarly, Fourier components $n_{\mathbf{k}a\alpha}$, $n_{\mathbf{k}a}$ and $\mathbf{S}_{\mathbf{k}a}$ of the density and spin operators are defined.

At $U = 0$, the tight-binding Hamiltonian has a linear dispersion near the Dirac points (K, K'), where the conduction and valence bands touch at half-filling (c.f. Fig. 1c), and correspondingly, the density of state vanishes at the Fermi energy (c.f. Fig. 1d), rendering the non-interacting system a

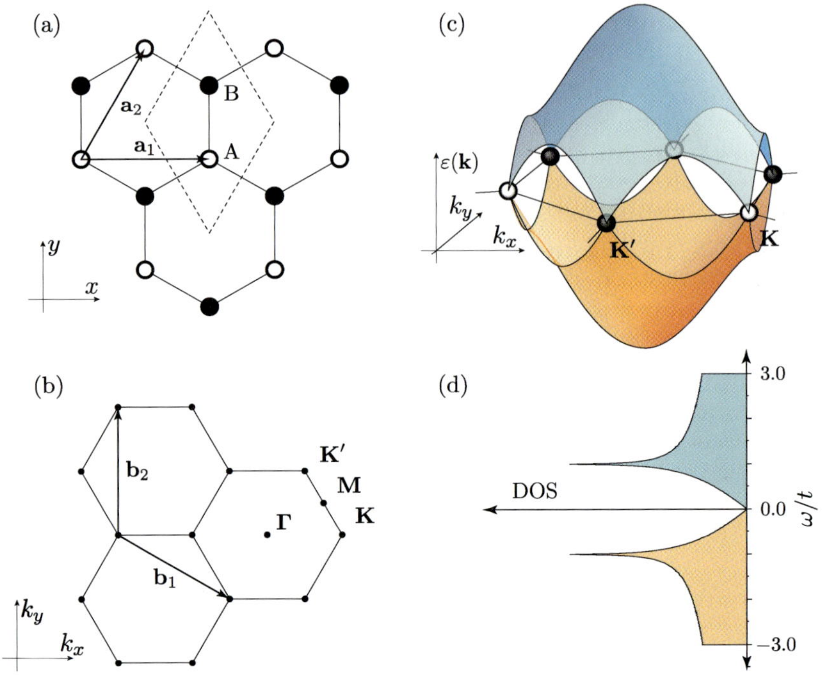

Fig. 1. Honeycomb lattice in real (**a**) and momentum space (**b**). In **a**, the unit cell is indicated, and the lattice vectors $\mathbf{a}_1 = \sqrt{3}a(1,0)$ and $\mathbf{a}_2 = \sqrt{3}a/2(1,\sqrt{3})$, with a equal to the distance between neighboring lattice sites. Open (full) circles indicate the sublattice A (B). In **b**, the Dirac points K and K', the M and the Γ point are indicated, and the reciprocal lattice vectors $\mathbf{b}_1 = 2\pi/(3a)(\sqrt{3},-1)$ and $\mathbf{b}_2 = 4\pi/(3a)(0,1)$. The free dispersion relation $\epsilon(\mathbf{k})$ is shown in **c**, where the π and π^* bands touch each other at K and K', this leads to a linearly vanishing density of states (DOS) at the Fermi level at half-filling (**d**)

semimetalic ground state. At half-filling, the finite-U region can be studied using projective (temperature $T = 0$) auxiliary field QMC simulations in the canonical ensemble. To obtain the ground state expectation values of physical observables A, one calculates

$$\langle A \rangle = \lim_{\Theta\to\infty} \frac{\langle \Psi_T | e^{-\Theta H/2} A e^{-\Theta H/2} | \Psi_T \rangle}{\langle \Psi_T | e^{-\Theta H} | \Psi_T \rangle}, \tag{3}$$

where the trial wavefunction $|\Psi_T\rangle$ must be non-orthogonal to the ground state. For details on the algorithm, see Refs. [37]. We study lattices of $N = 2L^2$ sites with periodic boundary conditions, and linear sizes up to $L = 18$. We found the projection parameter $\Theta = 40/t$ and an imaginary-time step $\Delta\tau = 0.05/t$ in the second-order Trotter decomposition to lead to converged quantities.

4 Results

In this section, we present a detailed analysis of our QMC data, in which we focus in particular on the region near the Mott transition. The analysis leads to the conclusion, that a gapped SL phase exists in the Hubbard model on the honeycomb lattice, separating the paramagnetic SM from the AF MI.

To monitor the electronic properties of the system upon increasing U, we extracted the single-particle excitation gap $\Delta_{sp}(\mathbf{k})$ from the imaginary-time displaced Green's function,

$$G(\mathbf{k},\tau)=\frac{1}{2}\sum_{a}\langle c^{\dagger}_{\mathbf{k}a\uparrow}(\tau)c_{\mathbf{k}a\uparrow}(0)\rangle=\frac{1}{2}\sum_{a}\langle c^{\dagger}_{\mathbf{k}a\downarrow}(\tau)c_{\mathbf{k}a\downarrow}(0)\rangle,$$

with $c^{\dagger}_{\mathbf{k}a\alpha}(\tau)=e^{\tau H}c^{\dagger}_{\mathbf{k}a\alpha}e^{-\tau H}$. At large imaginary time τ,

$$G(\mathbf{k},\tau)\propto\exp(-\tau\Delta_{sp}(\mathbf{k})),$$

and $\Delta_{sp}(\mathbf{k})$ corresponds to the lowest-lying particle (or hole) excitation energy. At $U=0$, the single-particle gap vanishes at the Dirac points K and K', and we thus consider $\Delta_{sp}(K)$ in the following. To obtain a robust estimate of $\Delta_{sp}(K)$ for each value of U and each linear system size L, we diagonalized the covariance matrix to minimize the correlation among QMC data before fitting the lowest exponential decay of the QMC data (main panel of Fig. 2). Then we extrapolated $\Delta_{sp}(K)$ for various fitting ranges in terms of varying the starting point τ_{start}, as seen in the inset of Fig. 2. Finally, we took the converged values of $\Delta_{sp}(K)$ and perform the finite size scaling to obtain the thermodynamic limit estimation of the gaps for different U values. Eventually, we also performed the same procedure to obtain the spin excitation gaps $\Delta_s(\Gamma)$ and $\Delta_u(\Gamma)$. Figure 3 shows the finite size scaling of the single-particle gap for different U values. We also performed the bootstrapping analysis as shown in the inset. Clearly, a single-particle gap opens beyond $U/t\approx 3.5$, signaling the break-down of the weak-coupling SM.

From previous investigations of the model, one expects long-range antiferromagnetic correlations beyond this point. The AF order resides within the unit cell of the honeycomb lattice, and we therefore measured the spin structure factor related to the staggered spin correlations at the Γ point

$$S_{AF}=\langle[\sum_{\mathbf{x}}(\mathbf{S}_{\mathbf{x}A}-\mathbf{S}_{\mathbf{x}B})]^2/N\rangle. \tag{4}$$

Figure 4 shows the QMC results together with a finite size extrapolation. AF order appears beyond $U/t\approx 4.3$, a value that is consistent with previous estimates for the onset of long-ranged AF order [18, 21]. This leaves an extended window $3.5<U/t<4.3$, within which the system is neither a SM, nor an AF MI.

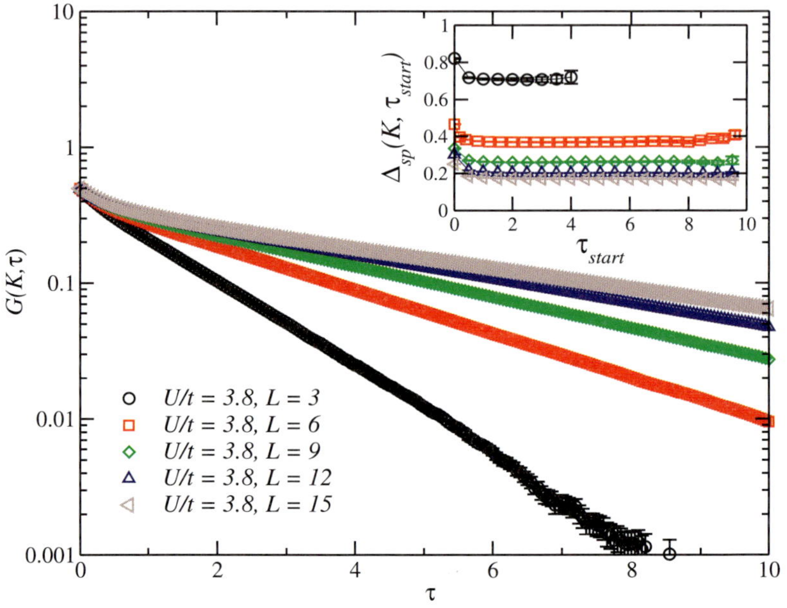

Fig. 2. Green's function at the Dirac point $G(K, \tau)$ for different system sizes at $U/t = 3.8$. The inset shows the determination of $\Delta_{sp}(K)$ as a function of the starting point of the fitting range τ_{start} with the covariance matrix of the Green's function taken into account

Further details on the nature of this intermediate region are obtained by examining the spin excitation gap, obtained from the long-time behavior of the imaginary-time displaced spin-spin correlation function. In the staggered sector at $\mathbf{k} = 0$, the correlation function is

$$S_s(\mathbf{k}, \tau) = \langle (\mathbf{S}_{\mathbf{k}A}(\tau) - \mathbf{S}_{\mathbf{k}B}(\tau)) \cdot (\mathbf{S}_{\mathbf{k}A}(0) - \mathbf{S}_{\mathbf{k}B}(0)) \rangle. \tag{5}$$

When τ is large enough, the spin gap $\Delta_s(\Gamma)$ is obtained from

$$S_s(\Gamma, \tau) \propto \exp(-\tau \Delta_s(\Gamma)).$$

The gap vanishes inside the AF phase due to the emergence of a Goldstone mode, as well as in the gapless SM phase. After performing the same analyzing procedures of QMC raw data as that of the single-particle gap, Fig. 5 shows finite size estimates of $\Delta_s(\Gamma)$ for different values of U/t, along with an extrapolation to the thermodynamic limit. A finite value of $\Delta_s(\Gamma)$ persists within an intermediate parameter regime $3.5 < U/t < 4.3$, while it vanishes both within the SM and the AF phase. This dome in the spin gap is also seen in the inset of Fig. 5, which shows both the finite-size data and the extrapolated values of $\Delta_s(\Gamma)$ as functions of U/t. We also calculated the uniform spin gap $\Delta_u(\Gamma)$ by extrapolating the spin gap observed at the smallest finite $\mathbf{k}$-vector

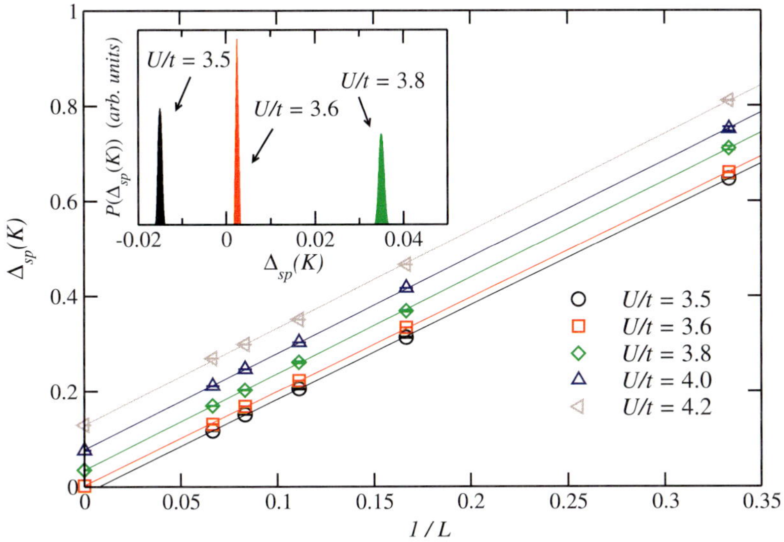

Fig. 3. Finite size extrapolation of the single-particle gap at the Dirac point $\Delta_{sp}(K)$ for different values of U/t, linear in $1/L$. A finite gap opens beyond $U/t \approx 3.5$, as seen from the histograms of the bootstrapping analysis (inset)

on each cluster to the thermodynamic limit. $\Delta_u(\Gamma)$ is found to be even larger than $\Delta_s(\Gamma)$ inside the intermediate region (e.g. $\Delta_s(\Gamma) = 0.023 \pm 0.007$ and $\Delta_u(\Gamma) = 0.099 \pm 0.001$ at $U/t = 4$), and vanishes in the SM and the AF phase ($\Delta_u(\Gamma)$ cannot be measured directly at $\mathbf{k} = 0$, because the uniform magnetization is a conserved quantity).

The observation of finite spin gaps rules out an algebraic SL, as well as triplet superconductivity, as proposed e.g. in Ref. [25, 34]. Candidate states consistent with the above results include (i) singlet superconductivity, (ii) an ordered state with a valence bond crystal (VBC), (iii) a charge density wave order, and (iv) a quantum Hall state. To discern between these possibilities, we turn to further QMC results.

In order to assess if superconductivity arises in the vicinity of the Mott transition, we use the method of flux quantization which probes the superfluid density and is hence independent of the specific symmetry of the pair wave function [38, 39]. Let Φ correspond to the magnetic flux of traversing the center of a torus on which the electronic system lies and $E_0(\Phi/\Phi_0)$ the total ground state energy, Φ_0 being the flux quanta. A superconducting state of Cooper pairs is present if in the TDL, the macroscopic energy difference $E_0(\Phi/\Phi_0) - E_0(\Phi/\Phi_0 = 1/2)$ is a function of period 1/2 [40]. In contrast, a metallic (or insulating) phase is characterized by an vanishing of $E_0(\Phi/\Phi_0) - E_0(\Phi/\Phi_0 = 1/2)$ as a function of system size. Figure 6a, b plots the macroscopic energy difference in the semi-metallic state at $U = 0$ and at $U/t = 4$ in the intermediate phase. In both cases the QMC data is

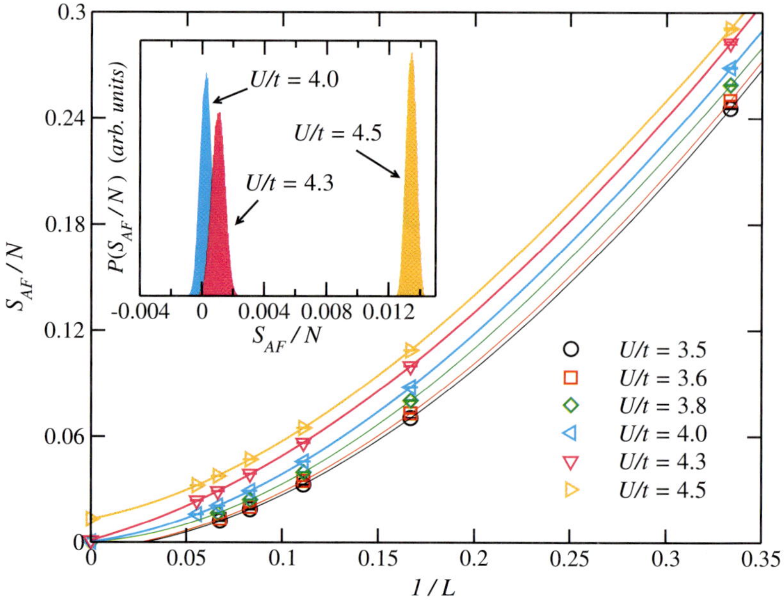

Fig. 4. Finite size extrapolation of the structure factor S_{AF} for various values of U/t using 3rd order polynomials in $1/L$. AF order sets in beyond $U/t \approx 4.3$, as seen from the histograms of the bootstrapping analysis (inset)

consistent with the vanishing of this quantity in the TDL. In addition, we measured singlet and triplet superconducting order parameters of (extended) s-, (complex) p-, and f-wave symmetry, which turn out to all vanish in the TDL. Hence, both flux quantization as well as a direct measurement of pairing correlations in some symmetry sectors lead to no sign of superconductivity.

Both the CDW and QHE trigger a breaking of the sub-lattice symmetry and thereby open a mass gap at the mean-field level. A detailed analysis of the charge-charge correlation functions rules out a CDW. Equivalently, we have found no signature of staggered current loops around next nearest neighbor sites [30]. This rules out the breaking of sublattice and time reversal symmetries as required for the QHE.

To examine the occurrence of a VBC, we probe for dimer-dimer correlations between dimers formed by nearest neighbor bonds $\langle ij \rangle$ and $\langle kl \rangle$ and separated by a distance $|i-k|$. $D_{ij,kl} = \langle O_{ij} O_{kl} \rangle - \langle O_{ij} \rangle \langle O_{kl} \rangle$. We have found no VBC, neither in the charge, $O_{ij} = \mathrm{Re} \sum_\alpha c^\dagger_{i\alpha} c_{j\alpha}$ and $O_{ij} = \mathrm{Im} \sum_\alpha c^\dagger_{i\alpha} c_{j\alpha}$, nor the spin sector, $O_{ij} = \mathbf{S}_i \cdot \mathbf{S}_j - 1/4$. Figure 7 shows the results of this measurement in the spin sector, i.e. the correlation between singlet dimers at $U/t = 4.0$. The bond in the center is the one with respect to which correlations were determined. They are found to be short-ranged, and consistent with the dominance of a resonating valence bond (RVB) state within the hexagons

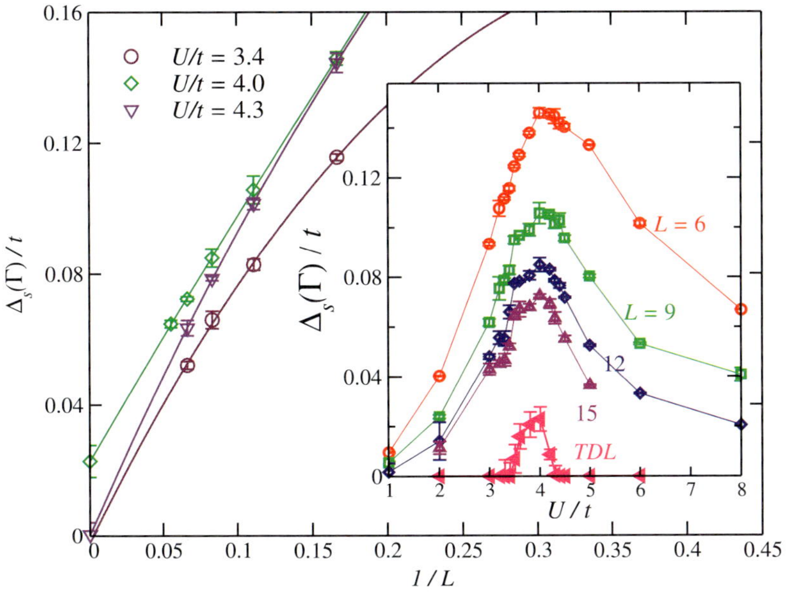

Fig. 5. Finite size extrapolation of spin gap $\Delta_s(\Gamma)$ at different values of U/t, using 2nd order polynomials in $1/L$. The inset shows a pronounced dome in the finite size data, that sharpens to a region with a finite spin gap between $U/t \approx 3.5$ and $U/t \approx 4.3$ in the thermodynamic limit (TDL)

of the honeycomb lattice. Consistently, we find no long-ranged order from the dimer-dimer structure factors in Fourier space. Our results thus provide strong indications for a SL RVB state in the intermediate coupling regime, stabilized in the vicinity of the Mott transition by the enhanced quantum fluctuations in this region, in spite of the honeycomb lattice being a bipartite one.

From analyzing the U-dependence of the kinetic energy,

$$E_{kin} = \langle -t \sum_{\langle i,j \rangle,\alpha} (c^\dagger_{i\alpha} c_{j\alpha} + c^\dagger_{j\alpha} c_{i\alpha})/N \rangle,$$

we obtain further insight into these different regimes and the emergence of local moments. As shown in Fig. 8, the curvature $d^2 E_{kin}/dU^2$ changes sign at $U/t \approx 4.3$. This marks a characteristic change from the weak-coupling region of positive curvature with de-localized electrons to the strong-coupling AF region with negative curvature. In the later region, localized spins from and order in an AF state. In the intermediate SL region, fluctuations are large enough to still prevent the formation of well-localized magnetic moments. Note, that around $U/t \approx 3.5$, a change in the curvature can be observed, that adds to the indications for an intermediate phase.

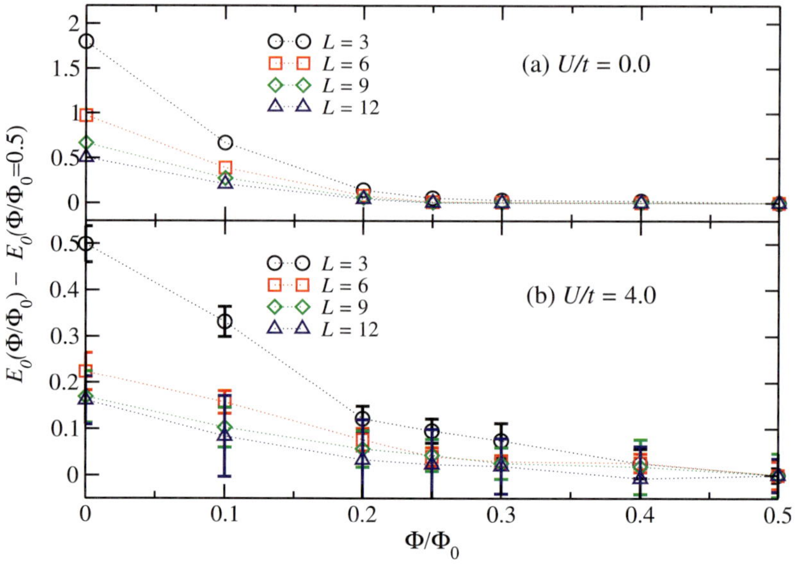

Fig. 6. Magnetic flux Φ dependence of the energy difference $E_0(\Phi/\Phi_0) - E_0(\Phi/\Phi_0 = 0.5)$ for different system sizes at $U = 0$ (**a**) and $U/t = 4$ (**b**)

5 Discussion

Finding a spin-liquid in the Hubbard model on the bipartite honeycomb lattice leads to two remarkable differences with other correlated systems. On the one hand, fluctuations close to the quantum critical point for SU(2) symmetry breaking do not lead to superconductivity as e.g. in heavy fermion systems [41], a fact that we can understand to be a consequence of the vanishing density of states at the Fermi energy. In this case, a finite coupling strength is needed, at least in the BCS-frame [35]. The SL-state suggests rather, that close to the Mott transitions corrections to the strong-coupling nearest-neighbor Heisenberg model induce efficient frustrations to the spin degrees of freedom in the Mott insulating region close to the Mott transition. In fact, the J_1-J_2 Heisenberg model on the honeycomb lattice was suggested to exhibit a RVB phase near $J_2/J_1 \approx 0.3$–0.35 [42]. Furthermore, a Klein Hamiltonian for a spin liquid state on the honeycomb lattice was constructed, including extended exchange interactions [43]. On the other hand, the presence of a spin liquid phase on the honeycomb lattice at half-filling, emerges as an unexpected realization of the RVB state, as proposed by Anderson [9] and Kivelson et al. [10] in connection with high temperature superconductors. It would be therefore highly interesting to explore the consequences of the RVB state on the honeycomb lattice on the appearance of superconductivity upon doping, in a spirit rather close to the original scenario proposed for the cuprates. Although such studies are beyond the power of the quantum Monte Carlo approach due to the sign

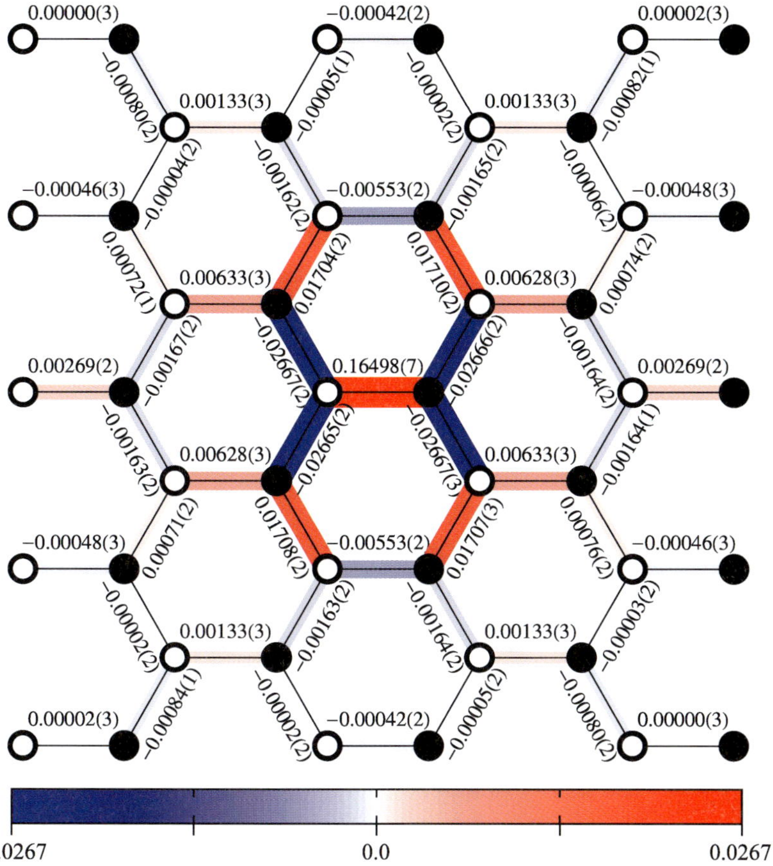

Fig. 7. Real space plot of the dimer-dimer correlation function $D_{ij,kl}$ in the spin section for an $L = 6$ system at $U/t = 4$. The reference bond is located in the center

problem, they could open promising perspectives e.g. in future experiments with ultra-cold atoms on a honeycomb optical lattice, or with honeycomb lattice based on group IV elements like expanded graphene (to enhance the ratio of U/t) or Si, where is the nearest neighour distance is expected to be approximately 50% larger than in graphene [44], such that correlation effects are enhanced. In fact, first attempts succeeded in synthesizing single-crystal silicon monolayers [45].

Acknowledgements. We wish to thank HLRS-Stuttgart (Project CorrSys) for the allocation of computer time. We also acknowledge financial support by the DFG programs SFB/TRR 21. We should like to thank L. Balents, S. Capponi, A.H. Castro Neto, A. Georges, M. Hermele, A. Läuchli, E. Molinari, Y. Motome, S. Sachdev, K.P. Schmidt and S. Sorella for fruitful discussions. We also thank NIC Jülich and the LRZ Munich for their allocation of CPU time.

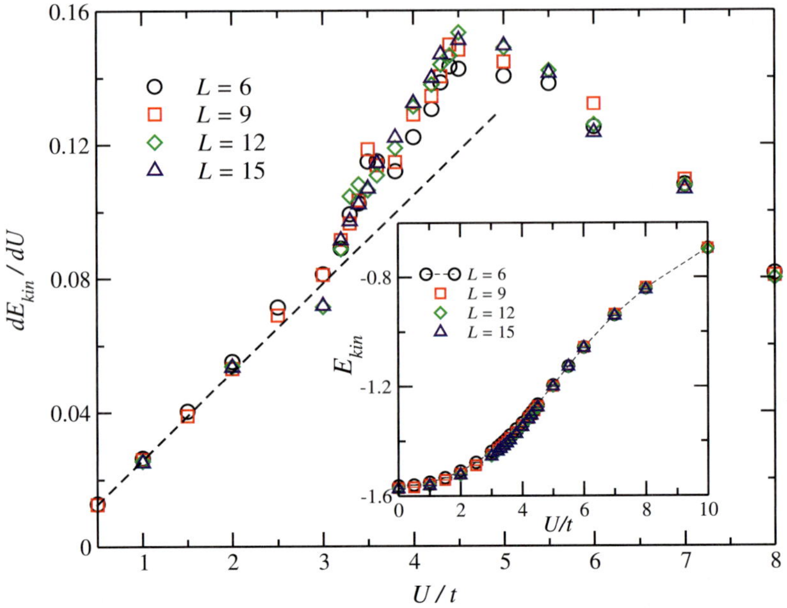

Fig. 8. Derivative dE_{kin}/dU of the kinetic energy as a function of U/t for systems of different sizes. The dashed line is a fit to the low-U behavior. The inset shows the QMC data for the kinetic energy E_{kin} from which the derivative is obtained by numerical differentiation

References

1. Z.Y. Meng et al., Nature **464**, 847 (2010).
2. J. Engel and S. Wessel, Phys. Rev. B **80**, 094404 (2009).
3. T. Fabritius, N. Laflorencie, and S. Wessel, subm. to Phys. Rev. B (2009).
4. A. Honecker and S. Wessel, Cond. Mat. Phys. **12**, 399 (2009).
5. S. Wessel, Phys. Rev. B **81**, 052405 (2010).
6. F. Heidrich-Meisner et al., Phys. Rev. A **80**, 041603(R) (2009).
7. P.W. Anderson, Mater. Res. Bull. **8**, 153 (1973).
8. P. Fazekas and P.W. Anderson, Philos. Mag. **30**, 423 (1974).
9. P.W. Anderson, Science **235**, 1196 (1987).
10. S.A. Kivelson, D.S. Rokhsar, and J.P. Sethna, Phys. Rev. B **35**, 8865 (1987).
11. E. Manousakis, Rev. Mod. Phys. **63**, 1 (1991).
12. B. Bernu, P. Lecheminant, C. Lhuillier, and L. Pierre, Phys. Rev. B **50**, 10048 (1994).
13. J.D. Reger, J.A. Riera, and A.P. Young, J. Phys.: Condens. Matter **1**, 1855 (1989).
14. J.E. Hirsch, Phys. Rev. B **31**, 4403 (1985).
15. S.R. White et al., Phys. Rev. B **40**, 506.
16. T. Mizusaki and M. Imada, Phys. Rev. B **74**, 014421 (2006).
17. T. Yoshioka, A. Koga, and N. Kawakami, Phys. Rev. Lett. **103**, 036401 (2009).
18. S. Sorella and E. Tosatti, Europhys. Lett. **19**, 699 (1992).

19. L.M. Martelo, M. Dzierzawa, L. Siffert, and D. Baeriswyl, Z. Phys. B **103**, 335 (1997).
20. N. Furukawa, J. Phys. Soc. Jpn. **70**, 1483 (2001).
21. T. Paiva et al., Phys. Rev. B **72**, (2005).
22. M. Bercx, T.C. Lang, and F.F. Assaad, Phys. Rev. B **80**, 045412 (2009).
23. S.-S. Lee and P.A. Lee, Phys. Rev. Lett. **95**, 036403 (2005).
24. I.F. Herbut, Phys. Rev. Lett. **97**, 146401 (2006).
25. M. Hermele, Phys. Rev. B **76**, 035125 (2007).
26. S. Hands and C. Strouthos, Phys. Rev. B **78**, 165423 (2008).
27. J.E. Drut and T.A. Lähde, Phys. Rev. Lett. **102**, 026802 (2009).
28. J.E. Drut and T.A. Lähde, Phys. Rev. B **79**, 165425 (2009).
29. C. Honerkamp, Phys. Rev. Lett. **100**, 146404 (2008).
30. S. Raghu, X.-L. Qi, C. Honerkamp, and S.-C. Zhang, Phys. Rev. Lett. **100**, 156401 (2008).
31. C.L. Kane and E.J. Mele, Phys. Rev. Lett. **95**, 226801 (2005).
32. X.-L. Qi, Y.-S. Wu, and S.-C. Zhang, Phys. Rev. B **74**, 085308 (2006).
33. A.M. Black-Schaffer and S. Doniach, Phys. Rev. B **75**, 134512 (2007).
34. B. Uchoa and A.H.C. Neto, Phys. Rev. Lett. **98**, 146801 (2007).
35. N.B. Kopnin and E.B. Sonin, Phys. Rev. Lett. **100**, (2008).
36. P. Sahebsara and D. Senechal, arXiv:0908.0474 (2009).
37. F.F. Assaad and H.G. Evertz, *Computational Many-Particle Physics*, *Lecture Notes in Physics 739* (Springer-Verlag, Berlin, 2008), p. 277.
38. F.F. Assaad, W. Hanke, and D.J. Scalapino, Phys. Rev. Lett. **71**, 1915 (1993).
39. F.F. Assaad, W. Hanke, and D.J. Scalapino, Phys. Rev. B **49**, 4327 (1994).
40. N. Byers and C.N. Yang, Phys. Rev. Lett. **7**, 46 (1961).
41. P. Gegenwart, Q.M. Si, and F. Steglich, Nature Physics **4**, 186 (2008).
42. J.B. Fouet, P. Sindzingre, and C. Lhuillier, Eur. Phys. J. B **20**, 241 (2001).
43. J.T. Chayes, L. Chayes, and S.A. Kivelson, Commun. Math. Phys. **123**, 53 (1989).
44. S. Cahangirov et al., Phys. Rev. Lett. **102**, 236804 (2009).
45. H. Nakano et al., Angew. Chem. **118**, 6451 (2006).

Ligand Protected Gold Alloy Clusters as Superatoms

Michael Walter

Freiburger Materialforschungszentrum, Albert-Ludwigs-Universität Freiburg, 79104 Freiburg i.Br., Germany, Michael.Walter@fmf.uni-freiburg.de

Summary. Density functional study of the experimentally observed ligand-protected gold alloy clusters reveal the same stabilization mechanism as in ligand protected pure AuN: the delocalized s-electron subsystem of a high symmetry metal core exhibits a shell closing. On the basis of this observation it is predicted that the substitution of a single Au atom in the well-known Au25(SR)18 compound with Pd, Ag, and Cd will produce stable clusters resulting in a method to tune redox properties in such a nanoscale building block. Similar shell closings are shown to stabilize the cores of experimentally known carbonyl protected nickel-gold and nickel-silver clusters. These species can be understood as structurally as well as electronically separated, weakly interacting gold/silver and nickel-carbonyl subsystems.

1 Introduction

The considerable scientific and technological interest in Ligand protected (LP) metal clusters is driven by the fact that these clusters can be synthesized in macroscopic quantities as well as by their pronounced stability in different environments and over a broad temperature range [6]. Currently, only few well characterized nanoscopic building blocks exist that possess these important prerequisites for a future use in large scale application such as sensors, markers or catalysts [7].

Prominent representatives of the LP metal clusters are the thiol, phosphine or carbonyl protected gold clusters. Interestingly, the wet chemical production of these systems results in mono-disperse samples with specific cluster sizes - so called magic numbers, whose origin was unclear until recently. Fortunately, there has been a remarkable progress in the last two years in this field both experimentally as well as theoretically. Experimentally the crystal structure of two thiol-protected species that waited structural characterization for a long time was determined [11, 13, 33]. Theoretically it was demonstrated that the traditionally separated fields of thiol- and phosphine-protected gold clusters can be joined under a unified view and that the magic stability of these clusters

W.E. Nagel et al. (eds.), *High Performance Computing in Science and Engineering '10*, DOI 10.1007/978-3-642-15748-6_3,

can be explained by delocalized electron shell closings (DESC) suggesting a super-atom picture for the clusters [28]. All the magic clusters fulfill the simple counting rule

$$N_{\mathrm{del}} - M - q = n_s \;, \tag{1}$$

where N_{del} is the number of delocalized electrons, M the number of electron localizing ligands, q the cluster charge and n_s the DESC number (2, 8, 18, 34, ...) [8]. Based on this concept the structure of a cluster was predicted [1] and independently found in the experiment [11, 33].

Recently, it was suggested to substitute some of the atoms in the LP gold cluster by more reactive transition metal atoms [9, 10]. This could be a versatile route to tailor the functionality (e.g. catalytic, magnetic, electronic or optical properties) of LP gold clusters. However, for the time being it is not evident at all, that LP gold alloy clusters obey the same stabilization mechanism as pure Au clusters.

The above mentioned studies concentrated on clusters protected by either phosphines/halogens or thiolates. Protected transition metal clusters like Ni or Pd clusters are usually stabilized by carbonyls however [19]. There exits large variety of mixed Ag/Au and Ni/Pd compounds. In particular the clusters $[Au_{16}Ni_{24}(CO)_{40}]^{-4}$ [32], $[Au_6Ni_{12}(CO)_{24}]^{-2}$ [14, 31] and $[Au_6Ni_{32}(CO)_{44}]^{-6}$ [26] and were completely characterized by x-ray diffraction of single crystals. All of these clusters contain a compact gold or silver core surrounded by a more or less closed nickel-carbonyl shell. Although the geometry of the cluster is well characterized there exist very limited information about the electronic structure of these compounds.

Here we report an explorative density functional theory (DFT) study of LP gold alloy clusters that was published already [30]. We also cover yet unpublished results about the electronic properties of transition metal clusters with coinage metal cores.

2 Methods

Our DFT calculations were performed with the real-space grid code GPAW [17] using a generalized gradient approximation [20]. The Kohn-Sham states were represented via the projector-augmented wave method [4]. The smooth wavefunctions are represented on a real space grid with 0.2 Å spacing. A frozen core approximation was used for the core electron states. Spin-polarization is considered if there are unpaired electrons in the calculation. Structures were considered to be relaxed [3] when the atomic forces were below 0.05 eV/Å.

3 Doped Gold Clusters

First, let us consider $PdAu_{12}(PR_3)_8Cl_4$—a cluster synthesized and characterized by Laupp and Strähle (LS) [16]. In this work the entire LS cluster is

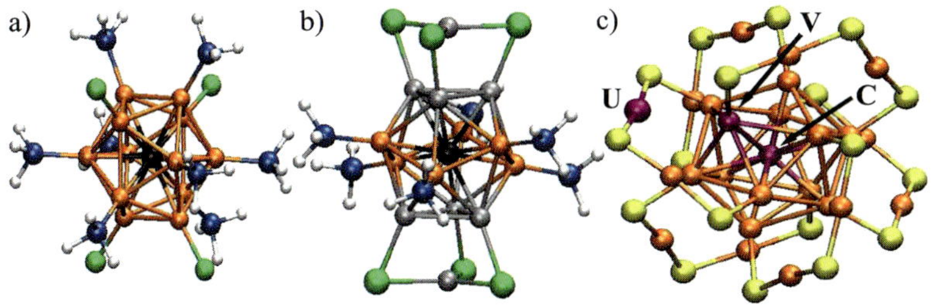

Fig. 1. The relaxed structures of **a** $PdAu_{12}Cl_4(PH_3)_8$, **b** $PtAu_6Ag_6(AgI_3)_2(PH_3)_6$ and **c** $Au_{25}(SR)_{18}$. In **c** the R groups are omitted for clarity. The purple atoms in **c** indicate the replacement positions for the substitutional metal atoms (X=Pd, Ag and Cd) resulting in the three different $XAu_{24}(SR)_{18}$ isomers **C**, **V** and **U**. Pd and Pt atoms are drawn in black, Au in orange or purple, Ag in grey, Cl or I in green, S in yellow, P in blue and H in white

investigated [1] in contrast to earlier calculations that disregarded the influence of the ligands completely [2]. The fully relaxed structure of $PdAu_{12}(PH_3)_8Cl_4$ is shown in Fig. 1a. The icosahedral metal core has a central Pd atom and each of the 12 gold vertex atoms bind to exactly one ligand.

The electronic density of states (DOS) of $PdAu_{12}(PR_3)_8Cl_4$ exhibits a remarkably large gap of 1.77 eV between the highest occupied molecular orbital (HOMO) and the lowest unoccupied molecular orbital (LUMO) in agreement with the observed high stability of this cluster. An inspection of the local DOS (LDOS) in Fig. 2a on the metal atoms reveals that the states around the Fermi energy (ϵ_F) are dominated by the metal s and p contributions (note, that the DOS is folded for better visibility in Fig. 2, which hinders the direct determination of the HOMO-LUMO gap from the plot). Pd contributes to the HOMO while the LUMO is located on gold atoms mainly. Similar as in pure ligand protected Au clusters, also the LS cluster's orbitals around ϵ_F are a delocalized combination of metal s and p orbitals.

These states are further analyzed as follows. The Kohn-Sham orbitals $\phi_n(\mathbf{r})$ are expanded in spherical harmonics $Y_{lm}(\hat{\mathbf{r}})$ centered at the central atom of the icosahedral cage (here the Pd atom) to obtain the radial weights

$$\phi_{nlm}(r) = \int d\hat{\mathbf{r}}\; Y^*_{lm}(\hat{\mathbf{r}})\, \phi_n(\mathbf{r}). \tag{2}$$

The orbitals weight for a particular global angular momentum l is then calculated as

$$c_{nl}(R_0) = \sum_m \int_0^{R_0} dr\, r^2\, |\phi_{nlm}(r)|^2 \tag{3}$$

where $R_0 = 4$ Å was used representing a cut-off roughly half way between the 12 Au atom shell and the ligands. In the following the global angular momenta

[1] The experimental PR_3 groups where replaced by PH_3 groups for computational convenience.

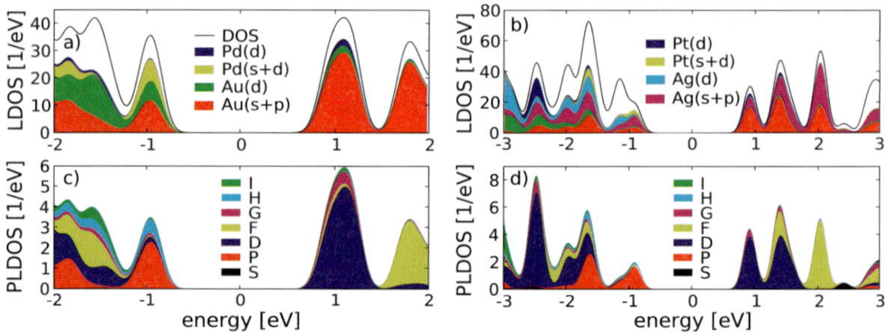

Fig. 2. Electronic properties of $Au_{12}Pd(PH_3)_8Cl_4$ (panels **a** and **c**) and $PtAu_6Ag_6(AgI_3)_2(PH_3)_6$ (panels **b** and **d**). Panels **a** and **b** show the DOS projected on the metal atoms while panels c and d exhibit the DOS projected on global spherical harmonics (see text for further details). The energy is given relative to Fermi energy and all density of states are broadened by Gaussians of 0.1 eV width

will be denoted by capital letters ($l = S, P, D, \ldots$) to distinguish them from the local angular momenta in the LDOS.

The global angular momentum analysis for the LS cluster is presented in Fig. 2b. Three states with P symmetry are located below ϵ_F and five states with D symmetry above. The appearance of these symmetries is a confirmation of the simple DESC model. The cluster closes the P shell, hence having 8 valence electrons effectively. Therefore the formula (1) perfectly works for this structure. There are 12 delocalized Au(s) electrons, from which 4 electrons are localized by the Cl atoms and the cluster is neutral. The phosphine ligands play the role of a Lewis base, not contributing to the delocalized electron count—as in the case of pure Au clusters.

Another experimentally completely characterized gold alloy cluster is the tri-metallic $PtAu_6Ag_6(AgI_3)_2(PR_3)_6$ synthesized by Teo and Zhang (TZ) [25]. The relaxed structure of the TZ cluster is shown in Fig. 1b. This cluster contains a central icosahedral core, which now consists of a Pt atom in the center and a 12 metal atom cage that is formed by a Au_6 ring sandwiched between two Ag_3 caps. Each silver cap is decorated with a AgI_3 unit while the gold ring is terminated with 6 phosphines.

Also this compound has a large HOMO-LUMO gap of 1.82 eV as can be seen in the LDOS shown in Fig. 2b. Similar to the LS cluster the HOMOs are located at the metal atoms with a major contribution of Au and Ag $s+p$. Applying the simple counting rule in (1), the icosahedral cage has 12 delocalized electrons originating from the Ag and Au atoms. The Platinum atom does not contribute to the delocalized electron count as its 6s electron is transfered to the 5d shell upon binding to the other metal atoms—a tendency that is also observed in the Pd bulk phase [5] [2]. The AgI_3 units localize 2 electrons each

[2] Opening the Pd 5d shell by forcing a spin splitting of the $PtAu_6Ag_6(AgI_3)_2(PH_3)_6$ cluster destroys the large HOMO-LUMO gap and results in an overall energy cost of 1.77 eV.

and again the PH_3 groups act as Lewis base ligands. Consequently, also the TZ cluster exhibits a DESC with $n_s = 8$. This finding is again confirmed by a global angular momentum analysis (Fig. 2d).

We now turn to the well studied thiol protected gold cluster $(Au_{25}(SR)_{18})^-$. This compound was shown to be exceptionally stable and even survived an excess thiol environment [24]. Only last year the structure of this cluster was elucidated [1, 11, 33]. There is experimental evidence that a similar cluster also might be stable if one Au atom is substituted by a Pd [10]. In the following the stability of $XAu_{24}(SCH_3)_{18}$ with X=Pd, Ag and Cd will be investigated. The idea behind this study is that the substitutional atoms Pd, Ag and Cd contribute with 0, 1 and 2 delocalized electrons respectively, hence modifying the super-atom nature of the cluster substantially.

The structure of the $Au_{25}(SR)_{18}$ cluster is shown in Fig. 1c. It comprises a 13 Au atom core and 6 SR-Au-SR-Au-SR units. The Au_{13} core itself consists of a 12 atom cage around a central atom, where the cage atoms are all in nearly equivalent icosahedral positions, each of them connecting to a sulfur atom from the six units. The units are placed in nearly octahedral symmetry only slightly distorted by the asymmetric direction of the CH_3 ligand molecules (not shown). There are three distinct positions for a replacement of a single Au by another metal atom resulting in three different isomers (**C**, **V** and **U**). In **C** the central atom is replaced, in **V** one of the 12 vertex atoms and in **U** one of the Au atoms from the thiol-gold units. According to this scheme, we have replaced one gold atom of the icosahedral $Au_{25}(SCH_3)_{18}$ and relaxed the $(XAu_{24}(SCH_3)_{18})^{(q)}$ (X=Pd, Ag, Cd, $q = 1, 0, -1$) to the next local minimum. In the case of Cd and Ag, the GS is **V** whereas the Pd atom prefers isomer **C** as GS. This finding is in line with the **C** position of the Pd or Pt atoms found experimentally in the LS and TZ clusters.

Significant HOMO-LUMO gaps of 1.23 eV, 1.18 eV and 1.18 eV are found for X=Pd, Ag, Cd in the cluster charge states $q = -2, -1, 0$ respectively. These values compare very well to the 1.2 eV gap of $Au_{25}(SCH_3)_{18}^-$ [1]. To study the electronic properties, we will concentrate on the GS and the charge states with the largest HOMO-LUMO gap, i.e. isomers **C**, **V**, **V** and charge states $q = -2, -1, 0$ for X=Pd, Ag and Cd respectively, in the following. The angular momentum projected DOS of all three compounds (Fig. 3) is similar to Fig. 2b and Fig. 2d and resemble the rare-gas super-atom picture known from $Au_{25}(SCH_3)_{18}^-$ also shown in the figure. The LUMO is of D and the HOMO of P symmetry showing again a DESC with $n_s = 8$ and explaining the large HOMO-LUMO gaps. The splitting seen in the D shell is a consequence of the octahedral symmetry of the $SR(AuSR)_2$ units around the cage. In the case of $Au_{25}(SCH_3)_{18}^-$ or X=Pd, where the Pd atom is located in the cluster center, the P-shell shows a single peak only. This is not the case anymore for X=Ag and X=Cd since here the alloy atom is located on a vertex leading to symmetry breaking and hence to the small splitting of the P-shell.

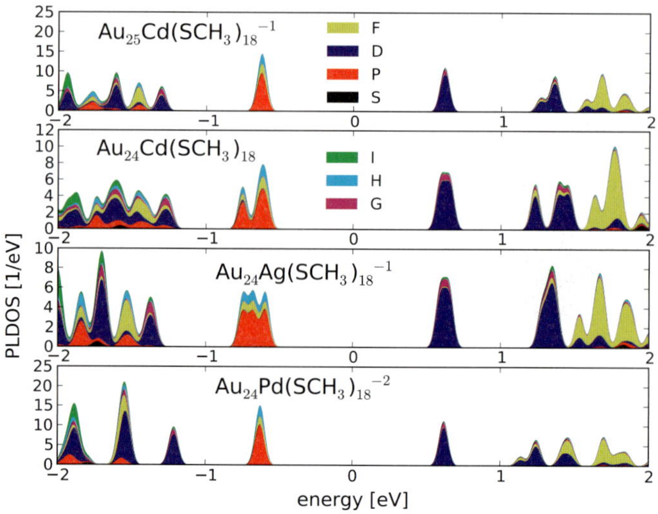

Fig. 3. The global Y_{lm} projected density of states (PDOS) for the core region of $Au_{25}(SCH_3)_{18}^-$, $Au_{24}Cd(SCH_3)_{18}$, $Au_{24}Ag(SCH_3)_{18}^-$ and $Au_{24}Pd(SCH_3)_{18}^{-2}$. The energy axis is defined relative to Fermi energy and the PLDOS is broadened by Gaussians of 0.03 eV width

The DESC for the different charge states shown in Fig. 3 can be understood by a simple s-electron counting as given in (1): Cd donates two, Ag one and Pd no electron to the number of delocalized electrons in the cluster, hence $N_{\mathrm{del}} = 12, 13, 14$ in the case of X=Pd, Ag, Cd respectively. All clusters contain $M = 6$ electron localizing units and therefore a charge of $q = -2, -1, 0$ is needed to fulfill the 8 electron shell closing for Pd, Ag, Cd respectively.

4 Nickel-Carbonyl Protected Superatoms

The first nickel containing cluster to analyze is $[Ag_{16}Ni_{24}(CO)_{40}]^{-4}$ reported by Zang and Dahl [32]. The structure of this cluster is shown in Fig. 4 and can be described as a central Ag_{16} core is protected by 4 $Ni_6(CO)_{10}$ units. These units are not directly connected to each other. The separation between these structural elements already suggests that the elements gain their stability mainly by themselves even so that they are not stable alone. This suggestion will be confirmed for the electronic properties in the following.

In our calculations the structure of $[Ag_{16}Ni_{24}(CO)_{40}]^{-4}$ was taken directly from the experimental crystal structure [32]. Relaxation of the experimental structure did not change the atomic positions much. The bond lengths changed less than 2% and the gap changes from 0.90 eV to 0.91 eV due to relaxation.

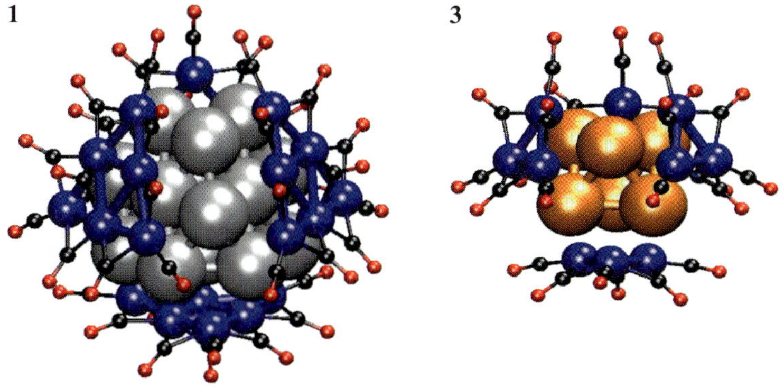

Fig. 4. The structures of $[Ag_{16}Ni_{24}(CO)_{40}]^{-4}$ and $[Au_6Ni_{12}(CO)_{24}]^{-2}$. Ag atoms are shown in silver, Au in gold, Ni in blue, C in black and O in red

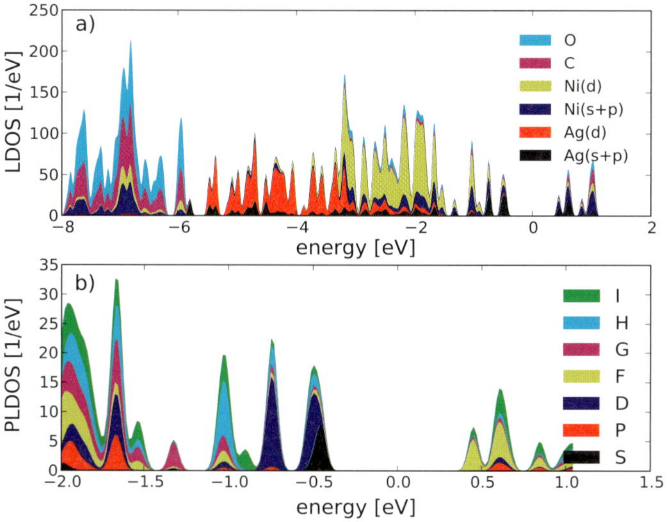

Fig. 5. **a** The atom projected DOS and **b** the angular expansion of the electronic states of $[Ag_{16}Ni_{24}(CO)_{40}]^{-4}$. The energy axis is defined relative to Fermi energy and the density of states are broadened by Gaussians of 0.03 eV width

First we ask the question about the contribution of the different structural elements to the spectrum of electronic single particle states. For this purpose, the local density of states projected on the different elements is shown in Fig. 5a. In the important region for stability, around Fermi-energy, the states have very small Ni(d) character and are mainly located on the silver part of the cluster. Therefore we can assign the silver part of the cluster to be mainly responsible for the electronic and hence chemical stability of this cluster.

The analysis of the global angular momentum character is presented in Fig. 5b. The occupied states below Fermi energy are of D and S delocal-

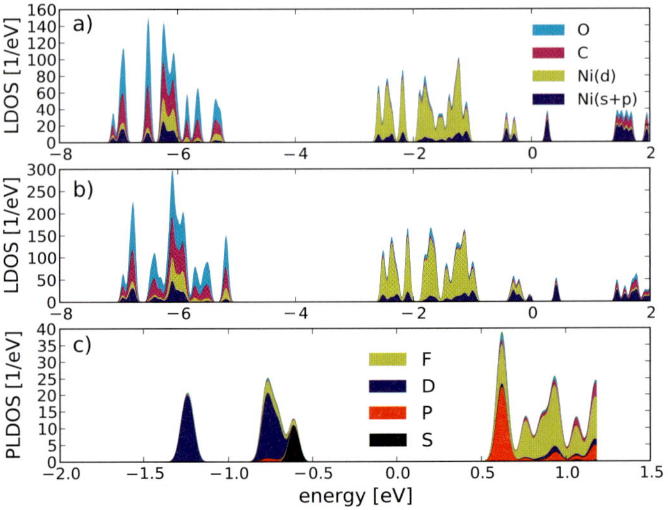

Fig. 6. The atom projected DOS (LDOS) of **a** the $Ni_6(CO)_{10}$ unit and **b** the $Ni_{24}(CO)_{40}$ shell. **c** Angular expansion of the electronic states (PLDOS) around Fermi energy for the Ag_{16} core. The energy axis is defined relative to Fermi energy and the density of states are broadened by Gaussians of 0.03 eV width

ized angular symmetry, whereas the unoccupied states above Fermi energy showThis finding is in perfect agreement with a Jellium shell closing of 20 delocalized electrons ($1S^2 1P^6 1D^{10} 1S^2 1F^0$) [8] which would be the preferred electron configuration of a filled silver ball with 20 delocalized electrons.

To verify our hypothesis of the separability of the structural components of **1**, we determine the electronic properties of the components alone. I.e. we study $[Ag_{16}]^{-4}$ without the Ni-carbonyl shell and the $Ni_{24}(CO)_{40}$ shell and the single $Ni_6(CO)_{10}$ unit without the silver core separately. The LDOS of the $Ni_{24}(CO)_{40}$ unit is shown in Fig. 6a. The unit is not magnetic whereas the pure Ni_6 cluster obtained by stripping away the CO ligands contains a large magnetic moment. We have found that $S_z = 6\ \hbar/2$ has the lowest energy suggesting a total spin of $S \geq 6\ \hbar/2$ [3] in flat triangular Ni_6. The CO shell hence quenches the magnetic moment, an effect which is well known for nickel and palladium containing clusters in the literature [18, 23, 27] Its generality is still under debate however [22]. The unit has a medium gap in the neutral state, but shows a large gap for the charge -2. The LDOS of the full $Ni_{24}(CO)_{40}$ shell is very similar to the LDOS of the $Ni_6(CO)_{40}$ unit as shown in Fig. 6b. It is also non-magnetic as can be suggested by the construction principle via the four $Ni_6(CO)_{10}$ units.

The PDOS of the $[Ag_{16}]^{-4}$ core is shown in Fig. 6b. In accordance to the Jellium model, one finds the occupied states just below the Fermi level to be

[3] In DFT only the projection of the total spin S_z can be specified.

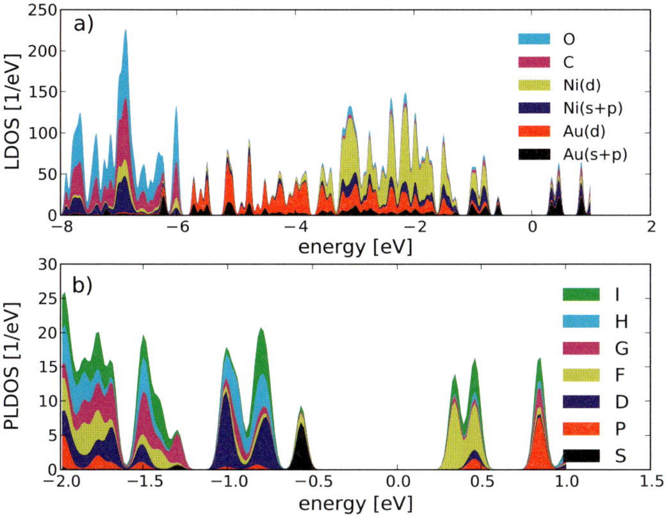

Fig. 7. **a** The atom projected DOS and **b** the angular expansion of the electronic states of $[Au_{16}Ni_{24}(CO)_{40}]^{-4}$. The energy axis is defined relative to Fermi energy and the density of states are broadened by Gaussians of 0.03 eV width

of S and D symmetry. These peaks can be found in very similar positions in the PLDOS of the full cluster in Fig. 5b. A similar situation is given for the unoccupied states of $[Ag_{16}]^{-4}$ just above the Fermi level: these show clear P and F symmetry. The lowest energy peak consists of three states with mixed P and F symmetry due to the non-spherical shape of the $[Ag_{16}]^{-4}$ core. Also these states can be found in the PLDOS of the full compound in Fig. 5b.

To summarize, we can note the following points: (i) The electronic structure of $[Ag_{16}Ni_{24}(CO)_{40}]^{-4}$ shows a rather large gap which can be understood by a shell closing of the delocalized electrons in the $[Ag_{16}]^{-4}$ subsystem. (ii) The cluster is non-magnetic, due to the influence of both the CO shell and the presence of the $[Ag_{16}]^{-4}$ core.

In the same paper where $[Ag_{16}Ni_{24}(CO)_{40}]^{-4}$ was characterized, the corresponding isostructural gold nickel analogue $[Au_{16}Ni_{24}(CO)_{40}]^{-4}$ was mentioned. This was repeated in Ref. [12] as well as in an abstract of papers of the American chemical society before [15]. Nevertheless the characterisation never appeared in the literature.

We have derived the structure of the gold containing cluster from the experimental crystal structure of the silver containing cluster by replacing the silver atoms by gold and subsequent relaxation. The atom sizes and hence bond lengths of Ag and Au are very similar due to relativistic effects [21]. Therefore the bond lengths changed by less than 4% and the electronic gap is also 0.90 eV. In line with the similar structure, the compounds posses also very similar electronic properties as shown in Fig. 7. The LDOS is dominated by Au(s) contributions around Fermi energy and the angular momentum expan-

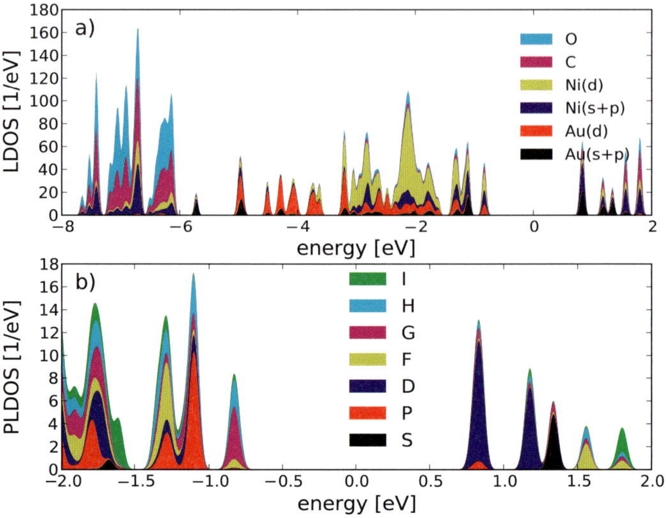

Fig. 8. **a** The atom projected DOS and **b** the angular expansion of the electronic states of $[Au_6Ni_{12}(CO)_{24}]^{-2}$. The energy axis is defined relative to Fermi energy and the density of states are broadened by Gaussians of 0.03 eV width

sion reveals the 20 delocalized electron closing in agreement with the Jellium picture of the $[Au_{16}]^{-4}$ core.

Finally, we consider the cluster $[Au_6Ni_{12}(CO)_{24}]^{-2}$. The synthesis and the structure of this cluster was reported already in the year 1991 [31]. Later this structure was better characterized and its electronic properties were studied using the Fenske-Hall MO model [14]. We have constructed the structure from the information given in Refs. [14, 31] and relaxed it. The structure of $[Au_6Ni_{12}(CO)_{24}]^{-2}$ is shown in Fig. 4 and consists of an Au_6 core with four $Ni_3(CO)_6$ units attached.

The LDOS of $[Au_6Ni_{12}(CO)_{24}]^{-2}$, shown in Fig. 8a, appears more sparse, but rather similar to the LDOS of $[Au_{16}Ni_{24}(CO)_{40}]^{-4}$ in Fig. 7. The only exception is the three-fold degenerate HOMO which, in contrast to the case of $[Au_{16}Ni_{24}(CO)_{40}]^{-4}$, does not contain significant $Au(s + p)$ weight. If we suppose the Au_6 core to be a superatom, we would expect it to be twofold negatively charged to close the 8 electron $1S^2 1P^6$ shell. Indeed, around Fermi energy one finds states of strong P symmetry on the occupied and nearly pure D symmetry on the unoccupied side as shown in Fig. 8b. However the HOMO of the cluster does not show an expected delocalized symmetry. This observation fits to the missing $Au(s + p)$ weight of these states. These states can therefore be associated to the nickel-carbonyl subsystem.

Note that this finding is in disagreement to the result of the Fenske-Hall (FH) analysis presented in Ref. [14]. There the HOMO of the cluster is threefold degenerate and of mainly $Au(6s)$ character corresponding to the P symmetry states found in our calculation. The FH model predicts a threefold degenerate LUMO originating from the nickel-carbonyl part of the cluster. DFT as higher level theory can clearly be expected to provide more accurate results and this disagreement shows the difficulties of lower level theories to describe the complex interactions in these type of clusters properly.

5 Conclusions

In conclusion we have shown for two experimentally characterized clusters that the super-atom-picture found for pure Au clusters applies equally well to protected Au alloy clusters. The stability of these clusters is a consequence of the 8-electron shell closing, where the elements Ag and Au donate one and the elements Pd and Pt donate no electron to set of delocalized electrons.

Moreover, we have proposed the stability of clusters similar to the exceptionally stable thiol protected $Au_{25}(SR_3)_{18}$ via the replacement of one of the Au atoms by X=Pd, Ag and Cd. Whereas the replacement by Cd or Ag is nearly neutral in energy, the replacement by Pd is found to be exothermic. We have shown, that the underlying energetics even for different charge states of the clusters can be understood in a simple model that splits structural and electronic contributions. The replacement energetics and the differences between the elements are so clear that we do not expect substantial changes by using a different exchange-correlation functional.

Finally, we have shown, that the stability of the well known carbonyl protected nickel-silver/gold clusters is governed by the same delocalized electronic shell closings. The clusters not only separate sterically, but also electronically into nickel-carbonyl and silver/gold subsystems.

Acknowledgements. M. W. acknowledges collaborations with M. Moseler, H. Häkkinen and R. Whetten on the reported topics. The computational ressources of RZ Karlsruhe are kindly acknowledged. These ressources where also used for an exploration of STM signatures of supported palladium clusters [29] not reported here.

References

1. J. Akola, M. Walter, R. Whetten, H. Häkkinen, and H. Grönbeck. On the structure of thiolate-protected Au_{25}. *J. Am. Chem. Soc.*, 130(12):3756–3757, 2008.
2. H.-A.L. Arratia-Perez R. Relativistic electronic structure of an icosahedral $Au_{12}Pd$ cluster. *Chem. Phys. Lett.*, 303:641–648, 1999.
3. E. Bitzek, P. Koskinen, F. Gähler, M. Moseler, and P. Gumbsch. Structural relaxation made simple. *Phys. Rev. Lett.*, 97(17):170201, 2006.
4. P.E. Blöchl. Projector augmented-wave method. *Phys. Rev. B*, 50(24):17953–17979, Dec 1994.
5. G.C. Bond. The electronic structure of platinum-gold alloy particles. *Platinum Metals Rev.*, 51(2):63–68, 2007.

6. C. Burda, X. Chen, R. Narayanan, and M.A. El-Sayed. Chemistry and properties of nanocrystals of different shapes. *Chem. Rev.*, 105:1025–1102, 2005.
7. M.-C. Daniel and D. Astruc. Gold nanoparticles: Assembly, supramolecular chemistry, quantum-size-related properties, and applications toward biology, catalysis, and nanotechnology. *Chem. Rev.*, 104:293–346, 2004.
8. W.A. de Heer. The physics of simple metal clusters: experimental aspects and simple models. *Rev. Mod. Phys.*, 65:611–676, 1993.
9. D. en Jiang and S. Dai. From superatomic Au25(SR)18- to superatomic M@Au24(SR)18q core-shell clusters. *Inorg. Chem.*, 48:2720–2722, 2009.
10. C. Fields-Zinna, M. Crowe, A. Dass, and R.W. Murray. Mass spectrometry of small bimetal monolayer-protected clusters. *Langmuir*, 25:7704–7710, 2009.
11. M. Heaven, A. Dass, P. White, K. Holt, and R. Murray. Crystal structure of the gold nanoparticle [N(C8H17)4][Au25(SCH2CH2Ph)18]. *J. Am. Chem. Soc.*, 130(12):3754–3755, 2008.
12. S.A. Ivanov, M.A. Kozee, W.A. Merrill, S. Agarwal, and L.F. Dahl. Cyclo-[Ni(μ2-SPh)2]9 and cyclo-[Ni(μ2-SPh)2]11: new oligomeric types of toroidal nickel(ii) thiolates containing geometrically unprecedented 9- and 11-membered ring systems. *J. Chem. Soc., Dalton Trans.*, pages 4105–4115, 2002.
13. P.D. Jadzinsky, G. Calero, C.J. Ackerson, D.A. Bushnell, and R.D. Kornberg. Structure of a thiol monolayer-protected gold nanoparticle at 1.1 Å resolution. *Science*, 318:430–433, 2007.
14. A.J.W. Johnson, B. Spencerb, and L.F. Dahl. Synthesis and experimental/theoretical investigation of the high-nuclearity cubic td [au6ni12(co)24]2-cluster, an initial example of a discrete gold-nickel bimetallic-bonded species: comparative analysis of the results of electron-counting methods and the fenske-hall mo model in rationalizing the bonding interactions of its au6ni12 core consisting of five face-fused metal octahedra. *Inorg. Chim. Ac.*, 227:269–283, 1994.
15. M.A. Kozee, J.M. Zhang, and L.F. Dahl. Isostructural T-D [Au16Ni24(CO)(40)](4-) and [Ag16Ni24(CO)(40)](4-) clusters: Stabilization of a microscopic ccp chunk of gold or silver metal. *Abstracts of Papers of Am. Chem. Soc.*, 219:U832–U832, 2000.
16. M. Laupp and J. Strähle. [(ph3pau)6(dppeau2)(aucl)4pd], an icosahedral au12 cluster with a central pd atom. *Ang. Chem. Int. Ed.*, 33:207–209, 1994.
17. J.J. Mortensen, L.B. Hansen, and K.W. Jacobsen. Real-space grid implementation of the projector augmented wave method. *Phys. Rev. B*, 71:035109, 2005.
18. G. Pacchioni, L. Ackermann, and N. Rösch. Chemical bonding in low-, medium- and high-nuclearity nickel carbonyl clusters. *Gazz. Chim. Ital.*, 122:205–214, 1992.
19. G. Pacchioni and N. Rösch. Carbonylated nickle clusters: From molecules to metals. *Acc. Chem. Res.*, 28:390–397, 1995.
20. J.P. Perdew, K. Burke, and M. Ernzerhof. Generalized gradient approximation made simple. *Phys. Rev. Lett.*, 77:3865–3868, 1996.
21. P. Pyykkö. Theoretical chemistry of gold. *Ang. Chem. Int. Ed.*, 43:4412–4456, 2004.
22. M. Ricco, T. Shiroka, S. Carretta, F. Bolzoni, C. Femoni, M.C. Iapalucci, and G. Longoni. The magnetic behaviour of [NnBu4]4[Ni16Pd16(CO)40]: An even-electron homoleptic carbonyl–metal cluster anion displaying a J=2 ground state. *Chem. Eur. J.*, 11:2856–2861, 2005.

23. N. Rösch, L. Ackermann, and G. Pacchioni. Developing magnetic and metallic behavior in high-nuclearity nickel cluster carbonyls. *J. Am. Chem. Soc.*, 114:3549–3555, 1992.
24. Y. Shichibu, Y. Negishi, H. Tsunoyama, M. Kanehara, T. Teranishi, and T. Tsukuda. Extremely high stability of glutathionate-protected au25 clusters against core etching. *Small*, 3:835–839, 2007.
25. B.K. Teo and H. Zhang. Molecular and crystal structure of an iodo-gold-silver-platinum cluster, [(ph3p)6au6ag6pt(agi3)2](thf)2, with a trimetallic icosahedral core capped with two planar [agi3] units. *J. Organomet. Chem.*, 614-615:66–69, 2000.
26. N.T. Tran, M. Kawano, D.R. Powell, R.K. Hayashi, C.F. Campana, and L.F. Dahl. Isostructural [Au6Pd6(Pd6-xNix)Ni20(CO)44]6- and [Au6Ni32(CO)44]6- Clusters Containing Corresponding Nonstoichiometric Au6Pd6(Pd6-xNix)Ni20 and Stoichiometric Au6Ni32 Nanosized Cores: Substitutional Pd/Ni Crystal Disorder (Coloring Problem) at Only Six Specific Nonadjacent Pseudoequivalent Metal Sites in the 38-Atom Trimetallic Close-Packed Framework. *J. Am. Chem. Soc.*, 121:5945–5952, 1999.
27. D.A. van Leeuwen, J.M. van Ruitenbeek, L.J. de Jongh, A. Ceriotti, G. Pacchioni, O.D. Häberlen, and N. Rösch. Quenching of magnetic moments by ligand-metal interactions in nanosized magnetic metal clusters. *Phys. Rev. Lett.*, 73(10):1432–1435, Sep 1994.
28. M. Walter, J. Akola, O. Lopez-Acevedo, P.D. Jadzinsky, G. Calero, C.J. Ackerson, R.L. Whetten, H. Grönbeck, and H. Häkkinen. A unified view of ligand-protected gold clusters as superatom complexes. *Proc. Nat. Acad. Sc.*, 105(27):9157–9162, 2008.
29. M. Walter and M. Moseler. How to observe the oxidation of magnesia supported Pd clusters by scanning tunnelling microscopy. Accepted by Phys. Stat. Solidi, 2009.
30. M. Walter and M. Moseler. Ligand protected gold alloy clusters: doping the superatom. *J. Chem. Phys. C*, 113:15834, 2009.
31. A.J. Whoolery and L.F. Dahl. Synthesis and structural-bonding analysis of the [Au6Ni12(CO)24]2- dianion containing an unprecedented 18-vertex cubic td metal core composed of five face-fused octahedra: the first example of a discrete gold/nickel bimetallic-bonded species. *J. Am. Chem. Soc.*, 113:6683–6685, 1991.
32. J. Zhang and L.F. Dahl. First-known high-nuclearity silver–nickel carbonyl cluster: nanosized [Ag16Ni24(CO)40]4- possessing a new 40-atom cubic Td closed-packed metal-core geometry. *J. Chem. Soc., Dalton Trans.*, page 1269–1274, 2002.
33. M. Zhu, C.M. Aikens, F.J. Hollander, G.C. Schatz, and R. Jin. Correlating the crystal structure of a thiol-protected au25 cluster and optical properties. *J. Am. Chem. Soc.*, 130(18):5883–5885, 2008.

The Chiral Critical Surface of QCD

Ph. de Forcrand[1,2] and O. Philipsen[3]

[1] Institut für Theoretische Physik, ETH Zürich, CH-8093 Zürich, Switzerland
[2] Physics Department, CERN, CH-1211 Geneva 23, Switzerland
[3] Institut für Theoretische Physik, Westfälische Wilhelms-Universität Münster, 48149 Münster, Germany

We calculate the critical surface bounding the region featuring chiral phase transitions in the quark mass and chemical potential parameter space of QCD with three flavours of quarks. Our calculations are valid for small to moderate quark chemical potentials, $\mu \lesssim T$. In a first step, we compute the leading Taylor coefficients of an expansion in μ/T for the theory with three degenerate flavours in two different ways and demonstrate, that such an expansion is controllable. Next we extend our calculations to the case of non-degenerate quark masses. These calculations are done on coarse $N_t = 4$ lattices. Finally, we present first results for the $N_f = 3$ QCD phase diagram at zero and finite density on $N_t = 6$ lattices, corresponding to a lattice spacing of $a \sim 0.2$ fm.

1 Introduction

The fundamental theory describing the strong interactions is Quantum Chromodynamics (QCD) with two light quark flavours, the $u-$ and $d-$quarks, and a heavier $s-$quark. Based on the fact that the interaction weakens at asymptotically large energy scales, QCD predicts at least three different forms of nuclear matter: the usual hadronic matter at low temperature and baryon density, a quark gluon plasma at high temperature and low baryon density, and colour superconducting nuclear matter at low temperatures and high baryon density. The theoretical determination of the QCD phase diagram at finite temperature and density is thus of paramount significance for many fields in nuclear and astroparticle physics, and, most importantly, for current and future heavy ion collision programs at RHIC (Brookhaven), LHC (CERN Geneva) and FAIR (GSI Darmstadt), where these phases are to be investigated experimentally. A conjectured phase diagram based on simplified models is shown in Fig. 1, where the mentioned regions are separated by phase transitions. In particular, the transition between the hadronic phase and the quark gluon plasma is expected to end in a chiral critical point.

W.E. Nagel et al. (eds.), *High Performance Computing in Science and Engineering '10*, DOI 10.1007/978-3-642-15748-6_4,

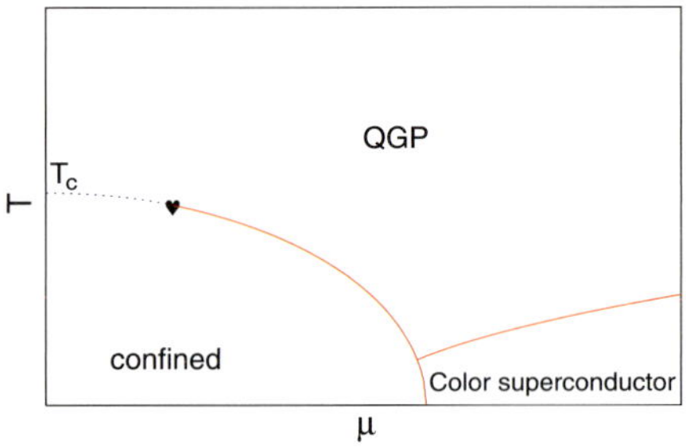

Fig. 1. Possible QCD phase diagram, based on theoretical expectations and models

Direct Monte Carlo simulations of the finite density QCD phase diagram are impossible because of the so-called sign problem. For finite chemical potentials μ_B for baryon number (or $\mu = \mu_B/3$ for quark number), the fermionic part of the theory becomes complex and cannot be used for importance sampling. Therefore indirect methods need to be employed which work for small enough μ/T only (for an overview and references, see [1]). Moreover, because simulations with dynamical fermions cannot yet be performed for physically light quark masses and on sufficiently fine lattices, Monte Carlo investigations have to proceed step by step in quark masses and lattice spacings in order to achieve an eventual extrapolation to the physically interesting case.

At zero chemical potential, the nature of the quark-hadron phase transition depends on the quark masses, as summarised in Fig. 2. In the limits of zero and infinite quark masses, order parameters for the breaking of the global chiral and centre symmetry, respectively, can be defined, and one finds numerically that first order phase transitions take place at some finite temperature T_c. On the other hand, for intermediate quark masses the transition is an analytic crossover. Hence, each corner of first order phase transitions is bounded by a second-order critical line as in Fig. 2. The physical quark masses are light, so our interest is in the lower left boundary, which is called the chiral critical line, as opposed to the deconfinement critical line in the heavy mass region.

When a chemical potential is switched on, the chiral critical line will sweep out a surface, as shown in Fig. 3. According to standard expectations, based on various models with a dynamics simpler than that of QCD [2], for small but non-zero $m_{u,d}$ the chiral critical line should continuously shift with μ to larger quark masses until it passes through the physical point at μ_E, corresponding to the endpoint of the QCD phase diagram. This is depicted in Fig. 3 (left), where the critical point is part of the chiral critical surface. Note, however, that there is no a priori reason for this. In principle it is also possible for the chiral critical surface to bend towards smaller quark masses, cf. Fig. 3 (right), in which case there would be no chiral critical point or phase transition at moderate densities. For definiteness, let us specialise to the theory with three degenerate quarks, which lives on the diagonal in the quark mass plane. The

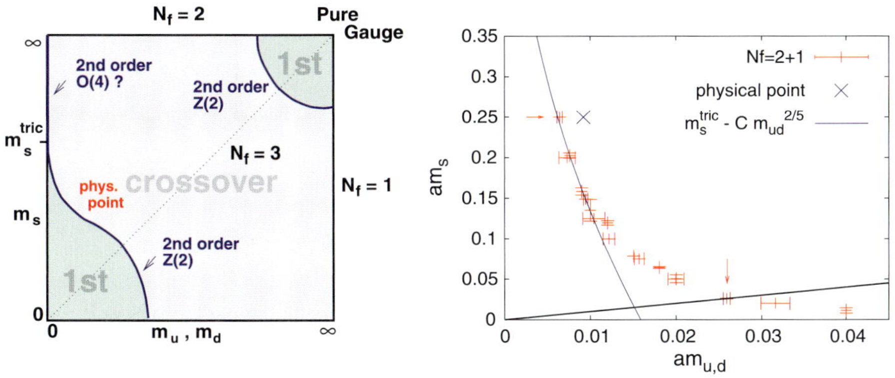

Fig. 2. Left: Schematic phase transition behaviour of $N_f = 2+1$ QCD for different choices of quark masses $(m_{u,d}, m_s)$ at $\mu = 0$. Right: The chiral critical line in the bare mass plane at $\mu = 0$. $N_f = 3$ corresponds to the solid line. Also shown is an extrapolation using conjectured tri-critical scaling, giving a tri-critical point at $m_s^{tric} \sim 2.8T$ [5]. Arrows mark the points for which the curvature of the chiral critical surface was calculated

critical quark mass corresponding to one point on the chiral critical line can be expanded as function of chemical potential,

$$\frac{m_c(\mu)}{m_c(0)} = 1 + \sum_{k=1} c_k \left(\frac{\mu}{\pi T}\right)^{2k} . \tag{1}$$

A strategy to learn about the chiral critical surface is to tune the quark mass to $m_c(0)$ and evaluate the leading coefficients of this expansion. In particular, the sign of the curvature c_1 will tell us which of the scenarios in Fig. 3 is realised.

In previous work the location of the boundary line has been determined for the case of degenerate quark masses, $N_f = 3$ [3, 4], where it was also shown that it belongs to the 3d Ising, or 3d $Z(2)$, universality class. On the lattice, temperature and lattice spacing are related by $T = 1/(aN_t)$, i.e. larger N_t corresponds to finer lattices for a fixed physical temperature. We have used $N_t = 4$ lattices, corresponding to a lattice spacing $a \sim 0.3$ fm, to map out how this line changes i) for $N_f = 3$ as a function of chemical potential μ [4, 5] and ii) for $\mu = 0$ in the case of non-degenerate quark masses $m_{u,d} \neq m_s$ [5]. This is shown in Fig. 2 (right). It was found that the physical point is located very close to the boundary line on the crossover side.

2 The Binder Cumulant and Universality

In order to investigate the critical behaviour of the theory, we use the Binder cumulant [9] as an observable. It is defined as

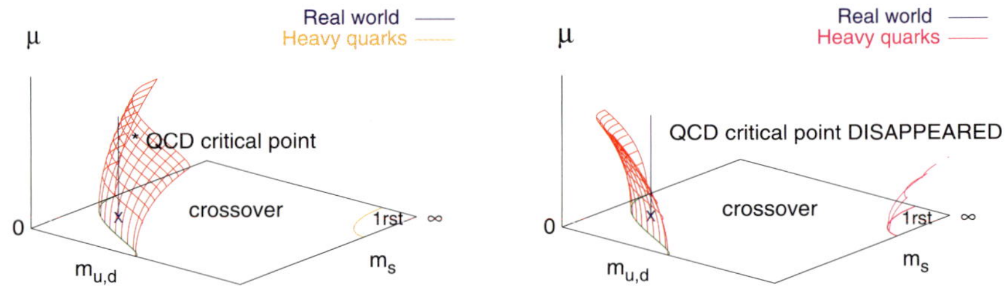

Fig. 3. Same as Fig. 2, but with chemical potential for quark number as an additional parameter. The chiral critical line sweeps out a critical surface as μ is turned on. Depending on the curvature, a QCD chiral critical point is present or absent. Also shown is the deconfinement critical surface for heavy quark masses

$$B_4(m,\mu) = \frac{\langle(\delta X)^4\rangle}{\langle(\delta X)^2\rangle^2}, \tag{2}$$

with the fluctuation $\delta X = X - \langle X \rangle$ of the order parameter of interest. Since we investigate the region of chiral phase transitions, we use the chiral condensate, $X = \bar{\psi}\psi$. For the evaluation of the Binder cumulant it is implied that the lattice gauge coupling has been tuned to its pseudo-critical value, $\beta = \beta_c(m,\mu)$, corresponding to the phase boundary between the two phases. In the infinite volume limit the Binder cumulant behaves discontinuously, assuming the values 1 in a first order regime, 3 in a crossover regime and the critical value 1.604 reflecting the 3d Ising universality class at a chiral critical point. On a finite volume the discontinuities are smeared out and B_4 passes continuously through the critical value. This is sketched in Fig. 4 (left). In the neighbourhood of the chiral critical point at zero chemical potential it can be expanded linearly

$$B_4(m,\mu) = A + B\,(am - am^c(0)) + C(a\mu)^2 \ldots, \tag{3}$$

with $A \to 1.604$ for $V \to \infty$.

The curvature of the critical surface in lattice units is directly related to the behaviour of the Binder cumulant via the chain rule,

$$\frac{dam_c}{d(a\mu)^2} = -\frac{\partial B_4}{\partial (a\mu)^2}\left(\frac{\partial B_4}{\partial am}\right)^{-1}. \tag{4}$$

While the second factor is sizeable and easy to evaluate in a simulation, the μ-dependence of the cumulant is excessively weak and requires enormous statistics to extract. In order to guard against systematic errors, we compute the derivative directly and without recourse to fitting via the finite difference quotient [6]

$$\frac{\partial B_4}{\partial (a\mu)^2} = \lim_{(a\mu)^2 \to 0} \frac{B_4(a\mu) - B_4(0)}{(a\mu)^2}. \tag{5}$$

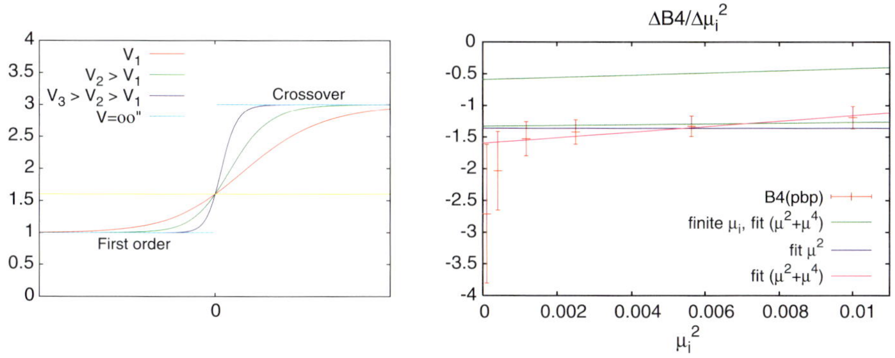

Fig. 4. Left: Schematic behaviour of the Binder cumulant as a function of quark mass for $\mu = 0$. First order transitions and crossovers correspond to $B_4 = 1, 3$, respectively, whereas a second order 3d Ising transition is characterised by $B_4 = 1.604$. On finite volumes the step function gets smeared out. Right: Curvature of B_4 from fits to imaginary μ results (error band) and from finite differences (data points)

Because the required shift in the couplings is very small, it is adequate and safe to use the original Monte Carlo ensemble for $am_0^c, \mu = 0$ and reweight the results by the standard Ferrenberg-Swendsen method. Moreover, by reweighting to imaginary μ the reweighting factors is real positive and close to 1.

In addition we cross-checked our methodology for $N_f = 3$ by a second independent calculation. We also computed the chiral critical surface for imaginary chemical potential and extracted the first two coefficients in the series Eq. 1 by fitting a polynomial to the data at imaginary μ. The two approaches give fully consistent results as shown in Fig. 4 (right).

3 Chiral Critical Surface, $N_f = 3$, $N_t = 4$

Based on our evaluations of the Binder cumulant, we thus conclude that the chiral critical point of the $N_f = 3$ theory (lower right arrow in Fig. 2) recedes as in Fig. 3 (right) [6, 7],

$$\frac{m_c(\mu)}{m_c(0)} = 1 - 3.3(3)\left(\frac{\mu}{\pi T}\right)^2 - 47(20)\left(\frac{\mu}{\pi T}\right)^4 - \dots . \tag{6}$$

As a consequence, the physical point remains in the crossover region also for moderate baryon densities. This is in contrast to the scenario reported by earlier lattice simulations using the reweighting method [8]. Moreover, it is fundamentally different from the expectations sketched above, and therefore a careful examination of the systematic errors in such simulations needs to be performed.

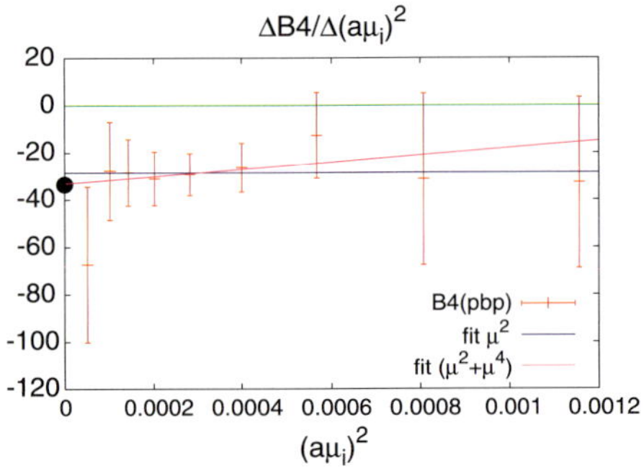

Fig. 5. Curvature (in μ) of the Binder cumulant on the chiral critical line for $N_f = 2+1$ with physical strange quark mass

Here, we perform two checks of the systematics. First, we reconsider the theory with non-degenerate masses with more statistics and compute the curvature of the chiral critical surface at the point marked by the upper left arrow in Fig. 2 (right). Second, we present results for $m_c(0)$ as well as the curvature of the $N_f = 3$ theory (lower right arrow in Fig. 2) on a finer $N_t = 6$ lattice, corresponding to $a \sim 0.2$ fm.

4 Chiral Critical Surface, $N_f = 2+1$, $N_t = 4$

Having established that the leading Taylor coefficients of the chiral critical surface can indeed be controlled for $N_f = 3$, it is important to check the behaviour for non-degenerate quark masses at a point nearer to physical QCD. For this purpose we have fixed the strange quark mass to its physical value and scanned for the critical surface in the $m_{u,d}$-direction. This calculation is more expensive than the previous ones, because now quark masses lighter than physical have to be simulated. This quantity was already the subject of an earlier investigation, in the meantime we have accumulated more statistics, by also utilising computer time on the EGEE Grid as explained below. The μ^2-derivative of the Binder cumulant for this case is shown in Fig. 5. Intercept and slope of the curve tell us that the first two coefficients of the series Eq. 1 are negative and the chiral critical line is receding towards smaller quark masses also in this region of the parameter space. Our final result is published in [10],

$$\frac{m^c_{u,d}(\mu)}{m^c_{u,d}(0)} = 1 - 39(8)\left(\frac{\mu}{\pi T}\right)^2 - \dots . \qquad (7)$$

5 Results for $N_f = 3$ and $N_t = 6$

Having ascertained that the calculational method for the Taylor coefficients is under control and the curvature is negative for both $N_f = 3$ and $N_f = 2+1$, there remains another source of considerable systematic errors. On coarse

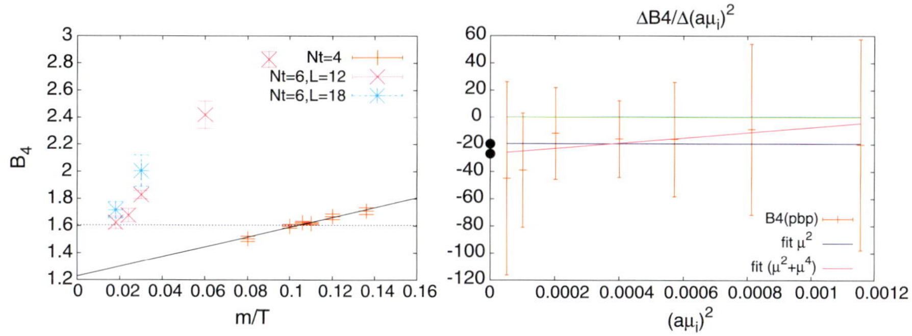

Fig. 6. First results on $N_t = 6$ for $N_f = 3$. Left: critical quark mass $m_c(0)$ from the intersection of B_4 with the Ising value shrinks compared to $N_t = 4$. Right: Curvature of the Binder cumulant. The dots on the left give the value of the curvature when the difference quotient is fitted as a constant or with a subleading term

lattices the discretisation effects can be larger than the finite density effects. This has been demonstrated recently by repeating the zero density studies on $N_t = 6$ lattices. For the case of $N_f = 3$, it was found that the critical quark mass shrinks by almost a factor of five on $N_t = 6$ compared to $N_t = 4$ [6], as shown in Fig. 6 (left). Again, this effect corresponds to a weakening of the transition. However, being much larger than the actual finite density effects, the question now is what happens to the curvature of the critical surface as the continuum is approached. First results are shown in Fig. 6 (right). The leading and next-to-leading order fits consistently give negative values for the curvature. The result needs to still be improved in accuracy and then be extended to $N_f = 2 + 1$. A continuum extrapolation would require the same computations for at least one additional lattice spacing.

6 Simulation Details

For our Monte Carlo simulations we use the standard Wilson gauge and Kogut-Susskind fermion actions. Configurations are generated using the Rational Hybrid Monte Carlo algorithm [11]. Our numerical procedure to compute the Binder cumulant is as follows. For each set of fixed quark mass and chemical potential, we interpolate the critical coupling β_c from a range of typically 3-5 simulated β-values by Ferrenberg-Swendsen reweighting [12]. For each simulation point 50–100 k RHMC trajectories have been accumulated, measuring the gauge action, the Polyakov loop and up to four powers of the chiral condensate after each trajectory. Thus, the estimate of B_4 for one set of mass values consists of at least 200 k trajectories, and the estimate of a critical point at least 500 k trajectories. The derivatives on the $N_t = 4$ lattices are based on $\sim$ 500 k trajectories, on $N_t = 6$ we have so far also collected $\sim$ 500 k trajectories.

The simulations are performed on the NEC SX-8 at the HLRS in Stuttgart and the EGEE Grid with support from the IT group at CERN. A scan in parameter space involves simulations of many parameter sets. For such a problem, parallelisation is achieved trivially by running one set of couplings per node, each node running in vector mode. This way of parallelising allows to explore large regions of the parameter space at the same time, which is necessary when mapping out a phase diagram. At the same time, there is no overhead for parallelisation and communication, ensuring maximal computing efficiency and one-to-one scaling of compute power with the number of processors. Moreover, this procedure permits efficient use of Grid-computing at CERN, which we have employed for the more expensive $N_f = 2+1$ and part of the $N_t = 6$ calculations. The NEC in Stuttgart is used to thermalise multiple decorrelated copies for the configurations, which are then transferred to the Grid for production runs.

7 Conclusions

We have performed systematic scans in the parameter space spanned by the quark masses, chemical potential and temperature in order to determine the order of the QCD phase transition as a function of those parameters. In order to do so we have employed a combination of a fast vector machine and Grid computing, which is well suited to the specific nature of this thermodynamics problem.

On coarse $N_t = 4$ lattices, we have found that the chiral phase transition for light quarks weakens as a real chemical potential $\mu \lesssim T$ for fermion number is switched on. This implies that for physical QCD, whose transition is an analytic crossover at zero chemical potential, it remains a crossover also at finite chemical potential. Those findings persist on finer lattices, $N_t = 6$. We stress, however, that we have only investigated the *chiral* critical surface. Our findings do not exclude the existence of a critical point at larger chemical potentials or one pertaining to non-chiral dynamics.

References

1. O. Philipsen, Eur. Phys. J. ST **152** (2007) 29 [arXiv:0708.1293 [hep-lat]].
2. A.M. Halasz et al., Phys. Rev. D **58** (1998) 096007 [arXiv:hep-ph/9804290]. M.A. Stephanov, K. Rajagopal and E.V. Shuryak, Phys. Rev. Lett. **81** (1998) 4816 [arXiv:hep-ph/9806219].
3. F. Karsch, E. Laermann and C. Schmidt, Phys. Lett. B **520** (2001) 41 [arXiv:hep-lat/0107020].
4. P. de Forcrand and O. Philipsen, Nucl. Phys. B **673** (2003) 170 [arXiv:hep-lat/0307020].
5. P. de Forcrand and O. Philipsen, JHEP **0701** (2007) 077 [arXiv:hep-lat/0607017].

6. P. de Forcrand, S. Kim and O. Philipsen, PoS **LAT2007** (2007) 178 [arXiv:0711.0262 [hep-lat]].
7. P. de Forcrand and O. Philipsen, JHEP **0811** (2008) 012 [arXiv:0808.1096 [hep-lat]].
8. Z. Fodor and S.D. Katz, JHEP **0404** (2004) 050 [arXiv:hep-lat/0402006].
9. K. Binder, Z. Phys. B **43** (1981) 119.
10. J.T. Moscicki, M. Wos, M. Lamanna, P. de Forcrand and O. Philipsen, arXiv:0911.5682.
11. M.A. Clark, A.D. Kennedy and Z. Sroczynski, Nucl. Phys. Proc. Suppl. **140** (2005) 835 [arXiv:hep-lat/0409133].
12. A.M. Ferrenberg and R.H. Swendsen, Phys. Rev. Lett. **63** (1989) 1195.

Mesoscopic Simulations of Polyelectrolyte Electrophoresis in Nanochannels

Jens Smiatek[1] and Friederike Schmid[2]

[1] Institut für Physikalische Chemie, Westfälische Wilhelms-Universität Münster, D-48149 Münster, Germany, jens.smiatek@uni-muenster.de
[2] Institut für Physik, Johannes-Gutenberg Universität-Mainz, D-55699 Mainz, Germany, friederike.schmid@uni-mainz.de

Summary. We present the results of mesoscopic dissipative particle dynamics (DPD) simulations of coupled electrohydrodynamic phenomena on the micro- and nanoscale. The effects of electroosmotic flow and slippage combined with polyelectrolyte electrophoresis are investigated in detail, taking full account of hydrodynamic and electrostatic interactions. Our numerical results are in excellent agreement with analytical calculations.

1 Introduction

Microfluidic devices like bio-MEMS (micro-electronical-mechanical-systems) and bio-NEMS (nano-electronical-mechanical-systems) have attracted broad interest over the last years due to their huge potential in biotechnology. The flow profiles in such micro- or nanosized devices are strongly influenced by the properties of the boundaries due to the large surface-to-volume ratio in these systems. Surface characteristics like the wetting behaviour and/or slippage have a dramatic effect on the microscopic flow, leading to sometimes unexpected behaviour.

One particularly important mechanism is electroosmotic transport: in contact with a liquid, many materials commonly used in nanotechnology (e.g., polydimethylsiloxane (PDMS)) become charged due to ionizations of surface groups [1]. As a consequence, surfaces are often covered by a compensating counterion layer [2]. In an external electric field, the ions are driven in one direction, dragging the surrounding solvent with them. As a result, a flow is induced in the fluid, the electroosmotic flow (EOF). This electrokinetic effect has numerous consequences. For example, it alters drastically the migration dynamics of mesoscopic objects like polyelectrolytes or colloids [3]. In microchannels, the EOF generated at the channel walls induces a total net flow, which is technologically attractive because it can be controlled and manipulated more easily on the submicrometer scale than pressure- or shear-driven flow.

W.E. Nagel et al. (eds.), *High Performance Computing in Science and Engineering '10*, DOI 10.1007/978-3-642-15748-6_5,

One important application of microchannels is to separate different fragments of biological matter like DNA by their length for sequencing or further manipulation. High molecular weight polyelectrolytes cannot be separated by naive electrophoresis in free salt solution due to the fact that the electrophoretic mobility becomes length independent for long chains [3]. In many separation methods, the polyelectrolytes are driven through micro- or nanostructured environments to overcome this problem, e.g., a disordered gel (in gel electrophoresis), or structured microchannels. The presence of boundaries alters the dynamical behaviour of the macromolecules drastically [4], and their dynamical behaviour results from a complex interplay of electrostatics, hydrodynamics, and confinement effects on the molecules.

Our research focuses on the investigation of the explicit coupling of electrohydrodynamic effects in the nanometer scale in confined geometries by coarse-grained mesoscopic simulations [5, 11]. We use Dissipative Particle Dynamics (DPD), which is a popular mesoscopic method in mesoscopic simulations. The results for polyelectrolyte electrophoresis in small microchannels are related to the experiments published by Mathé et al. in [4]. In particular, we investigate in detail the influence of the electroosmotic flow on the total polyelectrolyte mobility. We find an excellent agreement between theory and numerical results.

2 Dissipative Particle Dynamics

Dissipative Particle Dynamics (DPD) was originally developed by Hoogerbrugge and Koelman [6] as a combination of Lattice Gas Automata characteristics and Molecular Dynamics methods. Compared to atomistic Molecular Dynamics simulations, this method gives access to much longer time- and length scales and is therefore suited to study the long-time behaviour of soft matter systems and transport phenomena. It is coarse-grained, momentum-conserving, and creates a well-defined canonical ensemble.

The basic DPD equations are given by the forces on one particle, which involve two-particle interactions that are given by

$$\mathbf{F}_i^{DPD} = \sum_{i \neq j} \mathbf{F}_{ij}^C + \mathbf{F}_{ij}^D + \mathbf{F}_{ij}^R \tag{1}$$

with the conservative force $\mathbf{F}_{ij}^C$

$$\mathbf{F}_{ij}^C = -\boldsymbol{\nabla}_{ij} U_{ij}(r_{ij}), \tag{2}$$

which is often of the form [7]

$$\mathbf{F}_{ij}^C = a_{ij}\left(1 - \frac{r_{ij}}{r_c}\right) \cdot \hat{r}_{ij} \tag{3}$$

with the repulsion parameter a_{ij}, the cutoff radius r_c and the distance r_{ij} between interacting particles. The dissipative force $\mathbf{F}_{ij}^D$ reads

$$\mathbf{F}_{ij}^D = -\gamma_{DPD}\omega_D(r_{ij})(\hat{r}_{ij}\cdot\mathbf{v}_{ij})\hat{r}_{ij} \tag{4}$$

with the friction coefficient γ_{DPD}, whereas the additional random force $\mathbf{F}_{ij}^R$ is given by

$$\mathbf{F}_{ij}^R = \sigma\omega_R(r_{ij})\check{\zeta}_{ij}\hat{r}_{ij}. \tag{5}$$

where the weighting function is given by

$$\omega_D(r_{ij}) = [\omega_R(r_{ij})]^2 \equiv \omega_{DPD}(r_{ij}), \tag{6}$$

such that a canonical ensemble is sampled at equilibrium [7, 8]. The random number $\check{\zeta}$ has zero mean and unit variance and is symmetric, $\check{\zeta}_{ij} = \check{\zeta}_{ji}$, to ensure the conservation of momentum and the weighting function is arbitrary and often chosen linear [7]. It depends on the interparticle distance r_{ij} and the cut-off radius r_c. The strength of the interaction is steered by the distance of the particles with

$$\omega_{DPD}(r_{ij}) = \begin{cases} 1-\frac{r_{ij}}{r_c} & : r_{ij} < r_c \\ 0 & : r_{ij} \geq r_c \end{cases} \tag{7}$$

while the amplitude of the Gaussian white noise in (5) is given by

$$\sigma^2 = 2\gamma_{DPD}k_BT \tag{8}$$

with the Boltzmann constant k_B and the temperature T. (1) can be integrated by an ordinary Molecular Dynamics integration scheme like the Velocity-Verlet algorithm [9] which is used in the ESPResSo package.

3 The Software Package ESPResSo

All simulations in this work have been carried out using extensions of the software package ESPResSo (An **E**xtensible **S**imulation **P**ackage for **Res**earch on **So**ft matter) [10]. ESPResSo was mainly developed for coarse-grained mesoscopic simulation approaches. One of the advantages of this program is its high performance MPI-parallelization implemented for simulations on supercomputers. Users can furthermore change and extend the program code to adopt it for their own purposes. ESPResSo incorporates several simulation techniques like Lattice-Boltzmann, Dissipative Particle Dynamics, Stochastic Dynamics as well as pure Molecular Dynamics techniques. Another feature are the several implemented electrostatic algorithms like MMM1D, MMM2D, MMM3D, P3M, ELC and screened Debye-Hückel potentials, which allow to choose between the fastest calculation methods available. In summary, ESPResSo provides a well founded basis code for high performance computing on parallel clusters. The steering of the simulations is based on a TCL (Tool Command Language) script. For starting a simulation, no explicit knowledge of implementation details is needed. Even newcomers in the methods of computer simulations can successfully run a simulation after a short time.

Several tools for analysis are additionally included in the program. ESPResSo is under public license and free to download [10]. Users that develop new ideas are invited to submit their source code, written in the programming language C to be incorporated after testing in the newest release version. The development of ESPResSo continues and a number of extensions is already planned.

We ran our simulations on the NEC SX-8 Cluster at the High Performance Computing Center Stuttgart. The number of computing nodes has normally been chosen to 4 which corresponds to 32 CPUs. Each job has got a typical runtime of 4 hours by using the concept of setting checkpoints.

4 Polyelectrolyte Electrophoresis in Microchannels

4.1 General Theory

We consider for simplicity a planar slit channel with identical walls at $z = \pm L/2$, exposed to an external electric field E_x in the x direction. The electrostatic potential $\Phi(x, y, z)$ then takes the general form $\Phi(x, y, z) = \psi(z) + E_x\, x + \text{const.}$ where we can set $\psi(0) = 0$ for simplicity. The electrolyte in the channel is taken to contain n different ion species i with local number density $\rho_i(z)$ and valency Z_i, which results in a net charge density $\rho(z) = \sum_{i=1}^{n} (Z_i e)\rho_i(z)$. The electric field then generates a force density $f_x(z) = \rho(z)E_x$ in the fluid. Comparing the Poisson equation for the electrostatic potential ψ,

$$\frac{\partial^2 \psi(z)}{\partial z^2} = -\frac{\rho(z)}{\epsilon_r} \tag{9}$$

(where ϵ_r is the dielectric constant), with the Stokes equation

$$\eta_s \frac{\partial^2 v_x(z)}{\partial z^2} = -f_x(z) = -\rho(z)E_x \tag{10}$$

(with the shear viscosity η_s), one finds immediately $\partial_{zz} v_x(z) = \partial_{zz}\psi(z)\ (\epsilon_r\ E_x/\eta_s)$. For symmetry reasons, the profiles v_x and ψ must satisfy the boundary condition $\partial_z v_x|_{z=0} = \partial_z \psi|_{z=0} = 0$ at the center of the channel. This gives the relation

$$v_x(z) = \frac{\epsilon_r\, E_x}{\eta_s} \psi(z) + v_{\text{EOF}}, \tag{11}$$

where we have used $\psi(0) = 0$ and identified the fluid velocity at the center of the channel with the EOF velocity, $v_x(0) = v_{\text{EOF}}$. We further define $\psi_B := \psi(\pm z_B)$ (for no-slip boundaries, ψ_B is the so-called Zeta-Potential [2, 11]).

The roughness of the channel boundaries is included in the partial slip boundary condition.

$$\delta_B \ \partial_z v(x)|_{z_B} = v_x(z)|_{z_B}, \tag{12}$$

where $v_x(z)$ denotes the component of the velocity in x-direction evaluated at the position z_B of the so-called "hydrodynamic boundary". This boundary condition is characterised by two effective parameters, namely (i) the slip length δ_B and (ii) the hydrodynamic boundary z_B. We note that the latter does not necessarily coincide with the physical boundary. Inserting the partial-slip boundary condition for the flow, (12), we finally obtain the following simple expression for the electroosmotic mobility,

$$\mu_{\text{EOF}} = v_{\text{EOF}}/E_x = \mu^0_{\text{EOF}} \ (1 + \kappa \, \delta_B), \tag{13}$$

where we have defined the inverse 'surface screening length'

$$\kappa := \mp \frac{\partial_z \psi}{\psi}|_{z=\pm z_B}, \tag{14}$$

and μ^0_{EOF} is the well-known Smoluchowski result [2] for the electroosmotic mobility at sticky walls,

$$\mu^0_{\text{EOF}} = -\epsilon_r \ \psi_B/\eta_s. \tag{15}$$

The remaining task is to determine the screening parameter κ. If the surface charges are very small and the ions in the liquid are uncorrelated, it can be calculated analytically within the linearised Debye-Hückel theory [1]. The Debye-Hückel equation for the evolution of the potential ψ in an electrolyte solution reads $\partial_{zz}\psi = \kappa_D^2 \psi$ with the inverse Debye-Hückel screening length

$$\kappa_D = \sqrt{\frac{\sum_{i=1}^n (Z_i e)^2 \rho_{i,0}}{\epsilon_r \, k_B T}}, \tag{16}$$

where $\rho_{i,0}$ is the density of ions i far from the surface. It is solved by an exponentially decaying function,

$$\psi(z) \propto (e^{\kappa_D z} + e^{-\kappa_D z} - 2). \tag{17}$$

Inserting that in (14), one finds $\kappa = \kappa_D$, *i.e.*, the surface screening length is identical with the Debye screening length.

Unfortunately, the range of validity of the Debye-Hückel theory is limited, it breaks down already for moderate surface potentials ψ_B and/or for highly concentrated ion solutions. Nevertheless, the exponential behaviour often persists even in systems where the Debye-Hückel approximation is not valid. For high ion concentrations a Debye-Hückel-type approximation can still be used in a wide parameter range, if κ_D is replaced by a modified effective screening length [12]. For high surface charges, analytical solutions are again available in the so-called 'strong coupling limit', where the profiles are predicted to decay exponentially with the Gouy-Chapman length [13]. This limit is very special and rarely encountered. At intermediate coupling regimes, the decay length must be obtained empirically, e.g., by fitting the charge distribution $\rho(z)$ to an exponential behaviour, which is characterised by the same exponential behaviour than $\psi(z)$ by virtue of the Poisson equation,

$$\sum_{i=q}^{n}(Z_i e)\,\rho_i(z) \propto \frac{\partial^2 \psi(z)}{\partial z^2} \propto (e^{\kappa z} + e^{-\kappa z}). \tag{18}$$

Assuming that the electrophoretic velocity of the polyelectrolyte

$$v_p(x) = \mu_e E_x \tag{19}$$

is influenced by the electroosmotic mobility of the counterions μ_{EOF}, a total mobility has to be defined $\mu_t = \mu_{\text{EOF}} + \mu_e$ which describes the overall mobility of the electrophoretic object. Putting everything together, the total net electrophoretic mobility μ_t of a polyelectrolyte in the channel can be expressed in terms of the electroosmotic mobility μ_{EOF} as

$$\frac{\mu_t}{\mu_{\text{EOF}}} = 1 + \frac{\mu_e}{\mu^0_{\text{EOF}}(1 + \kappa\,\delta_B)}, \tag{20}$$

where the ratio μ_e/μ^0_{EOF} depends only weakly on the ionic strength of the electrolyte and the slip length of the surface. The main effect of slippage is incorporated in the factor $(1 + \kappa\,\delta_B)^{-1}$i [11].

4.2 Simulation Details

We have studied the electrophoresis of a single charged polymer of length $N = 20$ in electrolyte solutions, confined by a planar slit channel with charged walls. All particles, polymer, solvent and ions, are modeled explicitly. We use a simulation box of size $(12\sigma \times 12\sigma \times 10\sigma)$ which is periodic in x- and y-direction and confined by impermeable walls in the z-direction. The walls repel the particles *via* a soft repulsive WCA potential of range σ and amplitude ϵ. (Hence the accessible channel width for the particles is actually $L_z = 8\sigma$). Ions and monomers repel each other with the same WCA potential. In addition, chain monomers are connected by harmonic springs

$$U_{harmonic} = \frac{1}{2}k(r_{ij} - r_0)^2 \tag{21}$$

with the spring constant $k = 25\epsilon/\sigma^2$ and $r_0 = 1.0\sigma$. Neutral solvent particles have no conservative interactions except with the walls. The wall contains immobilized, negatively charged particles at random positions. Every second monomer on the polyelectrolyte with 20 beads carries a negative charge resulting in 10 charged monomers. The solvent contains the positive counterions for the walls and the polyelectrolyte, and additional (positive and negative) salt ions. All charges are monovalent, and the system as a whole is electroneutral. In addition to their other interactions, charged particles interact *via* a Coulomb potential with the Bjerrum length $\lambda_B = e^2/4\pi\epsilon_r k_B T = 1.0\sigma$, and they are exposed to an external field $E_x = -1.0\epsilon/e\sigma$. Specifically, we have studied systems with a surface charge density of $\sigma_A = -0.208e\sigma^{-2}$ which corresponds to 30 charged particles per wall. In a recent publication [5], we have shown that this corresponds to the 'weak-coupling regime', *i.e.*, the regime where the Poisson-Boltzmann theory is valid. The electrostatic coupling constant [13] $\Xi = 2\pi Z^3 \lambda_B^3 \sigma_A$ ($Z = 1$ is the valency of the cations), which gives the strength of electrostatic interactions between the surface and the ions compared to thermal energy, is close to unity, $\Xi \sim 1.3$. In this earlier work, we have also studied the effect of using homogeneously charged wall instead of discrete embedded charge, and found the differences to be negligible [5]. The use of discrete embedded charges has practical advantages in our simulation code, which is why we use them here. The total counterion density was $\rho = 0.06\sigma^{-3}$ and the salt density varied between $\rho_s = 0.05625$, 0.0375, 0.03, 0.025, and $0.015\sigma^{-3}$. In molar units, this corresponds to 0.272, 0.181, 0.145, 0.121 and 0.072 mol/l, if we identify $\lambda_B \approx 0.7$ nm, *i.e.*, the Bjerrum length in water at room temperature [3].

We use DPD simulations with a friction coefficient $\gamma_{\text{DPD}} = 5.0\sigma^{-1}(m\epsilon)^{1/2}$. (More precisely, we only use the DPD *thermostat*, not the soft conservative DPD forces. All conservative forces in our system are either WCA forces, spring forces, or Coulomb forces as explained above). The density of the solvent particles was $\rho = 3.75\sigma^{-3}$, and the temperature of the system was $T = 1.0\epsilon/k_B$. For these parameters, the shear viscosity of the DPD fluid – as determined by fitting the amplitude of Plane Poiseuille flows [15] – is given by $\eta_s = (1.334 \pm 0.003)\sigma^{-2}(m\epsilon)^{1/2}$. The DPD timestep was $\delta t = 0.01\sigma(m/\epsilon)^{1/2}$.

Tunable-slip boundary conditions were used with friction coefficients $\gamma_L =$ 0.1, 0.25, 0.5, 0.75, 1.0, and $6.1\sigma^{-1}(m\epsilon)^{1/2}$. The range of the viscous layer was $z_c = 2.0\sigma$. Only the solvent particles interact with the tunable-slip boundaries. By performing Plane Poiseuille and Plane Couette flow simulations with the above given parameters, the slip length δ_B and the hydrodynamic boundary positions z_B can be determined independently [15]. The hydrodynamic boundary position is found at $|z_B| = (3.866 \pm 0.266)\sigma$ in all simulations.

4.3 Results

Polyelectrolytes in Absence of External Fields

To understand the behaviour of polyelectrolytes in salty solution we first studied the dynamics in absence of external fields. The general theory [3] implies that hydrodynamic interactions between the monomers are important to describe the dynamic behaviour adequately. This results in the so called Zimm dynamics [3]. If external electric fields are present, hydrodynamic interactions between the charged monomers are screened due to the presence of salt. The two rivalling effects which are the electrophoresis of the polyelectrolyte as well as the electroosmosis of the mobile counterions in opposite direction lead to a crucial screening of hydrodynamic interactions which is described by Rouse dynamics [3].

A powerful tool to investigate the underlying dynamics in experiments as well as in computer simulations is given by the dynamic version of the structure factor which is defined by

$$S(\mathbf{k}, t) = \frac{1}{N} \sum_{i,j} < e^{i\mathbf{k}(\mathbf{R}_i(t) - \mathbf{R}_j(t_0))} > \tag{22}$$

with the actual monomer position $\mathbf{R}_i$ respectively $\mathbf{R}_j$. For the inverse length scale $1/R_g \ll k \ll 1/a_0$ with the gyration radius R_g [16] and the smallest microscopic length a_0 and the finite time interval $t_b \ll t \ll \tau$ after the ballistic time t_b and the longest relaxation time τ, the dynamic structure factor obeys the following scaling relation [17]

$$S(k, t) = S(k, 0) f(k^z t) \tag{23}$$

which depends on the parameter z. This parameter differentiates between the different regimes. For the Zimm dynamics it is given by

$$z = 3 \quad \text{(Zimm-Regime)} \tag{24}$$

and for Rouse dynamics

$$z = 2 + 1/\nu \quad \text{(Rouse-Regime)} \tag{25}$$

with the Flory parameter ν which describes the inverse fractal dimension of the polyelectrolyte [16].

We have compared an uncharged polymer and a half charged polyelectrolyte in a salt solution of concentration $\rho_s = 0.05\sigma^{-3}$ and a solvent density of $\rho = 3.0\sigma^{-3}$ with free periodic boundary conditions in all dimensions. All the other parameters are in agreement to the above described simulation details. Figure 1 presents our simulation results for an uncharged chain and the polyelectrolyte. Our results indicate that although external fields are absent, polyelectrolytes can be described best by Rouse dynamics whereas uncharged polymers can be described by Zimm dynamics. More results on that topic can be found in [18, 19].

Polyelectrolyte Electrophoresis in Microchannels

In this section our numerical results are presented which indicate the drastic influence of the electroosmotic flow on the total mobility of the polyelectrolyte. Larger slip lengths even enhance this effect.

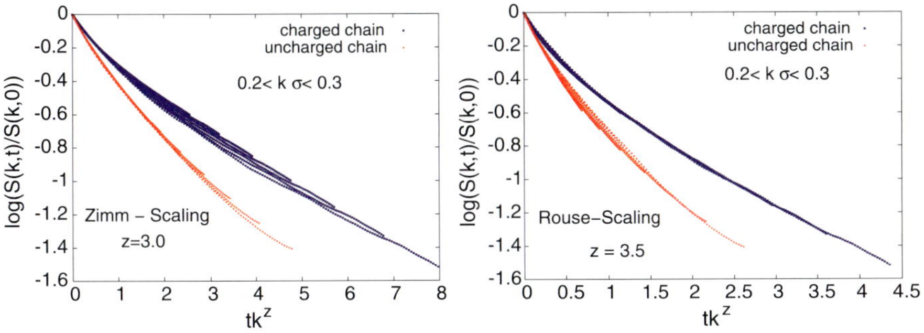

Fig. 1. Dynamic structure factor S(k,t) for an uncharged chain (red) in correspondence to a half charged chain (blue) for $0.2 < k\sigma < 0.3$ and salt concentration $\rho_s = 0.05\sigma^{-3}$. The timescale for the uncharged chain is $0 < t/\tau < 180$ in contrast to $0 < t/\tau < 300$ for the half charged polyelectrolyte. Both chains consist of $N = 50$ monomers. Left side: Zimm-scaling with $z = 3$. Right side: Rouse-scaling with $z = 3.5$

Figure 2 presents the results of the ionic distribution for a salt concentration of $\rho_s = 0.05625\sigma^{-3}$ and a counterion density of $\rho = 0.0525\sigma^{-3}$ with a surface ion density of $\sigma_s = 0.208\sigma^{-2}$ in presence of the polyelectrolyte. Due to the large number of cations in comparison to the anions, the cation density in the middle of the channel is increased (inset of Fig. 2). Calculating the ionic difference $\Delta\rho = \rho_c - \rho_a$ as the difference of the cationic and anionic density yields

$$\frac{\partial^2}{\partial z^2}\psi(z) = -\sum_i^N \frac{Z_i e}{\epsilon}\rho_i(z) = -\frac{Ze}{\epsilon}\Delta\rho \tag{26}$$

with the unique valency $|Z|$ [11]. As we have mentioned in Sect. 4, the application of the Debye-Hückel theory for non electroneutral bulk systems is not valid and has to be replaced by an effective inverse screening length κ. Thus we have used the following fit function

$$\Delta\rho = \Delta\rho_0(e^{-\kappa z} + e^{\kappa z}) + c \tag{27}$$

to determine the effective inverse screening length κ, with the fit parameters $\Delta\rho_0, c$ and κ_D. The red line in Fig. 2 shows that the fit describes the ionic difference adequately. The fitted parameter values of κ for the different salt concentrations are presented in Table 1. Significant differences between the inverse screening lengths are only observed for salt concentrations $\rho_s \geq 0.0375\sigma^{-3}$. Thus, for lower salt concentrations, the screening of electrostatic interactions is mainly effectuated by the much larger number of counterions instead of the salt ions [11].

The corresponding electroosmotic flow profiles for an exemplary salt concentration of $\rho_s = 0.05625\sigma^{-3}$ are shown in Fig. 3. The different magnitudes correspond to varying slip lengths. Larger slip lengths enhance the flow profile drastically. All points are in good agreement to the integrated analytical

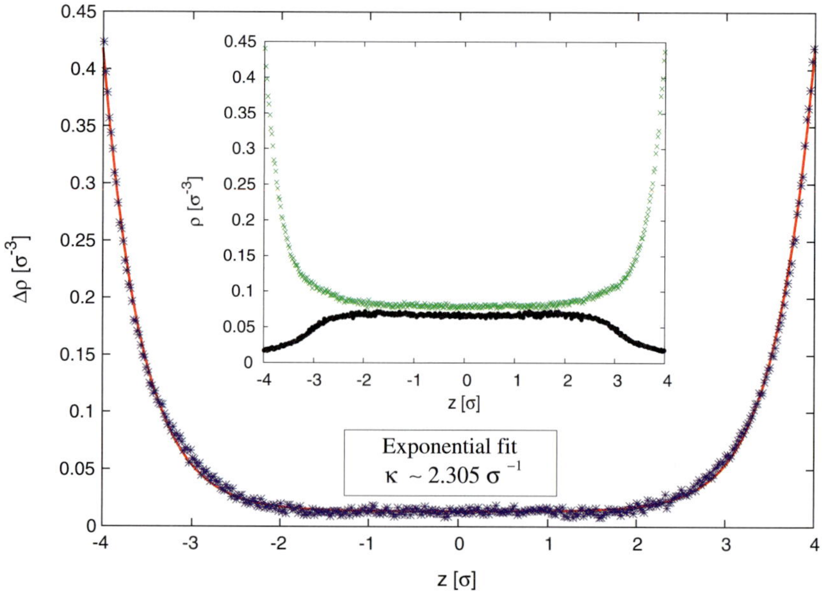

Fig. 2. Distribution of the ionic difference $\Delta\rho = \rho_c - \rho_a$ for an exemplary salt concentration of $\rho_s = 0.05625\sigma^{-3}$ with a surface ion density of $\sigma_s = 0.208\sigma^{-2}$. The red line corresponds to the exponential fit of (27) with an effective inverse screening length of $\kappa = 2.305 \pm 0.025\sigma^{-1}$. **Inset:** Distribution of cations (salt cations and counterions) with green symbols and anions (salt anions) with black symbols for an the exemplary salt concentration for the above given parameters

Table 1. Fitted inverse screening lengths κ for the different salt concentrations ρ_s and a fixed counterion density of $\rho = 0.0525\sigma^{-3}$

$\rho_s[\sigma^{-3}]$	$\kappa[\sigma^{-1}]$
0.015	1.996 ± 0.041
0.0225	2.011 ± 0.049
0.03	1.983 ± 0.041
0.0375	2.182 ± 0.047
0.05625	2.305 ± 0.025

expression of (27) in terms of the Stokes equation (10) with partial-slip boundary conditions. Thus the description of the electroosmotic flow in terms of the Stokes theory is valid.

Figure 4 compiles our numerical results for the electroosmotic mobility for all salt concentrations and slip lengths. They are in very good agreement to the theoretical prediction of (13), where $\mu_{0,EOF}$ has been determined indepen-

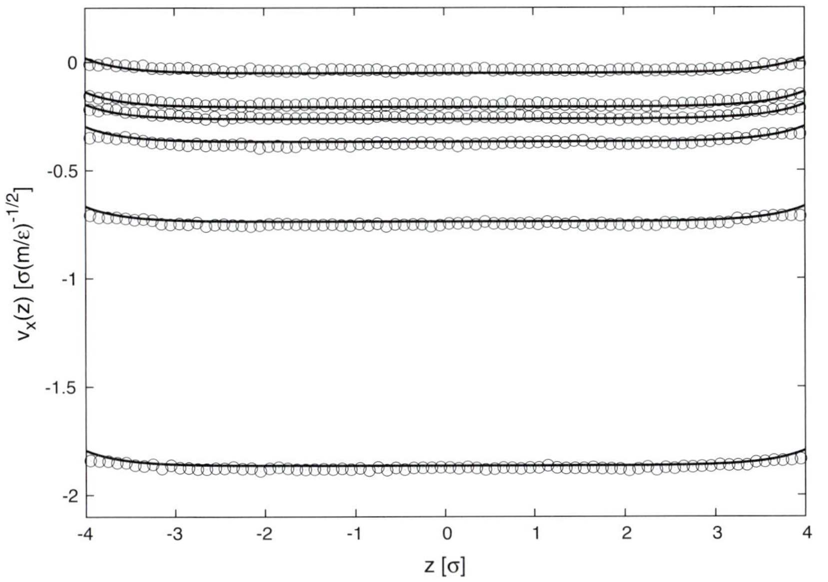

Fig. 3. Exemplary flow profiles for a salt concentration $\rho_s = 0.05625\sigma^{-3}$ for varying slip lengths (from bottom to top: $\delta_B = (14.98, 5.66, 2.63, 1.77, 1.29, 0.00)\sigma$.) The black lines are the integrated theoretical predictions in terms of the Stokes equation (10) with a fitted inverse screening length of $\kappa = 2.305\sigma^{-1}$

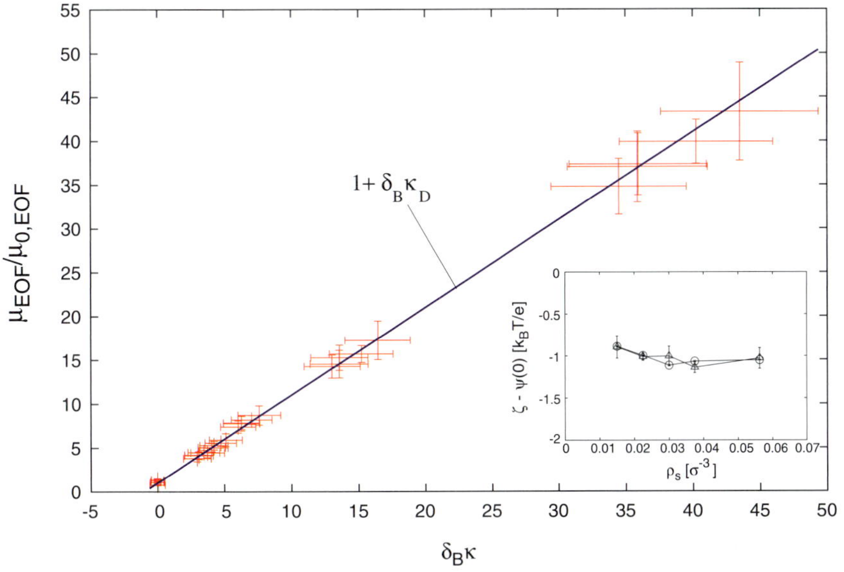

Fig. 4. Ratio $\mu_{EOF}/\mu_{0,EOF}$ plotted against $\delta_B\kappa$ for the different salt concentrations and screening lengths given in Table 1 with counterions of density $\rho = 0.0525\sigma^{-3}$. The blue line is the theoretical prediction of (13) with slope $1 + \delta_B\kappa$

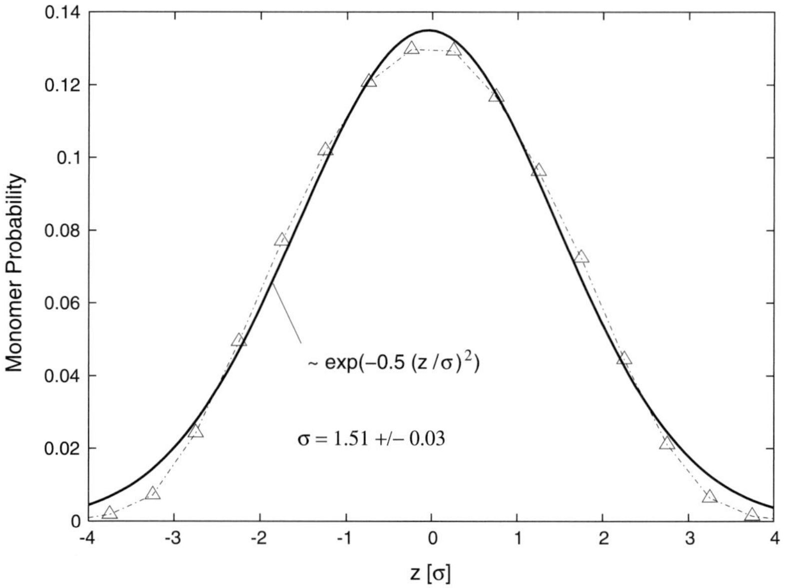

Fig. 5. Monomer appearance probability inside the channel for a salt concentration of $\rho_s = 0.05625\sigma^{-3}$

dently by a linear regression for each salt concentration. It is worth noting that the presence of the polyelectrolyte does not perturb the amplitude of the electroosmotic flow.

After investigating the electroosmotic flow of the solvent, we focus on the dynamics of the polyelectrolyte. The monomer distribution for a salt concentration of $\rho_s = 0.05625\sigma^{-3}$ is presented in Fig. 5. The fit shows that the distribution is dominated by a peak in the middle of the channel with a variance of $Var \sim 2.28\sigma$. Thus the main drag of the electroosmotic flow on the polyelectrolyte is exerted in the middle of the channel, and the rapid variations of the electroosmotic flow in close vicinity to the boundaries (Fig. 3) have little influence on the polyelectrolyte mobility. Estimating the total mobility by assuming a constant plug-like flow profile for the fluid in the middle of the channel from $z = -1.52\sigma$ to $z = 1.52\sigma$ due to the variance of the monomer distribution therefore seems reasonable.

The influence of the electroosmotic flow on the polyelectrolyte can be investigated by regarding static properties like the radius of gyration $R_g^2 = (1/2N^2) \sum_{i,j=1}^{N} < (\mathbf{R}_i - \mathbf{R}_j)^2 >$ and the end-to-end radius with $R_e^2 =< (\mathbf{R}_N - \mathbf{R}_1)^2 >$ [16]. The values of these parameters are shown in Table 2. Both properties decrease with larger salt concentration due to a more pronounced screening of electrostatic interactions which is in accordance to standard theories [3]. Although the values for the end to end radius are quite large, in all cases the box dimensions in the unconfined directions with $b_{x,y} = 12\sigma$ are larger than

Table 2. Radius of gyration R_g and end to end radius R_e for a polyelectrolyte with $N = 20$ monomers for different salt concentrations ρ_s

$\rho_s[\sigma^{-3}]$	$R_g[\sigma]$	$R_e[\sigma]$
0.015	3.2218 ± 0.047	10.6480 ± 0.0314
0.0225	3.1661 ± 0.0041	10.2736 ± 0.0266
0.03	3.1486 ± 0.0451	10.1777 ± 0.0292
0.0375	3.1279 ± 0.0045	10.0819 ± 0.0287
0.05625	3.0825 ± 0.0045	9.8331 ± 0.0280

the maximal average extension of the chain. Possible explanations for these large ratios between the end-to-end and gyration radii could be shear-induced elongation [12] or direct squeezing due to the presence of the channel walls.

The total mobility of the polyelectrolyte for varying boundary conditions is finally presented in Fig. 6. The theoretical prediction of (20) is in good agreement with the numerical results with the fitted ratio $\mu_e/\mu^0_{EOF} = -3.778 \pm 0.128$. With no-slip boundaries ($\delta_B \approx 0$) one obtains ordinary behaviour where the polyelectrolyte migrates in opposite direction to the electroosmotic flow. In the presence of slip, the absolute mobility may become negative if the amplitude of the electroosmotic flow exceeds a critical value given by a combination of the inverse screening and slip lengths, and if the sign of the wall charges is identical to the net charge of the polyelectrolyte. If the wall is oppositely charged to the polyelectrolyte, slippage effects should even enhance the total velocity of the polyelectrolyte.

The right inset of Fig. 6 shows the total displacement of the chains center of mass for various slip lengths and a salt concentration of $\rho_s = 0.05625\sigma^{-3}$. In nearly all cases except for $\delta_B \approx 0$, the total mobility of the polyelectrolyte is negative with $|\mu_p| \ll |\mu_{EOF}|$, indicating negative values of μ_t/μ^0_{EOF}. This can only be explained in terms of the drag force of the electroosmotic flow and the surplus of cations in close vicinity to the boundaries (Fig. 2) and is in agreement to the results derived above.

To summarise, the assumptions and calculations of Sect. 4 are in good agreement to the presented numerical results in this section. The total mobility of the polyelectrolyte can therefore be adequately described by (20).

5 Summary

We have presented mesoscopic DPD simulations of polyelectrolyte electrophoresis in narrow microchannels, taking full account of hydrodynamic and electrostatic interactions. We have shown that the product of the inverse screening length κ and the slip length δ_B massively influences the electroosmotic flow

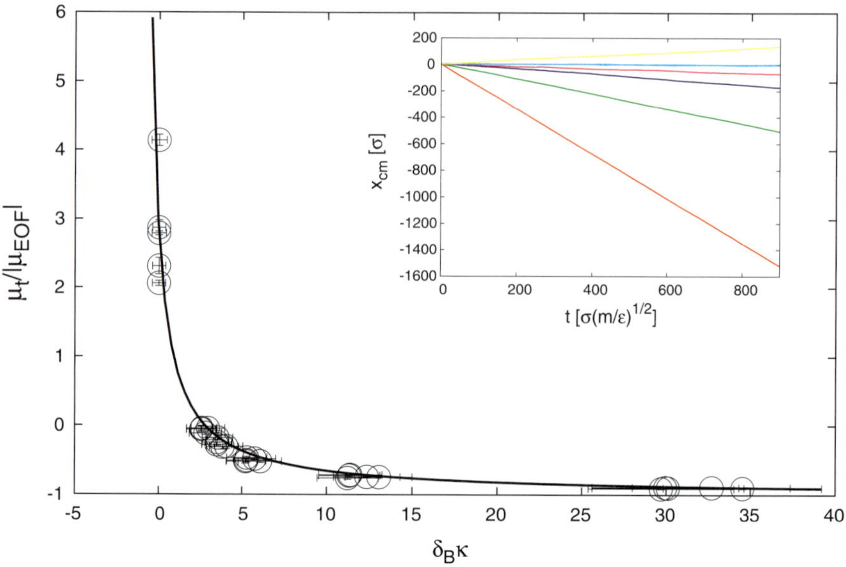

Fig. 6. Ratio $\mu_t/|\mu_{EOF}|$ plotted against $\delta_B\kappa$ for all salt concentrations. The black line is the theoretical prediction of (20) with absolut values of $|\mu_{EOF}|$. In the limit $\delta_B\kappa \to \infty$, the total mobility of the polyelectrolyte is equal to the electroosmotic flow. The ratio μ_p/μ^0_{EOF} has been fitted to -3.778 ± 0.128. Negative values of $\mu_t/|\mu_{EOF}|$ indicate absolute negative total mobilities of the polyelectrolyte. **Inset:** Total displacement of the polyelectrolytes center of mass for different boundary conditions and a salt concentration of $\rho_s = 0.05625\sigma^{-3}$. The total mobility becomes negative if the relation $|\mu_e| \ll |\mu_{EOF}|$ is fulfilled. The lines correspond from top to bottom to the slip lengths $\delta_B \approx (0.00, 1.292, 1.765, 2.626, 5.664, 14.98)\sigma$. Thus larger slip lengths indirectly enhance the total mobility of the polyelectrolyte

and therefore the total mobility of the polyelectrolyte. Thus the characteristics of the boundaries have to be taken into account for a proper description of the polyelectrolyte migration dynamics. For certain parameter sets, even a negative mobility can be achieved. All our numerical results are in good agreement to the analytical derived results.

In summary, only a combination of electroosmotic, electrophoretic, electrostatic and slippage effects does describe the total mobility of polyelectrolytes in microchannels adequately. Our simulations indicate and explain total negative mobilities due to boundary effects which also have been observed in recent experiments [4].

The characteristics of the channel walls could be used to significantly enhance flow profiles which offers the possibility to reduce the time needed for polymer migration or separation techniques. This could be an important aspect for future applications in microchannels or micropumps to accelerate the measuring time in experiments.

Acknowledgements. We thank Christian Holm, Burkhard Dünweg, Ulf D. Schiller, Marcello Sega, Michael P. Allen and Kai Grass for nice and fruitful discussions. Furthermore we thank the HLRS in Stuttgart for computing time and the Volkswagen Stiftung for funding.

References

1. Israelachvili, J.: Intermolecular and Surface Forces. Academic Press, London (1991)
2. Hunter, R.J.: Foundations of Colloid Science, Vol. 1. Clarendon Press, Oxford (1991)
3. Viovy, J.-L.: Electrophoresis of DNA and other polyelectrolytes: Physical mechanisms. Rev. Mod. Phys., **72**, 813–872 (2000)
4. Mathe, J., Di Meglio, J.-M., Tinland, B.: Electrophoretic separation of large DNAs using steric confinement. J. Colloid Interface Sci., **316**, 831–835 (2007)
5. Smiatek, J., Sega, M., Holm, C., Schiller, U.D., Schmid, F.: Mesoscopic simulations of the counterion-induced electroosmotic flow: A comparative study. J. Chem. Phys., **24**, 244702–244710 (2009)
6. Hoogerbrugge, P.J., Koelman, J.M.V.A.: Simulating microscopic hydrodynamic phenomena with dissipative particle dynamics. Europhys. Lett., **19**, 155–160, (1992)
7. Groot, R.D., Warren, P.B.: Dissipative particle dynamics: Bridging the gap between atomistic and mesoscopic simulation. J. Chem. Phys., **107**, 4423–4435 (1997)
8. Español, P., Warren, P.B.: Statistical mechanics of dissipative particle dynamics. Europhys. Lett., **30**, 191–195 (1995)
9. Frenkel, D., Smit, B.: Understanding Molecular Simulation, 2nd Edition. Academic Press, San Diego CA (2008)
10. Arnold, A., Mann, B.A., Limbach, H.-J., Holm, C.: ESPResSo: An extensible simulation package for research on soft matter systems. Comp. Phys. Comm., **174**, 704–727 (2005)
11. Smiatek, J., Schmid, F.: Polyelectrolyte electrophoresis in nanochannels: A dissipative particle dynamics simulation. J. Phys. Chem. B., **114**, 6266–6272 (2010)
12. Boroudjerdi, H., Kim, Y.-W., Naji, A., Netz, R.R., Schlagberger X., Serr A.: Statics and dynamics of strongly charged soft matter. Phys. Rep., **416**, 129–199 (2005)
13. Moreira, A.G., Netz, R.R.: Binding of similarly charged plates with counterions only. Phys. Rev. Lett., **87**, 078301–078305 (2001)
14. Hockney, R.W., Eastwood, J.W.: Computer Simulation using Particles. McGraw-Hill, New York (1981)
15. Smiatek, J., Allen, M.P., Schmid, F.: Tunable-slip boundaries for coarse-grained simulations of fluid flow. Eur. Phys. J. E, **26**, 115–123 (2008)
16. Doi, M., Edwards S.F.: The Theory of Polymer Dynamics. Oxford Press, Oxford (1986)
17. Dünweg, B., Kremer, K.: Molecular dynamics simulation of a polymer chain in solution. J. Chem. Phys., **99**, 6983–6998 (1993)
18. Smiatek, J.: Mesoscopic simulations of electrohydrodynamic phenomena. PhD thesis, Bielefeld University (2009)
19. Smiatek, J., Scherer, C., Schmid, F., manuscript in preparation

The SuperN-Project: An Update on Core-Collapse Supernova Simulations

B. Müller, L. Hüdepohl, A. Marek, F. Hanke, and H.-Th. Janka

Max-Planck-Institut für Astrophysik, Karl-Schwarzschild-Strasse 1, Postfach 1317, D-85741 Garching bei München, Germany, bjmuellr@mpa-garching.mpg.de

Summary. We give an overview of the challenges and the current status of our two-dimensional (core collapse) supernova modelling, and present the system of equations and the algorithm for its solution that are employed in our code, and report on our continuing efforts to improve the physics in our supernova code VERTEX as well as its the computational efficiency. We also discuss recent results of simulations performed on the NEC SX-8 at the HLRS, which include the first multi-dimensional general-relativistic neutrino transport simulations—conducted with a new extension of the VERTEX code—as well as simulations of neutron star cooling over several seconds for different nuclear equations of state.

1 Introduction

A star more massive than about 8 solar masses ends its live in a catastrophic explosion, a supernova. Its quiescent evolution comes to an end, when the pressure in its inner layers is no longer able to balance the inward pull of gravity. Throughout its life, the star sustained this balance by generating energy through a sequence of nuclear fusion reactions, forming increasingly heavier elements in its core. However, when the core consists mainly of iron-group nuclei, central energy generation ceases. The fusion reactions producing iron-group nuclei relocate to the core's surface, and their "ashes" continuously increase the core's mass. Similar to a white dwarf, such a core is stabilised against gravity by the pressure of its degenerate gas of electrons. However, to remain stable, its mass must stay smaller than the Chandrasekhar limit. When the core grows larger than this limit, it collapses to a neutron star, and a huge amount ($\sim 10^{53}$ erg) of gravitational binding energy is set free. Most ($\sim 99\%$) of this energy is radiated away in neutrinos, but a small fraction is transferred to the outer stellar layers and drives the violent mass ejection which disrupts the star in a supernova.

Despite 40 years of research, the details of how this energy transfer happens and how the explosion is initiated are still not well understood. Observational

W.E. Nagel et al. (eds.), *High Performance Computing in Science and Engineering '10*, DOI 10.1007/978-3-642-15748-6_6,

evidence about the physical processes deep inside the collapsing star is sparse and almost exclusively indirect. The only direct observational access is via measurements of neutrinos or gravitational waves. To obtain insight into the events in the core, one must therefore heavily rely on sophisticated numerical simulations. The enormous amount of computer power required for this purpose has led to the use of several, often questionable, approximations and numerous ambiguous results in the past. Fortunately, however, the development of numerical tools and computational resources has meanwhile advanced to a point, where it is becoming possible to perform multi-dimensional simulations with unprecedented accuracy. Therefore there is hope that the physical processes which are essential for the explosion can finally be unravelled.

An understanding of the explosion mechanism is required to answer many important questions of nuclear, gravitational, and astro-physics like the following:

- How do the explosion energy, the explosion timescale, and the mass of the compact remnant depend on the progenitor's mass? Is the explosion mechanism the same for all progenitors? For which stars are black holes left behind as compact remnants instead of neutron stars?
- What is the role of the—poorly known—equation of state (EoS) for the proto neutron star? Do softer or stiffer EoSs favour the explosion of a core collapse supernova?
- How do neutron stars receive their natal kicks? Are they accelerated by asymmetric mass ejection and/or anisotropic neutrino emission?
- What are the generic properties of the neutrino emission and of the gravitational wave signal that are produced during stellar core collapse and explosion? Up to which distances could these signals be measured with operating or planned detectors on earth and in space? And what can one learn about supernova dynamics or nuclear and particle physics from a future measurement of such signals in case of a Galactic supernova?
- How do supernovae contribute to the enrichment of the intergalactic medium with heavy elements? What kind of nucleosynthesis processes occur during and after the explosion? Can the elemental composition of supernova remnants be explained correctly by the numerical simulations? Does the rapid neutron capture process (r-process), which produces e.g. gold and the actinides, take place in supernovae?

2 Numerical Models

2.1 History and Constraints

According to theory, a shock wave is launched at the moment of "core bounce" when the neutron star begins to emerge from the collapsing stellar iron core. There is general agreement, supported by all "modern" numerical simulations, that this shock is unable to propagate directly into the stellar mantle and envelope, because it looses too much energy in dissociating iron into free nucleons while it moves through the outer core. The "prompt" shock ultimately stalls. Thus the currently favoured theoretical paradigm needs to exploit the fact that a huge energy reservoir is present in the form of neutrinos, which are abundantly emitted from the hot, nascent neutron star. The absorption of electron neutrinos and anti-neutrinos by free nucleons in the post shock layer is thought to reenergize the shock, and lead to the supernova explosion.

Detailed *spherically symmetric* hydrodynamic models, which recently include a very accurate treatment of the time-dependent, multi-flavour, multi-frequency neutrino transport based on a numerical solution of the Boltzmann transport equation [1, 2], reveal that this "delayed, neutrino-driven mechanism" does not work as simply as originally envisioned. Although in principle able to trigger the explosion (e.g., [3–5]), neutrino energy transfer to the post-shock matter turned out to be too weak. For inverting the infall of the stellar core and initiating powerful mass ejection, an increase of the efficiency of neutrino energy deposition is needed.

A number of physical phenomena have been pointed out that can enhance neutrino energy deposition behind the stalled supernova shock. They are all linked to the fact that the real world is multi-dimensional instead of spherically symmetric (or one-dimensional; 1D) as assumed in the work cited above:

(1) Convective instabilities in the neutrino-heated layer between the neutron star and the supernova shock develop to violent convective overturn [6]. This convective overturn is helpful for the explosion, mainly because (a) neutrino-heated matter rises and increases the pressure behind the shock, thus pushing the shock further out, and (b) cool matter is able to penetrate closer to the neutron star where it can absorb neutrino energy more efficiently. Both effects allow multi-dimensional models to explode easier than spherically symmetric ones [7–9].
(2) Recent work [10–13] has demonstrated that the stalled supernova shock is also subject to a second non-radial low-mode instability, called standing accretion shock instability or "SASI" for short, which can grow to a dipolar, global deformation of the shock [12, 14, 15].
(3) Convective energy transport inside the nascent neutron star [16–19] might enhance the energy transport to the neutrinosphere and could thus boost the neutrino luminosities. This would in turn increase the neutrino-heating behind the shock.

This list of multi-dimensional phenomena awaits more detailed exploration in multi-dimensional simulations. Until recently, such simulations have been performed with only a grossly simplified treatment of the involved microphysics, in particular of the neutrino transport and neutrino-matter interactions. At best, grey (i.e., single energy) flux-limited diffusion schemes were employed. Since, however, the role of the neutrinos is crucial for the problem, and because previous experience shows that the outcome of simulations is indeed very sensitive to the employed transport approximations, studies of the explosion mechanism require the best available description of the neutrino physics. This implies that one has to solve the Boltzmann transport equation for neutrinos.

2.2 The Mathematical Model

As core-collapse supernovae involve such a complex interplay of hydrodynamics, self-gravity and neutrino heating and cooling, numerical modellers face a classical "multiphysics" problem. Although the overall problem can still be formulated as a system of non-linear partial differential equations, rather dissimilar methods—sometimes with conflicting requirements on the computer architecture and the parallelization strategy—need to be applied to treat individual subsystems. In the case of our code, the system of equations that needs to be solved consists of the following components:

- The multi-dimensional Euler equations of (relativistic) hydrodynamics, supplemented by advection equations for the electron fraction and the chemical composition of the fluid, and formulated in spherical polar coordinates;
- equations for the space-time metric (or in the Newtonian case, the Poisson equation) for calculating the gravitational source terms in the Euler equations;
- the Boltzmann transport equation which determines the (non-equilibrium) distribution function of the neutrinos;
- the emission, absorption, and scattering rates of neutrinos, which are required for the solution of the Boltzmann equation;
- the equation of state of the stellar fluid, which provides the closure relation between the variables entering the Euler equations, i.e. density, momentum, energy, electron fraction, composition, and pressure.

In what follows we will briefly summarise the neutrino transport algorithms, thus focusing on the major computational kernel of our code. For a more complete description of the entire code we refer the reader to [20, 21], and the references therein.

2.3 "Ray-by-Ray Plus" Method for the Neutrino Transport Problem

The crucial quantity required to determine the source terms for the energy, momentum, and electron fraction of the fluid owing to its interaction with the neutrinos is the neutrino distribution function in phase space, $f(r, \vartheta, \phi, \epsilon, \Theta, \Phi, t)$. Equivalently, the neutrino intensity $I = c/(2\pi\hbar c)^3 \cdot \epsilon^3 f$ may be used. Both are seven-dimensional functions, as they describe, at every point in space (r, ϑ, ϕ), the distribution of neutrinos propagating with energy ϵ into the direction (Θ, Φ) at time t (Fig. 1).

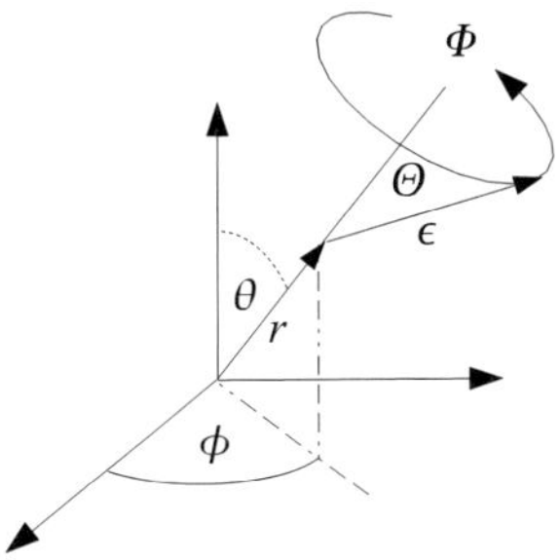

Fig. 1. Illustration of the phase space coordinates (see the main text)

The evolution of I (or f) in time is governed by the Boltzmann equation, and solving this equation is, in general, a six-dimensional problem (as time is usually not counted as a separate dimension). A solution of this equation by direct discretization (using an S_N scheme) would require computational resources in the PetaFlop range. Although there are attempts by at least one group in the United States to follow such an approach, we feel that, with the currently available computational resources, it is mandatory to reduce the dimensionality of the problem.

Actually this should be possible, since the source terms entering the hydrodynamic equations are *integrals* of I over momentum space (i.e. over ϵ, Θ, and Φ), and thus only a fraction of the information contained in I is truly required to compute the dynamics of the flow. It therefore makes sense to consider angular moments of I, and to solve evolution equations for these moments, instead of dealing with the Boltzmann equation directly. The 0th to 3rd order moments are defined as

$$J, \boldsymbol{H}, \mathbf{K}, \mathbf{L}, \ldots (r, \vartheta, \phi, \epsilon, t) = \frac{1}{4\pi} \int I(r, \vartheta, \phi, \epsilon, \Theta, \Phi, t) \, \boldsymbol{n}^{0,1,2,3,\ldots} \, \mathrm{d}\Omega \qquad (1)$$

where $\mathrm{d}\Omega = \sin\Theta \, \mathrm{d}\Theta \, \mathrm{d}\Phi$, $\boldsymbol{n} = (\cos\Theta, \sin\Theta\cos\Phi, \sin\Theta\sin\Phi)$, and exponentiation represents repeated application of the dyadic product. Note that the moments are *tensors* of the required rank.

This leaves us with a four-dimensional problem. So far no approximations have been made. In order to reduce the size of the problem even further, one needs to resort to assumptions on its symmetry. At this point, one usually employs azimuthal symmetry for the stellar matter distribution, i.e. any dependence on the azimuth angle ϕ is ignored, which implies that the hydrodynamics of the problem can be treated in two dimensions. It also implies $I(r, \vartheta, \epsilon, \Theta, \Phi) = I(r, \vartheta, \epsilon, \Theta, -\Phi)$. If, in addition, it is assumed that I is even independent of Φ, then each of the angular moments of I becomes a *scalar*, which depends on two spatial dimensions, and one dimension in momentum space: $J, H, K, L = J, H, K, L(r, \vartheta, \epsilon, t)$. Thus we have reduced the problem to three dimensions in total.

The System of Equations

With the aforementioned assumptions it can be shown [20], that in the Newtonian approximation the following two transport equations need to be solved, in order to compute the source terms for the energy and electron fraction of the fluid:

$$
\begin{aligned}
&\left(\frac{1}{c}\frac{\partial}{\partial t}+\beta_r\frac{\partial}{\partial r}+\frac{\boldsymbol{\beta_\vartheta}}{\boldsymbol{r}}\frac{\boldsymbol{\partial}}{\boldsymbol{\partial\vartheta}}\right)J+J\left(\frac{1}{r^2}\frac{\partial(r^2\beta_r)}{\partial r}+\frac{\mathbf{1}}{\boldsymbol{r}\,\mathbf{sin}\,\boldsymbol{\vartheta}}\frac{\boldsymbol{\partial}(\mathbf{sin}\,\boldsymbol{\vartheta}\boldsymbol{\beta_\vartheta})}{\boldsymbol{\partial\vartheta}}\right)\\
&+\frac{1}{r^2}\frac{\partial(r^2H)}{\partial r}+\frac{\beta_r}{c}\frac{\partial H}{\partial t}-\frac{\partial}{\partial\epsilon}\left\{\frac{\epsilon}{c}\frac{\partial\beta_r}{\partial t}H\right\}-\frac{\partial}{\partial\epsilon}\left\{\epsilon J\left(\frac{\beta_r}{r}+\frac{\mathbf{1}}{\mathbf{2}\boldsymbol{r}\,\mathbf{sin}\,\boldsymbol{\vartheta}}\frac{\boldsymbol{\partial}(\mathbf{sin}\,\boldsymbol{\vartheta}\boldsymbol{\beta_\vartheta})}{\boldsymbol{\partial\vartheta}}\right)\right\}\\
&\qquad-\frac{\partial}{\partial\epsilon}\left\{\epsilon K\left(\frac{\partial\beta_r}{\partial r}-\frac{\beta_r}{r}-\frac{\mathbf{1}}{\mathbf{2}\boldsymbol{r}\,\mathbf{sin}\,\boldsymbol{\vartheta}}\frac{\boldsymbol{\partial}(\mathbf{sin}\,\boldsymbol{\vartheta}\boldsymbol{\beta_\vartheta})}{\boldsymbol{\partial\vartheta}}\right)\right\}\\
&\qquad+J\left(\frac{\beta_r}{r}+\frac{\mathbf{1}}{\mathbf{2}\boldsymbol{r}\,\mathbf{sin}\,\boldsymbol{\vartheta}}\frac{\boldsymbol{\partial}(\mathbf{sin}\,\boldsymbol{\vartheta}\boldsymbol{\beta_\vartheta})}{\boldsymbol{\partial\vartheta}}\right)\\
&\quad+K\left(\frac{\partial\beta_r}{\partial r}-\frac{\beta_r}{r}-\frac{\mathbf{1}}{\mathbf{2}\boldsymbol{r}\,\mathbf{sin}\,\boldsymbol{\vartheta}}\frac{\boldsymbol{\partial}(\mathbf{sin}\,\boldsymbol{\vartheta}\boldsymbol{\beta_\vartheta})}{\boldsymbol{\partial\vartheta}}\right)+\frac{2}{c}\frac{\partial\beta_r}{\partial t}H=C^{(0)}, \qquad (2)
\end{aligned}
$$

$$
\begin{aligned}
&\left(\frac{1}{c}\frac{\partial}{\partial t}+\beta_r\frac{\partial}{\partial r}+\frac{\boldsymbol{\beta_\vartheta}}{\boldsymbol{r}}\frac{\boldsymbol{\partial}}{\boldsymbol{\partial\vartheta}}\right)H+H\left(\frac{1}{r^2}\frac{\partial(r^2\beta_r)}{\partial r}+\frac{\mathbf{1}}{\boldsymbol{r}\,\mathbf{sin}\,\boldsymbol{\vartheta}}\frac{\boldsymbol{\partial}(\mathbf{sin}\,\boldsymbol{\vartheta}\boldsymbol{\beta_\vartheta})}{\boldsymbol{\partial\vartheta}}\right)\\
&\qquad+\frac{\partial K}{\partial r}+\frac{3K-J}{r}+H\left(\frac{\partial\beta_r}{\partial r}\right)+\frac{\beta_r}{c}\frac{\partial K}{\partial t}-\frac{\partial}{\partial\epsilon}\left\{\frac{\epsilon}{c}\frac{\partial\beta_r}{\partial t}K\right\}\\
&\qquad-\frac{\partial}{\partial\epsilon}\left\{\epsilon L\left(\frac{\partial\beta_r}{\partial r}-\frac{\beta_r}{r}-\frac{\mathbf{1}}{\mathbf{2}\boldsymbol{r}\,\mathbf{sin}\,\boldsymbol{\vartheta}}\frac{\boldsymbol{\partial}(\mathbf{sin}\,\boldsymbol{\vartheta}\boldsymbol{\beta_\vartheta})}{\boldsymbol{\partial\vartheta}}\right)\right\}\\
&-\frac{\partial}{\partial\epsilon}\left\{\epsilon H\left(\frac{\beta_r}{r}+\frac{\mathbf{1}}{\mathbf{2}\boldsymbol{r}\,\mathbf{sin}\,\boldsymbol{\vartheta}}\frac{\boldsymbol{\partial}(\mathbf{sin}\,\boldsymbol{\vartheta}\boldsymbol{\beta_\vartheta})}{\boldsymbol{\partial\vartheta}}\right)\right\}+\frac{1}{c}\frac{\partial\beta_r}{\partial t}(J+K)=C^{(1)}. \qquad (3)
\end{aligned}
$$

These are evolution equations for the neutrino energy density, J, and the neutrino flux, H, and follow from the zeroth and first moment equations of the comoving frame (Boltzmann) transport equation in the Newtonian, $\mathcal{O}(v/c)$ approximation. The quantities $C^{(0)}$ and $C^{(1)}$ are source terms that result from the collision term of the Boltzmann equation, while $\beta_r = v_r/c$ and $\beta_\vartheta = v_\vartheta/c$, where v_r and v_ϑ are the components of the hydrodynamic velocity, and c is the speed of light. The functional dependencies $\beta_r = \beta_r(r, \vartheta, t)$, $J = J(r, \vartheta, \epsilon, t)$, etc. are suppressed in the notation. This system includes four unknown moments (J, H, K, L) but only two equations, and thus needs to be supplemented by two more relations. This is done by substituting $K = f_K \cdot J$ and $L = f_L \cdot J$, where f_K and f_L are the variable Eddington factors, which for the moment may be regarded as being known, but in our case is indeed determined from a separate simplified ("model") Boltzmann equation.

The moment equations (2) and (3) are very similar to the $\mathcal{O}(v/c)$ equations in spherical symmetry which were solved in the 1D simulations of [22] (see (7),

(8), (30), and (31) of the latter work). This similarity has allowed us to reuse a good fraction of the one-dimensional version of VERTEX, for coding the multi-dimensional algorithm. The additional terms necessary for this purpose have been set in boldface above.

Finally, the changes of the energy, e, and electron fraction, Y_{e}, required for the hydrodynamics are given by the following two equations

$$\frac{\mathrm{d}e}{\mathrm{d}t} = -\frac{4\pi}{\rho} \int_0^\infty \mathrm{d}\epsilon \sum_{\nu \in (\nu_{\mathrm{e}}, \bar{\nu}_{\mathrm{e}}, \dots)} C_\nu^{(0)}(\epsilon), \tag{4}$$

$$\frac{\mathrm{d}Y_{\mathrm{e}}}{\mathrm{d}t} = -\frac{4\pi\, m_{\mathrm{B}}}{\rho} \int_0^\infty \frac{\mathrm{d}\epsilon}{\epsilon} \left(C_{\nu_{\mathrm{e}}}^{(0)}(\epsilon) - C_{\bar{\nu}_{\mathrm{e}}}^{(0)}(\epsilon) \right) \tag{5}$$

(for the momentum source terms due to neutrinos see [20]). Here m_{B} is the baryon mass, and the sum in (4) runs over all neutrino types. The full system consisting of (2)–(5) is stiff, and thus requires an appropriate discretization scheme for its stable solution.

Method of Solution

In order to discretize (2)–(5), the spatial domain $[0, r_{\mathrm{max}}] \times [\vartheta_{\mathrm{min}}, \vartheta_{\mathrm{max}}]$ is covered by N_r radial, and N_ϑ angular zones, where $\vartheta_{\mathrm{min}} = 0$ and $\vartheta_{\mathrm{max}} = \pi$ correspond to the north and south poles, respectively, of the spherical grid. (In general, we allow for grids with different radial resolutions in the neutrino transport and hydrodynamic parts of the code. The number of radial zones for the hydrodynamics will be denoted by N_r^{hyd}.) The number of bins used in energy space is N_ϵ and the number of neutrino types taken into account is N_ν.

The equations are solved in two operator-split steps corresponding to a lateral and a radial sweep.

In the first step, we treat the boldface terms in the respectively first lines of (2), (3), which describe the lateral advection of the neutrinos with the stellar fluid, and thus couple the angular moments of the neutrino distribution of neighbouring angular zones. For this purpose we consider the equation

$$\frac{1}{c}\frac{\partial \Xi}{\partial t} + \frac{1}{r \sin\vartheta} \frac{\partial(\sin\vartheta\, \beta_\vartheta\, \Xi)}{\partial \vartheta} = 0\,, \tag{6}$$

where Ξ represents one of the moments J or H. Although it has been suppressed in the above notation, an equation of this form has to be solved for each radius, for each energy bin, and for each type of neutrino. An explicit upwind scheme is used for this purpose.

In the second step, the radial sweep is performed. Several points need to be noted here:

- terms in boldface not yet taken into account in the lateral sweep, need to be included into the discretization scheme of the radial sweep. This can be

done in a straightforward way since these remaining terms do not include derivatives of the transport variables J or H. They only depend on the hydrodynamic velocity v_ϑ, which is a *constant* scalar field for the transport problem.
- the right hand sides (source terms) of the equations and the coupling in energy space have to be accounted for. The coupling in energy is non-local, since the source terms of (2) and (3) stem from the Boltzmann equation, which is an integro-differential equation and couples all the energy bins
- the discretization scheme for the radial sweep is *implicit* in time. Explicit schemes would require very small time steps to cope with the stiffness of the source terms in the optically thick regime, and the small CFL time step dictated by neutrino propagation with the speed of light in the optically thin regime. Still, even with an implicit scheme $\gtrsim 10^5$ time steps are required per simulation. This makes the calculations expensive.

Once the equations for the radial sweep have been discretized in radius and energy, the resulting solver is applied ray-by-ray for each angle ϑ and for each type of neutrino, i.e. for constant ϑ, N_ν two-dimensional problems need to be solved.

The discretization itself is done using a second order accurate scheme with backward differencing in time according to [22]. This leads to a non-linear system of algebraic equations, which is solved by Newton-Raphson iteration with explicit construction and inversion of the corresponding Jacobian matrix with the Block-Thomas algorithm.

2.4 Parallelization

The ray-by-ray approximation readily lends itself to parallelization over the different angular zones. In order to make efficient use of the NEC SX-8 with its small shared-memory units (8 CPUs per node), distributed memory parallelism is indispensable. For this reason, a cooperation between MPA and the Teraflop Workbench at HLRS was set up in 2007 to initiate the development of an MPI version of the VERTEX code, which was completed in 2008. VERTEX has recently been re-integrated into the production version of our code, and now provides MPI functionality not only on the NEC SX-8, but also on other platforms used by our group.

Scaling tests demonstrate the great potential of the new code version. Figure 2 shows the parallel efficiency of strong scaling as a function of the number of processors for a test run with 256 angular rays on the NEC machine SX-8 (8 CPUs per node) at HLRS, on the JUROPA cluster at the Jülich Supercomputing Centre (JSC), on the SGI Altix 4700 at the Leibniz Rechenzentrum (LRZ), and on the IBM Power6 575 (32 CPUs/node) at the Rechenzentrum Garching (RZG) of the Max-Planck-Gesellschaft. Unfortunately, only a few tests could be conducted on the NEC SX-9 at HLRS so far, and no statements concerning the scaling on more than two nodes can be made yet.

of the neutron star and hence reduces the efficiency of heat conduction and lepton number transport by neutrino diffusion. The net lepton number loss rate (cp. Fig. 5) is therefore considerably different for these two equations of state, and the deleptonization timescale (i.e. the e-folding time-scale of the lepton number flux after ~ 2 s) is therefore about 80% larger in the case of the EoS of Lattimer&Swesty (3.4 s compared to 1.9 s) in the case of Shen). A similar dependence on the equation of state, albeit somewhat less pronounced, was also found for the total neutrino luminosities and the cooling timescale. In the event of galactic supernova, these observations may provide important clues for the analysis of the prospective neutrino signal and help to obtain constraints on neutron star properties and the nuclear equation of state.

4 Conclusions and Outlook

We continued to simulate 1D and 2D models of core collapse supernovae with detailed neutrino transport at the HLRS. With our new general relativistic neutrino hydrodynamics code, we confirmed earlier results for an $11.2\,M_{\odot}$ and an $15\,M_{\odot}$ star: For the $11.2\,M_{\odot}$ we found a rather early explosion, developing already ~ 150 ms after bounce, while the $15\,M_{\odot}$ does not explode during the first ~ 400 ms. However, compared to earlier Newtonian simulations conducted by our group we found quantitative differences in the gravitational wave signals of these two models, which deserve further analysis. At the same time, we continued our work on explosions of low mass progenitors with O-Ne-Mg cores: Using the progenitor model of Nomoto [28, 29], we performed several long-time simulations covering the explosion and ~ 10 s of the neutron star cooling phase, and conducted a detailed analysis of the neutrino signal, the nucleosynthesis conditions and equation of state dependencies.

Currently, our studies are limited to a axisymmetric (2D) models so far because the prohibitively long turn-around time for supernova simulations with our sophisticated ray-by-ray-plus method still precludes 3D simulations. For this reason, we are currently extending the 2D MPI version of the VERTEX code (which was developed by K. Benkert at HLRS in cooperation with MPA) to the 3D case. Test runs conducted on different high-performance computing platforms, including NEC SX-8 and SX-9, show that very good scaling can be expected even on a large number of cores. However, in order to efficiently use several thousand CPUs as required for 3D models and to avoid exceedingly large memory demands for temporary storage of output data, it is also necessary to optimize parallel I/O, and we are therefore working on Parallel HDF5 support in the VERTEX code. Once these technical issues are taken care of, we intend to begin pilot studies in 3D.

Acknowledgements. Support by the Deutsche Forschungsgemeinschaft through the SFB/TR27 "Neutrinos and Beyond" and the SFB/TR7 "Gravitational Wave Astronomy", and by the Cluster of Excellence EXC 153 "Origin and Structure of the Universe" (http://www.universe-cluster.de) are acknowledged, as well computer time grants of the HLRS, NIC Jülich, and Rechenzentrum Garching are acknowledged. We thank especially K. Benkert for her extremely valuable work on the MPI version of VERTEX.

References

1. Rampp, M., Janka, H.T.: Spherically Symmetric Simulation with Boltzmann Neutrino Transport of Core Collapse and Postbounce Evolution of a 15 $M_\odot$ Star. Astrophys. J. **539** (2000) L33–L36
2. Liebendörfer, M., Mezzacappa, A., Thielemann, F., Messer, O.E., Hix, W.R., Bruenn, S.W.: Probing the gravitational well: No supernova explosion in spherical symmetry with general relativistic Boltzmann neutrino transport. Phys. Rev. D **63** (2001) 103004–+
3. Bethe, H.A.: Supernova mechanisms. Reviews of Modern Physics **62** (1990) 801–866
4. Burrows, A., Goshy, J.: A Theory of Supernova Explosions. Astrophys. J. **416** (1993) L75
5. Janka, H.T.: Conditions for shock revival by neutrino heating in core-collapse supernovae. Astron. Astrophys. **368** (2001) 527–560
6. Herant, M., Benz, W., Colgate, S.: Postcollapse hydrodynamics of SN 1987A – Two-dimensional simulations of the early evolution. Astrophys. J. **395** (1992) 642–653
7. Herant, M., Benz, W., Hix, W.R., Fryer, C.L., Colgate, S.A.: Inside the supernova: A powerful convective engine. Astrophys. J. **435** (1994) 339
8. Burrows, A., Hayes, J., Fryxell, B.A.: On the nature of core-collapse supernova explosions. Astrophys. J. **450** (1995) 830
9. Janka, H.T., Müller, E.: Neutrino heating, convection, and the mechanism of Type-II supernova explosions. Astron. Astrophys. **306** (1996) 167–+
10. Thompson, C.: Accretional Heating of Asymmetric Supernova Cores. Astrophys. J. **534** (2000) 915–933
11. Blondin, J.M., Mezzacappa, A., DeMarino, C.: Stability of Standing Accretion Shocks, with an Eye toward Core-Collapse Supernovae. Astrophys. J. **584** (2003) 971–980
12. Scheck, L., Plewa, T., Janka, H.T., Kifonidis, K., Müller, E.: Pulsar Recoil by Large-Scale Anisotropies in Supernova Explosions. Phys. Rev. Letters **92** (2004) 011103–+
13. Foglizzo, T., Galletti, P., Scheck, L., Janka, H.T.: Instability of a Stalled Accretion Shock: Evidence for the Advective-Acoustic Cycle. Astrophys. J. **654** (2007) 1006–1021
14. Scheck, L., Kifonidis, K., Janka, H.T., Müller, E.: Multidimensional supernova simulations with approximative neutrino transport. I. Neutron star kicks and the anisotropy of neutrino-driven explosions in two spatial dimensions. Astron. Astrophys. **457** (2006) 963–986
15. Scheck, L., Janka, H.T., Foglizzo, T., Kifonidis, K.: Multidimensional supernova simulations with approximative neutrino transport. II. Convection and the advective-acoustic cycle in the supernova core. Astron. Astrophys. **477** (2008) 931–952

16. Keil, W., Janka, H.T., Müller, E.: Ledoux Convection in Protoneutron Stars—A Clue to Supernova Nucleosynthesis? Astrophys. J. **473** (1996) L111
17. Burrows, A., Lattimer, J.M.: The birth of neutron stars. Astrophys. J. **307** (1986) 178–196
18. Pons, J.A., Reddy, S., Prakash, M., Lattimer, J.M., Miralles, J.A.: Evolution of Proto-Neutron Stars. Astrophys. J. **513** (1999) 780–804
19. Marek, A.: Multi-dimensional simulations of core collapse supernovae with different equations of state for hot proto-neutron stars. PhD thesis, Technische Universität München (2007)
20. Buras, R., Rampp, M., Janka, H.T., Kifonidis, K.: Two-dimensional hydrodynamic core-collapse supernova simulations with spectral neutrino transport. I. Numerical method and results for a 15 Mȯ star. Astron. Astrophys. **447** (2006) 1049–1092
21. Müller, B., Janka, H.T., Dimmelmeier, H.: A new multi-dimensional general relativistic neutrino hydrodynamics code for core-collapse supernovae. I. Method and code tests in spherical symmetry. ArXiv e-prints **999** (2010), submitted to Astrophys. J.
22. Rampp, M., Janka, H.T.: Radiation hydrodynamics with neutrinos. Variable Eddington factor method for core-collapse supernova simulations. Astron. Astrophys. **396** (2002) 361–392
23. Marek, A., Janka, H.T.: Delayed Neutrino-Driven Supernova Explosions Aided by the Standing Accretion-Shock Instability. Astrophys. J. **694** (2009) 664–696
24. Hüdepohl, L., Müller, B., Janka, H.T., Marek, A., Raffelt, G.G.: Neutrino Signal of Electron-Capture Supernovae from Core Collapse to Cooling. ArXiv e-prints **999** (2009), submitted to Phys. Rev. Lett.
25. Dimmelmeier, H., Font, J.A., Müller, E.: Relativistic simulations of rotational core collapse I. Methods, initial models, and code tests. Astron. Astrophys. **388** (2002) 917–935
26. Isenberg, J.A.: Waveless approximation theories of gravities. *University of Maryland Preprint*, *ArXiv e-print* gr-qc/0702113 (1978)
27. Buras, R., Janka, H.T., Rampp, M., Kifonidis, K.: Two–dimensional hydrodynamic core–collapse supernova simulations with spectral neutrino transport. II. Models for different progenitor stars. Astron. Astrophys. **457** (2006) 281–308
28. Nomoto, K.: Evolution of 8-10 solar mass stars toward electron capture supernovae. I – Formation of electron-degenerate O + NE + MG cores. Astrophys. J. **277** (1984) 791–805
29. Nomoto, K.: Evolution of 8-10 solar mass stars toward electron capture supernovae. II – Collapse of an O + NE + MG core. Astrophys. J. **322** (1987) 206–214
30. Lattimer, J.M., Swesty, F.D.: A generalized equation of state for hot, dense matter. Nuclear Physics A **535** (1991) 331–+
31. Shen, H., Toki, H., Oyamatsu, K., Sumiyoshi, K.: Relativistic Equation of State of Nuclear Matter for Supernova Explosion. Progress of Theoretical Physics **100** (1998) 1013–1031

Higgs Boson Mass Bounds from a Chirally Invariant Lattice Higgs-Yukawa Model

Philipp Gerhold[1], Karl Jansen[2], and Jim Kallarackal[1]

[1] Humboldt-Universität Berlin, Newtonstr. 15, D-12489 Berlin, Germany, gerhold@physik.hu-berlin.de, Jim.Kallarackal@physik.hu-berlin.de

[2] DESY-Zeuthen, Platanenallee 6, D-15738 Zeuthen, Germany, karl.jansen@desy.de

Summary. We consider a chirally invariant lattice Higgs-Yukawa model based on the Neuberger overlap operator $\mathcal{D}^{(ov)}$. The model will be evaluated using PHMC-simulations and we will present final results on the upper and lower Higgs boson mass bound. The question of a fourth generation of heavy quarks has recently gained attention and we will illustrate preliminary results of the Higgs boson mass bounds within this framework. Finally we will discuss our progress on properties of the Higgs boson with respect to its unstable nature, such as the decay width and the resonance mass of the Higgs boson.

1 Introduction

In the early 1990's there has been a large activity of investigating lattice Higgs- and Higgs-Yukawa-models driven by the desire for a non-perturbative determination of lower and upper Higgs boson mass bounds as well as its decay properties, see e.g. Refs. [1–7] for reviews. These earlier investigations were, however, blocked by the lack of a consistent formulation of chiral symmetry on the lattice, which is, of course, indispensable for the lattice construction of chiral theories like the Higgs-Sector of the standard model.

There are two main developments which warrant to reconsider these questions: firstly, with the advent of the LHC, we are to expect that properties of the standard model Higgs boson, such as the mass and the decay width, will be revealed experimentally. Secondly, there is, in contrast to the situation of the earlier investigations, a consistent formulation of a lattice Higgs-Yukawa model with an exact lattice chiral symmetry [8] based on the Ginsparg-Wilson relation [9], allowing thus to go beyond those earlier models. Based on this development, the interest in lattice studies of Higgs-Yukawa models has recently been renewed [10–14].

Establishing non perturbative bounds for the Higgs boson mass was the main subject of research in the last three years and lead to the successful

W.E. Nagel et al. (eds.), *High Performance Computing in Science and Engineering '10*, DOI 10.1007/978-3-642-15748-6_7,

completion of a PhD thesis of PG [15]. Within this time it was possible to establish bounds for the lower as well as the upper Higgs boson masses using dynamical fermions obeying an exact lattice chiral symmetry. The final results were published in [15–17] and will be presented in Sect. 4.

Recently, we focused our interest on the possibility of a fourth generation of heavy quarks and leptons. Earlier analysis of electroweak precision measurements of the Z resonance peak stated that the number of fermion generations shall be three. This analysis though, was based on the assumption that the fourth generation neutrino is as light as the other three. An extension of the standard model of particle physics with heavy fermions, i.e. a heavy doublet in the quark sector, denoted by (t', b') as well as a heavy lepton doublet, denoted by (τ', ν_τ'), is not excluded [18]. Beyond the mere possibility of a heavy fermion generation, this extension has appealing properties. The extension of the standard model with a fourth generation of heavy quarks and leptons (**SM4**) can provide a sufficient source for CP violation in order to explain the observed asymmetry between matter and antimatter [18]. Furthermore, the heavy fermions have a significant impact on the running of the coupling constants such that a unification of the gauge couplings is possible without imposing super symmetry [19]. The electroweak precision data is well compatible with more than three generations of quarks and leptons [18]. Investigating such heavy fermions implies a strong Higgs-Yukawa coupling, which in turn necessitates a genuine non-perturbative study. We will present some first data on the Higgs boson mass bounds in the presence of heavy quarks.

Finally, we report on our progress on computing the resonance parameters of the standard model Higgs boson and address its decay properties. The standard model Higgs boson can decay into two weak gauge bosons $(W^{\pm}, Z)$ or a pair of fermion and anti-fermion $(\bar{\mathrm{t}}\mathrm{t}, \bar{\mathrm{b}}\mathrm{b})$. The Goldstone equivalence theorem states that the decay in to the weak gauge bosons can be determined by computing the decay of the Higgs boson into two Goldstone bosons and allows us to associate our computations with upcoming experimental results. Due to the fact that the Higgs boson can decay into lighter particles the computation of the Higgs boson mass by means of the two point correlation function is not anymore reliable. It has been shown [20, 21], that the volume dependence of two-particle energy states are connected to scattering phases as long as the energy levels are within the purely elastic region. The scattering phases will allow us to compute the resonance mass as well as its decay width.

2 The $\mathrm{SU}(2)_L \times \mathrm{SU}(2)_R$ Invariant Higgs-Yukawa Model

The model, we consider here, is a four-dimensional, chirally invariant $SU(2)_L \times SU(2)_R$ Higgs-Yukawa model discretized on a finite lattice with L_s lattice sites in space directions and L_t lattice sites in time direction. We set the lattice spacing a to unity throughout this paper. The model contains one four-component, real Higgs field Φ (equivalent to the complex doublet notation used in the standard model) and N_f degenerate fermion doublets represented by eight-component spinors $\psi^{(i)}$, $\bar{\psi}^{(i)}$, $i = 1, \ldots, N_f$ with the total action being decomposed into the Higgs action S_Φ, and the fermion action S_F. The partition function can hence be written as

$$Z = \int D\Phi\, D\psi\, D\overline{\psi}\ \exp\left(-S_\Phi[\Phi] - S_F[\Phi, \psi, \overline{\psi}]\right), \tag{1}$$

where N_f is the number of fermion doublets. It should be stressed here that *no gauge fields* are included within this model. The reason for neglecting the gauge fields is that we are interested in the Higgs and fermion masses here, which are dominated by the quartic self coupling and Yukawa coupling, whereas the coupling to the gauge bosons is strongly suppressed due to the renormalized gauge coupling constant being much smaller than the Yukawa coupling constant at the energy scale considered here which is given by the Higgs mass in this setup. Furthermore, the Goldstone equivalence theorem states, that the obtained results can be related to the standard electroweak model including gauge fields. Furthermore, we restrict ourselves to mass degenerate fermion doublets such that the Yukawa coupling of the scalar field to the fermions are equal.

The fermion action S_F is based on the Neuberger overlap operator $\mathcal{D}^{(ov)}$ [22] and can be written in terms of the fermionic matrix $\mathcal{M}$ according to

$$S_F = \sum_{i=1}^{N_f} \bar{\psi}^{(i)} \mathcal{M} \psi^{(i)}, \qquad \psi^{(i)} = \begin{pmatrix} t \\ b \end{pmatrix} \tag{2}$$

$$\mathcal{M} = \mathcal{D}^{(ov)} + y_N B[\Phi]\left(1 - \frac{1}{2\rho}\mathcal{D}^{(ov)}\right), \tag{3}$$

$$B[\Phi] = \Phi^\mu \left\{P_+ \theta^\dagger_\mu \mathbb{1}_2 + P_- \theta_\mu \mathbb{1}_2\right\} \tag{4}$$

$$\theta_\mu = (\mathbb{1}_2, -i\boldsymbol{\tau}) \tag{5}$$

$\boldsymbol{\tau}$ denotes the three Pauli matrices. The fermion matrix $\mathcal{M}$ describes the propagation of the fermion fields as well as their coupling to the scalar field ϕ.

The (doublet) Dirac operator $\mathcal{D}^{(ov)} = \hat{\mathcal{D}}^{(ov)} \otimes \hat{\mathcal{D}}^{(ov)}$ is given by the Neuberger overlap operator $\hat{\mathcal{D}}^{(ov)}$, which is related to the Wilson operator $\hat{\mathcal{D}}^{(W)} = \gamma^E_\mu \frac{1}{2}(\nabla^f_\mu + \nabla^b_\mu) - \frac{r}{2}\nabla^b_\mu \nabla^f_\mu$ by

$$\hat{\mathcal{D}}^{(ov)} = \rho\left\{1 + \frac{\hat{A}}{\sqrt{\hat{A}^\dagger \hat{A}}}\right\}, \qquad \hat{A} = \hat{\mathcal{D}}^{(W)} - \rho, \qquad 1 \leq \rho < 2r \tag{6}$$

with ∇^f_μ, ∇^b_μ denoting the forward and backward difference quotients and ρ is within its bounds a free parameter of the Neuberger Dirac operator. Note that in absence of gauge fields this kinetic part corresponds to the one of free fermions which will be exploited in the numerical construction of the overlap operator later. In particular, the eigenvalues and eigen vectors of $\hat{\mathcal{D}}^{(ov)}$ can be computed analytically. In momentum space with the allowed four-component momenta

$$p \in \mathcal{P} = \begin{cases} (-\pi, \pi]^{\otimes 4} & : \text{for } L_s, L_t = \infty \\ \{2\pi n / L_{s,t} \, : \, n \in \mathbb{N}_0, n < L_{s,t}\}^{\otimes 4} & : \text{for } L_s, L_t < \infty \end{cases} \tag{7}$$

the eigenvalues of the doublet operator $\mathcal{D}^{(ov)}$ are given by

$$\nu^{\epsilon}(p) = \rho + \rho \cdot \frac{\epsilon i \sqrt{\tilde{p}^2} + \frac{1}{2} r \hat{p}^2 - \rho}{\sqrt{\tilde{p}^2 + (\frac{1}{2} r \hat{p}^2 - \rho)^2}}, \quad \tilde{p}_\mu = \sin(p_\mu), \quad \hat{p}_\mu = 2 \sin\left(\frac{p_\mu}{2}\right), \tag{8}$$

and the corresponding eigen vectors of $\hat{\mathcal{D}}^{(ov)}$ are explicitly known to be the plane waves

$$\Psi_n^{p,\zeta\epsilon k} = e^{ip \cdot n} \cdot u^{\zeta\epsilon k}(p), \quad u^{\zeta\epsilon k}(p) = \sqrt{\frac{1}{2}} \begin{pmatrix} u^{\epsilon k}(p) \\ \zeta u^{\epsilon k}(p) \end{pmatrix}, \quad \zeta, \epsilon = \pm 1, k \in \{1, 2\} \tag{9}$$

with $u^{\epsilon k}(p)$ denoting the usual four-component spinor structure

$$u^{\epsilon k}(p) = \sqrt{\frac{1}{2}} \begin{pmatrix} \xi_k \\ \epsilon \frac{\tilde{p}^\mu \theta_\mu^\dagger}{\sqrt{\tilde{p}^2}} \xi_k \end{pmatrix} \text{ for } \tilde{p} \neq 0 \quad \text{and} \quad u^{\epsilon k}(p) = \sqrt{\frac{1}{2}} \begin{pmatrix} \xi_k \\ \epsilon \xi_k \end{pmatrix} \text{ for } \tilde{p} = 0. \tag{10}$$

Here $\xi_k \in \mathbb{C}^2$ are two orthonormal vectors and the four component quaternionic vector θ_μ is defined as $\theta_\mu = (\mathbb{1}, -i\boldsymbol{\tau})$.

The model then obeys an exact, but lattice modified, global $SU(2)_L \times SU(2)_R$ chiral symmetry according to

$$\begin{aligned} \psi \rightarrow \Omega_L \hat{P}_- \psi + \Omega_R \hat{P}_+ \psi \quad & \bar{\psi} \rightarrow \bar{\psi} P_+ \Omega_L^\dagger + \bar{\psi} P_- \Omega_R^\dagger \\ \phi \rightarrow \Omega_R \phi \Omega_L^\dagger \quad & \phi^\dagger \rightarrow \Omega_L \phi^\dagger \Omega_R^\dagger \end{aligned}$$

with $\Omega_L, \Omega_R \in \mathrm{SU}(2)$ recovering the chiral symmetry in the continuum limit [8].

Finally, the lattice Higgs action is given by the usual lattice notation

$$S_\Phi = -\kappa_N \sum_{n,\mu} \Phi_n^\dagger \left[\Phi_{n+\hat{\mu}} + \Phi_{n-\hat{\mu}}\right] + \sum_n \Phi_n^\dagger \Phi_n + \lambda_N \sum_n \left(\Phi_n^\dagger \Phi_n - N_f\right)^2 \tag{11}$$

with the only peculiarity that the fermion generation number N_f appears in the quartic coupling term which was a convenient convention for the large N_f analysis. However, this version of the lattice Higgs action is equivalent to the usual continuum notation

$$S_\varphi = \sum_n \left\{ \frac{1}{2} \left(\nabla_\mu^f \varphi\right)_n^\dagger \nabla_\mu^f \varphi_n + \frac{1}{2} m_0^2 \varphi_n^\dagger \varphi_n + \lambda_0 \left(\varphi_n^\dagger \varphi_n\right)^2 \right\}, \tag{12}$$

with the bare mass m_0 and the bare quartic coupling constant λ_0 corresponding to the continuum notation. The connection is established through a rescaling of the Higgs field and the involved coupling constants according to

$$\begin{aligned} \varphi_x &= \sqrt{2\kappa_N} \begin{pmatrix} \Phi_x^2 + i\Phi_x^1 \\ \Phi_x^0 - i\Phi_x^3 \end{pmatrix} \\ \lambda_0 &= \frac{\lambda_N}{4\kappa_N^2} \\ m_0^2 &= \frac{1 - 2N_f\lambda_N - 8\kappa_N}{\kappa_N} \\ y_0 &= \frac{y_N}{\sqrt{2\kappa_N}}. \end{aligned}$$

3 Implementation, Performance, and Parallelization

The simulation algorithm and the software components which were essential to produce the space time configurations were completed in 2009. The following chapter will describe the numerical challenge and the implemented algorithm. We will refer to the typical performance achieved on the XC4000 at the computing center in Karlsruhe.

The first step towards a numerical treatment of the considered Higgs-Yukawa model is to integrate out the fermionic degrees of freedom leading to the appearance of the determinant of the fermionic matrix $\mathcal{M}$ in the partition function

$$Z = \int D\Phi \, [\det(\mathcal{M})]^{N_f} \cdot \exp(-S_\Phi[\Phi]) \,. \tag{13}$$

where the fermionic matrix $\mathcal{M}$ was given in (3).

We remark here, that $\mathcal{M}$ is neither hermitean nor does it possess an orthogonal eigen basis. The fermion matrix $\mathcal{M}$ is, in contrast to QCD, not even γ_5-hermitean.

For the numerical treatment of the fermion dynamics the determinant in (13) is rewritten in terms of Gaussian integrations over N_f complex, so-called pseudo-fermionic fields ω_i, $i = 1, \ldots, N_f$. This step, however, requires the positivity and hermiticity of the underlying fermion matrix, which can be guaranteed by rewriting the partition function according to

$$Z = \int D\Phi \, [\text{sign} \det(\mathcal{M})]^{N_f} \cdot \left[\det\left(\mathcal{M}\mathcal{M}^\dagger\right)\right]^{\frac{N_f}{2}} \cdot \exp(-S_\Phi[\Phi]) \tag{14}$$

$$\begin{aligned} &= \int D\Phi \, D\omega \, D\omega^\dagger \, [\text{sign} \det(\mathcal{M})]^{N_f} \\ &\times \exp\left(-S_\Phi[\Phi] - \sum_{i=1}^{N_f} \frac{1}{2}\omega_i^\dagger \left[\mathcal{M}\mathcal{M}^\dagger\right]^{-\frac{1}{2}} \omega_i\right) \end{aligned} \tag{15}$$

leading to the numerically demanding problem of calculating the inverse square root of $\mathcal{M}\mathcal{M}^\dagger$. Note that (14) does not involve a general complex phase of the determinant but only its sign since the determinant of $\mathcal{M}$ is real as one can show.

In this reformulation the model can finally be numerically evaluated by using Monte-Carlo techniques. Here we have implemented a PHMC-algorithm (see [23] for a general review), which evaluates the extremely high-dimensional, i.e. $4 \cdot 16^{N_f} \cdot L_s^{3(N_f+1)} \cdot L_t^{N_f+1}$*-dimensional*, functional integral in (15) stochastically and solves the problem of determining the inverse square root of $\mathcal{M}\mathcal{M}^\dagger$ acting on a vector ω by applying polynomial approximations $P(\mathcal{M}\mathcal{M}^\dagger)\omega$.

The basic numerical task for performing this Monte-Carlo simulation is therefore the computation of $\mathcal{M}\mathcal{M}^\dagger\omega$. However, since $\mathcal{M}$ is a $8V \times 8V$ matrix for a lattice with volume $V = L_s^3 \times L_t$, this is a very time-consuming operation. In fact, $\mathcal{M}$ is too large to be held in computer memory. Instead, only the prescription how $\mathcal{M}$ acts on a given vector is implemented in the software code.

At this point we exploit the fact that there are no gauge fields included within our model. The Dirac operator $\mathcal{D}^{(ov)}$ is thus block-diagonal in momentum space. The fermion matrix $\mathcal{M}$ is therefore composed of the coupling matrix ϕ, being diagonal in position space, and the Dirac operator, being diagonal in momentum space. In our approach we use a Fast Fourier Transform for the computation of $\mathcal{M}\mathcal{M}^\dagger\omega$ to switch between position and momentum representations, such that all operator applications can be trivially performed due to their actual block-diagonal structure. This is particularly advantageous for the overlap operator, since the standard construction of this operator would be based on much more demanding computations and moreover would be constructed only approximately. A second advantage of this approach is that the applied Dirac operator can easily be replaced by other operators simply by adopting the corresponding eigenvalues.

The current program, which is running on the XC2 has been improved in many ways, including algorithmic as well as technical issues, which will be illustrated below. Table 1 shows the typical performance on the XC2 and a comparison to the performance achieved in 2008. The auto-correlation times τ are calculated according to the Γ-strategy [24] and refer to the vacuum expectation value of the scalar field.

When examining the differences in the running speeds in Table 1 one notices that the speed does not exactly scale proportional to the volume. Although a Fast Fourier Transform belongs in general to the complexity class $V \cdot \log(V)$ one needs to keep in mind that a FFT runs only optimal on lattices with sizes being a power of two. If other prime factors appear in the lattice size, the algorithm looses its efficiency. This is clearly seen on the $20^3 \times 32$ lattice containing a 5 as prime factor.

This observation also reveals the fact that the biggest portion of the used computer time is spent on performing the Fast Fourier Transforms. The im-

Table 1. Typical speeds, auto-correlation times τ for the vacuum expectation value of the scalar field and configuration file sizes for typical lattice sizes. The main improvement which enables us to perform calculations on lattice sizes up to 40^4, was to exploit the multi processor environment and algorithmic efforts which helped to reduce the maximum polynom degree down to 20

Lattice size $L_s^3 \times L_t$	Speed [generated configurations per day]	Auto-correlation time τ for Higgs vev	File size for single configuration
Typical performance in 2008:			
$12^3 \times 32$	700 to 900	5 to 6	1.7 MBytes
$16^3 \times 32$	300 to 400	5 to 6	4.0 MBytes
$20^3 \times 32$	80 to 120	5 to 6	7.8 MBytes
Current typical performance:			
$32^3 \times 32$	80 to 150	3 to 5	32.0 MBytes
$40^3 \times 40$	80 [(1)]	3 to 5	78.0 MBytes

(1) This lattice has been computed on the XC2-FAT-NODE, while the other numbers refer to computations on the THIN-NODES. Due to the high demands in memory, the 40^4 lattices can only run on the FAT-NODES.

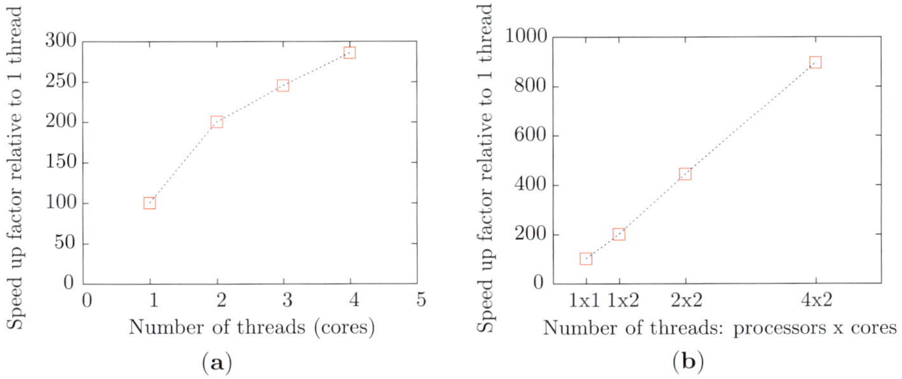

Fig. 1. a Scaling behavior of FFTW using several threads (cores) on a single processor. **b** Scaling performance using one or two cores and up to four processors. The number of threads is given as: $n_p \times n_c$ where n_p denotes the number of processors and n_c denotes the number of cores. Both performance plots were performed on a 32^4 lattice

plemented parallelization techniques therefore mainly focus on exploiting the available multi processor system for calculating the FFT. Figure 1a shows a straight forward use of the multi-threaded FFTW package. The reason for this bad scaling behavior lies in the fact, that FFTs are mostly limited by mem-

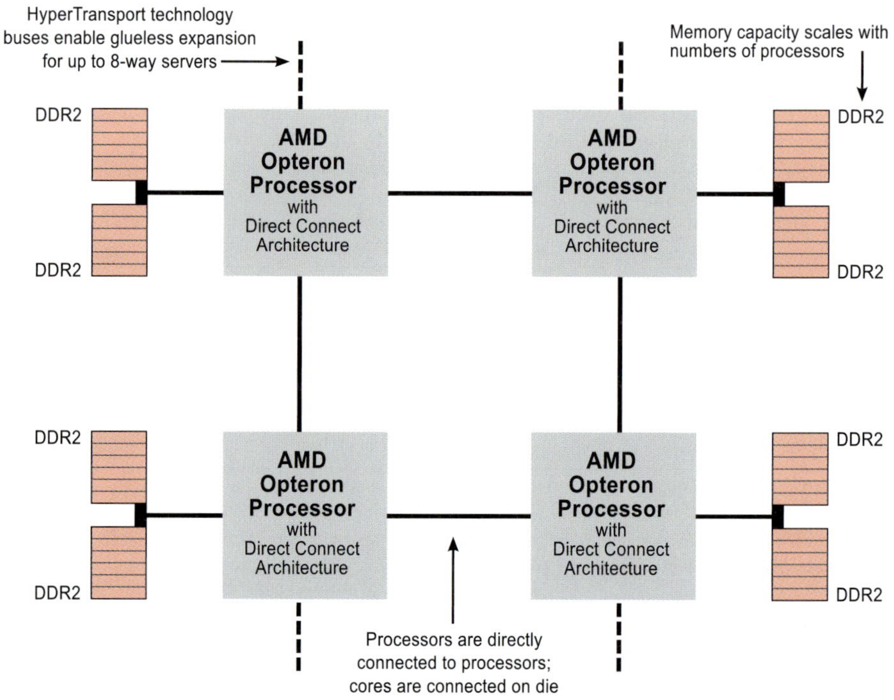

Fig. 2. ccNUMA architecture. Image taken from *4P Server Comparison* at http://www.amd.com/us-en/assets/content_type/DownloadableAssets/

ory access speeds, i.e. bus speeds, and not by CPU power. Figure 2 shows the *ccNUMA-architecture* underlying an AMD Opteron system. Each processor is connected to its neighboring unit through a hyper transport link which has a throughput of 4 GB/s. The figure also shows that each processor has its own memory segment. A thread running on one node but accessing memory that is resident on a different node, will have to make use of the hyper transport link. The idea was to align processes to memory segments where the memory had been allocated. The eight components of the fermionic vector can then be distributed to each of the four segments, such that each thread can access its data through its own bus leading to a very good scaling behavior. Figure 1b shows the performance in scaling with respect to the number of threads by aligning the processes to a specific memory segment.

Currently, we are running 32^4 and 40^4 lattices on the *fat* nodes having 8 cores each. The simulation program will need approximately 20 GB of main memory on each node for the 40^4-lattice. The fat nodes with their 128 GB of memory are therefor perfectly suited for our computations. There are about 200 Simulations which are stored in archive space and which occupy 8.4 TB.

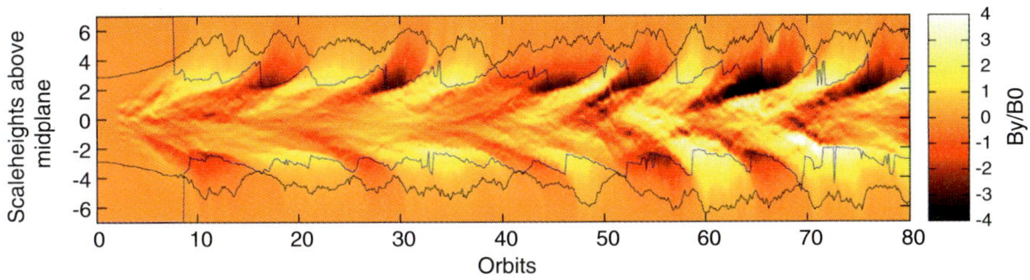

(a) Azimuthal magnetic field component. The green and blue curves denote the location of the magnetosphere and the photosphere, respectively.

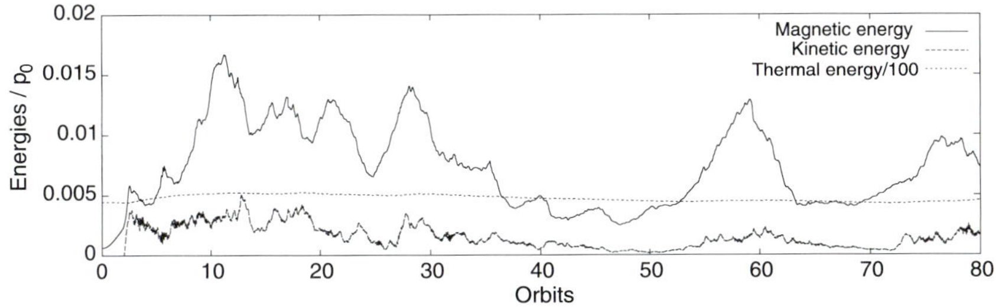

(b) Volume averaged magnetic, turbulent kinetic and thermal energy (note the different scaling for the thermal energy).

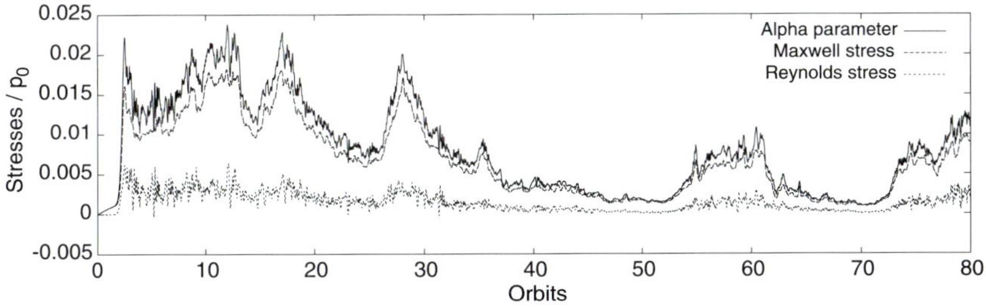

(c) Volume averaged turbulent stresses normalised to the mean gas pressure.

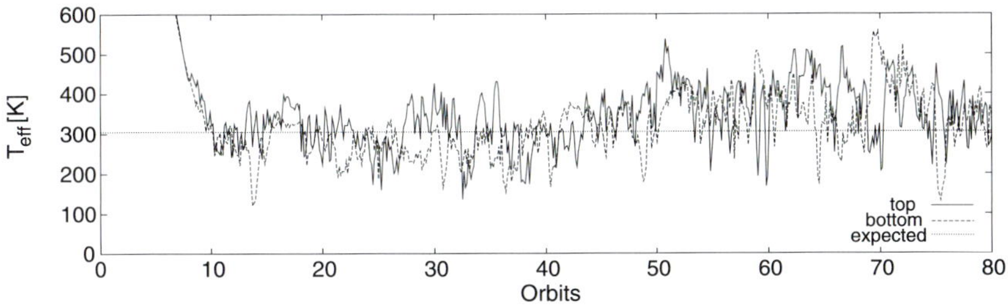

(d) Photospheric temperature at top and bottom boundary.

Fig. 2. Time history of various physical variables for model R7 illustrating the changes in turbulent activity and the correlations between stresses, magnetic field strength and luminosity of the disc. For further explanation see the text

In Fig. 2a we also plot the location of the photosphere (defined as the location where the optical depth equals unity) and the position of the magnetosphere (defined as the location where the magnetic pressure starts to exceed the gas pressure). During active phases, the photosphere is pushed outwards because of the expansion of the disc, and at the same time the corona is pushed inwards due to the higher magnetisation caused by stronger turbulence, which means that for most of the time the discs corona is for a large part optically thick.

We plot the time history of the energies (thermal energy, turbulent kinetic energy and magnetic energy) in Fig. 2b. During active periods the magnetic energy is much larger then during a quiet period. The turbulent kinetic energy (that is the total kinetic energy minus the kinetic energy of the background shear flow) is also larger during active phases, but here the variations are significantly smaller than the ones of the magnetic energy. The long-term changes in the thermal energy are much smaller than the up-and-down variations, which shows that we have indeed reached thermal equilibrium.

The turbulent stresses that determine the angular momentum transport in the disc are shown in Fig. 2c. The angular momentum transport is dominated by the Maxwell stress which is about four to five times larger than the Reynolds stress. This ratio is consistent with previous results for stratified boxes [for example [26]]. From a comparison between Fig. 2b and c it is evident that the magnetic energy and the alpha parameter are strongly correlated, with the time-dependence of the alpha paramater following that of the magnetic field. Indeed, it is known that in shearing-box calculations, the time-dependence of the alpha parameter can be fitted by a formula of the form $\langle \alpha \rangle \propto \langle B^2 \rangle + \text{const.}$ [see [5]].

Finally, we may ask how the observational appearance of the disc will change due to phases of varying turbulent activity. To address this question, we plot on Fig. 2d the effective temperature (i.e. the temperature at the location where the optical depth equals unity). A comparison of Fig. 2d with Fig. 2a and Fig. 2c, shows that in general a high degree of turbulent activity in one half of the disc leads to a higher flux of radiation through the corresponding boundary. The increased flux balances the increased turbulent heating, so that the thermal energy content in the disc remains approximately constant. The lag between photospheric temperature and stress as inferred from the cross-correlation between these two quantities is about 2.5 orbits. The time-scale for photons to diffuse away from the midplane is also 2.5 orbits, so these two time-scales are nicely consistent.

4.3 Vertical Structure

One of the most interesting quantity that our models are able to provide is the self-consistently calculated temperature profile, which has not been calculated before for a protoplanetary disc. Plots of the mean temperature profiles for the radiative models can be found in Fig. 3. Near the disc midplane, the

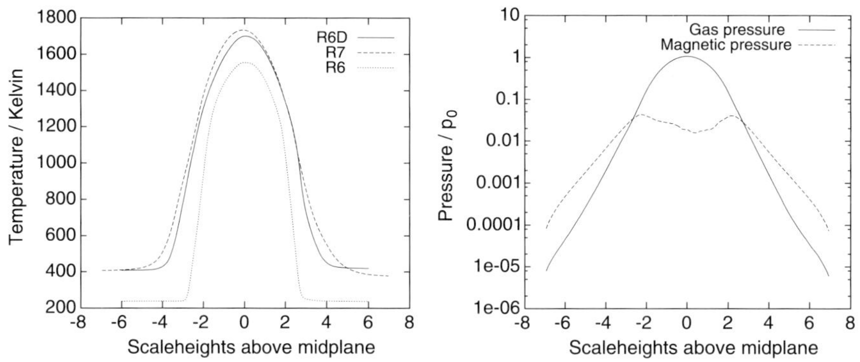

Fig. 3. *Left:* Temperature profiles for various radiative models. *Right:* Gas pressure and magnetic pressure for model R7

temperature profile resembles an inverted parabola, while in the optically thin upper layers it becomes flat. For the small resolution model R6 the temperatures are lower than in the higher resolution models, since this model is not numerically converged (see the discussion in Sect. 4.4). The higher resolved models yield similar temperatures, indicating convergence with respect to the heating rates.

For our model setup, the forces arising from radiation pressure are small compared to the forces arising from gas pressure, magnetic pressure and magnetic stresses. In the right panel of Fig. 3 we plot gas and magnetic pressures as a function of height. In the gas pressure dominated midplane region, the magnetic pressure is roughly constant, while in the magnetically dominated upper layers it decreases sharply.

The turbulence constantly creates new magnetic field and heats the disc gas, balancing the radiative and resistive losses and preventing the collapse of the disc. The strength of the turbulence can be measured in terms of the alpha parameter, that is the total stress divided by the pressure. In Fig. 4, left panel, the alpha parameter is plotted as a function of height for different models. We encounter a similar picture as in Fig. 3 for the magnetic pressure: Inside the gas-pressure dominated midplane region the stress profiles are roughly constant, while outside this region they decline exponentially. This means that most of the angular momentum transport happens in the midplane region.

We now look at the velocity distribution in the disc. In Fig. 4, right panel, we plot the turbulent velocity normalised to the gas sound speed. In the midplane region, the turbulence is subsonic, while in the corona it becomes highly supersonic, exceeding Mach 5 at the boundaries. For comparison with other works we also plot the velocity normalised to the initial isothermal sound speed at the midplane and the velocity normalised to the total sound speed c_{tot}, where $c_{\text{tot}}^2 = (p + B^2/8\pi)/\rho$. Our results are in agreement with the isothermal simulations done by [20] who report Mach numbers of about two

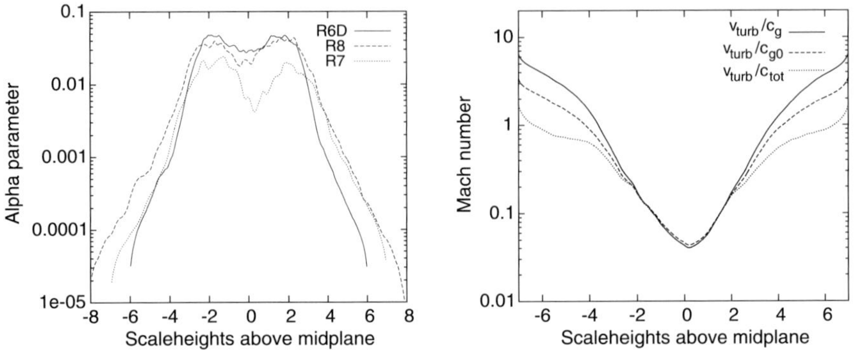

Fig. 4. *Left:* Stress profiles for various models. *Right:* Mach numbers of the turbulent velocities for model R7, where c_g denotes the gas sound speed, c_{g0} is the initial gas sound speed at the mid-plane and c_{tot} is the total sound speed (including gas and magnetic pressure)

and with HKS, who report Mach numbers (with respect to the total sound speed) between one and two.

In observations, the turbulence will show itself in the form of a broadening of spectral lines due to the velocity dispersion induced by it. At the mean location of photosphere the turbulent velocities are about Mach 2-3 which implies a significant effect on the line widths. The detection of CO overtone emission in young stellar objects (YSOs) provides a useful diagnostic tool to detect turbulent line broadening. The reason for this is that the near overlap of CO transitions near the $v = 2 - 0$ band allows a separation of the local broadening (e.g. turbulence) from the macrobroadening (caused for example by a disc wind). In the case of several YSOs, there is indeed substantial empirical evidence for supersonic turbulent line broadening of the magnitude that we find in our simulations [7, 12, 21].

4.4 Turbulent Saturation Level

As has already been remarked in the introduction, in numerical simulations of MRI turbulence, the turbulent saturation level is influenced by a number of numerical factors. By performing a suite of simulations with different box sizes an at different resolutions, we are able to gauge the strength of the influence of these numerical parameters.

In Table 2 we have tabulated the values of Reynolds stress, Maxwell stress and the alpha parameter for the models that were simulated. When increasing the resolution from $32 \times 64 \times 256$ (models I6 and R6), to the double resolution of $64 \times 128 \times 512$ (models I6D and R6D), the values of the stresses increase considerably. A similar increase of the stresses is observed also when only the resolution in the vertical direction is increased (models I5, I7, R7 and R8), which is analogous to what has been found in the radiative simulations of

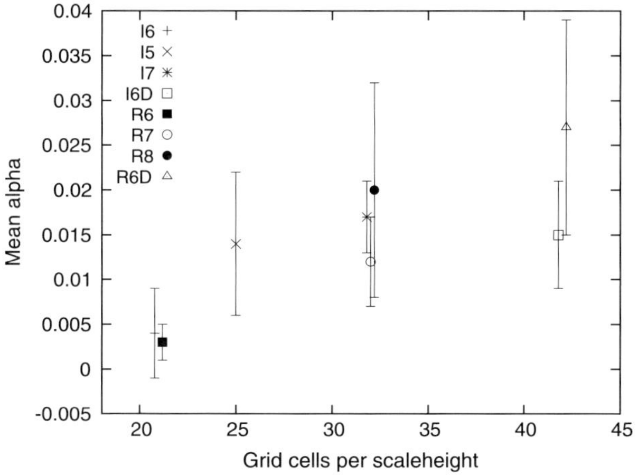

Fig. 5. Time-averaged value of the alpha parameter as a function of the number of grid cells in the *vertical* direction per scale height for all models. The "error bars" correspond to the deviation from the mean value

SKH. A look at Fig. 4, left panel, suggests that it is especially the region near the midplane that is not sufficiently resolved, since the stress drops there noticeably, while in the better resolved models it does not.

The results depicted in Fig. 2 suggest that once a resolution of about 30 grid cells per scaleheight in the vertical direction has been achieved, the results will no longer significantly depend on the numerical resolution. All models except `I6` and `R6` are consistent with an α-value of about $\alpha \sim 0.015 - 0.02$, with the isothermal models yielding somewhat lower stresses that are closer to the lower end of $\alpha \sim 0.015$, while the stresses found in the radiative models are grouped around the value $\alpha \sim 0.02$. No tendency is found for the turbulent saturation level to change with respect to neither the vertical box size nor the resolution in the radial and azimuthal directions.

As a consequence of the lower stress, the heating in the low-resolution model `R6` is also smaller than compared to the other radiative models, leading to significantly lower temperatures in this model. Apart from this, the temperature profile for model `R6` is qualitatively similar to the other models.

As has already remarked before, the heating does intrinsically depend on numerical effects. However, as can be seen from Fig. 3, left panel, the temperature profiles for the better resolved models (`R7`, `R8` and `R6D`) are very similar, suggesting that the heating rates are also converged.

5 Ressources Used

Table 3 provides an overview of the typical resources that were needed for some selected runs. The isothermal runs (starting with letter `I`) need much

Table 3. Computational resources needed by the simulations reported in the present paper

Name	Runtime	CPU hours	Processors	Memory	Data produced
`I6`	3 days	1 200	16	24 MB	20 GB
`I7`	6 days	4 600	32	48 MB	20 GB
`I6D`	11 days	34 000	128	192 MB	20 GB
`R6`	18 days	6 900	16	26 MB	21 GB
`R7`	20 days	15 300	32	52 MB	32 GB
`R6D`	31 days	95 200	128	208 MB	98 GB

less CPU time as compared to the radiative runs (starting with letter `R`) since they do not include the expensive radiation transport step.

6 Conclusion & Outlook

We have performed 3D radiative MHD simulations of MRI-turbulent protoplanetary discs. Including radiation transport allows us to obtain a self-consistent picture of the vertical structure which results from a dynamic balance between turbulent heating and radiative cooling. Such simulations do thus contain a greater level of realism as compared to previous isothermal simulations. Furthermore, radiative simulations are an important step towards bringing numerical simulations into contact with observations.

We are currently pursuing the following two lines of research: On the one hand we add new physics to our existing radiative models by including a realistic physical resistivity which is calculated by solving a simplified chemical network. On the other hand we extend the local shearing box setup to a global scale by switching to cylindrical/spherical coordinates. Example snapshots

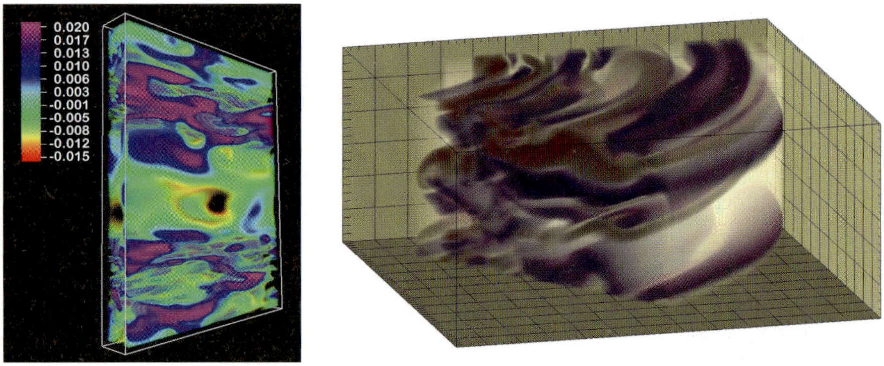

Fig. 6. *Left:* Initial stage of a non-ideal calculation including a chemical network. In the middle of the disc, a MRI-inactive "dead zone" starts to form. *Right:* Development of the magnetorotational instability in a cylindrical disc

from such simulations are shown in Fig. 6. The ultimate aim is to set up a global, radiative protoplanetary disc model with a self-consistently calculated dead zone.

References

1. S.A. Balbus. Enhanced Angular Momentum Transport in Accretion Disks. *Ann. Rev. A& A*, 41:555–597, 2003.
2. S.A. Balbus. Magnetohydrodynamics of Protostellar Disks. *ArXiv e-prints*, June 2009.
3. Steven A. Balbus and John F. Hawley. Instability, turbulence, and enhanced transport in accretion disks. *Rev. Mod. Phys.*, 70(1):1–53, Jan 1998.
4. D.S. Balsara and C. Meyer. Evaluating the Magnetorotational Instability's Dependence on Numerical Algorithms and Resolution. *ArXiv e-prints*, February 2010.
5. A Brandenburg. Turbulent protostellar discs. *Physica Scripta*, page 014016, 2008.
6. S.M. Cisowski and L.L. Hood. The RELICT magnetism of meteorites. In C.P. Sonett, M.S. Giampapa, & M.S. Matthews, editor, *The Sun in Time*, pages 761–784, 1991.
7. J.S. Carr, A.T. Tokunaga, and J. Najita. Hot H2O Emission and Evidence for Turbulence in the Disk of a Young Star. *Astrophysical J.*, 603:213–220, March 2004.
8. S.W. Davis, J.M. Stone, and M.E. Pessah. Sustained Magnetorotational Turbulence in Local Simulations of Stratified Disks with Zero Net Magnetic Flux. *ArXiv e-prints*, September 2009.
9. M. Flaig, R. Kissmann, and W. Kley. Growth of the MRI in accretion discs - the influence of radiation transport. *Monthly Notices Roy. Astr. Soc.*, 394:1887–1896, April 2009.
10. S. Fromang and J. Papaloizou. MHD simulations of the magnetorotational instability in a shearing box with zero net flux. *Astron. & Astrophys.*, 476(3):1113–1122, 2007.
11. S. Fromang, J. Papaloizou, G. Lesur, and T. Heinemann. MHD simulations of the magnetorotational instability in a shearing box with zero net flux. II. The issue of transport coefficients. *Astron. & Astrophys.*, 476:1123–1132, December 2007.
12. S.D. Hügelmeyer. *Multidimensional radiative transfer in circumstellar disks.* PhD thesis, Georg-August-Universität Göttingen, Göttingen, 2009.
13. L. Hartmann, N. Calvet, E. Gullbring, and P. D'Alessio. Accretion and the Evolution of T Tauri Disks. *Astrophysical J.*, 495:385–+, March 1998.
14. J.F. Hawley, C.F. Gammie, and S.A. Balbus. Local Three-dimensional Magnetohydrodynamic Simulations of Accretion Disks. *Astrophysical J.*, 440:742–+, February 1995.
15. M. Ilgner and R.P. Nelson. On the ionisation fraction in protoplanetary disks. The effect of turbulent mixing on gas-phase chemistry. *Astron. & Astrophys.*, 445:223–232, January 2006.

16. A. Johansen, A. Youdin, and H. Klahr. Zonal Flows and Long-lived Axisymmetric Pressure Bumps in Magnetorotational Turbulence. *Astrophysical J.*, 697:1269–1289, June 2009.
17. R. Kissmann, J. Kleimann, H. Fichtner, and R Grauer. Local turbulence simulations for the multiphase ISM. *Monthly Notices Roy. Astr. Soc.*, 391:1577–1588, December 2008.
18. Alexander Kurganov, Sebastian Noelle, and Guergana Petrova. Semidiscrete central-upwind schemes for hyperbolic conservation laws and Hamilton–Jacobi equations. *SIAM J. Sci. Comput.*, 23(3):707–740, 2001.
19. A.R. King, J.E. Pringle, and M. Livio. Accretion disc viscosity: how big is alpha? *Monthly Notices Roy. Astr. Soc.*, 376:1740–1746, April 2007.
20. K.A. Miller and J.M. Stone. The Formation and Structure of a Strongly Magnetized Corona above a Weakly Magnetized Accretion Disk. *Astrophysical J.*, 534:398–419, May 2000.
21. J. Najita, J.S. Carr, A.E. Glassgold, F.H. Shu, and A.T. Tokunaga. Kinematic Diagnostics of Disks around Young Stars: CO Overtone Emission from WL 16 and 1548C27. *Astrophysical J.*, 462:919–+, May 1996.
22. M. Oppenheimer and A. Dalgarno. The Fractional Ionization in Dense Interstellar Clouds. *Astrophysical J.*, 192:29–32, August 1974.
23. J. Pomoell, R. Vainio, and R. Kissmann. MHD Modeling of Coronal Large-Amplitude Waves Related to CME Lift-off. *Sol. Phys.*, 253:249–261, December 2008.
24. J.B. Simon, J.F. Hawley, and K. Beckwith. Simulations of Magnetorotational Turbulence with a Higher-Order Godunov Scheme. *Astrophysical J.*, 690:974–997, January 2009.
25. A. Sicilia-Aguilar, L. Hartmann, N. Calvet, S.T. Megeath, J. Muzerolle, L. Allen, P. D'Alessio, B. Merín, J. Stauffer, E. Young, and C. Lada. Disk Evolution in Cep OB2: Results from the Spitzer Space Telescope. *Astrophysical J.*, 638:897–919, February 2006.
26. J.M. Stone, J.F. Hawley, C.F. Gammie, and S.A. Balbus. Three-dimensional Magnetohydrodynamical Simulations of Vertically Stratified Accretion Disks. *Astrophysical J.*, 463:656–+, June 1996.
27. J.-M. Shi, J.H. Krolik, and S. Hirose. What Is The Numerically Converged Amplitude of MHD Turbulence in Stratified Shearing Boxes? *ArXiv e-prints*, September 2009.
28. N.I. Shakura and R.A. Syunyaev. Black holes in binary systems. Observational appearance. *Astron. & Astrophys.*, 24:337–355, 1973.
29. N.J. Turner, T. Sano, and N. Dziourkevitch. Turbulent Mixing and the Dead Zone in Protostellar Disks. *Astrophysical J.*, 659:729–737, April 2007.

Solid State Physics

Prof. Dr. Werner Hanke

Institut für Theoretische Physik und Astrophysik, Universität Würzburg,
Am Hubland, 97074 Würzburg, Germany

The solid state projects in 2009 at the HLRS can be categorized essentially into 3 fields: the first topic is concerned with first-principle, typically density functional theory (DFT) type of calculations. They range from organic metal interface physics over the ab initio characterization of the electronic properties of quantum dots to the optical response of semi-conductor nanowires. The second topic is somewhat related to this first one, namely laser ablation of metals and the third topic is concerned with transport properties of strongly interacting quantum systems.

Starting with the first topic ab initio calculation of electronic properties, there are two projects carried out by the F. Bechstedt group in Jena. The first one with B. Höffling et al. is concerned with interface physics, namely the adsorption of organic molecules on crystal surfaces, especially to metal/organic interfaces. These interface systems are of high technological importance, for example, for the development of molecular electronics. Theoretical ab initio investigation using high performance computers, such as, in particular, the HLRS, have become a indispensable tool for the investigation of the corresponding adsorption processes. The complexity of bonding mechanisms together with a huge number of atoms and configurations creates an enormous demand for massively parallel machines and codes, that run efficiently on those supercomputing machines. These simulations, like the example of the Jena group, can result in atomistic bonding models on top of which one proceeds with calculations of spectroscopic properties, which, in turn, maybe measured experimentally.

Also the second topic by the Jena group has potential technological interest, namely as light sources in the mid-infrared spectral region. It is concerned with the ab initio study of electronic properties of a specific quantum dot (PbTe) embedded in a specific matrix (CdTe). This and related systems are crucial for many applications, e.g. for gas sensors or medical diagnostic systems. In particular, these arrays of embedded quantum dots show quantum confinement effects for electrons and holes, and, thus, are promising candidates for light-emitting diodes and lasers. These systems and their electronic properties have been studied in the Jena-project, using state of the art DFT. One challenge in these transition-metal quantum dot systems embedded also in a transition metal matrix is the inclusion of d-states. Furthermore, in the case of the description of heavy elements, like Pb, the inclusion of relativistic effects such as the spin-orbit coupling is needed. A nice result of the Jena study is that the resulting spatial separation of electrons and holes inside the nano-crystal region leads to a quantum confined Stark effect.

The Si(111)-In nano wire optical response, again using ab initio calculations, has been studied by W.G. Schmidt and a group of co-workers of the University of Paderborn. The main idea of this work is to use the first-principle type of calculation of optical response as a type of sensor to detect the ground-state geometry of an important prototype quasi one-dimensional electronic system. A project related to a complicated process of removing material with very intensive pulse laser radiation, namely laser ablation of metals, has been carried out with the help of the HLRS in the Institute for Physics, University of Stuttgart, by J. Roth et al. The main difficulty concerning the laser ablation is the time scales of the processes which are involved and which are in the range from a several hundred femto-seconds (laser pulse) to a few pico-seconds (electron-lattice-relaxation time) up to many nano-seconds (e.g. cooling of the sample). These different time scales can most effectively be accounted for using molecular dynamics (md) simulations, which are the method of choice to get a deeper insight into the issue. The results, especially for aluminum, clearly demonstrate that md-calculations are not limited by memory, but by the speed of processes and the communication network. Future efforts to make progress have, therefore, to be directed towards these issues.

The Karlsruhe group around P. Schmitt-Eckert has already a long standing tradition at the HLRS, performing simulation of transport properties of strongly interacting quantum systems. Somewhat more specifically, they typically apply density matrix renormalization group methods (DMRG) to study transport properties of quantum devices attached to metallic leads. The probably most important result of this study is, that due to the HLRS simulation insights have been gained in this topic at a fundamental level. The results show that a simple picture of non-interacting electrons passing through the transport region, "one after the other", gives a wrong result compared to the realistic one, including electron-electron interactions. In this latter case, the Karlsruhe group proposed a type of Kondo-effect as the explanation for remarkable line shaped effects observed.

The project by S. Körbel and Ch. Elsässer from the Fraunhofer Institute for Mechanics and Materials in Freiburg is concerned with density functional simulation of piezo-electric devices, like actors and sensors. Here, a lead-free ferroelectric potassium-sodium compound is studied using up-to-date density functional schemes. The density functional calculations of this group can be used as a database for optimizing empirical interatomic potentials, which can be used in future studies of complex lattice structure behavior (domain wall motion) in e.g. Cu-doped $KNbO_3$ systems.

Organic-Metal Interface: Adsorption of Cysteine on Au(110) from First Principles

B. Höffling, F. Ortmann, K. Hannewald, and F. Bechstedt

European Theoretical Spectroscopy Facility (ETSF) and Institut für Festkörpertheorie und -optik, Friedrich-Schiller-Universität Jena, Max-Wien-Platz 1, 07743 Jena, Germany, benjamin.hoeffling@uni-jena.de

Summary. We present *ab initio* studies of the adsorption of the amino acid cysteine on the Au(110) surface. We perform density functional theory calculations using the repeated-slab supercell method to investigate the molecule-surface interaction which is driven by two functional groups: the deprotonized thiolate head group and the amino group. The interaction of these functional groups with the surface is studied analyzing bonding site, bonding energy, charge redistribution, and changes in the density of states for single bond molecules registering to the surface via only one of the functional groups. For the Au-amino bond we find that positions close to the top-Au-site are energetically favorable, leading to strong bonds that are largely electrostatic in nature. The covalent Au-thiolate bond is strongest for a bonding position at bridge and off-bridge sites and the bonding energy is found to be very sensitive to changes in the bonding geometry.

1 Introduction

In recent years, much attention has been devoted to the adsorption of organic molecules on crystal surfaces, especially to metal/organic interfaces [1–9], which are of particular interest for the development of molecular electronics [10–12]. The existence of different functional groups leads to numerous practical configurations for the organic functionalization of the substrates, each of which is characterized by a unique electronic signature as evident from single molecule conductance experiments [13] or from reflectance anisotropy spectroscopy [14]. Moreover, the underlying substrate gives rise to additional degrees of freedom, which may result in a change of the reconstruction upon adsorption [15].

Theoretical *ab-initio* investigations using high performance computers have become an indispensable tool for the investigation of adsorption processes. The complexity of bonding mechanisms together with a huge number of atoms and configurations creates a high demand for massively parallel machines and codes that run efficiently on those machines. These simulations

W.E. Nagel et al. (eds.), *High Performance Computing in Science and Engineering '10*, DOI 10.1007/978-3-642-15748-6_9,

result in atomistic bonding models on top of which one proceeds with calculations of spectroscopic properties which, in turn, may be measured experimentally.

In researching surface functionalization, one resorts to a large pool of functional groups such as the amino group which, for instance, allows for modifications of the metal work functions. From this viewpoint, cysteine is of particular interest. Its interaction with metal surfaces has been the subject of intense investigations both experimental [14–19] and theoretical [20–23]. Since cysteine not only contains a -SH thiol head group, but also a NH_2 amino group, its adsorption behavior on Au surfaces is rather complex.

While the bonding of cysteine on Au(111) has been investigated in terms of bonding configuration and changes in the density of states (DOS) [20, 21], prior to our work [22, 23] only rudimentary studies have been conducted for the Au(110) surface. Especially the possible contribution of a bonding via the amino group and its interplay with the thiolate bond in the formation of a flat or more vertical adsorption geometry needs a deeper understanding. The focus of this article lies on the detailed analysis of the properties of the molecule-substrate bonds.

After the introduction of the methodology in Sect. 2, we present the results in Sect. 3. We study the Au-thiolate bond and investigate it in terms of preferred bonding site, influence of the bond geometry, changes in the density of state and charge redistribution in Sect. 3.1. In Sect. 3.2 we investigate the amino-gold interaction using the same tools. Finally, in Sect. 4, we conclude with a short summary and give an outlook.

2 Computational Methods

The calculations are performed using the *Vienna ab initio simulation package* (VASP) [24, 25] within the framework of density functional theory (DFT) in the generalized gradient approximation (GGA) for exchange and correlation (XC) as parametrized by Perdew and Wang [26, 27]. GGA treatment has been shown by Maul et al. [28] to be superior to the local density approximation for the description of amino acid bonds. The explicit inclusion of long-range correlations as in previous adsorption studies [29, 30] has not been considered because such contributions are expected to be of minor importance in the present case. The pseudopotentials and wave functions are represented within the projector-augmented wave (PAW) method [31]. The electron wave functions in the regions between the cores are expanded in a plane wave basis set up to a cutoff energy of 500 eV. This allows for an accurate treatment of the first-row elements as well as the Au $5d$ electrons. The Brillouin zone (BZ) integration is represented by a sum over Monkhorst-Pack points [32]. The supercells consisted of material slabs of five atomic layers followed by a vacuum region equivalent to ten atomic layers. The lateral cell size is set as a 4×4 surface to minimize molecule-molecule interaction. Energy convergence with

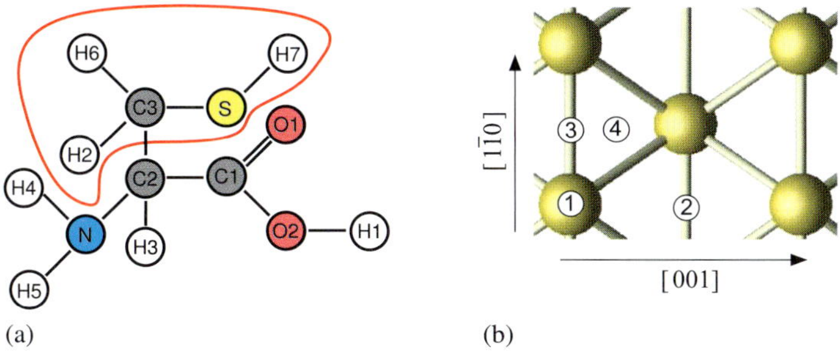

Fig. 1. **a** Used notation for atoms of cysteine. The red enclosed area indicates the side group that is replaced by -H for calculations with glycine (see text). **b** Unreconstructed Au(110) surface unit cell. High symmetry positions are indicated as follows: top site (1), [001] bridge site (2), [1$\bar{1}$0] bridge site (3), and [1$\bar{1}$0] off-bridge site (4)

respect to all these parameters has been checked carefully and the slabs have been found to reproduce the surface energy of the Au(110) surface accurately [33].

3 Results and Discussion

3.1 The Thiolate-Gold Bond

The interaction of the thiol head group -SH with the Au-(110)-(1×1) surface is analyzed calculating the adsorption energies for a series of molecular positions in deprotonized configuration. The cysteine molecule is placed at sites on a regular mesh of points over the surface unit cell. For each adsorption site we optimize the height of the sulfur atom by total energy minimization. The orientation of the molecule is kept fixed throughout all calculations and chosen to maximally restrict the interaction with the surface to the thiolate head group. The atoms are not allowed to relax. One set of calculations is performed with the C3 carbon atom located directly above the thiolate S atom and a second series of calculations with the S-C3 bond tilted with respect to the surface normal (nomenclature is presented in Fig. 1a). The results are presented in Fig. 2a and b respectively.

We find that the preferred bonding site in both scans is the [1$\bar{1}$0] off-bridge site (cf. Fig. 1b), with S located 1.6 Å above the Au surface layer. However, the strong asymmetry in the energy distribution in Fig. 2b shows that the bonding geometry has a major impact on the adsorption energy. Besides the [1$\bar{1}$0] off-bridge site strong Au-S bonds may also be established at the long [001] bridge site. At both these positions S binds to two Au atoms. The energy gain with respect to the non-bonded radical and clean surface varies strongly with the bonding geometry, ranging from 1.8 eV at the right [1$\bar{1}$0] off-bridge

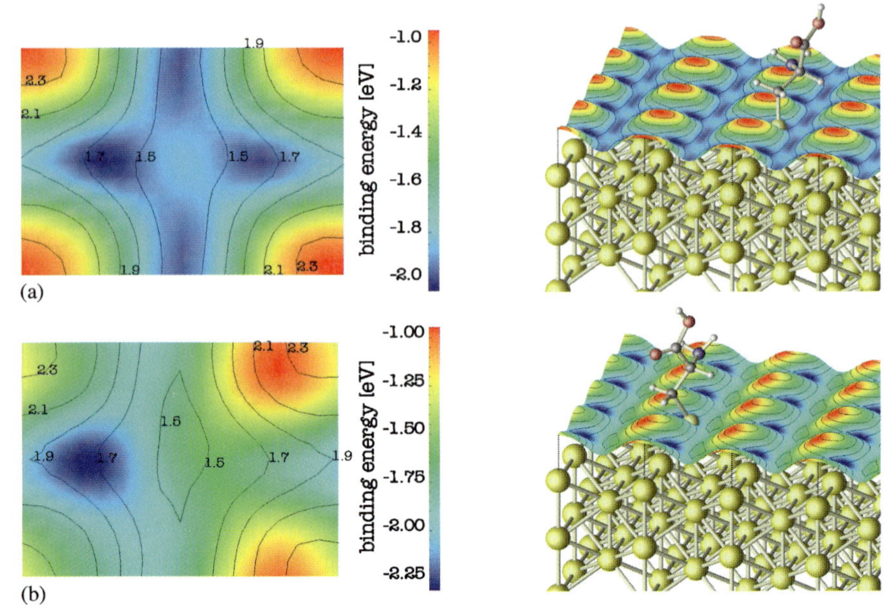

Fig. 2. Position dependent adsorption energy of the Au-thiolate bond. The height profile and contour lines denote the optimal molecule-surface distance for the given lateral position (values in Å). The colors indicate the binding energy. **a** Scan performed with the S-C3 bond in surface normal direction, **b** scan performed with S-C3 tilted

site in Fig. 2b to 2.3 eV at the left $[1\bar{1}0]$ off-bridge site. The bonding geometry seems to favor a local tetrahedron geometry as also suggested by calculations of S-Cu bonds on a Cu(110) substrate [34].

To quantify the dependence on the bonding angle, we did a series of calculations, in which the position of the S atom relative to the surface was kept fixed and the orientation of the rigid molecule was varied. For bridge and off-bridge S adsorption sites the tilted molecular geometries can be described by just one angle α (cf. Fig. 3). The two nearest gold atoms serve as reference points and the angle is measured with respect to the mid position between these two atoms. For the definition of α we introduce two vectors. The vector **b** points from the sulfur atom to the mid position between the two Au atoms (bridge) and the vector **c** points from S to C3. Both vectors have no component in $[1\bar{1}0]$ direction if S is located at $[1\bar{1}0]$ bridge or off-bridge site. The tilt angle is given by

$$\alpha = \arccos\left(\frac{\mathbf{b}\cdot\mathbf{c}}{|\mathbf{b}||\mathbf{c}|}\right). \tag{1}$$

In a perfect tetrahedron this angle would be $\alpha_{\mathrm{tetr}} = \arccos(-1/\sqrt{3}) \approx 125.26°$. This tetrahedron geometry implies an Au-S-Au angle of 109.5°. Given an Au-Au distance of $a_0/\sqrt{2} = 2.95$ Å between the two gold atoms, however, this

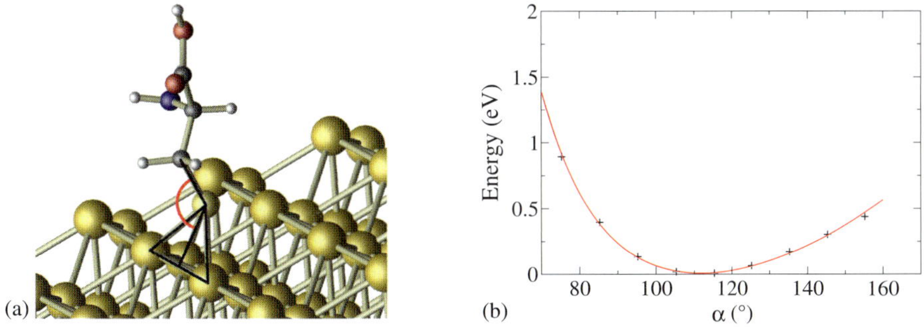

Fig. 3. **a** Illustration of the tilt angle α (in red) between the Au bridge atoms and the S-C3 bond in the thiolate bonding configuration. **b** Energy variation versus the tilt angle α of the thiolate bond at the $[1\bar{1}0]$ off-bridge site

results in a S-Au bond length of 1.81 Å, significantly smaller than the sum of the covalent radii 2.36 Å [35]. The calculated equilibrium bond length is in fact 2.45 Å, so such an ideal tetrahedron geometry is not possible. The Au-S-Au angle only amounts to 77.8°. Different to an ideal tetrahedron where the minimum angle of all pairs out of four vectors is maximized at 109.5°, in the distorted tetrahedron, two of the vectors are moved closer to each other, and the angle Au-S-Au is reduced to 77.8°. Consequently, the other angles can be larger and, α will be smaller than the optimal value of 125.26°. We therefore expect the optimal value for our bond angle α to be smaller than the ideal 125.26°.

Figure 3b shows the change in the systems free energy with varying α. The energy minimum is found at $\alpha_{\mathrm{opt}} = 111.8°$. As can be seen, the influence of the thiolate bonding geometry on the system energy is quite strong. Deviations of about 40° from α_{opt} result in energy changes up to 0.8 eV. The optimal angle α_{opt} decreased, as expected, from the ideal tetrahedron value 125.26° to 111.8°. Our observation is in contrast to the common belief in literature that the optimization of the bonding geometry is not important for finding the lowest energy configuration [36].

Electron Transfer

The S-Au bond length of about 2.45 Å corresponding to the strongest bond is very close to the sum of the covalent radii 2.36 Å of sulfur (1.02 Å) and gold (1.34 Å) [35]. Therefore, a strong covalent contribution to the thiolate bonding can be expected. Another indication for covalent bonding was found from the analysis of the optimal angle α and the resulting tetrahedron geometry. To gain insight into the nature of the thiolate-gold bond beyond indirect indications we investigate the redistribution of the electron density. A molecule adsorbed at the [001] bridge site with $\alpha = \alpha_{\mathrm{opt}}$ is used to study the bonding characteristics of the thiolate bond. The spatially resolved charge density

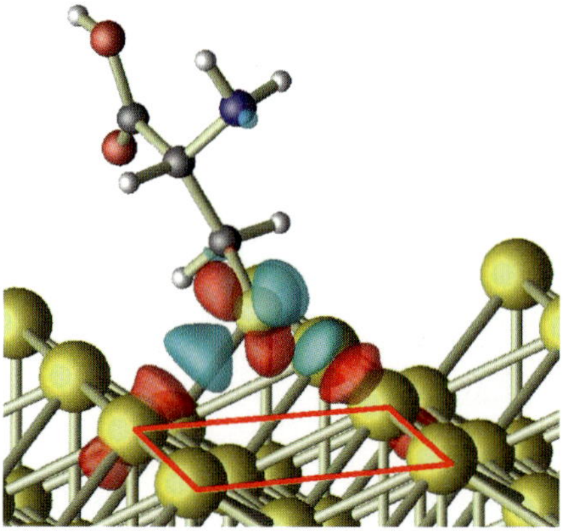

Fig. 4. Charge density difference plots for the thiolate bond, S at [001] bridge site. Regions of electron accumulation/depletion are marked in blue(+)/red(-). Isosurface value: $\pm 0.025\, e\, \text{Å}^{-3}$. The surface unit cell is indicated

difference is computed according to

$$\Delta\rho(\mathbf{x}) = \rho_{\text{ads/sub}}(\mathbf{x}) - \rho_{\text{ads}}(\mathbf{x}) - \rho_{\text{sub}}(\mathbf{x}), \tag{2}$$

where $\rho_{\text{ads/sub}}$ is the charge density of the adsorbate-substrate system and ρ_{ads} and ρ_{sub} are the charge densities of the adsorbate without surface and the clean surface, respectively. Thereby, the isolated systems have the same geometry as in the combined structure. The charge density difference is plotted in Fig. 4 and exhibits dominating contributions from sulfur-centered orbitals and the gold atoms involved in the bonding. In the molecule we observe charge redistribution mainly in the regions of the sulfur sp^3 hybrids in tetrahedron geometry. There is electron depletion in the area corresponding to the Au d lobes and electron accumulation along the bond axis, which corresponds to a σ-type bonding orbital. The accumulation/depletion regions indicate that the thiolate-gold bond is due to localized interaction, i.e., electronic rearrangement occurs largely at the sulfur atoms and the contributing gold atoms. Symmetrical positions of S and C3 with respect to the surface structure, particularly to the Au bonding partners, give rise to largely symmetric charge redistribution patterns, which shows that the contribution of the asymmetrical tailgroup to the bond is small.

The complicated charge redistribution plotted in Fig. 4 does not allow for an estimate of the net transfer of charges between the gold substrate and the deprotonized cysteine radical. Such an electron transfer would be important for an electrostatic contribution to the bonding energy. Moreover it is highly relevant for considering geometries with a simultaneous bond of the sulfur head group and the amino group to the surface. Electrostatic effects may also strongly influence the molecule-molecule interaction and lead to chiral recognition effects [37].

In order to analyze the total charge transfer, in a first step we laterally integrate over $\Delta\rho(\mathbf{x})$, i.e., we calculate the total charge-difference in a plane of thickness dz at given height z

$$\Delta Q(z, dz) = \int_{z}^{z+dz} d^3x \Delta\rho(\mathbf{x}). \tag{3}$$

In a second step we determine the total charge transfer into the Au(110) substrate

$$Q(z) = \int_{z_0}^{z} \Delta Q dz, \tag{4}$$

where z_0 is a position in the vacuum far from the surface. Charge difference $\Delta Q(z, dz)$ and total charge transfer $Q(z)$ are shown in Fig. 5a.

We observe electron accumulation ($\Delta Q > 0$ and Q increasing) at the clean lower surface (at $z \approx 5$ Å) and a region of electron depletion ($\Delta Q < 0$ and Q decreasing) at the upper surface to which the cysteine is attached (at $z \approx 12$ Å). The fact that the integrated charge $Q(z)$ returns to zero at the uppermost Au layer means that there is no charge transfer into the gold slab and we only observe charge redistribution from the covered surface to the clean surface underneath. This charge redistribution amounts to a net dipole moment in the gold slab, which is induced by the adsorbate. The overall dipole moment of the supercell containing adsorbate and substrate is calculated to $|\mathbf{p}_{\text{ads/sub}}| = 2.74$ D in the direction of the surface normal. Most of this dipole moment, however, is not due to adsorption induced electron rearrangement. It is caused by the deprotonization of the cysteine radical, as can be seen by investigating the polarization of the supercell containing only the free cysteine radical. Here, the calculated dipole moment is $|\mathbf{p}_{\text{ads}}| = 2.37$ D. Fig. 5a shows considerable charge transfer within the molecule (between 13 and 17 Å) with aggregation regions at the upper atoms and electron reduction in the central areas of the adsorbed radical. Consequently, multipole moments are likely to play a role in this bond. The zero net charge transfer from molecule to surface rules out the possibility of an ionic contribution to the bonding energy.

Electronic State Changes

The adsorbate bonding due to S-Au interaction and the accompanying electron redistribution have consequences for the electronic structure of the adsorbate system. We investigate these changes in terms of the Kohn-Sham (KS) eigenvalues of DFT and the corresponding DOS of the adsorbate system, the free molecule and the metal surface. In doing so, one should keep two points in mind. First, the KS eigenvalues do not account for the excitation aspect and quasiparticle character of single-particle electronic excitations [38]. However,

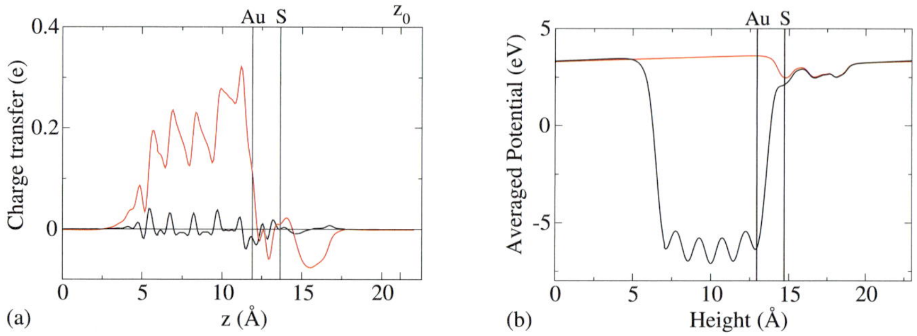

Fig. 5. a Total charge difference for the thiolate bond, S at [001] bridge site: total charge difference $\Delta Q(z)$ at given height z (black) and total charge transfer from above z to below $Q(z)$ (red). The position of the Au top layer, the position of S, and z_0, the starting point of the integral in (4), are marked. **b** Plane-averaged potential of adsorbate-substrate system (black) and free deprotonized cysteine radical (red). The position of the top-row Au layer and the S atom are indicated. The vacuum levels (plateaus) outside the systems are aligned

these effects mainly influence the energetical distance of occupied and unoccupied states [28] and occur in the adsorbed as well as the gas phase molecule. For the metal, these effects vanish close to the Fermi level [38]. Second, comparing the DOS of two different systems—adsorbate and molecule—one has to deal with three different energy scales which are significantly influenced by the local electrostatics. We solve this problem by aligning the energy scales by means of the electrostatic potentials from the KS equation. More precisely, we use plane-averaged potentials where the planes are parallel to the surface. The vacuum level in the potentials of the three systems are used for the energy-scale alignment (see Fig. 5b).

There are however two problems associated with the vacuum-level-alignment method. The first is that due to surface dipoles and electron density tailing the vacuum level close to interfaces or molecules is not equal to the vacuum level in larger distances [2]. Since we are using the vacuum level right above the cysteine radical for the alignment, this problem does not apply to our system. The second is that both our systems possess an overall dipole moment that—due to the periodic boundary conditions of the DFT supercell approach—adds a sawtooth shape to the electrostatic potential. But since the dipole moment is almost the same for both adsorbate system and free radical, a reasonably good alignment can nevertheless be obtained.

Results for the total DOS and its projection onto several sets of atoms, i.e., the projected DOS (PDOS), are shown in Fig. 6. We project to the metal substrate and the molecule separately by summing up over the corresponding metal slab and molecule atoms. In comparison, the DOS of the free cysteine radical is also plotted. A refined analysis also considers the character of the

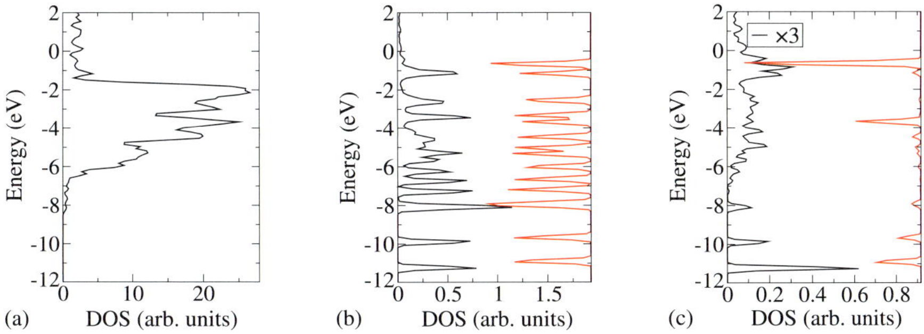

Fig. 6. a Au PDOS of the adsorbate-substrate system. **b** Cysteine-radical PDOS of the adsorbate-substrate system (black) and of the free cysteine radical (red). **c** S PDOS of the adsorbate-substrate system (black) and of the free cysteine radical (red). The S PDOS of the adsorbate-substrate system has been magnified by a factor of three to allow for better visibility of the lineshape. Energy alignment was obtained using the vacuum level of the electrostatic potential. The energy zero is set to the Fermi level of the metal

wave function by projecting onto the *s*-, *p*-, and *d*-states of the considered atom as well.

The band extending from about 2 to 6 eV below the Fermi level in Fig. 6a is characteristic for the Au *d* orbitals. The Au PDOS is only marginally affected by the single molecule adsorption. Significant changes, however, occur for the adsorbate (cf. Fig. 6b). The most noticeable difference between molecule-projected density of states in the gas phase and the adsorbed state is the disappearance of the sharp, distinct HOMO peak at −0.8 eV of the gas phase radical upon adsorption. The HOMO of the free deprotonized cysteine radical is localized at the sulfur atom as evident from Fig. 6c. Therefore, the most pronounced changes are observed in the DOS projected on the S atom. The S PDOS shows considerable broadening for all S-localized states above −7 eV. In particular we find a rather broad distribution over the energy interval between 0 and −6 eV, which is the range of the gold *d* orbitals. There are small peaks at the upper and at the lower end of this distribution, indicating a hybridization into bonding and antibonding states. Sulfur PDOS peaks below the main region of the Au *d* bands and all other molecular orbitals are unaltered in shape and slightly shifted to lower energies. The weak modifications of molecular states which are either not localized on sulfur or which, if localized on sulfur, are not in the energy range of the Au *d* states lead to the following final conclusions: (i) The molecule-substrate bond is strongly localized at the sulfur atom. (ii) Rehybridization of orbitals leads to a bond with sp^3 character. (iii) Electrostatic effects do not play a significant role.

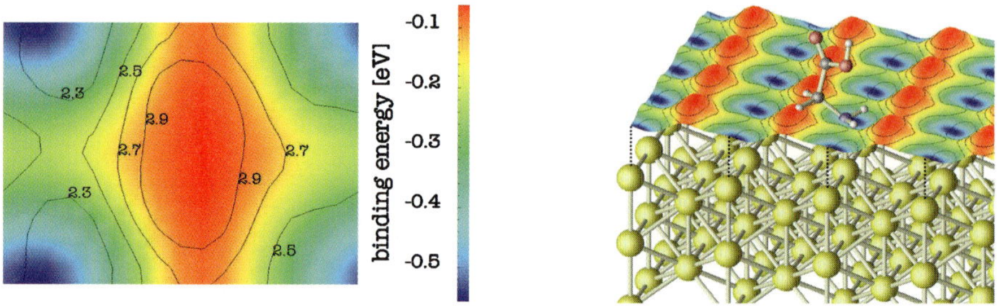

Fig. 7. Position dependent adsorption energy for the Au-amino bond

3.2 The Amino-Gold Bond

Geometry and Energetics

To probe the bonding properties of cysteine via the NH_2 amino group on the Au(110) surface, a series of calculations analogous to the calculations describing the thiolate-gold bond is carried out. The molecule is placed at different locations on a threedimensional mesh and the corresponding adsorption energies are determined. To investigate the Au-amino bond in isolation, it is important to restrict the molecule-gold interaction to the amino group nitrogen. Such a study of an isolated amino bond to a metal is possible for a flat molecule like adenine [9, 39] or ammonia [40], where no additional sidegroups give rise to bonding contributions interfering with the Au-amino bond. In the case of cysteine, however, this is not possible due to steric effects. In order to avoid additional interactions besides the amino group-Au bond which may be caused by the remaining parts of the molecule, we reduce the molecule by replacing CH_2-SH (see Fig. 1a) by H which gives the closely related amino acid glycine. The resulting structure closely resembles the most stable glycine conformation, with the amino group pointing away from the C1-C2 axis at almost a tetrahedron angle [41]. We place the thus obtained glycine model on a regular mesh of possible positions above the surface. The orientation of the molecule is chosen so that the amino group is symmetrical in $[1\bar{1}0]$-direction, and the N-H bonds form an angle of 17° with the surface plane. The results are displayed in Fig. 7.

The energy is minimal for a position very close to the top site (cf. Fig. 1b). In this off-top position the height of the amino group nitrogen is 2.2 Å above the topmost Au layer. The high symmetry of the energy distribution in Fig. 7, both in [001] and $[1\bar{1}0]$-direction, indicates that the influence of the bonding geometry and the orientation of the molecular tail group is negligible. The energy gain due to bonding with respect to the free molecule and surface amounts to 0.53 eV. This is significantly larger than for amino group bonding on Cu(110) [39]. Similar to the findings of Ref. [39], although maybe not so pronounced, the energy can be assumed to be a lower limit due to the incorrect description of the van der Waals interaction (see also Refs. [6] and [29]).

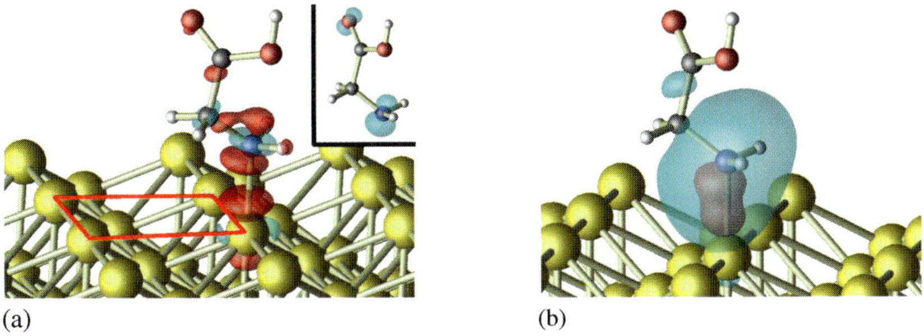

Fig. 8. a Charge density difference for the amino bond with N at top position. Regions of electron accumulation/depletion are indicated in blue(+)/red(-). Isosurface value: $\pm 0.025\, e\, \text{Å}^{-3}$. The surface unit cell is indicated. The inset depicts the free glycine HOMO, isosurface value: $0.3\, e\, \text{Å}^{-3}$. **b** Isosurfaces of the spatially-resolved electrostatic potential difference $\Delta V_{el}(\mathbf{x})$ (5). Isosurface values are -1 (blue) and -2 (red) eV

Electron Transfer

The interpretation of the amino-Au bond is difficult. On the one hand, the amino group and the N-C bond remain intact. On the other hand, the N-Au distance of 2.26 Å approaches the sum of the covalent radii of Au (1.35 Å) and N (0.75 Å) [35], indicating a strong bond. For a better understanding we calculate the charge density difference $\Delta\rho$ according to (2).

Figure 8a shows isosurfaces of $\Delta\rho$ for N at top site, i.e. slightly displaced from the most favorable site. The size of the isosurface-enclosed regions indicates the amount of charge transfer. The redistribution of the electrons is concentrated around the amino nitrogen and the nearest gold atom. The regions of electron depletion above and below the interacting Au atom show the characteristic shape of d electron orbitals which were already observed in the charge redistribution due to the Au thiolate bond in Sect. 3.1. Smaller contributions are found in the region of the N-C2 bond. Charge redistribution features in the molecule are similar to the glycine HOMO, which is also depicted in Fig. 8a. It is predominantly located at the amino group nitrogen and coincides with the region of electron depletion. So far, the emerging picture of the bond consists of electron rearrangement due to interaction of the HOMO state with the Au d orbitals. In this regard, the bonding of the amino group seems to be rather similar to the bonding of the sulfur head group discussed in Sect. 3.1. In the following however, we discuss why it is not. The complexity of the depletion and accumulation regions in Fig. 8a does not clearly indicate covalent nor ionic bonding, but the above findings might have an influence on the total charge transfer.

In order to determine the total charge transfer from adsorbate to substrate we calculate $\Delta Q(z, dz)$ and $Q(z)$ according to (3) and (4). The results of this procedure are plotted in Fig. 9. We observe charge accumulation at the lower

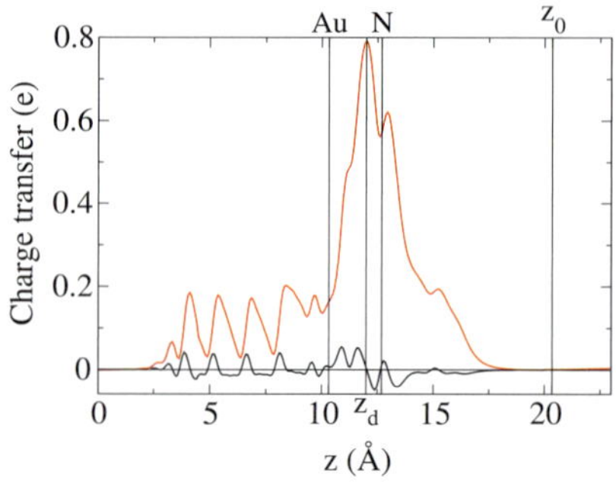

Fig. 9. Total charge difference for the amino bond, N at top site: total charge difference $\Delta Q(z)$ at given height z (black) and total charge transfer from above z to below $Q(z)$ (red). The position of the Au top layer, the position of N, the starting point of the integral z_0 from (4) and the dividing plane z_d are marked

surface (at $z \approx 4$ Å) and, especially, at the upper surface (at $z \approx 11$ Å) of the slab. As can be seen, the total charge transfer depends strongly on where we draw the dividing plane z_d between slab and molecule. However, since the maximum total charge transfer lies approximately halfway between the topmost Au layer and the amino nitrogen atom, it seems appropriate to place z_d at the height of maximum charge transfer. The total charge transfer from molecule to surface $Q(z_d)$ is 0.79 electrons for N at top site. The picture that presents itself here is that electrons are transfered from the adsorbate into the substrate. The excess charge in the slab, however, is not spread out through the whole gold slab but transferred to its surfaces. The non-vanishing charge transfer indicates an electrostatic contribution to the bond and should have visible consequences on the electronic structure of the adsorbed molecule.

Such an electrostatic effect can be visualized by computing the spatially resolved adsorption-induced potential-difference

$$\Delta V_{\mathrm{el}}(\mathbf{x}) = V_{\mathrm{el}}^{\mathrm{ads/sub}}(\mathbf{x}) - V_{\mathrm{el}}^{\mathrm{ads}}(\mathbf{x}) - V_{\mathrm{el}}^{\mathrm{sub}}(\mathbf{x}). \tag{5}$$

This quantity is presented in Fig. 8b. We find a region of considerable potential shift centered at the N-Au bond axis and reaching up to the molecule.

The above findings are not particular to the chosen adsorption site. $\Delta\rho$, $\Delta Q(z)$ and $Q(z)$ were determined and investigated for other bonding sites as well. The charge density difference isosurfaces show patterns very similar to those displayed in Fig. 8a, but the enclosed regions are smaller for other adsorption sites, which indicates less charge transfer for other sites. We find that the configuration with the largest charge transfer coincides with the strongest molecule-substrate bond, which supports the description of the bond between amino group and transition metal in terms of Coulomb interaction, as proposed previously for adenine on Cu(110) [39].

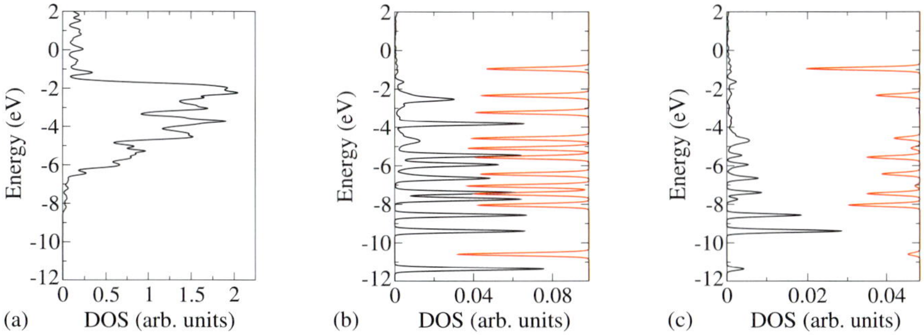

Fig. 10. **a** Au-projected DOS of the adsorbate-substrate system. **b** Glycine-projected DOS of the adsorbate-substrate system (black) and of the free glycine molecule (red). **c** N-projected DOS of adsorbed (black) and gas phase (red) system. The Fermi level of the metal is used as energy zero

Electronic State Changes

The above-described charge redistribution of nearly one electron upon adsorption with N at top site implies considerable changes in the electronic DOS. This is analyzed by means of the site projected density of states for free and adsorbed molecule as well as the Au slab, shown in Fig. 10.

In the gas phase the glycine HOMO peak appears at -1 eV (see Fig. 10b). This peak is not present in the glycine-projected DOS of the adsorbed system. The projection onto the amino group nitrogen (Fig. 10c) reveals that the glycine HOMO is mainly localized at N in the gas phase. The interaction with the surface, however, either depopulates this orbital or shifts it to another energy, possibly accompanied by a broadening. In view of the findings from the charge transfer analysis, which revealed a considerable electron depletion at the molecule, all molecular states should undergo such a shift. Indeed, Fig. 10b shows that the HOMO-10 (at -11 eV) is shifted by -1 eV upon adsorption. Similarly, higher energy states experience also a downward shift which is consistent with a loss of electrons in the molecule. The potential difference depicted in Fig. 8b gives rise to the downwards shift in energy for all amino group localized states as observed in Fig. 10b. The new HOMO state is considerably broadened and slightly decreased in energy.

The picture we get of the Au-amino bond is that through interaction with the Au *d* orbitals the glycine HOMO state is rearranged and charge is transfered from the area of the amino group into the slab. This charge transfer is accompanied by considerable potential shifts close to the amino group. It also introduces multipole moments in both adsorbate and substrate. A strong monopole attraction is visible in the adsorption-induced shift of molecule-localized states to smaller energies. This causes considerable attraction and a strong contribution to the total bonding energy. From the symmetry and localization of the electron redistribution depicted in Fig. 8a as well as the conservation of the lineshape of the glycine-projected DOS of states which are not localized at the amino group (as discussed above) we conclude that the contribution of the remaining molecule to the Au-amino bond is small. Therefore our findings regarding the adsorption of glycine are highly relevant for cysteine adsorption.

4 Summary and Outlook

We have studied the adsorption of the amino acid molecule cysteine on the Au(110)-(1×1) surface in detail by means of density functional theory calculations within the framework of a semilocal exchange-correlation description and the projector-augmented wave method. The total-energy optimizations with respect to the atomic positions have been accompanied by investigations of the electron transfers and the changes in the electronic structure of both molecule and substrate upon adsorption. To understand the driving forces for the adsorption and the formation of adsorbate geometries we have investigated the possible docking of cysteine through isolated sidegroups.

For the isolated thiolate-Au bond, fragmentation of the molecule in a radical and a hydrogen atom is important. The $[1\bar{1}0]$ off-bridge and the [001] bridge adsorption sites were found to be the energetically most favorable positions of the thiolate head group. We observed localized electronic interaction and hybridization of S-derived states in the S-projected density of states of the thiolate bonded system. The bonding energy was demonstrated to be strongly determined by the bonding geometry. The favored local geometry was conclusively shown to have nearly a tetrahedron structure with consequences for a sp^3 hybridization.

In order to study the bonding via the amino group, we have dropped the functionality of the thiol group and studied the adsorption of the smaller amino acid molecule glycine. We found an off-top position of the N atom of the amino group as the most favored adsorption site. However, the precise bonding geometry was less important, indicating a certain flexibility of the amino-Au bond. The studies of the charge redistribution revealed similarities to the amino bonding on the Cu(110) surface. The bond was shown to be largely electrostatic, and multipole terms beyond the ionic attraction seem to play a considerable role. The filling and energetics of electronic states located at the nitrogen atom have been shown to be important for the bonding behavior. For instance, the adsorption-induced charge transfer was shown to be due to interaction of the N-localized glycine HOMO state with the Au *d* bands. The electron transfer from molecule to surface causes a potential shift in the area of the amino group, thus shifting amino localized states to lower energies.

Having worked out basic bonding principles of that interaction of the cysteine functional groups, namely the thiol and amino group, then simultaneous bonding via the amino group nitrogen and the thiolate group sulfur remains to be studied. As next steps we are planning to use our results on favored bonding sites and preferred bond geometries to predict several likely molecular adsorption geometries and perform full ionic relaxation calculations for these molecule-substrate systems. This will allow us to determine the energetically most stable configuration and investigate wether the local character of the bonds will conserve the characteristics of the isolated functional group-metal interactions. The interplay of both bonding mechanisms is another interesting question.

Acknowledgements. We gratefully acknowledge fruitful discussions with J. Furthmüller and M. Preuss. The work was financially supported by the EU NANO-QUANTA Network of Excellence and the German Federal Government (BMBF Project No. 13N9669). Grants of computer time from the Höchstleistungsrechenzentrum Stuttgart are gratefully acknowledged.

References

1. A. Nilsson and G.M. Petterson, Surf. Sci. Rep. **55**, 49 (2004).
2. H. Ishii, K. Sugiyama, I. Eisuke, and K. Seki, Adv. Mater. **11**, 605 (1999).
3. G. Heimel, L. Romaner, J.-L. Brédas, and E. Zojer, Phys. Rev. Lett. **96**, 196806 (2006).
4. H. Vásquez, Y.J. Dappe, J. Ortega, and F. Flores, J. Chem. Phys. **126**, 144703 (2007).
5. I.G. Hill, A. Rajagopal, A. Kahn, and Y. Hu, Appl. Phys. Lett. **73**, 662 (1998).
6. W.G. Schmidt, K. Seino, M. Preuss, A. Hermann, F. Ortmann, and F. Bechstedt, Appl. Phys. A **85**, 387 (2006).
7. C. Vericat, M.E. Vela, and R.C. Salvarezza, Phys. Chem. Chem. Phys. **7**, 3258 (2005).
8. V. De Renzi, R. Rousseau, D. Marchetto, R. Biagi, S. Scandolo, and U. del Pennino, Phys. Rev. Lett. **95**, 046804 (2005).
9. E. Rauls, S. Blankenburg, and W.G. Schmidt, Surf. Sci. **602**, 2170 (2008).
10. C. Joachim, J.K. Gimzewski, and A. Aviram, Nature **408**, 541 (2000).
11. H.B. Akkerman, P.W.M. Blom, D.M. de Leeuw, and B. de Boer, Nature **441**, 69 (2006).
12. C.P. Collier, E.W. Wong, M. Belohradsky, F.M. Raymo, J.F. Stoddart, P.J. Kuekes, R.S. Williams, and J.R. Heath, Science **285**, 391 (1999).
13. S.Y. Quek, L. Venkataraman, H.J. Choi, S.G. Louie, M.S. Hybertsen, and J.B. Neaton, Nano Lett. **7**, 3477 (2007).
14. R. LeParc, C.I. Smith, M.C. Cuquerella, R.L. Williams, D.G. Fernig, C. Edwards, D.S. Martin, and P. Weightman, Langmuir **22**, 3413 (2006).
15. A. Kühnle, T.R. Linderoth, B. Hammer, and F. Besenbacher, Nature **415**, 891 (2002).
16. A. Kühnle, L.M. Molina, T.R. Linderoth, B. Hammer, and F. Besenbacher, Phys. Rev. Lett. **93**, 086101 (2004).
17. A. Kühnle, T.R. Linderoth, and F. Besenbacher, J. Am. Chem. Soc. **128**, 1076 (2005).
18. A. Kühnle, T.R. Linderoth, and F. Besenbacher, J. Am. Chem. Soc. **125**, 14680 (2003).

19. R.R. Nazmutdinov, J.D. Zhang, T.T. Zinkicheva, I.R. Manyurov, and J. Ulstrup, Langmuir **22**, 7556 (2006).
20. R. Di Felice, A. Selloni, and E. Molinari, J. Phys. Chem. B **107**, 1151 (2003).
21. R. Di Felice and A. Selloni, J. Chem. Phys. **120**, 4906 (2004).
22. B. Höffling, F. Ortmann, K. Hannewald, and F. Bechstedt, Phys. Rev. B **81**, 045407 (2010).
23. B. Höffling, F. Ortmann, K. Hannewald, and F. Bechstedt, Phys. Stat. Solidi C **7**, 149 (2010).
24. G. Kresse and J. Furthmüller, Comp. Mater. Sci **6**, 15 (1996).
25. G. Kresse and J. Furthmüller, Phys. Rev. B **54**, 11169 (1996).
26. J.P. Perdew, *Electronic Structure of Solids '91*, p. 11, Akademie-Verlag, Berlin (1991).
27. J.P. Perdew and Y. Wang, Phys. Rev. B **45**, 13244 (1992).
28. R. Maul, M. Preuss, F. Ortmann, K. Hannewald, and F. Bechstedt, J. Phys. Chem. A **111**, 4370 (2007).
29. F. Ortmann, W.G. Schmidt, and F. Bechstedt, Phys. Rev. Lett. **95**, 186101 (2005).
30. F. Ortmann, W.G. Schmidt, and F. Bechstedt, Phys. Rev. B **73**, 205101 (2006).
31. G. Kresse and D. Joubert, Phys. Rev. B **59**, 1758 (1999).
32. H.J. Monkhorst and J.D. Pack, Phys. Rev. B **13**, 5188 (1976).
33. B. Höffling, F. Ortmann, K. Hannewald, and F. Bechstedt, in: W.E. Nagel, D.B. Kröner, and M.M. Resch, eds., *High Performance Computing in Science and Engineering '09*, p. 53, Springer, Heidelberg (2009).
34. S. D'Agostino, L. Chiodo, F. Della Sala, R. Cingolani, and R. Rinaldi, Phys. Rev. B **75**, 195444 (2007).
35. Sargent-Welch, *Table of Periodic Properties of the Elements*, Sargent-Welch, Skokie (1980).
36. A. Bilić, J.R. Reimers, and N.S. Hush, J. Chem. Phys. **122**, 094708 (2005).
37. S. Blankenburg and W.G. Schmidt, Phys. Rev. Lett. **99**, 196107 (2007).
38. W.G. Aulbur, L. Jonson, and J.W. Wilkins, Solid State Phys. **54**, 1 (2000).
39. M. Preuss, W.G. Schmidt, and F. Bechstedt, Phys. Rev. Lett. **94**, 236102 (2005).
40. A. Bilić, J.R. Reimer, N.S. Hush, and J. Hafner, J. Chem. Phys. **116**, 8981 (2002).
41. R. Maul, F. Ortmann, M. Preuss, K. Hannewald, and F. Bechstedt, J. Comp. Chem. **28**, 1817 (2007).

Ab-initio Characterization of Electronic Properties of PbTe Quantum Dots Embedded in a CdTe Matrix

R. Leitsmann[a] and F. Bechstedt

European Theoretical Spectroscopy Facility (ETSF) and Institut für Festkörpertheorie und -optik, Friedrich-Schiller-Universität Jena, Max-Wien-Platz 1, 07743 Jena, Germany
[a] Present address: GWT-TUD GmbH, Annabergerstr. 240, 09125 Chemnitz, Germany, leitsmann@matcalc.de

Summary. The electronic properties of embedded PbTe nanocrystals (NCs) are investigated using an *ab initio* pseudopotential method and a repeated supercell approximation. The occurrence of polar dot matrix interfaces lead to the occurrence of an electrostatic field within the NC region. The consequences of this field on the electronic properties of the dot matrix system will be discussed in detail. The resulting spatial separation of electrons and holes inside the nanocrystal region lead to a quantum confined Stark effect.

1 Introduction

Light sources in the mid-infrared spectral region are crucial for many applications, e.g. for gas sensor or medical diagnostic systems. Arrays of embedded quantum dots (QDs), which show quantum confinement effects for electrons and holes, are promising candidates for light-emitting diodes and lasers. In the last years, huge efforts towards the development of QD-laser systems in this spectral region has been made using PbTe QDs embedded in a CdTe matrix. [1, 2] The highly symmetric QDs (see Fig. 1) with rhombo-cubo-octahedron shape exhibit a high mid-infrared luminescence yield [3–5]. The polar QD-matrix interfaces lead to an internal electrostatic field, which strongly influences the structural properties of the QDs [6]. However, its influence on electronic properties like the highest occupied nanocrystal orbital (HONO) or the lowest unoccupied nanocrystal orbital (LUNO) is not clarified.

Typically such questions can be addressed using state of the art density functional theory (DFT). However, the *ab initio* description of Pb- and Cd-salts is a challenging task for several reasons. First of all Cd exhibits very shallow *d*-states. [7] Therefore, one has to treat (beside the s, p states) the

W.E. Nagel et al. (eds.), *High Performance Computing in Science and Engineering '10*, DOI 10.1007/978-3-642-15748-6_10,

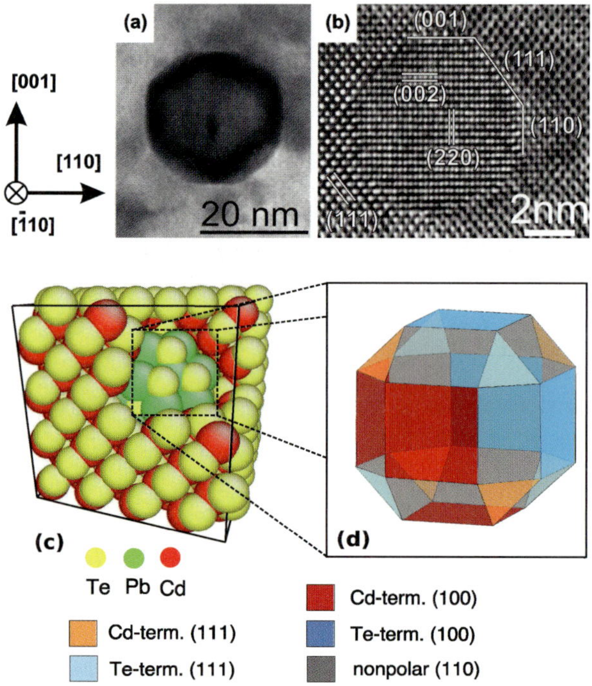

Fig. 1. a TEM image of a typical PbTe QD in a CdTe matrix [11]. **b** HRXTEM image of a similar but smaller QD with indications of the symmetry-allowed lattice planes and interface orientations [11, 13]. **c** Schematic model of a PbTe QD embedded in CdTe matrix [4]. **d** Schematic representation of the occurring dot-matrix interface facets

outermost Pb- and Cd-*d* states as valence electrons. On the other hand, for a precise description of heavy elements like Pb the inclusion of relativistic effects (e.g. spin-orbit coupling—SOC) is needed. [8] This increases the computational costs of those calculations considerable. In addition we have to use hugh supercells ($\approx$ 1000 atoms) to be able to compare our results to experimental observations. In former publications [9–12] we have already analyzed the geometry and electronic properties of PbTe/CdTe interfaces and there influence on the equilibrium crystal shape of embedded PbTe QDs as well as the atomistic geometry of the PbTe/CdTe QD-matrix system [6].

In the present study we focus on the electronic properties of PbTe QDs embedded in a CdTe matrix. In particular, the influence of the internal electric field, different supercell sizes, and the PbTe/CdTe band offset on the energetic and the spatial position of the electron and hole wavefunctions will be discussed in detail.

2 Computational Method

2.1 Theoretical Background

We are using total energy calculations based on DFT [14] within the local density approximation (LDA) [15]. In this framework not only questions of academic interest but also real-life systems can be studied. According to the Hohenberg-Kohn theorems the ground state properties of the investigated structures can be obtained by the minimization of the total energy, which is a unique functional of the ground-state electron density. To simplify the computation of such a many electron problem one can map it onto a system of non-interacting electrons $\{\Psi_i\}_i^N$ that has the same ground state density as the system of interacting electrons. This leads to the so-called Kohn-Sham (KS) equations [16]:

$$\left\{-\frac{\hbar^2}{2m}\nabla^2 + V_{\mathbf{ion}}(\mathbf{r}) + V_{\mathbf{H}}[n](\mathbf{r}) + V_{\mathbf{XC}}[n](\mathbf{r})\right\}\Psi_i(\mathbf{r}) = \varepsilon_i\Psi_i(\mathbf{r}), \qquad (1)$$

where $V_{\mathbf{H}}$ is the classical Hartree potential, while the exchange-correlation (XC) potential is defined as the density variation of the XC energy functional $\delta E_{\mathbf{XC}}/\delta n(\mathbf{r}) = V_{\mathbf{XC}}[n](\mathbf{r})$. For the numerical solution of the KS equation we expand the electronic wave functions into plane waves

$$\Psi_i(\mathbf{r}) \equiv \Psi_{n\mathbf{k}}(\mathbf{r}) = \langle \mathbf{r} \,|\, n\mathbf{k} \rangle = \sum_{\mathbf{G}} c_{n\mathbf{k}}(\mathbf{G}) e^{i(\mathbf{k}+\mathbf{G})\mathbf{r}} \qquad (2)$$

with the band index n and the wave vector $\mathbf{k}$, which is an element of the first Brillouin zone. This leads to very efficient algorithms for periodic systems because it renders the kinetic energy operator diagonal, and symmetries in reciprocal space can be exploited. The convergence is easily controlled by adjusting the energy cutoff of the plane-wave components in the basis. The representation of the rapid oscillations of the wave functions near the nuclei demands many plane waves. However, the valence electrons determine most of the interesting physical properties, and in the interstitial region the wave functions are rather smooth and can be described by comparatively few plane waves. We therefore employ the projector-augmented wave (PAW) method [17] which establishes a one-to-one relation between the exact wave functions and a set of smooth pseudo wave functions that match the exact ones outside a certain radius around each nucleus. The application of the plane wave expansion and the PAW pseudopotential method to the KS equation results in a generalized eigenvalue problem that can be efficiently solved with iterative algorithms like the Residual Minimization Method with Direct Inversion in Iterative Subspace (RMM-DIIS) [18] as implemented in the Vienna Ab-initio Simulation Package (VASP) [19, 20]. Parallelization is done using the Message Passing Interface (MPI).

For the description of heavy elements, for example lead atoms, relativistic effects have to be taken into account [8]. For this purpose (1) have to be generalized [21]:

$$\sum_{s'} \left\{ -\frac{\hbar^2}{2m} \nabla^2 \delta_{ss'} + V_{\mathbf{ion}}^{ss'}(\mathbf{r}) + V_{\mathbf{H}}[n](\mathbf{r}) \delta_{ss'} + V_{\mathbf{XC}}^{ss'}[n](\mathbf{r}) \right\} \Psi_i(\mathbf{r}, s') \qquad (3)$$
$$= \varepsilon_i \Psi_i(\mathbf{r}, s).$$

In this case the effective KS potentials $V_{\mathbf{XC}}^{ss'}$ and $V_{\mathbf{ion}}^{ss'}$ are not spin-diagonal any more. The generalized KS equation allows the treatment of external magnetic fields and SOC.

To obtain the atomic equilibrium positions we perform an ionic relaxation until the interatomic Hellmann-Feynman forces are smaller than 20 meV/Å in each x, y, z component. The resulting geometries are used to derive bond length (or nearest-neighbour distances) and interface bonding configurations. For the equilibrium positions and the electron density that minimized the total energy the single-particle potential can be obtained from the solution of the KS equation (1) or its generalized form (3). Its local part without the exchange-correlation part (the electron-ion interaction), will be herein referred to as electrostatic potential. The energy eigenvalues are used to discuss the electronic structure of the embedded nanocrystals (NCs).

2.2 Computational Cost

Since the KS matrix is diagonal in the index n of the eigenstate ("inter-band-distribution") the diagonalization of (1) and (3) can be efficiently parallelized. If there are enough nodes available, the diagonalization for the n-th state may be parallelized as well ("intra-band-distribution"). A limiting factor is, however, the communication overhead required for the redistribution of the wavefunctions between all nodes, which is necessary during the orthogonalization procedure of the eigenstates. Therefore, the scalability of the VASP-code depends on the size of the system as can be seen in Fig. 2a. In this graph we compare two systems containing 512 (circles) and 1000 (squares) atoms, which correspond to KS matrices of the size $(2810 \times 108383)^2$ and $(5445 \times 211441)^2$, respectively. A comparsion between the NEC SX-8 and the NEC SX-9 performance is shown in Fig. 2b. These benchmark tests demonstrate the very good scaling behavior of the code, which is in few cases biased by different numbers of electronic iteration cycles per ionic step. In addition the performance of a singe CPU on the SX-9 system is enhanced by about 60 % compared to the SX-8 machine. Our calculations are dominated by complex matrix-matrix multiplications (CGEMM). The sustained iteration performance for both cases exceeds 1 TFLOPS already on 16 nodes NEC SX-8. The sustained efficiency is between 79 and 50 % [22]. Compared to other high-computation facilities and local machines the implementation on the NEC SX-8/9 system

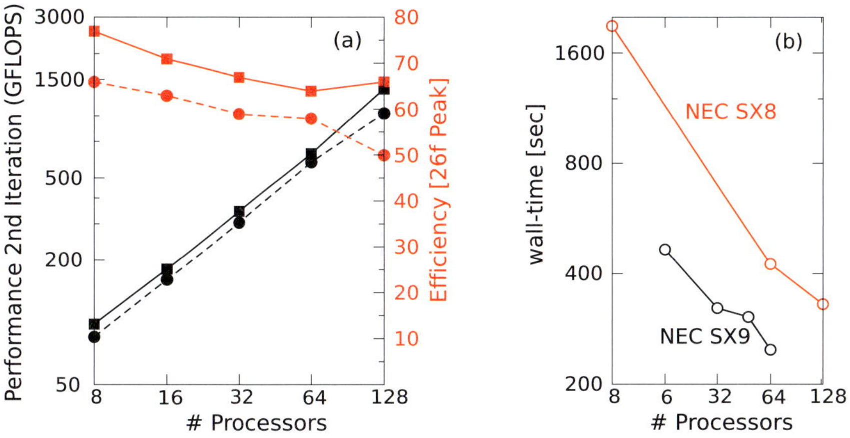

Fig. 2. **a** Scaling behavior of differently sized systems (circles 512 atoms, squares 1000 atoms) on the NEC SX-8; **b** Comparsion of the SX-8 and SX-9 performance for a 76 atom system

is the most efficient one. This fact is a consequence of the vector architecture, which can handle CGEMM operations in a very efficient way.

3 Nanocrystal Construction Using Supercells

The PbTe NCs embedded in CdTe matrix are modeled using the so-called supercell approach. [23] That means periodic boundary conditions are applied with a three-dimensional arrangement of nanocrystals embedded in a matrix material. For simplicity we use a simple-cubic arrangement of supercells. Each of them contains a PbTe QD and a certain amount of matrix material (CdTe). The size of the simple-cubic supercells vary with the diameter of the PbTe QDs and the size of the CdTe matrix.

The constructed supercell systems with different QD and matrix sizes are denoted as shown in Table 1. The denotation nm characterizes the number n of nearest-neighbour atom shells counted from the NC center and the edge length of the supercell ma_0 in units of the lattice constant a_0. All generated structures contain a central Te anion and exhibit a C_{3v} symmetry with the symmetry axis along the [111] direction. This symmetry reduction of the original O_h (PbTe) point group is related to the inequality of oppositely terminated but for reasons of electrostatic neutrality pairwise occurring dot-matrix interfaces. [6] Interestingly, those facet orientations lead to a dominance of the cation-terminated facets at the $[\bar{1}\bar{1}\bar{1}]$ halves of the NCs and vice versa of the anion-terminated facets at the [111] halves.

Table 1. Structural parameter of the constructed supercells: O-shell—outermost nanocrystallite shell, SC-size—supercell size in a_0, d_{NC}—nanocrystallite diameter in Å, d_{matrix}—distance between neighboring nanocrystallites in Å

Notation	33	43	44	45	46	55
O-shell	3	4	4	4	4	5
SC-size	3	3	4	5	6	5
d_{NC}	6.41	12.82	12.82	12.82	12.82	19.23
d_{matrix}	12.82	6.41	12.82	19.23	25.64	12.82

4 Results and Discussion

4.1 Induced Electrostatic Fields

Due to the pairwise occurring polar QD-matrix interfaces, one expects a strong electrostatic field along the [111] direction, which originates from the interface charges at the either Te- or Cd-terminated {111} or {100} interface facets (see Fig. 1d). To visualize this effect we have calculated the electrostatic potential acting on the KS-electrons. We are mainly interested in the variations of the electrostatic potential induced by the interface charges, i.e. variations at the length scale of the order of the QD size. The variation resulting from the distribution of the electrons around individual atoms (e.g. from the *d* electrons of the cations) or form chemical bonds between adjacent atoms should not be discussed. For that reason we spatially average the electrostatic potential using a Fourier-filter, which cuts of all variations at length scales smaller than the Pb-Te bond length (6.41 Å). The result is illustrated in Fig. 3a. The offset of the electrostatic potential inside the QD region and in the matrix material is clearly visible. It is of the same order ($\sim$ 3 eV) as observed at flat PbTe/CdTe interfaces [12]. It nicely illustrates the confinement of the KS electrons in the QD region.[1] The asymmetric behavior along the [111] direction is also clearly visible. As have been demonstrated in Ref. [6] this is a consequence of the interface charge distribution. Inside the QD region a linear slope of the potential can be estimated (see Fig. 3b) which corresponds to a electric field of 4.4 MV/cm. The influence of this field on the electronic properties of the QD-matrix system will be discussed in the following two sections.

4.2 Electronic Properties

To study the influence of the supercell size on the obtained electronic properties, we have varied the matrix size, i.e. the QD-QD distance, for a 1.28 nm QD. The position of several characteristic states close to the band gap have been plotted in Fig. 4a. Thereby, the highest occupied state in the supercell

[1] The elctrostatic potential shown here is not equivalent to the potential used in effective-mass potential well models.

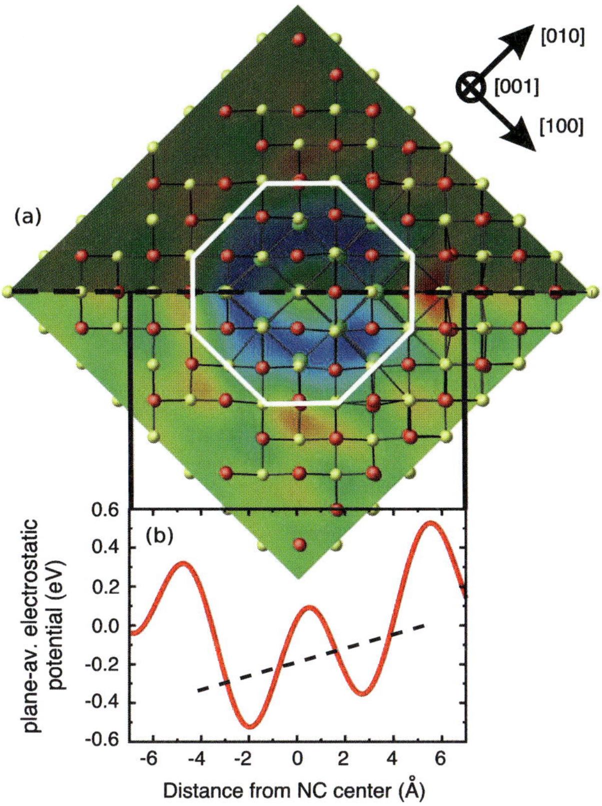

Fig. 3. **a** Fourier-filtered electrostatic potential (arb. units) of a 44 QD shown within the $(\bar{1}10)$ and $(0\bar{1}1)$ planes. Blue colors correspond to negative and red colors to positive values. The atomic positions of the PbTe QD (inside the white line) and the CdTe matrix are indicated by a stick and ball model. **b** The red solid line shows the spatially averaged electrostatic potential along the [111] direction. The estimated slope is indicated by a black dashed line

$\epsilon_{\mathrm{H}}^{\mathrm{SC}}$ have been used as energy zero. While the second lowest unoccupied supercell state $\epsilon_{\mathrm{C}_2}^{\mathrm{SC}}$ remains almost constant, the energetic position of the lowest unoccupied supercell state $\epsilon_{\mathrm{C}_1}^{\mathrm{SC}}$ decreases with increasing matrix size. The reason for this behavior is related to the spatial localization of these states. As can be seen in Fig. 4b, the states $\epsilon_{\mathrm{V}}^{\mathrm{SC}}$ and $\epsilon_{\mathrm{C}_2}^{\mathrm{SC}}$ are localized at the PbTe QD. Therefore, they can be interpreted as the PbTe-QD HONO and LUNO states. On the other hand, the $\epsilon_{\mathrm{C}_1}^{\mathrm{SC}}$ state is localized in the whole matrix region and, hence, corresponds the lowest unoccupied CdTe matrix state. This explains the variation of its energetic position with the matrix size. Its appearance inside the QD band gap region can be qualitatively explained with the aid of a simple potential well model.

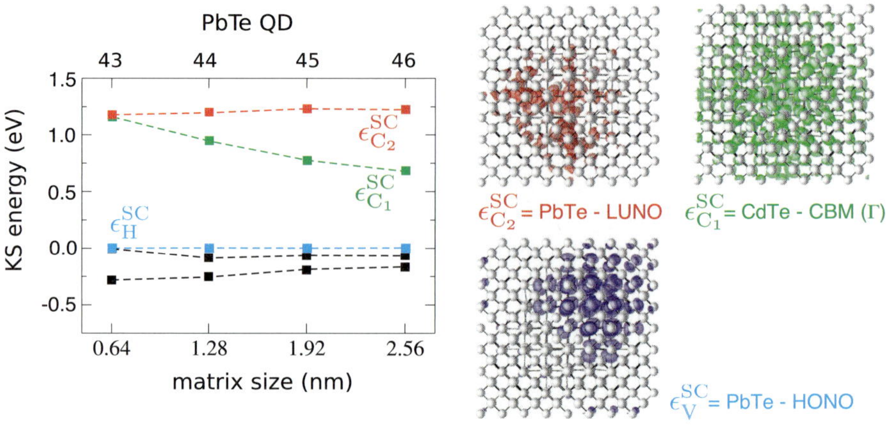

Fig. 4. Influence of the matrix size on the position of the electronic near-gap states of a 1.28 nm QD (left side). Spatial localization of the considered electronic states (right side)

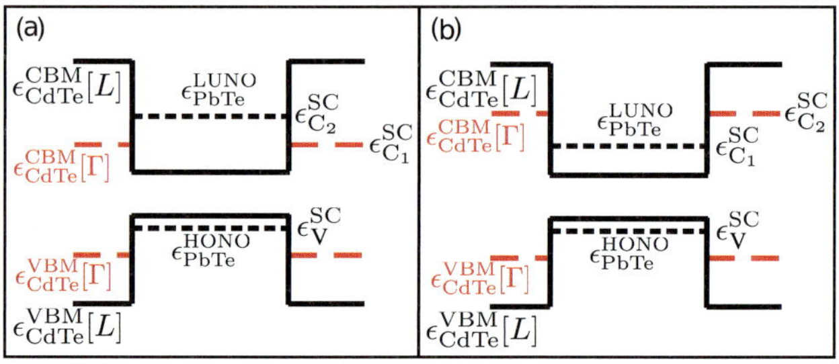

Fig. 5. Potential well model of the embedded PbTe QDs. The Cd 4*d* electrons are either treated as valence electrons (**a**) or as core electrons (**b**)

In Fig. 5 the effective confinement potentials of the electrons and holes are illustrated for an embedded QD. The depth of the potentials is determined by the relative position of the valence band maximum (VBM) and the conduction band minimum (CBM) of the corresponding bulk materials. However, it is a well known fact that the used KS approximation leads to an underestimation of the band gap values [24]. In the case of CdTe the calculated band gap of 0.34 eV underestimates the experimental value of 1.6 eV [25] drastically. This is at least partially the result of the to flat lying Cd4*d* electrons compared to the experiment because of the *pd*-repulsion [7, 26]. To study the influence of this large band gap error on the electronic properties we consider two different cases. In the first case we treat the Cd4*d* electrons as valence electrons (a) and in the second case the Cd4*d* electrons are treated as core electrons (b). In the latter situation the *pd*-repulsion is neglected and, hence, the energetic position of the bulk CdTe VBM is decreased, which leads to an overall band gap opening.

For the modelling of optical experiments, i.e. energetic transitions without momentum transfer, the depth of the effective potential wells has to be determined at one single k point. Since, bulk PbTe is a direct semiconductor with a band gap located at the L point, we have to use the L point to setup the potential well model for the PbTe/CdTe QD-matrix system. The state $\epsilon_{\mathrm{C}_1}^{\mathrm{SC}}$, however, corresponds to energetically deeper lying CdTe CBM located at the Γ point. Due to the strong under estimation of the CdTe bandgap this state lies only slightly above the bulk PbTe conduction band edge (see Fig. 5a). Due the strong quantum confinement effect the PbTe QD states are shifted to higher energies. As a consequence the state $\epsilon_{\mathrm{C}_1}^{\mathrm{SC}}$ lies below the PbTe LUNO level $\epsilon_{\mathrm{C}_2}^{\mathrm{SC}} = \epsilon_{\mathrm{PbTe}}^{\mathrm{LUNO}}$. As have been allready discussed the treatment of the Cd $4d$ electrons as core electrons opens the CdTe band gap. Therefore, in this case the state $\epsilon_{\mathrm{C}_1}^{\mathrm{SC}}$ lies above the PbTe LUNO level (see Fig. 5b).

To prove that this simple picture corresponds to the situation in our supercell calculations we have plotted the mean probability density $|\overline{\varphi}_n(\mathbf{r})|^2$ of the states $\epsilon_{\mathrm{V}}^{\mathrm{SC}}$, $\epsilon_{\mathrm{C}_1}^{\mathrm{SC}}$, and $\epsilon_{\mathrm{C}_2}^{\mathrm{SC}}$ versus the distance from the NC center along the [111] direction in Fig. 6. The treatment of the Cd $4d$ electrons as valence electrons leads to a $\epsilon_{\mathrm{C}_1}^{\mathrm{SC}}$ delocalized in the whole supercell, i.e. it corresponds

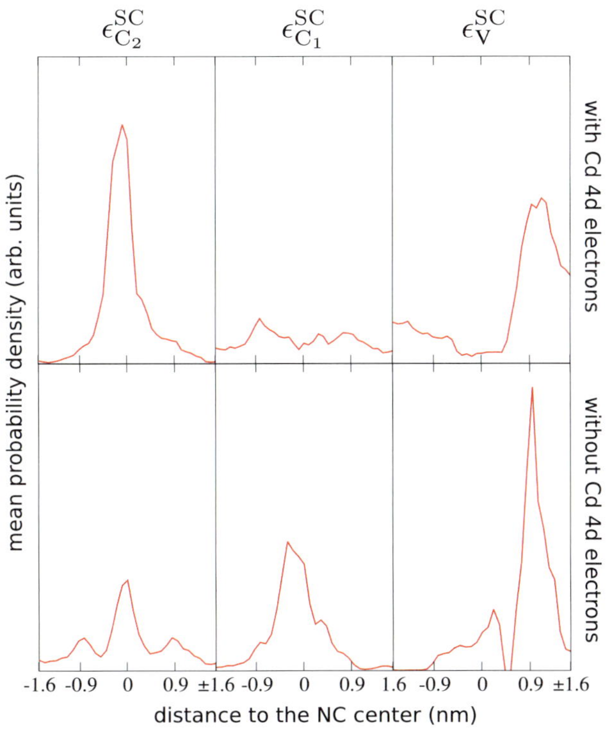

Fig. 6. Mean probability density $|\overline{\varphi}_n(\mathbf{r})|^2$ of the states $\epsilon_{\mathrm{C}_1}^{\mathrm{SC}}$, $\epsilon_{\mathrm{C}_2}^{\mathrm{SC}}$ and $\epsilon_{\mathrm{V}}^{\mathrm{SC}}$ of a 44 QD versus the distance from the NC center along the [111] direction

to the CdTe CBM(Γ). However, putting the Cd $4d$ electrons in the core yields a PbTe HONO state below the CdTe CBM(Γ), i.e. in this case the $\epsilon_{C_1}^{SC}$ state is localized within the NC region and corresponds to the PbTe HONO. The latter case most likely agrees with the experimental conditions.

Furthermore, we have investigated the influence of the CdTe band gap value (i.e. the treatment of the Cd $4d$ electrons) on the PbTe-QD gap. Independent of the QD size the PbTe-QD gap is increased when the Cd $4d$ electrons are frozen in the core. This is a result of the increased CdTe band gap and the consequently deeper confinement potentials, which leads to larger confinement energies. Hence, for the comparison of our theoretical predictions (concerning the electronic properties) with experimental observations one has to keep in mind, that in addition to the well known DFT band gap underestimation also the confinement energies are expected to be too small (according to the underestimation of the depth of the potential wells).

4.3 Spatial Localization of Electrons and Holes: Quantum Confined Stark Effect (QCSE)

Having a closer look at the states in Fig. 6 one realizes that the probability density of the PbTe HONO and LUNO are slightly shifted with respect to the NC center. While the HONO is shifted in positive [111] direction the LUNO is shifted in negative [111] direction. This is an effect of the interface charge induced electrostatic field with in the QD region (see Fig. 3). The effect of such a field on the electronic states can be easily understood using again the simple potential well model. The electric field and the periodic boundary conditions lead to an triangular shaped potential, which is superimposed to the original potential well model. The result is shown in Fig. 7.

The electric field leads to a lowering of the deepest electron-like QD states and vice versa to a shift of the hole-like QD states to higher energies. This yields a red shift of the HONO-LUNO transition energy. In addition the wave functions of the corresponding states are localized at opposite NC sites.

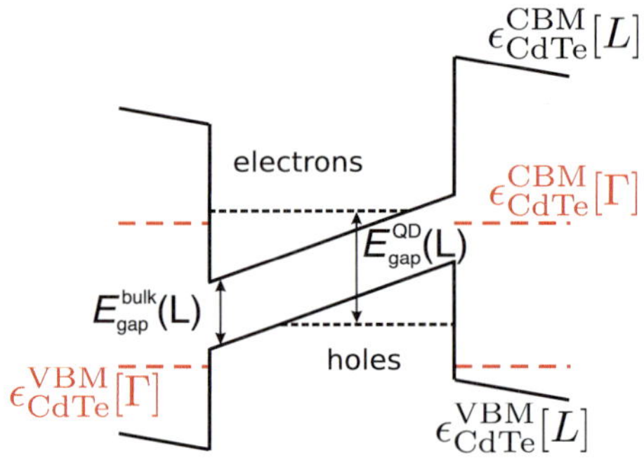

Fig. 7. Effect of an electrostatic field on the spatial localization of the electrons and holes within a simple potential well model

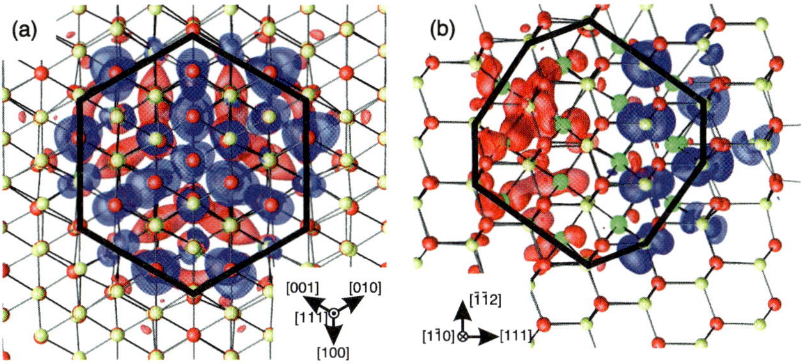

Fig. 8. Spatial localization of the PbTe HONO (blue) and LUNO (red) in a 44 QD supercell: **a** in a plane perpendicular to the [111] symmetry axis; **b** in a plane parallel to the [111] symmetry axis

In the literature such a behavior is called Quantum Confined Stark Effect (QCSE) [27].

In Fig. 8 we have plotted the isosurfaces of the PbTe 44 HONO (blue) and LUNO (red). The projection onto a plane perpendicular to the [111] symmetry axis (a) shows clearly the C_{3v} symmetry of the QD-matrix system. In Fig. 8b the projection onto a plane parallel to the [111] symmetry axis is shown. Obviously the HONO and LUNO are localized at opposite NC sites. This is an impressive prove of the applicability of the simple potential well model to our supercell calculations. One possible consequence of the predicted QCSE is an intensity reduction of the photoluminescence signal at very low temperatures, due to the reduced wave-function overlap of the HONO and LUNO states, instead of the expected increase. However, at higher temperatures also higher lying states contribute to the PL signal. Due to screening effects those states are less influenced by the QCSE. Therefore, at higher temperatures the PL intensity should be not reduced. Experimentally such a behavior have indeed been observed for PbTe QDs embedded in a CdTe matrix [4].

5 Summary and Outlook

We have investigated nanocrystals consisting of heavily ionic materials (PbTe) embedded in a matrix of polar semiconductor (CdTe) by means of ab initio methods. The influence of the band gap underestimation, inherent in the applied density functional theory within the local density approximation, on the obtained results have been discussed in detail. To identify the electronic states and to relate them either to the quantum dot or to the matrix region a simple potential well model have been used.

In this way we could demonstrate the large influence of the interface charge induced electrostatic field on the electronic properties. The electrons and holes inside the quantum dot are localized at opposite nanocrystal sides, i.e. a quantum confined Stark effect have been observed. This prediction is in good agreement with recent photoluminescence measurements.

Acknowledgements. We acknowledge valuable discussions with colleagues of our group J. Furthmüller and F. Ortmann. In addition we thank Prof. Schäffler from the University of Linz for providing us the experimental HRXTEM data. The work was financially supported through the Fonds zur Förderung der Wissenschaftlichen Forschung (Austria) in the framework of SFB25, Nanostrukturen für Infrarot-Photonik (IR-ON) and the EU e-I3 ETSF (GANo. 211956). Grants of computer time from the Höchstleistungsrechenzentrum Stuttgart are gratefully acknowledged.

References

1. H. Groiss, E. Kaufmann, G. Springholz, T. Schwarzl, G. Hesser, F. Schäffler, W. Heiss, K. Koike, T. Itakura, T. Hotei, M. Yano, and T. Wojtowicz, Appl. Phys. Lett. **91**, 222106 (2007).
2. W. Heiss, H. Groiss, E. Kaufmann, H. Hesser, M. Böberl, G. Springholz, F. Schäffler, R. Leitsmann, F. Bechstedt, K. Koike, H. Harada, and M. Yano, J. Appl. Phys. **101**, 081723 (2007).
3. W. Heiss, H. Groiss, E. Kaufmann, M. Böberl, G. Springholz, F. Schäffler, K. Koike, H. Harada, and M. Yano, Appl. Phys. Lett **88**, 192109 (2006).
4. R. Leitsmann and F. Bechstedt, Phys. Rev. B **80**, 165402 (2009).
5. R. Leitsmann and F. Bechstedt, ACS Nano **3**, 3505 (2009).
6. R. Leitsmann and F. Bechstedt, Phys. Rev. B **78**, 205324 (2008).
7. S.-H. Wei and A. Zunger, Phys. Rev. B **37**, 8958 (1988).
8. E.A. Albanesi, C.M.I. Okoye, C.O. Rodriguez, E.L.P. y Blanca, and A.G. Petukhov, Phys. Rev. B **61**, 16589 (2000).
9. R. Leitsmann, L.E. Ramos, and F. Bechstedt, Phys. Rev. B **74**, 085309 (2006).
10. H. Groiss, G. Hesser, W. Heiss, F. Schäffler, R. Leitsmann, F. Bechstedt, K. Koike, and M. Yano, Phys. Rev. B **79**, 235331 (2009).
11. R. Leitsmann, L.E. Ramos, F. Bechstedt, H. Groiss, F. Schäffler, W. Heiss, K. Koike, H. Harada, and M. Yano, New J. Phys. **8**, 317 (2006).
12. R. Leitsmann and F. Bechstedt, Phys. Rev. B **76**, 125315 (2007).
13. W. Heiss, H. Groiss, E. Kaufmann, M. Böberl, G. Springholz, F. Schäffler, K. Koike, H. Harada, and M. Yano, Appl. Phys. Lett. **88**, 192109 (2006).
14. P. Hohenberg and W. Kohn, Phys. Rev. **136**, B864 (1964).
15. D.M. Ceperley and B.J. Adler, Phys. Rev. Lett. **45**, 566 (1980).
16. W. Kohn and L.J. Sham, Phys. Rev. **140**, A1133 (1965).
17. P.E. Blöchl, Phys. Rev. B **50**, 17953 (1994).
18. P. Pulay, Chem. Phys. Lett. **73**, 393 (1980).
19. G. Kresse and J. Furthmüller, Comp. Mat. Sci. **6**, 15 (1996).
20. G. Kresse and J. Furthmüller, Phys. Rev. B **54**, 11169 (1996).
21. U. von Barth and L. Hedin, J. of Phys. C: Solid State Phys. **5**, 1629 (1972).
22. S. Haberhauer, NEC—High Performance Computing Europr GmbH, SHaberhauer@hpce.nec.com (2006).

23. M.C. Payne, M.P. Teter, D.C. Allan, T.A. Arias, and J.D. Joannopoulos, Rev. Mod. Phys. **64**, 1045 (1992).
24. J.P. Perdew and M. Levy, Phys. Rev. Lett. **51**, 1884 (1983).
25. P. Dziawa, B. Taliashvili, W. Domuchowski, L. Kowalczyk, E. Lusakowska, A. Mycielski, V. Osinniy, and T. Story, Phys. Stat. Sol. (c) **2**, 1167 (2005).
26. N.J. Shevchik, J. Tejeda, M. Cardona, and D.W. Langer, Phys. Stat. Sol. (b) **59**, 87 (1973).
27. Y.-H. Kuo, Y.K. Lee, Y. Ge, S. Ren, J.E. Roth, T.I. Kamins, D.A.B. Miller, and J.S. Harris, Nature **437**, 1334 (2005).

Si(111)-In Nanowire Optical Response from Large-scale Ab Initio Calculations

W.G. Schmidt, S. Wippermann, E. Rauls, U. Gerstmann, S. Sanna, C. Thierfelder, M. Landmann, and L.S. dos Santos

Lehrstuhl für Theoretische Physik, Universität Paderborn, 33095 Paderborn, Germany

Summary. The anisotropic optical response of Si(111)-(4×1)/(8×2)-In in the mid-infrared, where significant changes in the band structure between competing models of this important quasi-1D system are expected, has been calculated from *first principles.* Two characteristic peaks are calculated for the hexagon model of the (8×2) structure, but not for the trimer model. The comparison with recent infrared reflection anisotropy spectroscopy (RAS) data—showing the replacement of the anisotropic Drude tail of the (4×1) phase by two peaks at 0.50 eV and 0.72 eV—gives strong evidence for the hexagon model. Our calculations thus settle decades of intense debate about the ground-state geometry of this important prototype for quasi one-dimensional electronic systems.

1 Introduction

One-dimensional (1D) electronic systems are attracting considerable interest for both fundamental and technological reasons. Fascinating phenomena, such as spin-charge separation [1], and charge-density wave (CDW) formation due the Peierls instability may appear, while quasi-1D structures are being investigated as atomic-scale interconnects in devices. Such structures can be grown by self-assembly on Si surfaces [2], where the existence of an energy gap prevents the coupling of the electronic states of the nanostructure with the substrate in the vicinity of the Fermi level, thus preserving the quasi-1D character of the nanostructure states.

An intensively studied model system is the ordered array of In nanowires that self-assembles at the Si(111) surface to form the Si(111)-(4×1)-In phase, which has quasi-1D metallic properties at room temperature (RT). Results from many different experiments, as well as *ab initio* band structure calculations, agree that the structure is metallic, with the In atoms forming two parallel zigzag chains, which are separated by zigzag Si chains that resemble the π-bonded chains of Si(111)-(2×1) [3–14]. The structure has three quasi-1D surface state bands, which disperse strongly and cross the Fermi level along

W.E. Nagel et al. (eds.), *High Performance Computing in Science and Engineering '10*, DOI 10.1007/978-3-642-15748-6_11,

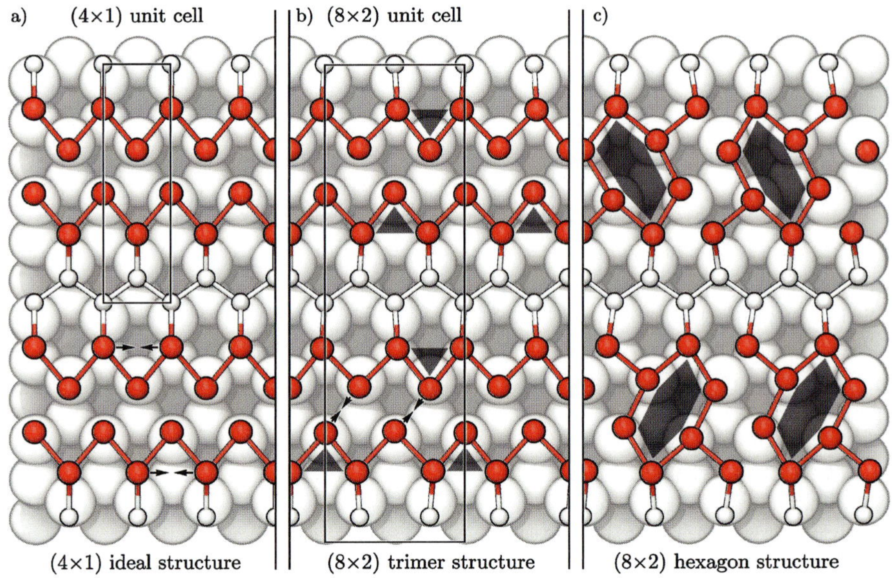

Fig. 1. Schematic top views of the room temperature Si(111)-(4×1)-In phase (**a**) and the trimer (**b**) hexagon models (**c**) proposed to describe the low-temperature structure of the Si(111)-In nanowire array. Red and grey balls indicate In and Si atoms, respectively

the chain direction, and show only a weak dispersion perpendicular to the chain.

Yeom et al. [15] reported that Si(111)-(4×1)-In undergoes a reversible phase transition below 120 K to an (8×2) structure, with a strong reduction of the density of states at the Fermi energy. From photoemission and scanning tunnelling microscopy (STM) studies they concluded that this phase formed a 1D CDW system driven by a Peierls instability. These intriguing results provoked much experimental and theoretical work. However, the origin of the phase transition remains controversial. The phase transition cannot be based on a simple CDW model because only one of the metallic bands nests properly [3, 15–19]. A triple-band Peierls instability has been suggested, where an interband charge transfer modifies the Fermi surface to improve nesting [17], while a periodic lattice distortion that lowers the energy has also been suggested [6, 9, 10, 20]. Recently, it has been proposed that, while the Peierls transition is important, other structural transitions are cooperatively involved [21].

Not only the mechanism of the (4×1) $\rightarrow$ (8×2) phase transition, but also the LT ground-state and its properties remain controversial. While the RT (4×1) phase is a quasi-1D metal [12, 16, 22], it has been variously suggested that the LT (8×2) phase is metallic, but with a lower density of states at the Fermi level [19, 20, 23], semimetallic [19], and semiconducting with a fundamental energy gap of 0.1–0.3 eV [15–18, 21]. Most *ab initio* calculations predict the nanowire ground state to be characterized by the formation of In trimers (cf. Fig. 1b) and find no gap opening [6, 7, 9, 10]. Recently, however, a structure with In hexagons, resulting from shear distortions in neighbouring chains (Fig. 1), has been predicted, which is semiconducting [24]. In Ref. [24] it was argued that the metallic RT phase arises from dynamic fluctuations between degenerate ground states, but recent photoemission [21] and Raman spectroscopy [11] results have cast doubt on this model. Total-energy calculations [13] of the hexagon and the trimer model for the LT phase concluded that an unambiguous identification of the internal structure of the ground state on energetic arguments is problematic. The energy differences between the competing structures are very small and depend on the approximations made in the calculations, e.g., the treatment of the In *4d* states, and exchange and correlation effects. Cho and Lee [25] state that the hexagon model is not stable, but the model has been supported recently by positron diffraction studies [26].

Very recently, optical spectroscopic information from a new spectral region has become available. Optical excitations around the Fermi level, in the mid-IR spectral range from 0.31 eV to 0.99 eV were probed with infrared spectroscopic ellipsometry (IRSE) and reflection anisotropy spectroscopy (RAS) [27]. As we will show in the following, the comparison of the measured data with spectra calculated from *first principles* provides strong evidence that excludes the trimer model and supports the hexagon model of the In nanowire ground state.

2 Computational Method

The theoretical data are obtained using density functional theory (DFT) within the local density approximation (LDA) for exchange and correlation as implemented in VASP [28]. Within the DFT calculations, the system of Kohn-Sham equations

$$\left\{-\frac{\hbar^2}{2m}\triangle + V_{ext}(\mathbf{r}) + \int \frac{n(\mathbf{r}')}{|\mathbf{r}-\mathbf{r}'|}d\mathbf{r}' + V_{xc}(\mathbf{r})\right\}\psi_{n\mathbf{k}}(\mathbf{r}) = \varepsilon_{n\mathbf{k}}\psi_{n\mathbf{k}} \tag{1}$$

$$n(\mathbf{r}) = \sum_{n,\mathbf{k}} f_{n\mathbf{k}}|\psi_{n\mathbf{k}}|^2 \tag{2}$$

is solved iteratively for the external potential $V_{ext}(\mathbf{r})$ until self-consistency in the total electron density $n(\mathbf{r})$ is reached. Plane waves serve as basis set for the Kohn-Sham orbitals $\psi_{n\mathbf{k}}(\mathbf{r})$. The ground-state DFT calculations were parallelized for different bands and sampling points in the Brillouin zone using the message passing interface (MPI). Parallelization over bands and plane wave coefficients at the same time reduces the communication overhead significantly.

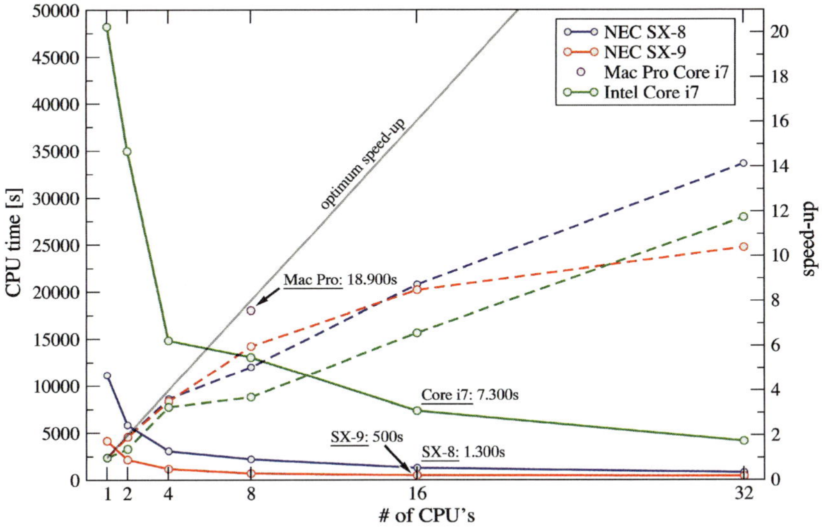

Fig. 2. Wall clock time and speedup for DFT calculations for the hexagon model of the Si(111)-In nanowire array containing around 200 atoms. The calculations were performed with the Stuttgart optimized VASP version on the HLRS NEC SX-8 and SX-9 machines. In comparison, we show data for a local Linux cluster (Intel Core i7, 24 Twin-nodes with 4 CPU's 2.5 GHz Quad Core Xeon each) and Mac Pro workstations (Intel Core i7)

Figure 2 shows benchmark calculations to determine the electronic ground state of the 200 atom cell used for surface modeling in our project. The calculations within this project were performed on the NEC SX-8 and SX-9 of the Höchstleistungs-Rechenzentrum Stuttgart. As can be seen, a reasonable scaling is achieved for using up to 32 CPUs.

Reflection anisotropy spectroscopy (RAS) is a non-destructive optical probe of surfaces that is capable of operation within a wide range of environments [29–31]. For a long time, however, a lack of adequate theoretical treatments presented a substantial barrier to the understanding and full exploitation of RAS spectra. This situation changed at the end of the nineties, when powerful computers allowed for the realistic modelling of the surface optical response from *first principles* [32–34]. This computational progress could build on substantial earlier efforts to understand the light-surface interaction, see, e.g., Refs. [35, 36]. Del Sole obtained an expression for the surface contribution [37] to the reflectance, $\Delta R/R$, where R is the reflectance according to the Fresnel equation. For s-light polarized along i and normal incidence it holds

$$\frac{\Delta R_i}{R}(\omega) = \frac{4\omega}{c}\Im\left\{\frac{\Delta\epsilon_{ii}(\omega)}{\epsilon_b(\omega)-1}\right\}, \tag{3}$$

where ϵ_b is the bulk dielectric function, and

$$\Delta\epsilon_{ij} = \int dzdz' \left[\epsilon_{ij}(\omega; z, z') - \delta_{ij}\delta(z - z')\epsilon_0(\omega; z)\right] - \int dzdz'dz''dz''' \epsilon_{iz}(\omega; z, z')\epsilon_{zz}^{-1}(\omega; z', z'')\epsilon_{zj}(\omega; z'', z'''). \quad (4)$$

Here $\epsilon_{ij}(\omega; z, z')$ is the non-local *macroscopic* dielectric tensor of the solid-vacuum interface accounting for all many-body and local-field effects [38]. (4) can be evaluated by replacing the semi-infinite crystal by an artificial super-cell, large enough to represent the vacuum as well as the surface and bulk regions of the crystal under investigation. Provided that (i) the material slab is large enough to properly describe the surface region of the crystal, i.e., the surface as well as surface-modified bulk wave-functions and (ii) the off-diagonal terms of the dielectric tensor are small compared to the diagonal ones, a simple expression for the surface contribution to the reflectivity can be derived [39]

$$\frac{\Delta R_i}{R}(\omega) = \frac{4\omega}{c}\Im\left\{\frac{\alpha_{ii}^{hs}(\omega)}{\epsilon_b(\omega) - 1}\right\}. \quad (5)$$

Here $\alpha_{ii}(\omega)$ with $i = x, y$ is the diagonal tensor component of the slab polarizability.

Here we calculate $\alpha_{ij}(\omega)$ in the independent-particle approximation from the Kohn-Sham conduction/valence orbitals $|c\mathbf{k}\rangle/|v\mathbf{k}\rangle$ and eigenvalues $\varepsilon_{c/k}$

$$\alpha_{ij}(\omega) = \frac{4e^2\hbar^2}{\Omega}\sum_{\mathbf{k}}\sum_{c,v}\frac{\langle c\mathbf{k}|v_i|v\mathbf{k}\rangle\langle v\mathbf{k}|v_j|c\mathbf{k}\rangle}{\left[\varepsilon_c(\mathbf{k}) - \varepsilon_v(\mathbf{k})\right]\left(\left[\varepsilon_c(\mathbf{k}) - \varepsilon_v(\mathbf{k})\right]^2 - \hbar^2\left[\omega + i\eta\right]^2\right)}, \quad (6)$$

where v_i denotes the velocity operator. The calculation of the polarizability tensor for the structure shown in Fig. 1 typically requires 96 GB memory and 21 hours wall clock time using 16 CPU's of the SX-8.

3 Results

Figure 3 shows measured and calculated RAS spectra for the In nanowire array. The measured data reveal a smooth increase to low energy for the RT (4×1) phase, while the LT (8×2) phase shows two sharp positive peaks at 0.50 eV and 0.72 eV. Positive anisotropy indicates that optical transitions parallel to the chains are dominant in this spectral region. Both phases show the broader, negative 1.9 eV feature, which splits below the metal-insulator transition [7, 40, 41].

The calculated anisotropy of both the trimer and hexagon models of the (8×2) structure agree well with the experiment above 0.7 eV, as has been reported previously [41]. Below 0.7 eV, only the hexagon model looks similar to the experimental results. In particular, two positive peaks are predicted,

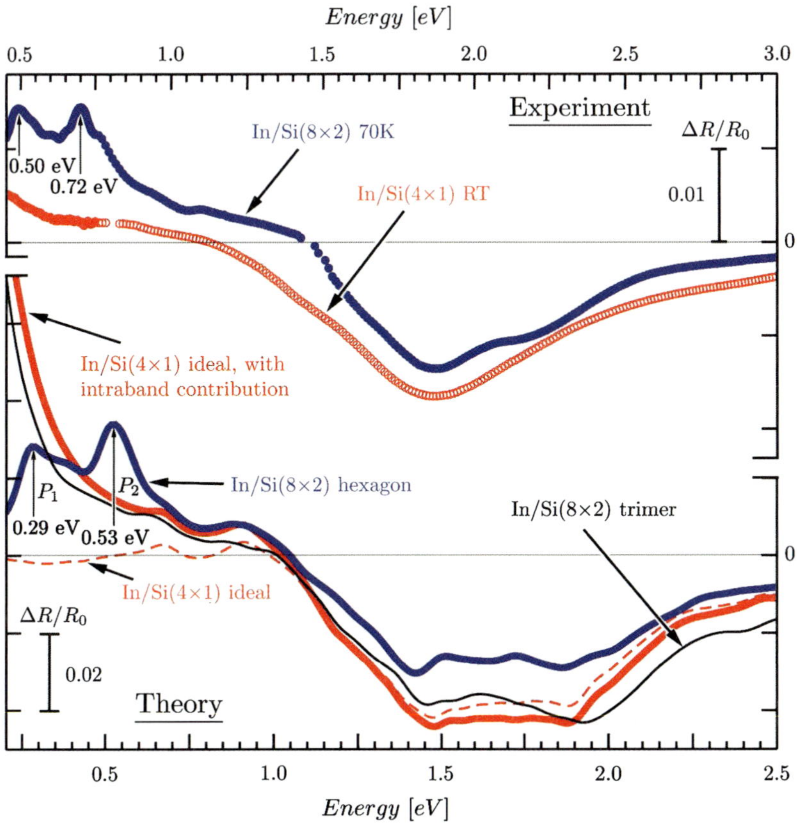

Fig. 3. RAS spectra of Si(111)-(4×1)-In at RT (300 K) and Si(111)-(8×2)-In at LT (70 K): upper, experiment; lower, theory. Note the different scales

separated by 0.24 eV. This splitting agrees very well with the experimental splitting of 0.22 eV. The origin of the two peaks in the mid IR can be traced to optical transitions close to the M point of the surface Brillouin zone, indicated by P_1 and P_2/P_2' in Fig. 4. Around M, nearly parallel valence and conduction bands close to the Fermi level give rise to a high joint density of states. From the orbital character of states (not shown here) we can assign P_1 to transitions between bonding and non-bonding In chain states within the single In zigzag chains, while P_2 and P_2' involve in addition In-In bonds between the two parallel zigzag chains. These bonds are exclusively formed for the hexagon model (Fig. 1). The spectra measured in the mid-IR are thus directly related to the hexagon structure of the In nanowire array.

In addition to the optical transitions in the region of the $\overline{XM}$ high-symmetry line we observe significant contributions to the optical anisotropy from transitions within the Brillouin zone. To visualize these transitions a 3-dimensional representation of the surface Brillouin zone is shown in Fig. 5.

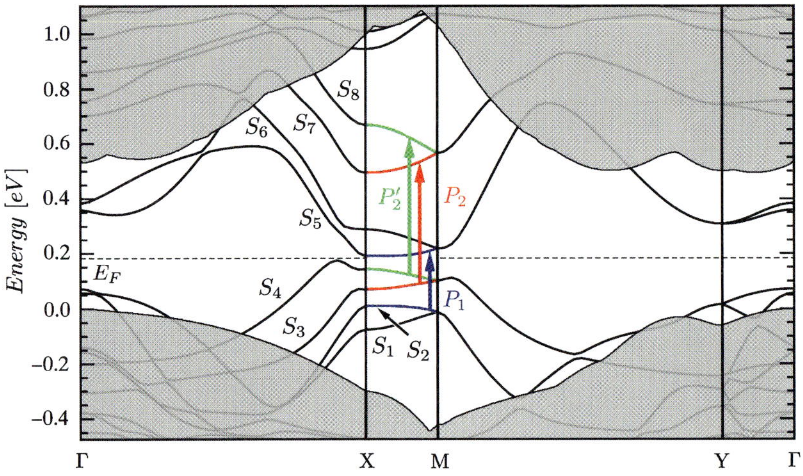

Fig. 4. Band structure of the hexagon model for Si(111)-(8×2)-In calculated within DFT-LDA. Pronounced optical transitions showing up in the RAS spectra are marked. Gray regions correspond to the projected Si bulk bands. The bulk valence band maximum is chosen as energy zero. The Fermi level is indicated

The notation of the surface bands $S_1 - S_8$ and the peaks P_1, P_2/P_2' refers to Figs. 4 and 3, respectively. The transitions indicated in Fig. 5b, c take place over the whole width of the Brillouin zone, while the transitions in Fig. 5d, e occur either near the $\overline{\Gamma X}$ or $\overline{MY}$ high-symmetry lines only. In some cases transitions from and to the same bands contribute to different peaks: Transitions from $S_2 \rightarrow S_5$ contribute to either P_1 or P_2, depending on the exact location of the transition inside the surface Brillouin zone.

Detailed comparison of the simulated spectra with the measured data reveals that the calculated mid-IR peaks are redshifted by 0.25 eV (note the different scales in Fig. 3). The underestimation of excitation energies is typical for DFT calculations where self-energy effects are neglected. The complexity and size of the In nanowire structure prevents the calculation of optical spectra using many-body perturbation theory that includes self-energy and excitonic effects [42]. Quasiparticle calculations for the high-symmetry points of the hexagon model surface band structure found self-energy effects to increase the lowest transition energies by 0.26 eV on average [13]. A larger shift of 0.5 eV, typical for Si excitation energies [42], applies to the higher energy negative optical anisotropies, because the optical transitions involve Si states [41]. Allowing for these energy shifts, the agreement between the calculated and measured RAS spectra is truly impressive.

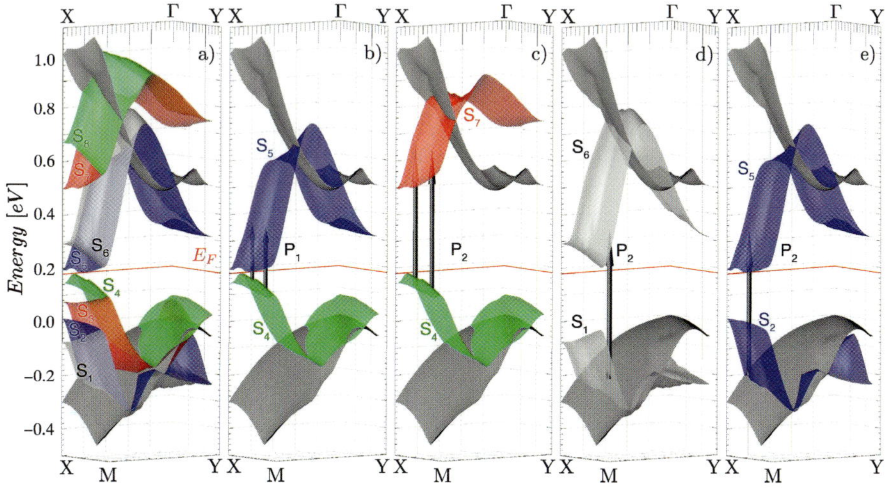

Fig. 5. **a** 3-D band structure of the hexagon model for Si(111)-(8×2)-In calculated within DFT-LDA. The notation of bands is consistent with Fig. 4. Pronounced optical transitions from *within* the Brillouin zone causing the peaks P_1 and P_2 in the RAS spectra are shown in **b** and **c**–**e**, respectively. The Fermi level is indicated by E_F

4 Summary

In conclusion, *ab-initio* optical response calculations have been reported for the In-Si nanowire system. The calculated spectra support the picture of a metal-insulator phase transition between a metallic Si(111)-(4×1)-In phase formed by zig-zag chains of In atoms at room temperature and a semiconducting (8×2) phase characterized by In atoms forming hexagons. The present computational results in conjunction with recent experimental data are likely to settle an intense debate on the ground-state geometry of one of the arguably most important model systems for quasi-1D metallic nanowires.

Acknowledgements. Generous grants of computer time from the Höchstleistungs-Rechenzentrum Stuttgart (HLRS) and the Paderborn Center for Parallel Computing (PC^2) are gratefully acknowledged. We thank the Deutsche Forschungsgemeinschaft for financial support.

References

1. Elliott H. Lieb and F.Y. Wu, Phys. Rev. Lett. **20**, 1445 (1968).
2. F.J. Himpsel, K.N. Altman, R. Bennewitz, J.N. Crain, A. Kirakosian, J.L. Lin, and J.L. McChesney, J. Phys.: Condens. Matter **13**, 11097 (2001).
3. T. Abukawa, M. Sasaki, F. Hisamatsu, T. Goto, T. Kinoshita, A. Kakizaki, and S. Kono, Surf. Sci. **325**, 33 (1995).

4. I.G. Hill and A.B. McLean, Phys. Rev. Lett. **82**, 2155 (1999).
5. O. Bunk, G. Falkenberg, J.H. Zeysing, L. Lottermoser, R.L. Johnson, M. Nielsen, F. Berg-Rasmussen, J. Baker, and R. Feidenhans'l, Phys. Rev. B **59**, 12228 (1999).
6. J.-H. Cho, D.-H. Oh, K.S. Kim, and L. Kleinman, Phys. Rev. B **64**, 235302 (2001).
7. S. Wang, W. Lu, W.G. Schmidt, and J. Bernholc, Phys. Rev. B **68**, 035329 (2003).
8. K. Fleischer, S. Chandola, N. Esser, W. Richter, and J.F. McGilp, Phys. Rev. B **67**, 235318 (2003).
9. J.-H. Cho, J.-Y. Lee, and L. Kleinman, Phys. Rev. B **71**, 081310(R) (2005).
10. X. Lopez-Lozano, A. Krivosheeva, A.A. Stekolnikov, L. Meza-Montes, C. Noguez, J. Furthmüller, and F. Bechstedt, Phys. Rev. B **73**, 035430 (2006).
11. K. Fleischer, S. Chandola, N. Esser, W. Richter, and J.F. McGilp, Phys. Rev. B **76**, 205406 (2007).
12. T. Kanagawa, R. Hobara, I. Matsuda, T. Tanikawa, A. Natori, and S. Hasegawa, Phys. Rev. Lett. **91**, 036805 (2003).
13. A.A. Stekolnikov, K. Seino, F. Bechstedt, S. Wippermann, W.G. Schmidt, A. Calzolari, and M. Buongiorno Nardelli, Phys. Rev. Lett. **98**, 026105 (2007).
14. S. Wippermann, N. Koch, and W.G. Schmidt, Phys. Rev. Lett. **100**, 106802 (2008).
15. H.W. Yeom, S. Takeda, E. Rotenberg, I. Matsuda, K. Horikoshi, J. Schaefer, C.M. Lee, S.D. Kevan, T. Ohta, T. Nagao, and S. Hasegawa, Phys. Rev. Lett. **82**, 4898 (1999).
16. T. Tanikawa, I. Matsuda, T. Kanagawa, and S. Hasegawa, Phys. Rev. Lett. **93**, 016801 (2004).
17. J.R. Ahn, J.H. Byun, H. Koh, E. Rotenberg, S.D. Kevan, and H.W. Yeom, Phys. Rev. Lett. **93**, 106401 (2004).
18. S.J. Park, H.W. Yeom, S.H. Min, D.H. Park, and I.W. Lyo, Phys. Rev. Lett. **93**, 106402 (2004).
19. J. Guo, G. Lee, and E.W. Plummer, Phys. Rev. Lett. **95**, 046102 (2005).
20. K. Sakamoto, H. Ashima, H.W. Yeom, and W. Uchida, Phys. Rev. B **62**, 9923 (2000).
21. Y.J. Sun, S. Agario, S. Souma, K. Sugawara, Y. Tago, T. Sato, and T. Takahashi, Phys. Rev. B **77**, 125115 (2008).
22. T. Uchihashi and U. Ramsperger, Appl. Phys. Lett. **80**, 4169 (2002).
23. H.W. Yeom, K. Horikoshi, H.M. Zhang, K. Ono, and R.I.G. Uhrberg, Phys. Rev. B. **65**, 241307(R) (2002).
24. C. Gonzalez, F. Flores, and J. Ortega, Phys. Rev. Lett. **96**, 136101 (2006).
25. Jun-Hyung Cho and Jung-Yup Lee, Phys. Rev. B **76**, 033405 (2007).
26. Y. Fukaya, M. Hashimoto, A. Kawasuso, and A. Ichimiya, Surf. Sci. **602**, 2448 (2008).
27. S. Chandola, K. Hinrichs, M. Gensch, N. Esser, S. Wippermann, W.G. Schmidt, F. Bechstedt, K. Fleischer, and J.F. Mcgilp, Phys. Rev. Lett. **102**, 226805 (2009).
28. G. Kresse and J. Furthmüller, Comp. Mat. Sci. **6**, 15 (1996).
29. W. Richter, Philos. T. Roy. Soc. A **344**, 453 (1993).
30. D.E. Aspnes, Solid State Commun. **101**, 85 (1997).

31. P. Weightman, D.S. Martin, R.J. Cole, and T. Farrel, Rep. Prog. Phys. **68**, 1251 (2005).
32. W.G. Schmidt, F. Bechstedt, and J. Bernholc, J. Vac. Sci. Technol. B **18**, 2215 (2000).
33. W.G Schmidt, K. Seino, P.H. Hahn, F. Bechstedt, W. Lu, S. Wang, and J. Bernholc, Thin Solid Films **455/456**, 764 (2004).
34. W.G. Schmidt, F. Fuchs, A. Hermann, K. Seino, F. Bechstedt, R. Paßmann, M. Wahl, M. Gensch, K. Hinrichs, N. Esser, S. Wang, W. Lu, and J. Bernholc, J. Phys.: Condens. Matter **16**, S4323 (2004).
35. J.D.E. McIntyre and D.E. Aspnes, Surf. Sci. **24**, 417 (1971).
36. A. Bagchi, R.G. Barrera, and A.K. Rajagopal, Phys. Rev. B **20**, 4824 (1979).
37. R. Del Sole, Solid State Commun. **37**, 537 (1981).
38. R. Del Sole and E. Fiorino, Phys. Rev. B **29**, 4631 (1984).
39. F. Manghi, R. Del Sole, A. Selloni, and E. Molinari, Phys. Rev. B **41**, 9935 (1990).
40. K. Fleischer, S. Chandola, Norbert Esser, W. Richter, J.F. Mcgilp, W.G. Schmidt, S. Wang, W. Lu, and J. Bernholc, Appl. Surf. Sci. **234**, 302 (2004).
41. S. Wippermann and W.G. Schmidt, Surf. Sci. **603**, 247 (2009).
42. P.H. Hahn, W.G. Schmidt, and F. Bechstedt, Phys. Rev. Lett. **88**, 016402 (2001).

Laser Ablation of Metals

Johannes Roth, Carolina Trichet, Hans-Rainer Trebin, and Steffen Sonntag

Institut für Theoretische und Angewandte Physik, Universität Stuttgart, Stuttgart, Germany, johannes@itap.physik.uni-stuttgart.de

1 Introduction

Laser ablation is the process of removing material with very intensive, pulsed laser radiation. It is a technology which gains increasingly greater importance for drilling, eroding, welding, structuring and marking of all kinds of materials. With the advent of ultrashort pico- and femtosecond pulses especially, a large number of new processing methods were developed. A deeper understanding of the ablation process itself still lacks behind.

The time scales of the processes involved in laser ablation are in the range from a several hundred femtoseconds (laser pulse) through a few picoseconds (electron-lattice-relaxation time) up to many nanoseconds (extrusion of the melt, cooling of the sample). Thus molecular dynamics (MD) simulations are the method of choice to get a deeper inside into the issue. The time steps applied are of the order of a few femtoseconds; on current computer architectures a few million time steps can be simulated. Thus the whole scope of interesting phenomena can be covered with atomistic resolution. Such a detailed treatment of the time steps and particle coordinates cannot be handled on common institute workstations, neither the wall clock time nor the storage of the results. Thus for most of the computations we rely heavily on the supercomputers of the HLRS.

1.1 Physical Challenge

The challenge for these kind of MD simulations is that the laser beam interacts with the conduction electrons of the metal. But the heat conduction of the electrons is governed by quantum mechanics and cannot be simulated directly with classical molecular dynamics. To treat the lattice ions and the electrons separately a two-temperature model (TTM) [1] has been implemented in our existing molecular dynamics package `IMD`. The interaction of the laser light with the electrons is described by a continuum equation and the transmission of the heat to the lattice by a coupling term added to the equations of motion.

W.E. Nagel et al. (eds.), *High Performance Computing in Science and Engineering '10*, DOI 10.1007/978-3-642-15748-6_12,

Up to now simulations of simple metals like Al, Cu and Fe and the complex metallic alloy $Al_{13}Co_4$ have been carried out. In the following we will present results of these simulations obtained on the Nehalem cluster of the HLRS.

2 Report

The starting point for this project was the expertise of our institute in simulations of mechanical properties like dislocation motion, crack propagation or shock waves in metals, complex metallic alloys (CMA) and quasicrystals (QC). In the SFB 382, terminated in 2006, we started already to develop a molecular dynamics simulation package called `IMD` which is especially suitable for simulations of the previously mentioned materials and their properties up to sample sizes of several billions of atoms [2, 3]. In general, the program supports a large repertoire of options and short range potentials for all kinds of solids. Therefore the upgrade of `IMD` for laser ablation simulations was quite natural. `IMD` has been adapted to new computer architectures and optimized continuously. Thus it is very flexible and powerful and can be run on simple PCs as well as massively-parallel supercomputers. `IMD` is freely available at www.itap.physik.uni-stuttgart.de/˜imd and is currently used by about 250 scientists worldwide.

2.1 Interaction Parameters for Aluminium

As a prototype material for our studies we chose pure aluminium since embedded atom (EAM) potentials for this metal have been derived by Mishin et al. [4] and by Ercolessi and Adams [5]. A second criterion was that aluminium shows no temperature- and pressure-dependent phase transitions for example from fcc to bcc like other metals. It is necessary to derive the phase diagram in advance since all states of aggregation can occur during ablation. Therefore melting and boiling simulations have been carried out. First problems occurred at the boiling point since it was not possible to find it by further heating of the melt. To find a reasonable estimate of the boiling point anyhow, gaseous samples have been cooled at constant pressure. By the observation of drop formation the boiling point could then be determined. Figure 1 displays the pressure density phase diagram for aluminium as obtained from the simulations. The phases important for ablation are obviously represented correctly. Only the potentials from Mishin et al. show some unphysical behavior at high pressures and temperatures like a positive compression modulus. Thus all subsequent simulations are based on the potentials of Ercolessi and Adams. The phase diagram of copper and iron could also be obtained with the experience gained for aluminium. Just for iron a rather large deviation for the melting temperature (about 1000 K) was observed. A possible explanation is a weakness of the applied potentials together with the complicated phase diagram of iron. The potentials reflect the underlying data set on which the potential

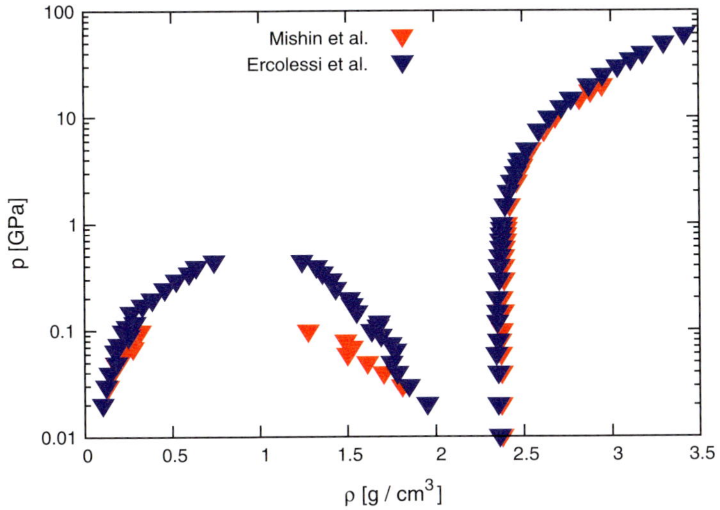

Fig. 1. Calculated pressure-density phase diagram for the two embedded atom potentials from Mishin et al. and Ercolessi and Adams

was trained. Usually these are rather low pressure and most often also low temperature equilibrium data. If insufficient solid-liquid transition data are present then the potentials cannot represent them accurately enough. With the help of the program `potfit`, which has also been developed at our institute [6, 7] we are able to derive better embedded atom potentials directly from ab-initio quantum mechanics calculations. The procedure, however, is very time consuming and requires a lot of experience.

Physical observables like melting temperature or velocity of sound of a material are properties governed by the lattice atoms and are therefore given by the applied potential. Thermal properties like heat conduction, heat capacity, and a proper coupling of the external laser field to the atoms are not directly determined by the lattice atoms. The conduction electrons actually lead to the high thermal conductivity of the metals. The thermal conductivity K_e, the heat capacity C_e and the electron-phonon coupling constant κ have decisive influence on the ablation process. While K_e and C_e can be determined rather easily from experiment, there is no standard experimental procedure to obtain the electron-phonon coupling constant. One has to take a detour via the measurement of phonon frequencies, electrical conductivity and thermal power to fix the size of the coupling constant.

2.2 Heat Propagation and Two-Temperature Model

There are several possibilities in `IMD` to couple the laser field from outside into the simulation sample. First a phenomenological ansatz has been implemented where at each time step the kinetic energies of the atoms are rescaled.

The laser power decays exponentially within the metal as given by the law of Lambert and Beer. The velocities are adjusted accordingly. The advantage of the method lies in its simplicity, the drawback is that in metals the heat is transported faster into the interior by the electrons. This behavior cannot be simulated by a simple adaption of the velocity distribution near the surface. Compared to the experiment more or less severe deviations occur depending on the size of the material parameters. The simple rescaling works well if the heat conductivity of the lattice is large. A much more realistic heat conduction process could be achieved by the implementation of the two-temperature model, namely the solution of a generalized heat conduction equation for the electrons and the ions separately:

$$C_e(T_e)\frac{\partial T_e}{\partial t} = \nabla[K_e \nabla T_e] - \kappa(T_e - T_i) + S(\mathbf{x}, t) \tag{1}$$

$$C_i(T_i)\frac{\partial T_i}{\partial t} = \nabla[K_i \nabla T_i] + \kappa(T_e - T_i) \tag{2}$$

(1) and (2) describe the time evolution of the electronic (T_e) and the ionic (T_i) or lattice temperature within the metal. $C_{e,i}$ are the heat capacities, $K_{e,i}$ the heat conductivities, κ the electron-phonon coupling constant and $S(\mathbf{x}, t)$ the external laser field. With these equations the laser field is coupled physically meaningfully into the system: first the energy is brought into the electronic system via a source term $S(\mathbf{x}, t)$. Then the electronic system transports the heat diffusively into the bulk and at the same time interacts with the atom cores. To work on an atomic scale, (2), which is a continuum description of the temperature field, has to be replaced by molecular dynamics. This means that instead of (2) the following equations of motion have to be solved together with (1):

$$m_j \frac{d^2\mathbf{x}_j}{dt^2} = -\nabla_{\mathbf{x}_j} U(\{\mathbf{x}_k\}) - \frac{\kappa}{C_l}\frac{(T_i - T_e)}{T_i} m_j \frac{d\mathbf{x}_j}{dt}. \tag{3}$$

Macroscopic observables like pressure, density or lattice temperature can be determined by averaging from coordinates $\mathbf{x}_j(t)$ and momenta $\mathbf{p}_j(t)$, respectively. The last velocity-dependent friction term in the equation regulates the temperature of the system.

The differential equation for the electronic heat conduction is solved on a fixed lattice by a finite-difference scheme. The local temperature is determined by averaging over about 100 atoms. The number of atoms in the surface cells decreases when ablation has started and the temperature determination gets inaccurate. A better method would be a smooth-particle-hydrodynamics (SPH) type atom-based solution of the differential equation. First approaches along this line have appeared in literature, however, they have not been pursued meanwhile and are rather complicated anyhow. For our purpose the problem has been solved by switching of the two-temperature model if the atom density in the finite-difference cells became too low.

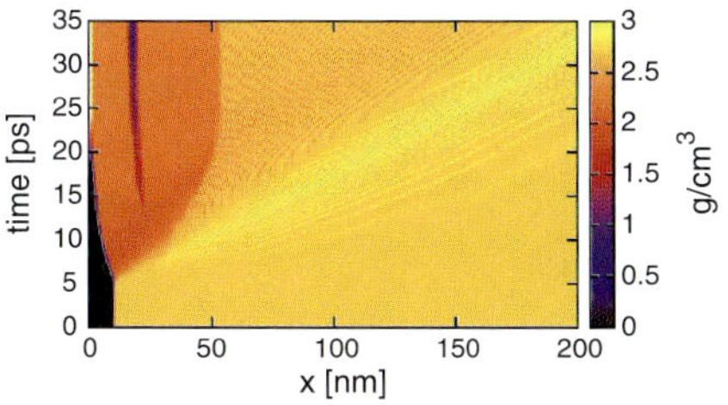

(a) Time evolution of the mass density after a 100 fs laser pulse

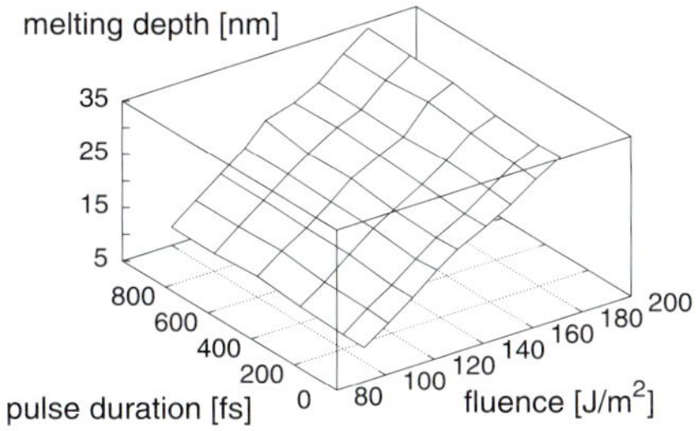

(b) Melting depth depending on different pulse durations and pulse strengths

Fig. 2. Ablation experiment for simple heating of pure aluminium

2.3 Simulation of Laser Ablation in Aluminium

Different types of ablation experiments can be simulated depending on the shape of the source term $S(\mathbf{x}, t)$ in (1). For the case of simple heating (as it occurs in the center of an experimentally applied Gaussian laser beam as compared to simulated sizes) a source term is chosen with a spatially homogeneous energy density. The result shall be discussed by means of Fig. 2a: Two picoseconds after the start of the simulation the 100 fs laser beam hits the surface. The melting front immediately starts to propagated into the bulk of the material. The solid-liquid interface is clearly represented by the density difference. In this experiment the laser fluence (laser power per area) lies above the ablation threshold which means that material is removed. A complete layer is split off with the full width of the sample—an effect which is also observed in experiment. Figure 2b displays a parameter study of the two laser parameters pulse duration and laser fluence. For the melting depth a linear dependency with the laser energy is observed.

A further possibility for ablation simulations is an inhomogeneous power density S as it is found in TEM00 modes. Today the smallest beam diameters are approximately 1 μm—a length with has not yet been reached in molecular dynamics simulations. To obtain size relations between penetration depth (a few nm) and beam geometry as realistic as possible a large simulation box is required. The standard box size for inhomogeneous heating was $80 \times 80 \times 180$ nm^3 and contained 48 million atoms. All simulations were carried out on the Nehalem cluster of the HLRS on 384 compute cores. The beam diameter was 65 nm. No visualization tools existed for simulations of this size (a snapshot of all coordinates and velocities requires 5 GB storage) which could be run on standard institute hardware. In collaboration with project D.3 of the SFB 716 the program `MegaMol` [11] was developed which permits to extract the interesting information from the pile of data. Similar to the homogeneous case a linear dependence between the melting depth and the laser power was obtained in these simulations (see Fig. 3b).

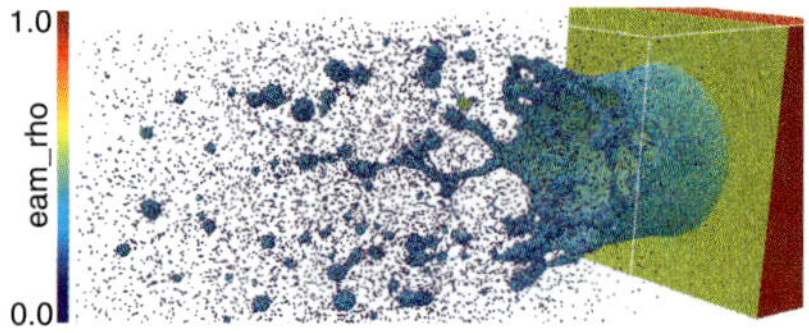

(a) Snapshot of a simulation 80 ps after the laser pulse. The picture has been generated with the program `MegaMol`

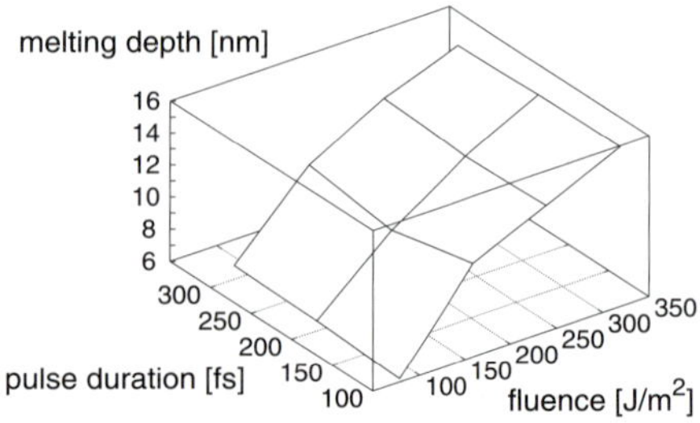

(b) Melting depth in dependence of different pulse durations and pulse strengths

Fig. 3. Ablation experiment with inhomogeneous heating. The simulation size is 48 million atoms in a box of $80 \times 80 \times 180\ \mathrm{nm}^3$. The gas phase of the plume extends up to 1.5 μm above the surface

2.4 Active Boundary Conditions

The laser radiation causes not only ablation but isochoric modification of the aggregate state and a pressure wave moving through the probe. Without further precaution the pressure wave will be reflected at the end of the simulation box due to the finite sample size in a simulation. The back-moving rarefaction wave leads to new effects which are not present in experiment, for example a contraction of the surface. The backward motion of the pressure wave can be avoided in the simulation by longer samples—an approach at the cost of longer run times, or—much more elegantly—by active boundary conditions at the end of the simulation box. While periodic boundary conditions are applied perpendicular to the direction of the laser beam a damping ramp is introduced into the equations of motions at the end of the sample. A coordinate-dependent friction of the form $-\xi(\mathbf{x})\dot{\mathbf{x}}_j$ leads to a smooth dying out of the pressure wave. Intensive tests have shown that this procedure is sufficiently accurate for the simulation of laser ablation, such that more complex methods like a dynamically computed compensating force at the end of the probe are not required.

2.5 Cluster Analysis of the Gas Phase

The inhomogeneous ablation process leads to a cluster-rich gas phase as can bee seen in Fig. 3a. Thus not only the events in the bulk below the surface are interesting. For the study of the cluster size distribution a set of new analysis tools had to be developed and tested. Currently the DBScan algorithm for cluster detection is used which works rather slowly. For larger sample sizes it is necessary to make the tools available which are already present in

MegaMol. The cluster analysis of laser ablation is very important for the rating of the effectivity of the material removal. It turns out that this work is very tedious. Many single atoms and small clusters are emitted and thus a meaningful statistics is possible. But large clusters and drops of all sizes up to a maximal drop size are also generated. Most often, the largest drop falls back into the ablation crater. The number of these larger objects is small (ablated total volume per pulse 1–5%). Thus many simulations are required for good statistics especially if the ablation behavior is to be evaluated as a function of laser fluence.

2.6 Anisotropic Materials

Up to now only the ablation behavior of metals with homogeneous and isotropic thermal properties has been studied. It was self-evident to investigate the ablation behavior of complex metallic alloys since our institute has a great expertise in their physics. As a prototype system we started with the approximant $Al_{13}Co_4$ of the decagonal quasicrystal phase Al-Ni-Co. The structure of this alloy features properties of periodic and quasiperiodic solids: it consists of quasiperiodic planes stacked periodically with a period of 14 Å. Therefore the heat conductivity K_e and the electron-phonon coupling possess a direction dependency. For the simulation of simple heating we used embedded atom potentials developed at our institute with the help of potfit for Al-Ni-Co. The size of the sample was $18 \times 22 \times 79$ nm^3 (2.3 million atoms). Figure 4 displays the ablation behavior for the different crystal directions in comparison with the results for aluminium. Different melting depths exist as expected

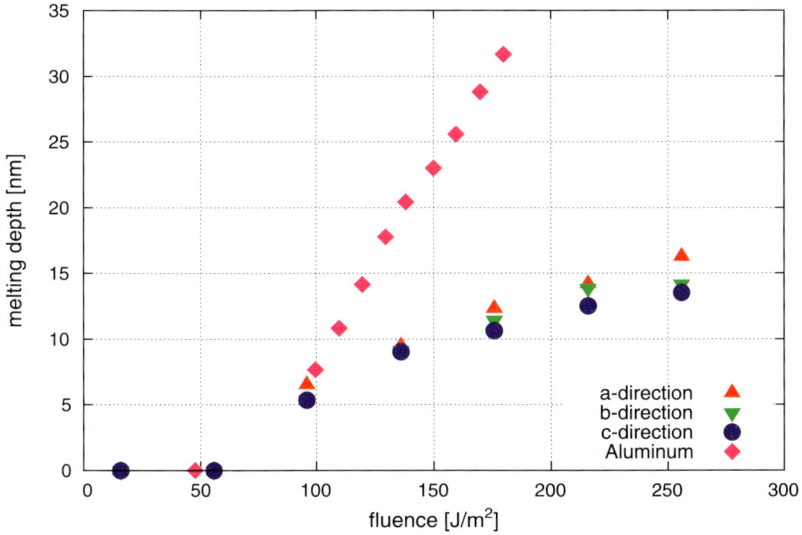

Fig. 4. Melting depth of $Al_{13}Co_4$ in comparison to pure aluminium

due to the anisotropic heat conduction—although the effect is small. An explanation for the weak direction dependence is given by the relatively high electron-phonon coupling of Al-Co (five times larger than that of aluminium). This leads to a very short electron-lattice relaxation time ($\tau_{el-ion} \approx C_e/\kappa$) which shortens the diffusive heat conduction of the electrons considerably.

3 Performance

3.1 General Considerations

Originally, `IMD` has been developed for largely homogeneous materials. Its strength lies in the simulations of phase diagrams, cracks, dislocations, or diffusion.

`IMD` distributes the volume of a freely chooseable simulation box evenly among the requested processors. The box is a cuboid and is divided into an array of $n_1 \times n_2 \times n_3$ equal blocks. Each block is treated independently by its own processor and is supposed to require the same wall time. If the probe is not cuboidal or has an inhomogeneous density, for example in a gas-fluid equilibrium state or a sphere in vacuum, then the blocks need a largely different amount of computing power which leads to an unequal load balancing and bad performance.

In the simulations of shock waves and laser ablation one open surface is required and above that a volume which is empty at the beginning. Thus the processors allocated to this part have nothing to do and spoil the performance. In a crack simulation a new open surface develops, but this has a minor effect since the volume in between is comparatively small. In the case of laser ablation the performance decay can be reduced considerably if the processor distribution is adapted in a smart way. Namely the longitudinal dimension is split up in as few blocks as possible while the transversal dimension is divided in as many blocks as possible. Hence each processor has a volume which contains vacuum, damaged and pristine material and thus has comparable work load. This trick works well for laser ablation and can be applied for increasing samples since we are interested in broad, flat samples. It does not perform so well for shock waves since in that case the samples should be rather long in order to be able to study the shock wave for a time interval as long as possible.

In future we plan to make the volume assignment of `IMD` more flexible. This should be rather easy to achieve. The optimal solution would be a completely dynamic load balancing which also works for complex sample geometries. This, however, requires very deep modifications to the existing code.

3.2 Benchmarks

General benchmark data for `IMD` have been given by Stadler et al. [2]. The data show that `IMD` scales almost linearly in weak scaling (same number of

Table 1. Timings of IMD on the BlueGene/L in microseconds per step and atom, for pair and EAM interactions, and for various system sizes and CPU numbers

#proc	#nodes	atoms/proc	#atoms	pair	EAM
1	1	2k	2k	19.00	33.99
1	1	16k	16k	18.22	33.67
1	1	54k	54k	20.01	33.47
1	1	128k	128k	20.01	33.97
64	32	2k	128k	21.05	36.80
64	32	16k	1m	19.36	35.18
64	32	54k	3.4m	20.82	34.59
64	32	128k	8m	20.52	35.01
256	128	4k	1m	20.71	36.11
256	128	32k	8m	20.79	35.07
2048	1024	512	1m	31.47	38.33
2048	1024	2k	4m	20.40	36.42
2048	1024	4k	8m	19.63	35.75
2048	1024	16k	32m	19.02	34.92
2048	1024	128k	256m	20.04	34.67

atoms per processor) and fairly well for strong scaling (total number of atoms constant, thus communication load growing). Some more recent data have been obtained on the BlueGene/L at the NIC in Jülich (see Table 1). This shows that the performance has not been worsened by recent extensions of the code.

For comparison with Table 1 we give the performance numbers obtained for a typical run with 512 cores on the Nehalem cluster of the HLRS: A timestep of the two-temperature model takes $1.52 \cdot 10^{-5}$ s/step/atom without the option stress and $1.85 \cdot 10^{-5}$ s/step/atom with this option which is used for more detailed analysis. Without two-temperature model a time step requires $6 \cdot 10^{-6}$ s/step/atom.

Most of the simulations reported here were obtained for samples of size $50 \times 50 \times 100$ nm^3 which contained of the order of 60 million atoms and were run for 18–20 ps. Longer simulation runs had to be restarted. The simulations fitted into a 12 h queue with 512 compute cores. A configuration file needs 6 GBytes of storage, thus a 160 ps movie of a sample which melts after 40–50 ps requires about 800 frames and a storage of 4.8 TBytes.

A sample length of 100 nm is sufficiently long since the laser beam intensity is virtually zero at a depth of 40 nm due to the penetration depth in aluminium of 8 nm. Thus we run currently new simulations with sample size $200 \times 200 \times 100$ nm^3 and 200 million atoms. A configuration file requires 20 GByte of storage in this case and a 16 TByte for a full movie.

To give an impression of the overall computing source requirements: A simulation of the cluster size distribution usually requires simulations for 5 different fluences with 5 different starting conditions to improve statistics for a single 100 fs pulse length.

4 Summary

We have presented results of laser ablation especially of aluminium and $Al_{13}Co_4$. Similar studies for Cu and Fe are in progress. We found that the new Nehalem cluster of the HLRS is very suitable for our molecular dynamics simulation code `IMD`. In future we plan to scale up our simulations up to the maximal size that can be run on the Nehalem cluster. A recent short test at our institute showed that we could achieve 1.3 billion atoms on 2 quadcore Nehalems with a main memory of 192 GB. But a time step would take "forever". This clearly demonstrates that molecular dynamics is not limited by memory but by the speed of the processors and the communication network.

References

1. Anisimov, S.I., Kapeliovich, B.L., Perel'man, T.L., Electron emission from metal surfaces exposed to ultra short laser pulses. J. Exp. Theo. Phys. Lett. **39** (1974)
2. Stadler, J., Mikulla, R., Trebin, H.-R., IMD: A software package for molecular dynamics studies on parallel computers. Int. J. Mod. Phys. C **8**, 1131–1140 (1997)
3. Roth, J., Gähler, F., Trebin, H.-R., A molecular dynamics run with 5.180.116.000 particles. Int. J. Mod. Phys. C **11**, 317–322 (2000)
4. Mishin, Y., Farkas, D., Mehl, M.J., Papaconstantopoulos, D.A., Interatomic potentials for monoatomic metals from experimental data and *ab initio* calculations. Phys. Rev. B **59**, 3393–3407 (1999)
5. Ercolessi, F., Adams, J.B., Interatomic potentials from first-principles calculations: The force-matching method. Europhys. Lett. **26**, 583–588 (1994)
6. Brommer, P., Gähler, F., Potfit: effective potentials from ab-initio data. Mod. Sim. Mat. Sci. Eng. **15**, 295–304 (2007)
7. Brommer, P., Development and Test of Interaction potentials for complex metallic alloys. PhD Thesis, Universität Stuttgart, Stuttgart (2009)
8. Sonntag, S., Roth, J., Gähler, F., Trebin, H.-R., Femtosecond laser ablation of aluminum. Appl. Surf. Sci. **255**, 9742–9744 (2009)
9. Sonntag, S., Roth, J., Trebin, H.-R., Molecular dynamics simulations of laser ablation in orthorhombic $Al_{13}Co_4$. Appl. Phys. **A** (2010). http://www.springerlink.com/index/10.1007/s00339-010-5762-5
10. Sonntag, S., Roth, J., Trebin, H.-R., Molecular dynamics simulations of laser ablation in Aluminum. In preparation
11. Grottel, S., Reina, G., Vrabec, J., Ertl, T., Visual verification and analysis of cluster detection for molecular dynamics. IEEE Trans. on Visual. and Comp. Graph. **13**, 1624–1631 (2007)

Conductance and Noise Correlations of Correlated Nanostructures

Alexander Branschädel[1] and Peter Schmitteckert[2]

[1] Institut für Theorie der Kondensierten Materie, Karlsruhe Institute of Technology, Universität Karlsruhe, Wolfgang Gaede Straße 1, D-76128 Karlsruhe, Germany
[2] Institut für Nanotechnologie, Research Center Karlsruhe, Karlsruhe Institute of Technology, D-76021 Karlsruhe, Germany

Transport properties of strongly interacting quantum systems are a major challenge in todays condensed matter theory. In our project we apply the density matrix renormalization group (DMRG) method [1–6] to study transport properties [7–11] of quantum devices attached to metallic leads. To this end we have developed two complementary approaches to obtain conductance of a structure coupled to left and right leads. First we use the Kubo approach [12] to obtain linear conductance. Combined with leads described in momentum space [13, 14] we have obtained high resolution in energy. In this report we extend the results based on the Kubo approach to systems with degenerate orbitals. The second approach is based on simulating the time evolution [15–17] of an initial state with a charge imbalance [18]. In a cooperation with Edouard Boulat and Hubert Saleur we have been able to show that our approach is in excellent agreement with analytical calculations in the framework of the Bethe ansatz [19]. This agreement is remarkable as the numerics is carried out in a lattice model, while the analytical result is based on field theoretical methods in the continuum. Therefore we have to introduce a scale T_B to compare the field theoretical result to our numerics. Remarkably, at the so called self-dual point the complete regularization can be expressed by a single number, even for arbitrary contact hybridization t'. Most strikingly we proved the existence of a negative differential conductance (NDC) regime even in this simplistic model of a single resonant level with interaction on the contact link. In an extension of this approach we present results for current-current correlations, including shot noise, based on our real time simulations.

1 Shot Noise in the Interacting Resonant Level Model

The study of current fluctuations in nanodevices such as quantum point contacts and tunnel junctions is deeply connected with some of the most important physical questions. These include the nature of fundamental excitations

W.E. Nagel et al. (eds.), *High Performance Computing in Science and Engineering '10*, DOI 10.1007/978-3-642-15748-6_13,

in strongly interacting electronic systems [20, 21], the possibility of fluctuation theorems out of equilibrium [22], and the time evolution of many body entanglement [23]. Experimental progress in this area has been swift - second and third cumulants have been measured in several systems [24, 25], shot noise of single hydrogen molecules has been measured [26], and even the full counting statistics has been obtained in semi conductor quantum dots [27].

In [11] we reported on our attempt to calculate shot noise from the simulation of the time evolution of a one-dimensional system with two leads coupled to a single impurity. We described a method that allows to extract the noise power spectrum by means of a fourier transform of the current fluctuations. To remove finite size effects we could show that it may be sufficient to apply a linear extrapolation to infinite system size. Therefore, we calculated shot noise for the Resonant Level Model analytically for the thermodynamic limit using the Landauer–Büttiker approach in order to be able to compare to our numerical results. For the numerical simulation we have been able to use exact diagonalisation methods, based on a single particle representation and therefore limited to non-interacting systems. We found very nice agreement of analytical and numerical results. However, the real challenge begins when interaction is included, since then the full many-particle problem has to be treated. Therefore one has to resort to approximative methods. The td-DMRG approach allows for the simulation of the time evolution of interacting 1D systems. In particular we calculated the time dependent current fluctuations

$$S(t,t') = \langle \Delta\hat{I}(t)\Delta\hat{I}(t')\rangle_{\Psi}, \quad \Delta\hat{I}(t) = \hat{I}(t) - \langle \hat{I}(t)\rangle_{\Psi}, \tag{1}$$

$$\hat{I}(t) = \mathrm{e}^{\mathrm{i}\hat{H}t}\,\hat{I}\,\mathrm{e}^{-\mathrm{i}\hat{H}t}, \quad \langle\cdot\rangle_{\Psi} = \langle\Psi|\cdot|\Psi\rangle, \tag{2}$$

for the Interacting Resonant Level Model (IRLM)

$$\hat{H} = \hat{H}_0 + \hat{H}_{\mathrm{B}}, \quad \hat{H}_0 = -J\Big\{\sum_{m=-M_{\mathrm{L}}}^{-2} \hat{c}^{\dagger}_{m+1}\hat{c}_m + \sum_{m=1}^{M_{\mathrm{R}}-1} \hat{c}^{\dagger}_{m+1}\hat{c}_m\Big\} + \text{H.c.}, \tag{3}$$

$$\hat{H}_{\mathrm{B}} = \sum_{m=\pm 1} \Big\{U\Big(\hat{n}_m - \frac{1}{2}\Big)\Big(\hat{n}_{\mathrm{d}} - \frac{1}{2}\Big) - J'(\hat{c}^{\dagger}_m\hat{d} + \hat{d}^{\dagger}\hat{c}_m)\Big\} + V_{\mathrm{g}}\hat{n}_{\mathrm{d}}, \tag{4}$$

with the creation and annihilation operators as well as the density operators in the leads ($\hat{c}^{\dagger}_j$, $\hat{c}_j$, $\hat{n}_j = \hat{c}^{\dagger}_j\hat{c}_j$) and on the level ($\hat{d}^{\dagger}$, $\hat{d}$, $\hat{n}_{\mathrm{d}} = \hat{d}^{\dagger}\hat{d}$). The size of the leads has to be finite due to the operation principle of the method. The number of lattice sites in the left (right) lead is denoted by M_{L} (M_{R}). The fluctuations then allow for the calculation of the noise power spectrum by means of a fourier transform

$$S(\omega) = 4\mathrm{Re}\int_0^{\infty} \mathrm{d}t\,\mathrm{e}^{\mathrm{i}\omega t} S(t,t') \tag{5}$$

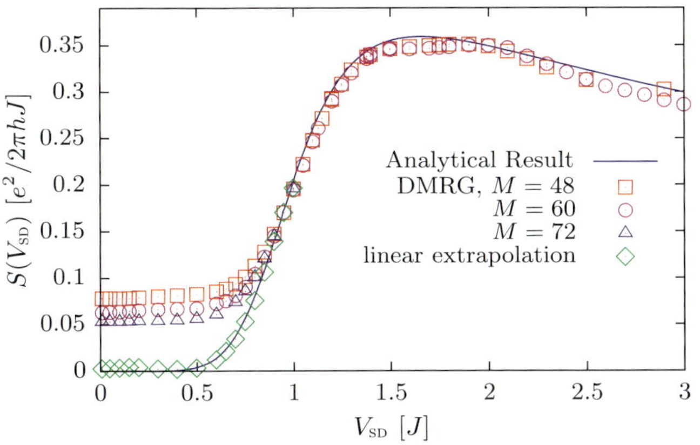

Fig. 1. Linear extrapolation of the numerical shot noise data. The linear scaling of the low voltage finite size error is exploited to perform a linear extrapolation $1/M \to 0$. We find nice agreement of numerical and analytical results

where we restricted ourselves to the low frequency limit $\omega = 0$. The initial state $|\Psi\rangle$ is obtained as the quantum mechanical ground state of the Hamiltonian $\hat{H} + V_{\text{SD}}(\hat{N}_{\text{L}} - \hat{N}_{\text{R}})/2$. The particle number operators in the left ($\hat{N}_{\text{L}}$) and right ($\hat{N}_{\text{R}}$) lead enter with opposite sign, which generates a charge imbalance in the initial state corresponding to a difference of the chemical potential in the left and the right lead.

The details of the numerical simulation have been discussed in [11, 28]. At this point we want to mention that the calculations have been carried out mainly at the XC2 of the SCC Karlsruhe consuming about one year of computation time with a typical load of 200–500 CPU cores. In contrast to [11], where we have been able to present results for the non-interacting RLM and preliminary results for the IRLM only, we now have nice results for the IRLM, Fig. 1, in the resonant tunneling case $V_{\text{g}} = 0$ at the so-called self-dual point $U = 2.0J$ [29]. The analytical result has been provided by Edouard Boulat and Hubert Saleur allowing for a verification of our approach. The points correspond to results obtained numerically, with the coupling of the structure to the leads set to $J' = 0.4J$. The figure shows results for different system sizes with a total number of lattice sites $M = M_{\text{L}} + M_{\text{R}} + 1 = 48 \ldots 72$ ($M_{\text{R}} = M_{\text{L}} + 1$), as well as the thermodynamic limit obtained by a linear extrapolation $1/M \to 0$. The comparison with the analytical result shows that the correction of finite size errors works very nicely and leads to good agreement of the both approaches. In contrast to the analytical solution the numerical approach is not restricted to a specific value of the interaction.

2 Transport in the Presence of Degenerate Levels

Here we consider a model consisting of a (hexagonal) ring of sites, through which is threaded a single flux quantum. The ring is then coupled symmetrically to leads in a two-terminal setup, as indicated in Fig. 2. While penetrating a real benzene molecule with a flux quantum would need magnetic field of the order of $\sim 10^5$T, we have chosen this model as the non-interacting ring has now the interesting property that its conductance vanishes for all gate voltages [30].

The flux quantum through the ring is modeled as a phase on the hopping-matrix elements on the ring using Peierls substitution, such that the relative phase of the hopping elements on the ring changes by π going once around the ring. The distribution of this phase over the ring is insignificant as it can always be gauged into a single bond. In this work we choose to modify a single hopping element only, denoted J_{BC}, as indicated in red in Fig. 2. Including a particle-hole symmetric nearest-neighbor density-density interaction on the ring the Hamiltonian of the model is

$$\mathcal{H} = \mathcal{H}_{\mathrm{Ring}} + \mathcal{H}_{\mathrm{lead}} + \mathcal{H}_{\mathrm{Contact}}, \tag{6}$$

$$\begin{aligned} \mathcal{H}_{\mathrm{Ring}} = & \sum_{i=1}^{M_{\mathrm{S}}-1} \left(J_i d_i^\dagger d_{i+1} + J_i^* d_{i+1}^\dagger d_i \right) + J_{\mathrm{BC}} d_{M_{\mathrm{S}}}^\dagger d_1 + J_{\mathrm{BC}}^* d_1^\dagger d_{M_{\mathrm{S}}} \\ & + \sum_{i=1}^{M_{\mathrm{S}}-1} U_i \left(n_i - \frac{1}{2} \right) \left(n_{i+1} - \frac{1}{2} \right) + U_{M_{\mathrm{S}}} \left(n_{M_{\mathrm{S}}} - \frac{1}{2} \right) \left(n_1 - \frac{1}{2} \right) \\ & + \sum_{i=1}^{M_{\mathrm{S}}} V_{\mathrm{gate}} n_i \end{aligned} \tag{7}$$

$$\begin{aligned} \mathcal{H}_{\mathrm{lead}} = & \sum_{i=1}^{M} \sum_{\alpha=L,R} \left(J_{i,\alpha} c_{i,\alpha}^\dagger c_{i+1,\alpha} + J_{i,\alpha}^* c_{i+1,\alpha}^\dagger c_{i,\alpha} \right) \\ & + \sum_{k} \sum_{\alpha=L,R} \left(\epsilon_{k,\alpha} n_{k,\alpha} + J_{k,\alpha} c_{M,\alpha}^\dagger c_{k,\alpha} + J_{k,\alpha}^* c_{k,\alpha}^\dagger c_{M,\alpha} \right) \end{aligned} \tag{8}$$

$$\mathcal{H}_{\mathrm{Contact}} = J_{\mathrm{L}} d_1^\dagger c_{1,L} + J_{\mathrm{L}}^* c_{1,L}^\dagger d_1 + J_{\mathrm{R}} d_{M_{\mathrm{S}}/2+1}^\dagger c_{1,R} + J_{\mathrm{R}}^* c_{1,R}^\dagger d_{M_{\mathrm{S}}/2+1}, \tag{9}$$

where $n_\ell = d_\ell^\dagger d_\ell$ is the local density operator of site ℓ, and $n_{k,\alpha} = c_{k,\alpha}^\dagger c_{k,\alpha}$ is the density operator for momentum level k in lead α. $M_{\mathrm{S}} = 6$ denotes the number of sites in the ring, M the number of real-space sites in the leads, and k labels the momentum-space sites of the leads. In this work we use the values $J_i = J_{\mathrm{Dot}}$ on the dot and give values of J_{BC} in units of J_{Dot}. Mostly we consider the case $J_{\mathrm{BC}} = -J_{\mathrm{Dot}}$ corresponding to a single flux-quantum through the ring, $\Phi = \Phi_0$. Further, we use the values $J_{\mathrm{L}} = J_{\mathrm{R}} = 0.5 J_{\mathrm{Dot}}$ for the coupling to the leads. Additionally we use a combination of a logarithmic discretization

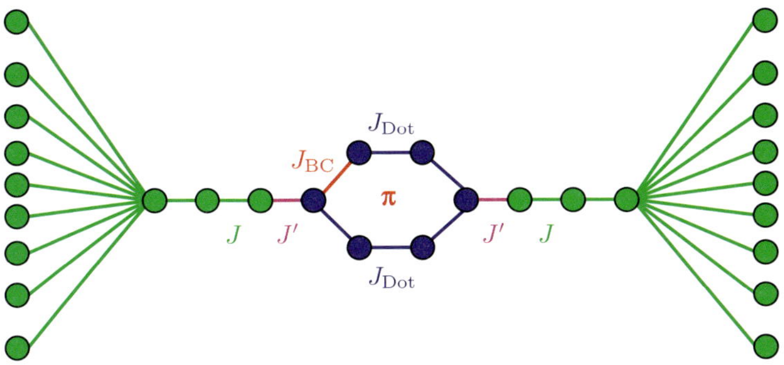

Fig. 2. Illustration of the finite size setup used in the DMRG calculations. The sites in the ring are shown in blue, and the implementation of the flux, $J_{\rm BC}$, is shown as a red hopping between sites 1 and 6 on the ring. Each lead (shown in green) is described by a real-space part coupled to a momentum-space part. The real-space part of the leads ensures a proper representation of local physics, whereas the momentum-space part ensures that the low-energy spectrum is properly represented. Note that the outermost real-space site in each lead is coupled to *all* momentum-space sites of that lead

to cover a large energy-scale of the band, and switch to a linear discretization for the low-energy sector close to the Fermi edge [13]. We model each lead by a real-space tight-binding chain coupled to a momentum-space part [13], an advantageous setup for representing all relevant energy scales of a given problem. Within this setup we evaluate the Kubo formula for conductance, explicitly the two correlators [12]

$$g_{I_i I_j} = \frac{e^2}{h} \langle\psi_0| \, I_i \frac{8\pi\eta(\mathcal{H}-E_0)}{\left((\mathcal{H}-E_0)^2+\eta^2\right)^2} I_j \, |\psi_0\rangle \,, \tag{10}$$

$$g_{I_i N} = -\frac{e^2}{h} \langle\psi_0| \, I_i \frac{4\pi i\eta}{(\mathcal{H}-E_0)^2+\eta^2} N \, |\psi_0\rangle \,, \tag{11}$$

where $I_\ell = i(J_\ell^* c_{\ell-1}^\dagger c_\ell - J_\ell c_\ell^\dagger c_{\ell-1})$ denotes the current operator on site ℓ and $N = \frac{1}{2}(N_L - N_R)$ is the rigid shift of the levels in the leads corresponding to the applied voltage perturbation. Note that the Hamiltonian $\mathcal{H}$ in (10), (11) contains all interactions and couplings but not the voltage perturbation, and $|\psi_0\rangle$ is the ground state of this Hamiltonian.

2.1 Kubo Approach for Degenerate Ground-States

In the case where the ground-state $|\psi_0\rangle$ is (near-) degenerate the evaluation of the Kubo formula sketched above breaks down. Rather than a single ground-state correlator at zero temperature, a finite temperature multiple 'ground-state' average must be used, since in general it cannot be decided numerically whether a finite gap is a physical gap, or caused by numerical inaccuracies. The evaluation of the Kubo-formula in this case is thus

$$g = \frac{1}{Z} \sum_{n=1}^{N} e^{-\beta(E_n - E_0)} g_n, \tag{12}$$

where g_n is the conductance of the n'th (near-) degenerate ground-state level calculated using (10), (11), β is the inverse temperature, E_n is the energy of the n'th level, and $Z = \sum_{n=1}^{N} e^{-\beta(E_n - E_0)}$ is the partition sum. Here we set β of the order of the inverse level spacing of the leads, thus averaging over low lying states having an excitation gap smaller than the finite size resolution of the leads. In addition N is chosen sufficiently large to cover all relevant (near-) degenerate states.

2.2 Results

It was shown in [31, 32] that a continuum model of a 1D ring of non-interacting spinless fermions threatened by a flux π shows antiresonances at flux $\phi = \pi$ leading to new phenomena in the high temperature limit when interaction is turned on [33]. In the non-interacting limit the transport properties of the benzene-like ring-structure sketched in Fig. 2 can be calculated exactly and in the infinite lead limit the conductance is identical to zero for all gate voltages, for all couplings J' and at all temperatures due to a perfect interference between the two paths through the ring. Note that on a lattice this property does not hold for a ring consisting of four or eight sites. Using DMRG to evaluate the Kubo formula for conductance we calculate the conductance for different values of the strength of the interaction. We typically use 400–800 states per block in the DMRG procedure, and the momentum-space part of the lead is described by 40 logarithmically discretized levels to cover the broad energy-range of the band, and additionally 10 linearly discretized levels close to the Fermi-edge to ensure a good discretization here.

The obtained conductances are shown in Fig. 3. In the non-interacting limit we do indeed find a vanishing conductance with only minor finite-size deviations, originating from the finite size of the lead used in the DMRG setup.

Increasing the strength of the interaction, U, the value of the conductance also increases, and eventually a resonance is formed at zero gate-potential. For the interaction strength $U = 2$ the 'resonant' value of the conductance is thus found to be $g \approx 0.75$. The shape of this resonance, i.e. the exponential decay of the resonance with the gate voltage, differs significantly from the Lorentzian shape usually found in simple resonant systems [12, 13], indicating that a more complicated mechanism is at play.

Increasing the strength of the interaction further, $U = 2.25$, a broad 'plateau' in the conductance is formed around zero gate-potential. The plateau

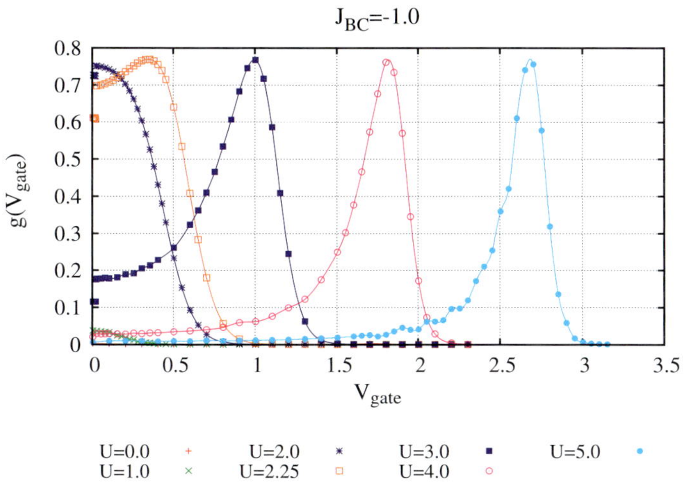

Fig. 3. Conductance versus gate-potential for the ring-structure varying the nearest-neighbor interaction strength U. In the non-interacting limit the conductance is zero to the precision of our finite-size setup, whereas the values for moderate interaction strengths approaches the unitary limit. Lines are added to the DMRG data as guides to the eye

is significantly wider than a Lorentzian of the corresponding height, and resembles somewhat a split Kondo resonance, the splitting introduced by the hopping to the leads that also allows for transport. To show the similarity to the single impurity Anderson model (SIAM) we label in Fig. 4 the left (right) lead as up (down) electrons. Setting $J_{\mathrm{Dot}} = 0$ no mixing of 'up' and 'down' states occurs, and strong interaction forbids adding/removing an additional particle. In order to have transport it is necessary to switch on the hybridization between the dot sites, which acts like a magnetic field supressing the proposed geometric Kondo effect. In our case, due to the added flux, the degeneracy of the single-particle levels is not lifted leaving room for Kondo physics. Nevertheless, the hybridization of the dot provides a mixing of the up and down states proposed in Fig. 4, which is necessary to enable transport through the ring. However, by increasing U a charge density wave (CDW) ordering again becomes preferred when the interaction reestablishes two well separated states. It is interesting to note, that the effect is most dominant for interaction values close to where a phase transition to a CDW ordered state appears in the thermodynamic limit at $U_{\mathrm{c}} = 2J$.

In order to test this idea we have performed similar calculations on slightly asymmetric rings, reducing the magnitude of a single hopping slightly from unity, breaking the degeneracy between the single particle levels, and obtained the conductances shown in Fig. 5. As the figure shows for the interaction strength $U = 2.25$, the introduced asymmetry rapidly destroys the effect, as expected for Kondo physics.

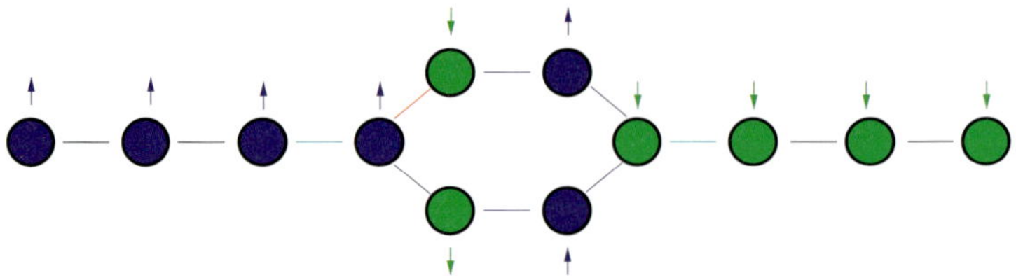

Fig. 4. Conjectured Kondo setup

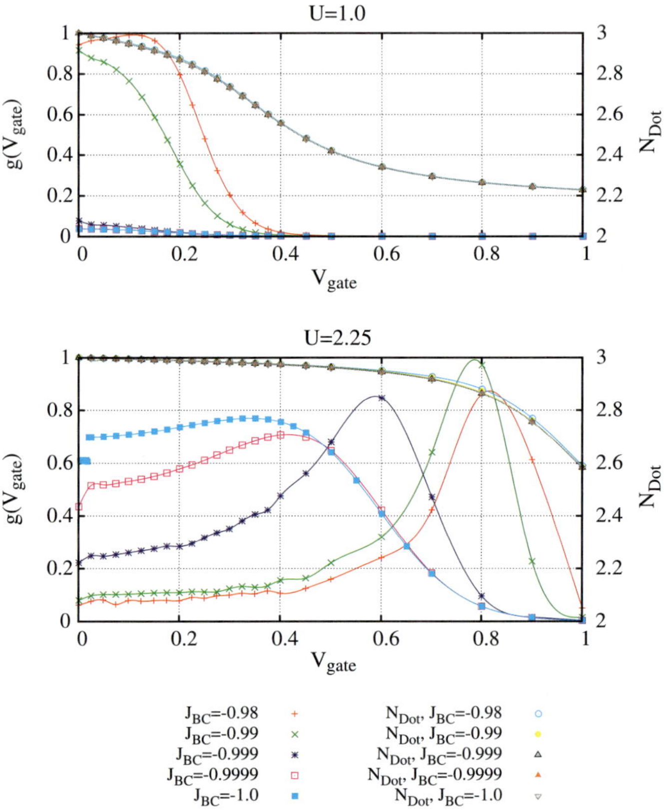

Fig. 5. An asymmetry of the hopping in the ring rapidly destroys the geometric Kondo-effect. This figure shows the effect on the conductance and the occupation of the ring when modifying a single hopping by 0.01-2%. When the geometric Kondo-effect is destroyed a normal resonant structure pattern is rediscovered, and the resonance moves towards the normal position where the occupation is half-integer. Lines are added to the DMRG data as guides to the eye

In [30] a Kondo temperature of the order of $T_{\mathrm{K}} \sim 10^{-4}$ for the parameter in this study was estimated based on Schrieffer–Wolff transformation and ignoring the levels far away from the Fermi surface of the leads, which is in nice agreement with the results of Fig. 5, since the Kondo peak gets destroyed by a perturbation of the order of T_{K}.

Also plotted in Fig. 5 is the total density of the ring for the various parameter choices. Remarkably, the electron-density of the ring remains virtually unchanged when the asymmetry is varied, although the conductance of the ring changes significantly. This clearly demonstrates that the observed effect is an interference effect. Furthermore, increasing the asymmetry of the ring and thereby destroying the interference effect, the resonance is pushed towards the usual location for resonant structures in the Coulomb blockade regime, where the particle number on the structure is half-integer, and at the same time the line-shape becomes increasingly Lorentzian, although an asymmetry with a long tail persists.

In the case $U = 1.0$, where no geometric Kondo-effect is present, reducing the symmetry in the ring also results in changes. From the symmetric case, $J_{\mathrm{BC}} = -J_{\mathrm{Dot}}$, where only a very small resonance is found, to the most asymmetric case considered here, $J_{\mathrm{BC}} = -0.98 J_{\mathrm{Dot}}$, a clear resonance develops.

We explain this new resonance as the result of a population blocking mechanism, similar to [34]: When applying a gate voltage the upper level of the two levels close to the Fermi surface of the leads is pushed out of resonance first. The second level remains occupied since emptying the level would cost interaction energy due to the particle-hole symmetric interaction. Therefore adding or removing a particle costs interaction energy and the lower level remains occupied although it is pushed above the Fermi level. In that way the degeneracy of the levels is lifted and therefore the destructive interference is removed, details are explained in [30].

A note is in order about the calculations for small V_{gate}: In a region around $V_{\mathrm{gate}} = 0$ the combined lead and ring system is effectively degenerate, and thus the degenerate method is applied in this parameter range. However, even using this expression the evaluation of the conductance for small gate-potentials remains difficult due to numerically difficult resolvent equations, and the results obtained there are not expected to be accurate. In future work we aim at resolving this problem by considering spectral functions in combination with the Meir-Wingreen formula, rather than evaluating the Kubo formula for conductance.

At the fundamental level the results shown in this work clearly demonstrate that the simple picture of individual electrons passing through the transport region one after the other gives a significantly different result as opposed to the situation where moderate electron-electron interaction is included. Rather a complicated many-body interference effect is formed, and we have proposed/conjectured a Kondo-effect as the explanation for the remarkable line-shape observed. The proposed Kondo-effect lies in the geometrical degree of freedom, and hence differs from the standard Kondo-effect in the spin-degree of freedom. Introducing an asymmetry in the ring clearly destroys the effect, in a manner similar to the effect of a magnetic field on the standard Kondo-effect. In addition we have shown that due to interaction effects new conductance peaks can appear which are based on a population blocking mechanism and which are not related to Kondo physics and it will be intresting to see how these interaction effects could be reflected in an exact density functional theory approach [35].

Acknowledgements. The DMRG calculations were performed on the HP XC4000 at the Steinbuch Center for Computing (SCC) Karlsruhe under project RT-DMRG, with support through project B2.10 of the DFG Centre for Functional Nanostructures. The section on the Kubo approach was a joined project with Dan Bohr [30].

References

1. S.R. White. *Phys. Rev. Lett.*, 69:2863, 1992.
2. S.R. White. *Phys. Rev. B*, 48:10345, 1993.
3. I. Peschel, X. Wang, M. Kaulke, and K. Hallberg, editors. *Density Matrix Renormalization*, 1999.
4. Reinhard M. Noack and Salvatore R. Manmana. Diagonalization- and numerical renormalization-group-based methods for interacting quantum systems. In Adolfo Avella and Ferdinando Mancini, editors, *Lectures on the Physics of Highly Correlated Electron Systems IX: Ninth Training Course in the Physics of Correlated Electron Systems and High-Tc Superconductors*, volume 789, pages 93–163, Salerno, Italy, 2005. AIP.
5. Karen A. Hallberg. New trends in density matrix renormalization. *Adv. Phys.*, 55(5):477–526, 2006.
6. U. Schollwöck. The density-matrix renormalization group. *Rev. Mod. Phys.*, 77(1), 2005.
7. Peter Schmitteckert. Nonequilibrium electron transport using the density matrix renormalization group. *Phys. Rev. B*, 70:121302(R), 2004.
8. P. Schmitteckert and G. Schneider. Signal transport and finite bias conductance in and through correlated nanostructures. In Wolfgang E. Nagel, Willi Jäger, and Michael Resch, editors, *High Performance Computing in Science and Engineering '06*, pages 113–126. Springer, Berlin, 2006.
9. P. Schmitteckert. Signal transport in and conductance of correlated nanostructures. In Wolfgang E. Nagel, Dietmar B. Kröner, and Michael Resch, editors, *High Performance Computing in Science and Engineering '07*, pages 99–106. Springer, Berlin, 2007.
10. Tobias Ulbricht and Peter Schmitteckert. Signal transport in and conductance of correlated nanostructures. In Wolfgang E. Nagel, Dietmar B. Kröner, and Michael Resch, editors, *High Performance Computing in Science and Engineering '08*, pages 71–82. Springer, Berlin, 2008.
11. Alexander Branschädel, Tobias Ulbricht, and Peter Schmitteckert. Conductance of correlated nanostructures. In Wolfgang E. Nagel, Dietmar B. Kröner, and Michael Resch, editors, *High Performance Computing in Science and Engineering '09*, pages 123–137. Springer, Berlin, 2009.
12. D. Bohr, P. Schmitteckert, and P. Wölfle. DMRG evaluation of the Kubo formula – conductance of strongly interacting quantum systems. *Europhys. Lett.*, 73:246, 2006.

13. Dan Bohr and Peter Schmitteckert. Strong enhancement of transport by interaction on contact links. *Phys. Rev. B*, 75(24):241103(R), 2007.
14. Peter Schmitteckert. Calculating green functions from finite systems. *J. Phys.: Conf. Ser.*, 220:012022, 2010.
15. Tobias Ulbricht and P. Schmitteckert. Is spin-charge separation observable in a transport experiment? *EPL*, 86(5):57006+, 2009.
16. Tobias Ulbricht and Peter Schmitteckert. Tracking spin and charge with spectroscopy in spin-polarised 1d systems. *EPL*, 89:47001, 2010.
17. Tobias Ulbricht, Rafael A. Molina, Ronny Thomale, and Peter Schmitteckert. Color-charge separation in trapped su(3) fermionic atoms. *Phys. Rev. A*, 82(1):011603, Jul 2010.
18. A. Branschädel, Guenter Schneider, and P. Schmitteckert. Conductance of inhomogeneous systems: Real-time dynamics. *Ann. Phys. (Berlin)*, 522:657, 2010.
19. E. Boulat, H. Saleur, and P. Schmitteckert. Twofold advance in the theoretical understanding of far-from-equilibrium properties of interacting nanostructures. *Physical Review Letters*, 101(14):140601, 2008.
20. R. de Picciotto, M. Heiblum, H. Shtrikman, and D. Mahalu. *Phys. Rev. Lett.*, 75:3340, 1995.
21. A. Kumar, L. Saminadayar, D.C. Glattli, Y. Jin, and B. Etienne. *Phys. Rev. Lett.*, 76:2778, 1996.
22. M. Esposito, U. Harbola, and S. Mukamel. *Rev. Mod. Phys.*, 81:1665, 2009.
23. I. Klich and L. Levitov. *Phys. Rev. Lett.*, 102:100502, 2009.
24. B. Reulet, J. Senzier, and D.E. Prober. *Phys. Rev. Lett.*, 91:196601, 2003.
25. Y. Bomze, G. Gershon, D. Shovkun, L.S. Levitov, and M. Reznikov. *Phys. Rev. Lett.*, 95:176601, 2005.
26. D. Djukic and J.M. van Ruitenbeek. Shot noise measurements on a single molecule. *Nano Letters*, 6(4):789, 2006.
27. S. Gustavsson, R. Leturcq, B. Simovic, R. Schleser, T. Ihn, P. Studerus, and K. Ensslin. *Phys. Rev. Lett.*, 96:076605, 2006.
28. A. Branschädel, E. Boulat, H. Saleur, and P. Schmitteckert. Numerical evaluation of shot noise using real time simulations. arXiv:1004.4784.
29. A. Branschädel, E. Boulat, H. Saleur, and P. Schmitteckert. Shot noise in the self-dual interacting resonant level model. *Phys. Rev. Lett.*, page n./a. (accepted for publication).
30. Dan Bohr and Peter Schmitteckert. The dark side of benzene: interference vs. interaction. (unpublished).
31. M. Büttiker, Y. Imry, and M.Ya. Azbel. Quantum oscillations in one-dimensional normal-metal rings. *Phys. Rev. A*, 30(4):1982, 2007.
32. Y. Gefen, Y. Imry, and M.Ya. Azbel. Quantum oscillations and the Aharonov-Bohm effect for parallel resistors. *Phys. Rev. Letters*, 52(2):129, 2007.
33. A.P. Dmitriev, I.V. Gornyi, V.Yu. Kachorovskii, and D.G. Polyakov. Aharonov-Bohm conductance through a single-channel quantum ring: Persistent-current blockade and zero-mode dephasing. *Phys. Rev. Lett.*, 105:036402, 2010.
34. Moshe Goldstein, Richard Berkovits, Yuval Gefen, and Hans A. Weidenmueller. Transmission phase of a quantum dot: Testing the role of population switching. *Phys. Rev. B*, 79(12):125309, 2009.
35. Peter Schmitteckert and Ferdinand Evers. Exact ground state density-functional theory for impurity models coupled to external reservoirs and transport calculations. *Phys. Rev. Lett.*, 100(8):086401, Feb 2008.

Cu Substitutionals and Defect Complexes in the Lead-Free Ferroelectric KNN

Sabine Körbel and Christian Elsässer

Fraunhofer Institute for Mechanics of Materials IWM, Wöhlerstraße 11, 79108 Freiburg, Germany, sabine.koerbel@iwm.fraunhofer.de

1 Introduction

Piezoelectric devices like actors and sensors are often made of lead zirconate titanate ($PbTi_xZr_{1-x}O_3$, PZT) because of its superior ferroelectric properties as compared to other available materials. Since efforts are made to remove toxic metals, for instance lead, from industrial materials, the lead-free ferroelectric potassium sodium niobate ($K_xNa_{1-x}NbO_3$, KNN) has become an object of intensive research. Like PZT, KNN crystallizes in the perovskite crystal structure and exhibits an approximately temperature-independent morphotropic phase boundary at the experimentally easily accessible composition of K:Na≈50/50. The Curie temperature of KNN is comparatively high (approximately 400° C), allowing operation in a wide temperature range.

The properties of ferroelectric materials are commonly tailored by doping with aliovalent metals, i.e. by generating point defects. It is assumed that acceptor dopants form defect dipoles with O vacancies for charge compensation, which impedes domain wall motion and thus causes ferroelectric hardening. It has been shown with density functional theory that defect dipoles of Cu or Fe substitutionals and O vacancies in $PbTiO_3$ tend to align parallel to the spontaneous polarization [4], and the reversed effect has been observed experimentally [7, 16]: In an aged system, the defect dipoles can act as a restoring force for the polarization after the external electric field is switched off, enabling reversible switching by 90° which is accompanied by a large strain already at comparatively low electric fields. This could therefore be utilized for actuator applications. For memory applications on the other hand, such a polarization restoring mechanism is undesirable.

Simulating the microscopic effects of doping on the macroscopic ferroelectric material properties can help to identify potentially advantageous doping elements and therefore possibly to reduce experimental efforts. Density functional theory (DFT) is a widely employed method for accurately calculating structural and energetical properties [22]. DFT has been extensively applied

W.E. Nagel et al. (eds.), *High Performance Computing in Science and Engineering '10*, DOI 10.1007/978-3-642-15748-6_14,

to perovskites and/or defects in numerous studies documented in the literature. In this study, we performed DFT calculations in order to determine which Cu defects and defect complexes of Cu substitutionals and vacancies are stable in KNN.

The lattice site on which a dopant is incorporated determines which types of defect complexes can occur in the material. It can be influenced to some extent by the chemical processing conditions, i. e. the partial pressure of oxygen and the temperature. We determined the most stable lattice site for Cu dopants as function of the oxygen partial pressure, the temperature, the alkali stoichiometry and the chemical potential for electrons (the Fermi level) [9]. For Cu dopants on the stable lattice sites, we then analyzed the stability of several conceivable defect complexes consisting of Cu dopants and oxygen or alkali vacancies [3].

2 Physical Approach

The most stable lattice site for a defect, e. g. a dopant, is determined by the defect formation enthalpy, for which in most cases of crystalline systems the defect formation energy E^f is a good approximation [22]:

$$E^f[X^q] = E_{\mathrm{tot}}[X^q] - E_{\mathrm{tot}}[\mathrm{bulk}] - \sum_i \mu_i n_i + q \cdot E_F, \qquad (1)$$

where X is the defect, q its charge, E_{tot} the total energy, bulk corresponds to an undoped reference cell, n_i is the number of atoms of species i exchanged with a reservoir when incorporating the dopant, and μ_i is the chemical potential. In the case of charged defects, electrons are exchanged with an electronic reservoir with the electronic chemical potential (the Fermi energy) E_F. Because the material is an insulator, the position of the Fermi energy in the band gap is not determined and is treated as a variable here.

Stability requirements confine the possible values of the chemical potentials to certain ranges [22]. Experimentally, the chemical potentials can be varied inside the stability area via, e.g., the temperature and/or the oxygen partial pressure.

The temperature and oxygen partial pressure can be related to the chemical potential of oxygen by, e.g., the ideal gas equation [18]:

$$\mu_{\mathrm{O}}(T,p) = \frac{1}{2}\left(H_{\mathrm{O}_2}(T,p_0) - TS_{\mathrm{O}_2}(T,p_0)\right) + \frac{1}{2}k_B T \ln\left(\frac{p}{p_0}\right),$$

where p_0 is the oxygen partial pressure in air at 1 bar, T is the temperature, and G, H and S are the Gibbs free energy, the enthalpy and the entropy per oxygen molecule. The latter two can be found in thermochemical tables and were taken from [10].

The stability of a defect complex is determined by its binding energy, which was calculated as the difference in defect formation energies of the defect complex and the isolated defects.

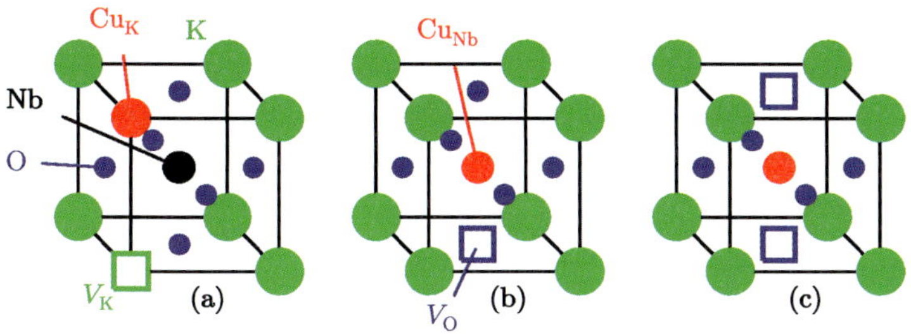

Fig. 1. Defect configurations: **a** ($Cu_K - V_K$), **b** ($Cu_{Nb} - V_O$), and **c** ($V_O - Cu_{Nb} - V_O$)

3 Computational Method

3.1 Details of the DFT Calculations

Density functional theory (DFT) in the local density approximation (LDA) was applied to determine the defect formation energies of substitutional Cu atoms on A or B sites in the perovskite ABO_3, for A or O vacancies, and for defect complexes consisting of substitutional Cu atoms and neighboring vacancies in $KNbO_3$ and KNN 50/50. The LDA may be suitable for modelling ferroelectric perovskites [19]. We considered the three defect complexes ($Cu_K - V_K$), ($Cu_{Nb} - V_O$), and ($V_O - Cu_{Nb} - V_O$), as schematically depicted in Fig. 1.

The defect formation energies were calculated for atomistic supercell models containing $2 \times 2 \times 2$ simple cubic (sc) unit cells (40 atoms) for isolated Cu substitutionals in $NaNbO_3$, KNN 50/50 and $KNbO_3$ and $2 \times 2 \times 3$ sc unit cells (60 atoms) for defect complexes. The cubic high-temperature structure and the LDA lattice constants of the perfect crystals were assumed throughout.

Optimally smooth norm-conserving pseudopotentials as proposed by Vanderbilt [20], the Ceperley-Alder [2] LDA exchange-correlation functional as parametrized by Perdew and Zunger [15], and Monkhorst-Pack [14] $4 \times 4 \times 4$ (40-atoms cells) and $4 \times 4 \times 3$ (60-atoms cells) k-point meshes with Gaussian broadening [6] of 0.2 eV were applied. The calculations were performed with the mixed-basis pseudopotential (MBPP) method [5, 8, 11–13], employing a basis of plane waves up to a maximum energy of 340 eV combined with atom-centered basis functions for alkali $s+p$ semicore states, oxygen p valence states, Nb $s + p$ semicore and d valence states, and Cu d valence states. The atomic positions were optimized using the BFGS algorithm until the maximum force was smaller than 10 meV/Å. In the case of charged defects a compensating homogeneous background charge density was included in the calculations of electrostatic energies and potentials.

The mixed perovskite KNN ($(K, Na)NbO_3$) was modelled with a so called virtual crystal approximation (VCA) for the alkali atoms, a method that replaces two species by a third, virtual species with intermediate properties. The VCA was applied to e.g. PZT by others before [1, 17].

The LDA bandgaps of bulk $KNbO_3$ (about 1.6 eV) and KNN (about 1.7 eV) were used throughout. We did not correct for the LDA bandgap error (the experimental bandgap of $KNbO_3$ is 3.3 eV [21]) because there exists no unique method for correcting the LDA bandgap which is computationally feasible for a considerable number of calculations as were necessary for this study (different defect configurations, each in several different charge states).

Spin-polarization of the substitutional Cu impurities was neglected. Test calculations including spin-polarization yielded energy differences of about 0.2 eV between the polarized and the unpolarized state, thus the spin-polarization of the Cu defects has only a small effect on the defect formation energies.

3.2 HPC Resources Used for the Present Study

In the DFT code used for the present calculations (MBPP), the k-loop is paralellized using MPI. The shortest computation time is therefore achieved for one processor per k-point. The number of k-points (and processors) used in the present calculations lies between 4 and 32, depending on the symmetry of the structure. The typical runtime for one SCF (self-consistent field) calculation varies between a few hours and about two days. In some cases, it took more than 50 SCF calculations until the forces fell below the given limit. The program stores binary restart data that can amount to about 800 MB, the maximum memory required by any process was in the same range (up to about 700 MB). The largest part of the memory needed is used for storing the eigenvalues of the Hamilton matrix.

Figure 2 shows the time needed for calculating the total energy of a test structure (a grain boundary in Fe) for different numbers of processors, and

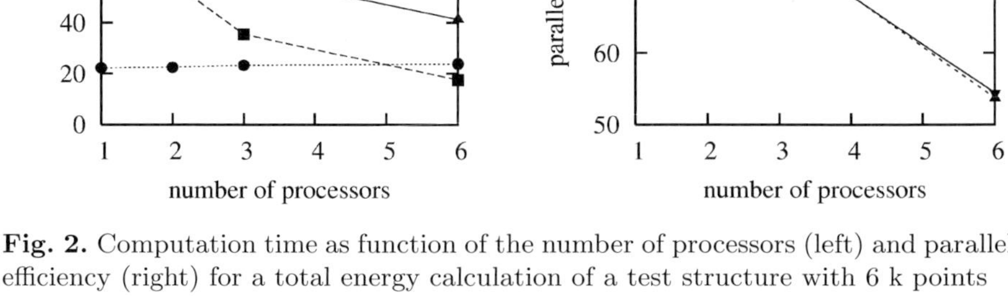

Fig. 2. Computation time as function of the number of processors (left) and parallel efficiency (right) for a total energy calculation of a test structure with 6 k points

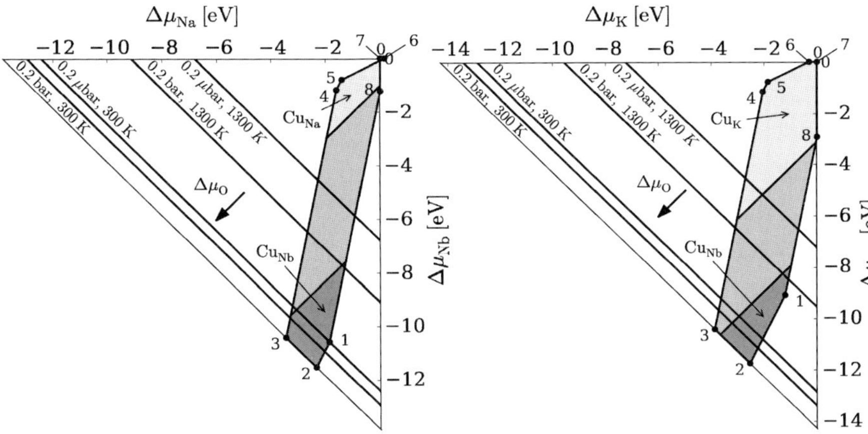

Fig. 3. Phase stability diagram for $NaNbO_3$ (left) and $KNbO_3$ (right). The areas of chemical potentials for which the perovskite phase is stable with respect to other competing phases is shaded in grey. The regions for which Cu favors the *A* or *B* site ($Cu_{Na/K}$ and Cu_{Nb}, respectively) are indicated by arrows

the parallel efficiency as compared to Amdahl's law. The present MPI implementation follows Amdahl's law very closely, due to good load balancing (every processor performs the same task for a different *k*-point) and very little communication.

4 Results

4.1 Energetically Preferred Lattice Sites for Cu Dopants

Figure 3 shows the phase stability diagrams of $NaNbO_3$ and $KNbO_3$. The shaded areas are the allowed regions for the chemical potentials. For ambient temperatures and pressures, the defect formation energy (cf. (1)) for Cu substitutionals is lower on a Nb than on an alkali site for all Fermi levels in the LDA bandgap (indicated by the dark grey area). If the air pressure is reduced to about 1 μbar and the temperature approaches 1300 K (close to the sintering temperature), the alkali substitution becomes more favorable (light gray area). In the region in-between, the defect formation energies as function of the Fermi level cross inside the bandgap, so that the Fermi level determines which site is more favorable.

Increasing the K content (from 0 in $NaNbO_3$ to 1 in $KNbO_3$) slightly stabilizes Cu on an alkali site. A calculation for KNN 50/50 (not shown) gives intermediate results. However, the influence of the alkali stoichiometry is small as compared to those of temperature and oxygen partial pressure.

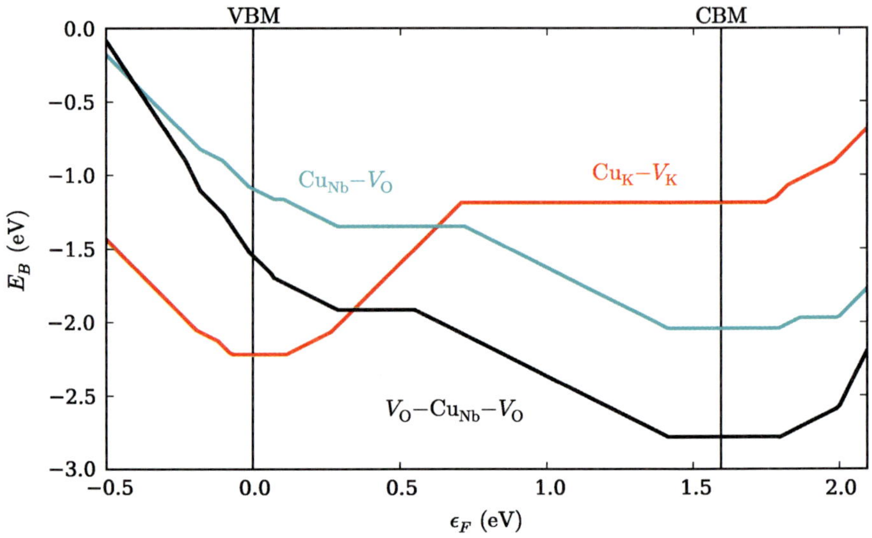

Fig. 4. Binding energies of the defect complexes ($Cu_K - V_K$), ($Cu_{Nb} - V_O$), and ($V_O - Cu_{Nb} - V_O$) versus Fermi energy in $KNbO_3$. The LDA valence band maximum (VBM) and conduction band minimum (CBM) are indicated by vertical lines

4.2 Stability of Defect Complexes of Cu Substitutionals and O Vacancies

Figure 4 shows the binding energies of the defect complexes ($Cu_K - V_K$), ($Cu_{Nb} - V_O$), and ($V_O - Cu_{Nb} - V_O$) in $KNbO_3$. All three considered defect complexes in $KNbO_3$ have negative binding energies and are therefore stable at least at low temperatures for all Fermi levels in the LDA bandgap. For KNN (not shown), qualitatively the same behavior is observed. In KNN, the complexes are less strongly bound than in $KNbO_3$ (e.g., for a Fermi level at the VBM, E_B of ($V_O - Cu_{Nb} - V_O$) is about −1.5 eV in $KNbO_3$ and about −0.8 eV in KNN).

5 Discussion

The presented DFT calculations show that Cu dopants in KNN may substitute either on A or B sites of the perovskite structure ABO_3, depending on temperature and oxygen partial pressure. Therefore the processing conditions can determine if Cu acts as a donor on the A site or an acceptor on the B site.

Like in $PbTiO_3$, associates of dopants and vacancies in KNN are more stable than the isolated defects and therefore likely to form. ($V_O - Cu_{Nb} - V_O$) is energetically more favorable than ($Cu_{Nb} - V_O$), but with respect to defect concentrations at finite temperatures, ($V_O - Cu_{Nb} - V_O$) should be compared to ($Cu_{Nb} - V_O$) and an isolated O vacancy (V_O). The latter configuration has a much higher configurational entropy because of the many possible sites for the isolated V_O, and may therefore become the prevailing defect type at elevated temperatures.

Two of the three defect complexes of Cu in $KNbO_3$ studied in this work, ($Cu_K - V_K$) and ($Cu_{Nb} - V_O$), should exhibit a defect dipole moment, which could lead to ferroelectric hardening and to the aging effect described in [7], where a restoring force on the polarization, presumably caused by defect dipoles, was experimentally observed in doped $BaTiO_3$, $PbTiO_3$, and KNN crystals [7, 16].

6 Summary and Outlook

If processed at ambient temperature and pressure, Cu in KNN seems to occupy more likely a Nb site, but it could be driven to substitute on the alkali sites by a reducing atmosphere during processing. The Cu substitutionals can form the stable defect complexes ($Cu_A - V_A$) , ($Cu_{Nb} - V_O$) and ($V_O - Cu_{Nb} - V_O$), where A stands for K or Na. To simulate the reorientation of the polarization in the presence of these defect complexes, and the reorientation of the defect complexes themselves, requires large computational resources and will be addressed in a future work. DFT results for Cu doped $KNbO_3$ supercells will be used to optimize an empirical interatomic potential, and the DFT simulations of switching in Cu doped KNN will be complemented by atomistic simulations to reduce the computational effort.

Acknowledgements. This work is funded by the German Research Foundation (projects EL 155/21-1,2). The provided computation time and storage on the HP XC4000 computer at the Karlsruhe Institute of Technology is gratefully acknowledged.

References

1. Bellaiche L., Vanderbilt D. Virtual Crystal approximation revisited: Application to dielectric and pietoelectric properties of perovskites. Phys. Rev. B, **61**, 7877 (2000).
2. Ceperley, D.M., Alder, B.J. Ground State of the Electron Gas by a Stochastic Method. Phys. Rev. Lett., **45**, 566 (1980).
3. Erünal, E., Eichel, R.-A., Körbel, S., et al. Defect structure of copper doped potassium niobate ceramics, Functional Materials Letters, **3** 19–24 (2010).
4. Erhart, P., Eichel, R.A., Träskelin, P., Albe, K. Association of oxygen vacancies with impurity metal ions in lead titanate. Phys. Rev. B, **76**, 174116 (2007).
5. Elsässer, C., Takeuchi, N., Ho, K., Chan, C., Braun, P., Fähnle, M. Relativistic effects on ground state properties of 4d and 5d transition metals. J. Phys. Cond. Matter, **2**, 4371 (1990).

6. Fu, C.L., Ho, K.M. First-principles calculation of the equilibrium ground-state properties of transition metals: Applications to Nb and Mo. Phys. Rev. B, **28**, 5480 (1983).
7. Feng Z., Ren X. Striking sinilarity of ferroelectric aging effect in tetragonal, orthorhombic and rhombohedral crystal structures. Phys. Rev. B, **77**, 134115 (2009).
8. Ho, K., Elsässer, C., Chan, C., Fähnle, M. First-principles pseudopotential calculations for hydrogen in 4d transition metals. I. Mixed-basis method for total energies and forces. J. Phys. Cond. Matter, **4**, 5189 (1992).
9. Körbel, S., Marton, P., Elsässer, C. Formation of vacancies and copper substitutionals in potassium sodium niobate under various processing conditions. Phys. Rev. B, **81**, 174115 (2010).
10. Lide, D.R. (ed.): CRC Handbook of Chemistry and Physics, 89th ed. CRC Press, Boca Raton (2008).
11. Lechermann, F., Welsch, F., Elsässer, C., Ederer, C., Fähnle, M., Sanchez, J.M., Meyer, B. Density-functional study of Fe_3Al: LSDA versus GGA. Phys. Rev. B, **65**, 132104 (2002).
12. Meyer, B., Hummler, K., Elsässer, C., Fähnle, M. Reconstruction of the true wavefunctions from the pseudowavefunctions in a crystal and calculation of electric field gradients. J. Phys. Cond. Matter, **7**, 9201 (1995).
13. Meyer, B., Lechermann, F., Elsässer, C., Fähnle, M. Fortran90 Program for Mixed-Basis Pseudopotential Calculations for Crystals. Max-Planck-Institut für Metallforschung, Stuttgart.
14. Monkhorst, H.J., Pack, J.D. Special points for Brillouin-zone integrations. Phys. Rev. B, **13**, 5188–5192 (1976).
15. Perdew, J.P., Zunger, A. Self-interaction correction to density-functional approximations for many-electron systems. Phys. Rev. B, **23**, 5048 (1981).
16. Ren X. Large electric-field-induced strain in ferroelectric crystals by point-defect-mediated reversible domain switching. Nature Mater., **3**, 91 (2004).
17. Ramer N.J., Rappe A.M. Application of a new virtual crystal approach for the study of disordered perovskites. J. Phys. Chem. Sol., **61**, 315 (2000).
18. Reuter, K., Scheffler, M. Composition, structure, and stability of $RuO_2(110)$ as a function of oxygen pressure. Phys. Rev. B, **65**, 035406 (2001).
19. Umeno, Y., Meyer, B., Elsässer, C., Gumbsch, P. Ab initio study of the critical thickness for ferroelectricity in ultrathin $Pt/PbTiO_3/Pt$ films. Phys. Rev. B, **74**, 060101 (2006).
20. Vanderbilt, D. Optimally smooth norm-conserving pseudopotentials. Phys. Rev. B, **32**, 8412 (1985).
21. Wiesendanger, E. Dielectric, mechanical and optical properties of orthorhombic $KNbO_3$. Ferroelectrics, **6**, 263 (1974).
22. Van de Walle, C.G., Neugebauer, J. First-principles calculations for defects and impurities: Applications to III-nitrides. J. Appl. Phys., **95**, 3851 (2004).

Reacting Flows

Prof. Dr. Dietmat Kröner

Abteilung für Angewandte Mathematik, Universität Freiburg,
Hermann-Herder-Str. 10, 79104 Freiburg, Germany

Spray combustion and simulations of combustion in combination with steam generation in utility boilers are the main topics in the three contributions of this section. All of them have been run on the NEC-platform at the HLRS. At least the codes in Matschke et al. and Lempke et al. are developed for improving the design process for real devices. These papers prove that the software development in this field has reached a level such that "the influence of a component failure ... is predictable in early design stage. The large acceleration that is gained from employing supercomputers like the NEC SX-9 allows engineers to keep in time with development and design changes."

In the contribution of Kronenburg et al. simulations of evaporating 2D methanol droplets in laminar and turbulent environments are analyzed. A level-set technique is combined with the ghost fluid method in order to track the liquid interphase and to simulate the surface tension. The mathematical model is based on the system of the compressible Navier-Stokes-equations and transport equations for the mass fractions of the species including diffusion and advection. The model is implemented in an in-house-low-Mach-number CFD code (BOFFIN), which is second order in space and time. The level set equation is solved by a 5th order WENO-scheme. Several numerical experiments in a laminar or turbulent environment are performed.

The main focus of the contribution of Lempke et al. is the utilization of an Euler Lagrange approach to simulate the interaction of liquid and gaseous phases in a model rocket combustor. The mathematical model consists also of the full system of the compressible Navier Stokes equations, transport equations for the species and two additional equations for the q-ω-low-Reynolds-number-turbulence model. To take into account turbulence chemical interactions a probability-density-function approach can be switched on. Eight species for the hydrogen oxidation in a set of 18 reactions are taken into account. The discrete equations are solved by an implicit lower-upper-symmetric-Gauss-Seidel-finite-volume-scheme. The chemistry is treated fully coupled with a fluid motion.

The liquid droplet spray is considered as a dispersed phase and treated by a Lagrangian approach. The dispersed phase is represented by a large number of computational particles, driven by the velocity from the Navier Stokes equations. For each droplet an ODE for the diameters of the droplets, the 3D velocities and the temperatures are taken into account. The source terms for this ODEs control in particular the vaporization rates and the heat fluxes through the surface of the droplets. Secondary break up is also included. The coupling between the spray and the surrounding gas is coded in the source terms of the compressible Navier Stokes equations and the ODEs for the spray.

The results are considered as a proof for the general applicability of the Euler-Lagrange approach. Comparisons with other results indicate that some aspects have to be improved.

In the paper of Matschke et al. the coupled simulation of the entire water/steam cycle for large utility boilers is studied. In particular the simulation of the combustion and the steam generation as well as the interaction between furnace and the steam cycle is considered. The new concept in this paper is a study of a three flux heat exchanger which exchanges heat between the flue gas, the super heater and the reheater.

The underlying mathematical model consists of the system of the compressible Navier Stokes equations and a global reaction scheme for the combustion. The authors mainly use direct numerical simulations. A first order upwind scheme is used for the discretization of all transport equations, a 3D-code for the furnace simulation (AIOLOS) and a 1D-code (DYNAMIC) for the water/steam simulation in the tubes. Numerical results concerning the tube temperature in the reheater are obtained and are in good agreement with measurements.

Scalar Mixing in Droplet Arrays in Stagnant and Convective Environments

A. Kronenburg[1], M.R.G. Zoby[2], S. Navarro-Martinez[2], and A.J. Marquis[2]

[1] Institut für Technische Verbrennung, University of Stuttgart, Stuttgart, Germany, kronenburg@itv.uni-stuttgart.de
[2] Department of Mechanical Engineering, Imperial College London, London SW7 2AZ, UK, m.zoby06@imperial.ac.uk, s.navarro@imperial.ac.uk, a.marquis@imperial.ac.uk

1 Introduction

Modelling sprays and spray combustion is a challenging task due to the wide range of the associated length and time scales and the complex interaction of turbulence and the physico-chemical processes, such as heat and mass transfers across interfaces. Most computational studies of evaporating sprays do not resolve the liquid phase nor the near field and the droplets are treated as point sources of mass, momentum, energy and species. Even if evaporation, and to some degree combustion, are directly dependent on local conditions, these studies neglect the conditions in the immediate neighbourhood of the individual droplet. Large differences will be found between local and cell-averaged fuel concentrations in the inter-droplet region [1]. These unresolved inhomogeneities can have profound effects on the accuracy of mixture fraction based combustion models that rely on strong dependencies of scalar dissipation and reactive species on mixture fraction within one CFD cell. Schroll et al. [2] pointed out that satisfactory closures may not be obtained using the source point approximation due to lack of resolution in the near liquid field.

In the present work, two-dimensional DNS of methanol droplets are analysed, assessing scalar mixing in terms of mixture fraction PDF and dissipation. Several droplet loadings in static, laminar and turbulent environments are simulated and the sensitivity of the local fuel mixing field to these parameters is quantified. The droplets are organised in infinite, regular, planar layers with thicknesses of one, two, three and four droplets. The droplets and interdroplet spaces are fully resolved. The relative velocity between droplets and surroundings varies from 0 m/s to 35 m/s.

The model used in the present work combines the one-fluid and two-fluid formulations for multiphase flows where energy is solved based on a one-fluid formulation while species, velocities and pressure are solved with a two-fluid

W.E. Nagel et al. (eds.), *High Performance Computing in Science and Engineering '10*, DOI 10.1007/978-3-642-15748-6_15,

formulation. In addition, a Level Set technique is combined with the Ghost Fluid method in a mass conserving approach in order to track the liquid interface and model surface tension. The simulations are computationally expensive and high performance computers are needed. This work has been developed using the NEC Nehalem Cluster Platform.

The next section describes the methodology applied, describing the equations used and the computational approach. Then, the scalar mixing and droplet interactions test cases are described and the results analysed.

2 Methodology

The equations that describe a fluid flow are derived from the principles of mass, momentum and energy conservation.

The continuity equation is written as

$$\frac{\partial \rho}{\partial t} + \frac{\partial \rho u_i}{\partial x_i} = 0 \tag{1}$$

where ρ is the fluid density, t is time, u_i is the velocity in i-direction and x_i is the cartesian coordinate in i-direction.

The momentum equation can be expressed as

$$\frac{\partial \rho u_k}{\partial t} + \frac{\partial \rho u_i u_k}{\partial x_i} = F_{ext} - \frac{\partial p}{\partial x_k} + \frac{\partial \sigma_{ik}}{\partial x_i} \tag{2}$$

where F_{ext} is the sum of the forces acting in the fluid (gravity, etc) and σ_{ik} is the viscous stress tensor.

Considering a Newtonian fluid and neglecting the gravity, the Navier-Stokes equations are written as

$$\frac{\partial \rho u_k}{\partial t} + \frac{\partial \rho u_i u_k}{\partial x_i} = F_{ext} - \frac{\partial p}{\partial x_k} + \frac{\partial}{\partial x_i}(\mu(\frac{\partial u_k}{\partial x_i} + \frac{\partial u_i}{\partial x_k} - \frac{2}{3}\delta_{ik}\frac{\partial u_j}{\partial x_j})) \tag{3}$$

where δ_{ik} is the Kronecker delta and μ is the viscosity of the fluid.

The species conservation, based on the continuity equation, is given by

$$\frac{\partial \rho Y_\alpha}{\partial t} + \frac{\partial \rho Y_\alpha u_i}{\partial x_i} = \omega_\alpha - \frac{\partial J_{\alpha,i}}{\partial x_i} \tag{4}$$

where Y_α is the mass fraction of the specie α in the fluid, ω_α is the mass source term and $J_{\alpha,i}$ is the diffusional flux with contribution of concentration gradients, mass diffusion due to temperature, external forces and pressure gradients. If diffusion can be described by Fick's law, valid for binary diffusion, then

$$J_{\alpha,i} = -\rho D_\alpha \frac{\partial Y_\alpha}{\partial x_i} \tag{5}$$

where D_α is the diffusion coefficient of species α.

The enthalpy, h, is a function of species mass fractions and temperature.

$$h = h(Y_1, Y_2, \ldots, Y_n, T) = \sum Y_\alpha (h_{0\alpha} + \int c_{p\alpha} dT) \tag{6}$$

where $h_{0\alpha}$ is the enthalpy of formation and $c_{p\alpha}$ is the heat capacity of species α.

At low Mach number, the transport equation is

$$\frac{\partial \rho h}{\partial t} + \frac{\partial \rho h u_i}{\partial x_i} = \frac{\partial}{\partial x_i}(\rho D_h \frac{\partial h}{\partial x_i}) + \frac{\partial p}{\partial t} - \rho \omega_{rad} \tag{7}$$

where D_h is the enthalpy diffusion coefficient, ω_{rad} is the heat loss rate due to radiation and the pressure derivative is often neglected in combustion modelling.

In the energy equation, the contribution of pressure variations due to the movement of the liquid-vapour interface is relatively small and can be neglected. Therefore the temperature equation, derived from the energy conservation equation, can be written as

$$\frac{\partial \rho c_p T}{\partial t} + \frac{\partial \rho c_p T u_i}{\partial x_i} = \frac{\partial}{\partial x_i}(\rho D_h \frac{\partial c_p T}{\partial x_i}) + \omega_T \tag{8}$$

where c_p is the heat capacity of the fluid, T is the temperature, ω_T is the heat source term.

Two concepts are used in two-phase flow modelling: one-fluid and two-fluid formulations. In the one-fluid approach one set of equations for the variables (velocities and scalars) is valid in the entire flow field, including interfaces. The two-fluid formulation uses a set of equations for each phase in the flow field. A phase indicator is used in both approaches to separate the phases.

In this work, the model uses a new hybrid method where energy is solved based on a one-fluid formulation while species, velocities and the pressure are solved with a two-fluid formulation. In the energy equation, the properties at the interface cells are defined depending on the volume fractions of gas and liquid. In the species, momentum and pressure equations, source terms are included to take into account the transfer terms between the phases. Mass transfer terms and subsequently all other transfer terms can be deduced from the temperature gradients at the interface whose position is known by solving (9) (see below). The approach avoids the introduction of interface thickness and the smoothing of fluid properties. The physical properties (density, diffusivity, specific heat and conductivity) are defined in each cell according to the phase indicator, local composition and temperature using standard kinetic theory. The equation of state of the gas phase assumes ideal gas and as the temperature of the liquid phase is assumed constant and at saturation value, the liquid properties are assumed to be constant.

In addition, a Level Set method [3] has been combined with the Ghost Fluid method [4] in a mass conserving approach for the liquid interface tracking. The Level Set method is based on the definition of a smooth function, such as the minimum distance of a point to the interface, in the entire computational domain. The use of a smooth function, ϕ, avoids steep gradients and numerical instabilities. Negative values correspond to one phase while positive values to the other and the interface is represented by the zero level of the function. The phase indicator is defined by the sign of the level set function. The function evolves in time, t, according to the following equation.

$$\frac{\partial \phi}{\partial t} + u_{iI}\frac{\partial \phi}{\partial x_i} = 0 \tag{9}$$

where u_{iI} is the local interface velocity in the ith direction.

(9) is solved explicitly for the cells close to the interface ($|\phi| \leq 1.5\Delta x$, where Δx is the grid spacing), using a second-order accurate method in time [5] and a first order upwind method in space. The interface velocity, u_{iI}, is calculated [6] by

$$u_{iI} = u_{i,d} - \frac{k}{\rho_l \cdot h_{fg}}\frac{\partial T}{\partial x_i} \tag{10}$$

where $u_{i,d}$ is the droplet velocity, k is the thermal conductivity, ρ_l is the liquid density, h_{fg} is the enthalpy of evaporation and T is temperature whose gradient is approximated by a fourth-order central difference scheme. It is important to maintain ϕ as a distance function so the interface can be easily reconstructed. In order to keep the Level Set as a distance function, it is necessary to reinitialise ϕ periodically. This procedure is computationally expensive especially in three-dimensional simulations. The reinitialisation [3] is achieved by calculating the new Level Set function in the entire domain by

$$\frac{\partial \phi}{\partial \tau} + sign(\phi)(|\frac{\partial \phi}{\partial x_i}| - 1) = 0 \tag{11}$$

where τ is time normalised by an arbitrary time scale. (3) ensures that $|\nabla\phi| = 1$. Integration in time is achieved by a third-order Runge-Kutta scheme with an arbitrary time step equal to $0.8\Delta x$ and a defined maximum residual in a band close to the interface ($|\phi| \leq 2.5\Delta x$). The absolute partial derivatives of the Level Set function gradient are approximated by a fifth-order WENO scheme [7]. The interfacial curvature,κ, is easily calculated within all cells that contain an interface by,

$$\kappa = \frac{\partial}{\partial x_i}\Big(\frac{\frac{\partial \phi}{\partial x_i}}{|\frac{\partial \phi}{\partial x_i}|}\Big). \tag{12}$$

Despite being efficient and allowing accurate interface reconstruction, the Level Set method is not inherently mass conserving. Some recent works [8] combine it with mass conserving methods in order to overcome this problem. Interface tracking methods have also been coupled to the Ghost Fluid method to better handle jump conditions [9] and to reduce the parasitic currents generated by surface tension modelling. Following this idea, the method developed in this work combines the Level Set method with a mass conserving approach and the Ghost Fluid method. Mass conservation is ensured by checking the error and redistributing the liquid mass loss/gain among the droplets.

The Ghost Fluid method proposes that derivatives in each phase use only values of that specific phase. For that, values must be extrapolated for each phase when jumps occur. The formulation of a second order central difference derivative for pressure, p, for example, is

$$\frac{\partial p}{\partial x} = \frac{p_{i+1} - p_{i-1}}{x_{i+1} - x_{i-1}} \tag{13}$$

In the Ghost Fluid method, the derivatives become

$$\frac{\partial p}{\partial x} = \frac{p^{+}_{i+1} + Ju - p_{i-1}}{x_{i+1} - x_{i-1}} \tag{14}$$

The location and value of the jump, Ju, in the case of surface tension is $Ju = \sigma\kappa$. Then, just the value of pressure in the ghost cell, p^{+}_{i+1}, must be set.

The model is implemented in an in-house low-Mach number CFD code (BOFFIN). The code is second order accurate in time (Crank-Nicholson scheme) and in space away from discontinuities. The model has been validated as presented in [10].

2.1 Code Parallelisation

The simulations are computationally expensive and high performance computers are needed. This work has been developed using the NEC Nehalem Cluster Platform. A computational domain is split into several sub-domains which can be calculated in different processors. Each sub-domain has halo cells which overlap its neighbour domains cells as shown in Fig. 1. These halo cells keep the information necessary for the domain to compute derivatives, check interfaces and check level set errors. However, the information is calculated only in the domain where the cells belong to and not in the halo cells. Therefore, once the solution of the domains is calculated, information have to be exchanged to the halo cells so the neighbour domains have the information they need to calculate the next solutions.

All the scalars (temperature, energy and species), velocities and pressure are exchanged between domains as indicated in Fig. 1. As TVD schemes are used for the scalars, the number of halo cells rows passed is 2. For velocities and pressure only one row of halo cells is passed. The lateral parallelisation is explained in detail in [11].

The level set, however, requires 3 rows of halo cells and corners exchanges (see Fig. 2) because of the WENO scheme in the reinitialisation procedure and in order to check the existence of interfaces. Figure 2 shows the way the corners exchanges are done. Extra information about diagonal neighbours must be passed so these exchanges can be done.

The communication between processors is implemented using Message Passing Interface (MPI) routines.

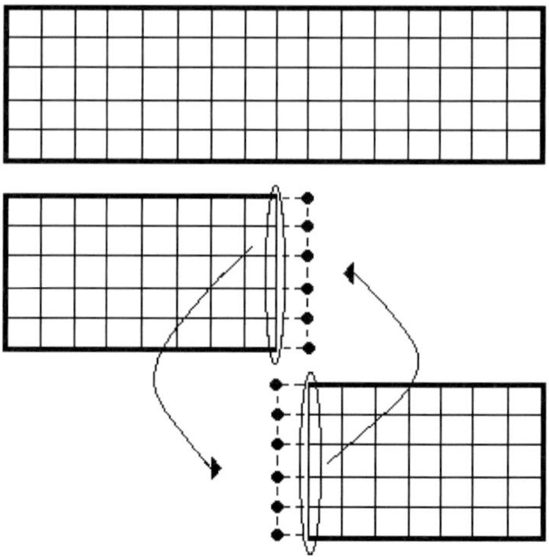

Fig. 1. Basic scheme of halo cells for the left-side sub-domain

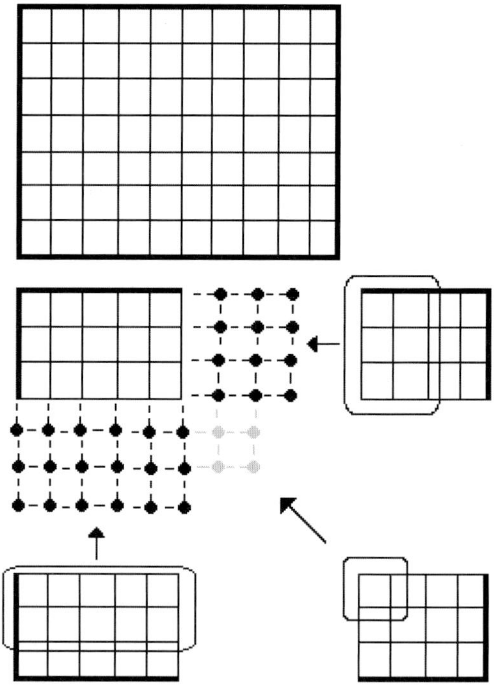

Fig. 2. Scheme of halo cells with level set information for the upper left-side sub-domain

3 Test Cases

The simulations concerned the study of evaporation and scalar mixing in droplet arrays [12]. The domains consist on a 2D space that extends infinitely in x-direction and 3 mm in y-direction, populated with one, two, three or four infinite rows of equidistant droplets. Inflow/outflow boundary conditions perpendicular to the droplet layers are used with periodic boundary conditions in the remaining direction. Scalar dissipation and the mixture fraction field are investigated for arrays of droplets of diameter of 100 μm. The grid mesh is $\Delta x = \Delta y = 5$ μm and the simulations run until 90% of the droplet mass has evaporated. Different regions need to be defined to allow a zone dependent analysis of the local conditions of the combustible mixture. Three regions are defined: the global region comprises all of the computational domain; the inner and outer regions refer to inter-droplet spaces and regions outside the droplet cloud. In addition, in the four-droplet case, different interdroplet regions are also identified (see Fig. 3) and analysed. Liquid is assumed at 338 K with $\rho_l = 750$ kg/m^3, $\sigma = 1.85 \times 10^{-2}$ N/m, $\mu = 3.5 \times 10^{-4}$ Pa.s and $h_{fg} = 1.097$ MJ/kg.

In the laminar flow test case, hot air (2530 K) is injected at 5 m/s and in the turbulent flow test the mean velocity is 10 m/s with maximum fluctuations of 25 m/s. Furthermore, in the turbulence test, the flow field is initialised with velocities scaled from a 2D turbulent field and the inflow is given by 2D turbulent DNS data. In order to sustain the turbulence through the domain, the gas phase viscosity was fixed at a lower value ($\mu = 10^{-6}$ kg/m.s) and pressure was increased from 1 atm to 5 atm. The Kolmogorov length scale is of the order of the droplet radius. A typical vorticity field is presented in Fig. 4, and it is evident from the figure that the flow is strongly affected by the droplets and develops differently for each of the cases.

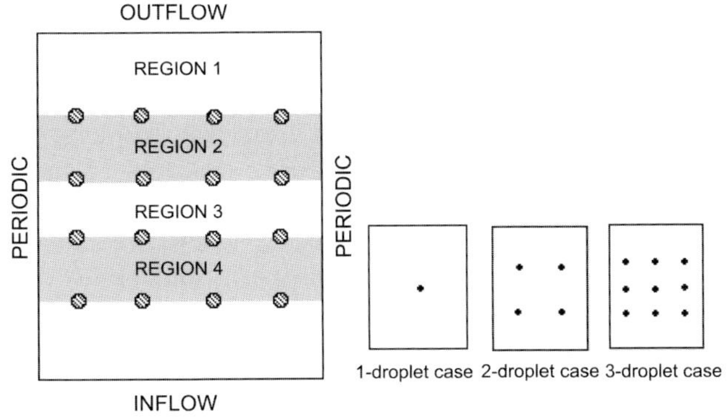

Fig. 3. Geometry of the 4-droplet layer test case domain and defnition of the regions (left figure). The right figure illustrates the droplet configurations for the 1-, 2- and 3-droplet layer test cases

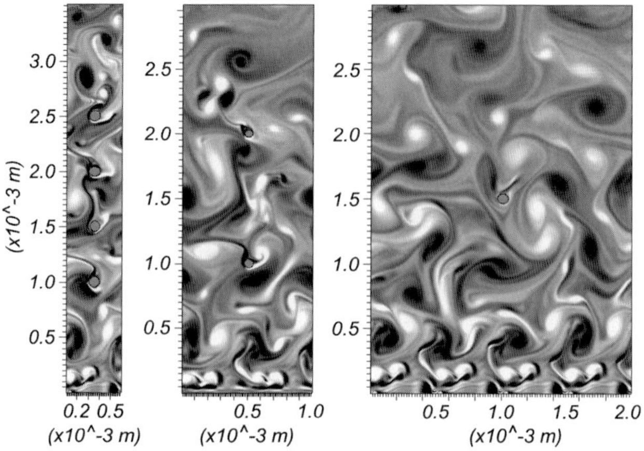

Fig. 4. Vorticity field after 5 ms for the 4- (left), 2- (centre) and 1- (right) droplet layer test cases

4 Scalar Mixing in Droplet Arrays

A first quantitative analysis of the spatial evolution of the mixture fraction field involves a comparison of the computational results with an analytic description of the near droplet field for inertial droplets. Mixture fraction can be obtained from the a linear combination of species mass fractions, Y_α, and the solution of the system of equations as described in Sect. 2 can be conveniently condensed into one scalar, namely mixture fraction, that is then compared to the results of the analytical solution. Following [13], mixture fraction, f, along a transversal line perpendicular to the mean flow field in the near droplet region is given by (15).

$$f = f_\infty + \frac{J_m(f_d - f_\infty)}{4\pi\rho D \frac{l}{u_{mean}}} e^{\left(\frac{-r^2}{4D\frac{l}{u_{mean}}}\right)} \tag{15}$$

f_∞ is the value of mixture fraction far from the droplets, f_d is the value of mixture fraction inside the droplet initially centered at (x_d, y_d), r is the transverse distance from the droplet centre, $r = |x - x_d|$, l is the distance from the line to the droplet, $l = y - y_d$, u_{mean} is the flow mean velocity, J_m is the mass evaporation rate of a single droplet, ρ and D are the density and diffusivity in the interdroplet space, respectively. The results for the laminar case present good agreement with (15) as depicted in Fig. 5, where results from the one- and two-droplet layer cases are shown. In the two-droplet layer case, one line per droplet is assessed.

With respect to the probability distribution, it has been shown in [10] that in stagnant environments, where expansion of the gas phase occurs only due to the Stefan flow, the presumed β-PDF does not capture two dominant peaks

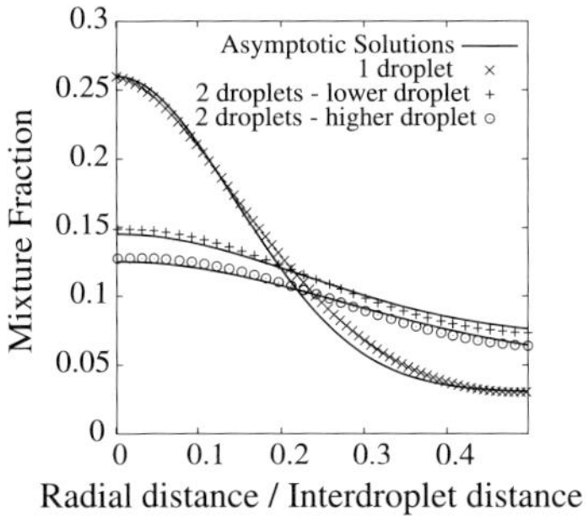

Fig. 5. Mixture fraction along a transversal line ($l = 500\ \mu$m) in the laminar flow test—comparison with the asymptotic solution, (15), from [13]

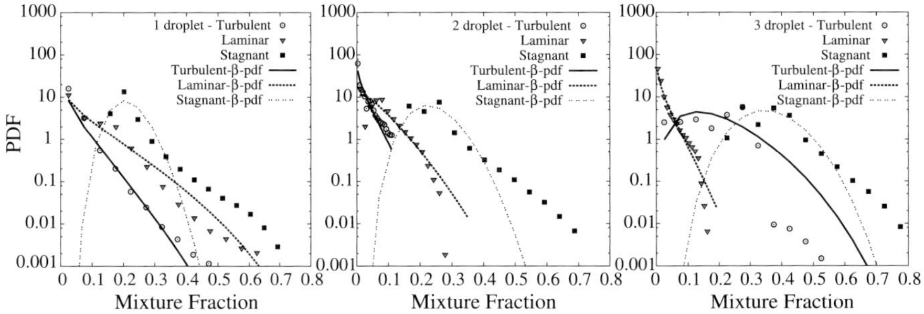

Fig. 6. Comparison of DNS data and a presumed β-shape for the mixture fraction PDF for various flow fields and various droplet loadings

that develop at two different mixture fraction values in the three-droplet layer case. The β-PDF is defined as

$$P(f) = \frac{f^{\alpha-1}(1-f)^{\beta-1}}{\Gamma(\alpha)\Gamma(\beta)}\Gamma(\alpha+\beta) \tag{16}$$

with $\alpha = \mu\gamma$, $\beta = (1-\mu)\gamma$ and $\gamma = \frac{\mu(1-\mu)}{\sigma}$. The parameter μ is the mean value of f, σ is the variance of f and Γ is the Gamma function.

As shown in Fig. 6, in the stagnant environment, the double peak does not occur when the distance between the droplets is large (one-droplet case). However, once the interdroplet distance decreases (two- and three-droplet cases) the bimodal nature of the PDF appears. For the laminar flows, the β-PDF seems to capture better the behaviour. In the turbulent environments, the β-PDF represents reasonably well the probability for the lower droplets density cases but it does not capture the profile for the higher density case (Fig. 6). It is clear that structural information must be taken into account when modelling the mixture fraction PDF in two-phase flows, however, defining the appropriate parameters is not evident.

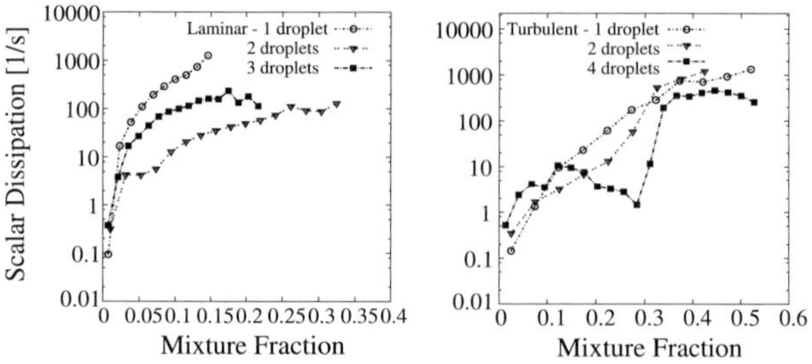

Fig. 7. Global conditionally averaged scalar dissipation for laminar (left) and turbulent (right) convective environments

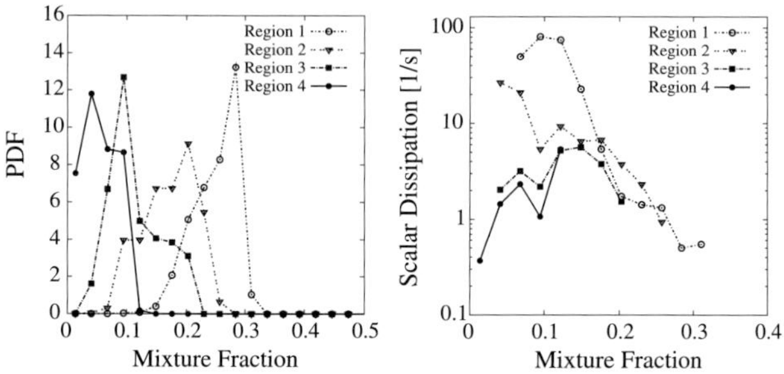

Fig. 8. Mixture fraction and Scalar dissipation in the 4-droplet turbulent case

Figure 7 compares scalar dissipation for the different droplet arrays. It can be seen that the higher the number of droplets (reduced interdroplet space) the lower is the scalar dissipation in the laminar cases. However, for turbulent flows, this trend is not repeated, and it cannot be assumed that a higher droplet concentration necessarily decreases scalar dissipation in all the mixture fraction range.

Different regimes are identified when the global computational domain is split into different areas depending on their position relative to the droplets (as defined in Fig. 3). In Fig. 8, the PDF and the scalar dissipation of the four-droplet case with turbulent flow are presented. Unsurprisingly, we can observe a clear shift of the PDF peak towards higher mixture fraction values with downstream distance due to accumulation of the fuel vapour as the flow passes a succession of droplet layers. Less obvious are the results for scalar dissipation. Scalar dissipation in region 1 is higher than in the interdroplet regions (regions 2, 3 and 4) for most of the mixture fraction range.

Figure 9 shows that the stagnant, laminar and turbulent flows have completely different scalar dissipation in the two-droplet cases. While in the stagnant case the inner and outer region's branches are well defined, in the laminar and specially in the turbulent tests, it is possible to have the same value of scalar dissipation over a wide range of radial locations. It is clear that the flow affects significantly the vapour distribution and in terms of scalar dissi-

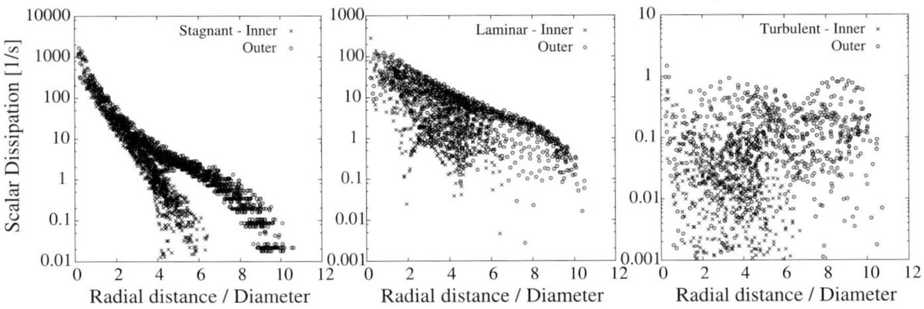

Fig. 9. Scatter plot of global Scalar dissipation in the 2-droplet cases

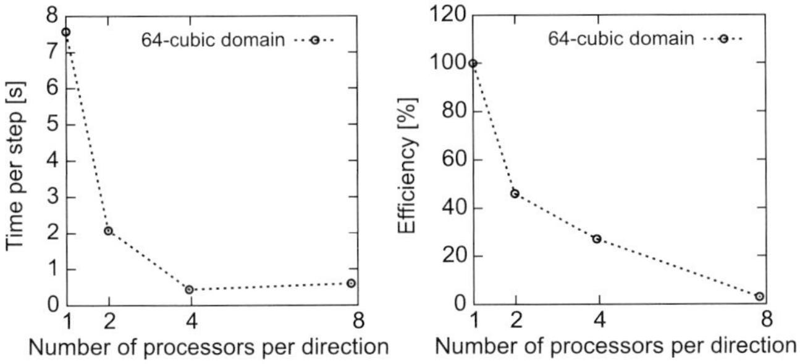

Fig. 10. Time per step as function of number of processors used in each direction

pation, a clear distinction between inner and outer region is no longer possible once turbulent mixing occurs and the radial distance does not seem to give sufficient structural information needed for accurate modelling.

5 Computational Resources

The efficiency of the parallelisation is tested in order to evaluate the code's performance. A 64^3 domain with a centered droplet under hot laminar environment is used as test case. The tests are performed using $1 \times 1 \times 1$, $2 \times 2 \times 2$, $4 \times 4 \times 4$ and $8 \times 8 \times 8$ processors in x, y and z directions, respectively. Figure 10 shows that there is a significant reduction in the simulation time when the number of processors is increased from 1 to 4 in each direction. However, an increase to 8 processors per direction does not really reduce the time per step. This is because the cost of communication is high enough to compensate the decrease in the size of the domain per processor. In this case, it is not efficient to increase the number of processors for more than 4 per direction.

The domains used in this work have constant and very fine mesh ($\Delta x = 5\ \mu m$) and are composed of $600 \times 400 \times 1$ cells in x, y and z directions, respectively. The time steps are of the order of 0.1 μs and the simulations run for approximately 100000 steps. The domains are split into 16 or 24 processors and the average computational time per simulation is 4000 cpu hours. Twelve simulations (4 configurations and 3 environments in total) are performed for this work.

6 Conclusions

A novel computational method is used to simulate evaporating droplet arrays in laminar and turbulent environments using DNS. The simulations are computationally expensive and high performance computers are used. Mixture fraction PDFs and scalar dissipation are assessed in order to investigate their dependence on the position relative to dense droplet layers. As in stagnant environments, different dependencies of PDF and scalar dissipation can be identified and large inhomogeneities exist for all test cases within the computational domain. The β-PDF seems to capture the global behaviour for laminar environment test cases well, but not for higher droplet densities under stagnant and turbulent conditions. Radial distance to the droplets and droplet density do not provide sufficient structural information for accurate modelling. The analysis of a suitable parameterisation for the modelling of the PDF and scalar dissipation as well as 3D simulations are subject to ongoing work at Imperial College and the University of Stuttgart.

References

1. R.I. Imaoka, W.A. Sirignano, *Proc. Combustion Institute* (30) (2005) 1981–1989.
2. P. Schroll, A.P. Wandel, R.S. Cant, E. Mastorakos, *Proc. Combust. Inst* (1) (2008) 2275–2282.
3. M. Sussman, P. Smereka, S. Osher, *J. Comp. Phys* (114) (1994) 146–159.
4. R. Fedkiw, T. Aslam, B. Merriman, S. Osher, *J. Comp. Phys* (152) (1999) 457–492.
5. N.N. Yanenko, *Springer, Berlin.*
6. R.P. Selvam, L. Lin, R. Ponnappan, *Int. J. Heat and Mass Transfer* (49) (2006) 4265–4278.
7. G.S. Jiang, D. Peng, *SIAM J. Sci. Comp.* (21) (2000) 2126–2143.
8. E.R.A. Coyajee, M. Herrmann, B.J. Boersma, *Proc. Summer Program - Center for Turbulence Research* (1) (2004) 315–327.
9. S. Tanguy, T. Menard, A. Berlemont, *J. Comp. Phys.* (221) (2006) 837–853.
10. M.R.G. Zoby, S. Navarro-Martinez, A. Kronenburg, *Proc. of the 4th European Combustion Meeting.*
11. A.J. Marquis, M. Wille, *BOFFIN for parallel Computers, Technical Memorandum*, Imperial College, London (2008).
12. M.R.G. Zoby, S. Navarro-Martinez, A. Kronenburg, A.J. Marquis, *Notes on Numerical Fluid Mechanics and Multidisciplinary Design (in press).*
13. A.Y. Klimenko, R.W. Bilger, *Progress in Energy and Combustion Science* 25 (1999) 595–688.

Euler-Lagrange Simulation of a LOX/H2 Model Combustor with Single Shear Coaxial Injector

Markus Lempke[1], Peter Gerlinger[1], Michael Rachner[2], and Manfred Aigner[1]

[1] Institut für Verbrennungstechnik der Luft- und Raumfahrt, Universität Stuttgart, Pfaffenwaldring 38-40, 70569 Stuttgart, Germany
markus.lempke@dlr.de
[2] Institut für Verbrennungstechnik, DLR Köln, Linder Höhe, 51147 Köln, Germany

Summary. In this paper a mixed Euler-Lagrange approach is used for a 3D simulation of a LOX/H2 model rocket combustor with a single shear coaxial injector. The specific test case presented is the MASCOTTE combustor at 10 bar pressure in the so called A-10 configuration. The simulation of the gas phase is conducted with the scientific code TASCOM3D which works in an Eulerian mode while the liquid fuel droplets are treated by the scientific code SPRAYSIM in a Lagrangian framework. The two codes and the coupling mechanisms are explained and results of a preliminary simulation will be presented. At the end an outlook is given focusing on how to obtain an even more accurate representation of the experiment in subsequent simulations. Finally some comments on the computational costs of the calculations and the performance of the two codes on the NEC SX-9 are given.

1 Introduction

Although computational tools have become an important part of the design processes in many technical fields the usage of numerical simulations to assist the design of rocket combustors is still fairly low [1]. Even today the development is mainly based on experiences gained from trial-and-error methodologies as well as expensive ground testing. Taking into account the growing competition on the space transport market and the invariably rising demands with respect to mission flexibility, reliability and payload weight [2] an increase in the employment of numerical tools seems worthwile. Especially the potential of CFD calculations to reduce the amount of needed ground tests for qualification is attractive [3]. However, the application of computational modeling procedures as design tool involves numerous challenges like the appropriate descriptions of chemical kinetics, turbulence and turbulent combustion effects, as well as a meaningful description of the two phases that exist in spray combustion systems [4].

W.E. Nagel et al. (eds.), *High Performance Computing in Science and Engineering '10*, DOI 10.1007/978-3-642-15748-6_16,

With respect to these challenges the focus of this work is the utilization of an Euler-Lagrange approach to address the coexistence and interaction of liquid and gaseous phases. Both phases are treated separately, the gas phase in an Eulerian and the liquid droplet spray in an Lagrangian framework. The spray code tracks the droplet through the domain and calculates their evaporation along the trajectories using the gas phase properties delivered by the CFD code. The source terms due to evaporating droplets are then mapped to the cells of the computational domain and incorporated in the gas phase computations. Sophisticated simulations of the combustion process with finite-rate chemistry and detailed reaction mechanisms guarantee an appropriate description of chemical kinetics. Turbulence-chemistry interaction is not yet taken into account but will be considered in the future via the PDF-approach that is briefly introduced in Sect. 2.1.

Whatever the approach, a detailed experimental database of model combustors is required for model validation. In this respect rocket combustion studies of hydrogen/oxygen propellants are particularly challenging since they require high propellant flow rates. Moreover, the safety issues of the propellants demand for extensive facilities development and operation capabilities [5]. Consequently comparatively few experimental data is available in this field. One of the most sophisticated approaches to improve this situation is the MASCOTTE program [6]. This cryogenic combustion test facility has been established at ONERA (Office National d'Études et de Recherches Aérospatiales) in the 1990s and has progressively been extended towards operation at higher propellant flow rates and thus higher pressures. Since experimental investigations [7] imply that the atomization of coaxial jets is mainly dependent on the momentum flux ratio J, this number was kept approximately constant in the different versions (V01-V03) to enable a meaningfull comparison between low and high pressure behavior. A detailed overview over the realized operation conditions can for example be found in [3]. Numerous experimental investigations including Phase Doppler Particle Analysis (PDPA), laser-induced fluorescence (LIF) and Coherent Anti-Stokes Raman Scattering (CARS) measurements have been carried out and offer a profound basis for the validation of numerical codes. The specific test case examined numerically in this work is the subcritical configuration MASCOTTE A-10.

2 Governing Equations and Numerical Schemes

2.1 Gas Phase

For the simulation of the gas phase the scientific in-house code TASCOM3D (Turbulent All Speed Combustion Multigrid 3D) was used. This code computes reacting flows by solving the full compressible Navier-Stokes and species transport equations. The turbulence closure problem requires two more transport equations for variables of the two-equation low-Reynolds-number q-ω

turbulence-model of Coakley et al. [8]. Additionally an assumed PDF (probability density function) approach can be activated to take turbulence chemistry interaction into account. In this case transport equations for the variance of the temperature and for the variance of the sum of species mass fractions are solved as well. This set of equations in three-dimensional conservative form can be written as

$$\frac{\partial \mathbf{Q}}{\partial t} + \frac{\partial(\mathbf{F} - \mathbf{F}_\nu)}{\partial x} + \frac{\partial(\mathbf{G} - \mathbf{G}_\nu)}{\partial y} + \frac{\partial(\mathbf{H} - \mathbf{H}_\nu)}{\partial z} = \mathbf{S}\,, \tag{1}$$

where

$$\mathbf{Q} = \left[\bar{\rho}, \bar{\rho}\tilde{u}, \bar{\rho}\tilde{v}, \bar{\rho}\tilde{w}, \bar{\rho}\tilde{E}, \bar{\rho}q, \bar{\rho}\omega, \bar{\rho}\sigma_T, \bar{\rho}\sigma_Y, \bar{\rho}\mathbf{Y}\right]^T \ . \tag{2}$$

The variables in the conservative variable vector $\mathbf{Q}$ are the density $\bar{\rho}$ (averaged), the velocity components (Favre averaged) $\tilde{u}$, $\tilde{v}$ and $\tilde{w}$, the total specific energy $\tilde{E}$, the turbulence variables $q = \sqrt{k}$ and $\omega = \epsilon/k$ (where k is the kinetic energy and ϵ the dissipation rate of k), the variance of the temperature σ_T and the variance of the sum of the species mass fractions σ_Y. The vector $\mathbf{Y}$ contains the species mass fractions $\tilde{Y}_i$ for $i = 1, 2, \ldots, N_k - 1$ and N_k being the total number of gaseous species. Since the set of equations includes the conservation of mass $N_k - 1$ species transport equations have to be solved, only. In (1) the vectors $\mathbf{F}$, $\mathbf{G}$ and $\mathbf{H}$ specify the inviscid fluxes in x-, y- and z-direction, $\mathbf{F}_\nu$, $\mathbf{G}_\nu$ and $\mathbf{H}_\nu$ the viscous fluxes, respectively. The source vector $\mathbf{S}$ results from turbulence and chemistry and is given by

$$\mathbf{S} = \left[0, 0, 0, 0, 0, \bar{S}_q, \bar{S}_\omega, \bar{S}_{\sigma_T}, \bar{S}_{\sigma_Y}, \mathbf{S}_Y\right]^T \ , \tag{3}$$

where $\bar{S}_q$ and $\bar{S}_\omega$ are the averaged production rates of the turbulence variables, $\bar{S}_{\sigma_T}$ and $\bar{S}_{\sigma_Y}$ the source terms of the variance variables (σ_T and σ_Y) and $\mathbf{S}_Y$ the source term vector of the species mass fractions $\bar{S}_{Y_i}$.

For the work presented here the hydrogen/oxygen reaction mechanism according to Ó Conaire [9] was employed which comprises 8 species for hydrogen oxidation in a set of 18 reactions. This choice was influenced by the fact that this reaction mechanism has been most extensively validated for the high pressure regime. The unsteady set of equations in (1) is solved using an implicit Lower-Upper-Symmetric Gauss-Seidel (LU-SGS) [10–13] finite-volume algorithm, where the finite-rate chemistry is treated fully coupled with the fluid motion. The code is fully parallelized employing MPI and optimized for vector CPUs. For more details on TASCOM3D the interested reader is referred to [12–16].

2.2 Liquid Droplet Spray

The liquid droplet spray is simulated by the research code SPRAYSIM. This code is based on the Lagrangian treatment of the dispersed phase. In contrast to the Eulerian treatment the dispersed phase is represented by a large number of computational particles. Each of these computational particles is tracked numerically across the gasfield, which has to be supplied by the CFD code, until the particle is evaporated or has left the computational domain. The set of equations for each droplet can be written in vector form as

$$\frac{\mathrm{d}\mathbf{Q}_d}{\mathrm{d}t} = \mathbf{H}_d \tag{4}$$

with the variable vector

$$\mathbf{Q}_d = [D_d, u_d, v_d, w_d, T_d]^T \tag{5}$$

where D_d is the diameter of the droplet, u_d, v_d and w_d are the velocities of the droplet in x-, y- and z-direction and T_d is the temperature of the droplet. As for the right hand side vector

$$\mathbf{H}_d = [H_D, H_x, H_y, H_z, H_T]^T \tag{6}$$

the correlations

$$H_D = -\frac{1}{\rho_{liq}}\left(\frac{\dot{m}_{vap}}{\frac{\pi}{2}D_d^2} + \frac{D_d}{3}\frac{\mathrm{d}\rho_{liq}}{\mathrm{d}t}\right) \tag{7}$$

$$H_i = \frac{F_{drag,i}}{m_d} = \frac{A_d\, c_{drag}\, \rho_{gas}\, u_{rel,i}^2}{m_d} \quad , \quad i = x, y, z \tag{8}$$

$$H_T = -\frac{1}{c_{p,liq}\,\rho_{liq} V_d}\left(\dot{m}_{vap}\,\Delta h_{vap} + \dot{Q}\right) . \tag{9}$$

are used to calculate the sourceterms, where ρ_{liq} is the density and $c_{p,liq}$ the heat capacity of the liquid. The variables m_d and V_d are the mass and the volume of the droplet, respectively and $\dot{m}_{vap}$ denotes the instantaneous vaporization rate. The drag forces $F_{drag,i}$ are calculated by standard drag expressions for spherical particles with the droplet crossection A_d, the drag coefficient c_{drag} and the relative velocity between droplet and ambient gas field $u_{rel,i}$. The energy balance at the surface of the droplet has to consider the energy losses due to evaporating liquid $\dot{m}_{vap}\,\Delta h_{vap}$ and the heatflux through the surface $\dot{Q}$. It has to be noted that besides the drag force also $\dot{m}_{vap}$ and $\dot{Q}$ are functions of the ambient gas field properties.

There are two ways to generate spray initial conditions in SPRAYSIM. Either they can be calculated by the code based on state-of-the-art models for primary atomization or they can be defined as distributions obtained from experimental data. The implemented models for the secondary breakup include the Taylor Analogy Breakup (TAB), enhanced TAB (ETAB) and Cascade Atomization and Drop Breakup (CAB) approaches. With respect to the evaporation model there are three choices available. In the order of computational effort these are the uniform-temperature model, the conduction-limit model and the effective-conductivity model. At present SPRAYSIM is not parallelized and doesn't account for droplet interaction.

2.3 Coupling

Along its trajectory across the gasfield each particle exchanges mass, momentum and energy with the surrounding gas. By accumulation of all contributions from the computational particles that met a finite volume the spray feeds sourceterms to each cell of the computational domain. Trajectories and property variations during each integration timestep of the computational particles (parcels) are calculated using gas field properties provided by the gas flow solver. Then source terms stemming from the parcels are fed back to the flow solver (two-way-coupling) again. Accordingly the sourceterm vector for the gas phase computation in (1) has to be expanded by the spray source terms according to

$$\mathbf{S}_{tot} = \mathbf{S} + \mathbf{S}^{spray} \tag{10}$$

with the spray sourceterm vector due to evaporation

$$\mathbf{S}^{spray} = \left[\sum_{i=1}^{N_k} \bar{S}_{Y_i}^{evap}, \bar{S}_x, \bar{S}_y, \bar{S}_z, \bar{S}_E, \bar{S}_q, \bar{S}_\omega, 0, 0, \mathbf{S}_Y^{evap} \right]^T . \tag{11}$$

Details regarding the calculation of the sourceterms in (11) can be found in [17]. At the moment the exchange of sourceterms and flowfield variables between the two codes is realized through file I/O.

3 Boundary Conditions and Computational Procedure

The inflow conditions for the simulations of the MASCOTTE test case A-10 are given in Table 1. The model rocket combustor with single element shear coaxial injector has been extensively investigated experimentally for these operating conditions [18–22]. Moreover, this testcase has been subject for many numerical investigations and validation attempts [23–26]. The setup of the presented computations are based on recommendations of the 2nd Workshop on Rocket Combustion Modeling [27].

Figure 1 shows a sketch of the computational domain which represents a quarter of the real geometry. This simplification was introduced to reduce

Table 1. Thermophysical properties of the MASCOTTE A-10 test case

	H_2	O_2
pressure [MPa]	1	1
temperature [K]	287	85
massflow [g/s]	23.7	50
density [kg/m^3]	0.84	1170
velocity [m/s]	319	2.18
surface tension [N/m]	–	$1.44 \cdot 10^{-2}$

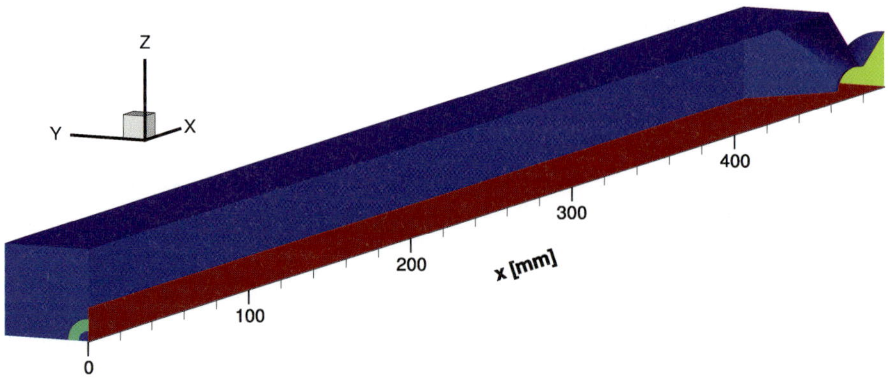

Fig. 1. Computational representation of MASCOTTE combustor with color coded boundary conditions

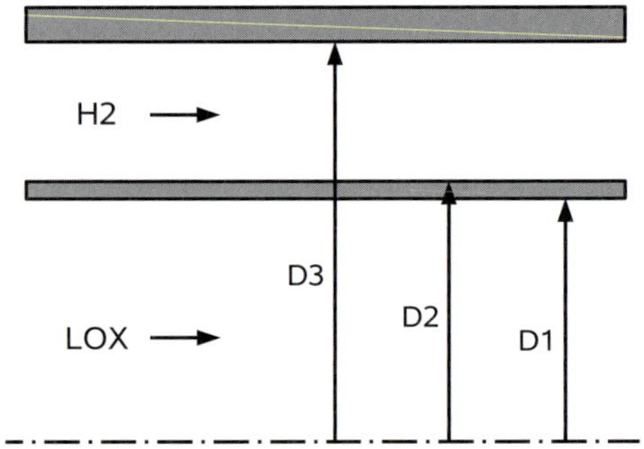

Fig. 2. Injector geometry for the 10 bar testcase. The values for the diameters D_1-D_3 are 5.0 mm, 5.6 mm and 12.0 mm, respectively

the computational cost. The boundary conditions are color coded with "blue" denoting solid walls, "red" for symmetry planes and "green" and "yellow" representing inflow and outflow boundaries, respectively. In the presented simulation the domain has been discretized with a little more than 750 000 volumes. Near solid walls the structured grid is strongly refined to meet the condition $y^+ \leq 1$. The combustion chamber has a square cross section of 50 mm × 50 mm to enable optical access via side windows. The windows are cooled with a helium film that has no detectable effect on the main flow [21] and therefore is neglected in the computations. The distance from the injector face plane to the nozzle exit plane is 497 mm and the nozzle throat has a diameter of 15 mm. The geometry of the coaxial injector is described in Fig. 2. Liquid oxygen is fed to the combustion chamber through the inner tube with a diameter $D_1 = 5.0$ mm. The annular hydrogen feeding has an inner diameter of $D_2 = 5.6$ mm and an outer diameter of $D_3 = 12$ mm.

For spray modeling 36 000 numerical particles are used, each representing a varying number of real droplets. Up to 4.5 million actual droplets per particle and second ensure an accurate reproduction of the spray. The spray is initialized in form of a Rosin-Rammler-Distribution

$$Q_3 = 1 - \exp\left[-\left(\frac{D}{D_e}\right)^m\right] \tag{12}$$

that was fitted to experimental data, where $D_e = 130\,\mu\text{m}$ and $m = 2.25$. It has to be noted that the experimental data was not obtained in the injector plane but 30 mm downstream of the injector. In the computation, however, the spray is injected in the injector face plane with a uniform axial velocity of 10 m/s. The radial velocity component is calculated depending on the radial injection position of the parcel by (13) for the angle of injection that was derived from experimental observations [28]:

$$\tan\theta\,(r) = \frac{r}{L_c\left(1 - \frac{r}{R_i}\right) + \frac{R_i}{\tan\theta_0}} \tag{13}$$

with

$$\tan\theta_0 = \tan\theta\,(R_i) = 0.68\left(\frac{U_{gas}}{U_{liq}} - 1\right)\sqrt{\frac{\rho_{gas}}{\rho_{liq}}}\ . \tag{14}$$

The variable L_c denotes the liquid core length (7.8 mm) that was calculated according to the observations of Rehab et al. [29] while the radius R_i corresponds to $R_i = D_1/2$. In (14) the variables U_{gas} and U_{liq} are the velocity of H_2 and liquid oxygen, respectively. Accordingly ρ_{gas} and ρ_{liq} are the densities of propellant and oxidator. This correlations result in injection angles with respect to the axial direction of 0° for $r = 0$ mm to approximately 70° at $r = R_i$. The evaporation of the droplets along their trajectories is calculated by the effective conductivity model of Abramzon and Sirignano [30]. In the presented simulation secondary breakup has not yet been taken into account.

To generate an initial solution of the problem first a pure gas phase simulation was conducted. This was done in such a way that gaseous oxygen was injected at a mass flow rate corresponding to those of the liquid phase. Ignition was then realized by artificially increasing the temperature in the mixing zone near the injector plane since the propellant temperatures are by far to low for auto-ignition. After achieving a steady solution for the gas phase the coupled computations are started and the gaseous oxygen mass flow was subsequently reduced as the liquid oxygen mass flow is increased over time. Finally the inflow boundary for gaseous oxygen was replaced by a solid wall as denoted in Fig. 1 and the fully coupled simulations are started. Hereby spray computations were typically performed every 10 000 to 50 000 CFD iterations.

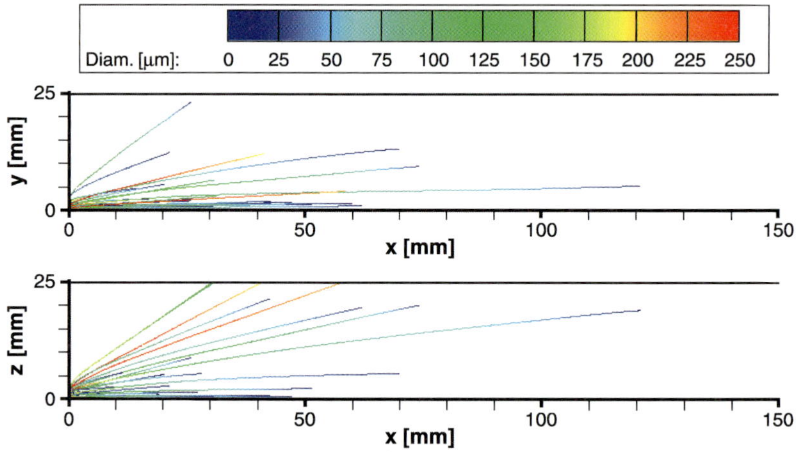

Fig. 3. Projection of fifty arbitrarily chosen particle trajectories on xy-plane (top) and xz-plane (bottom)

4 Results and Discussion

Figure 3 shows fifty arbitrarily chosen particle paths calculated by the spray code. Hereby the top illustration shows the projection of the trajectories on the xy-plane and the bottom one the projection on the xz-plane. The color represents the diameter of the droplets in μm. It has to be noted that not the complete length of the combustion chamber is shown but the first 150 mm downstream the injector only. Accordingly it can be seen that the majority of the droplets evaporates within the first 50 mm of the combustion chamber. A small number of very large droplets which are injected under very high angles of up to 70° don't fully evaporate before they reach the combustion chamber walls. However, their trajectories are clearly bend in the region where they encounter the hydrogen co-flow. In the presented calculation droplets which hit walls are treated as if they had left the computational domain. However, the mass flow of drops impinging on walls is less than 2% of the total oxygen mass flow.

The gas phase calculations resulting from the source terms of the spray computation according to Fig. 3 are illustrated in Fig. 4. From top to bottom contour plots of temperature, oxygen mass fraction and OH mass fraction are shown. As can be seen from the temperature as well as from the OH distribution the flame is anchored directly at the injector post and reaches far into the combustion chamber. The oxygen contour plot indicates the existence of gaseous oxygen in the near wall region. This has to stem from liquid oxygen droplets with high injection angles traversing the hydrogen co-flow. This oxygen mixes and reacts with the hydrogen in the recirculation zone causing the high temperatures and hydroxyl mass fractions that can be seen in the respective contourplots. Since there is no experimental evidence for this strong

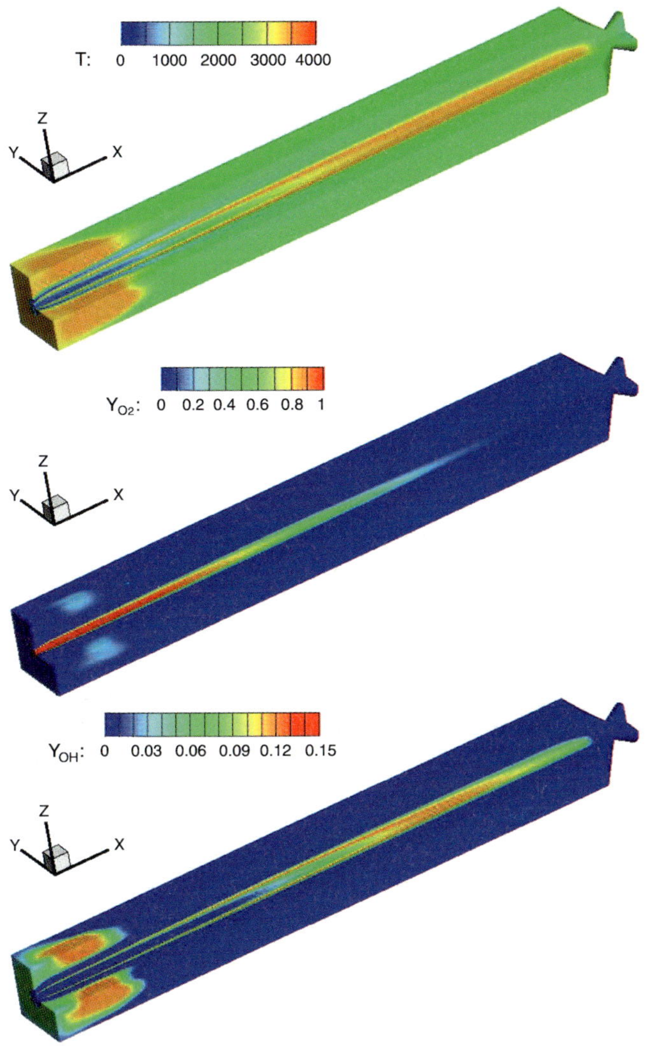

Fig. 4. Calculated contour plots of temperature (top) and gaseous oxygen (middle) and hydroxyl (bottom) mass fractions

reactions in the recirculation zone it seems necessary to critically reinvestigate the spray initial conditions.

It has to be kept in mind that the results presented here are of preliminary character. They are proof for the general applicability of the Euler-Lagrange approach and represent a basis for the ongoing investigations of the MASCOTTE A-10 testcase. Only after grid independence has been verified meaningful quantitative results can be given.

Subsequently there a variety of aspects that are worth to be investigated. According to [27] flow visualizations have shown that the atomization process is far from being complete at the position 30 mm downstream the injector. Thus simulations employing different models for secondary breakup seem worthwile to obtain a more accurate description of the atomization process. To include an adequate representation of the liquid core the computational domain and the spray injection position should be changed. Rather than originating from a circular plane the spray should be injected on the surface of a cone reaching into the combustion chamber with the height of $L_c = 7.8$ mm. This cone then has to be cut out of the computational domain. Other effects with strong influence that are not yet taken into account in this work are turbulence-chemistry interaction and unsteadyness of the flow. The utilization of the assumed PDF-approach implemented in TASCOM3D will surely augment the accuracy of the result. However, also the computational cost will get significantly higher.

5 Performance

The vectorization capability of SPRAYSIM on the NEC SX-9 has proven to be very poor as could be expected for a Lagrange code. Despite some optimization efforts the vector operation ratio didn't even exceed 60% . Thus the Spray computations were performed on the frontend ONTAKE with a typical CPU time of around 30–60 minutes. Since SPRAYSIM is not yet parallelized, only a single CPU of one of the 4 Intel Xeon X7350 quadcore processors was used. However, the neglected parallelization is not critical because only a small fraction of the required total CPU time is needed for the Lagrange solver.

In previous contributions [31–33] the good vectorization properties of the gas phase code TASCOM3D on the precursor NEC SX-8 have been shown in detail. These findings are also valid for the performance on the NEC SX-9. The performance analysis shows a vector operation ratio of 99.2%, an averaged vector length of 216 and a performance of 7.1 GFLOPS. The CFD computations have been performed using one single node of the NEC SX-9. Therefore the computational domain was divided into 16 blocks with an equal number of cells per block to ensure good load balancing. Overall around 800 hours of CPU-time were needed for the gas field calculations including the preliminary calculations without liquid phase coupling and the transition calculations.

Acknowledgements. This work was performed within the framework of the SFB-TR 40 'Technological foundations for the design of thermally and mechanically highly loaded components of future space transportation systems' and thus funded by the *Deutsche Forschungsgemeinschaft (DFG)*. This support is greatly appreciated. Further information on this program and its goals can be found on http://www.sfbtr40.de/.

The presented simulations were performed on the national super computer NEC SX-9 and its frontend ONTAKE at the High Performance Computing Center Stuttgart (HLRS) under the grant number *scrcomb*. The authors wish to thank for the computing time and the technical support.

References

1. Yang, V., Habiballah, M., Popp, M., Hulka, J., Liquid rocket thrust chambers: aspects of modeling, analysis and design, Progress in Astronautics and Aeronautics, **200**, 2004.
2. Bundesministerium für Bildung und Forschung (BMBF), Deutsches Raumfahrtprogramm, 2001.
3. Haidn, O.J., Habiballah, M., Research on high pressure cryogenic combustion, Aerospace Science and Technology, **7**, pp. 473–491, 2003.
4. Deshpande, M., Grenda, J.M., Merkle, C.L., CFD Modeling for LOX/H2 Coaxial Combustor Flowfields, AIAA 96-2855, 1996.
5. Moser, M.D., Merenich, J.J., Pal, S., Santoro, R.J., OH-Radical Imaging and Velocity Field Measurements in a Gaseous Hydrogen/Oxygen Rocket, AIAA 93-2036, 1993.
6. Habiballah, M., Vingert, L., Traineau, J.C., Vuillermoz, P., MASCOTTE: a test bench for cryogenic combustion research, 47th International astronautical congress, Beijing, China, October 7.-11., 1996.
7. Hopfinger, E., Lasheras, J.C., Breakup of a water jet in high velocity co-flowing air, in: Proceedings of the 6th International Conference on Liquid Atomization, Rouen, France, 1994.
8. Coakley, T.J., Huang, P.G., Turbulence Modeling for High Speed Flows, AIAA paper 92-0436, 1992.
9. Ó.M. Conaire, Curran H.J., Simmie J.M., Pitz W.J., and Westbrook C.K., A comprehensive modeling study of hydrogen oxidation, International Journal of Chemical Kinetics, **36**, pp. 603–622, 2004
10. Shuen, J.S.: Upwind Differencing and LU Factorization for Chemical Non-Equilibrium Navier-Stokes Equations, Journal of Computational Physics, **99**, pp. 233–250, 1992.
11. Jameson, A., Yoon, S.: Lower-Upper Implicit Scheme with Multiple Grids for the Euler Equations, AIAA Journal, **25**, pp. 929–937, 1987.
12. Gerlinger, P., Brüggemann, D.: An Implicit Multigrid Scheme for the Compressible Navier-Stokes Equations with Low-Reynolds-Number Turbulence Closure, Journal of Fluids Engineering, **120**, pp. 257–262, 1998.
13. Gerlinger, P., Möbus, H., Brüggemann, D.: An Implicit Multigrid Method for Turbulent Combustion, Journal of Computational Physics, **167**, pp. 247–276, 2001.
14. Gerlinger, P.: Investigations of an Assumed PDF Approach for Finite-Rate-Chemistry, Combustion Science and Technology, **175**, pp. 841–872, 2003.
15. Stoll, P., Gerlinger, P., Brüggemann, D.: Domain Decomposition for an Implicit LU-SGS Scheme Using Overlapping Grids, AIAA-paper 97-1896, 1997.
16. Stoll, P., Gerlinger, P., Brüggemann, D.: Proceedings of the 4th ECCOMAS Conference, **1**, pp. 205–212, John Wiley & Sons, 1998.

17. Le Clercq, P., Doué, N., Rachner, M., Aigner, M., Validation of a Multicomponent-Fuel Model for Spray Computations, AIAA 2009-1188, Proceedings of the 47th Aerospace Sciences Meeting, Orlando, FL, USA, January 5.-8., 2009.
18. Herding, G., Snyder R., Scouflaire, P., Rolon, C., Candel, S., Emission and Laser Induced Fluorescence Imaging of Cryogenic Propellant Combustion, Conference on Propulsive Flows in Space Transportation Systems, Bordeaux, France, 1995.
19. Brummund, U., Cessou, A., Oschwald, M., Vogel, A., Grisch, F., Bouchardy, P., Péalat, M., Vingert, L., Habiballah, M., Snyder, R., Laser diagnostics for cryogenic propellant combustion studies, Proceedings of the 2nd International Symposium on Liquid Rocket Propulsion, Châtillon, France, 1995.
20. Bazile, R., Guerre, S., Stepowski, D., "Planar Laser Induced Fluorescence of hot O2 in MASCOTTE", Second French-German Colloquium on Research on Liquid Rocket Propulsion, Aachen, Germany, 1996.
21. Candel, S., Herding, G., Snyder, R., Scouflaire, P., Rolon, C., Vingert, L., Habiballah, M., Grisch, F., Péalat, M., Bouchardy, P., Stepowski, D., Cessou, A., Colin, P., Experimental Investigation of Shear Coaxial Cryogenic Jet Flames, Journal of Propulsion and Power, **14**, 1998.
22. Habiballah, M., Orain, M., Grisch, F., Vingert, L., Gicqel, P., Experimental studies of high-pressure cryogenic flames on the MASCOTTE facility, Combustion Science and Technology, **178**, pp. 101–128, 2006.
23. Pourouchottamane, M., Dupoirieux, F., Vingert, L., Habiballah, M., Burnley, V., Numerical Analysis of the 10 bar MASCOTTE Flow Field, in: Proceedings 2nd International Workshop on Rocket Combustion Modeling, Lampoldshausen, Germany, March 25.-27., 2001
24. Farmer, R., Cheng, G., Chen, Y.-S., CFD Simulation of Liquid Rocket Engine Injectors, Part 2: Simulations of the RCM-2 Experiment, in Proceedings 2nd International Workshop on Rocket Combustion Modeling, Lampoldshausen, Germany, March 25.-27., 2001
25. Nicole, A., Dupoirieux, F., Vingert, L., Habiballah, M., Theron, M., Simulation of a subcritical LOX/GH2 MASCOTTE test case (10 bar), 5th International Spacecraft Propulsion Conference, Heraklion, Greece, May 5.-8., 2008.
26. Nicole, A., Ordonneau, G., Theron, M., 3D Simulation of LOX/GH2 MASCOTTE test case at 10 bar, Proceedings of the 3rd European conference for aero-space sciences (EUCASS), Versailles, France, July 6.-9., 2009.
27. Vingert, L., Habiballah, M., Test case RCM 2: Cryogenic spray combustion at 10 bar at Mascotte, in: Proceedings 2nd International Workshop on Rocket Combustion Modeling, Lampoldshausen, Germany, March 25.-27., 2001
28. Careé, I., Etude d'un injecteur coaxial assisté, PhD thesis, Université de Rouen, 1990.
29. Rehab, H., Villermaux, E., Hopfinger, E., Flow Regimes of Large Velocity ratio Coaxial Jets, Journal of Fluid Mechanics, **345**, pp. 357–381, 1997
30. Abramzon, B., Sirignano, W.A., Droplet vaporization model for spray combustion calculations, International Journal of Heat and Mass Transfer, **32**, pp. 1605–1618, 1989.
31. Kindler, M., Rust, B., Gerlinger, P., Aigner, M., Numerical Investigations of NO_X-Formation in Scramjet Combustors, In: Nagel, W.E., Kröner, D.B., Resch, M.M. [Hrsg.]: High Performance Computing in Science and Engineering '09, Transactions of the High Performance Computing Center, Stuttgart (HLRS), pp. 197–207, ISBN 978-3-642-04664-3, 2010.

32. Kindler, M., Blacha, T., Lempke, M., Gerlinger, P., Aigner, M., Numerical Investigations of Model Scramjet Combustors, In: Nagel, W.E., Kröner, D.B., Resch, M.M. [Hrsg.]: High Performance Computing in Science and Engineering '08, Transactions of the High Performance Computing Center, Stuttgart (HLRS), pp. 153–166, ISBN 978-3-540-88301-2, 2009.
33. Kindler, M., Gerlinger, P. and Aigner, M.: Assumed PDF Modeling of Turbulence Chemistry Interaction in Scramjet Combustors, In: Nagel, W.E., Kröner, D.B., Resch, M.M. [Hrsg.]: High Performance Computing in Science and Engineering '07, Transactions of the High Performance Computing Center, Stuttgart (HLRS), pp. 203–213, 2008.

Simulation of Triflux Heat Exchangers in Utility Boilers

Alan Matschke, Michael Müller, Uwe Schnell, and Günter Scheffknecht

Institute of Combustion and Power Plant Technology—IFK, University of Stuttgart, Pfaffenwaldring 23, 70569 Stuttgart, Germany

Summary. In this paper the recent development of the coupled simulation of large utility boilers is described. The coupled simulation is an advanced tool to investigate the interaction between the furnace and the steam cycle of power plants. Gained knowledge is useful both for design of future facilities and optimisation of operational performance of existing power plants. In this article the detailed simulation of a Triflux Heat Exchanger (THX) is presented. In utility boilers THX are used to exchange heat between the flue gas, the Superheater (SH) and the Reheater (RH) steam cycle. The presented methodology allows to calculate the steam and tube temperatures of all heated tubes of an utility boiler including the THX. Results and a comparison between measured and calculated values are presented for a lignite-fired boiler.

1 Introduction

Today the production of heat and power is still predominantly based on fossil fuels. The fastest option to reduce the production of carbon dioxide and other emissions is to increase the efficiency of the energy conversion process. In utility boilers the heat value of the fuel is transferred to the water/steam cycle by radiation, convection, and conduction. In the furnace of a coal-fired steam-generator radiation is the predominant mechanism of heat transfer, whereas in the area of tube bundles the convection dominates. CFD-based simulation tools have been applied to the combustion process with fixed thermal boundary conditions from the water/steam cycle. The coupling of a utility boiler simulation with a numerical code for the water/steam cycle leads to a new dimension for the analysis of thermal power plants as the mutual influences between flue gas and water/steam side are considered.

It is a well known fact that the heat absorption of the bundle heat exchangers varies over the continuous operation period. Due to slagging the heat absorption of the SH decreases and the heat absorption of the RH increases. However, the steam parameters before the turbine have to be controlled.

W.E. Nagel et al. (eds.), *High Performance Computing in Science and Engineering '10*, DOI 10.1007/978-3-642-15748-6_17,

There are several ways to deal with this problem. First, the slag adherent to the tube bundles is continuously removed by soot blowers. The soot blowing equipment, however, is uncapable of removing all slag from the tube bundles.

Second, most of the heat exchangers are equipped with spray water injections to cool down the steam in the connecting steamlines between the heat exchangers. Spray water injection into the SH steam cycle has no influence on the overall efficiency of the power plant. Injected spray water into the reheater steam cycle leads to a reduced power plant efficiency [13]. Regarding the necessary effort to raise the overall efficiency a reduction of efficiency due to injected spray water into the RH steam cycle has to be avoided.

A third way to control the steam parameters before the turbine without the necessity to inject spray water into the RH is the THX. According to its name, heat is not only exchanged between the flue gas and either the SH or RH steam cycle but between all three flows. This can be achieved by leading all tubes of a SH through the tubes of a RH. Depending on the temperature level heat is transferred from the flue gas to the reheater tubes, which then heat or cool the inner superheater tubes. The amount of exchanged heat is regulated by the mass flow through the inner superheater tube.

2 Models

Apart from many auxiliary systems a large utility boiler consists of two major components: the furnace, changing fuel energy to heat, and the steam generator absorbing this heat to superheat steam. The combustion process takes place in the furnace where the distribution of velocity, temperature, and chemical species are of interest. In order to reproduce the conditions inside the furnace appropriately, 3D-methods have to be applied. The superheating of steam inside the tubes proceeds at high Reynolds numbers and so it is assumed that the flow is evenly distributed along the cross-section of the tube. In order to simulate the entire water/steam cycle with each single tube within reasonable time, only 1D-codes can be applied. Both, the 3D-code AIOLOS for the furnace simulation [15], and the 1D-code DYNAMIC [10, 11] for the water/steam side simulation have been developed at the IFK and will be described in the following sections.

2.1 The 3D-CFD-Code AIOLOS

In AIOLOS all transport equations are implemented by the general transport equation for a value Φ:

$$\frac{\partial(\rho\Phi)}{\partial t} + \frac{\partial(\rho u_j \Phi)}{\partial x_j} = \frac{\partial}{\partial x_j}\left(\Gamma\frac{\partial\Phi}{\partial x_j}\right) + S_{\Phi} \tag{1}$$

with the diffusion coefficient Γ and the density ρ. The four terms of this equation correspond to a transient, convective, diffusive, and source term. In the present work, only steady state conditions are considered, the transient term is neglected. The computational domain is discretised into a finite number of control volumes in which all values are assumed constant. Integration over a control volume V and applying the Gauss integration rule, the general transport equation is written as:

$$\int_A \rho u_j \Phi n_j dA = \int_A \Gamma \frac{\partial \Phi}{\partial x_j} n_j dA + \int_V S_\Phi dV \tag{2}$$

where A is the total surface area of the control volume, n_j the normal unit vector of the infinitesimal surface area dA and u_j the velocity component in the direction of n_j. Hence, the balance over a control volume with a finite number of faces f with surface area A_f is given as:

$$\sum_f (\rho u_j \Phi n_j A)_f = \sum_f \left(\Gamma \frac{\partial \Phi}{\partial x_j} n_j A \right)_f + S_\Phi V \tag{3}$$

The values for ρ_f and Γ_f on the cell faces are derived via central differencing, whereas the velocity u_f can be interpolated using one of several available discretisation schemes, e.g. central differencing by deferred correction, upwind, higher upwind, monotorised linear upwind, and Quick. The different schemes have previously been discussed by many authors [3, 6, 14]. In the present work, a first order upwind scheme is used for all transport equations. The velocity on a face is replaced by the velocity of the upwind adjacent cell node. This method generally yields a very stable solution with reasonable accuracy.

In order to fulfil both the continuum and momentum equations, a SIMPLE pressure correction scheme [1] is applied to calculate an updated velocity and pressure field. However, linear interpolation of the pressure gradient in the momentum equation may lead to decoupling of pressure and velocity in the cell centre. The coupling of pressure and velocity is achieved by the method of Rhie and Chow [8]. The final discretised form of the general transport equation can be reduced to the following form:

$$a_P \Phi_P = \sum_N a_N \Phi_N + b \tag{4}$$

where P and N denote values of the actual control volume and its neighboring cells, respectively. To solve (4) for the entire computational domain several algorithms have been developed. According to the recommendations of Schneider [14] the SIP method [16] is applied for the solution of the pressure field, and a Gauss-Seidel algorithm with underrelaxation [9] is used for all other variables.

Turbulence modelling can be done using a standard k, ϵ approach [4] or a higher Reynolds-Stress-Model [7, 18]. With an additional set of analytical equations the viscosity, density, and specific heat capacity are calculated.

The combustion processes inside industrial furnaces are very complex since a large number of chemical reactions are involved even for natural gas flames.

The complexity rises for solid fuels, e.g. coal, due to their inhomogeneous structure. Hence, the reaction mechanisms have to be simplified in order to be calculated within a CFD framework. In the present work, coal combustion is described by a global reaction scheme covering two heterogeneous reactions for coal pyrolysis and char combustion and two homogeneous reactions for the oxidation of hydrocarbons and carbon monoxide. A more detailed description of the reaction model can be found in literature [5, 15, 19].

2.2 The 1D Representation of Tubes

The water/steam cycle is discretised using a one dimensional tube segment of the finite length Δz. The governing equations for this segment are the one dimensional continuity, momentum and enthalpy conservation equations:

$$\frac{\partial \dot{m}}{\partial z} = -A \cdot \frac{\partial \rho}{\partial t} \tag{5}$$

$$\frac{\partial p}{\partial z} = -\rho \left(\frac{\partial w}{\partial t} + w \cdot \frac{\partial w}{\partial z} \right) - \rho g \sin \beta - \left| \frac{\partial p}{\partial z} \right|_{fric} \tag{6}$$

$$\rho \cdot \left(\frac{\partial h}{\partial t} + w \cdot \frac{\partial h}{\partial z} \right) = \frac{\partial p}{\partial t} + \frac{U}{A} \cdot \dot{q} + w \cdot \left(\frac{\partial p}{\partial z} + \left| \frac{\partial p}{\partial z} \right|_{fric} \right) \tag{7}$$

with pressure p, enthalpy h, velocity w, and, gravitational acceleration g inside the segment, the inner cross-sectional area A and the inner circumference U. As for the 3D-part, transient terms will be neglected. The pressure drop due

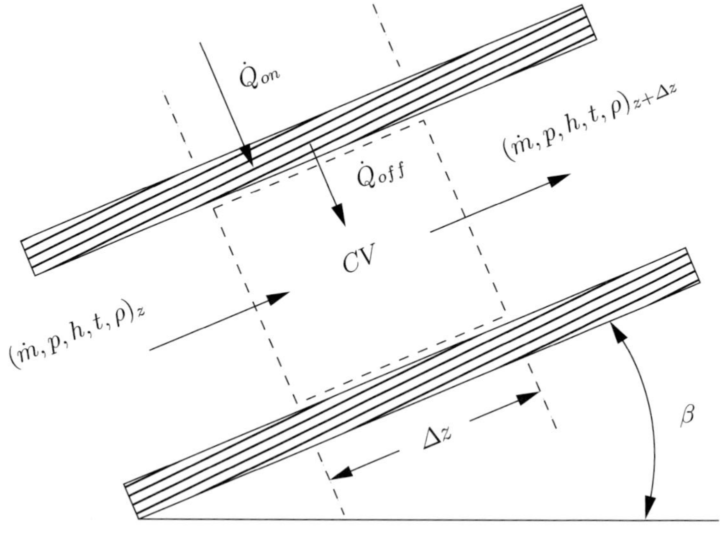

Fig. 1. Control volume of a heated tube segment

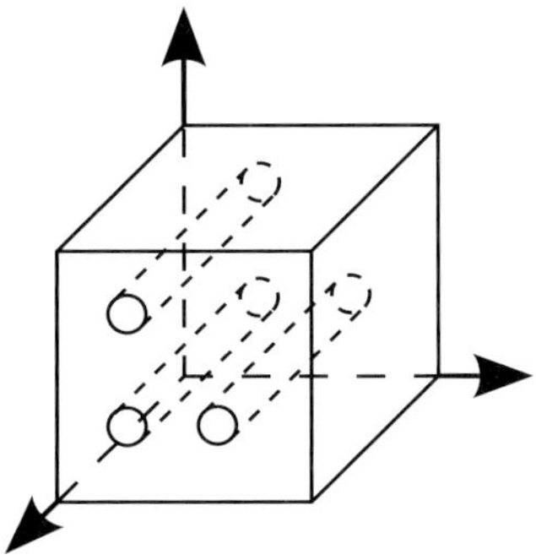

Fig. 2. Tubes in a CFD-Cell

to friction $|\frac{\partial p}{\partial z}|_{fric}$ is calculated from Colebrook's equations representing the Moody-Diagram [2]. The Nusselt Number, necessary to calculate the tube temperatures is calculated from Gnielinski's approach [2]. In a steady-state approach the tube wall does not accumulate heat. All heat from the surrounding flue gas ($\dot{Q}_{on}$) is directly transferred to the water/steam side ($\dot{Q}_{off}$). In large utility boilers the feedwater pump generates supercritical pressures. Therefore, no phase change between water and steam occurs.

2.3 Interaction Between 3D CFD-Cells and 1D Tube Segments

For the interaction between 3D Cells and the 1D tube segments some of the 3D CFD-Cells (i.e. the ones containing tubes) have to be treated as porous cells. Porous cells do have additional sources/sinks for enthalpy, radiation and momentum. The source/sink term for momentum depends on the geometry of the bundle heat exchangers [12].

Each tube segment absorbs a certain amount of energy either by convection or radiation from the surrounding flue gas. After calculating the entire water/steam cycle the surface temperature of each tube is passed back as a new boundary information to the 3D CFD-Cell. After multiple iterations between the 3D simulation of the furnace and the 1D representation of the water/steam cycle a converged solution is reached.

2.4 Triflux Heatexchangers

In Fig. 3 a small segment of a THX is shown. The outer tube belongs to the reheater steam cycle, the inner tube to the superheater steam cycle. The outer tube is heated by the surrounding flue gas. Depending on the temperature difference between the SH-steam in the inner tube and the RH-steam in the outer tube a certain amount of heat is exchanged between the superheater and the reheater steam cycle. Starting with an assumed amount of transferred heat the mean tube temperature of the inner tube is calculated from both sides.

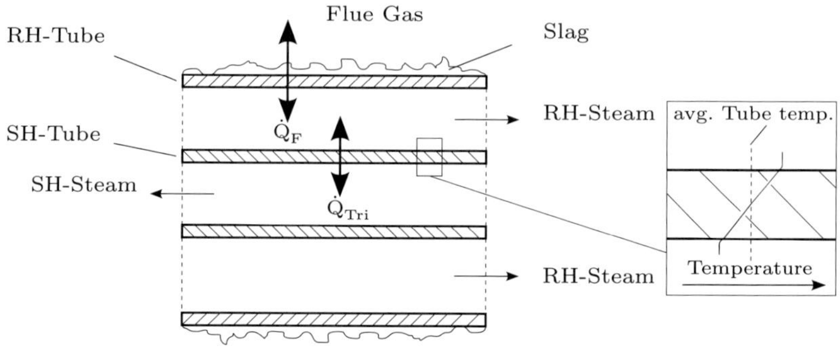

Fig. 3. Segment of a Triflux tube

During each iteration the amount of transfered heat is adjusted until the mean inner tube temperature is equal calculated from the reheater side as well as from the superheater side.

2.5 Water and Steam Properties

The water/steam properties are of great importance for the appropriate representation of supercritical water/steam. The IAPWS-97 [17] is a standard that describes the numerical implementation of routines to calculate water/steam properties. To avoid the time consuming reoccurant calculation of water/steam properties its values at discrete points are stored in large arrays in the memory of the computer. Explicit values are determined by bilinear interpolation between discrete points.

3 Results

For the validation of the coupled simulation several simulations of a lignite-fired utility boiler were carried out. Figure 4 shows the entire numerical grid in the centre. The porous cells in the shape of the bundle heat exchangers of the convective part interacting with the water/steam cycle are shown on the left. On the right hand side details of the main burners are shown. The overall dimensions are 150 m × 23 m × 23 m. Blue faces are either inlets for coal, primary, and secondary air or outlets for the recirculated flue gas; red cells are interface cells for domain decomposition. For the furnace 2.2 million cells are used, 1.9 million cells for the coal burners, and the convective part consists of 0.6 million grid cells. This boiler is qualified for model evaluation because of its good instrumentation with measuring points and the associated data base for storing measured data. Figure 5 shows results of the AIOLOS furnace simulation. The 1200 °C iso-surface reaches slightly into the tube bundles of the convective part (left). On the right, the predicted temperature distribution

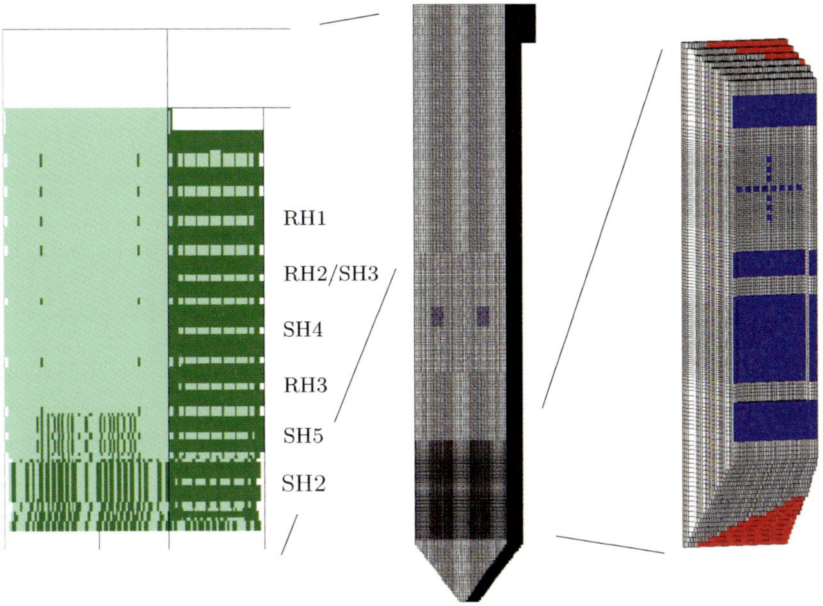

Fig. 4. Grid of a large utility boiler with porous cells of the convective part (left) and details for coal burners (right)

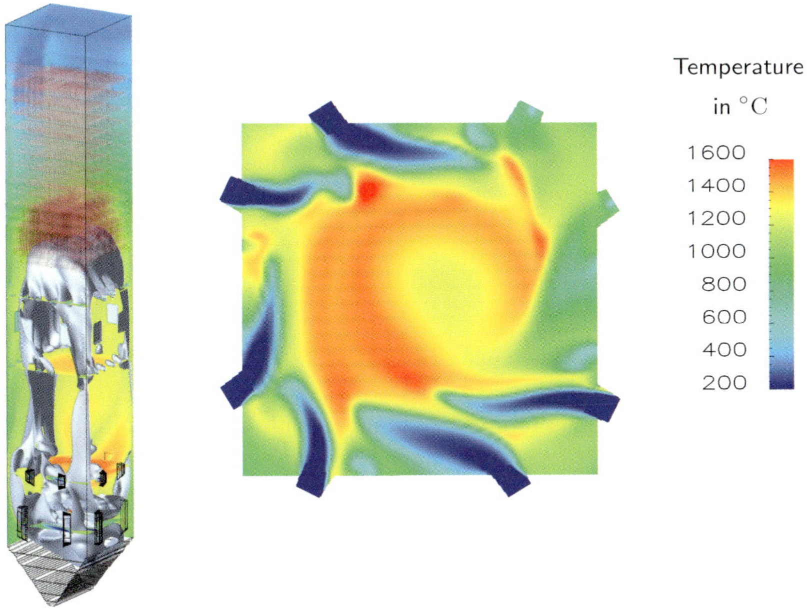

Fig. 5. 1200 °C Temperature iso-surface in furnace (left) and temperature distribution on cross-section 25 m above bottom (right)

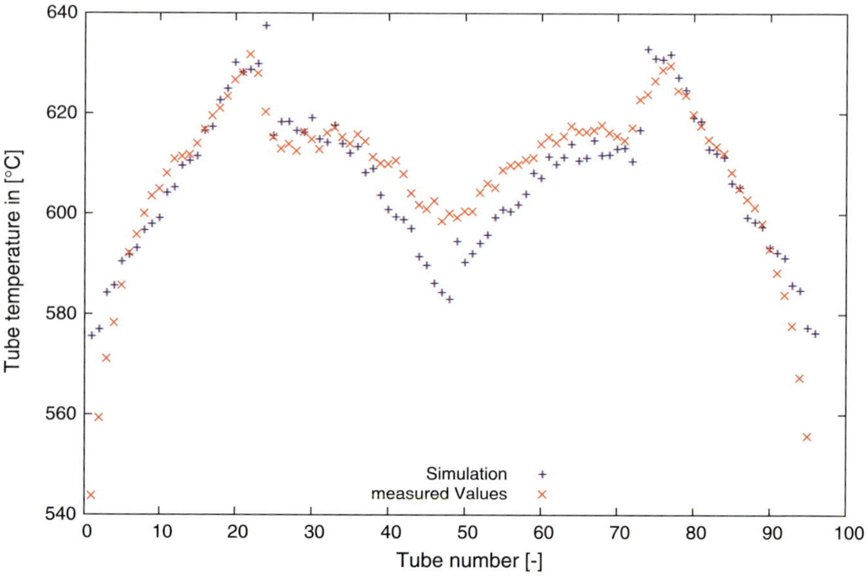

Fig. 6. Single tube temperatures of bank 7 of RH 3

is shown on the level of the main coal burners. With two coal burners of eight out of service, an unbalanced flue gas distribution across the boiler width is induced.

The third reheater (RH 3) consists of eighteen tube banks arranged horizontally and 96 sections arranged vertically in total 1728 tubes. At RH 3, the tube bank number seven is equipped with temperature sensors across the entire boiler width. Figure 6 shows the comparison between calculated and measured values. The distribution indicates that the real conditions are very well represented by the coupled simulation.

3.1 Triflux Heat Exchanger

Initially a fully operational power plant was considered. All bundle heat exchangers are connected via four steam lines. In Fig. 7 the steam temperatures of the four steam lines after steam injection 2, immediately before the SH 3 are shown. The large difference between simulated and measured steam temperature for the steam line 3 was not explainable. After consultation with the power plant operator it became obvious that the steam injection (SI) for steam line 3 was out of service. By adjusting this behaviour for the steam injection 3, the simulation results for the steam temperatures of all four steam lines are significantly better. Most notably, the high temperatures of steam line 3 could now be predicted. In Fig. 8 the corresponding mass flows for each steam line are depicted. The same effect occurs at the outlet of Triflux on the

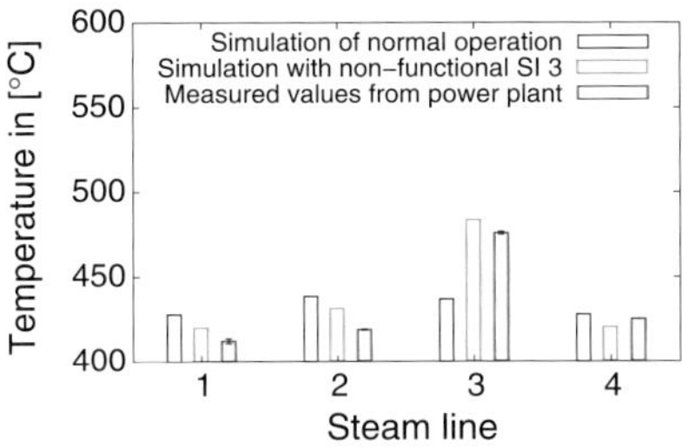

Fig. 7. Comparison of steam temperatures on SH after SI before Triflux

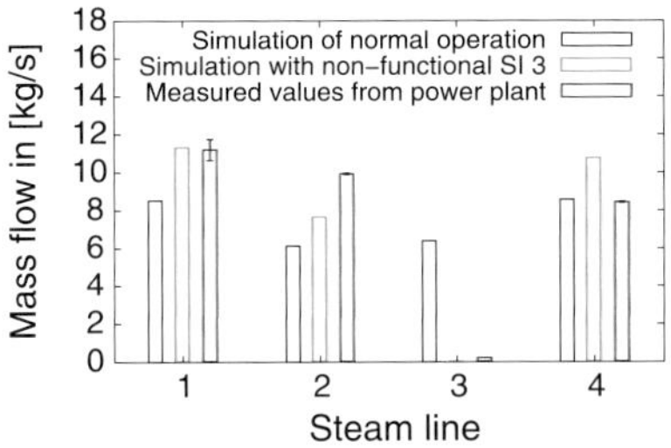

Fig. 8. Comparison of injected steam mass flow of SI 2, on SH before Triflux

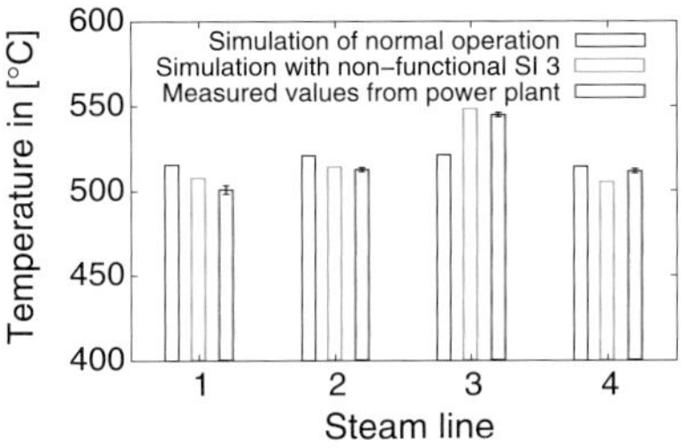

Fig. 9. Comparison of steam temperatures on RH after THX

RH side. In Fig. 9 the steam temperatures of the four steam lines immediately after the RH 2 are shown. Once more, with the assumption of normal operation the high temperatures of steam line 3 at the outlet of RH 2 could not be explained. The failure of the steam injection 3 on the SH side causes significantly higher steam temperatures for steam line 3 on the RH side. Figure 10 depicts three distinct steam temperatures of SH 3. The dotted lines stand for normal operation. The blue line is the steam temperature of the coldest tube of SH 3. The green line displays the average steam temperature of all tubes of SH 3. The red line is the hottest tube of SH 3. The continuous lines depict the same steam tubes but simulated with the failure of the steam injection 3. The hottest tube temperature rises significantly by approx. 16.5 °C. An additional 16.5 °C of steam temperature brings the tube material close to the maximum tolerable temperature. Therefore, the failure of a steam injection before the THX should be repaired as soon as possible after detection.

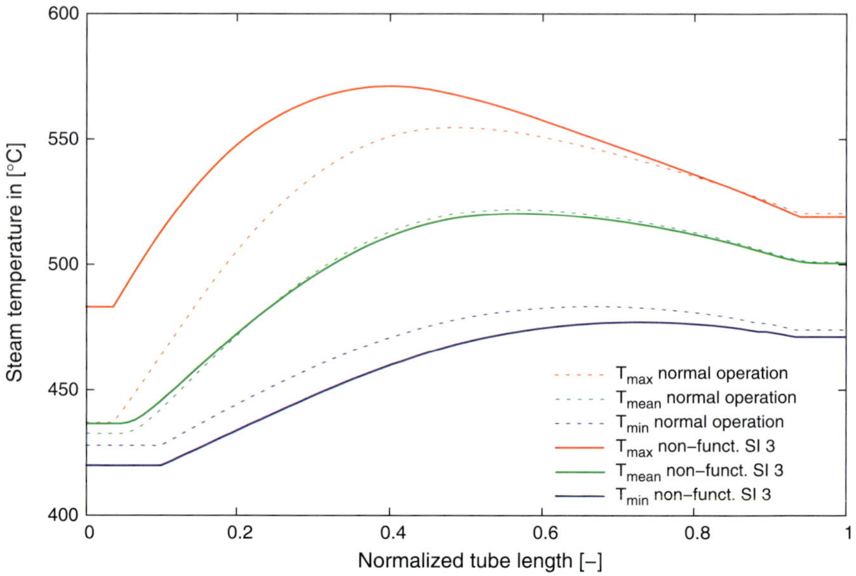

Fig. 10. Steam temperatures for selected tubes of SH 3 (inner tube of Triflux)

3.2 Performance

The simulations were performed on the NEC SX-9 installed at the HLRS in Stuttgart using one node with 16 CPUs. An average performance of 3.0 GFLOPS per processor and an average vector length of 248 is reported. The averaged cpu-time for one simulation is around 190 h. The total memory usage is 12.3 GB. The 1D-part of the coupled simulation is not parallelized yet whereas the 3D-part scales quite well with additional CPUs [9]. About 90% of the real time is spent for the 3D simulation of the furnace.

4 Conclusions

In utility boilers the heat absorption in parallel tubes of heat exchangers is influenced by the flue gas distribution as well as by the design of the heat exchangers. Since the local temperatures of the individual tubes are unknown, engineers apply surcharges to the mean steam temperature during the design process of boilers. These surcharges are either based on experience or taken from empirical formulae. The coupled simulation of the combustion and the steam generation process allows the prediction of single tube temperatures and uneven mass distribution between the different steam lines. Furthermore, with the presented model of a THX unknown and unmeasurable tube temperatures inside the TXH can be predicted. The main constructive and process parameters can be analysed with regard to deviations from mean temperatures. The influence of a component failure like an non-functional steam injection is predictable in early design stage. The huge acceleration that is gained from employing supercomputers like the NEC SX-9 allows engineers to keep in time with development and design changes.

References

1. Caretto, L.S., Gosman, A.D., Patankar S.V., Spalding, D.B.: Two calculation procedures for steady, three-dimensional flows with recirculation. Proc. Third Int. Conf. Numer. Methods Fluid Dyn., Paris (1972)
2. Çengel, Y.A.: Heat Transfer: A Practical Approach. McGraw-Hill Companies, New York (2002)
3. Knaus, H.: Simulation turbulenter reagierender Zweiphasenströmungen in industriellen Feuerungen mit komplexen Geometrien. VDI Verlag, Düsseldorf (2001)
4. Launder, B.E., Spalding D.B.: The Numerical Computation of Turbulent Flows, Computer Methods in Applied Mechanics and Engineering, 3:269–289 (1974)
5. Magnussen, B., Hjertager B.: On mathematical modelling of turbulent combustion with special emphasis on soot formation and combustion. 16th Symp. (Int.) on Comb., pp. 719–729, The Combustion Institute (1976)
6. Noll, B.: Numerische Strömungsmechanik. Springer Verlag, Berlin (1993)
7. Rotta, J.: Statistische Theorie der Turbulenz, Zeitschrift für Physik, Vol. 129, S. 547–572 (1951)
8. Rhie, C., Chow, W.: Numerical study of turbulent flow past an airfoil with trailing edge separation. AIAA Journal, 21(11):1525–1532 (1983)
9. Risio, B.: Effizienz und Verlässlichkeit numerischer Vorhersagen zur rechnergestützten Optimierung von Grosskraftwerksbrennkammern. VDI Verlag, Düsseldorf (2001)
10. Rolf, A.: Simulation des nichtlinearen, dynamischen Verhaltens von Wärmetauschern sowie ihrer komplexen Schaltung im Kraftwerksbau mit einem semianalytischen Berechnungsverfahren. PhD thesis University of Stuttgart (1984)
11. Sauer, C.: Detaillierte gekoppelte Simulation von Kraftwerksfeuerung und Dampferzeuger. PhD thesis, University of Stuttgart (2007)
12. Sauer, C., Berreth, A., Risio, B., Schnell, U., Hein, K.R.G.: Application of a Porous Cell Model for the Simulation of Tube Bundles in Convective Parts of Industrial Utility Boilers. 6th European Conference on Industrial Furnaces Boilers (INFUB), Estoril - Lisbon (Portugal) (2002)
13. Scheffknecht, G.: Vorlesungsmanuskript Dampferzeugung (2006)
14. Schneider, R.: Beitrag zur numerischen Berechnung dreidimensionaler reagierender Strömungen in industriellen Brennkammern. PhD thesis, University of Stuttgart (1998)
15. Schnell, U.: Numerical modelling of solid fuel combustion process using advanced CFD-based simulation tools. Progress in Computational Fluid Dynamics, 1 (4), 208–218 (2002)

16. Stone, H.L.: Iterative solution of implicit approximations of multidimensional partial differential equations. SIAM J. Numer. Anal., 5, 530–558 (1968)
17. Wagner, W., Kreztschmar, H.-J.: International Steam Tables. Springer Verlag, Berlin, (2008)
18. Wilcox, D.C.: Turbulence Modelling for CFD. DCW Industries, California (1993)
19. Zinser, W.: Zur Entwicklung mathematischer Flammenmodelle für die Verfeuerung technischer Brennstoffe. VDI Verlag, Düsseldorf (1985)

Computational Fluid Dynamics

Prof. Dr.-Ing. Siegfried Wagner

Institut für Aerodynamik und Gasdynamik, Universität Stuttgart, Pfaffenwaldring 21, 70550 Stuttgart, Germany

The majority of papers within CFD deals with Direct Numerical Simulation (DNS) and Large Eddy Simulation (LES) or with hybrid methods, i.e. a combination of DNS or LES with RANS (Reynolds Averaged Navier-Stokes) methods because they require high performance computers (HPCs) with big capability both with respect to memory and performance.

The increasing demand of memory and performance since the last workshop at HLRS in 2009 can be demonstrated by a few highlights. The Direct Numerical Simulations of the project by Friedrich and Kloker required a memory of 0.4 Tbyte RAM since their computational grid contained 10^9 grid points. The DNS of turbulent plane Cuette flow by Garcia-Villalba, Azagro and Uhlmann needed $6.06 \cdot 10^8$ grid cells on the HP XC4000 of SCCK. The big demand on more capable computers was also demonstrated by von Terzi, Venema, Bauer and Rodi who studied unsteady heat transfer by DNS. They used 10^8 grid cells for their simulation but did foresee a demand of 10^9 grid cells in the future.

Another important point is the scaling behaviour of a code as the number of processors per platform increased steadily in the past and will do so in the future. Chan-Braun, Garcia-Villalba and Uhlmann started DNS with 64 processors on the HP XC4000 of SCCK and demonstrated a good scaling behaviour up to 256 processors. A good speed-up up to 128 processors was also experienced by Galeazzo, Habisreuther and Zazalis. It is hoped that they can preserve this behaviour up to 1000 processors and more in the future.

LES and even more DNS require HPCs with extremely high performance and big memory. But they are mandatory since many problems cannot be simulated accurately by eddy-viscosity based RANS. Von Terzi, Schneider and Bauer showed that RANS cannot inherently account for mean secondary vortices in corners of diffusers which remarkably influence the performance of these devices. Thus, the authors used LES, but needed up to $22 \cdot 10^6$ grid cells.

The development of advanced numerical methods and algorithms, i.e. the software, is equally important as the hardware. Friedrich and Kloker reached a speed-up of about 2 compared to the performance of the NEC SX-8 last year by improving communication and employing optimized FFT. They reached a performance of 1.1 TFlop/s on the NEC SX-8. Furthermore, they used for their DNS of pinpoint suction in the boundary layer of airfoils and wings sixth order finite differences to resolve the complex flow in the boundary layer. To further improve the performance of the code they replaced the fully explicit Runge-Kutta 4th order time integration by a three-stage Heune method, i.e. the standard Heune method plus an additional second corrector step. They applied a spatial filter necessary for the downstream direction since the alternating forward-backward differences did not provide enough numerical damping for high wave number modes. The computer code LESOCC2 of von Terzi, Venema, Bauer and Rodi was run on the HP XC3000 of SCCK for preliminary simulations. They could obtain a better performance on a given platform when using different implementations of the computationally most intensive algorithms. They gained this experience by testing their code on the NEC SX-8 and SX-9, SGI-Altix and on several Linux clusters besides the HP XC3000.

To simulate the industrial production of large silicon single crystals Raufeisen, Breuer, Botsch and Delgado added the explicit computation of the melt-crystal interface to the previous version of their code FASTEST and computed the dynamics of the three-phase boundary where the melt, the crystal and the surrounding atmosphere meet. They extended the code to an ALE (Arbitrary Lagrangian Eulerian) formulation and could that way explicitly compute the free surface that was necessary to more accurately simulate the process.

Bensing, Keßler and Krämer used for their simulation of flow-structure interactions of helicopter rotors also an ALE formulation. They used an unstructured code on the NEC Nehalem cluster where they reached a sustained performance of 12 per cent peak performance using 128 processors. The present mesh size of $6.09 \cdot 10^6$ will be increased up to $5 \cdot 10^8$ cells in the future.

Roidl, Meinke and Schröder used for their numerical investigation of shock wave boundary-layer interactions a zonal RANS-LES ansatz whereby the transition from RANS to LES took place in an overlapping zone using synthetic turbulence generation methods and a reconstruction of eddy viscosity for transition. Up to 31 million mesh points were used. A performance of 483.2 GFlop/s was reached on the NEC SX-9 of HLRS using 2 nodes.

A shortcoming in the reports turned out to be the usage of a relatively low number of CPUs. This situation has to be improved since some of the HPCs within GCS (Gauss Centre of Supercomputing) are already massively parallel platforms and some additional could come in the near future. Although the three centres of supercomputing in Jülich, Munich and Stuttgart do already a lot with respect to education and courses for the customers there is still a need to continue this effort and may be to even increase it. A change in the philosophy of running the job classes might also help. Roidl, Meinke and Schröder stated in their paper that they used a relatively low number of nodes to minimize the overall user time for the simulation. They took into account that a rather high number of nodes would increase the turnaround time as the scheduling system prefers jobs with fewer nodes.

Direct Numerical Simulation of Swept-Wing Laminar Flow Control Using Pinpoint Suction

Tillmann A. Friederich and Markus J. Kloker

Institut für Aerodynamik und Gasdynamik, Universität Stuttgart, Pfaffenwaldring 21, D-70550 Stuttgart, Germany, friederich@iag.uni-stuttgart.de, kloker@iag.uni-stuttgart.de

Summary. Recent investigations of laminar flow control for swept-wing boundary-layer flows provide promising results with respect to crossflow-transition delay using a technique called *pinpoint suction*. Strong, localized suction through holes accurately positioned with respect to the crossflow-vortex position can directly weaken the growth of secondary instabilities that are responsible for the final laminar breakdown. With our incompressible code *N3D* extremely resolved grids in wall-normal direction have to be used to obtain numerical convergence. In order to confirm the results with such strong, localized suction, a single suction hole has been simulated in a Blasius boundary layer with the *N3D* code and additionally with our compressible code *NS3D*. Results from grid studies carried out for both codes as well as a comparison of the compressible/incompressible results are provided. Incompressible simulations with various pinpoint suction scenarios in the swept-wing flow show the application of successful pinpoint suction. Performance data for both codes focusing on a comparison of the NEC SX-8/SX-9 conclude this report.

1 Introduction

Improving the fuel efficiency of aircrafts has become an important task within the last decades. Not only do airlines benefit from saving fuel by decreasing their direct operation costs regarding the oil price development but also the environmental aspect has gained growing interest and it will only be a matter of time until environmental laws limiting exhaust gases will be approved. Current commercials for newly designed aircrafts are inconceivable without sentences like "The 787 Dreamliner is using 20 percent less fuel than any other airplane of its size" (Boeing homepage).

Until today actually applied optimisations for new airplanes are limited to enhanced shaping, higher surface quality and engine improvement, but little potential is thought to be left in these research fields. New concepts have therefore to be envisaged. Laminar flow control (LFC) provides a total drag reduction potential of up to 16 % by e.g. realizing 40 % laminar flow on wings and control surfaces of a current airliner. Much LFC research has been

W.E. Nagel et al. (eds.), *High Performance Computing in Science and Engineering '10*, DOI 10.1007/978-3-642-15748-6_18,

carried out for two-dimensional flow scenarios: By methods like boundary-layer suction or active T-S wave control; triggering anti-phase disturbances; transition to turbulence caused by travelling disturbances can nowadays be successfully controlled, see, e.g. Peltzer et al. [14] and Liepmann et al. for the fundamentals [8] and Liepmann & Nosenchuck [9].

However, the physical mechanisms in a three-dimensional flow, present on every swept (airliner) wing, are more difficult to influence. Steady crossflow vortices (CFVs) typically trigger early transition whereas in a two-dimensional scenario unsteady T-S waves cause transition to turbulence. After the high-frequency secondary instability of CFVs had been fully eludicated by means of experiments, theory and DNS within the last decades (White & Saric [23], Koch et al. [7], Wassermann & Kloker [20–22]), several methods aiming at crossflow transition control have been developed since: Saric proposed the distributed roughness elements (DRE) method, see e.g. [15, 16] and for a more recent overview [17]. A one-time excitation of steady CFVs that are spaced narrower than the naturally most amplified ones creates a flow scenario with, at first, secondarily stable CFVs and thus delayed transition to turbulence. The upstream flow deformation concept (UFD), proposed by Wassermann & Kloker [18, 19] and also [20], pursues a similar goal, not necessarily connected with roughness.

Messing & Kloker [12, 13] and also [6] proposed a combination of UFD and suction leading to the general concept of distributed flow deformation (DFD) and in particular formative suction. By designing a slot-suction panel the narrower-spaced useful vortices are continuously excited and additionally suction is present. Also, succeeding panels can be adapted to the changing stability characteristics when proceeding downstream and thus continuously excite the locally optimal DFD mode.

A new idea of directly influencing already growing secondary instabilities, called *pinpoint suction* (see Friederich & Kloker [5] and [6] for a tentative study) is currently developed at our institute and topic of this report. The idea to use this specialized, pinpointed suction arose from the discovery that even a small velocity component normal to a local shear layer can substantially reduce the shear-layer instability (Bonfigli & Kloker [3], Friederich [4]). Localized, strong suction through a hole at the updraft side of the crossflow vortex, i.e. the position of the eigenfunction maximum of high-frequency secondary modes, directly reduces their growth while also reducing the vortex strength. For a recent overview including UFD, DFD and pinpoint suction see also Kloker [6].

2 Numerics

2.1 Incompressible Code N3D

Our code *N3D* solves the full 3-d unsteady incompressible Navier-Stokes equations using a disturbance formulation. A velocity-vorticity formulation is cho-

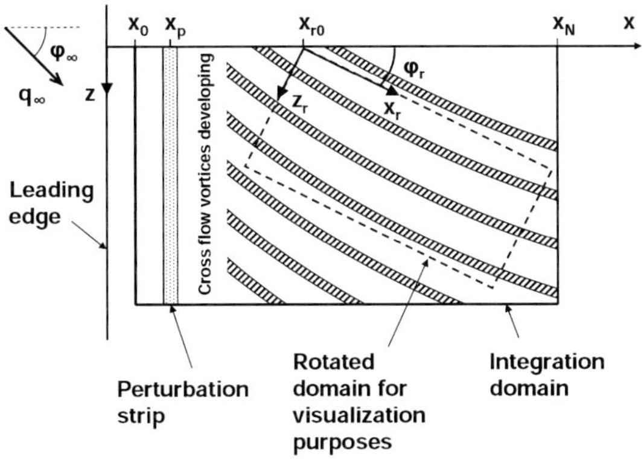

Fig. 1. Flat plate (top view) with integration domain and rotated reference system (x_r, z_r) used for visualization purposes

sen. Each flow quantity q is divided into its steady baseflow part q_B and its unsteady component q' to ease formulation of the boundary-conditions. For the downstream and wall-normal directions x and y sixth-order compact (splitted) finite differences are implemented whereas for the spanwise direction z we use a Fourier spectral representation with $K+1$ modes. A sketch of the considered domain is shown in Fig. 1, and for a detailed description of the numerical method see [20] and [3]. The necessity of a highly-resolved wall-normal coordinate for strong suction ($\bar{v}_{wall,max}/\bar{u}_\infty > 10\%$) demanded the development of a new semi-implicit time-integration scheme. Due to the inhibiting viscous time-step limit arising from the y-direction the standard fully explicit RK4-O4 time integrator was replaced by a three-stage Heun method—standard Heun method plus and additional, second corrector step—where the second y-derivatives are treated implicitly. The vorticity-formulation required then an iteration method for the equations at the wall. The still small time step forced by the convective terms in y-direction made a spatial filter necessary for the downstream direction since the alternating forward-backward differences did not provide enough numerical damping of high wavenumber modes.

2.2 Compressible Code NS3D

The DNS code *NS3D* (see Babucke, Linn, Kloker, and Rist [1]) is used for the compressible test simulations. This code is based on the complete 3-d unsteady compressible Navier-Stokes equations and a calorically perfect gas. The equation set is also solved in a rectangular integration domain on the flat plate. In streamwise (x-) and wall-normal (y-) direction, the discretization is realized by splitted compact finite differences of 6th order. In the spanwise (z-) direction, the flow is assumed to be periodic, thus a Fourier spectral representation is employed to compute the z-derivatives. In contrast to the incompressible code *N3D*, *NS3D* largely computes in physical space. After transformation to Fourier space and simple computation of the z-derivatives, the back transformation is done with de-aliasing using the 2/3-rule. For time integration the classical 4th-order Runge-Kutta method is employed as in the incompressible case. A detailed description of the discretization algorithm and boundary conditions is given in Babucke & al. [1] and Linn & Kloker [10].

3 Results

For the incompressible code *N3D* a strong dependence of the necessary wall-normal resolution at the wall, Δy_{wall}, and the maximum suction amplitude prescribed at the hole was found. In order to confirm the results with very high suction amplitudes, a grid-study test case was simulated with the incompressible and compressible codes *N3D* and *NS3D*, respectively, with varying Δy_{wall}. A single suction hole with 30% maximum suction velocity in a two-dimensional Blasius boundary layer flow (U_∞=15.0$\frac{m}{s}$) was chosen. All remaining flow parameters were matched to Meitz's case of Goldstein [11]. Since for the compressible code the allowable time step is roughly proportional to the Mach number, U_∞ of the compressible boundary layer was chosen to be 4.6 times higher in order to achieve Ma=0.2 and thus a reasonable time step. In order to allow for comparison, $Re_{\delta_1}(x_{hole})$, d_{hole}/δ_1 and d_{hole}/s_z have been kept constant for both codes, where d_{hole} is the hole diameter, δ_1 the local displacement thickness, and $s_z = \lambda_{z,0}$ is the spanwise spacing of the holes and equal to the fundamental spanwise wavelength. The parameters for both simulations can be found in Table 1. Figure 2 provides the displacement thickness for both cases and the corresponding x-locations for the suction hole centers can directly be extracted for $\delta_1 = 0.01224$.

Table 1. Parameters for the test case

	N3D	NS3D
U_∞	$15.0\ \frac{m}{s}$	$69.5\ \frac{m}{s}$
Ma_∞	–	0.2
T_∞	–	$270.0\,K$
p_∞	–	$1.0511\,bar$
ν_∞	$1.5\cdot 10^{-5}\ \frac{m^2}{s}$	$1.5\cdot 10^{-5}\ \frac{m^2}{s}$
L	$0.1\,m$	$22.7{\cdot}10^{-3}\,m$
Re	100000	100000
$Re_{\delta_1,hole}$	1224	1224
$\delta_{1,hole}$	0.01224	0.01224
d_{hole}	$0.84426\cdot\delta_1 = 0.0103$	$0.84426\cdot\delta_1 = 0.0103$
$\lambda_{z,0}$	$2\cdot d_{hole} = 0.0206$	$2\cdot d_{hole} = 0.0206$
x_{hole}	5.0797	4.9649

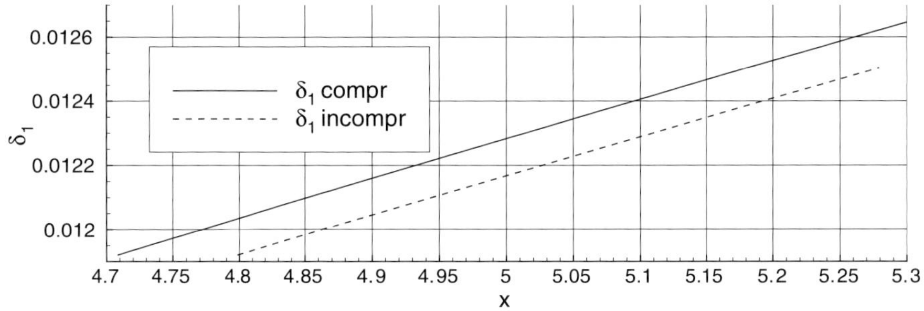

Fig. 2. Displacement thickness δ_1 for compressible and incompressible Blasius test cases of the corresponding, unperturbed baseflows

3.1 Incompressible Test Case

A two-dimensional self-similar solution with zero streamwise pressure gradient is chosen as baseflow for this test case. The extensions of the integration domain in streamwise and wall-normal direction are (short/long case) $L_x = 0.48/L_x = 0.72$ ($N_x = 642/962$, $\Delta x = 7.50 \cdot 10^{-4}$) and $L_y = 0.056$ ($N_y = 233$, step size decreases while approaching the wall, various Δy_{wall}). The suction hole in these cases is modeled by prescribing

$$v(x, 0, z) = -v_{max} \cos^3(\frac{\pi r}{d}) \tag{1}$$

where

$$d = d_{hole}, r = \sqrt{(x - x_{hole})^2 + (z - z_{hole})^2}, \quad r \leq \frac{d_{hole}}{2}. \tag{2}$$

Later on, also a different function for v(x,0,z) has been used.

Figure 3 provides results for three different values of Δy_{wall}. Convergence for 30% suction can be found for $\Delta y_{wall} = 1/1320 \cdot \delta_1$.

3.2 Compressible Test Case

A Blasius similarity solution is used as compressible baseflow as well where Ma=0.2 is chosen in order to achieve a reasonable time step limit as explained above. The pressure p_∞ is chosen such that the kinematic viscosity ν=$1.5 \cdot 10^{-5} \frac{m^2}{s}$ is equal to the incompressible case. Physical parameters are provided in Table 1. The correct suction hole x-location can be extracted from Fig. 2. The extensions of the integration domain in streamwise and wall-normal direction are $L_x = 0.60$ ($N_x = 598$, stretched grid before and after suction hole, $\Delta x_{min} = 6.68 \cdot 10^{-4}$) and $L_y = 0.105$ ($N_y = 142$, step size decreases while approaching the wall, various Δy_{wall}). The suction hole in these cases is modeled by

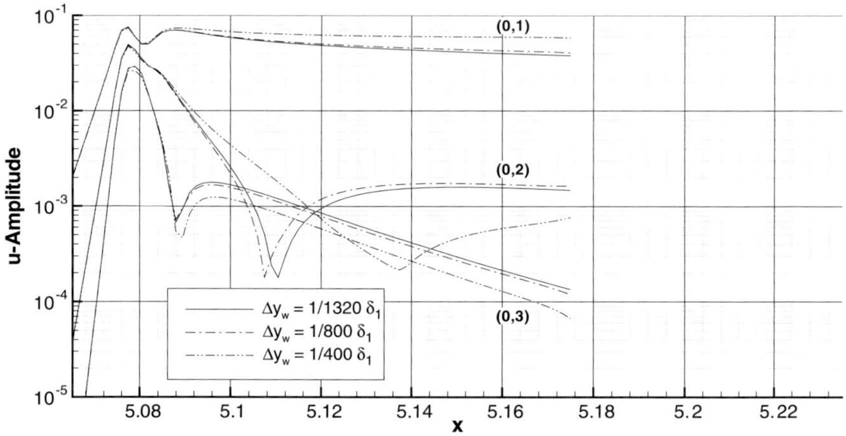

Fig. 3. Results from a grid study with the incompressible code *N3D*. Downstream modal development of stationary u-amplitudes for the short Blasius test case with one 30% suction hole. The suction hole is modeled using a $\cos^3$-function for the suction velocity. Mode designation (h,k): frequency $h\omega_0$, spanwise wavenumber $k\gamma_0$.

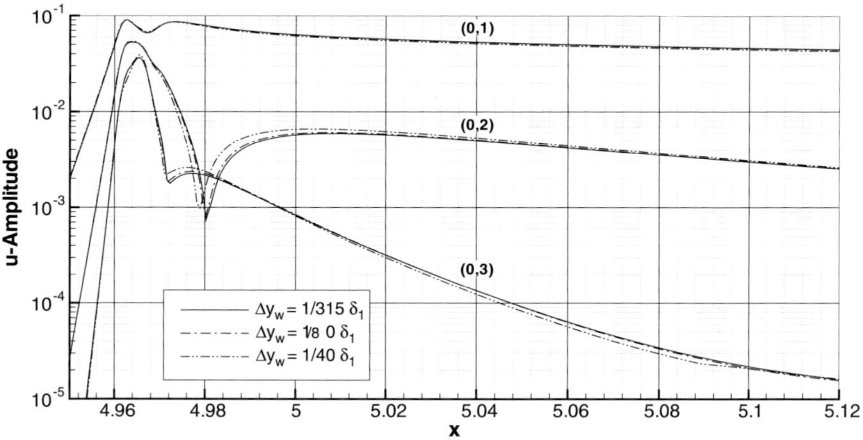

Fig. 4. Results from a grid study with the compressible code *NS3D*. Downstream modal development of stationary u-amplitudes for the Blasius test case with one 30% suction hole. The suction distribution is modeled using a 5th-order polynomial

$$v(x,0,z) = -v_{max} \cdot (1 - 6(\frac{r}{r_{hole}})^5 + 15(\frac{r}{r_{hole}})^4 - 10(\frac{r}{r_{hole}})^3) \quad (3)$$

where

$$r = \sqrt{(x - x_{hole})^2 + (z - z_{hole})^2}, \quad r \le r_{hole}. \quad (4)$$

Figure 4 shows that all resolutions provide useful results, convergence can be found for $\Delta y_{wall} = 1/80 \cdot \delta_1$ which is larger by more than an order of magnitude compared to the incompressible case.

3.3 Comparison

In order to directly compare results from both codes, one more test case was simulated with the incompressible code *N3D*. The suction distribution was changed from the $\cos^3$-function to the polynomial distribution in order to prescribe the identical boundary condition. Also, since the compressible code prescribes the quantity ρv, the value of $v_{hole,max}$ was extracted once a steady state was reached and fed into the incompressible simulation ($\rho v_{max,compr} = -0.295000$, $v_{max,incompr} = -0.295878$. Results from both simulations are displayed in Fig. 5. The highest wall-normal resolutions are used for the comparison. The important modes (0,1), (0,2) match perfectly. Slight deviations can be found for the higher modes. Figure 6 shows a comparison of the physical flow domain. The velocity distributions match almost perfectly and so do the $u = 0.0$ isocontours which mark areas of reversed flow. Only the wall-normal velocity component v shows some deviations in the outer flow field. The λ_2-isosurface reveals slight differences in the streamwise vortex extension. Velocity profiles at four downstream locations including the position of the suction hole are provided in Fig. 7. Obviously, the upper boundary condition produces minor differences in the far-field of the wall-

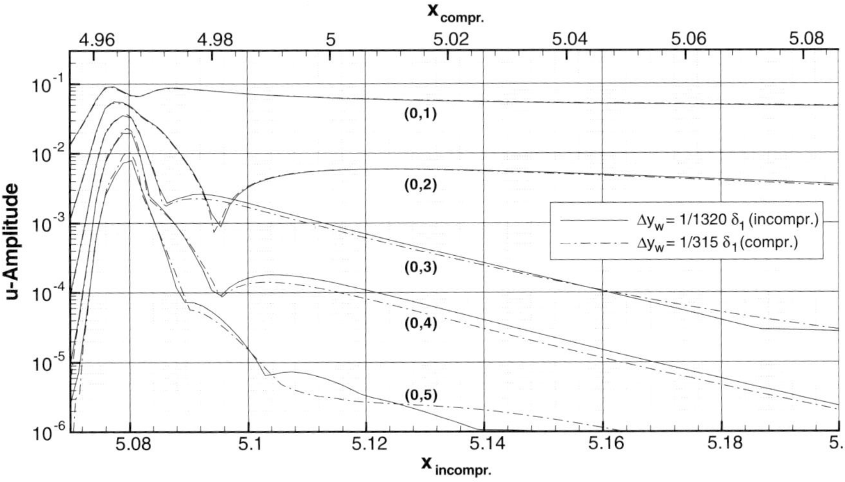

Fig. 5. Comparison of results from incompressible code *N3D* and compressible code *NS3D*. Downstream modal development of stationary u-amplitudes for the Blasius test case with one 30% suction hole. The suction distribution is modeled using a 5th-order polynomial in both codes

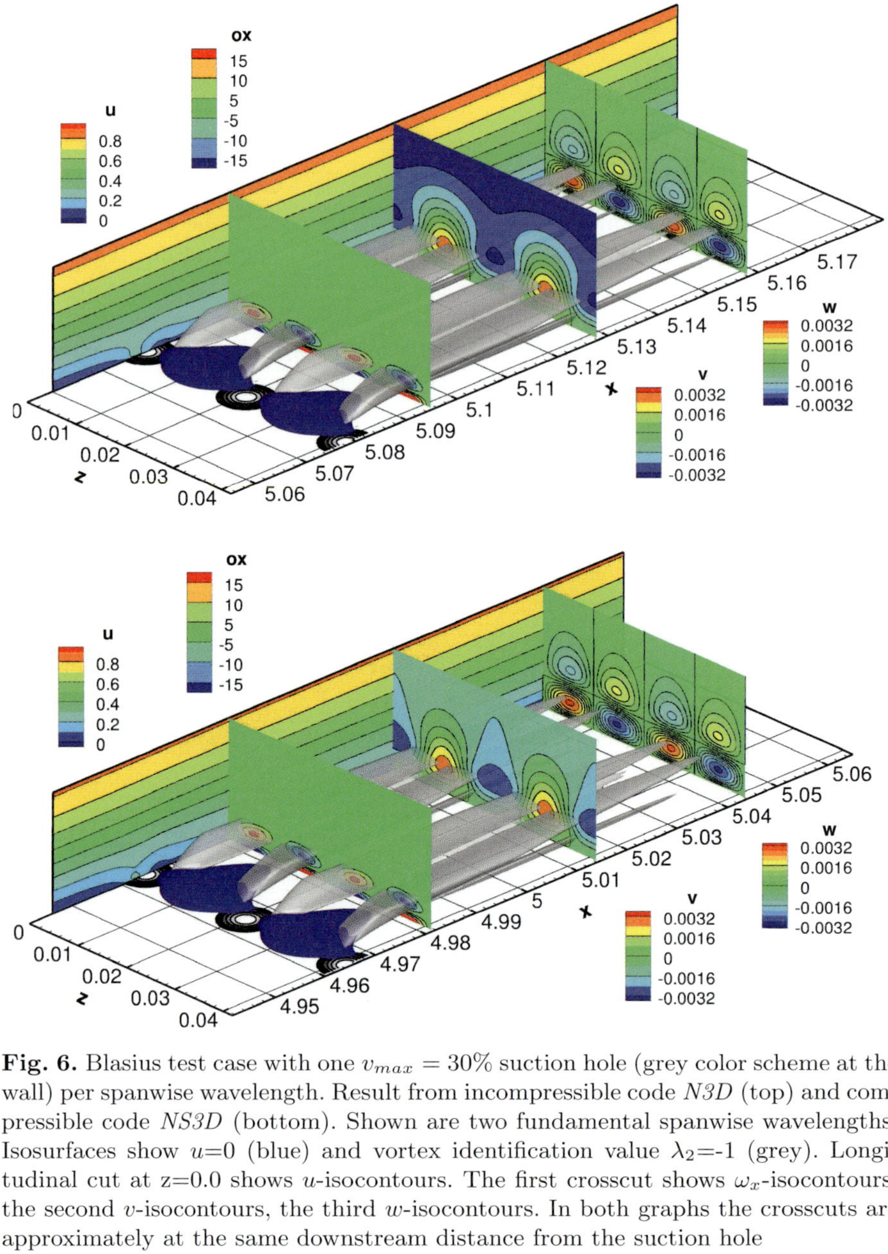

Fig. 6. Blasius test case with one $v_{max} = 30\%$ suction hole (grey color scheme at the wall) per spanwise wavelength. Result from incompressible code *N3D* (top) and compressible code *NS3D* (bottom). Shown are two fundamental spanwise wavelengths. Isosurfaces show u=0 (blue) and vortex identification value λ_2=-1 (grey). Longitudinal cut at z=0.0 shows u-isocontours. The first crosscut shows ω_x-isocontours, the second v-isocontours, the third w-isocontours. In both graphs the crosscuts are approximately at the same downstream distance from the suction hole

normal velocity component. For the incompressible code exponential decay $\partial v'/\partial y = v'_y = \alpha \cdot v'$ is prescribed for the spanwise modes $k > 0$ whereas for the 2-D mode k=0 $v_y = 0$ is used in order to allow mass to stream into the integration domain. For the baseflow $v_{B_y} = 0$ holds. In the compressible code

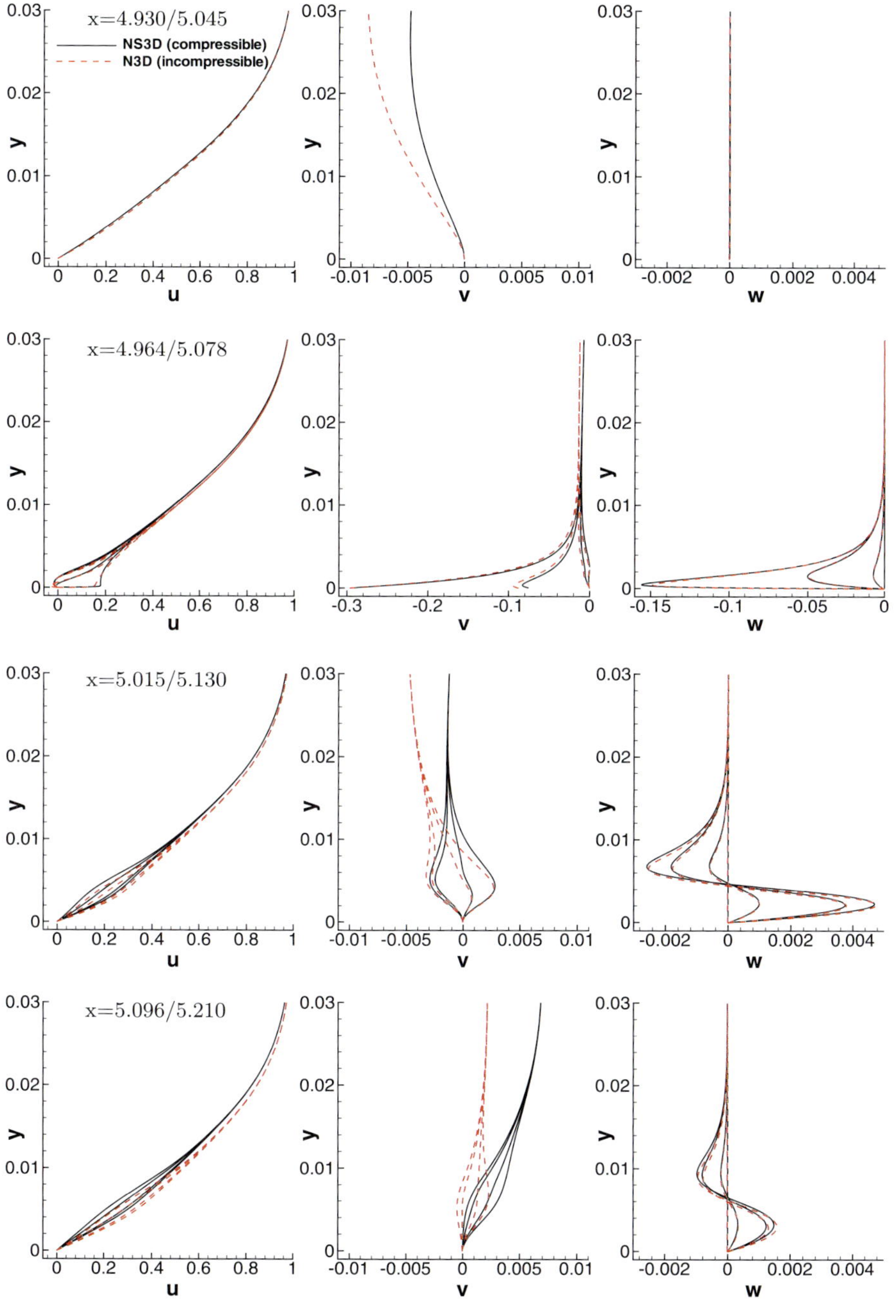

Fig. 7. u-, v-, and w-profiles at four downstream locations (given are compressible/incompressible coordinates). x=4.964/5.078 is the respective center of the suction hole. Note the different v- and w-scales. Each subfigure contains profiles at 4 equidistant spanwise locations ($z \in [0.00, \lambda_{z,0}/2]$)

a potential-flow boundary condition is used that keeps u from the baseflow and sets $v'_y = w'_y = 0$. Despite the minor differences in v these results confirm the fully converged, correct solution for both codes.

The very low Δy_{wall} value for the incompressible case—the compressible value is about a factor of 16 higher—is probably due to the incompressible code formulation in velocity-vorticity variables and due to the fact that the wall-vorticities have to be calculated using different equations. It turned out that up to maximum suction values $\bar{v}_{wall,max}/\bar{u}_\infty = 10\%$ the wall-normal resolution has not to be specifically refined. Note that the overall computational effort is still higher for the compressible code, see Sect. 3.5.

3.4 Pinpoint Suction Results

Here, we use the baseflow corresponding to the DLR-Göttingen Prinzip-Experiment [2, 3], where a flat-plate flow with negative streamwise pressure gradient was generated by a displacement body above. The free-stream velocity $q^*_\infty = 19\,\frac{m}{s}$ and a sweep angle of the plate $\Phi_\infty = 42.5°$ were chosen. Thus, the component $U^*_\infty = 14\frac{m}{s}$ and a reference length of $L^* = 0.1\,m$ are used for non-dimensionalization. The Reynolds number is $Re = 92000$ based on U^*_∞ and L^*. The extensions of the integration domain in streamwise and wall-normal direction are $L_x = 0.42\,m$ ($N_x = 3202$) and $L_y = 0.0076\,m$ ($N_y = 225$, step size decreases while approaching the wall). The fundamental wavelength in spanwise direction $\lambda_{z,0} = 0.012\,m$ is discretized with $K = 15$ harmonics and the fundamental reference frequency is $\omega_0 = 6$ ($f^* = 133\,Hz$).

For pinpoint-suction amplitudes of 50%, convergence was achieved with $\Delta y_{wall} = 1/2500 \cdot \delta_1$. The grid is continuously stretched up to the upper edge of the integration domain ($\Delta y_{edge} \approx 120 \Delta y_{wall}$).

Figure 8 shows an example of a pinpoint suction setup. The crossflow-vortex position is marked by the λ_2-isocontours. The eigenfunction of a high-frequency z-mode is located at the updraft side of the vortex. A crosscut through the suction hole (arrows) shows schematically the correct position of pinpoint suction. For all pinpoint cases, the $\cos^3$-function from (1) and (2) are used to prescribe the wall-normal velocity distribution inside the hole surface.

After a reference case *REF* has reached a quasi-steady state, the exact CFV position is determined and different suction scenarios can be set up. A case with a single 50%-suction hole (case *1-50p*) is setup to verify the pinpoint idea, i.e. strong single-hole suction is applied. To improve the pinpoint suction effect on the vortex without increasing the maximal suction amplitude (and thus dodging wall-normal resolution problems) various scenarios with consecutive 50%-suction holes have been considered. In case *2-50p* one additional suction hole has been added at some downstream distance. Cases *3-50p* and *9-50p* provide setups with 3 or 9 narrow-spaced holes in downstream direction. In order to compare the three-dimensional contribution of pinpoint suction, a "two-dimensional" case has been simulated (case *3-2D*) corresponding to case *3-50p*, i.e. only the spanwise average suction part is

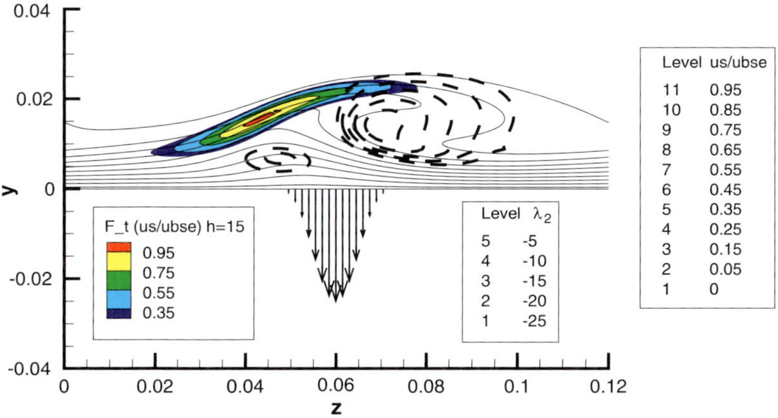

Fig. 8. Pinpoint concept: Crosscut through a crossflow vortex without suction (lines: $\tilde{u}_s = u_s/u_{b,s,e}$-isocontours; dashed lines: λ_2-isocontours; colored: modal $\tilde{u}_s$-amplitude distribution of high-frequency secondary instability mode h=15) and design example of pinpoint-suction position at the wall (arrows, crosscut through center of suction hole)

activated resulting in 3 consecutive 2-D slits with equivalent $c_q = \bar{v} \cdot P/\bar{U}_{s,e}$ ($\bar{v}$: average suction of the hole; P: porosity of the suction panel, $\bar{U}_{s,e}$: local streamwise edge velocity). As a last reference case, also the x-variation has been eliminated and perfectly homogeneous suction has been applied for case *hom*. Results are provided as follows: Figs. 9 and 10 show vortical structures (snapshots) for cases *REF*, *1-50p*, *2-50p*, *3-50p*, *3-2D*, and *9-50p*. The small top-view plots have been added to point out the respective suction hole position for each case. The color scheme represents the wall-normal distance of the vortex. The red, fuzzy structures represent transitional or already turbulent flow. The exact same unsteady background disturbance input is included in every simulation. Note the streamwise shift of transition for the pinpoint suction cases. The comparison between cases *3-50p* and *3-2D* shows that the 3-D part of pinpoint suction indeed helps transition delay by a stronger reduction of the CFV. For case *9-50p* no transitional flow can be observed.

For a quantitative representation the flow is Fourier-analysed in time over one fundamental period and the downstream development of the disturbance quantity $\tilde{u}'_s$ is analyzed and shown in Fig. 11 for all cases. The pure steady 3-D deformation h=0-(0,0), serving as a measure for the vortex strength, shows different attenuation effects of the different suction scenarios. The evaluation of unsteady, high-frequency secondary instability modes reveals transition delay for all suction scenarios compared to the reference case *REF*. The suction case *hom* reveals the weakest delay (c_q equivalent to case *3-50p*), followed by cases *1-50p* and *3-2D*. Cases *2-50p* and *3-50p* achieve approximately the same shifted transition position whereas for case *9-50p* transition and also secondary growth is eliminated completely.

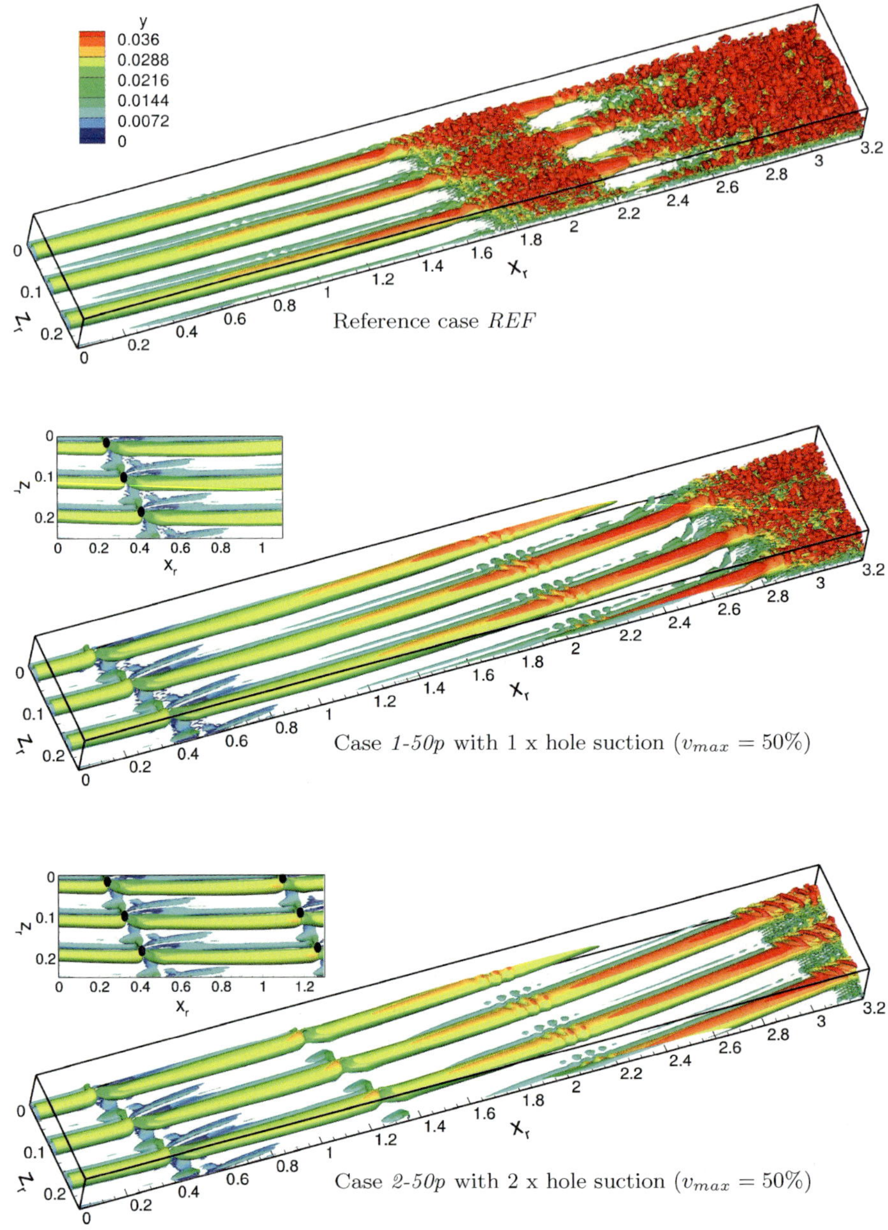

Fig. 9. Visualisation (snapshots) of vortical structures using the λ_2-criterion—color indicates the position in wall-normal coordinate y—for the reference case and two suction setups in a rotated reference system ($\Phi_r = 45°$). Note the compression of the x_r-axis. For each case, suction is marked in a small top-view (black orifices). Shown are three fundamental spanwise wavelengths

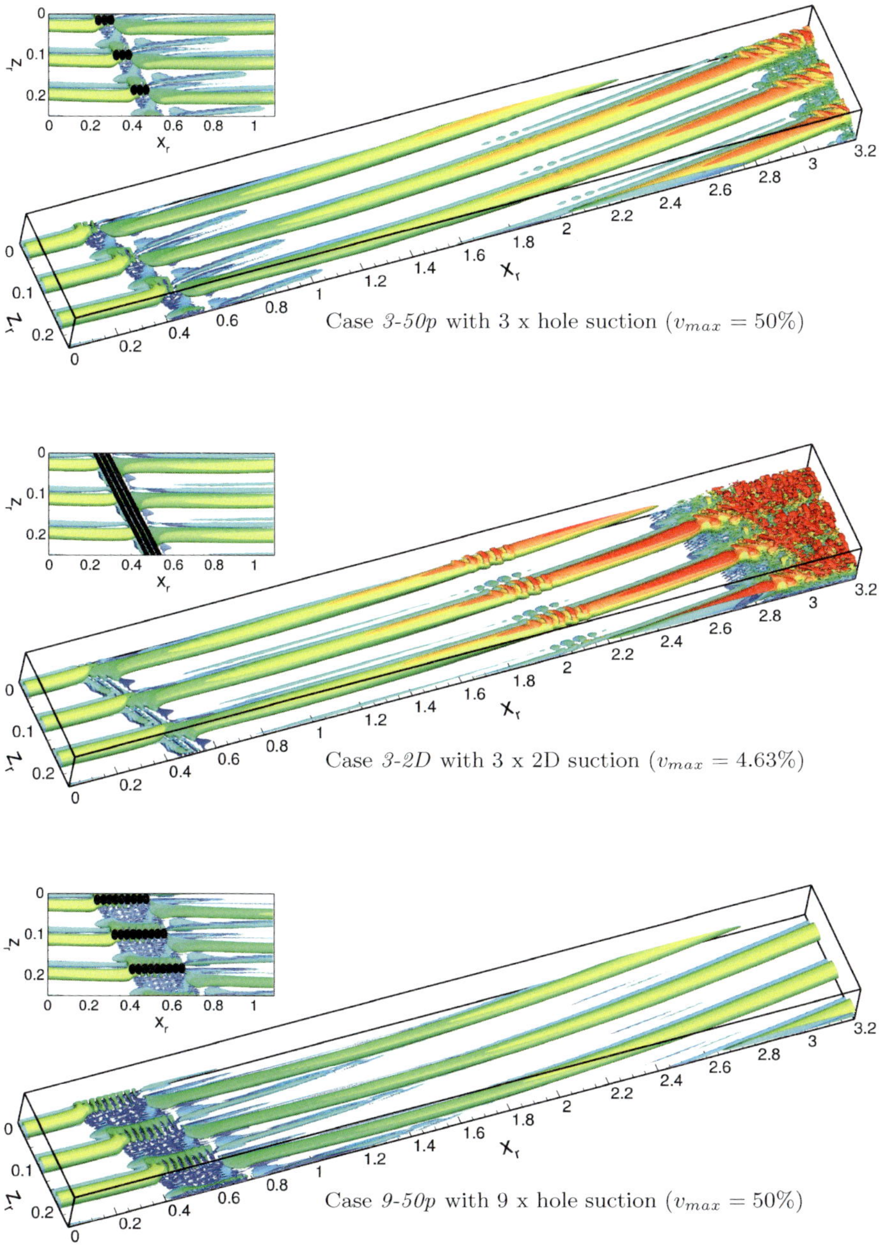

Fig. 10. For caption see Fig. 9. Three more suction cases

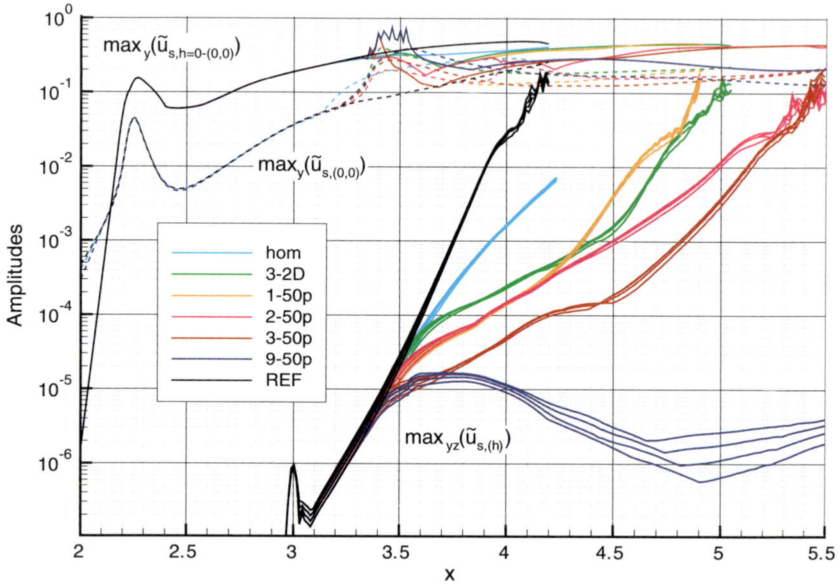

Fig. 11. Downstream development of modal $\tilde{u}_s = u_s/u_{b,s,e}$-amplitudes. Shown are the y-maxima of the mean flow deformation (0,0), the pure 3-D deformation h=0-(0,0), and of some important high-frequency modes h=16, 17, 18, 19 for the reference case and various suction scenarios

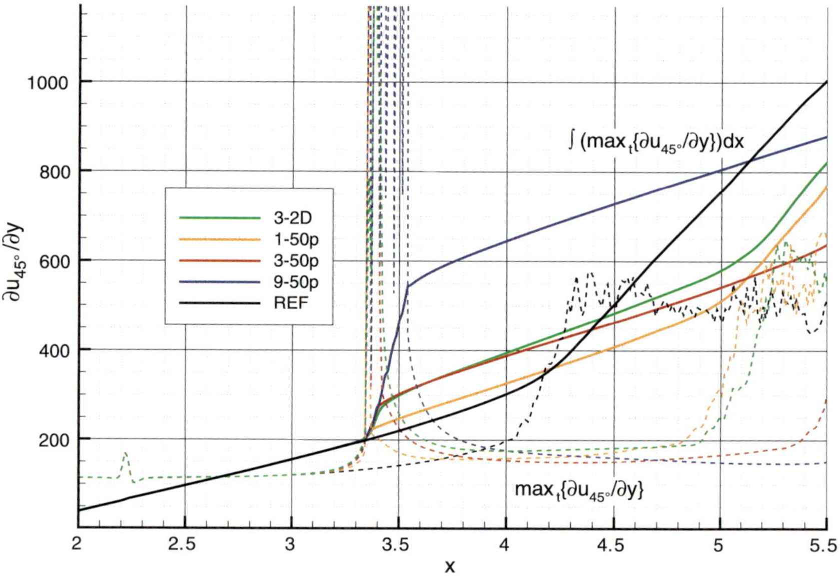

Fig. 12. Local and (in downstream direction) integrated wall-normal gradient of velocity component in direction of the oncoming flow ($\Phi = 45°$) serving as a measure of the skin friction coefficient. Reference case, three pinpoint suction cases, and the case *3-2D* with slits for comparison (cf. Figs. 9, 10)

Following the goal of skin-friction reduction by means of transition delay using suction one must keep in mind that suction generally produces "fuller" streamwise velocity profiles that result in higher skin friction. The wall-normal gradient of the velocity component in direction of the oncoming flow has been evaluated in order to compare the overall effect of the suction setups on the skin friction coefficient. For all cases, the maximum at each downstream location over one fundamental period in time has been used in order to compensate for intermittency effects. For the reference case this means that not single, travelling pulses are considered but a fully turbulent flow scenario. The same procedure is carried out for the pinpoint suction cases. Results of $max_t\{\partial u_{45^\circ}/\partial y\}$ and the integral value $\int max_t\{\partial u_{45^\circ}/\partial y\}dx$ are provided in Fig. 12. The local values at the suction holes achieve extremely high values up to 4000 (not shown). However, the integral effect for all suction cases is small enough to outperform the reference case. The slope of the integrated value for the reference case rises strongly starting at x=4.0 due to turbulent flow. The suction drawback is compensated for all cases as can be seen from the intersection between the curve $(\int max_t\{\partial u_{45^\circ}/\partial y\}dx)_{REF}$ and the corresponding suction curves. Starting at the intersection less overall skin friction is achieved.

Another illustration of the pinpoint suction concept is given in Fig. 13. Snapshots of vortex visualisations using the λ_2-criterion from three different simulations are included. The reference case *REF* (red) with oncoming CFVs and no suction exhibits early transition to turbulence. Note that values for $z_r > 1.4$ and $x_r > 1.9$ are blanked for visual clarity. The result for suction holes *without* oncoming vortices is provided by the green structures with blanking for $x_r > 1.0$. Evolving CFVs can be observed that develop in between the red ones. The final scenario (blue) with oncoming vortices and pinpointed hole suction (case *3-50p*, cf. Fig. 10) shows the successful transition delay where early stages of turbulent flow can be detected only at the end of the visualisation domain. Obviously, nonlinear superposition hinders the growth of the oncoming main vortex and its secondary instability.

3.5 Computational Aspects

General Overview

The incompressible code *N3D* has been optimized in the frame of a former Teraflop workbench project for the NEC SX-8 with its 8 CPUs per node (see also [12]). It primarily computes in Fourier space because of the Fourier-spectral ansatz in spanwise direction that decouples the (discretizedly huge) three-dimensional Poisson-type equations in $(K+1)$ independent two-dimensional problems, where K is the maximum Fourier-mode number. The code is parallelized using OpenMP (intranodal) and MPI (typically internodal), and a speed-up of a factor of about 2 has been achieved by improving the communication and employing optimized FFTs within the workbench project.

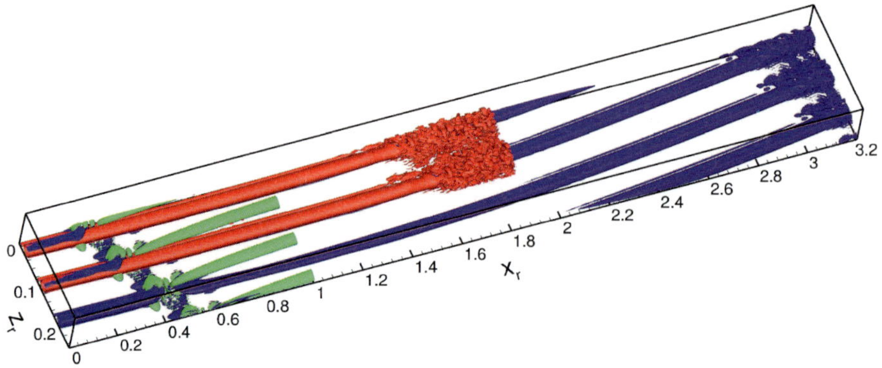

Fig. 13. Visualization of successful transition delay for unstable crossflow vortices using the λ_2-criterion in a rotated reference system ($\Phi_r = 45°$). Snapshot in time. Note the compression of the x_r-axis. Three scenarios are shown: Reference case without suction (red); case with suction hole only, i.e. no oncoming crossflow vortex (green); combined case, i.e. with oncoming crossflow vortex and successful *pinpoint suction* (blue), case *3-50p* is shown here. Shown are three fundamental spanwise wavelengths

Each node works on 8 spectral components in the optimal case for the Poisson-equation part, or 16 on the NEC SX-9. The nonlinear convective terms are computed pseudospectrally, i.e in physical space, and thus a transformation to physical space and back is performed using optimized FFTs. To avoid aliasing by the nonlinear generation of higher modes only 2/3 of the gained modes are used: Using $K = 10$ modes for example means adding 6 modes with zero values, transformation to 32 points ($2^{K_{exp}}$, $K_{exp} = 5$) in physical space, computation of the nonlinear terms, and back transformation keeping only 10 modes of the 16. The MPI parallelization is then done for blocks in the chordwise x-direction, without necessity for computing derivatives, and within each node the CPUs do slices of the wall-normal y-direction in parallel.

The compressible code *NS3D* primarily computes in physical space. A true domain decomposition (in the x-y plane) is implemented, i.e. derivatives by compact finite differences have to be computed over domain boundaries, and the number of domains typically limits the number of used nodes. Increasing the number of domains then can significantly decrease the turnaround time for a job by using many nodes, whereas for *N3D* in its present form K is decisive. The CPUs within a node compute longitudinal cuts of the flow field by working on a given number of z-positions. Optimal combinations of spanwise-mode numbers and domains exist, different for symmetric and non-symmetric flow fields. A grid transformation is embedded so that arbitrary bodies can be considered. Thus the code is more flexible coping also with aero-acoustic problems. However it needs much smaller time steps—about one order of magnitude—for subsonic flows than *N3D* because of the time step limit governed by the transport of fast sound waves. For a typical subsonic laminar-turbulent transition problem the compressible code would need about 20 times as long in CPU time.

Until now our typical supercomputing data, including simulations in the group of U. Rist at IAG, were for:

- N3D, laminar flow control of a swept-wing flow by suction or active control, turbulent separation control using inclined slot blowing, control of laminar separation bubbles, mechanisms/control of boundary-layer transition. Largest Simulations on NEC SX-8: 1 billion grid points, 0.4 Tb RAM (0.4 kb/point), 34 nodes, 4.0 Gflop/s per CPU, 1.1 Tflop/s, 100 h wall time.
- NS3D, mechanisms and control of shear-layer noise, hypersonic transitional boundary-layer flow on plates/cones: Largest Simulations on NEC SX-8: 283 million grid points, 0.307 Tb RAM (1.1 kb/point), 16 nodes, 6.1 Gflop/s per CPU, 0.8 Tflop/s, 100 h wall time.

Our plans for the near future are:

- *N3D*, more transition/turbulence/active control in two- and three-dimensional base flows, higher Reynolds numbers, larger domains for wings, possibly more complex geometry—needing 5 billion grid points and about 10.0 Gflop/s per CPU.
- *NS3D*, disturbance receptivity/transition/control in three-dimensional base flows for high subsonic Mach numbers and supersonic Mach numbers with shock layers, complete flow around wing profiles with disturbance feed back, high-temperature effects in hypersonic shear flows including ducts in the wall—needing 20 billion grid points, more nodes, and for a compressible biglobal stability eigenvalue solver 1 Tb RAM per node.

Performance Data for SX-8/SX-9

A performance analysis for the incompressible code *N3D* on the NEC SX-9 machine (1 node, 16 CPUs) was carried out for case *9-50p* ($N_x \times N_y \times N_z$: $4862 \times 225 \times 64$, KMAX=15, IMAX=2). With the new semi-implicit 3-stage Heun time integrator and activated iterative solver for the wall vorticity (5, 4, and 3 iterations for the first, second, and third Heun stage, respectively) the code achieved 8.8 Gflop/s per CPU and 0.181 μs per point and timestep. For comparison, a similar case with coarser wall-normal resolution was simulated using the original explicit RK-4 time integrator that has been used until now in our *N3D* code. The performance data showed 11.4 Gflop/s per CPU and 0.089 μs per point and timestep. The higher value of 0.181 μs (factor two) for the iterative solver is due to the iteration procedure in the semi-implicit scheme, where the expensive v-poisson equation has to be solved many times.

Executed on 1 node (8 CPUs) NEC SX-8, the case with the Heun-3 integrator achieved 3.8 Gflop/s per CPU and 0.457 μs per point and timestep. This means the same problem is simulated on NEC SX-9 roughly 2.5 times faster.

The incompressible 2-D test case from Sect. 3.1 ($N_x \times N_y \times N_z$: $642 \times 257 \times 64$, KMAX=15, IMAX=1) was executed on SX-8 and SX-9: The Heun-3 integrator is used and 4, 3, and 2 iterations for the successive Heun substages are chosen which results in 2.8 Gflop/s per CPU and 0.235 μs per point and timestep on 1 node (8 CPUs) SX-8. 5.4 Gflop/s per CPU and 0.127 μs per point and timestep on 1 node (16 CPUs) SX-9 are obtained. This means the same problem can be simulated on SX-9 roughly 2 times faster. The test case is obviously not large enough for a meaningful comparison of performance data on SX-8 and SX-9.

For the compressible *NS3D* code, a large symmetrical case with $16 \times 425 \times 325 \times 64$ points (no. of domains $\times N_x \times N_y \times N_z$) reached 5.5 Gflop/s per CPU on 16 nodes NEC SX-8 whereas the NEC SX-9 achieved 14.3 Gflop/s per CPU on 4 nodes which is a factor of 2.6 higher. Values of 1.792 μs per point and timestep (SX-8) compared to 0.677 μs per point and timestep (SX-9) confirm this factor.

Finally, a comparison of computation time between the incompressible and compressible code for the 2-D test case is provided. The only reasonable value to compare is the total computational time necessary for a specific physical result. Computations on SX-9 are compared:

- *N3D*: As mentioned above, the computational performance is 5.4 Gflop/s per CPU and 0.127 μs per point and timestep on 1 node (16 CPUs) SX-9. For the case with $\Delta y_{wall} = 1/1320 \cdot \delta_1$, resulting in $\Delta t = 2.0944 \cdot 10^{-4}$, 20000 timesteps have to be simulated to reach a steady state, where the amplitudes of unsteady modes from a Fourier analysis are less than a certain threshold value. This corresponds to a running length of approximately 341 δ_1. A total CPU time of *26897 sec* was necessary.
- *NS3D*: For the case with $\Delta y_{wall} = 1/315 \cdot \delta_1$ ($\Delta t = 5.236 \cdot 10^{-6}$) 390000 time steps were necessary corresponding to a running length of roughly 167 δ_1. Despite diminished necessary running length, a total CPU time of *1002608 sec* results, being a factor of 40 higher than using *N3D*. For the cases with $\Delta y_{wall} = 1/80 \cdot \delta_1$ ($\Delta t = 6.283 \cdot 10^{-5}$, 130000 time steps) and $\Delta y_{wall} = 1/40 \cdot \delta_1$ ($\Delta t = 3.1416 \cdot 10^{-5}$, 65000 time steps) this factor reduces to 10 or 5, respectively.

Acknowledgements. The provided supercomputing time and technical support of the Höchstleistungsrechenzentrum Stuttgart (HLRS) within the project *LAMTUR* is gratefully acknowledged.

References

1. Babucke, A., Linn, J., Kloker, M.J. and Rist, U.: Direct numerical simulation of shear flow phenomena on parallel vector computers. In High Performance Computing on Vector Systems 2005 (ed. M. Resch & al), Proc. High Performance Computing Center Stuttgart (HLRS), pp. 229-247, Springer (2006).

2. Bippes, H.: Basic experiments on transition in three-dimensional boundary layers dominated by crossflow instability, *Progress in Aerospace Sciences*, vol. 35, pp. 363–412, 1999.
3. Bonfigli, G., Kloker, M.J.: Secondary instability of crossflow vortices: validation of the stability theory by direct numerical simulation, *J. Fluid Mech.*, vol. 583, pp. 229–272, 2007.
4. Friederich, T.: Active control of the crossflow secondary instability in a 3-d boundary layer using steady blowing and suction. Master thesis, Institut für Aerodynamik und Gasdynamik, Universität Stuttgart, 2007.
5. Friederich, T., Kloker, M.J.: Localized blowing and suction for direct control of the crossflow secondary instability, Seattle AIAA-2008-4394.
6. Kloker, M.J.: Advanced Laminar Flow Control on a Swept Wing—Useful Crossflow Vortices and Suction, Seattle AIAA-2008-3835.
7. Koch, W., Bertolotti, F. P., Stolte, A. and Hein, S.: Nonlinear equilibrium solutions in a three-dimensional boundary layer and their secondary instability, *J. Fluid Mech.*, vol. 406, pp. 131–174, 2000.
8. Liepmann, H. W., Brown, G.L. and Nosenchuck, D.M.: Control of laminar-instability waves using a new technique, *J. Fluid Mech.*, vol. 118, pp. 187–200, 1982.
9. Liepmann, H.W., Nosenchuck, D.M.: Active control of laminar-turbulent transition, *J. Fluid Mech.*, vol. 118, pp. 201–204, 1982.
10. Linn, J., Kloker, M.J.: Numerical Investigations of Film Cooling. RESPACE - Key Technologies for Reusable Space Systems (ed. A. Gülhan), NNFM 98, pp. 151–169, Springer (2008).
11. Meitz, H.L.: Numerical investigation of suction in a transitional flat-plat boundary layer, Dissertation, Department of Aerospace and Mechanical Engineering, University of Arizona, Tucson, USA (1996).
12. Messing, R., Kloker, M.J.: Smart suction - an advanced concept for laminar flow control of three-dimensional boundary layers, in High Performance Computing on Vector Systems 2007 (ed. M. Resch & al), Proc. High Performance Computing Center Stuttgart, Springer (2008).
13. Messing, R., Kloker, M.J.: Investigation of suction for laminar flow control of three-dimensional boundary layers, *J. Fluid Mech.* (in press)
14. Peltzer, I., Wicke, K., Pätzold, A. and Nitsche, W.: In-flight experiments on active TS-wave control on a 2D-laminar wing glove. *Seventh IUTAM Symposium on Laminar-Turbulent Transition*, Stockholm, Sweden, 2009 (ed. P. Schlatter & D. Henningson). Springer.
15. Saric W.S., Carrillo R. and Reibert M.: Leading-edge Roughness as a Transition Control Mechanism. AIAA 98-0781
16. Saric W.S., Carrillo R. and Reibert M.: Nonlinear Stability and Transition in 3-D Boundary Layers. Meccanica 33: 469–487
17. Carpenter, A.L. Saric, W.S., and Reed, H.L.: Laminar Flow Control on A Swept Wing With Distributed Roughness, AIAA 2008-7335.
18. Wassermann, P., Kloker, M.: Direct numerical simulation of the development and control of boundary-layer crossflow vortices. In *New Results in Numerical and Experimental Fluid Dynamics II* (ed. W. G. Nitsche, H.-J. Heinemann & R. Hilbig). *Proc. 11. AG STAB/DGLR Symposium (1998).* Notes on Numerical Fluid Mechanics, vol. 72. Vieweg.

19. Wassermann, P., Kloker, M.: DNS-investigations of the development and control of crossflow vortices in a 3-D boundary-layer flow. In *Laminar-Turbulent Transition. Proc. IUTAM Symp., Sedona, AZ, USA 1999* (ed. H. Fasel & W. Saric). Springer.
20. Wassermann, P., Kloker, M.J.: Mechanisms and passive control of crossflow-vortex-induced transition in a three-dimensional boundary layer, *J. Fluid Mech.*, vol. 456, pp. 49–84, 2002.
21. Wassermann, P., Kloker, M.J.: Transition mechanisms induced by travelling crossflow vortices in a three-dimensional boundary layer, *J. Fluid Mech.*, vol. 483, pp. 67–89, 2003.
22. Wassermann, P., Kloker, M.J.: Transition mechanisms in a three-dimensional boundary layer with pressure-gradient changeover, *J. Fluid Mech.*, vol. 530, pp. 265–293, 2005.
23. White, E.B., Saric, W.S.: Secondary instability of crossflow vortices, *J. Fluid Mech.*, vol. 525, pp. 275–308, 2005.

A Numerical Study of Turbulent Stably-Stratified Plane Couette Flow

Manuel García-Villalba, Elena Azagra, and Markus Uhlmann

Institut für Hydromechanik, Karlsruher Institut für Technologie, Karlsruhe, Germany, villalba@ifh.uka.de

Summary. Direct numerical simulations (DNS) of stably-stratified, turbulent plane Couette flow are currently being performed on the XC-4000. The friction Reynolds number is kept approximately constant, $Re_\tau \simeq 540$ and the Richardson number Ri_w varies between 0 and 0.1. The flow is divided into two regions: the region close to the wall and the core region. The region close to the wall presents strong velocity and density gradients and its structure is similar to unstratified wall-turbulence. In the core region, the gradients of mean velocity and density are approximately constant, and the structure of the flow is similar to homogeneous stratified turbulence with shear. With increasing stratification the formation of layers is clearly observed. The layers are inclined with respect to the wall plane and their thickness decreases with increasing stratification.

1 Introduction

Stably-stratified turbulent shear flows are encountered in many applications in engineering and geophysics [1]. The atmospheric boundary layer is typically stably stratified at night [2, 3] while oceanic flows are almost always stably stratified [4, 5]. Due to the importance of such flows for the environment, there is a growing interest in improving the understanding of stratified flows. Topics of current research are, among others, the quantification of mixing [6] and the dynamics of strongly stratified turbulence [7, 8].

Previous work in this project was directed towards improving the understanding of stably-stratified pressure-driven plane channel flow [9, 10]. The gradient Richardson number $Ri_g = N^2/S^2$, where N is the buoyancy frequency and S is the shear rate, has been identified as the most important parameter for the characterization of the flow [11, 12]. Ri_g varies from low values close to the wall where the shear is maximum to very high values in the core of the channel where the shear vanishes due to symmetry. In Couette flow on the other hand, the shear never vanishes and therefore Ri_g does not have a singular region.

W.E. Nagel et al. (eds.), *High Performance Computing in Science and Engineering '10*, DOI 10.1007/978-3-642-15748-6_19,

There are many experimental and numerical studies of turbulent, plane Couette flow [13–17]. For the stably stratified case, we are only aware of theoretical studies using variational principles [18–20]. In this report we present preliminary DNS results for this configuration.

2 Computational Setup

In this study, stably-stratified turbulent plane Couette flow is simulated directly. A stable stratification is maintained by imposing a constant upper-wall density which is smaller than the constant bottom-wall density. We consider thermally-stratified air and therefore the Prandtl number is $Pr = 0.7$. The relevant non-dimensional parameters of this flow are the Reynolds and Richardson numbers. The bulk non-dimensional numbers are defined using the difference in velocity between both walls ΔU, the difference in density $\Delta\rho$ and the channel height $H = 2h$, where h is the half-height, so that

$$Re_w = \frac{\Delta U H}{\nu}, \quad Ri_w = \frac{\Delta\rho g H}{\rho_0 \Delta U^2}, \tag{1}$$

where ν is the kinematic viscosity, ρ_0 is a reference density and g is the gravitational acceleration. The Reynolds number can be also defined in terms of the friction velocity u_τ, $Re_\tau = u_\tau h/\nu$. This can be interpreted as the ratio between two length scales: the outer scale h and the inner scale $\delta_v = \nu/u_\tau$.

The governing equations are the Navier-Stokes equations under the Boussinesq approximation,

$$\frac{\partial u_i}{\partial x_i} = 0, \tag{2}$$

$$\frac{\partial u_i}{\partial t} + u_j \frac{\partial u_i}{\partial x_j} = -\frac{\partial p}{\partial x_i} + \frac{1}{Re_w} \frac{\partial^2 u_i}{\partial x_j \partial x_j} - Ri_w \rho' \delta_{i2}, \tag{3}$$

$$\frac{\partial \rho}{\partial t} + u_j \frac{\partial \rho}{\partial x_j} = \frac{1}{Re_w Pr} \frac{\partial^2 \rho}{\partial x_j \partial x_j}, \tag{4}$$

where u_i is the velocity field, ρ the density field, ρ' the density fluctuations and p the kinematic pressure field that remains after removing the component that is in hydrostatic balance with the mean density field.

The numerical code integrates the governing equations in the form of evolution problems for the wall-normal vorticity ω_y, the Laplacian of the wall-normal velocity $\nabla^2 v$ and the density ρ, following the formulation of Kim et al. [21]. After some algebra, (2)–(3) can be reduced to yield

$$\frac{\partial}{\partial t} \nabla^2 v = h_v + \frac{1}{Re_w} \nabla^4 v - Ri_w \left(\frac{\partial^2 \rho}{\partial x^2} + \frac{\partial^2 \rho}{\partial z^2} \right), \tag{5}$$

$$\frac{\partial}{\partial t} \omega_y = h_g + \frac{1}{Re_w} \nabla^2 \omega_y, \tag{6}$$

Table 1. Overview of the simulations performed

Case	Re_w	Ri_w	Line color
A1	44000	0	Black
A2	57000	0.03	Green
A3	72000	0.06	Red
A4	112000	0.1	Blue

where h_v and h_g are the nonlinear terms (see [21] for explicit expressions). (4) is not modified.

The spatial discretization uses de-aliased Fourier expansions in the wall-parallel planes and Chebyshev Polynomials in y. The streamwise and spanwise coordinates and velocity components are respectively x, z and u,w. The temporal discretization is third-order, semi-implicit Runge-Kutta [22]. The numerical code has been extensively validated for the unstratified case [23, 24]. The simulation parameters are summarized in Table 1.

In order to study the effects of stable stratification in turbulent plane Couette flow there are several possibilities. The most straightforward would be to fix Re_w and systematically increase Ri_w. This would lead to a systematic decrease in $Re_\tau = h/\delta_v$ as a consequence of the reduced friction observed when the stratification increases. Therefore, the ratio between the outer and inner length scales in the flow would be reduced, leading to a direct effect of viscosity in a bigger portion of the channel. To avoid this we have taken a different approach. Instead of fixing Re_w we have tried to keep Re_τ constant. This is very easy to do in Pouiseille flow since the flow is driven by a pressure-gradient that can be kept constant [12]. In Couette flow this is not so straightforward and a trial and error procedure has been used, in which we fix pairs of values (Re_w, Ri_w) until the desired Re_τ is obtained. Azagra [25] reported simulations with a $Re_\tau \sim 130$ and in this report we present simulations with $Re_\tau \sim 540$. For all cases we use the same computational box and resolution as in the channel flow simulation at $Re_\tau = 550$ of [24].

3 Computational Details

Computational aspects and performance of the employed code were reported in previous years, including the speed up of parallelization up to 128 processors [9]. In the present case a typical production run uses a computational box $(Lx, Ly, Lz) = (8\pi h, 2h, 4\pi h)$ with $(Nx, Ny, Nz) = (1536, 257, 1536)$ grid cells. First, for a given Ri_w, the flow is initialized with the flow field from the previous Ri_w. The flow evolves for a long transient period until it reaches the new steady state. Then, statistics can be collected. To reduce the computational effort, the transient period is run with a lower resolution. The second phase of a run consist of at least 200 time units $(H/\Delta U)$ or approximately 150000 time steps. This corresponds to a clock time of approximately 51 days

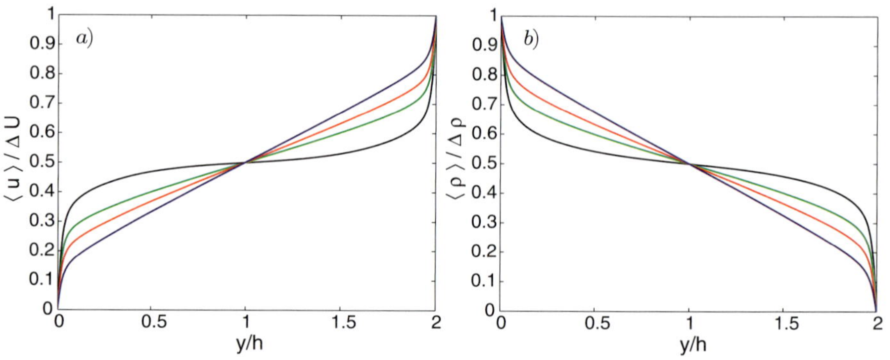

Fig. 1. **a** Mean streamwise velocity and **b** mean density. Line colors defined in Table 1

and a total number of CPU hours of about 160000. It should be mentioned that the higher the stratification, the smaller the time step, and therefore more time steps are required to accumulate the same amount of statistical information.

4 Results

Figure 1 shows the mean streamwise velocity and the mean density for all cases considered. With increased stable stratification the profiles come close to the laminar solution (a linear profile). This means that both gradients decrease near the wall and increase in the core of the channel.

There are several parameters which can be used to characterize the importance of stable stratification for shear flows. On unbounded situations, the gradient Richardson number Ri_g has been identified as the most important parameter. It represents the relative importance of buoyancy and shear. Linear analysis [26] gives $Ri_g \geq 0.25$ everywhere in the flow as a sufficient condition for stability in stratified shear flows. This value has been confirmed in various experiments. Figure 2a shows the vertical distribution of Ri_g. It is interesting to note that Ri_g does not reach values beyond 0.2, as was also observed at lower Reynolds numbers [25].

Another important non-dimensional number is the flux Richardson number, defined as

$$Ri_f = \frac{\frac{g}{\rho_0}\langle \rho' v' \rangle}{\langle u'v' \rangle \frac{\mathrm{d}\langle u \rangle}{\mathrm{d}y}}. \tag{7}$$

It represents the ratio of buoyant destruction to shear production in the turbulent kinetic energy equation. In Fig. 2b the distribution of Ri_f as a function of Ri_g is displayed. The values obtained are consistent with data from experiments in stably-stratified flows reviewed by Linden [27].

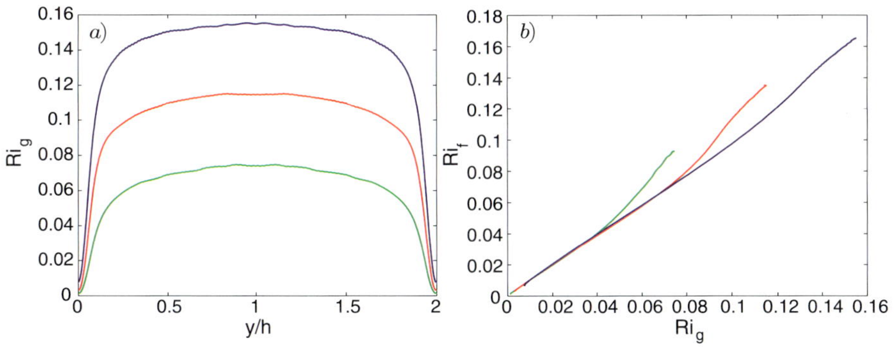

Fig. 2. **a** Vertical profile of the gradient Richardson number Ri_g. **b** Flux Richardson number Ri_f vs. Ri_g. Line colors defined in Table 1

The characterization of the state of the turbulence can be done in terms of the buoyancy Reynolds number $Re_B = \varepsilon/(\nu N^2)$ and the Froude number $Fr = u_c/(Nl_c)$ [6], where u_c and l_c are characteristic velocity and length scales, respectively. Note that because $\varepsilon \sim u_c^3/l_c$, it is possible to form a turbulent Reynolds number $Re_T = u_c l_c/\nu \simeq Re_B/Fr^2$. The buoyancy Reynolds number is typically used in the oceanographic literature. It has been proposed that active turbulence is present only when Re_B exceeds a value of about 20 [28].

Brethouwer et al. [7] performed a scaling analysis of strongly stratified turbulent flows, in terms of Re_B and a horizontal Froude number F_h. The latter was defined by using as velocity scale the root mean square horizontal velocity $u_c = u_{rms}$ and as length scale $l_c = u_{rms}^3/\varepsilon$, i.e. $F_h = \varepsilon/(Nu_{rms}^2)$. They distinguished two extreme regimes. First, the strongly stratified turbulence regime, for $Re_B >> 1$, $F_h << 1$, which is very relevant in a geophysical context [8]. Lindborg [29] made the estimate $F_h < 0.02$ for the limit under which strongly stratified turbulence can be observed. At the other extreme, Brethouwer et al. [7] defined the viscosity-affected stratified flow regime for $Re_B << 1$, $F_h << 1$. This regime might not be so important for geophysical flows, but it is relevant for numerical simulations and laboratory experiments. Note that $Re_B \sim Re_T F_h^2$, so that Re_T might be large even if $Re_B << 1$.

Therefore, in addition to Re_B, which is shown in Fig. 3a, an estimation of F_h for the present simulations is required to fully characterize the state of the turbulence. This has been done by using as velocity scale $(u_{rms}^2 + w_{rms}^2)^{0.5}$, and the profiles are shown in Fig. 3b. We obtain values of F_h between 0.3 (case A4) and 0.6 (case A2) in the core of the channel. This indicates that the turbulent state that we obtain cannot be considered strongly stratified. Figure 4 displays the various regimes in stably stratified flows [7]. In the present simulations the flow is in the weakly stratified turbulence regime. To reach the strongly stratified turbulence regime the Re_τ of the flow should be at least of the order of 2000.

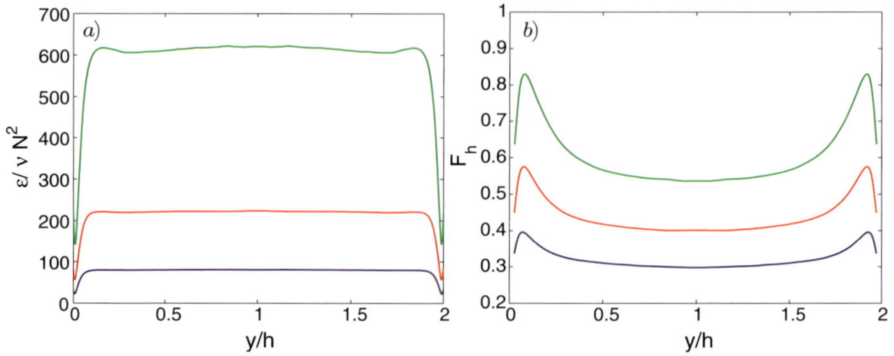

Fig. 3. a Vertical profile of the buoyancy Reynolds number $Re_B = \varepsilon/\nu N^2$. **b** Vertical profile of the horizontal Froude number F_h. Line colors defined in Table 1

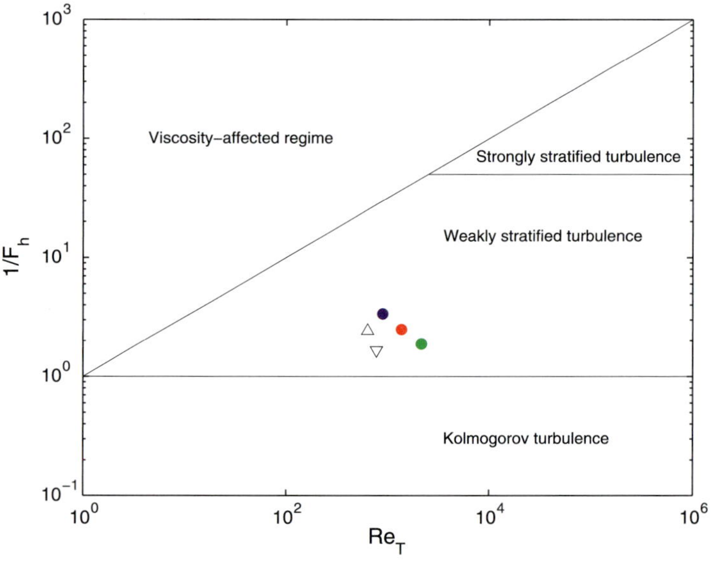

Fig. 4. Regimes in stably stratified flows, following [7]. $Re_T \simeq Re_B/F_h^2$ vs. $1/F_h$. Circle colors defined in Table 1. Triangles correspond to the lower Reynolds number simulations reported in [25]

Finally, Figs. 5–7 show various visualizations of the flow for cases A1, A3 and A4. In all cases the quantity displayed is the instantaneous density fluctuation normalized by the friction density ρ'/ρ_τ, the latter is defined as $\rho_\tau = q_w/u_\tau = -k(\partial_y\langle\rho\rangle)_w/u_\tau$, where q_w is the density flux at the wall and k is the molecular diffusivity of density. Figure 5 shows a horizontal plane in the middle of the domain, Fig. 6 shows a transverse vertical plane $y-z$ and Fig. 7 shows a streamwise vertical plane $x-y$. In the unstratified case very long and wide coherent motions are observed which are also correlated

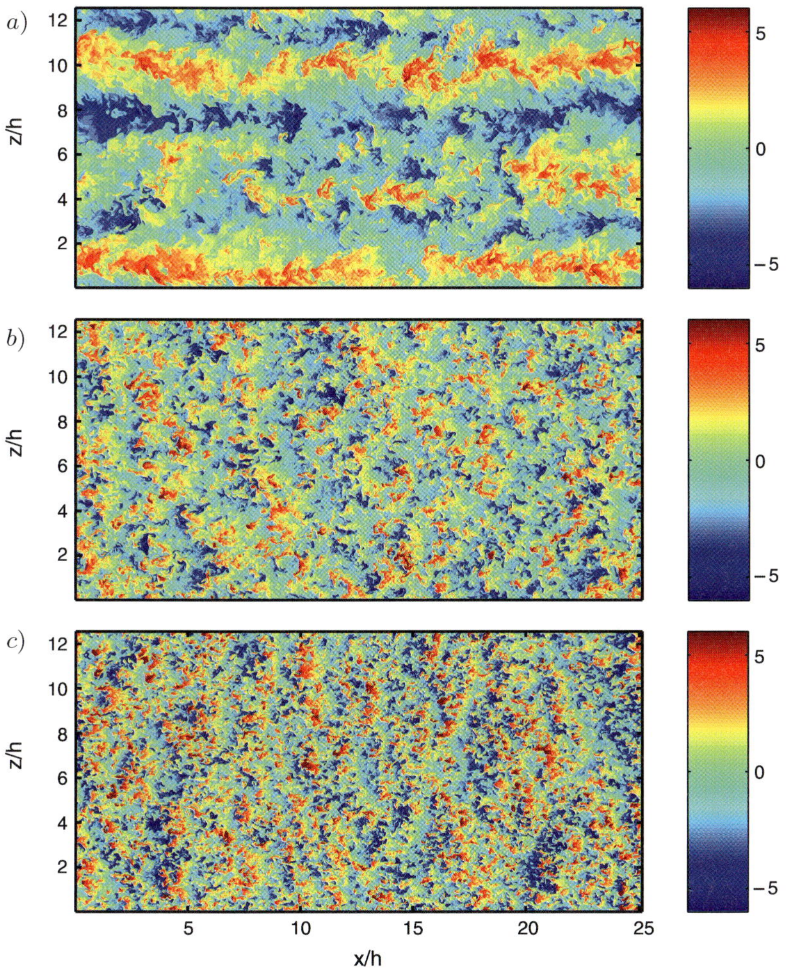

Fig. 5. Instantaneous density fluctuations ρ'/ρ_τ in a wall parallel plane in the middle of the domain $y = h$. **a** A1. **b** A3. **c** A4

across the full height of the domain. These structures were studied at a lower Reynolds number by Tsukahara et al. [17]. With increasing stratification these structures are not observed. Instead there is a clear formation of density layers. Formation of layers is a well-known phenomenon in the oceanic literature [1, 4]. The layers are inclined with respect to the wall plane and their thickness decreases with increasing stratification.

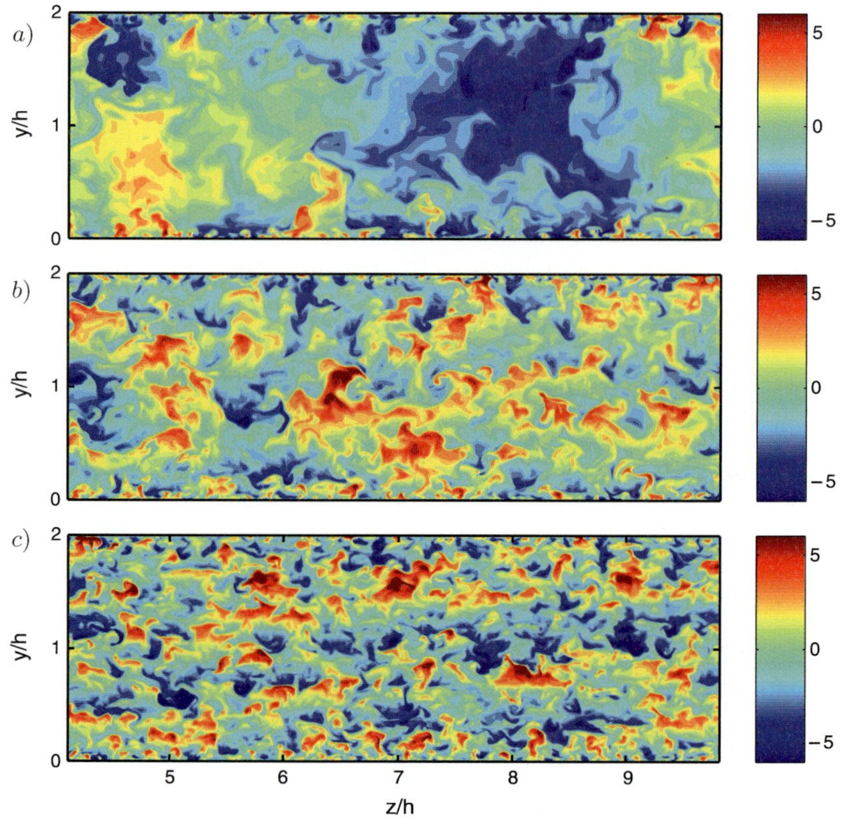

Fig. 6. Instantaneous density fluctuations ρ'/ρ_τ in a yz-plane. **a** A1. **b** A3. **c** A4

5 Conclusions

In this report, preliminary results from direct numerical simulations of turbulent, stably-stratified, plane Couette flow at $Re_w \sim 540$ and various Ri_w have been reported. We have tried to characterize the state of the flow using various non-dimensional quantities proposed in the literature. Furthermore, the state of the turbulence has been characterized following the scaling analysis reported by Brethouwer et al. [7]. Our simulations can be described as being in the weakly-stratified turbulence regime. Visualizations of the flow show that the stratification acts by suppressing the large scales of the flow and that density layers are formed. We plan to continue the project by trying to reach the strongly-stratified turbulence regime. In order to achieve this goal, it is necessary to increase significantly Re_w and Ri_w, which will require a significantly larger number of cells but this could be partially compensated by considering smaller computational boxes, since large structures are damped by the stratification.

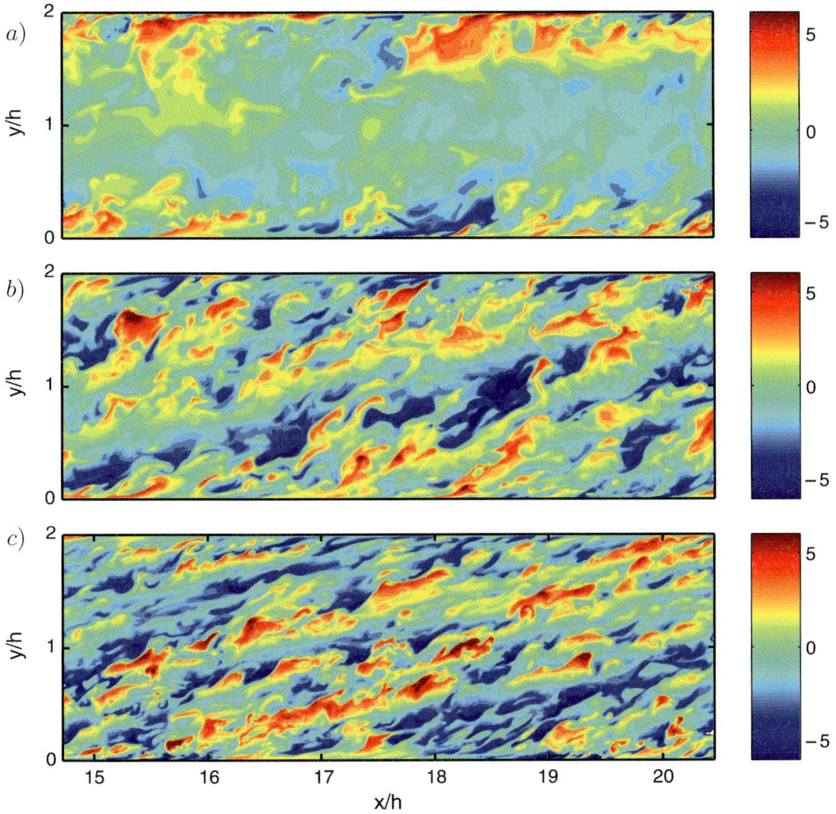

Fig. 7. Instantaneous density fluctuations ρ'/ρ_τ in a xy-plane. **a** A1. **b** A3. **c** A4

Acknowledgements. This work was funded by the German Research Foundation (DFG). The calculations were carried out on the national super computer HP XC4000 of the Steinbuch Centre for Computing, Karlsruhe.

References

1. J.S. Turner. *Buoyancy effects in fluids.* Cambridge Univ. Press, 1973.
2. F.T.M. Nieuwstadt. The turbulent structure of the stable, nocturnal boundary layer. *Journal of the Atmospheric Sciences*, 41(14):2202–2216, 1984.
3. R.B. Stull. *An introduction to boundary layer meteorology.* Kluwer Academic, 1999.
4. S.A. Thorpe. The turbulent ocean. Cambridge Univ. Press, Cambridge 2005.
5. C. Wunsch and R. Ferrari. Vertical mixing, energy, and the general circulation of the oceans. *Annu. Rev. Fluid Mech.*, 36:281–314, 2004.
6. G.N. Ivey, K.B. Winters, and J.R. Koseff. Density stratification, turbulence, but how much mixing? *Annu. Rev. Fluid Mech.*, 40:169–184, 2008.

7. G. Brethouwer, P. Billant, and E. Lindborg. Scaling analysis and simulation of strongly stratified turbulent flows. *J. Fluid Mech.*, 585:343–368, 2007.
8. J.J. Riley and E. Lindborg. Stratified turbulence: a possible interpretation of some geophysical turbulence measurements. *J. Atmos. Sci.*, 65:2416–2424, 2008.
9. M. García-Villalba and J.C. del Álamo. Turbulence and internal waves in a stably-stratified channel flow. In W.E. Nagel et al., editor, *High Performance Computing in Science and Engineering '08*, pages 217–227. Springer, 2008.
10. M. García-Villalba and J.C. del Álamo. Turbulence modification by stable stratification in channel flow. In preparation, 2010.
11. S. Komori, H. Ueda, F. Ogino, and T. Mizushina. Turbulent structure in stably stratified open-channel flow. *J. Fluid Mech.*, 130:13–26, 1983.
12. V. Armenio and S. Sarkar. An investigation of stably stratified turbulent channel flow using large-eddy simulation. *J. Fluid Mech.*, 459:1–42, 2002.
13. A. Lundbladh and A.V. Johansson. Direct simulation of turbulent spots in plane Couette flow. *J. Fluid Mech.*, 229:499–516, 1991.
14. K.H. Bech, N. Tillmark, P.H. Alfredsson, and H.I. Andersson. An investigation of turbulent plane Couette flow at low Reynolds numbers. *J. Fluid Mech.*, 286:291–325, 1995.
15. J. Komminaho, A. Lundbladh, and A.V. Johansson. Very large structures in plane turbulent Couette flow. *J. Fluid Mech.*, 320:259–285, 1996.
16. C.H. Liu. Turbulent plane Couette flow and scalar transport at low Reynolds numbers. *J. Fluid Mech.*, 286:291–325, 1995.
17. T. Tsukahara, H. Kawamura, and K. Shingai. DNS of turbulent Couette flow with emphasis on the large-scale structure in the core region. *J. Turbulence*, 7(19), 2006.
18. C.P. Caulfield and R.R. Kerswell. Maximal mixing rate in turbulent stably stratified Couette flow. *Phys. Fluids*, 13(4):894–900, 2001.
19. C.P. Caulfield, W. Tang, and S.C. Plasting. Reynolds number dependence of an upper bound for the long-time-averaged buoyancy flux in plane stratified Couette flow. *J. Fluid Mech.*, 498:315–332, 2004.
20. W. Tang, C.P. Caulfield, and R.R. Kerswell. A prediction for the optimal stratification for turbulent mixing. *J. Fluid Mech.*, 634:487–497, 2009.
21. J. Kim, P. Moin, and R. Moser. Turbulence statistics in fully developed channel flow at low Reynolds number. *J. Fluid Mech.*, 177:133–166, 1987.
22. P.R. Spalart, R.D. Moser, and M.M. Rogers. Spectral methods for the Navier-Stokes equations with one infinite and two periodic directions. *J. Comp. Phys.*, 96:297–324, 1991.
23. J.C. del Álamo and J. Jiménez. Spectra of the very large anisotropic scales in turbulent channels. *Phys. Fluids*, 15:L41–L44, 2003.
24. J.C. del Álamo, J. Jiménez, P. Zandonade, and R.D. Moser. Scaling of the energy spectra in turbulent channels. *J. Fluid Mech.*, 500:135–144, 2004.
25. E. Azagra. Mixing efficiency in stably-stratified turbulent Couette flow. Master thesis, Karlsruhe Institute of Technology, 2009.
26. J. Miles. On the stability of heterogeneous shear flows. *J. Fluid Mech.*, 10:496–508, 1961.
27. P.F. Linden. Mixing in stratified fluids. *Geophys. Astrophys. Fluid Dyn.*, 13:3–23, 1979.

28. W.D. Smyth and J.N. Moum. Length scales of turbulence in stably stratified mixing layers. *Phys. Fluids*, 12(6):1327–1342, 2000.
29. E. Lindborg. The energy cascade in a strongly stratified fluid. *J. Fluid Mech.*, 550:207–242, 2006.

DNS of Unsteady Heat Transfer Increase on a Curved Surface Due to Wake-Induced Turbulence

D. von Terzi[1], L. Venema[1], H.-J. Bauer[1], and W. Rodi[2]

[1] Institut für Thermische Strömungsmaschinen, Karlsruhe Institute of Technology (KIT), Kaiserstr. 12, D-76131 Karlsruhe, Germany, vonterzi@kit.edu, lars.venema@kit.edu, hans-joerg.bauer@kit.edu
[2] Institut für Hydromechanik, Karlsruhe Institute of Technology (KIT), Kaiserstr. 12, D-76131 Karlsruhe, Germany, rodi@kit.edu

Summary. From experimental studies it is well established that free-stream and wake-induced turbulence can increase the heat transfer on curved surfaces such as turbine blades or cylinders even if the mean flow matches corresponding laminar velocity profiles. Here, preliminary results of an investigation using (embedded) Direct Numerical Simulation (DNS) are reported that aim at identifying the responsible physical mechanisms for this phenomenon. The setup follows an existing experiment where a heated cylinder is subjected to the turbulent wake of a smaller cylinder placed upstream. The Reynolds number based on the diameter of the heated cylinder is $Re = 48,000$. A key parameter identified in an experimental correlation is the characteristic turbulence level Tu_a that includes a dependency on Re and the mean flow acceleration. For the present study, Tu_a is an order of magnitude higher than those reported in any previous DNS. The obtained Nusselt number is in good agreement with the experimental correlation. More and better resolved data is currently produced for further analysis.

1 Introduction

In jet engines and in modern gas turbines in power plants, the first turbine stages are subject to temperatures well beyond material limits requiring extensive cooling. The cooling amounts to losses in efficiency of the device and, therefore, it is of practical importance to understand the physical mechanisms responsible for the heat transfer. A critical region is the vicinity of the stagnation point of the turbine blade. Here, experiments documented an increase in heat transfer correlated to the passing of turbulent wakes of the preceding blades. However, the boundary layer remains laminar (in the mean) due to the strong acceleration caused by the curved surface. As a consequence, the Reynolds analogy does not hold and modeling of the heat transfer employing a turbulent Prandtl number concept cannot work. Moreover, the physical

W.E. Nagel et al. (eds.), *High Performance Computing in Science and Engineering '10*, DOI 10.1007/978-3-642-15748-6_20,

mechanisms causing this increase are still obscure making any kind of modeling difficult.

In the experiments of Magari & LaGraff [1], the turbine blades are simplified to cylinders: a large heated cylinder representing the leading edge of a turbine blade and a smaller one upstream acting as a wake generator. This geometric simplification allows for a controlled way to investigate the phenomenon taking place. For laminar flow around a single heated circular cylinder, Frössling [2] already obtained a semi-analytical relation describing the heat transfer. However, many investigations have demonstrated that free-stream and wake-induced turbulence causes a considerable increase in the heat transfer in the stagnation region. A correlation for the increase was proposed by Dullenkopf & Mayle [3]. The same effect was also noted in the experiment carried out in [1]. The increase in heat transfer is well documented in this work, but the mechanisms causing it are not clear. Here, a study employing Direct Numerical Simulation (DNS) could help to clarify these mechanisms as the results allow to extract all details of the flow development, the turbulent structures and also the heat flux caused by the turbulence. Wissink & Rodi [4] performed the first DNS of this kind. Inspired by the experiments in [1], they calculated by DNS the heat transfer from a cylinder placed in the wake of a smaller cylinder. However, in order to limit computational costs to match the available resources, the wake of this cylinder located upstream was calculated in a separate precursor DNS [5]. This data were then used as inflow conditions for the DNS of the flow around and heat transfer from the heated cylinder. Because of this procedure, the smaller, wake-generating cylinder had to be placed much farther upstream of the heated cylinder than in the experiments, and also a lower Reynolds number had to be chosen. As a consequence, the oncoming turbulence intensity in the stagnation region was drastically reduced so that the effect of the turbulence and, consequently, the increase in heat transfer is smaller than in the experiments.

The objective of the present study is to reproduce the increase of heat transfer due to wake turbulence using the setup in [1]. An important feature is that the wake generator is included in the computational domain. To accommodate for the high Reynolds number, embedded DNS is applied where the DNS is patched to Large Eddy Simulation (LES) downstream of the heated cylinder. In this report, results obtained from a preliminary DNS are presented. The data were compared with correlations in the literature and results are shown for the correlation found in [3]. Moreover, visualizations are used to give an impression of the flow structure and the instantaneous temperature field in the vicinity of the stagnation point. More comparisons with experimental data and additional results can be found in [6].

2 Setup

The numerical setup resembles the experiment in [1] as closely as possible. In Fig. 1 an overview is given. A large heated cylinder with diameter D rep-

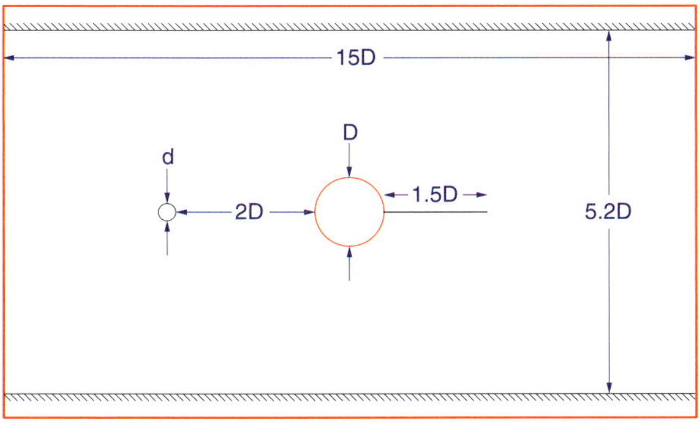

Fig. 1. Schematic overview of the computational setup

resents in its front part the leading edge of a turbine blade. To reduce the vortex shedding behind the large cylinder, a splitter plate with a length of $1D$ is placed at the rear. At a distance of $2D$ upstream, a smaller cylinder with a diameter of $0.25D$ is placed, acting as wake generator. The channel has a height of $5.2D$ and a width of $2.5D$. With a value of 19.2%, the blockage is relatively severe, hence all tunnel walls need to be included in the simulation. The length of the domain is chosen as $15D$, with the axis of the large cylinder at $7.5D$. The cylinder surfaces and the splitter plate are modeled as no-slip walls. Since the boundary layer thicknesses on the wind tunnel walls are unknown, all sidewalls are modeled as frictionless. The temperature is computed employing a scalar transport equation, since the flow is incompressible. On the large (heated) cylinder a Dirichlet boundary condition is applied with a temperature of $1.35T_0$, where T_0 is the free-stream temperature. The inflow is uniform and the Reynolds number (Re) based on D and the bulk velocity is 48,000.

2.1 Computational Grid

The computational domain is meshed with a block-structured grid, as displayed in Fig. 2. The baseline case has a total number of 100 millions cells distributed over 256 blocks. The preliminary results from this case are presented here and are available in more detail in [6]. Due to the high Reynolds number it is too costly to perform a DNS on the whole domain. The wake generator is resolved *"just well enough"*, such that the large coherent structures, which will reach the heated cylinder, are sufficiently well resolved. The smaller structures, which would be dissipated before reaching the stagnation point, are of little importance. Downstream of the large cylinder, a coupling with LES is realized to save further grid points and hence computational cost. The region of interest, the front part of the heated cylinder, has the highest cell density. Near the heated cylinder, at an angle of $\theta = 30°$, the radial distance from the wall in wall-units of the first computational point adjacent to

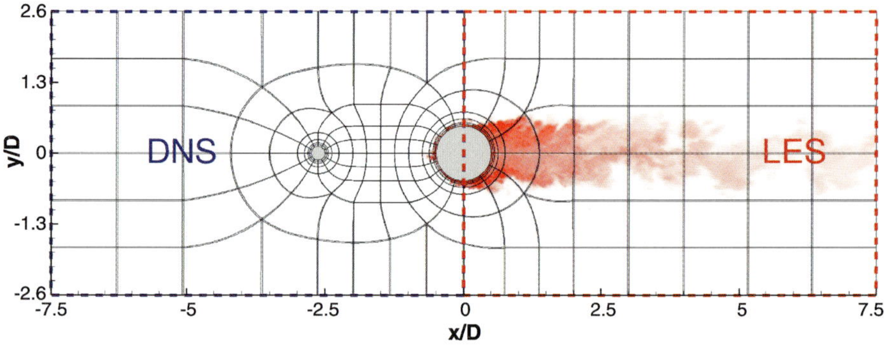

Fig. 2. Block boundaries. The contour gives an impression of the temperature field in the wake of the heated cylinder

the cylinder is $r^+ = 1.3$. In the circumferential and spanwise direction the grid-spacings in wall-units are $\Delta\theta^+ = 27$ and $\Delta z^+ = 34$, respectively. This resolution is not as good as one would like, but the results shown in [6] already justify some conclusions and indicate that the grid intended for the final simulation with a total number of 10^9 cells should be sufficient for an embedded DNS. In addition, a simulation with the baseline grid at a lower Reynolds number and, hence, a better resolution is in progress.

2.2 Reference Correlation

The DNS data were compared to several correlations in the literature describing the heat transfer increase for cylinder stagnation flow. In this report, for brevity, only the correlation of [3] is shown that appears to be the most accurate one. The correlation was obtained from a survey of experimental data:

$$Nu_a Pr^{-0.37} = \\ 0.571 + 0.0125\ Tu_a \left\{1 + \frac{1.8}{[1+(Tu_a/20)^3]}\right\} \tag{1}$$

with the characteristic Nusselt number $Nu_a = Nu/\sqrt{a_1 Re}$, the characteristic turbulence intensity $Tu_a = Tu\sqrt{Re/a_1}$, and the Prandtl number $Pr = 0.7$. The turbulence intensity is defined as $Tu = \sqrt{\frac{2K}{3}}/U_\infty$ with $K = \frac{1}{2}(\langle u'u'\rangle + \langle v'v'\rangle + \langle w'w'\rangle)$, where $\langle\cdot\rangle$ represents the time-average of a quantity. An important parameter is the coefficient a_1, the free-stream strain rate. Measured along the centerline, it describes the free-stream acceleration in tangential direction (s), with $a_1 = \frac{d(U_s/U_\infty)}{d(s/D)}$. For potential flow around a single cylinder in a free-stream, the theoretical value is $a_1 = 4$. However, depending on Reynolds number, turbulence level and wind tunnel blockage, values of 2.3–3 are reported in the literature, e.g. in [3]. In the present study,

a_1 is determined at the centerline, at a distance of $0.03D$ from the cylinder. From the DNS data a value of $a_1 = 1.43$ is obtained. This value is very low compared with those referenced above. The deviation is apparently caused by the large wake deficit at the centerline for the present tandem–cylinder setup. The turbulence level in the DNS was determined as $Tu = 0.22$, which is in fair agreement with the experimental value of $Tu = 0.21$ in [1].

3 Flow Solver and Code Performance

The simulations are performed with the Finite Volume code LESOCC2 [7] developed at KIT. LESOCC2 solves the incompressible, three-dimensional, time-dependent, Navier–Stokes equations and an arbitrary number of scalar transport equations using second-order accurate central differences for the discretization of the convective and viscous fluxes (and a second-order upwind scheme for the convective terms in the scalar transport equation for the temperature). Time advancement is accomplished by an explicit, low-storage Runge–Kutta method. Conservation of mass is achieved by the SIMPLE algorithm with the pressure–correction equation being solved with the strongly implicit procedure (SIP) of Stone. Velocity and pressure decoupling is avoided by a momentum interpolation technique. In the LES zone, the subgrid-scale model of Smagorinsky is used with $C_s = 0.1$.

The code has been used in numerous studies and substantial experience has been gained with respect to its numerical properties and its performance on various hardware platforms including the NEC SX-8, SGI-Altix, and several Linux clusters. Parallelization is accomplished via domain decomposition and MPI for the data transfer. To achieve better performance on a given hardware, different implementations of the computationally most intensive algorithms exist. The user can decide between highly vectorized versions and unvectorized versions of subroutines or can choose to trade off between memory requirements and number of operations in order to fit a run on a specific machine or possibly take advantage of cache effects.

For reference, part of the preliminary simulation on the baseline grid was carried out on the HP XC3000 at the Steinbuch Centre for Computing (SCC), an Intel Xeon Linux cluster. Using 256 processor cores, the baseline simulation yielded an acceptable turnaround time. However, this simulation is still relatively coarse, as stated in Sect. 2.1. The number of grid points for the final simulation will be a factor of ten larger. Since the number of blocks and, consequently, the maximum number of processors, cannot be increased accordingly, a switch to a parallel-vector machine is made. Previous simulations with LESOCC2 on the NEC SX-8 of the Bundeshöchstleistungsrechenzentrum Stuttgart (HLRS) delivered a good performance. A continuation of the preliminary simulation on the NEC SX-9 using 64 processors (the maximum available), however, resulted only in a speedup of a factor of two in comparison with the Linux cluster for the same number of processor cores. The lack of

Table 1. Performance of selected LESOCC2 routines on the NEC SX-9 with and without additional compiler directives for better vectorization

PROC. NAME	MFLOPS		V. OP. RATIO		AVER. V. LENGTH	
	standard	optimized	standard	optimized	standard	optimized
coeff3	40791.2	41295.7	99.60	99.60	256.0	256.0
sip3dp	65.0	2151.3	76.56	99.96	253.7	253.7

expected performance was analyzed using the profiler "ftrace". Key outputs of the profiler are compiled for two routines in Table 1. "coeff3" is an example for a routine that exhibited already the performance level consistent with past experience. On the other hand, "sip3dp", one of the most time consuming routines, yielded an abysmal performance. The analysis hinted to a lack of vectorization—even though compiler flags were set to optimal vectorization and the problem size was adjusted for an average vector length close to or at the optimal value of 256. As a consequence, additional compiler directives were implemented on a subroutine basis. The statements ensured that no dependencies were assumed by the compiler resulting in a scalar operation of key loops. The additional compiler directives produced a drastic performance enhancement with an additional speedup of the order of 30 for the simulation. This speedup makes the final DNS already feasible. However, compared with other routines, like "coeff3", a further speedup should still be possible. Current work is focusing on reducing the use of lookup tables in routine "sip3dp".

4 Preliminary Results

The baseline simulation, started on the HP XC3000 at the SCC and finished on the NEC SX-9 at the HLRS already show promising and novel results that are published in [6]. Below some of these findings are presented to highlight the significance of the results and the potential for the final simulation.

4.1 Mean Heat Transfer

The effect of wake-induced turbulence on the heat transfer at the cylinder stagnation region is the focus of the present study. Figure 3 shows a comparison of the stagnation point heat transfer obtained by the present DNS with the correlation in [3]. The heat transfer is represented by the characteristic Nusselt number and plotted versus the characteristic turbulence level. In addition, data from other simulations in the literature are plotted for reference (taken from [8]). These simulations exhibit a good agreement with the experimental data, but cover only a limited range of Tu_a at the lower end. In contrast, in the present study a significantly higher characteristic turbulence level is achieved. This is due to the relatively low free-stream strain rate

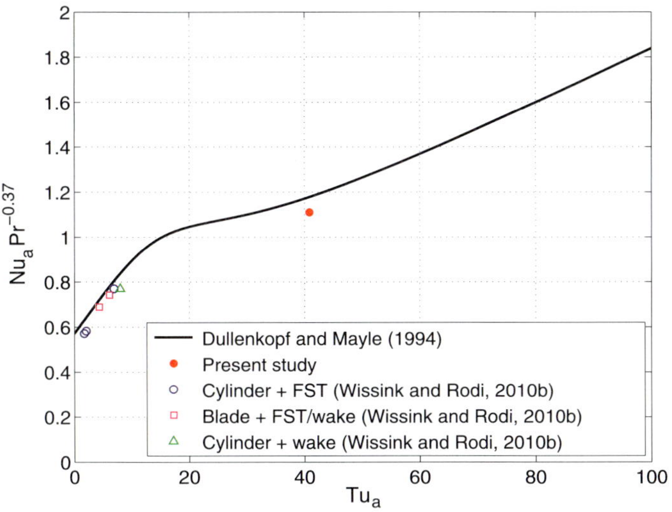

Fig. 3. Comparison of numerical simulations with the correlation of Dullenkopf and Mayle [3]. The symbols represent numerical simulations from the literature [8] (open symbols) and the present study (filled symbol); FST denotes free-stream turbulence

combined with the high turbulence intensity as described in Sect. 2.2. In accordance with the prediction in [2], an increase in heat transfer for all angles along the front circumferential area of the heated cylinder is found compared to fully laminar flow. However, presumably due to the presence of the wake, the maximum increase is shifted away from the centerline (see [6]). Considering the preliminary nature of the present simulation, a surprisingly good agreement with experimental correlations can be observed. Moreover, to the best of our knowledge, these results are the highest Tu_a levels obtained via DNS to date. This and the better resolved data of the final simulation can then be used for a detailed investigation of the physical mechanisms at play.

4.2 Instantaneous Temperature and Flow Field

In order to determine what causes the increase in heat transfer taking place in the vicinity of the stagnation point, detailed knowledge of the temperature and flow fields is of importance. In Fig. 4, the instantaneous temperature field adjacent to the heated cylinder is displayed. Many fine-structured temperature variations are visible and longitudinal hot and cold spots can be recognized. These spots are correlated with elongated flow structures that penetrate the boundary layer of the heated cylinder, as was already observed for a lower Reynolds number in [4]. The instantaneous flow field, displayed in Fig. 5, shows the appearance of such structures. These structures are visualized by means of pressure fluctuations ($p' = p - \langle p \rangle$) which is a common way to identify the larger vortical structures in a turbulent flow [9]. In between the

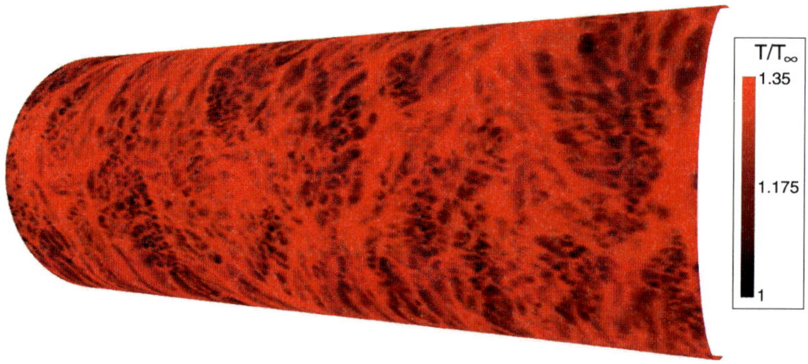

Fig. 4. Instantaneous temperature field adjacent to the heated cylinder

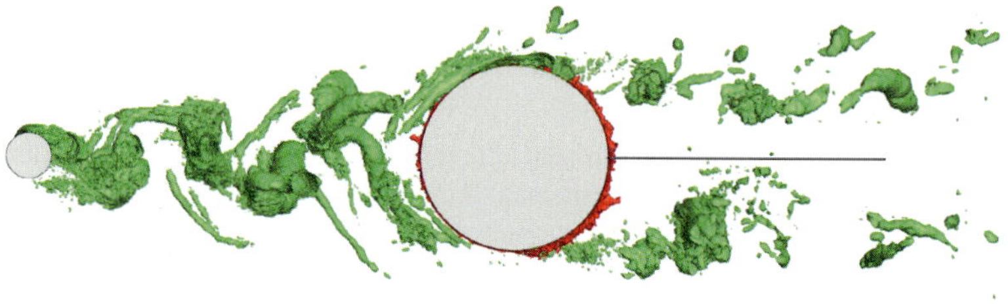

Fig. 5. Instantaneous pressure fluctuations and temperature iso-surfaces for $p' = -0.25$ and $T/T_\infty = 1.15$, respectively

large spanwise structures also streamwise structures impinge on the heated surface.

5 Conclusions

Heat transfer in the stagnation point area of a heated cylinder was investigated using embedded DNS. The heated cylinder is subjected to the turbulent wake of a smaller cylinder placed upstream. The Reynolds number based on the diameter of the heated cylinder is 48,000. The characteristic turbulence level for the present study is an order of magnitude higher than those reported in any previous DNS. The obtained Nusselt number is in good agreement with the experimental correlation in [3]. The increase in heat transfer was found for all angles along the front circumferential area, however, presumably due to the presence of the wake, the maximum increase is shifted away from the centerline. More details can be found in [6]. Although the baseline simulation presented here already provides some first insights, more resolution is needed for the detailed analysis envisaged for the current project.

Acknowledgements. The work reported here was carried out within the "Research Group Turbo-DNS" at the Institut für Thermische Strömungsmaschinen. Its financial support by means of the German Excellence Initiative and Rolls–Royce Deutschland is gratefully acknowledged. The authors are grateful for the provision of computer time by the Bundeshöchstleistungsrechenzentrum Stuttgart (HLRS).

References

1. P.J. Magari and J.E. LaGraff. Wake-induced unsteady stagnation-region heat transfer measurements. *Journal of Turbomachinery*, 116:29–38, Jan. 1994.
2. N. Frössling. Evaporation heat transfer and velocity distribution in two-dimensional and rotationally-symmetric laminar boundary layer flow. *NACA TM-1432*, 1958.
3. K. Dullenkopf and R.E. Mayle. The effects of incident turbulence and moving wakes on laminar heat transfer in gas turbines. *Journal of Turbomachinery*, 116:23–28, 1994.
4. J. Wissink and W. Rodi. Direct numerical simulation of heat transfer from the stagnation region of a heated cylinder affected by an impinging wake. *Submitted to J. Fluid Mech.*, 2010.
5. J.G. Wissink and W. Rodi. Large-scale computations of flow around a circular cylinder. In Michael Resch, Sabine Roller, Peter Lammers, Toshiyuki Furui, Martin Galle, and Wolfgang Bez. *High Performance Computing on Vector Systems 2007*. Springer-Verlag Berlin Heidelberg, 2008.
6. L. Venema, D. von Terzi, H.-J. Bauer, and W. Rodi. DNS of heat transfer increase in a cylinder stagnation region due to wake-induced turbulence. In *Proceedings of ETMM-8, 9–6 June 2010, Marseille, France*, 2010.
7. C. Hinterberger. *Dreidimensionale und tiefengemittelte Large-Eddy Simulation von Flachwasserströmungen*. PhD thesis, University of Karlsruhe, 2004.
8. J. Wissink and W. Rodi. Heat transfer from the stagnation area of a heated cylinder at $Re_d = 140000$ affected by free-stream turbulence. In *Proceedings of ETMM-8, 9–6 June 2010, Marseille, France*, 2010.
9. D.A. von Terzi, R.D. Sandberg, and H.F. Fasel. Identification of large coherent structures in supersonic axisymmetric wakes. *Computers & Fluids*, 38(8):1638–1650, 2009.

Application of a Novel Turbulence Generator to Multiphase Flow Computations

Christian Huber, Hassan Gomaa, and Bernhard Weigand

Institut für Thermodynamik der Luft- und Raumfahrt, Universität Stuttgart, Pfaffenwaldring 31, 70569 Stuttgart, Germany, christian.huber@itlr.uni-stuttgart.de

Summary. The application of a novel turbulent inflow generator is presented using a 3D Direct Numerical Simulation (DNS) method. For simulation the in-house 3D CFD program Free Surface 3D (FS3D) is applied, which solves the incompressible Navier Stokes Equations for flows with free surfaces using a Volume-of-Fluid (VOF) technique. Three different numerical setups are presented, demonstrating the wide range of application of the new inflow generator. For chosen cases qualitatively comparisons are made between the new and a former implemented turbulence generator. Another more general focus is put on the high influence of turbulent flow fields on simulation outcomes, which justifies the use of computationally intensive inflow generators. Further, information about the performance on the NEC SX-8 platform, where all the simulations were performed, are revealed.

1 Introduction

The application of artificial turbulent boundary conditions represents an essential part of today's numerical simulations. Examples in the field of multiphase flows are the investigation of rising air bubbles in a disturbed liquid, evaporating droplets in a turbulent free stream or the disintegration of turbulent liquid jets. In terms of accuracy it is especially important to reproduce the predominant physical disturbances at the inflow as accurate as possible. Therefore, an inflow generator is needed which can reliably and efficiently reproduce turbulent velocity fields, even on large inflow areas. In this paper the applicability of a novel approach for the generation of turbulent inflow conditions, proposed by Kornev and Hassel [1], is presented and for chosen cases qualitatively compared to a former method proposed by Klein et al. [2]. Both methods are currently implemented in the ITLR in-house DNS code FS3D, which is used to obtain the presented numerical results. Three different numerical applications have been investigated to demonstrate the wide range of application of the new inflow generator method. Those are the simulation of a turbulent free air stream, a disintegrating turbulent liquid jet and an evaporating water droplet in a hot turbulent air stream. Since Klein's method

W.E. Nagel et al. (eds.), *High Performance Computing in Science and Engineering '10*, DOI 10.1007/978-3-642-15748-6_21,

was specially optimized for the NEC SX-8 platform, all computations were performed there, to prevent a distortion of the performance comparisons that are also subject of this paper. The first section will serve as an introduction to FS3D and the implemented turbulent inflow generators. Subsequently, the applicability of the new inflow generator to the three aforementioned representative numerical setups will be discussed. The work will conclude with an overview on the achieved performance ratings on the NEC SX-8 platform.

2 Numerical Method

For simulation the in-house 3D CFD program FS3D was applied, that computes the Navier Stokes Equations and the energy equation for incompressible flows with free surfaces using DNS. The flow field is computed by solving the governing equations for mass, momentum and energy conservation

$$\rho_t + \nabla \cdot (\rho \mathbf{u}) = 0 \tag{1}$$
$$(\rho \mathbf{u})_t + \nabla \cdot ((\rho \mathbf{u}) \otimes \mathbf{u}) = \nabla \cdot (\mathbf{S} - \mathbf{I}p) + \rho \mathbf{g} + \mathbf{f}_\gamma \tag{2}$$
$$(\rho c_p T)_t + \nabla \cdot (\rho c_p T \mathbf{u}) = \nabla \cdot (k \nabla T) + T_0 \left[(\rho c_p)_t + \nabla \cdot (\rho c_p \mathbf{u}) \right] + \Delta h_v \dot{m}''' \tag{3}$$

Here $\mathbf{u}$ denotes the velocity vector, ρ the density, p the static pressure, $\mathbf{g}$ the external body force vector and $\mathbf{S}$ determines the viscous stress tensor that is defined by $\mathbf{S} = \mu[\nabla \mathbf{u} + (\nabla \mathbf{u})^T]$ for Newtonian fluids, where μ represents the dynamic viscosity. The surface tension, stated by the capillary stress tensor $\mathbf{f}_\gamma$, is handled as a volume force that acts on every cell containing an interface. It is computed by the conservative continuous surface stress (CSS) model suggested by Lafaurie et al. [3]. Regarding (3), c_p denotes the specific heat capacity at constant pressure, T the temperature, k the heat conductivity and T_0 is the reference temperature for zero enthalpy. Δh_v represents the latent heat of the evaporating fluid and $\dot{m}'''$ is the volumetric mass source of vapour. To track the interface,the VOF method, originally introduced by Hirt and Nichols [4], is applied. At this, the additional field variable f_1, describing the liquid volume fraction distinguishing between the liquid and gaseous phases, is introduced as follows:

$$f_1(\mathbf{x}, t) = \begin{cases} 0 & \text{in the gaseous phase} \\ (0;1) & \text{in cells containing the interface} \\ 1 & \text{in the liquid.} \end{cases} \tag{4}$$

The evaporation of the liquid phase is considered by a second VOF variable f_2 describing the volume fraction field of the vapour. It is defined in the same manner as f_1. The vapour is considered to be insoluble in the liquid phase but diffusion in the gaseous phase is allowed. The motion of the volume fraction fields is stated by the transport equations:

$$(f_1)_t + \nabla \cdot (f_1 \mathbf{u}_\Gamma) = -\dot{m}'''/\rho_l \tag{5}$$
$$(f_2)_t + \nabla \cdot (f_2 \mathbf{u}_{gp}) = \nabla \cdot (D_{vg} \nabla f_2) + \dot{m}'''/\rho_v \,, \tag{6}$$

where $\mathbf{u}_\Gamma$ and $\mathbf{u}_{gp}$ denote the advection velocity of the interface and of the gaseous phase. The generation of volume due to the evaporation of the liquid and the resulting production of vapour is considered by the ratio of $\dot{m}'''$ and the liquid density ρ_l respectively the vapour density ρ_v. The diffusion of the vapour is described by the binary diffusion coefficient of the vapour inside the gas D_{vg}. By knowledge of the volume fractions the local fluid properties (e.g. ρ, μ, k) are calculated in terms of a fluid formulation for the computational domain, e.g. for the density field

$$\rho(\mathbf{x}, t) = \rho_l f_1(\mathbf{x}, t) + \rho_v f_2(\mathbf{x}, t) + (1 - f_1(\mathbf{x}, t) - f_2(\mathbf{x}, t))\rho_g \,. \tag{7}$$

The presented equations are discretized by finite volume schemes with second order accuracy in space on a staggered grid. The calculation of the fluxes in (5), (6) is based on a piecewise linear reconstruction of the interface in each cell, the so called piecewise linear interface calculation (PLIC) method proposed by Rider and Kothe [5], which ensures a sharp interface and widely suppresses numerical dissipation. The mass and momentum convection, as well as the momentum diffusion and the occurring body forces are treated explicitly.

The Poisson equation for the pressure resulting from the assumption of incompressibility has to be solved implicitly. In FS3D this is carried out by a multigrid solver that uses Red-Black Gauss Seidel as smoothing algorithm. A detailed description of the implementation of the governing equations is given in [6–8].

2.1 Turbulent Inflow Generation

In FS3D two different methods are available, to generate turbulent inflow boundary conditions. In both cases the mean inflow velocity profile $\overline{u_i}(t, y, z)$ is superimposed by velocity fluctuations $u_i'(t, y, z)$ which is a usual approach for the generation of artificial turbulence.

The first method for generating the velocity fluctuation field with a prescribed autocorrelation function R_{ii} and Reynolds stresses r_{ij} is based on the digital filtering of random fluctuation fields [9, 2]. Here, at first a three dimensional field of random numbers is generated for each of the velocity components $U_j(t, y, z)$. According to Lund et al. [9] $U_j(t, y, z)$ have to satisfy the condition of zero mean ($\overline{U_j} = 0$) and unit variance as well as zero covariance ($\overline{U_i U_j} = \delta_{ij}$) which is achieved by applying the digital filtering technique proposed by Klein et al. [2]. The computation of the filtering coefficients is based on a multidimensional Newton method for the provided autocorrelation function of decaying homogeneous turbulence in a late stage

$$R_{ii}(r, 0, 0) = exp(\frac{-\pi r^2}{4L^2}) \qquad \text{with} \quad L = \sqrt{2\pi\nu(t - t_0)} \tag{8}$$

which satisfies the mentioned autocorrelation condition and moreover allows the prescription of the integral length scale L which, from an engineering point of view, is easier accessible than the knowledge of the full autocorrelation function. The generated fluctuation fields $U_j(t,y,z)$ possess the prescribed two-point spatial and temporal autocorrelation functions.

To take into account the cross correlations between the different velocity components in a second step the coordinate transformation $u_i(t,y,z) = \overline{u_i}(t,y,z) + a_{ij}U_j(t,y,z)$ is performed, where the amplitude tensor given by

$$a_{ij} = \begin{bmatrix} \sqrt{r_{11}} & 0 & 0 \\ r_{21}/a_{11} & \sqrt{r_{22} - a_{21}^2} & 0 \\ r_{31}/a_{11} & (r_{32} - a_{21}a_{31})/a_{22} & \sqrt{r_{33} - a_{31}^2 - a_{32}^2} \end{bmatrix} \tag{9}$$

is related to the Reynolds stresses r_{ij}. The final velocity fluctuations field $u_i'(t,y,z) = a_{ij}U_j(t,y,z)$ possesses the prescribed two-point spatial and temporal autocorrelation functions as well as the one-point correlations r_{ij}, hence the Reynolds stresses. In the following this method is referred to as *Klein's method.*

The second, recently implemented method for the synthesis of the turbulent inflow velocity field with a prescribed autocorrelation function, is based on the random spot method proposed by Kornev and Hassel [1, 10]. The method is based on the idea of the random placement of a collection of spots, with a certain inner velocity distribution in the outer front of the inflow plane. The spots then move through the inflow plane and the turbulent velocity field is gained by the superposition of the velocity oscillations that are induced by the spots at the inflow plane. It was shown by Kornev and Hassel [1] that under the consideration of certain symmetry conditions (e.g. spherical symmetry) of the carried vector fields it is possible to obtain an analytical solution for the spots inner velocity field from the shell-summed energy spectrum. Within this work the method was applied for the energy spectrum for decaying turbulence given in the Fourier transformed k-space by

$$E(k) = C_E k^4 exp(-k^2/w_0^2) \tag{10}$$

where $\omega_0 = \sqrt{\pi}/L$ and L denoting the longitudinal length scale. As shown by Kornev and Hassel [1] in the spot fixed cylindrical coordinate system (ρ,ϑ,z) the vorticity and velocity fields within each spot are obtained to

$$u_\rho = u_z = \omega_\vartheta = 0 \tag{11}$$

$$u_\vartheta = \pi\sqrt{2C_E/C_1}\omega_0^5 exp(-w_0^2r^2/2)\rho \tag{12}$$

$$\omega_z = \pi\sqrt{2C_E/C_1}\omega_0^5 exp(-w_0^2r^2/2)(2-\omega_0^2\rho^2) \tag{13}$$

$$\omega_\rho = \pi\sqrt{2C_E/C_1}\omega_0^7 exp(-w_0^2r^2/2)\rho\rho_3 \tag{14}$$

Here, ρ_i denotes the Cartesian spot fixed coordinate system, with $\rho = \sqrt{\rho_1^2+\rho_2^2}$ and $r = \sqrt{\rho_i^2}$. The Relation between the two spot-fixed coordinate systems is clarified in Fig. 1. In the technical application the amplitude factor C_E/C_1 is determined in a preprocessing step such that the desired root

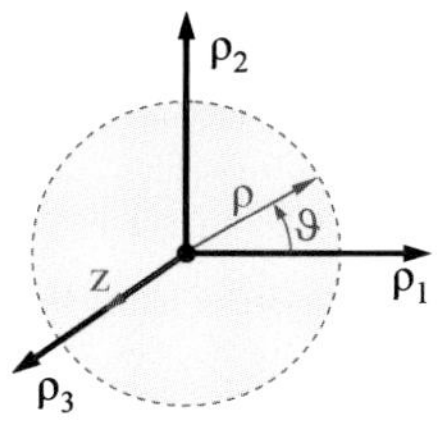

Fig. 1. Relation between spot-fixed cylindrical coordinate system (ρ, ϑ, z) in blue and spot-fixed Cartesian coordinate system (ρ_1, ρ_2, ρ_3) in black

mean square velocity values are satisfied throughout the inflow plane. In the following this method is referred to as *Kornev's method.*

For the current state of implementation an apparent advantage of the random spot method is given by the fact that time steps are variable and the inflow generation is realizable during runtime. Hence, the possibility to generate turbulent velocity fields for large inflow areas arises. However, the implemented random spot method supports only the homogeneous turbulent flow case of a freely decaying energy spectrum.

3 Applications

In the following, three different numerical setups, demonstrating the wide range of application of the new inflow generator, will be presented. The first setup, namely the simulation of a turbulent free air stream, will mainly serve as a validation case. The decay of the turbulent kinetic energy, as well as the decay of the normal stresses will be evaluated to prove the correct operation of the inflow generator. The latter two numerical setups represent more practical setups. A comparison between the two inflow generators will be made. Also the general influence of turbulent flow fields will be discussed, demonstrating the justification of using such computationally expensive methods to accurately reproduce physical fluctuations at the inlet. Also further information about the performance will be presented. For all simulations the integral length scale L is chosen in such a manner that the Kolmogorov length scale is resolved.

3.1 Turbulent Free Air Stream

The simulation of a free air stream serves as a validation case for the new turbulent inflow generator method. It will also allow to show the capabilities of the new method compared the former method of Klein, mainly regarding the performance on large inflow areas ($> 10^4$ *controll volumes*).

Numerical Setup

The 3D computational domain used for this setup is of quadratic shape and has an overall extension of $4 \cdot 10^{-3}$ m in each spatial direction. The resolution

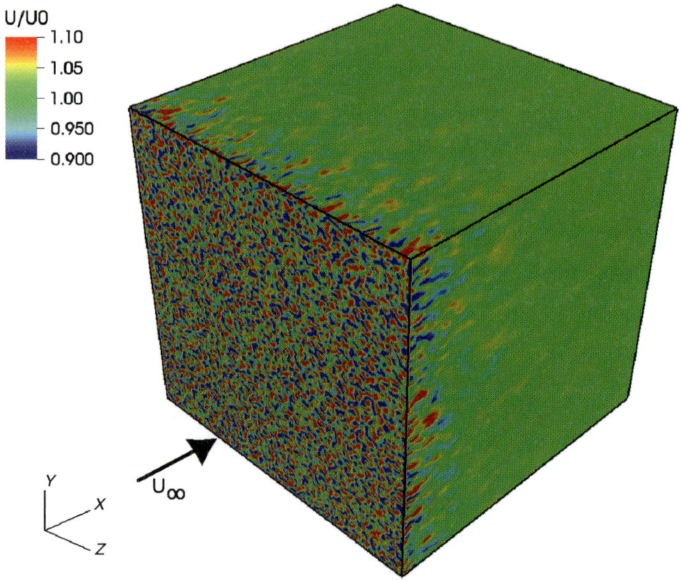

Fig. 2. Velocity fluctuations in the computational domain

is chosen to be 256 cells in each direction, which sums up to about $16.7 \cdot 10^6$ computational cells in the whole domain. On the left side of the computational domain a turbulent inflow condition is applied, were the fluid, in this case air at 20 $°C$ and 1 bar ambient pressure, enters the domain with a mean inflow velocity of 1 m/s and an intensity of turbulence of 10%. The integral length scale L is chosen to be $4 \cdot 10^{-5}$ m, one hundredth of the overall extension in one spatial direction. While on all lateral boundaries a symmetry condition is employed, on the right side of the computational domain, namely the outflow, a continuous (Neumann) condition is applied.

Results

In Fig. 2 the velocity fluctuations at the inlet and throughout the rest of the computational domain are exemplarily depicted for one timestep. While the intensity of the generated velocity fluctuations is highest at the inlet area, the decaying behaviour in the main flow direction is clearly visible.

For the considered case of decaying turbulence in a late stage, expressed by (10), the energy of a spot decays as $t^{-5/2}$ in a moving frame of reference [1]. Hence, in the laboratory fixed computational domain the turbulent kinetic energy decays proportional to $\xi^{-5/2}$, where ξ denotes the non-dimensional domain length in the spatial main flow direction. The good agreement between the theoretical power law and the numerical results is evident from Fig. 3. On the right side of Fig. 3 the decay of the normal Reynolds stresses is shown. The symmetry of the setup in y and z direction dictate that $\overline{u_2'^2}$ and $\overline{u_3'^2}$ are almost

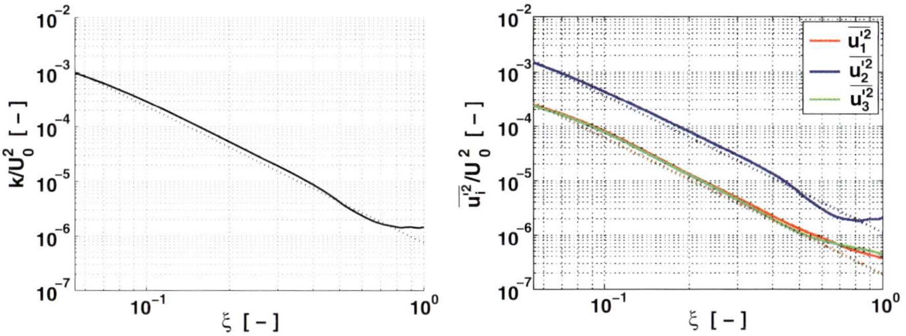

Fig. 3. Decay of turbulent kinetic energy (left) and normal Reynolds stresses (right) over the non-dimensional domain length. The solid lines show the numerically obtained data and the dotted lines the theoretical decay proportional to $\xi^{-5/2}$

equal and accordingly smaller than the axial root mean square $\overline{u_1'^2}$. Visible deviations at the outflow are explainable by the influence of the boundary conditions and their upstream effect due to the ellipticity of the governing equations. As is shown later, this effect is eliminated by introducing a damping zone right at the outlet.

In the present case a computational performance in the range of 2.0 $GFLOPS$ was attained on 8 CPUs. The total computational time required was 1 hour and 47 minutes. To be able to compare the two implemented turbulent inflow generators this setup was also performed applying the method of Klein. Thereby, the implemented algorithm was still occupied with the initialization step, filtering the random fluctuation fields, until the wall clock time was hit.

As the fluctuation fields could be reused, the difference would be essentially lesser in succeeding computations. Nevertheless, the capabilities and performance of the newly implemented approach for very large inflow areas is clear.

3.2 Disintegration of Turbulent Liquid Jets

In the following the disintegration of a turbulent liquid jet, for two different turbulence intensities, namely 0.5% and 5.0%, are presented. For the case of 0.5% turbulence intensity a qualitative comparison between the two available inflow generator methods is presented in addition.

Numerical Setup

A liquid decane jet is injected into a quiescent high pressure nitrogen environment of 51 bar, with a pressure drop of 1.5 bar across the nozzle. A jet diameter D of $2 \cdot 10^{-4}$ m is assumed for the chosen inflow velocity profile. The 3D computational domain depicted in Fig. 4 has an overall length of $12D$ in

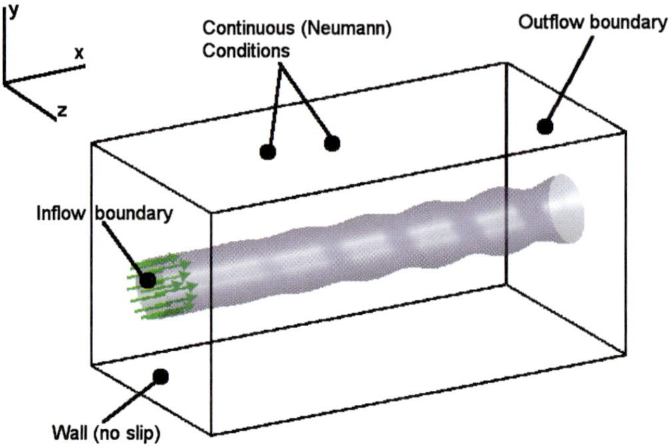

Fig. 4. Numerical setup for liquid jet disintegration

the direction of the main flow and $5D$ between the lateral boundaries. The liquid decane jet thereby enters the computational domain at the left side at a mean inflow velocity of $U_0 = 1.362\ m/s$ and a block velocity profile. This results in a Reynolds Number of $Re = \rho_l U_0 D/\mu_l = 2366$ and a Weber Number of $We = \rho_l U_0^2 D/\sigma = 1365$. An initial turbulence intensity of 0.5%, respectively 5%, is applied to it. A resolution of 768 cells in the main flow direction and 320 cells in the lateral directions is chosen, which results in a total of $78.6 \cdot 10^6$ million computational cells. At this resolution the jet diameter is resolved by 64 cells. The integral length scale L was chosen to be $2 \cdot 10^{-5}\ m$, one tenth of the nozzle diameter. On the inlet plane a no slip wall boundary condition is employed aside of the jet inflow area, while on all other boundaries a continuous (Neumann) condition is applied.

Despite the fact that this disintegration process is mainly dominated by high ambient pressure there is still a visible influence of turbulent fluctuations. A quantitative comparison and discussion on the influence of the related turbulence properties are given by Gomaa et al. [11] for liquid jets and by Sander and Weigand [12] for liquid sheets.

Results

In Figs. 5 and 6 the disintegration of a turbulent decane jet with an initial turbulence intensity of 0.5% is depicted over the whole computational domain of $12D$. While the results in Fig. 5 are computed by the inflow generator method of Klein, the results depicted in Fig. 6 were obtained using the new turbulent inflow generator method by Kornev and Hassel [1]. It can be seen that the wavy surface disturbances match each other very well. The surface waves are showing the same behaviour in both cases and have the same order of magnitude. Also both injection processes evolve in a similar way, showing the qualitative agreement between both turbulent generators.

Fig. 5. Jet disintegration of a round decane jet in a quiescent high pressure environment of 51 *bar*. $Re = 2366$, $We = 1265$, $L/D = 10$ and a turbulent intensity of 0.5%. Computed by the method of Klein

Fig. 6. Jet disintegration of a round decane jet in a quiescent high pressure environment of 51 *bar*. $Re = 2366$, $We = 1265$, $L/D = 10$ and a turbulent intensity of 0.5%. Computed by the method of Kornev

In Fig. 7 the same test case with an turbulence intensity of 5% instead of 0.5%, computed by the new method of Kornev, is shown. In comparison to the simulation results with a lower turbulence intensity a slight increase of the arising instability amplitudes along the core is visible. It can also be observed that the surface waves shift a little closer to the nozzle exit. The disintegration process taking place at the forefront of the liquid jet produces fewer but larger ligaments.

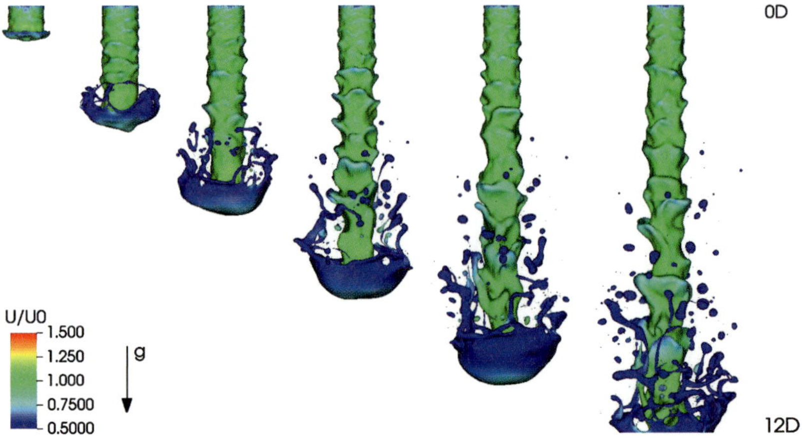

Fig. 7. Jet disintegration of a round decane jet in a quiescent high pressure environment of 51 *bar*. $Re = 2366$, $We = 1265$, $L/D = 10$ and a turbulent intensity of 5%. Computed by the method of Kornev

A performance up to 2.2 $GFLOPS$ was reached on 8 CPUs. It was additionally possible to compute twice as much timesteps with the new turbulent inflow generator method compared to the current implementation of Klein's method.

3.3 Drop Evaporation in a Turbulent Free Stream

Finally the evaporation of a water droplet at 8 $°C$ in a turbulent free air stream of 252 $°C$ is presented. Because of the lack of sufficient computational resources for applying Klein's method on large inflow areas, this demonstration case is only computed using the new inflow generator method by Kornev.

Numerical Setup

A water droplet with an initial diameter D of 10^{-3} m is initialized in the $3D$ computational domain depicted in Fig. 8. The overall length of the domain is $6D$ in the direction of the main flow and $4D$ between the lateral boundaries. The surrounding turbulent free air stream enters the computational domain at the left side at a mean inflow velocity of $U_0 = 1.362$ m/s, resulting in a Reynolds number of $Re = \rho_g U_0 D/\mu_g = 884$ and a Weber number of $We = \rho_g U_0^2 D/\sigma = 3$. An initial turbulence intensity of 0%, respectively 10%, is applied to the inflow. A resolution of 192 cells in the main flow direction and 128 cells in the lateral directions is chosen. At this resolution the jet diameter is resolved by 32 cells. The integral length scale L was chosen to be $D/10$ of the droplet diameter. Except the inflow plane, where an inflow boundary condition is employed, a continuous (Neumann) condition is applied to all

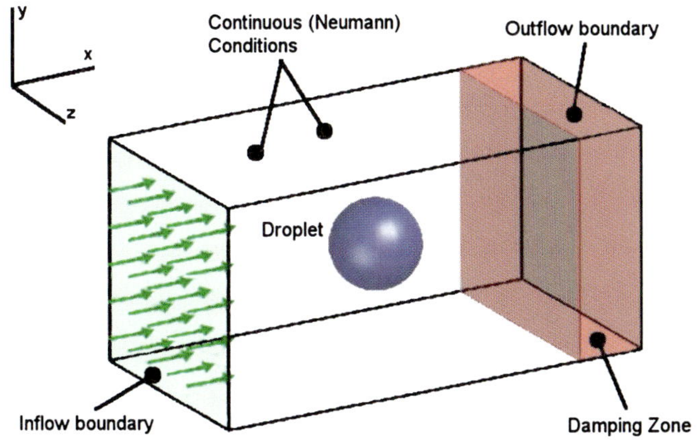

Fig. 8. Numerical setup for an evaporating water droplet

other boundaries. An additional damping zone at the end of the computational domain prevents backflow into the calculation domain.

In Fig. 9 the non-dimensional mass loss due to the evaporation of the droplet is depicted for two different inflow turbulent intensities. It is obvious that with turbulent boundary conditions applied, there is a difference in the evaporation rate. After 0.065 s there is already a variation of 1% regarding the mass loss, compared to the numerical simulation without a turbulent inflow condition. This difference may even increase when considering other, more volatile, fluids than water, or when larger simulation times are achieved.

One of the reasons why the influence of the turbulent velocity field on the evaporation process in this case lies within a single-digit percentage can be explained by the field of turbulent fluctuations surrounding the droplet. This is depicted in Fig. 10. As the approaching turbulent flow field is redirected by the droplet, the evolving boundary layer prevents most of the oncoming turbulent flow to directly interact with the droplet surface. As a consequence the turbulent velocity field can only directly interact with a limited area at the front of the droplet, resulting in a limited influence onto the evaporation process.

The performance reached in such simulations considering evaporation processes was up to 2.6 $GFLOPS$ on 8 CPUs of the NEC SX-8.

4 Architecture and Performance

All simulations were computed on the NEC SX-8 platform at HLRS, because the method of Klein et al. [2] was at the time of implementation especially optimized for the NEC SX-8 platform. Therefore, the comparison of the two inflow generators had to take place on this platform to not distort the comparison. A parallel version of FS3D using the (-P auto) option was used. All computations were conducted on an exclusive node using all 8 vector proces-

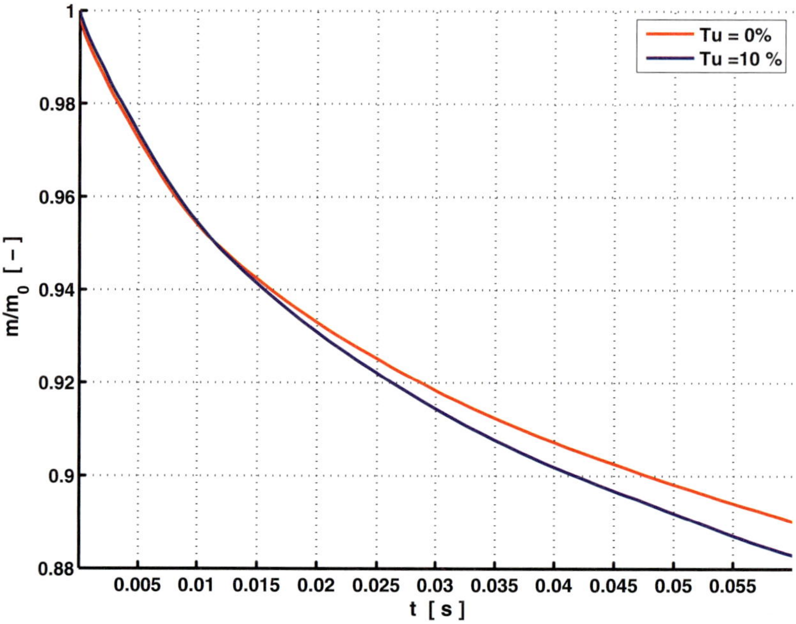

Fig. 9. Mass loss due to evaporation of a water droplet

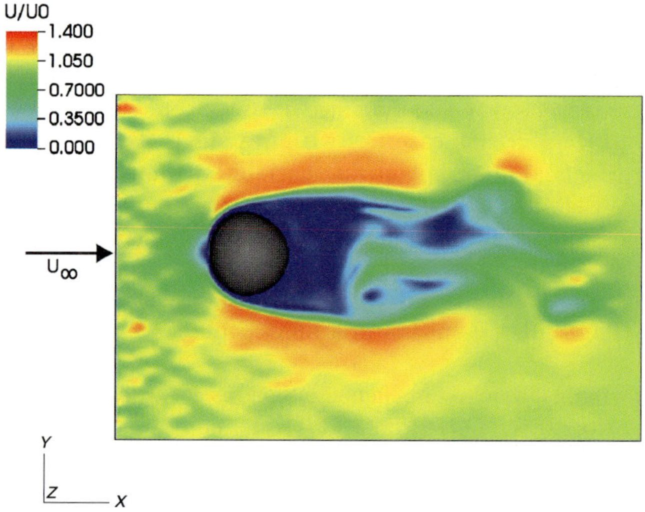

Fig. 10. Velocity fluctuations around an evaporating droplet

sor units. The total computational time was limited to 48 h per simulation. For additional optimization the program was compiled using the default optimization level (-C vopt), maximum inlining as well as a special code version which makes extensive use of unrolled loops. To gain meaningful performance information the environment variable F_PROGINF was set to *DETAIL* and the output was evaluated. The different numerical setups achieved average vector lengths up to 224 and vector operation ratios up to 92%. Also an overall computational performance in the range of 2.6 $GFLOPS$ was achieved. By optimization of the newly implemented turbulent inflow generator, mainly regarding the vectorization and parallelization of the source code, it is expected to further improve the attainable performance. Nevertheless a remarkable performance gain was achieved comparing the new turbulent inflow generator with the former implemented inflow generation method proposed by Klein et al. [2]. With the new method it is for the first time possible to calculate setups with turbulent inflow conditions on larger inlet areas in an acceptable amount of time, which marks a big milestone in the development of FS3D.

5 Conclusion

A novel turbulent inflow generator method was presented using 3D Direct Numerical Simulation to compute multiphase flow problems. The in-house code FS3D was used for the computations. The focus was particularly turned on the wide range of applicability of the novel turbulent inflow generator. Therefore, three different numerical setups were computed and discussed successively. Hereby, the influence of turbulent flow conditions was shown and where possible, a qualitative comparison was made versus the former implemented turbulent inflow generator. For the considered energy spectrum the advantages of the newly implemented method particularly lie in the ability to also handle large inlet areas, allowing for a wide range of new applications in the future. Also information about the performance of the new turbulence model was evaluated, showing a maximum computational performance of up to 2.6 $GFLOPS$, which will be further improved by optimizations regarding the vectorization and parallelization of the source code. In future work the method of Kornev will be extended to further allow other energy spectra rather than the decaying turbulence stated above. Moreover, the turbulence generator will be of great use for gaining a detailed physical understanding on the relevance of turbulence in multiphase flow processes.

Acknowledgements. The authors greatly appreciate the *High Performance Computing Center Stuttgart* (HLRS) for support and supply of computational time on the NEC SX-8 platform under the Grant No. FS3D/11142. In addition, the authors would like to thank Prof. N. Kornev and Prof. E. Hassel for the fruitful cooperation and all the helpful discussions.

References

1. N. Kornev and E. Hassel. Synthesis of homogeneous anisotropic divergence-free turbulent fields with prescribed second-order statistics by vortex dipoles. *Physics of Fluids*, Vol. 19, No. 6, pp. 068101, 2007.

2. M. Klein, A. Sadiki, and J Janicka. A digital filter based generation of inflow data for spatially developing direct numerical or large eddy simulations. *Journal of Computational Physics*, Vol. 186, pp. 652–665, 2003.
3. B. Lafaurie, C. Nardone, R. Scardovelli, S. Zaleski, and G. Zanetti. Modelling Merging and Fragmentation in Multiphase Flows with SURFER. *Journal of Computational Physics*, Vol. 113, pp. 134–147, 1994.
4. C.W. Hirt and B.D. Nichols. Volume of Fluid (VOF) Method for the Boundaries of Free Surfaces. *Journal of Computational Physics*, Vol. 39, pp. 201–225, 1981.
5. W.J. Rider and D.B. Kothe. Reconstructing volume tracking. *Journal of Computational Physics*, Vol. 141, pp. 112–152, 1998.
6. M. Rieber. *Numerische Modellierung der Dynamik freier Grenzflächen in Zweiphasenströmungen*. PhD thesis, Stuttgart, 2004.
7. M. Hase and B. Weigand. Transient heat transfer of deforming droplets at high Reynolds numbers. *International Journal of Numerical Methods for Heat and Fluid Flow*, Vol. 14, pp. 85–97, 2004.
8. J. Schlottke and B. Weigand. Direct numerical simulation of evaporating droplets. *Journal of Computational Physics*, Vol. 227, pp. 5215–5237, 2008.
9. T. Lund, X. Wu, and D. Squires. Generation of turbulent inflow data for spatially-developing boundary layer simulations. *Journal of Computational Physics*, Vol. 140, No. 1, pp. 233–258, January 1998.
10. N. Kornev and E. Hassel. Method of random spots for generation of synthetic inhomogeneous turbulent fields with prescribed autocorrelation functions. *Communications in Numerical Methods in Engineering*, Vol. 23, No. 1, pp. 35–43, January 2007.
11. H. Gomaa, C. Huber, W. Sander, and B. Weigand. Direkte numerische Simulation des primären Strahlzerfalls von Rundstrahlen. In *Proceedings 9. Workshop über Sprays, Techniken der Fluidzerstäubung und Untersuchung von Sprühvorgängen*, pp. 112–119, Heidelberg, May 2010.
12. W. Sander and B. Weigand. Direct numerical simulation and analysis of instability enhancing parameters in liquid sheets at moderate Reynolds numbers. *Physics of Fluids*, Vol. 20, No. 5, pp. 053301, May 2008.

Numerical Investigation on the Deformation of Droplets in High-Pressure Homogenizers

Kathrin Kissling, Steffen Schütz, and Manfred Piesche

Institute of Mechanical Process Engineering, University of Stuttgart, Böblinger Straße 72, 70199 Stuttgart, Germany, kissling@imvt.uni-stuttgart.de

1 Introduction

Emulsification is an important process both in pharmaceutical and cosmetical industry as well as in the food industry. In emulsions we find either aqueous droplets in oily matrices or vice versa oily droplets in an aqueous matrix. The characteristic of an emulsion is determined by the attributes of the system of dispersed droplets, which are the mean droplet size, the size distribution and the concentration. In this study macroemulsions are considered, with a mean droplet diameter in the range of 1 to 100 μm.

In high-pressure emulsification processes usually valves, nozzles or baffles are applied to force a strong elongational field in the gap. The matrix fluid ejects as a free jet in turbulent or transition flow regime, which causes vortices. Hence additional shear fields can be found downstream the gap. In high-pressure homogenizers different break-up phenomena are involved simultaneously, as there are cavitation, shear and turbulence. The order of importance of the effects is very difficult to rank.

In laminar flow regime the predominant deformation and break-up mechanisms are induced by viscous forces acting on the surface of the droplet. Walstra [10] and Mohr [8] describe the turbulence-induced break-up of the droplets as a two stage process depending on the size ranges of droplets and turbulence eddies. The primary break-up is induced by large vortex structures. The droplet is deformed in the shear field between two vortices and breaks up dependant on the local deformation rates. In analogy to the effects of laminar break-up, break-up only occurs, if the local Capillary number exceeds a flow type dependant critical Capillary number [4]. More recently investigations were carries out by [3] and [6], who focused on the flow field inside the homogenizing device.

Apart from macroscopic phenomena, the properties of the emulsions are influenced by microscopic effects. Due to the dissipation of small vortices, small pressure fluctuations are induced, which lead to a destabilization of the

W.E. Nagel et al. (eds.), *High Performance Computing in Science and Engineering '10*, DOI 10.1007/978-3-642-15748-6_22,

interface of droplets with larger diameters compared to the dimensions of the vortices [7].

Cavitation is defined as the formation of steam bubbles, which appear if the local pressure in the system drops below the saturation pressure. As soon as the local pressure exceeds the saturation pressure the steam bubbles collapse and intensive shear fields, pressure fluctuations, temperature gradients and so called micro jets appear in the vicinity of the cavitation bubbles, which cause destabilization of the interface resulting in the break-up of the droplets [5].

2 Characteristics of the Homogenizer

In this study a baffle high-pressure homogenizer in form of an orifice is examined (see Fig. 1). The diameter ratio rating the outer dimensions and the gap exceeds 10, hence the emerging jet can be treated as a free jet.

3 Governing Equations

The governing equations for the flow of the matrix fluid are the continuity equation and the momentum equation. In a first step incompressible Newtonian fluid flow is assumed. Hence the continuity equation is obtained as

$$\nabla \cdot \mathbf{U} = 0 \tag{1}$$

and the momentum equation as

$$\frac{\partial \rho \mathbf{U}}{\partial t} + \nabla \cdot (\rho \mathbf{U}\mathbf{U}) = -\nabla p + \nabla \cdot \boldsymbol{\tau} + \rho \mathbf{f}. \tag{2}$$

Contrary to other publications [3] no turbulence model is applied and a direct numerical simulation is carried out. Hence the dimensions of the computational control volumes need to be smaller than Kolmogorov's microscale [2]

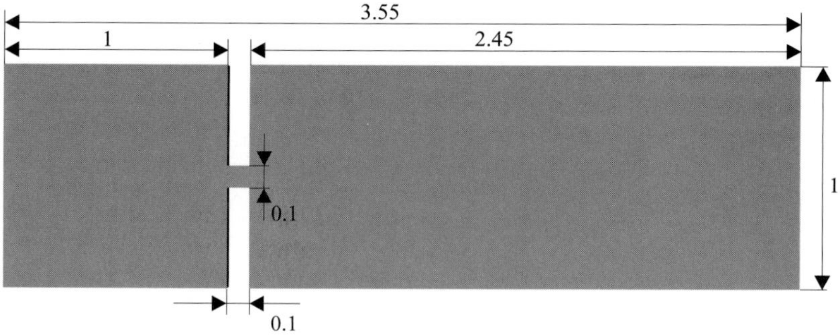

Fig. 1. Schematic representation of the high pressure homogenizer, dimensions in mm

$$l_{kol} = \left(\frac{\mu^3}{\rho^3 \epsilon} \right)^{1/4} \tag{3}$$

where ϵ describes the volume specific energy dissipation rate.

A Volume-of-Fluid method is applied to describe the free surface of the droplets. For each additional phase an additional state variable α is introduced, which describes the volume fraction. The interface is tracked by solving the transport equation for α_i

$$\frac{\partial \alpha_i}{\partial t} + \nabla \cdot (\alpha_i \mathbf{U}) = 0. \tag{4}$$

The volume fractions α_i are coupled by the closure equation:

$$\sum_{i=1}^{N} \alpha_i = 1, \qquad \alpha_i \in [0, 1]\,. \tag{5}$$

For the momentum equation a mixture approach is applied. Hence the phase averaged equation reads

$$\frac{\partial \rho \mathbf{U}}{\partial t} + \nabla \cdot (\rho \mathbf{U}\mathbf{U}) = -\nabla p + \nabla \cdot \boldsymbol{\tau} + \rho \mathbf{f} + \mathbf{F}_\sigma \tag{6}$$

with the volume averaged material properties density and viscosity

$$\rho = \sum_{i=1}^{N} (\alpha_i \rho_i) \tag{7}$$

and

$$\mu = \sum_{i=1}^{N} (\alpha_i \mu_i). \tag{8}$$

The effect of surface tension is implemented in the volume force $\mathbf{F}_\sigma$ according to the Continuous Surface Force approach of Brackbill [1]. Brackbill interpreted the surface tension as a continuous, three-dimensional effect across an interface. This approach is applied here since in interface-capturing methods like the Volume-of-Fluid method the interface is neither tracked explicitly nor shape or location are known. Therefore, an exact boundary condition cannot be applied to the interface. Hence, the influence of the surface tension formulated as volumetric force reads:

$$\mathbf{F}_\sigma = \sigma \kappa \left(\nabla \alpha \right). \tag{9}$$

The curvature is defined as

$$\kappa = \nabla \cdot \left(\frac{(\nabla \alpha)}{|(\nabla \alpha)|} \right), \tag{10}$$

where the unit normal vector $\mathbf{n}$ on the interface yields

$$\kappa = -\nabla \cdot \mathbf{n}. \tag{11}$$

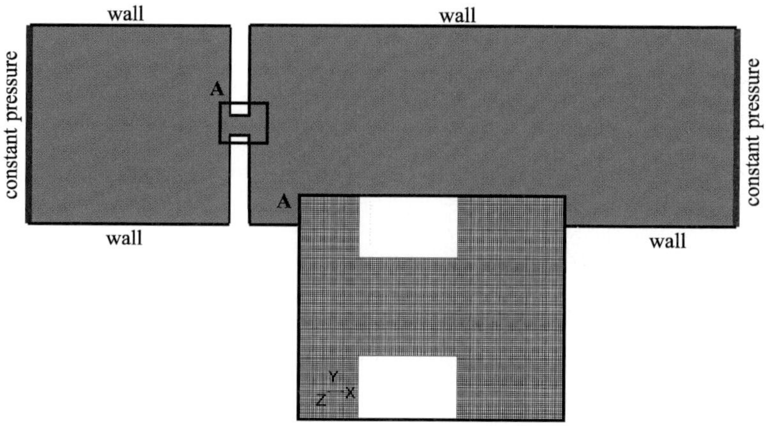

Fig. 2. Computational domain

4 Computational Domain, Grid and Boundary Conditions

The computational grid is shown in Fig. 2. The mesh is created by two to four million finite-volume elements. The dimensions of the cells correspond to the Kolmogorov microscale, dependent on the present pressure drop. The gap was resolved with at least 50 cells across the diameter. To avoid mesh dependency hexahedral, perfectly orthogonal cells were used. In this first step of the project the computational domain was reduced to a 2d model. The applied boundary conditions are shown in Fig. 2.

5 Flow Field

In this section the results of the investigation of the flow field in a high-pressure homogenizer are presented. In contrast to former investigations in literature no turbulence models were applied but a direct numerical simulation was carried out, since the assumptions of most turbulence models like isotropic turbulence are not valid in the narrow gap, where all turbulence is suppressed by the strong acceleration. Moreover if turbulence appears in the wake, it is not fully developed in most homogenizers. The influence of the local vortex structures on the deformation and break-up behaviour of the droplets is crucial. Hence the turbulent or transitional structures have to be resolved. The temporal development of the undisturbed flow field is shown in Fig. 3. While the fluid approaches the gap the flow pattern is predominantly laminar. The streamlines converge at the inlet of the gap. There the flow becomes strongly accelerative and shows strong elongational characteristics. Due to the strong acceleration the flow in the gap remains laminar, but unstable boundary layers can be

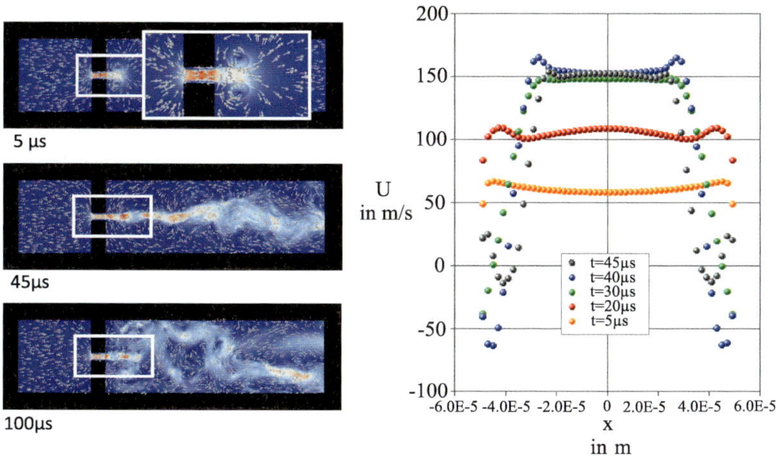

Fig. 3. Temporal development of the velocity field: Pressure difference 100 bar

identified at the walls. In the wake of the gap we find vortex rings, which emerge from the eddy structures in the boundary layers.

If we regard the flow inside the smallest gap at certain time steps velocity profiles can be found, which show two maxima and two or more inflection points. During the start up the profiles have two maxima close to the walls. Then the puls forms a third maximum in the center, with the other maxima still existing. During the next microseconds this behaviour changes rapidly. Now a local minimum can be found in the middle of the gap. Next to the wall unstable boundary layers can be found with strongly developed backflow and distinc shear layers. After 30 microseconds periodically oscillating flow profiles are reached. These flow profiles could not be found in former studies, since RANS turbulence models cannot capture these highly anisotropic phenomena.

A parameter study was performed to rate the influence of the pressure drop on the flow profiles. The pressure drop over the baffle varies from 5 to 800 bar. The corresponding velocity profiles are shown in Fig. 4. At a pressure drop of 5 bar, the profile is homogeneous over the gap width. At a pressure drop of 10 bar the backflow in the boundary layers develop, and velocity maxima appear. These effects get more distinct, the more the pressure drop rises. At high pressure drops the profiles are no longer axisymmetric, due to the highly transient behaviour of the jet. The jet is not statically directed along the gap axis, but starts to fluctuate in orthogonal direction.

Due to the high velocity gradients between jet and ambient fluid, transient or turbulent mixing and entrainment phenomena can be examined. To get a better understanding for these effects the flow type in the homogenizing orifice is evaluated. The corresponding flow parameter λ is defined as

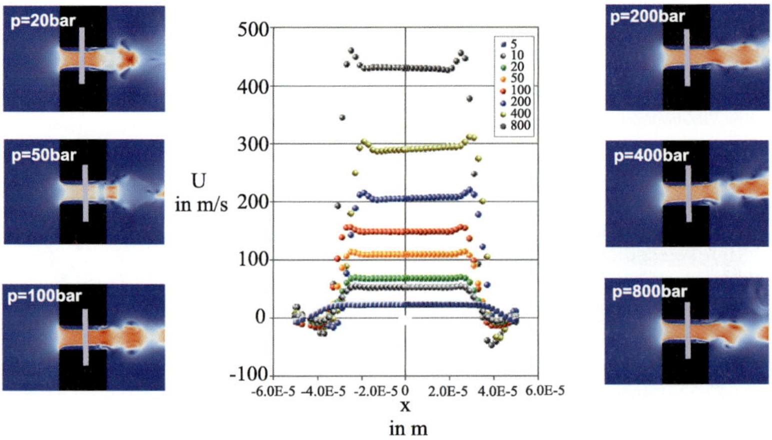

Fig. 4. Velocity profiles in the narrow gap at pressure drops of 5 to 800 bar

$$\lambda = \frac{|\nabla \mathbf{U}| - \left|\nabla \mathbf{U}^T - \nabla \mathbf{U}\right|}{|\nabla \mathbf{U}| + \left|\nabla \mathbf{U}^T - \nabla \mathbf{U}\right|}, \tag{12}$$

and describes the predominant flow type as there are elongational ($\lambda = 1$), shear ($\lambda = 0$) and rotational flow ($\lambda = -1$). For a homogenizing pressure of 100 bar the flow type in the high pressure homogenizer is shown in Fig. 5.

In the inner gap shear-layers can be examined, which result from unstable boundary layers. In the wake of the orifice clearly defined vortex rings can be found. The centers of the vortex rings show a dominant rotational flow pattern. In radial direction the flow field becomes increasingly shear dominated. According to the flow type characteristic deformation mechanisms can be predicted.

In the present paper the mechanisms of deformation and break-up of the droplets described by Walstra and Smulders in 1998 [10] are reconstructed in a qualitative manner. The investigations were carried out with a Volume-

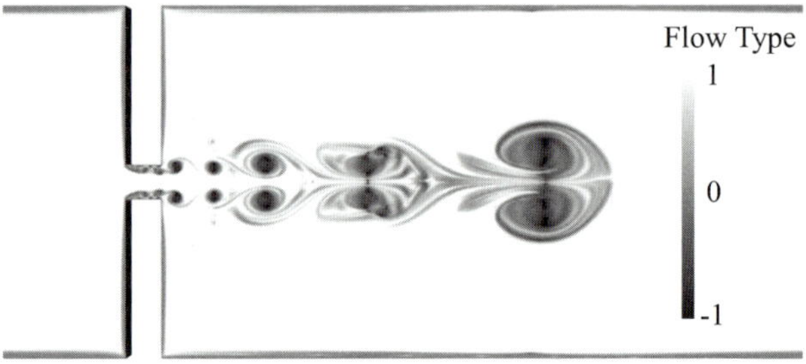

Fig. 5. Representation of the flow type in the orifice. Elongational flow ($\lambda = 1$); Shear flow ($\lambda = 0$); Rotational flow ($\lambda = -1$)

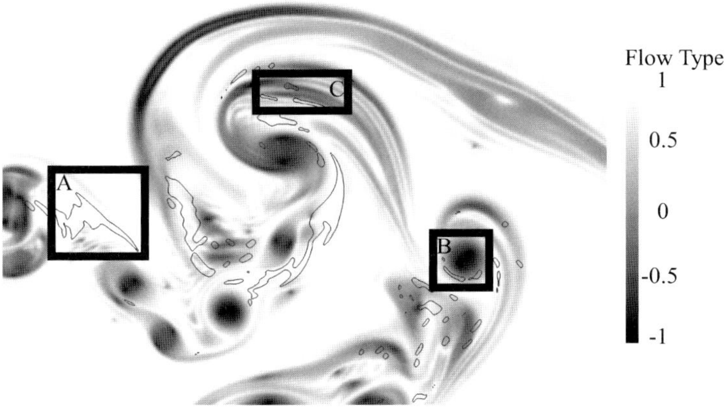

Fig. 6. Deformation and break-up mechanisms in the high-pressure homogenizer. A: Statistical deformation; B: deformation due to shear forces in a vortex system; C: deformation due to shear forces between two vortex systems

of-Fluid method. In Fig. 6 the deformation of a droplet is shown, where the described mechanisms can be distinguished. (A) shows the statistical deformation of the surface of a droplet caused by vortex structures, which have smaller dimensions compared to the diameter of the dispersed element. (B) and (C) show droplets in vortex structures or between two vortex structures, where they experience mainly shear induced deformation and develop the typical s-shape. The results show the strong influence of the vortex structures on the deformation and break-up of the droplets. Hence the importance of the representation of those structures is proved. The results show, that the smoothing behaviour of turbulence models applied in former studies in literature cannot capture these phenomena. Hence the high computational effort is justified.

6 Computational Effort

The computations are performed on the NEC Nehalem Cluster. Typically the computations are parallelized on 16 CPU. The computational effort is especially due to the low time step size of $1 \cdot 10^{-9}$ s, which is needed, to resolve the vortex structures in adequate precision. A typical job runs 72 hours on 16 processors.

7 Summary

In the present paper a direct numerical simulation of a high-pressure homogenizer is shown. Contrary to former investigations, the detailed flow mechanisms could be shown and their influence on the deformation and break-up could be proven. In the future the effects of cavitation need to be included in the model. Moreover the model needs to be enhanced to a 3d model. At the moment a Coupled-Level-Set-Volume-of-Fluid Method [9] is developed and implemented in OpenFOAM and will be applied to model the deformation and break-up of the droplets.

References

1. J.U. Brackbill, D.B. Kothe and C. Zemnach, A Continuum Method for Modelling Surface Tension. *Journal of Computational Physics* **100** (1992), pp. 335–354.
2. J.T. Davies, Turbulence Phenomena. *Academic Press Inc, London* (1972).
3. J. Floury, J. Bellettre, J. Legrand and A. Desrumaux, Analysis of a New Type of High Pressure Homogeniser. A: Study of the Flow Pattern. *Chemical Engineering Science* **59** (2004), pp. 843–853.
4. H.P. Grace, Dispersion Phenomena in High Viscosity Immiscible Fluid Systems and Application of Static Mixers as Dispersion Devices in Such Systems. *Chemical Engineering Communications* **14** (1982), pp 225–277.
5. E.S.R. Gopal, Principles of Emulsion Formation. *In: P. Sherman: Emulsion Science. Chapter 1, Academic Press Inc., London* (1968).
6. A. Hakansson, C. Trägardh and B. Bergenstahl, Studying the effects of adsorption, recoalescence and fragmentation in a high pressure homogenizer next term using a dynamic simulation model. *Food Hydrocolloids* **23** (2009), pp. 1177–1183.
7. J.O. Hinze, Fundamentals of the Hydrodynamic Mechanism of Splitting in Dispersion Processes. *AIChE Journal* **1** (1995), pp 289–295.
8. K.H. Mohr, High-Pressure Homogenization. Part I. Liquid-Liquid Dispersion in Turbulence Fields of High Energy Density. *Journal of Food Engineering* **6** (1987), pp. 177–186.
9. T. Menard, S. Tanguy and A. Berlemont, Coupling level set/VOF/ghost fluid methods: validation and application to 3D simulation of the primary break-up of a liquid jet, *Int. J. Multiphase Flow* **33** (2007), pp. 510–524.
10. P. Walstra, I. Smulders, Emulsion Formulation. *In: B. P. Binks: Modern Aspects of Emulsion Science, Chapter 2. The Royal Society of Chemistry, Cambridge* (1998).

Direct Numerical Simulation of Sediment Transport in Turbulent Open Channel Flow

Clemens Chan-Braun, Manuel García-Villalba, and Markus Uhlmann

Institute for Hydromechanics, Karlsruhe Institute of Technology, 76128 Karlsruhe, Germany, chan-braun@ifh.uka.de

Summary. Direct numerical simulations of interface-resolved sediment transport in horizontal open-channel flow are currently being performed on the XC-4000. The channel bottom boundary is roughened with a fixed layer of spheres and about 9000 particles are allowed to move within the computational domain. The density ratio of the solid and fluid phase is 1.7 and the bulk Reynolds number of the flow is 2880. In the present configuration, the particles tend to accumulate near the bed because of gravity, but due to the turbulent motions, a cycle of resuspension and deposition is produced. This leads to a particle concentration profile which decreases with the distance from the bed. The preliminary results show that the presence of particles strongly modifies the mean fluid velocity and turbulent fluctuation profiles. The dispersed phase lags the carrier phase on average across the whole channel height. Both observations confirm previous experimental evidence. The different observations suggest that particle inertia, finite-size and finite-Reynolds effects together with gravity play an important role in this flow configuration. Several potential mechanisms of turbulence-particle interaction are discussed.

1 Introduction

The erosion, transport and deposition of sediment by turbulent open-channel flow are important mechanisms for fluvial engineering applications. These phenomena are far from being completely understood due to the complex interactions between the turbulent flow and the sediment particles. Turbulent flow in an open-channel is statistically inhomogeneous in the wall-normal direction, and the Reynolds numbers of interest are typically high, which leads to a wide range of velocity and length scales in the flow. In addition, the presence of a range of sediment sizes, shapes and compositions further complicates the description. Sediment particles tend to accumulate near the bed of the channel due to the effect of gravity. However, under certain conditions, the turbulent motions might erode the bed and entrain sediments which can then be transported by the carrier fluid. On average this resuspension and deposition cycle

W.E. Nagel et al. (eds.), *High Performance Computing in Science and Engineering '10*, DOI 10.1007/978-3-642-15748-6_23,

leads to a concentration profile which decreases with the distance from the bed.

In order to improve the understanding of such flows, it is necessary to simplify the problem under consideration, while at the same time retaining the fundamental physics of the original problem. In this study, we consider fully-developed, particle-laden, horizontal, open-channel flow. The particles are spherical, all of them with the same diameter. For this configuration, we have performed direct numerical simulations (DNS) with interface resolution.

In spite of the difficulty of performing experiments in multi-phase flow, similar configurations as the one studied here have been investigated experimentally over the last two decades [1–5]. Most of these experiments were performed using dilute suspensions, with a volumetric concentration of order $\mathcal{O}(10^{-3})$, and various kinds of sediments. For example, in the study of [2] the density ratio between the solid and fluid phase was very high since they employed glass and PVC particles in air. The density ratio was lower in the rest of the studies, ranging from values as low as 1.05 in the experiments of [1] (polystyrene particles in water) to 2.6 in [3] and [4] (both used glass particles in water), while [5] covered the range 1.03-2.65 using natural sand and neutrally-buoyant sand (crushed nylon) in water. These experiments have shown the existence of a velocity lag (in the mean) between particle and fluid velocities, and modifications to the turbulence characteristics of the flow due to the presence of the suspended particles.

Computational studies are only now beginning to appear, with the exception of the pioneering work of [6], who conducted resolved DNS of turbulent particulate flow in a horizontal channel using 160 stationary and mobile particles.

Due to computational limitations, our present study does not exactly correspond to any of the experiments mentioned above. We have chosen parameter values that are comparable to those of the study of [3], with somewhat lower Reynolds number and density ratio, and in particular a substantially larger solid volume fraction. Details of our configuration are given below.

The objective of the present study is to investigate the effects observed in previous experiments with the aid of numerical simulations. In particular we are interested in the lag between average particle and fluid velocities, the modification of fluid turbulence properties due to the dispersed phase, the particle concentration profile and the particle velocity fluctuations.

2 Numerical Method

The present simulations have been carried out with the aid of a variant of the immersed boundary technique [7, 8] proposed by [9]. This method employs a direct forcing approach, where a localized volume force term is added to the momentum equations. The additional forcing term is explicitly computed at each time step as a function of the desired particle positions and velocities, without recurring to a feed-back procedure; thereby, the stability characteristics of the underlying Navier-Stokes solver are maintained in the presence of particles, allowing for relatively large time steps. The necessary interpolation of variable values from Eulerian grid positions to particle-related Lagrangian positions (and the inverse operation of spreading the computed force terms back to the Eulerian grid) are performed by means of the regularized delta function approach of [7, 8]. This procedure yields a smooth temporal variation of the hydrodynamic forces acting on individual particles while these are in arbitrary motion with respect to the fixed grid.

Since particles are free to visit any point in the computational domain and in order to ensure that the regularized delta function verifies important identities (such as the conservation of the total force and torque during interpolation and spreading), a Cartesian grid with uniform isotropic mesh width $\Delta x = \Delta y = \Delta z$ is employed. For reasons of efficiency, forcing is only applied to the surface of the spheres, leaving the flow field inside the particles to develop freely.

The immersed boundary technique is implemented in a standard fractional-step method for incompressible flow. The temporal discretization is semi-implicit, based on the Crank-Nicholson scheme for the viscous terms and a low-storage three-step Runge-Kutta procedure for the non-linear part [10]. The spatial operators are evaluated by central finite-differences on a staggered grid. The temporal and spatial accuracy of this scheme are of second order.

The particle motion is determined by the Runge-Kutta-discretized Newton equations for translational and rotational rigid-body motion, which are explicitly coupled to the fluid equations. The hydrodynamic forces acting upon a particle are readily obtained by summing the additional volume forcing term over all discrete forcing points. Thereby, the exchange of momentum between the two phases cancels out identically and no spurious contributions are generated. The analogue procedure is applied for the computation of the hydrodynamic torque driving the angular particle motion.

During the course of a simulation, particles can approach each other closely. However, very thin inter-particle films cannot be resolved by a typical grid and therefore the correct build-up of repulsive pressure is not captured which in turn can lead to possible partial 'overlap' of the particle positions in the numerical computation. In practise, we use the artificial repulsion potential of [11], relying upon a short-range repulsion force (with a range of $2\Delta x$), in order to prevent such non-physical situations.

The present numerical method has been submitted to exhaustive validation tests [9, 12–14], as well as grid convergence studies [15]. The computational code has been applied to the case of vertical plane channel flow [16].

Table 1. Parameters of the direct numerical simulation

Re_b	$= 2880$	Re_τ	$= 208$
U_{bh}/u_τ	$= 13.9$	D^+	$= 10$
ρ_p/ρ_f	$= 1.70$	H/D	$= 21$
St^+	$= 9.7$	St_b	$= 0.64$
Φ_s	$= 1.4\%$	gh/U_{bh}^2	$= 0.70$
$D/\Delta x$	$= 6$	Δx^+	$= 1.7$

3 Setup of the Simulation

The flow configuration consists of turbulent open-channel flow with suspended freely-moving spherical particles over a rough wall. The rough wall is formed by a layer of fixed spheres which are tightly packed in a cubic arrangement (see Fig. 1 left). The layer of fixed spheres consists of $Np_x = 192$ particles in streamwise and $Np_z = 48$ particles in spanwise direction. The spacing between the particle centres is $D + 2\Delta x$, where D is the sphere diameter and Δx is the grid spacing. Below this layer of spheres a rigid wall is located at $y = 0$. As can be seen in Fig. 1 left this rigid wall is roughened by spherical caps that can be defined by the part above $y = 0$ of spheres located at $y = D/2 - \sqrt{2}(D/2 - \Delta x)$ in a cubical arrangement, staggered to the fixed layer of spheres above.

Periodic boundary conditions are applied in streamwise and spanwise directions. At the upper boundary a free-slip condition is employed. At the bottom boundary and on the surface of the fixed spheres a no-slip condition is applied.

Table 1 summarises the numerical and physical flow parameters. The computational domain dimensions are $L_x/H \times L_y/H \times L_z/H = 12 \times 1 \times 3$, in streamwise, wall-normal and spanwise directions, respectively. An equidistant Cartesian grid was used throughout the entire domain using $1536 \times 129 \times 384$ grid points. The fixed as well as the moving particles are of spherical shape and have a diameter of $D/H = 4.69 \cdot 10^{-2}$. The global solid volume fraction, Φ_s, was set to $\Phi_s = 1.4\%$, which corresponds to 9216 particles. In order to scale the results, two quantities need to be specified: the friction velocity u_τ and the virtual wall location y_0, since for a rough wall, the real position of the wall cannot be unambiguously defined. In the present work we define the position of the virtual wall as $y_0 = 0.8D$, measured from the bottom of the computational domain. This value has been used often in the past [17] and the results are not very sensitive to small variations. Once y_0 was specified, u_τ was obtained using a linear fit of the total shear-stress profile. Knowing the position of the virtual wall the effective flow depth, h, can be defined as the distance from the virtual wall to the top boundary, $h = H - y_0$. The bulk velocity based on the domain height H is defined as $U_{bH} = 1/H \int_0^H \langle U \rangle dy$. Hereby, U is the time averaged streamwise velocity component and angle brackets denote spatial averaging along the homogeneous directions. The bulk Reynolds number,

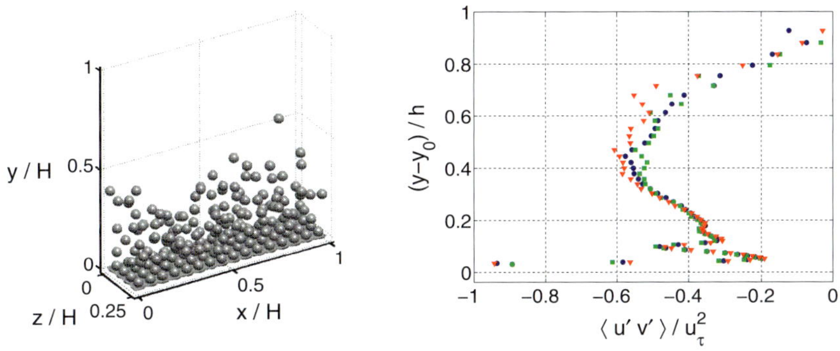

Fig. 1. Left: Close-up of the computational domain. Both fixed and moving particles are shown at a given instant; right: Wall-normal profile of the streamwise and wall-normal particle velocity cross-correlation $\langle u'_p v'_p \rangle / u_\tau^2$. Averaging was performed over different time intervals: dots, 8 flow through times; squares, 4 flow through times; triangles, 2 flow through times

$Re_b = U_{bH} H/\nu$, was kept constant at a value of 2880. This corresponds to a friction Reynolds number, $Re_\tau = u_\tau h/\nu \simeq 180$ in case of a smooth wall. In the present two-phase flow, the friction Reynolds number obtained $Re_\tau = u_\tau h/\nu$ is 208. Further quantities that can be derived are the particle diameter in wall-units $D^+ = u_\tau D/\nu = 10$, the Stokes number based on the particle time scale and the friction time scale, $St^+ = 9.7$, and the Stokes number based on the particle time scale and the outer time scale $t_b = h/U_{bh}$, $St_b = 0.64$. The non-dimensional gravity acceleration was set to $gh/U_{bh}^2 = 0.79$.

The grid resolution of the simulation in terms of wall units was $\Delta x^+ = 1.7$. This is a very high resolution away from walls, however it is marginal very close to the wall and very close to the surface of the spheres. Thus the present simulation should be considered as marginally resolved and a finer resolution is desirable. The time step was chosen such that the CFL number remains below 0.3.

The initial turbulent flow field was taken from a simulation of open-channel flow over a rough wall without moving particles. Then, particles were added at random positions and assigning them an initial translational velocity equal to the local fluid velocity. Subsequently, the simulation was run until the flow reached a statistically stationary state. The simulation was continued for 120 H/U_{bH}, which corresponds to 10 flow-through times, during which flow field statistics were sampled at every second time step and particle statistics were stored every tenth time step. This leads to 9000 samples of instantaneous flow and $4.9 \cdot 10^7$ samples of particle data from 1850 instantaneous fields. Based on this data the statistics were computed that are discussed in the following.

Due to the gravity effect, once the steady state is reached the concentration of particles is much higher close to the wall than in the upper part of the channel (see Fig. 1 left and Fig. 2). To provide a measure for the convergence of the disperse flow statistics, Fig. 1 right displays the wall-normal profile of the

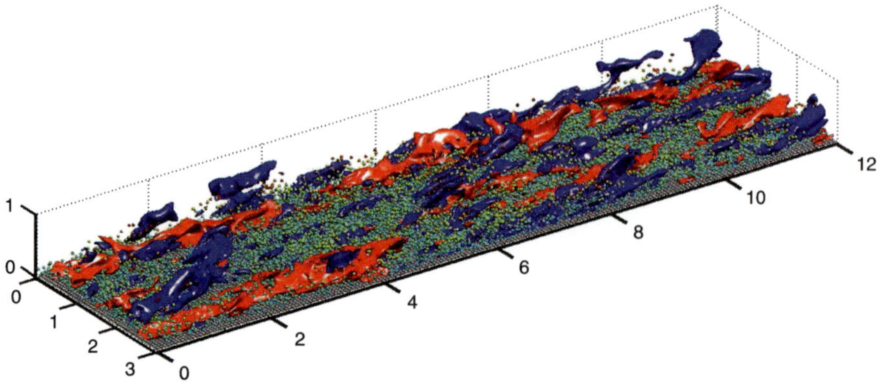

Fig. 2. Entire computational domain, moving particles clustered in the near wall region due to gravity, high-speed and low speed streaks in blue and red colour visualised by iso-contours of constant velocity fluctuation

cross-correlation between the streamwise and wall-normal particle velocities, $\langle u'_p v'_p \rangle$ computed over different time intervals. The statistics of the disperse phase below $y/h \approx 0.3$ can be regarded as steady and well converged. Above $y/h \approx 0.3$ the scatter of the data increases. The latter region is visited by few particles and therefore to obtain well-converged particle statistics would require significantly longer integration times.

4 Computational Costs and Efficiency

The simulation presented in this report was run on 64 processors and run for approximately $2 \cdot 10^5$ time steps. Each time step required in average 35 seconds of clock time, which corresponds to a total cost of approximately 120 000 CPU hours and a clock time of 3 month. As has been stated above a considerable longer integration time will be required for a full convergence of the statistics. Furthermore, a simulation with a higher spatial resolution is of great importance, as will be discussed later. The computational effort of such a simulation is reduced as the transient period will be considerably shorter. A simulation with a spatial resolution a factor of 2 higher in each direction is currently running on 256 processors and requires 80 seconds per time-step on average. This result is in agreement with the expected good scaling behaviour of the code found on several high-performance computers.

5 Results and Discussion

Figure 2 shows a snapshot of the flow field illustrating the complexity of the phenomena involved. Particles are coloured with the distance to the wall and the flow is visualized using iso-surfaces of positive (negative) streamwise

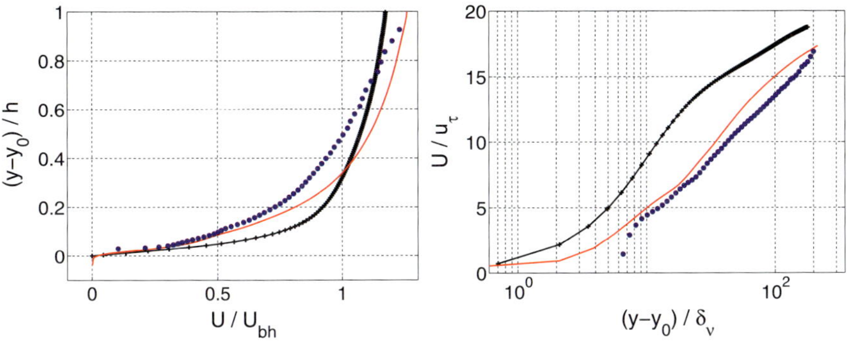

Fig. 3. Wall-normal profile of the mean streamwise velocity distribution of the fluid phase (solid line) and the solid phase (dots) in comparison with a single phase open channel flow over a smooth wall (solid line markers +); left: normalised by U_{bh}; right: normalised by u_τ

velocity fluctuations in red (blue). Particles accumulate close to the wall but occasionally, due to the large scale turbulent motion, particles might reach the upper part of the domain.

In the following discussion the results of the two-phase flow are compared with the results of a single-phase open channel over a smooth bed at the same bulk Reynolds number $Re_b = 2880$ which corresponds to a friction Reynolds number of $Re_\tau = 182$. This simulation has been performed with the in-house code LESOCC2 [18, 19].

The mean streamwise velocity profile of the fluid phase and the disperse phase are shown in Fig. 3 in comparison with the single-phase smooth-wall simulation. The first observation when comparing the profiles of the fluid phase is that the effect of the disperse phase is similar to the effect of roughness, i.e. the velocity profile in Fig. 3 right is shifted towards lower values. However, in addition to this, the gradient of the velocity profile decreases near the wall in the present situation, which is untypical for flow over roughness. A change of the gradient is also in contrast to the observations in previous experiments of two-phase flow in a horizontal channel [1, 3]. This effect might stem from the low Reynolds number considered, the lower ratio h/D used in the present simulation or the large concentration of particles in the near-wall region. Other explanations are also possible. Both Figures show that there is a mean velocity lag between solid and fluid phase over the whole flow depth. A lower velocity of the disperse phase has been observed by several previous experiments over a range of Stokes numbers from St_b of order 0.1 in [3] up to Stokes numbers of order 1 in [2]. It was also found to occur in flows with different density ratios; from nearly neutrally buoyant in the case of [1] to high density ratios in [2]. Furthermore, it has been observed also over a range of different solid volume fraction from as little as 0.023% in [1] up to 2.2% in the present case. There could be several explanations for this phenomenon. It

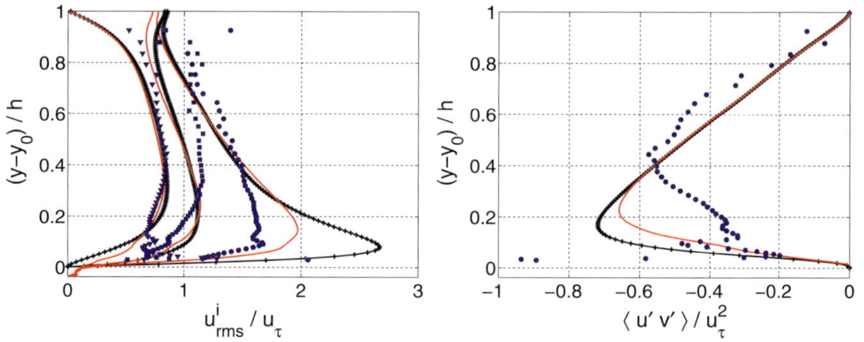

Fig. 4. Left: Wall-normal profile of root-mean square velocity fluctuations normalised by u_τ of the fluid phase (solid line) and the solid phase (solid markers) in comparison with a single phase open channel flow over a smooth wall (solid line markers +); from left to right: wall-normal, spanwise and streamwise fluctuations; right: Wall-normal profile of cross-correlation of streamwise and wall-normal velocities (Reynolds stress) normalised by u_τ^2 of the fluid phase (solid line) and the solid phase (solid markers) in comparison with a single phase open channel flow over a smooth wall (solid line with markers)

might be explained by the fact that particles close to the wall, and therefore with low momentum, are ejected by bursts to the outer region but retaining their low momentum for some time. Another possibility is that the driving pressure-gradient might act differently on fluid and particles.

Figure 4 left shows the velocity fluctuations of the two phases in comparison with the fluctuations in the smooth wall simulation. For the statistics of the fluid phase the strongest effect of the dispersed phase can be seen in the decrease of the near-wall peak of the streamwise component u_{rms}. The profiles of the fluctuations in wall-normal and spanwise direction differ only marginally from the profiles of the smooth wall simulation. This is in agreements with the findings of [3] who, however, observed a less pronounced drop of the near-wall peak in the streamwise fluctuations. This might be due to the smaller volume fraction in their experiment. Above $y/h = 0.4$ the single and two-phase fluid statistics overlap except for the region close to the upper boundary. Here, the streamwise and spanwise fluctuations are smaller compared to the single phase case while the wall-normal component again overlaps with the single phase case. This could indicate that the difference between the single-phase flow and the two-phase flow in this region can be found in the large-scale turbulent structures rather than in the small scale structures.

We turn now our attention to the statistics of the disperse phase. Near the wall, the fluctuations of the streamwise and spanwise components are smaller than the fluid counterparts, while the wall-normal component is larger, i.e. the particle fluctuation energy is more isotropic than the fluid turbulent kinetic energy. As mentioned before, in this region there is a higher concentration of particles, and therefore the collision rate is increased. As a consequence a more isotropic distribution is expected. [1] have also observed that the fluctuations of the wall-normal particle velocity are larger than the fluid fluctuations near the wall, while [3] did not observe this effect. Conversely, in the outer region (above $y/h \geq 0.4$) all particle velocity fluctuation components are larger than the fluid counterparts and exhibit smaller wall-normal gradients. This can possibly be explained by the increased inertia of the disperse phase.

In Fig. 4 right the profiles of the cross-correlation between streamwise and wall-normal velocity fluctuations (Reynolds stress) for both solid and fluid phases is shown. The statistics for the fluid phase above $y/h = 0.3$ collapse with the single-phase data. In this region the disperse phase seems to have only little effect on the distribution of the Reynolds stress. Below $y/h = 0.3$ the Reynolds stress of the two-phase flow is smaller. This could be due to some sort of interference between the particles and the buffer layer structures, the latter could be somewhat damped by the presence of the former leading to a lower Reynolds stress. Comparing the disperse phase cross-correlation and the fluid Reynolds stress, it is clear that a strong deviation can be found between $\langle u'v' \rangle$ and $\langle u'_p v'_p \rangle$ below $y/h \sim 0.4$. Visibly, the particle motion in the streamwise and wall-normal directions is less correlated than the fluid motion. Different explanations are possible, in particular the well-known crossing-trajectories effect of heavy particles in turbulent flow [20]. This observation deserves further consideration.

Finally, we discuss the vertical distribution of particles in the channel. It is customary in the literature to study equilibrium suspensions by using an equation for the concentration of the dispersed phase. This can be derived from a convection-diffusion equation of the concentration [21]. Further assumptions are the use of a logarithmic law of the wall for the mean streamwise velocity distribution of the fluid phase and an isotropic eddy viscosity model for the turbulent concentration flux. This later assumption is particularly questionable in light of the higher particle inertia. These hypotheses lead to the following law for the concentration of the dispersed phase

$$\frac{\langle C \rangle}{C_a} = \left[\frac{(h-y)/y}{(h-a)/a} \right]^{\beta} . \tag{1}$$

Hereby, $\beta = v_s/\kappa u_\tau$, where v_s is the 'nominal' settling velocity of the particle given by a balance between drag and immersed weight, using the standard drag formula [22], $\kappa \approx 0.41$ the von Kármán constant and u_τ the friction velocity. In (1) the quantity C_a is a near-bed reference concentration measured at a distance $y - y_0 = a$, where a cannot be zero because the expression becomes singular. Here we have selected a value of a which corresponds roughly to three particle diameters ($a/h = 0.15$), leading to $C_a = 0.0533$. (1) is often called the Rousean distribution for suspended sediment.

Figure 5 shows the concentration of particles versus the vertical coordinate normalized by the concentration C_a, together with the evaluation of the Rousean distribution for two different β values. The first one as given by its

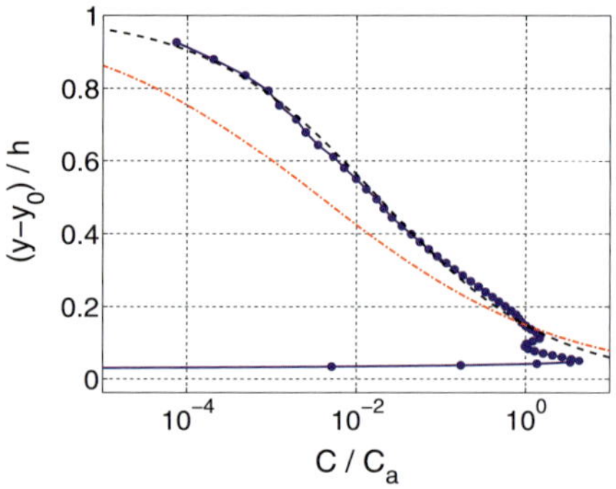

Fig. 5. Wall-normal profile of particle concentration (solid line with markers) in comparison with Rouse formula, (1), using the original definition of β (dash-dotted line) and using a 40% lower value of β (dashed line)

definition, $\beta = v_s/\kappa u_\tau$, does not provide a good approximation for the present particle distribution. This might be due to the high degree of simplification inherent in the formula, in particular the modeling of the vertical turbulent concentration flux by means of a simple eddy viscosity. A further evaluation with an optimized exponent (40 % lower value of β) leads to a remarkable fit with the present DNS data for $y/h \geq 0.12$.

6 Conclusions

We have performed interface-resolved direct numerical simulation of sediment transport in horizontal open channel flow. In our configuration the wall boundary was roughened by means of fixed spherical particles. The simulated parameter values were comparable to those of the study of [3], with the main exception of a substantially larger solid volume fraction.

Our results show that the presence of particles strongly modifies the mean fluid velocity profile, leading to lower gradients in the immediate vicinity of the wall and higher gradients somewhat outside of the region exhibiting the highest particle concentration. The dispersed phase, on the other hand, lags the carrier phase on average across the whole channel height (here we refer to the difference in the mean of the streamwise component of the velocities corresponding to the two phases). The fluid velocity fluctuations are similar to the single-phase reference data, except for the streamwise component, which is strongly damped in the buffer layer. Particle velocity fluctuations are smaller than the fluid counterpart near the wall and larger in the outer flow. The fluid Reynolds stress is found to be damped near the wall and essentially unchanged in the outer flow. The particle "Reynolds stress" is much lower than the fluid counterpart in the near-wall region, and both are similar outside.

Finally, we have compared the concentration profile with the Rousean distribution. Although the original formula already involves a free-to-choose parameter (the location of the reference point), we need to adapt the exponent (given as a function of the settling velocity, the friction velocity and the Karman constant) in order to obtain a good fit of the DNS data.

The different observations suggest that particle inertia, finite-size and finite-Reynolds effects together with gravity play an important role in this flow configuration. Several potential mechanisms of turbulence-particle interaction have been discussed in the text. It is our intention to gather the additional statistical data which is necessary to verify those hypotheses in the future. More urgently, however, we need to repeat the current simulation at increased spatial resolution, which is slightly marginal. Furthermore, it needs to be established whether the details of the numerical treatment of particle collisions affect the results significantly.

Acknowledgements. This work was supported by the German Research Foundation (DFG) under project JI 18/19-1. The computations have been carried out at the Steinbuch Centre for Computing (SCC) of Karlsruhe Institute of Technology.

References

1. D. Kaftori, G. Hetsroni, and S. Banerjee. Particle behavior in the turbulent boundary layer. II. Velocity and distribution profiles. *Phys. Fluids*, 7(5):1107–1121, 1995.
2. A. Tanière, B. Oesterlé, and J.C. Monnier. On the behaviour of solid particles in a horizontal boundary layer with turbulence and saltation effects. *Exp. Fluids*, 23(6):463–471, 1997.
3. K.T. Kiger and C. Pan. Suspension and turbulence modification effects of solid particulates on a horizontal turbulent channel flow. *J. Turbulence*, 3:019, 2002.
4. M. Righetti and G.P. Romano. Particle-fluid interactions in a plane near-wall turbulent flow. *J. Fluid Mech.*, 505:93–121, 2004.
5. M. Muste, K. Yu, I. Fujita, and R. Ettema. Two-phase flow insights into open-channel flows with suspended particles of different densities. *Environ. Fluid Mech.*, 9:161–186, 2009.
6. Y. Pan and S. Banerjee. Numerical investigation of the effects of large particles on wall-turbulence. *Physics of Fluids*, 9(12):3786–3807, 1997.
7. C. S. Peskin. *Flow patterns around heart valves: A digital computer method for solving the equation of motion.* PhD thesis, Albert Einstein College of Medicine, 1972.
8. C.S. Peskin. The immersed boundary method. *Acta Numerica*, 11:479–517, 2002.
9. M. Uhlmann. An immersed boundary method with direct forcing for the simulation of particulate flows. *J. Comput. Phys.*, 209(2):448–476, 2005.
10. R. Verzicco and P. Orlandi. A finite-difference scheme for three-dimensional incompressible flows in cylindrical coordinates. *J. Comput. Phys.*, 123:402–414, 1996.

11. R. Glowinski, T.W. Pan, T.I. Hesla, and D.D. Joseph. A distributed Lagrange multiplier fictitious domain method for particulate flows. *Int. J. Multiphase flow*, 25(5):755–794, Aug 1999.
12. M. Uhlmann. New results on the simulation of particulate flows. Technical Report 1038, CIEMAT, Madrid, Spain, 2003. ISSN 1135–9420.
13. M. Uhlmann. An improved fluid-solid coupling method for DNS of particulate flow on a fixed mesh. In M. Sommerfeld, editor, *Proc. 11th Workshop Two-Phase Flow Predictions*. 2005.
14. M. Uhlmann. Direct numerical simulation of sediment transport in a horizontal channel. Technical Report 1088, CIEMAT, Madrid, Spain, 2006. ISSN 1135-9420.
15. M. Uhlmann. Experience with DNS of particulate flow using a variant of the immersed boundary method. In P. Wesseling, E. Oñate, and J. Périaux, editors, *Proc. ECCOMAS CFD*. Egmond and Zee, The Netherlands, 2006.
16. M. Uhlmann. Interface-resolved direct numerical simulation of vertical particulate channel flow in the turbulent regime. *Phys. Fluids*, 20(5):053305, 2008.
17. I. Nezu and H. Nakagawa. *Turbulence in Open-Channel Flows*. IAHR/AIRH Monograph Series, Balkema Publishers, Rotterdam, 1993.
18. M. Breuer and W. Rodi. Large eddy simulation of complex turbulent flows of practical interest. In E.H. Hirschel, editor, *Flow simulation with high performance computers II*, volume 52 of *Notes on Numerical Fluid Mechanics*, pages 258–274. Vieweg, Braunschweig, 1996.
19. C. Hinterberger. *Dreidimensionale und tiefengemittelte Large-Eddy-Simulation von Flachwasserströmungen*. Ph.D. thesis, University of Karlsruhe, Institute for Hydromechanics, 2004.
20. M.I. Yudine. Physical consideration on heavy-particle diffusion. *Adv. Geophys.*, 6:185–191, 1959.
21. M. García. *Sedimentation engineering: processes, measurements, modeling, and practice*. American Society of Civil Engineers (ASCE), Reston, Va., 2008. ASCE Manual of Practice 110.
22. R. Clift, J.R. Grace, and M.E. Weber. *Bubbles drops and particles*. Academic Press, New York, 1978.

Grid Sensitivity of LES Heat Transfer Results of a Turbulent Round Impinging Jet

Sven Olaf Neumann[1], Naseem Uddin[1,2], and Bernhard Weigand[1]

[1] Institut für Thermodynamik der Luft- und Raumfahrt, Universität Stuttgart, Pfaffenwaldring 31, Stuttgart 70569, Germany
Olaf.Neumann@itlr.uni-stuttgart.de

[2] Mechanical Engineering Department, NED University of Engineering and Technology, Karachi 75270, Pakistan

Summary. Impinging jets are used in a variety of engineering applications like in chemical reactors, mixing devices, drying and cooling applications. Good quality simulations of this highly complex flow field is a challenging task. In this work, the flow field and heat transfer of turbulent jet impingement is investigated by Large Eddy Simulation (LES). The benchmark case of a turbulent impinging jet with out swirl and excitation, recommended by ERCOFTAC, is simulated on different grids. The jet's Reynolds number (Re) is 23000 and jet outlet-to-target wall distance (H/D) is 2. The agreement between experimental data and simulation results was improved by grid refinement in the free shear layer region behind the nozzle lip. The correlation between the heat transfer mechanism, flow kinematics and turbulence quantities was investigated.

1 Introduction

The high performance of gas turbines depends in part on the efficient cooling of the turbine blades. The leading edge of the blades is internally cooled through the impingement of cold turbulent jets. In order to control the heat transfer on the impingement wall, the knowledge of the interaction of the large-scale structures with each other and with the wall are key factors. In order to understand the role of vortical structures in heat transfer enhancement, [33–35] used a statistical technique with conditional sampling for the study of impingement phenomenon. They found that the large vortical structures increased the heat transfer. They proposed that the striking large scale structures cause the surface renewal effect, which results in the changes in the thermal field. [45] reported that the large vortical structures cause local augmentation of convective heat transfer. According to [13] it is the competition between the vortex rings and secondary vortices induced by them on the impinging wall that determines the wall surface temperature. [49] used Particle Image Velocimetery (PIV) and Laser Induced Fluorescence (LIF) to

W.E. Nagel et al. (eds.), *High Performance Computing in Science and Engineering '10*, DOI 10.1007/978-3-642-15748-6_24,

measure time dependent velocity and temperature distributions. It was found that the amplification of vorticity in the stagnation region was the main source of enhancement of heat transfer in impinging jets. An interesting inquiry in the impinging jet phenomenon is the reason of appearance of secondary peak in the Nusselt number radial distribution. It has been found that the heat transfer for an impinging jet is greatly affected by the jet outlet-to-wall distance (H/D). For H/D<4, prominent peaks appear in the radial distribution of Nusselt number. [27] found that the occurrence of secondary peaks depends not only on jet outlet-to-wall spacing but also on the turbulence level in the jet. He also suggested that the secondary peak is due to transition from free jet into wall jet region. [40] attributed the occurrence of the secondary peak to the higher turbulence in the boundary layer due to flow acceleration and intense shear between the radially ejecting jet and stagnant ambient. [22] have conducted an experimental study of the evolution of large scale structures in an impinging jet. They pointed out the difficulty in characterising this highly complex flow field by measurements and visualisation studies. The most important region for heat transfer is the near wall region, which unfortunately is difficult to approach via present experimental techniques. In the past there have been numerous numerical investigations of this phenomenon based on Reynolds Avergared Navier-Stokes Equations (RANS), however, computational resources have grown tremendously in the last years. Therefore, the characterisation of the complex turbulent flow through adequately resolved time dependent simulations of even the small scales at realistic Reynolds numbers became possible for distinct configurations. They can inherently enable a more universal representation of complex turbulent flows than calculations via RANS turbulence models. Nevertheless limited computational resources and the demand for an applicable time of turnover for numerical simulations entails often the use of RANS models at an acceptable or even good level of accuracy. Recently, [60] presented the review of turbulence models applied to impinging jet heat transfer. The traditional RANS simulation hardly capture all important flow features of turbulent jet very well, as they are limited due to representation of only the mean flow behaviour. Large Eddy Simulation studies provide both statistical and structural information of the flow field. [48] carried out a pioneering work of a LES of impinging jets. The focus of their work was on the dynamical interaction of flow structures in two jets impingement. [14, 15] investigated the mechanisms by which thermal eddies are generated and transported into the impingement region in order to understand the thermal interaction between the fluid and solid, by LES. The Reynolds number of the investigation was 6500. [44] used LES with two different subgrid scale (SGS) models in order to investigate the physical behaviour of a jet at Re=10000 in the wall region. [8] and [9] investigated the single and multiple plane impinging jets using LES at a Reynolds number of 5800. [51, 52] investigated plane and circular jets with periodic inlet conditions at a Reynolds number of 6000. Recently, [23] have done a very detailed Large Eddy Simulation of an impinging jet at Re=20000 for H/D=2, using unstructured grids, which can certainly serve as a benchmark case.

The fundamental goal of the present investigation is to understand the complex dynamics of passive scalar field in impingement and wall jet zones and to assess the grid sensitivity of the predictions. For this purpose, the Large Eddy Simulation (LES) of a single impinging turbulent cold jet has been done at a Reynolds number (Re $= DU_b/\nu$) of 23000 and H/D =2, using two different structured grids (Table 2). In the next section the numerical procedure, the problem setup and the resolution requirements are discussed. After that the results of flow field are presented, followed by the scalar-flux investigations. Understanding the basics of jet impingement phenomena with scalar transport creates the basis for subsequent investigations when adding further complexity to the turbulent impinging jet by pulsation or swirl at the inlet which can be found in engineering applications. It is important to understand the influence different jet modes on the heat transfer mechanism. In most of the experimental studies the jet issues from the converging nozzles, in which the issuing jet has a flat velocity profile with a thin boundary layer near the nozzle lip. The most important jet mode in this case is the shear layer mode. The most amplified frequency in the jet scaled with the initial momentum layer thickness (δ^{**}) and jet exit velocity, gives the range of Strouhal number for this mode as [28].

$$St_{\delta^{**}} = \frac{f^{'}\delta^{**}}{U_o} \tag{1}$$

where $f^{'}$ is the most amplified frequency and U_o is mean the velocity. However, if the jet emerges in a fully developed flow state from the pipe, the momentum thickness is of the order of the jet radius and the jet does not exhibit the shear-layer mode, instead it will attain the jet column mode. The mode represented by Strouhal number (scaled with diameter) is called jet column mode or preferred mode. The preferred mode of the jet is defined as:

$$St_D = \frac{f_o D}{U_o} \tag{2}$$

Gutmark and Ho [20] have found that the preferred mode depends upon the initial conditions and can occur in the range of Strouhal number from 0.3 to 0.85. The simulation conditions are tabulated below (Table 1), the Reynolds number corresponds to the axial bulk velocity at the inlet. In this report results are only presented for non-excited or non-swirling jets. However, Large Eddy Simulations for configurations with a specific inlet velocity exitation which are important in technical applications have been studied sucessfully in the past [53–57] but results will not be included in this report.

Table 1. LES simulations

Cases	Modulation	Re	Results included
Case-0-I	No-modulation	23000	yes
Case-0-II	No-modulation	23000	yes
Case-EV-I	Subharmonic excitation ($f_{n/2}$)	23000	no
Case-EV-II	Preferred mode excitation (f_n)	23000	no
Case-EV-III	Harmonic excitation (f_{2n})	23000	no
Case-EV-IV	Higher harmonic excitation (f_{5n})	23000	no
Case-S-I	Weak swirl (S=0.2)	23000	no
Case-S-II	Strong swirl (S=0.47)	23000	no

2 Numerical Procedure and Setup

A cylindrical domain is used for the investigation of the impinging jet. The jet issues from a pipe and strikes at the centre of the circular target wall. A top confinement wall is used with no slip boundary condition. The wall jet forms as a result of impingement spreads over the target wall and flow leaves the domain from the outlet circumferentially. The computational setup resembles closely with the experimental setup utilized in number of experiments (see [3, 38]). Figure 1 shows a schematic diagram of the impinging jet phenomenon. The temperature at the pipe inlet is T_o. The temperature attained by the target wall after the jet impingement is T_w. A constant heat flux (q_w) is applied at the target wall. The size of domain must be sufficiently large to capture all relevant large-scale structures. The computational domain consists of a cylindrical domain connected to a circular pipe as shown in Fig. 1. The length of the pipe is $2\pi D$, where, D is it's diameter. The pipe outlet-to-wall distance (H) is 2D. The cylindrical domain is $5\pi D$ in diameter. The local heat transfer coefficient is defined as:

$$h = \frac{q_w}{< T_w > - T_o} \tag{3}$$

where, T_o is jet inlet temperature, T_w is the target wall temperature and q_w is the heat flux at the wall, <> indicates averaging. The Nusselt number is defined as:

$$Nu = \frac{< h > D}{k} \tag{4}$$

3 Computational Details

3.1 Numerical Grid Resolution

The geometry of the computational domain used for the jet impingement simulation is shown in Fig. 1. The domain consists of a circular impingement zone and a circular pipe of length 6D. The target to wall distance (H) is 2D. The domain is confined by an adiabatic, no-slip wall at the top. Also the boundary

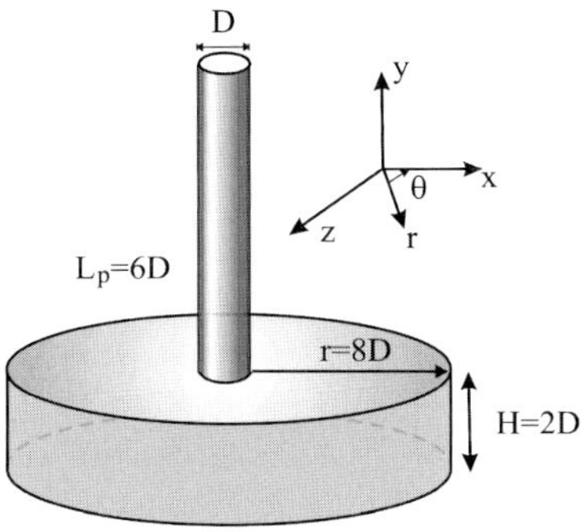

Fig. 1. Schematic diagram of the computational domain used for the simulation of the flow and temperature field

condition at the pipe wall is adiabatic. The domain is 16D in length. The heat flux of 1000 W/m^2 is applied at the impingement wall for all cases. The computational domain is modeled via hexahedral structured O-grids, developed via ICEM-CFD of ANSYS, Inc. The flow in the most important stagnation region and the wall jet differs considerably from a fully developed channel flow, however, following the convention of the grid spacing reporting, the grid resolution is reported in the dimensionless spacings. The values of Δr^+, $(r\Delta\theta)^+$ and $\Delta y^+ ta$ are reported in Table 2, where $()^+ = (.)u_\tau/\nu$. The mean grid spacing feasible for DNS simulation is $O(\eta)$, where, η is the Kolmogorov scale. In LES, however, the grid estimation is not very stringent as part of the flow field is modelled. For the present investigation, two different grids are used. One is comprised of 5 million control volumes (CV), and the other consists of 10 million control volumes. In our investigations, we have used a hexahedral structured grid. Both the pipe and cylindrical domain are modelled via O-grids. In the present simulation the streamwise direction in the pipe is 2πD which is equivalent to 2580 wall units. For both the grids the pipe has $\Delta L_p^+ = (\Delta L)_p u_\tau/\nu$=70. It has been found, that especially the results for the heat transfer were improved in the stagnation region (r/D<1) using the finer grid.

Table 2. Numerical grid – resolution

Grids	Total CV	CV (in impingement zone)	Δr^+	$(r\Delta\theta)^+$	Δy^+_{min}
Grid-I	5 M	2.2 M	36	20	0.9
Grid-II	10 M	4.5 M	29	17	0.2

In case of Grid-I, there are more than 5 grid points in within 10 wall units of the impingement wall, where as in case of Grid-II, there are more than

12 grid points within 10 wall units of the impingement wall. This ensures the good quality resolution of the near wall region and the presented results are primarily based on Grid-II, as with the finer resolution especially the simulation in the inner stagnation region, r/D $<$ 1, with respect to the heat transfer was improved.

3.2 Inflow Conditions

For inflow conditions, the digital-filter based turbulence generation procedure proposed by Klein et al. [36] was used. In the jet's axial direction, a fully developed turbulent pipe flow mean velocity profile, together with velocity fluctuations varying in time are prescribed at the inlet. The turbulence inflow conditions are generated, based on the procedure by Klein et al. [36]. The velocity fluctuations generated are based on digital filtering of random data. In this procedure, the inflow turbulent velocity generated is based on the relation:

$$u_i = < u_i > + a_{ij} u_j^{'}, \tag{5}$$

where $< u_i >$ is the mean axial velocity, $u_j^{'}$ are velocity fluctuations and a_{ij} is a tensor related to the Reynolds stress tensor (see [36]). According to this procedure, the prescribed auto-correlation function R_{uu} for homogeneous turbulence in later stages is used to describe the turbulence:

$$R_{uu}(\hat{r}) = exp\left(\frac{-\pi\hat{r}^2}{4L(t)^2}\right), \tag{6}$$

where $\hat{r}$ is the position vector and L(t) is the integral length scale at the inflow plane.

3.3 Outflow Conditions

At the outlet, the Convective boundary conditions are used. As this allow a negligible influence on the evolution of the flow structures in the finite-size computational domain. It is experienced that the use of the Neumann type boundary conditions with a full three-dimensional domain simulation, for the impinging jet case, generates a non-physical feedback, which strongly affects the global stability of the simulation. The convective boundary condition is spatially and temporally local, and defined as (e.g. [7, 11, 18])

$$\frac{\partial U}{\partial t} + \Lambda \frac{\partial U}{\partial r} = 0. \tag{7}$$

The quantity Λ is taken to be a constant convection velocity of the large-scale structures, which is set to a constant as per global mass conservation requirements.

4 Numerical Code & Its Performance

4.1 FASTEST

The computations were performed with the CFD code FASTEST (Flow Analysis Solving Transport Equations Simulating Turbulence). The code is based on a finite volume discretization with pressure velocity coupling done through a SIMPLE algorithm. For LES computations, a top hat filter was used. The three dimensional filtered Navier-Stokes equations along with the subgrid model proposed by Germano [17] were solved. The second order implicit Crank-Nicolson method is used for time discretization and for space discretization of the convective terms a second order central differencing scheme is used. The resulting set of equations is solved using a SIP solver [50].

4.2 Code Performance & Solution Control

Computations are done on two different computing clusters available at Höchstleistungsrechenzentrum (HLRS), Stuttgart, Germany (see Table 3). For first computations on grid-I the CRAY Opteron Cluster was used, then the FASTEST code was migrated to the NEC vector platform. The noteworthy improvements in the code are the vectorization of the subroutines related to LES subgrid scale model. The FASTEST has two different loop structures available in the code, which could be used for computations, depending on the architecture of the computing system. The routines related to LES had the three times nested loop version only. The use of a collapsed single loop is a natural choice for vector machines as it gives better vectorization. The single loop version is now also available for the LES-model. The code has attained 5.6 Gflops and high vectorization rates (>99.8%). For Case-0-II, the vectorized version of the code is used and computations are done on the NEC SX-8. The number of processors and computing platforms used are listed below: The computational domain for the jet is divided into 31 blocks. This domain

Table 3. Computational platforms

Cases	Re	Modulation	CPUs	Computing Platform
Case-0-I	23000	No-modulation	20	Cray Opteron
Case-0-II	23000	No-modulation	7	NEC SX-8

decomposition helps in computational load balancing on the processors. The inter-processor communications were performed with standard Message Passing Interface (MPI). The simulation of the jet requires approximately 5000 CPU hours for grid-II on NEC SX-8. The computations on NEC SX-8 are much faster and require three to four times less computational time than the one on the Cray Opteron Platform. On average total simulation time for each case is equivalent to 25 cycles, where one cycle corresponds to the natural frequency of the jet. The flow becomes statistical stationary after about 9 cycles. The simulation is controlled in such a way that the CFL number remains less than one. The dimensionless time step used is $\Delta t D/U_{ax}$ and equals to 7E-07.

5 Results

This sections contains the details of the results and discussion. First the frequencies present inside jet are investigated then the velocity distributions in the wall jet along with turbulent intensities are explored. In the following section U represents the mean velocity and, although the computations were done in a cartesian coordinate system, the $< u >$,$< v >$,$< w >$ in this paper represent the velocities in radial, vertical and azimuthal directions respectively, as indicated in Fig. 1. The subscripts b and CL represent the bulk and centre-line velocity at the pipe's exit.

5.1 Instabilities in an Axisymmetric Jet

The large scale structures originating from the shear layer instabilities of a Kelvin Helmhotz (KH) type, significantly affect the later development of the jet. The KH instabilities which exists between the fast moving jet flow and surrounding slow moving fluid, also enhance the entrainment of the surrounding fluid. The instabilities give rise to the formation of the Ring vortex, shown in Fig. 2. Figure 2 shows the development of initial shear layer. As can be noted, the fluid emanates from the pipe wall and forms the shear layer, see also Fig. 9, which indicates also the coherent structures extracted through

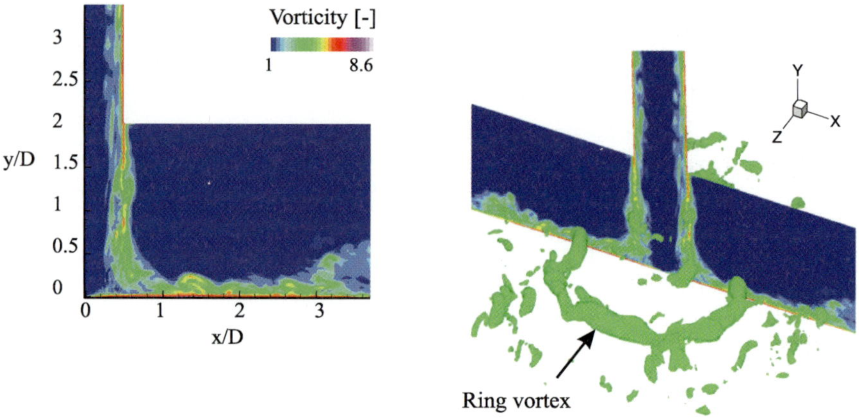

Fig. 2. Ring vortex visualized through iso-pressure surfaces (P/ρU_b^2 = -0.016) in the jet's shear layer

iso-Q surfaces (Definition (12)). In a free jet vortex rings are connected by these structures. The intermittently generated toroidal shaped (ring) vortices, lose their structure on striking the surface and the structures are stretched.

Velocity Frequency and Preferred Mode of Axisymmetric Jet

If a jet emerges from a converging nozzle and the boundary layer on the inner surface of nozzle is laminar, the initial region of a jet shear layer can be considered as parallel or two dimensional. In this case, the ratio of the jet diameter to the initial momentum thickness is large (ξ/D < 1–1.5), where, ξ is the distance from the nozzle or pipe outlet. The shear layer near the nozzle is initially dominated by a Kelvin Helmholtz instability mechanism. This would cause the formation of the first vortex at the nozzle lip. The formation of a vortex is attributed to most amplified frequency of small disturbances which grow exponentially ([28]). This mode represented by the Strouhal number $St_{\delta^*}=$ 0.01–0.018 ([20]), where, δ^*, is the momentum thickness of the boundary layer. The jet column mode is attained if the flow emerges in a fully developed state from a long pipe and the momentum thickness is of order of the jet radius. Then the frequency is scaled with diameter of the jet and the most amplified frequency is called the preferred mode of the jet, corresponding to the Strouhal number range $St_D=$ 0.25–0.85, [39]. They have found that the natural frequency of an impinging jet is the same for a free jet if ξ/D > 1. [44] have found that the mean velocity and turbulence intensity in an impinging jet starts to differ from those in a free jet at a distance of about 1.3D from the wall. [52] also found that the natural frequency in the impinging jet corresponds to St_D equal to 0.37. In order to investigate the most amplified frequencies in the jet, the data is monitored simultaneously inside the jet core, shear layer, impingement zone and wall jet region. The most amplified frequency is found inside the shear layer at the distance (ξ) equals one pipe diameter, downstream of the pipe outlet. This corresponds to a Strouhal number of $St_D = 0.328$ (preferred mode of the jet). This dimensionless frequency is very close to the natural frequency of the free jet.

5.2 Mean Flow Properties and Turbulent Flow Intensities

[16] have given the mean velocity distributions in a single impinging jet at r/D=0, i.e. stagnation line. Figure 3 shows a comparison between Particle Image Velocimetry (PIV) and Laser Doppler Anemometry (LDA) measurements with the current Large Eddy Simulation. A good match has been found in the results. The mean radial velocity $< u >$ is almost zero, which shows the axial symmetry. The velocity and turbulence were compared with the data by [6]. The Large Eddy Simulation agreed well with experiments. Figures 4, 5 and 6 show the comparison between simulation and experiments at r/D =2, 2.5 and 3.0. Note that although the trend of the turbulent flow intensities distribution are in agreement with the experimental data the magnitude of

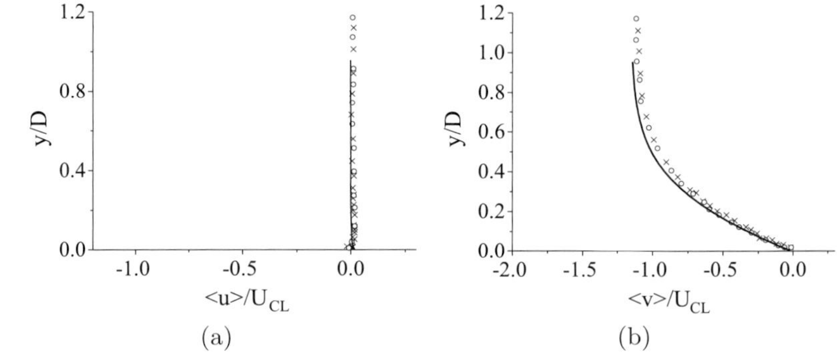

Fig. 3. The velocities **a** $< u > /U_{CL}$ (radial) and **b** $< v > /U_{CL}$ (vertical) at r/D = 0, ∘ PIV, × LDA measurements of [16], $< u >$ and $< v >$ are the components of mean velocity, U_{CL} is the velocity at the centreline at jet's exit

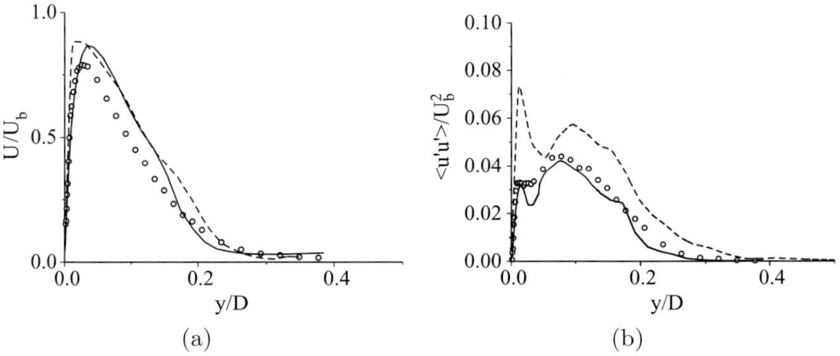

Fig. 4. The mean velocity **a** U/U_b and **b** $< u^{'}u^{'} > /U_b^2$ at r/D =2, solid line (LES-Grid-II), dashed line (LES-Grid-I), ∘ Hot-wire measurements of [6], U is mean velocity, U_b is the bulk velocity at jet's exit

the predicted turbulent intensities are higher than hot-wire measurements. Such measurements sometimes generate erroneous data in regions of high turbulence. Than they also under-predict the turbulence level, probably due to their intrusive flow measurement character. The distribution of turbulent stress in wall jet region is shown in Fig. 6. It is interesting to note that the turbulent stress does not vanish at a point where velocity gradient is zero. The same phenomenon is observed by [47] through experimental measurements. The eddy viscosity hypothesis based turbulence models assume that the turbulent shear stress is proportional to the local gradient of the mean velocities. The current simulation shows this assumption is invalid for the wall jet zone. Even velocity predictions on Grid-I represent fairly well the experimental finding they were improved running the simulation on Grid-II. The wall jet developing after impinging jet did not exhibit the conventional law

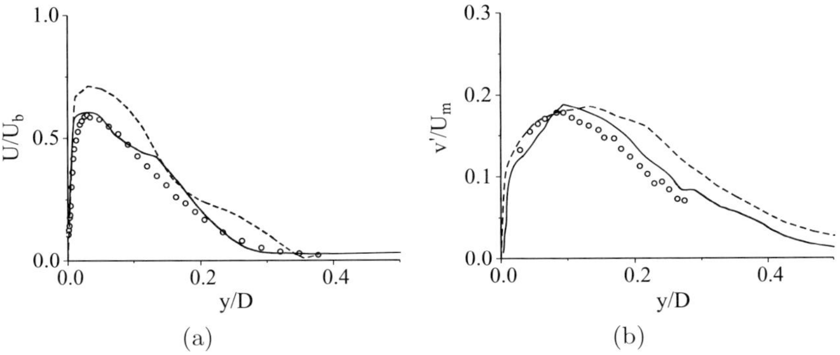

Fig. 5. The mean velocity **a** U/U_b and **b** $v^{'}/U_m$ at r/D =2.5, solid line (LES-Grid-II), dashed line (LES-Grid-I), ∘ Hot-wire measurements of [6], U_b is the bulk velocity at jet's exit, U_m is the local maximum velocity in wall jet zone

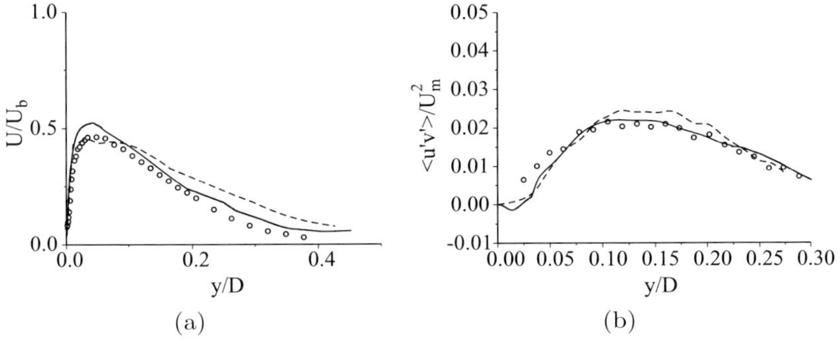

Fig. 6. The mean velocity **a** U/U_b and **b** $< u^{'}v^{'} > /U_m^2$ at r/D =3.0, solid line (LES-Grid-II), dashed line (LES-Grid-I), ∘ Hot-wire measurements of [6], U_b is the bulk velocity at jet's exit, U_m is the local maximum velocity in wall jet zone

of the wall behaviour. The following semi-logarithmic relation can be used to model the inner layer of the wall jet:

$$\frac{u}{u_\tau} = \frac{1}{\kappa} ln(\frac{y u_\tau}{\nu}) + A \quad (8)$$

where, u_τ is the friction velocity and κ is the von Karman constant. [58] have found that a logarithmic profile might be fitted to the data but the constant (A) present in law of the wall (8) must be adjusted. [45] have found that in case of an impinging jet, if the outer edge of the equilibrium layer is attached to the point of maximum velocity (U_m), which occurs close to wall, then this maximum velocity can be used as an appropriate scaling parameter. They proposed that the A in (8) should be treated as a function.

$$A = f(\frac{U_m}{u_\tau}) \quad (9)$$

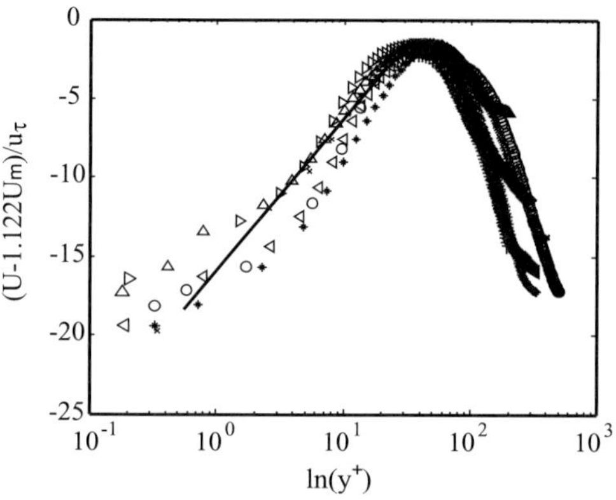

Fig. 7. ○ r/D=1, $*$ r/D=2, × r/D=2.5, ◁ r/D=3, ▷ r/D=3.5, △ r/D=4 Semilogarithmic velocity plot with subtraction of profile shift parameter, the line shows the characteristic of the equilibrium layer that it extends till the location of maximum velocity

Using the approach proposed by [21, 45] we have found the following relationship:

$$A = 1.121(\frac{U_m}{u_\tau}) - 10.53 \qquad (10)$$

The first term in the above equation is called profile shift parameter. If this shift parameter is subtracted from the law of the wall, the resulting curve shows the presence of an equilibrium layer that extend up to the point of maximum velocity. Figure 7 shows the radial velocity profiles with subtracting the shift parameter in semi-arithmic axes. The simulation thus confirms the previous experimental findings of [45] and [21] that the inner layer of the wall jet, which is formed after the jet impingement on the wall, can be cast in to scaling log-laws. As can be seen, the flow in an impinging jet and the subsequent wall jet is highly complex. It is found that the mean flow characteristics are influenced by the wall shear stress. It is important to analyse the nature of the turbulence in detail, which is the subject of the next section.

5.3 Heat Transfer and Coherent Structures

The radial distribution of the Nusselt number is ensemble-averaged in azimuthal direction and presented in Fig. 8. The Nusselt number from the simulation is plotted versus experimental data ranging from Reynolds number 20000 to 23300. As one can see, the first peak Nu_1 is only slightly off-set from the stagnation point and captured well by the simulation. The predicted heat transfer at the stagnation point is within the experimental scatter. The secondary peak is captured as well. However, the scatter in experimental data is becomes higher in the region of the secondary peak, indicating the complex

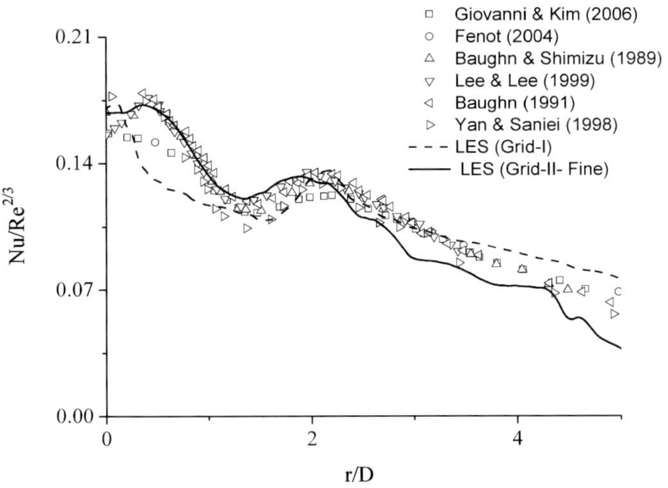

Fig. 8. Radial distribution of the Nusselt number in an impinging jet, Re=23000 and H/D=2., Nusselt number is normalised with $Re^{2/3}$ □ ([19], Re = 23000), ○([12], Re = 23000), △ ([2], Re = 23750), ▽ ([38], Re = 20000), ◁ ([3], Re = 23300), ▷([59], Re = 23000), Dashed line – (LES-Grid-I, Re = 23000), Solid line – (LES-Grid-II, Re = 23000)

dynamics of the flow in that region. Further outward at larger radii the simulation underpredicts the heat transfer compared to the experimental data, leading to the implication that the numerical scheme still performs slightly to dissipative in this region. The additional grid refinement which led to Grid-II had especially consequences in the predicted location of the first peak of the Nusselt number which is now in good agreement with experimental findings. For H/D equals 2, [38] have found that the following relationship between the local peaks hold:

$$Nu_2/Nu_1 = 0.613Re_D^{0.023} \tag{11}$$

From simulation it is found that the ratio between the two peaks Nu_2/Nu_1 is found to be 0.76, this agrees well with the correlation proposed by [38] according to which the ratio is 0.77, for Re=23000 and H/D=2. Through simulation it has been found that the ring vortices were found to be generated a-periodically and striking the surface. This causes constant oscillation of flow near the stagnation zone. [46] have found through the visualisation studies that the fluid in the vicinity of the stagnation point experiences a pulsation due to swelling and contraction of the jet core. It has been found that at r/D=1 the ring vortices are striking the surface which are causing a low pressure over there. At the same location, locally the velocity attains highest values in the impingement system. The periodic acceleration of the flow in this zone causes a boundary layer excitation, which decreases the boundary layer thickness and hence reduces the temperature over there. As the ring vortex breaks up due to radial stretching, the velocity increases and causes the local cooling effect on the wall. This causes an increase in heat transfer rate. The heat transfer rate reaches its maximum at r/D equals to two, which is normally attributed to the increase in turbulent kinetic energy. In order to understand the flow dynamics in this zone, we have utilized the Q-criterion ([30]) given by:

$$Q = \frac{1}{2}(\Omega_{ij}^2 - S_{ij}^2) \tag{12}$$

where, Ω_{ij}, and S_{ij} are the antisymmetric and symmetric parts of the velocity gradient tensor. This method uses the fact that Q has a positive value when vorticity is more predominant than shear strain. The condition also corresponds to the source term of the Poisson equation for the pressure ($\nabla^2 P = 2\rho Q$), so the condition shows that an area with $Q > 0$ has the minimum pressure and is a local low-pressure area, and that a larger value of Q represents a stronger vortex. The Q-criterion is used for the investigation of flow structures in the wall jet zone. Figure 9 shows the coherent structures in an impinging jet. It is found that there is a strong acceleration and stretching of structures in this region. The positive Q values are very high, not only above the stagnation region, but also in wall jet region. This shows the strong vortical nature of the flow. The accelerating boundary layer flow in the radial wall jet experiences the entrainment of the surrounding cold fluid. This whole scenario results in reduction in wall temperature. This reduction in temperature is found to be significant from r/D≈1 to r/D≈2. However, after $r/D>2$ the boundary layer becomes thin due to radial spreading and reduction in velocity, due to this reason the wall temperature starts increasing again, which results in reduction in Nusselts number. The Q surfaces colored by the mag-

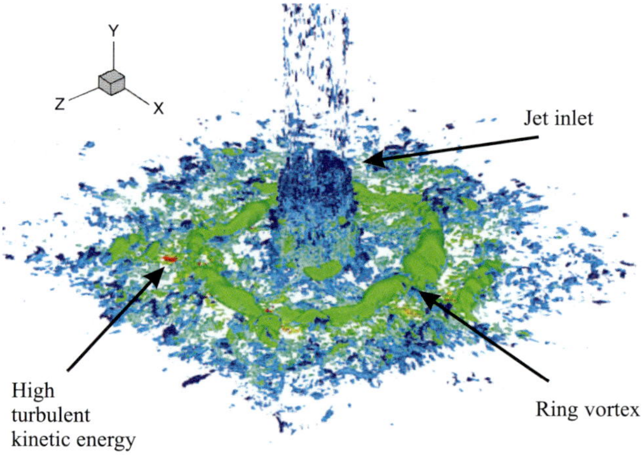

Fig. 9. The $Q > 0$ surfaces (Q= $+3.15U_b^2/D^2$) in an impinging jet case. The Q surfaces are colored by turbulent kinetic energy. The red Q-surfaces are the zone of high turbulent kinetic energy which coincides with the Ring vortex impingement zone

nitude of the turbulent kinetic energy are shown in Fig. 9. The regions of high turbulent kinetic energy coincides with the Ring vortex impingement location on the wall. This shows that the Ring vortex is affecting the flow field significantly and the common reason of the increase in turbulent kinetic energy and secondary peak in Nusselt number is closely linked to the coherent structures in the turbulent impinging jet. This reason of the secondary peak has been proposed earlier by [40] and [46]. [40] attributed the occurrence of the secondary peak to the higher turbulence in boundary layer due to flow acceleration and intense shear between the radially ejecting jet and stagnant ambient. Also, the so-called 'transition zone' ($r/D = 1 - 2$) suggested by [46], where, toroidal vortices and wall eddies developing and abruptly merging is in fact what was stated above. The Large Eddy Simulation confirms the previous experimental findings and it has been found that the seemingly different interpretations are the same.

6 Conclusions

A series of LES investigations of turbulent jet impingement are performed on the different HLRS computing platforms. It is found that:

1. In an Large Eddy Simulation of an impinging jet the grid resolution of the free shear layer zone and of the wall jet developing zone is a crucial issue. Whereas mean values of velocity and temperature are already quite reasonable predicted in the coarser grid-I simulation, correlations of the velocity fluctuations and heat transfer in the core of the impingement region can be improved by using the finer grid-II.
2. Through analysis of the coherent structures appearing in an single impinging jet system it has been found that the Ring vortex impingement location on the wall coincides with regions of highly time dependent flow pattern resulting locally high turbulent kinetic energy where the ring vortex interacts with the beginning wall jet. This shows that the Ring vortex is affecting the flow and temperature field significantly and the resulting boundary layer interaction phenomenon can be interpreted as the reason for the characteristic secondary peak in the local Nusselt number at these nozzle to wall distance impinging jet.

Acknowledgements. The authors would like to thank Prof. Dr. M. Schäfer and Dr.-Ing. D. Sternel, Fachgebiet für Numerische Berechnungsverfahren im Maschinenbau (FNB), Technische Universität, Darmstadt, Germany, for providing the FASTEST code and helpful discussions. The second author wants to thank the Higher Education Commission (HEC), Pakistan for financial support and the help of Dr.-Ing. P. Lammers, Höchstleistungs-Rechenzentrum (HLRS), Stuttgart is gratefully acknowledged.

References

1. Azevedo, L.F.A., Webb, B.W., Queiroz, M., Pulsed air jet impingement heat transfer, Experimental Thermal and Fluid Science **8**, 206–213 (1994)
2. Baughn, J.W., Shimizu, S., Heat transfer measurement from a surface with uniform heat flux and an impinging jet, Int. J. of Heat Transfer, **111**, 1096–1098 (1989)
3. Baughn, J.W., Hechanova, A.E., Yan, X., An experimental study of entrainment effects on the heat transfer from a flat surface to a heated circular impinging jet, Journal of Heat Transfer, **113**, 1023–1025 (1991)
4. Camci, C., Herr, F., Forced Convection heat transfer enhancement using a self-oscillating impinging planar jet, J. Heat Transfer, **124**, 770–782 (2002)
5. Chung, Y.M., Luo, K.H., Sandham, N.D., Numerical study of momentum and heat transfer in unsteady impinging jets, Int. J. of Heat and Fluid Flow, **23**, 592–600 (2002)
6. Cooper, D., Jackson, D.C., Launder, B.E., Liao, G.X., Impinging jet studies for turbulence model assessment – I. Flow field experiments, Int. J. Heat Mass Transfer, **36**, 2675–2684 (1993)
7. Colonius, T., Numerically non-reflecting boundary and interface conditions for compressible flow and aeroacoustic computations, AIAA, 7, **35**, 1126–1133 (1997)
8. Cziesla, T., Tandogan, E., Mitra, N.K., Large eddy simulation of heat transfer from impinging slot jets. Numerical Heat Transfer, Part A **32**, 1–17 (1997)
9. Cziesla, T., Biswas, G., Chattopadhyay, H., Mitra, N.K., Large eddy simulation of flow and heat transfer in an impinging slot jet. Int. J. Heat and Fluid Flow **22**, 500–508 (2001)
10. Didden, N., Ho, C., Unsteady separation in a boundary layer produced by an impinging jet, J. Fluid Mech, **160**, pp 235–256 (1985)
11. Enquist, B., Majda, A., Absorbing boundary conditions for the numerical simulation of waves, Math. of Computation, **31**, 139, 629–651 (1977)
12. Fenot, M., Etude du refroidissement par impact de jets. Application aux aubes de turbines, Universite de Poitiers, France (2004)
13. Fox, M.D., Kurosaka, M., Hedges, L., Hirano, K., The influence of vortical structures on the thermal fields of jets, J. Fluid Mech., **255**, 447–472 (1993)
14. Gao, S., Leslie, D.C., Voke, P.R., Large eddy simulation of thermal impinging jets, Rep ME-FD/91.02 Dep. Mech. Eng., University of Surrey, Guildford, UK (1991)
15. Gao, S., Voke, P.R., Large Eddy Simulation of turbulent heat transport in enclosed impinging jets, Int. J. Heat and Fluid Flow, **16**, 349–356 (1995)
16. Geers, L.F., Tummers, M.J., Hanjalić, K., Experimental investigation of impinging jet arrays. Experiments in Fluids, **36**, 946–958 (2004)
17. Germano, M., Piomelli, U., Moin, P., Cabot, W.H., A dynamic subgrid-scale eddy viscosity model, Phy. Fluid, **3**, 1760–1765 (1991)
18. Giles, M.B., Non-Reflecting Boundary Conditions for Euler Equation Computations, AIAA Journal, **28**, 12, 2050–2058 (1990)
19. Giovannini, A., Kim, N.S., Impinging jet: Experimental analysis of flow field and heat transfer for assessment of turbulence models, Annals of the Assembly for International Heat Transfer Conference 13, TRB-15 (2006)

20. Gutmark, E., Ho, C., Preferred modes and the spreading rates of jets, Phy. Fluids, **26**, No. 10, 2932–2938 (1983)
21. Guerra, D.R.S., Su, J., Freire, A.P.S., The near wall behaviour of an impinging jet, Int. J. Heat and Mass Transfer, **48**, 2829–2840 (2005)
22. Hall, J.W., Ewing, D., The development of the large-scale structures in round impinging jets exiting long pipes at two reynolds numbers, Experiments in Fluids, **38**, 50–58 (2005)
23. Hadžiabdić, M., LES, RANS and Combined Simulation of Impinging Flows and Heat Transfer, TU Delft, Netherlands (2005)
24. Ho, C.M., Nossier, N.S., Dynamics of impinging jet. Part 1, the feedback Phenomenon, J. Fluid Mechanics, **105**, 119–142 (1981)
25. Ho, C.M., Huang, L.S., Subharmonics and vortex merging in mixing layers, J. Fluid Mechanics, **119**, 443–473 (1982)
26. Hofmann, H.M., Wärmeübergang beim pulsierenden Prallstrahl, Dissertation, Universität Karlsruhe (2005)
27. Hoogendoorn, C.J., The effect of turbulence on heat transfer at a stagnation point, Int. J. Heat Mass Transfer, **20**, 1333–1338 (1997)
28. Hussain, A.K.M.F., Coherent structures - reality and myth, Phy. Fluids, **26**, No. 10, 2816–2850 (1983)
29. Hwang, S.D., Cho, H.H., Effects of acoustic excitation positions on heat transfer and flow in axisymmetric impinging jet: main jet excitation and shear layer excitation, Int. J. of Heat and Fluid Flow, **24**, 199–209 (2003)
30. Hunt, J.C.R., Wray, A.A., Moin, P., Eddies, stream and convergence zones in turbulent flows, Center for Turbulence Research, CTR-S88 (1988)
31. Jiang, P., Guo, Y.C., Chan, C.K., Lin, W.Y., Frequency characteristics of coherent structures and their excitations in small aspect-ratio rectangular jets using large eddy simulation, Computers Fluids, **36**, 611–621 (2007)
32. Jiang, X., Zhao, H., Luo, K.H., Direct computation of perturbed impinging hot jets, Computers & Fluids, **36**, 259–272 (2007)
33. Kataoka, K., Mihata, I., Maruo, K., Suguro, M., Chigusa, T., Quasi-periodic large- scale structure responsible for the selective enhancement of impinging jet heat transfer, Proceedings of 8th International Heat transfer Conference, San Francisco, CA, USA **3** (1986)
34. Kataoka, K., Suguro, M., Degawa, H., Maruo, K., Mihata, I., The effect of surface renewal due to large scale eddies on jet impingement heat transfer, Int. J. Heat and Mass Transfer, No. 3, **30**, 559–567 (1987)
35. Kataoka, K., Ase, H., Sako, N., Unsteady aspects of large-scale coherent structures and impingement heat transfer in round air jets with and without controlled excitations, Int. J. of Engg. Fluid Mech., **1**, 365–382 (1988)
36. Klein, M., Sadiki, A., Janicka, J., A digital filter based generation of inflow data for spatially direct numerical or large eddy simulations, Journal of Computational Physics, **18**, 652–665 (2003)
37. Landreth, C.C., Adrian, R.J., Impingement of a low Reynolds number Turbulent circular jet onto a flat plate at normal incidence, Experiments in Fluids **9**, 74–84 (1990)
38. Lee, J., Lee, S., Stagnation region heat transfer of a turbulent axisymmetric jet impingement, Experimental Heat Transfer, **12**, 137–156 (1999)
39. Liu, T., Sullivan, J.P., Heat transfer and flow structures in an excited circular impinging jet, Int. J. Heat and Mass Transfer, **17**, 3695–3706 (1996)

40. Lytle, D., Webb, B.W., Air jet impingement heat transfer at low nozzle-plate spacing, Int. J. Heat Mass Transfer, **137** (12), 1687–1697 (1994)
41. Martin, H., Wärmeübergang bei Prallströmung, VDI-Waermeatlas, VDI (1998)
42. Mladin, E.C., Zumbrunnen, D.A., Alterations to coherent flow structures and heat transfer due to pulsations in an impinging air-jet, Int. J. Therm. Sci., **39**, 236–248 (2000)
43. Nevins, R.G., Ball, H.D., Heat transfer between a flat plate and a pulsating impinging jet, National Heat Transfer Conference Boulder, CO, ASME (1961)
44. Olsson, M., Fuchs, L., Large eddy simulations of a forced semi-confined circular impinging jet, Phys. Fluid, **10** (2), 476–486 (1998)
45. Özdemir, I.B., Whitelaw, J.H., Impingement of an axis-symmetric jet on unheated and heated flat plates, J. Fluid Mech., **240**, 503–532 (1992)
46. Popiel, C.O., Trass, O., Visualisation of a free and impinging round jet, Experimental Thermal and Fluid Sciences, **4**, 253–264 (1991)
47. Poreh, M., Tsuei, Y.G., Cermak, J.E., Investigation of a turbulent radial wall jet, Journal of Applied Mechanics, **34**, 457–463 (1967)
48. Rizk, M.H., Menon, S., Large eddy simulations of axis-symmetric excitation effects on a row of impinging jets, Physics of Fluids, **31** (7), 1892–1903 (1988)
49. Sakakibara, J., Hishida, K., Maeda, M., Vortex structure and heat transfer in the stagnation region of an impinging plane jet (simultaneous measurements of velocity and temperature field by digital particle image velocimetry and laser induced fluorescence), Int. J. Heat Mass Transfer, **40** (13), 3163–3176 (1997)
50. Stone, H.L., Iterative solution of implicit approximations of multidimensional partial differential equations, SIAM J. Numer. Anal., 3, **5**, 530–558 (1988)
51. Tsubokura, M., Kobayashi, T., Taniguchi, N., A numerical study on the difference of the eddy structures between plane and round impinging jet. Engineering Turbulence Modelling and Experiments-5, W. Rodi and N. Fueyo, Elsevier, Amsterdam (2002)
52. Tsubokura, M., Kobayashi, T., Taniguchi, N., Jones, W.P., A numerical study on the eddy structures of impinging jets excited at the inlet, Int. J. of Heat and Fluid Flow, **24**, 500–511 (2003)
53. Uddin, N., Neumann, S.O., Weigand, B., Large eddy simulation of the heat transfer due to swirling and non-swirling jet impingement, Proc. of 2008 ASME Summer Heat Transfer Conference, number HT2008-56422, Jacksonville, Florida USA (2008)
54. Uddin, N., Neumann, S.O., Weigand, B., Investigation of the effect of inlet velocity field exitation of turbulent impinging jet on heat transfer using large eddy simulation, Proc. 2nd International Conference on Jets, Wakes and Separated Flows, ICJWSF-2008, 2008
55. Uddin, N., Neumann, S.O., Weigand, B., Lammers, P., Thermal & Flow Field Analysis of Turbulent Swirling Jet Impingement Using Large Eddy Simulation, volume 08 of High Performance Computing in Science and Engineering, pages 301–315, Springer, Berlin (2008). ISBN 978-3-540-88301-2
56. Uddin, N., Neumann, S.O., Weigand, B., Investigation of the effect of inlet velocity field exitation of turbulent impinging jet on heat transfer using large eddy simulation, 6th International Symposium on Turbulence, Heat and Mass Transfer, Rome (2009)
57. Uddin, N., Neumann, S.O., Weigand, B., Understanding the dynamics and control of a turbulent impinging jet via pulsation and swirl using Large Eddy

Simulation, High Performance Computing in Science and Engineering, vol. 09, Springer, Berlin (2009)
58. Wygnanski, I., Katz, Y., Horev, E., On the applicability of various scaling laws to the turbulent wall jet, J. Fluid Mech., **234**, 669–690 (1992)
59. Yan, X., Saniei, N., Heat Transfer Measurements From a Flat Plate to a Swirling Impinging Jets, Proceedings of 11th International Heat Transfer Conference, Kyonju, Korea (1998)
60. Zuckerman, N., Lior, N., Impingement heat transfer: correlations and numerical modeling, ASME J. Heat Transfer, **127**, 544–552 (2005)

Large Eddy Simulations of a Jet in Crossflow

Flavio Cesar Cunha Galeazzo, Peter Habisreuther, and Nikolaos Zarzalis

Engler-Bunte-Institute, Division of Combustion Technology – Karlsruhe Institute of Technology – Engler-Bunte-Ring 1, 76131 Karlsruhe, Germany
Flavio.Galeazzo@kit.edu

1 Introduction

The simple jet in crossflow (JIC) configuration, in which a jet is injected perpendicularly into a crossflow, is of major theoretical and practical importance. The phenomena associated with the JIC are illustrated in Fig. 1. The interaction between the two flows generate various turbulent vortices and complex coherent structures. Margason [8] reviewed the investigations in the '80s and in the beginning of the '90s. Despite of the huge progress in laser diagnostics and simulation tools, the JIC only received little attention in the recent years, mainly at weak turbulent conditions [1, 11].

The objective of the research project on which this work is included is to investigate the turbulent mixing of the JIC under high turbulent conditions [3, 4, 7]. This article deals with the efforts of using Large Eddy Simu-

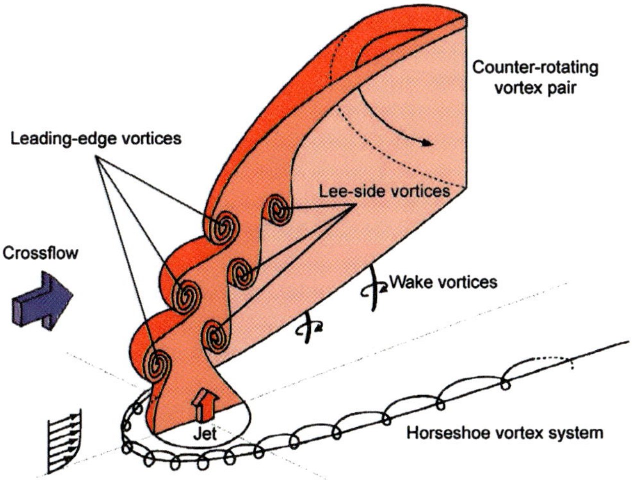

Fig. 1. Scheme of the jet in crossflow phenomenology (from [3, 4])

W.E. Nagel et al. (eds.), *High Performance Computing in Science and Engineering '10*, DOI 10.1007/978-3-642-15748-6_25,

lations (LES) to describe the phenomenology of the JIC; the results reported here were compared to highly sophisticated measurements using simultaneous combined Particle Image Velocimetry/Laser Induced Fluorescence laser diagnostics that can be seen in references [3, 4].

2 Boundary Conditions

The experimental facility consists of a channel with square cross section (108 × 108 mm) with a round jet (inner diameter D = 8 mm) mounted flush to the wall (see Fig. 2). The crossflow flows through a contraction cone, having a nearly perfect plug-flow profile when entering the channel, while the pipe flow is fully developed at the moment it flows into the channel. More information about the experimental facility can be seen at references [1, 7].

The most important parameter describing the jet penetration is the jet-to-crossflow velocity ratio, R. This is defined as the square root of the momentum ratio: $R = (\rho_{jet} U_{jet}^2 / \rho_{cross} U_{cross}^2)^{1/2}$, where U_{jet} represents the bulk jet velocity and U_{cross} represents the maximum crossflow velocity in a plane 1.5 D upstream of the jet inlet. This definition was chosen to fit to the experimental data set; more details can be seen at [1, 3, 4, 7]. The boundary conditions described in Table 1 yield a velocity ratio of $R = 4$. The Reynolds numbers

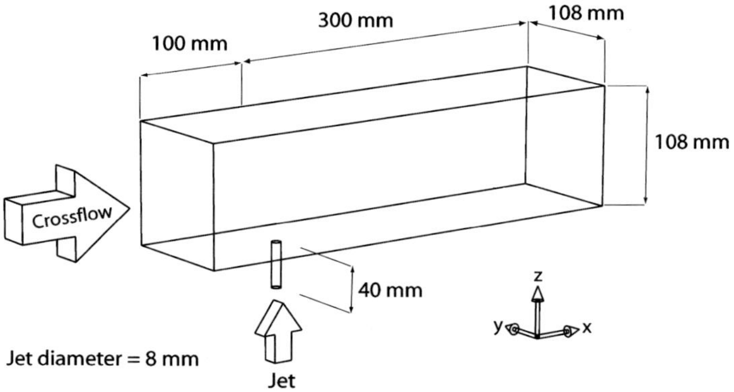

Fig. 2. Overview of the computational domain with dimensions

Table 1. Boundary conditions

Crossflow Inlet	Bulk velocity	9.08 m/s
	U_{cross}	9.43 m/s
	Turbulence intensity	1.5%
	Re	$6.24 \cdot 10^4$
Tube Inlet	Bulk velocity	37.72 m/s
	Turbulence intensity	7%
	Re	$1.92 \cdot 10^4$

were calculate using the channel cross section length and the pipe diameter as length scale for the crossflow and the jet flow, respectively. The turbulence intensities refer to the respective mean flows, away from walls.

3 Simulation Aspects

A solver coded using the open-source OpenFOAM CFD toolbox version 1.5 [9] was used to solve the modeled transport equations for mass, momentum and the passive scalar employing the finite volume method approach. The velocity equations were discretized using the filtered linear discretization scheme, while the scalar equation used a limited scheme to ensure the boundness of the results. The time was solved using the Euler implicit method.

The computational domain for the simulations was chosen in a way to save computational time, while capturing all important phenomena. The domain of the course grid extended 100 mm in the upstream crossflow direction and 300 mm in the downstream direction, as can be seen at Fig. 2. In order to further save computational time, the fine grid extended only 200 mm in the downstream direction. A pipe length of 40 mm was included into both domains.

The flow solver works with unstructured grids with elements of any shape. In this work the fine grid was composed of 7.5 million hexahedral elements, while the course grid contained 1.5 million elements. The fine and course grids have a resolution of 39 and 27 elements at the jet inlet, respectively. The time step was 2 microseconds long for the course grid and 1 microsecond long for the fine one, both yielding a maximum CFL number of approximately 0.3.

The simulation was parallelized via domain decomposition using METIS [5] as decomposition strategy and Open MPI [2] as message passing interface. The symmetric matrices were solved using the Preconditioned conjugate gradient (PCG) linear solver, while the asymmetric matrices used the Preconditioned bi-conjugated gradient (PBiCG) one. More details about the linear solvers can be found in [9].

The profiles of the mean velocity components at the inlets were computed using an extra RANS simulation with an extended domain and mapped to the inlets of the LES domain. The turbulent fluctuations were superimposed to the mean values following the work of Klein et al. [6]. The sub-grid scale turbulence for the LES was modeled using the standard Smagorinsky model [10]. The near-wall region was modeled using standard wall functions.

A passive scalar was used as a tracer for the mixing of the two flows. The sub-grid scale turbulent flux was solved using the gradient diffusion model and a turbulent Schmidt number of 1. The dimensionless scalar concentration has a value of 1 into the pure pipe flow and 0 into the pure crossflow.

Table 2. Comparison of the cluster systems

	In-house cluster	HP XC4000
Processors	Intel Core2 Q9550, 2.83 GHz	AMD Opteron, 2.6 GHz
Cores per node	1×4 cores	2×2 cores
Memory per node	8GB	16GB
Interconnect	Gigabit Ethernet	InfiniBand 4X
MPI version	OpenMPI	HPMPI

Computational Resources

Two systems were available to conduct the simulations: the in-house Linux cluster of the Division of Combustion Technology, and the HP XC4000 high performance computer of the federal state Baden-Württemberg, available trough the Steinbuch Centre for Computing of the Karlsruhe Institute of Technology. More information about the HP XC4000 system can be found at [12].

The in-house cluster consists of various subsystems:

- 1 12-core node with two Intel Xeon X7460 processors at 2.663 GHz.
- 19 4-core nodes with Intel Core2 Q9550 processors at 2.83 GHz.
- 17 2-core nodes with Intel Pentium Dual E2180 at 2 GHz.

The 12-core node was used to generate the grids, mainly because of the available 32 GB of memory. The simulations were conducted in the 4-core nodes.

Table 2 shows a comparison of the main aspects of the two systems.

Parallel Performance

The parallel performance of the computational systems was assessed by simulations using the course and fine grids. The course grid is relatively small, and fits well into the in-house cluster, while the fine grid is better suited for the HP XC4000 system.

For simulations using the course grid in the in-house cluster, the intra-node relative performance can be seen at Fig. 3. It is clear that the relative performance is far from a linear speedup; using four cores is only 49% faster than running at only one. It is also worth noting that there is no gain in performance increasing the number of cores from three to four, indicating that some bottleneck was achieved.

On the other hand, the inter-node performance is very good, as can be seen at Fig. 4. Using four nodes (summing to 16 cores) leads to a simulation time that is 3.81 times shorter than running the simulation on only one node, leading to a parallel efficiency of 95%. Running in four nodes, the mean computation time per time step is 3.2 seconds; one time step is 2 microseconds

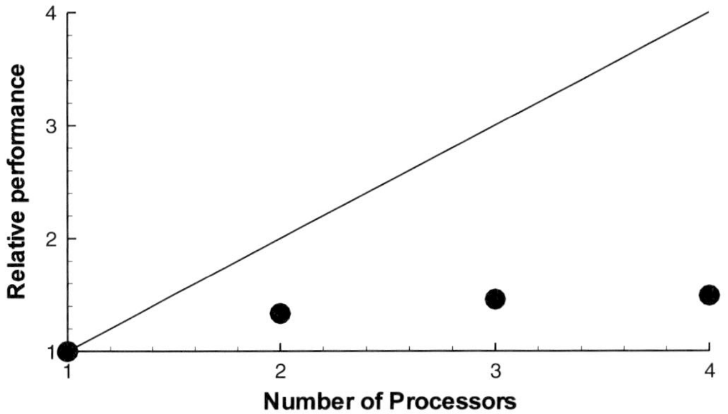

Fig. 3. Intra-node relative performance, in-house cluster, course grid

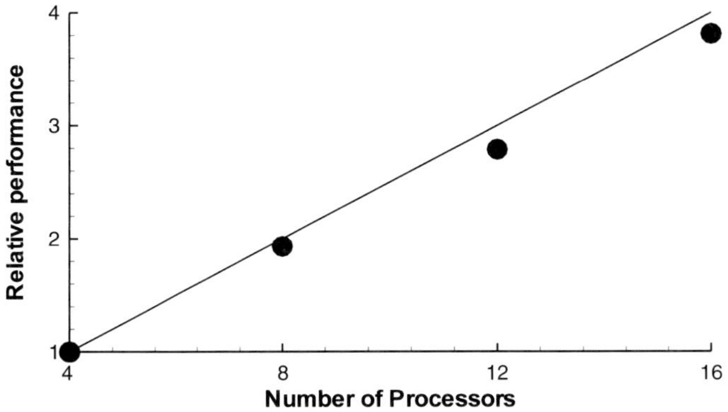

Fig. 4. Inter-node relative performance, in-house cluster, course grid

long. A complete simulation of 0.15 s, which corresponds to five residence times, can be simulated in less than 67 hours.

The same course grid was used as benchmark in the HP XC4000. The results for the intra-node performance can be seen at Fig. 5, and the behavior is different from the in-house cluster. The speedup from one to two processors is super linear, being the computation 2.19 times faster in two processors as in only one. This phenomenon can be explained by the different architectures. The nodes of the in-house cluster have one four-core processor, that shares all the cache memory and the main memory access channel between the cores. The nodes of the HP XC4000, on the other hand, have two sockets with two-core processors that can only take advantage in full extend of the cache memory when two cores on different physical processors are being used. When using the four cores the speedup is 2.72, which is far from the linear one but better than the in-house cluster nodes.

The inter-node performance is very good, as can be seen Fig. 6. The simulation was 27.57 times faster at 128 processors (32 nodes) than in only one

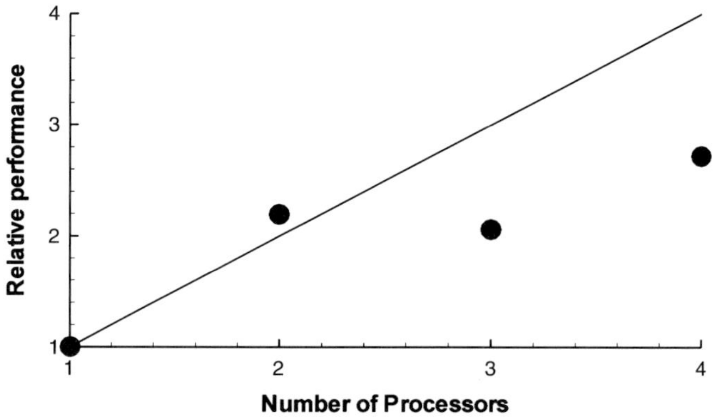

Fig. 5. Intra-node relative performance, HP XC4000, course grid

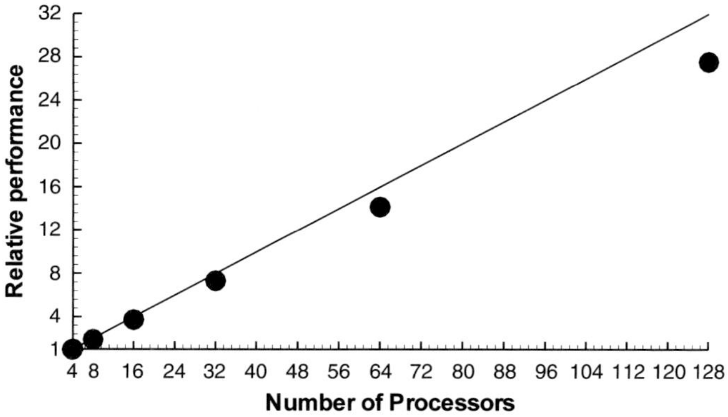

Fig. 6. Inter-node relative performance, HP XC4000, course grid

node, with a parallel efficiency of 86%. The efficiency is better with less nodes, as expected.

A direct comparison of the two system shows that, for one node, the in-house cluster needs 12.2 s and the HP XC4000 needs 9.7 s per iteration, a difference of 25%. The inter-node relative performance is comparable at 4 nodes, with a relative speedup of 3.81 for the in-house cluster and 3.73 for the HP XC4000, a difference of only 2%.

A more appropriate benchmark for the HP XC4000 is the simulation using the fine grid, as its requirements exceed the capacity of the in-house cluster. The Fig. 7 shows the inter-node performance using the fine grid, which is even better than using the course grid. Running in 64 processors (16 nodes) leads to a simulation 15.03 faster than using only one node, with 94% efficiency. The simulation have taken 3.2 seconds per time step using 16 nodes, which corresponds to less than 134 hours for 0.15 s of simulated flow, as the time step using the fine grid have to be shortened to 1 microsecond.

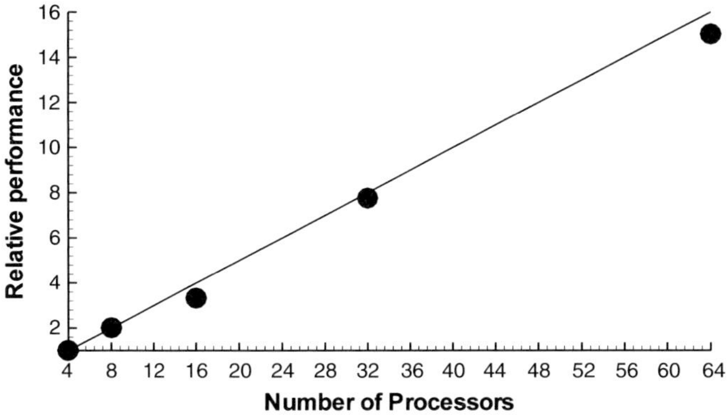

Fig. 7. Inter-node relative performance, HP XC4000, fine grid

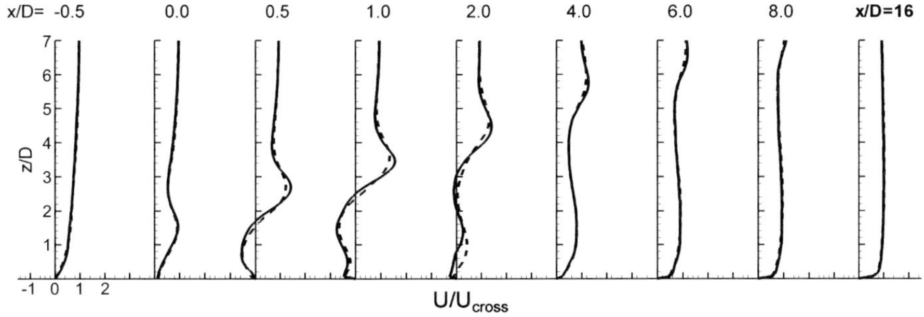

Fig. 8. Line plots of velocity component U/U_{cross} at the symmetry plane; $y/D = 0$. Solid line, fine grid; dashed line, course grid

4 Results

One of the most important features of the jet in crossflow is the jet trajectory, defined as the locus of maximum U velocity component in every velocity profile. Figure 8 shows a comparison between the U velocity component simulated using the course and fine grids at the symmetry plane (y/D = 0). The simulations show essentially the same jet trajectory, except in the region between x/D = 0.5 and x/D = 2, where the results deviates. The difference in the maximum value of the U velocity component is 14%, 13% and 7% for the profiles at x/D = 0.5, 1 and 2, respectively. It shows that the results of the course grid are not grid independent. To check if the resolution of the fine grid is already enough, a even finer grid have to be simulated, which is the next task of the project.

The comparison of the results using two dimensional maps of the various variables makes the differences between the grids even more clear. Figure 9 shows maps in the XY plane of the velocity component U and the specific Reynolds stress components $\overline{u'u'}$ and $\overline{v'v'}$ for z/D positions 1.5 and 3. At

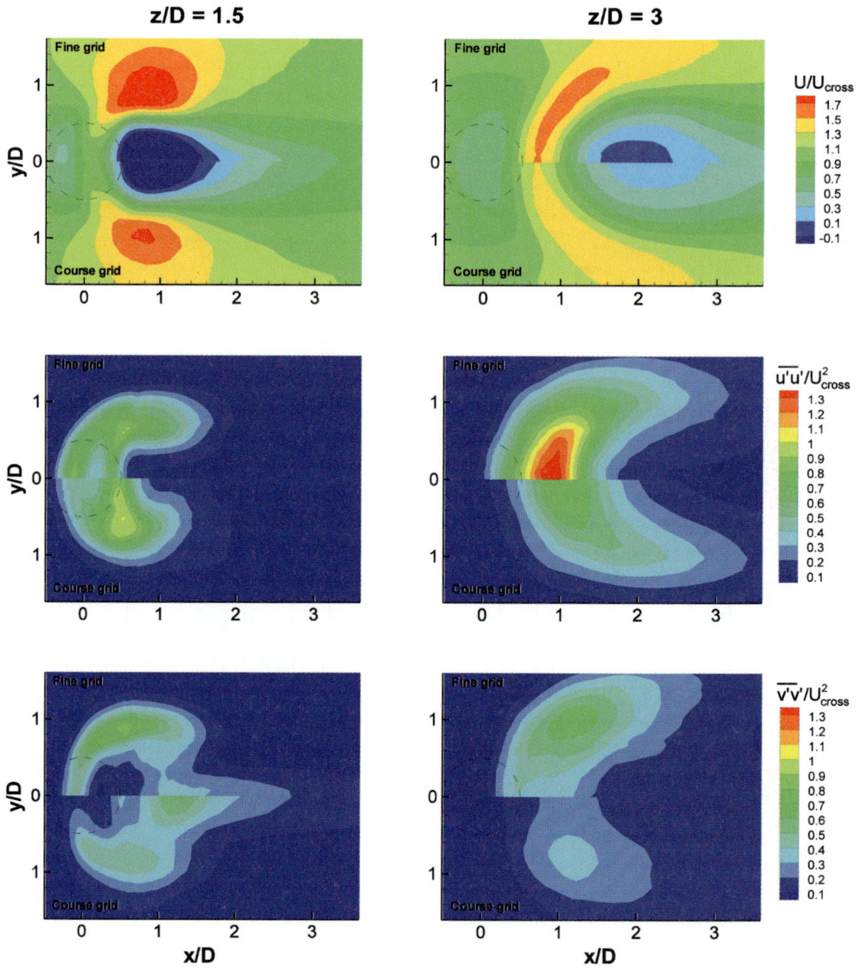

Fig. 9. Two dimensional maps of velocity component U/U_{cross} and specific Reynolds stress components $\overline{u'u'}/U^2_{cross}$ and $\overline{v'v'}/U^2_{cross}$ at z/D = 1.5 and 3.0. Upper half of the maps, fine grid; lower half, course grid

z/D = 1.5, there are small but visible differences in the velocity predictions away from the symmetry plane (y/D = 0). This difference is more evident at z/D = 3, where the peak magnitude of the U velocity component is 1.61 using the fine grid and 1.50 using the course one, a difference of 6%. For the Reynolds stress component $\overline{u'u'}$, the agreement is not satisfactory. While at z/D = 1.5 the fine grid produces two peaks, one directly in front of the jet and one at its side, the course grid predicts only one peak at the jet side. At z/D = 3 there is only one peak using both grids, but their position doesn't match. The peak value is also higher using the fine grid by 42% than using the course grid. The $\overline{v'v'}$ component exhibits the same behavior, with qualitative

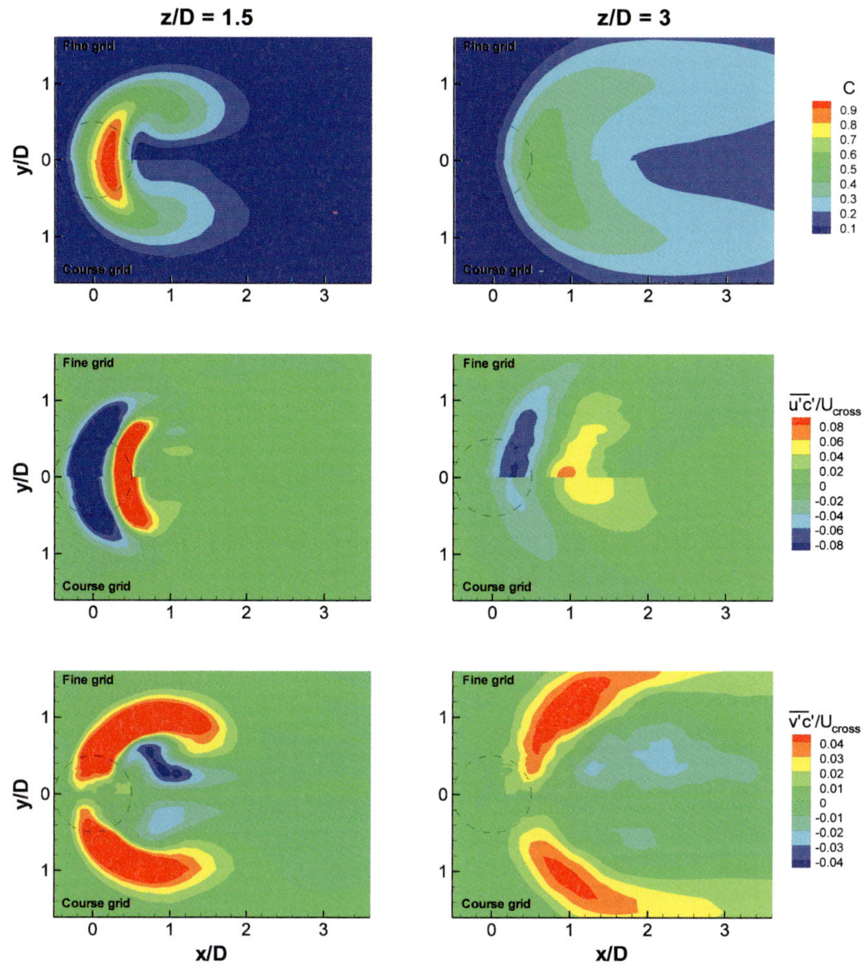

Fig. 10. Two dimensional maps of dimensionless concentration C and specific Reynolds flux components $\overline{u'c'}/U^2_{cross}$ and $\overline{v'c'}/U^2_{cross}$ at z/D = 1.5 and 3.0. Upper half of the maps, fine grid; lower half, course grid

differences between the simulations, and its value being 47% lower using the course grid in comparison to the fine one at z/D = 3.

The two dimensional maps for the mixing of the jet with the crossflow can be seen at Fig. 10, where the mean concentration C and the $\overline{u'c'}$ and $\overline{v'c'}$ components of the specific Reynolds flux vector are presented. The overall agreement for the mean concentration is good. There are, however, small quantitative differences. At z/D = 1.5, for example, the lee side of the jet presents sharper gradients using the fine grid than using the course one. The $\overline{u'c'}$ component of the specific Reynolds flux shows a good agreement at z/D = 1.5, however at z/D = 3 the peak values are more pronounced using the fine grid. The difference is 24% for the maximum values and 50% for the minimal ones. The same behavior can be seen at the $\overline{v'c'}$ maps, where the qualitative agreement is good, while the peak values differ by about 20% at z/D = 3.

5 Conclusion

The results show that the Large Eddy Simulations technique can predict well various features of the jet in crossflow. The comparison of the results obtained with the two grids shows that while only small differences are visible in the mean variables, the turbulence description is affected by the grid resolution. The simulation using the fine grid was used in the references [3, 4] in a comparison to experimental data obtained with highly sophisticated measurements using simultaneous combined Particle Image Velocimetry/Laser Induced Fluorescence laser diagnostics, with good agreement.

The parallel performance of the solver coded using the OpenFOAM CFD Toolbox was assessed in the in-house Linux cluster and in the HP XC400 system of the SCC. The intra-node performance in the in-house cluster was poor, with 4 processors being only 49% faster than a single one, while the HP XC4000 nodes have a speedup of 2.72 using four processors. The inter-node performance was good in both systems, being worth mentioning the very good speedup of the XC HP4000 system up to 128 processors.

Acknowledgements. One part of the computation time was kindly provided by the Steinbuch Centre for Computing (SCC) of the Karlsruhe Institute of Technology. The first author would like to acknowledge the financial support received from the "Conselho Nacional de Desenvolvimento Científico e Tecnológico"—CNPq—Brazil.

References

1. C. Cárdenas, R. Suntz, J.A. Denev, and H. Bockhorn. Two-dimensional estimation of reynolds-fluxes and -stresses in a jet-in-crossflow arrangement by simultaneous 2D-LIF and PIV. *Applied Physics B: Lasers and Optics*, 88(4):581–591, 2007.
2. E. Gabriel, G.E. Fagg, G. Bosilca, T. Angskun, J.J. Dongarra, J.M. Squyres, V. Sahay, P. Kambadur, B. Barrett, A. Lumsdaine, R.H. Castain, D.J. Daniel, R.L. Graham, and T.S. Woodall. Open MPI: Goals, concept, and design of a next generation MPI implementation. In *Proceedings, 11th European PVM/MPI Users' Group Meeting*, pages 97–104, Budapest, Hungary, September 2004.
3. F.C.C. Galeazzo, G. Donnert, P. Habisreuther, N. Zarzalis, R.J. Valdes, and W. Krebs. Measurement and simulation of turbulent mixing in a jet in crossflow. In *Proceedings of ASME Turbo Expo 2010: Power for Land, Sea and Air*. ASME, June 14-18 2010. Paper number GT2010-22709.

4. F.C.C. Galeazzo, G. Donnert, P. Habisreuther, N. Zarzalis, R.J. Valdes, and W. Krebs. Measurement and simulation of turbulent mixing in a jet in crossflow. *Journal of Engineering for Gas Turbines and Power*, accepted for publication, 2010.
5. G. Karypis and V. Kumar. A fast and high quality multilevel scheme for partitioning irregular graphs. *SIAM Journal on Scientific Computing*, 20(1):359–392, 1999.
6. M. Klein, A. Sadiki, and J. Janicka. A digital filter based generation of inflow data for spatially developing direct numerical or large eddy simulations. *Journal of Computational Physics*, 186:652–665, 2003.
7. T. Lischer, G. Donnert, F.C.C. Galeazzo, P. Habisheuther, N. Zarzalis, R. Valdes, and W. Krebs. Simultaneous velocity and concentration measurements using laser-optical measurement methods in comparison with Reynolds Averaged Navier-Stokes models. In *Proceedings of the 12th Internation Symposium on Transport Phenomena and Dynamics of Rotating Machinery*. ISROMAC, February 2008. Paper number ISROMAC12-2008-20112.
8. R.J. Margason. Fifty years of jet in cross flow research. In *AGARD, Computational and Experimental Assessment of Jets in Cross Flow*, chapter 1. AGARD CP-534, Winchester, UK, 1993.
9. OpenCFD Ltd. *OpenFOAM User Guide, Version 1.5*, 2008.
10. J. Smagorinsky. General circulation experiments with the primitive equations. I. The basic experiment. *Mon. Weather Rev.*, 91:99–164, 1963.
11. L.K. Su and M.G. Mungal. Simultaneous measurements of scalar and velocity field evolution in turbulent crossflowing jets. *Journal of Fluid Mechanics*, 513:1–45, 2004.
12. Universität Karlsruhe (TH), Rechenzentrum. *HP XC4000 User Guide*, 2009.

The Impact of Secondary Mean Vortices on Turbulent Separation in 3D Diffusers

Dominic von Terzi, Hayder Schneider, and Hans-Jörg Bauer

Institut für Thermische Strömungsmaschinen, Karlsruher Institut für Technologie (KIT), Kaiserstr. 12, D-76131 Karlsruhe, Germany, vonterzi@kit.edu, hayder.schneider@kit.edu, hans-joerg.bauer@kit.edu

Summary. In rectangular ducts with fully-developed turbulent flow, mean vortices in the corners form secondary flow patterns whose energy contents is orders of magnitude lower than that of the flow in the streamwise direction. In the present numerical experiments, it is demonstrated using Large Eddy Simulations (LES) that these Mean Secondary Vortices (MSV) exert a profound influence on flow separation in three-dimensional asymmetric diffusers following such a duct. By removing, enhancing or switching the sense of rotation of the MSV in the inlet duct of two diffusers the shape, location and extent of separation zones farther downstream were drastically altered and, hence, the performance of the device. These results provide an explanation why eddy-viscosity based Reynolds-Averaged Navier–Stokes (RANS) models, that inherently cannot account for MSV, fail in predicting even the location of the separated flow in such diffusers.

1 Introduction

Diffusers are commonly used in industrial applications where the flow needs to be decelerated. Their task is to recover in form of pressure as much of the kinetic energy as possible. A typical example is the turbulent flow in the connecting duct between a compressor and the combustion chamber in a jet engine. In general, the diffuser is three-dimensional (3D) and asymmetric in geometry. For aggressive designs or at off-design operation, it is prone to flow separation. The separation then reduces the amount of pressure that is recovered in the device and may also cause unwanted inhomogeneities in the exiting mean flow. Moreover, the separated flow is highly sensitive to geometric changes as was shown experimentally for two similar diffusers by Cherry et al. [1]. Both the reliable and accurate determination of the separated flow field in diffusers is therefore important in the design process. Unfortunately, this remains a major challenge for Computational Fluid Dynamics (CFD) based on Reynolds-Averaged Navier–Stokes (RANS) turbulence models as was demonstrated many times in the literature and, for the 3D asymmetric

W.E. Nagel et al. (eds.), *High Performance Computing in Science and Engineering '10*, DOI 10.1007/978-3-642-15748-6_26,

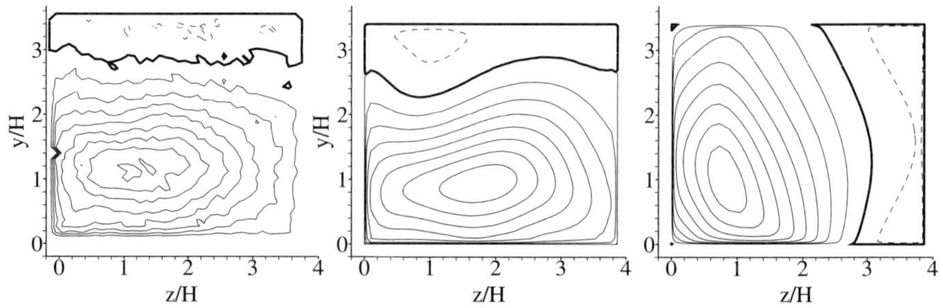

Fig. 1. Mean streamwise velocity U/U_b contours in a cross section at $x/H = 12$ for diffuser D1; from left to right: experiments of [1], LES and RANS of [4]; same U/U_b contours with interval 0.1 shown; thicker lines indicate zero velocity and dashed lines negative levels

diffusers of the experiments in [1], in two recent ERCOFTAC workshops [2, 3] and by Schneider et al. [4].

In [4], it was shown that RANS calculations with a standard two-equation eddy-viscosity turbulence closure fail in predicting the separated flow field in the 3D diffuser and cannot capture the geometric sensitivity. In contrast, Large Eddy Simulation (LES) was able to compute all major flow-field characteristics within measurement uncertainty for a given configuration. Figure 1 illustrates the magnitude of the failure of the RANS predictions. While, for the given geometry, the experiments and the LES revealed the location of the separation bubble close to the upper wall with predominantly horizontal alignment, the RANS calculation yielded a vertical alignment close to the right wall. Moreover, the RANS model overpredicted considerably the amount of separation in the diffuser. That these findings are independent of the chosen eddy-viscosity model, numerical method and computational grid can be easily verified by inspecting the data provided by different research groups in the ERCOFTAC workshops. But why the RANS models fail and what physics should be included in the modeling to remedy the situation are still in the realm of speculation.

By scrutinizing the workshop results one can recognize a marked difference between two classes of RANS models. Eddy-viscosity models always yielded results similar to those shown in Fig. 1 with misplaced separation bubbles. On the other hand, the use of a Reynolds Stress Model (RSM) resulted in very different flow fields between different groups. With respect to the alignment of the separation bubble, the RSM results sometimes, at least qualitatively, were close to the LES and experimental data. One difference between RSM and eddy-viscosity models is that the latter cannot account for mean secondary vortices (MSV) in the inlet duct at all. Although these secondary flow patterns contain only little kinetic energy, usually orders of magnitude less then for the mean flow in the streamwise direction, it is the premise of the present study that MSV have a tremendous impact on the separation dynamics within the diffuser and may explain the kind of failure of eddy-viscosity models shown in Fig. 1. To test our hypothesis, a set of numerical experiments was designed using LES. To this end, the MSV are artificially manipulated, e.g. by removing or enhancing these structures or even switching their sense of rotation. The hypothesis is then confirmed if the changes in MSV lead to the expected variations in the flow field.

2 Computational Setup

The reference experiments [1] that motivated the present study were conducted for two different diffuser geometries having the same expansion ratio, but different aspect ratios. The same development channel of length $l = 60H$ and constant rectangular cross-section was placed in front of the diffuser, thereby ensuring reproducible boundary conditions (BC). As the flow enters the diffuser (D1 or D2) it rapidly separates due to the geometric expansion. The location and extent of the separated flow in D1 and D2 are vastly different and can be attributed to the aspect ratio as the only changing parameter.

Our previous studies [4–6] included a thorough qualitative and quantitative comparisons between the numerical and experimental data. In [4], it was found that relatively coarse simulations with wall-functions using 1.6 million grid cells were sufficiently accurate to reproduce all major flow-field characteristics. In [6], a wall-resolving LES with 42.9 million grid cells was conducted and compared to coarser simulations with wall-functions and the experimental data. The wall-resolving LES delivered more accurate results, however, at significantly higher costs. For this simulation, the contribution of the subgrid-scale model was relatively small. The time-averaged fraction of turbulent to laminar viscosity ν_t/ν within the diffuser was only 0.03 in average with a maximum value of 0.17. More importantly, the study in [6] revealed that a long time-averaging of the order of $\mathcal{O}(10^2)$ flow-through times was required to obtain reliable turbulence statistics. In [5], more economical wall-resolving LES with 7.6 million cells were computed that were able to resolve secondary flow patterns in the inlet duct and yielded results in the diffuser within measurement uncertainties. The experience from the above studies allowed for deriving a computational setup that captured all physical aspects of importance for the present study while optimizing the computational costs.

2.1 Computational Domain and Boundary Conditions

Figure 2 illustrates a schematic of the computational domain representing the experimental setup [1] simulated here. Table 1 contains further information on the geometrical features of the two diffusers. In the following, all values reported here are made dimensionless using the bulk velocity $U_b = 1$ and a representative length such as the height (H), or the width (B) of the inlet channel, or the length of the diffuser (L). The origin of the Cartesian

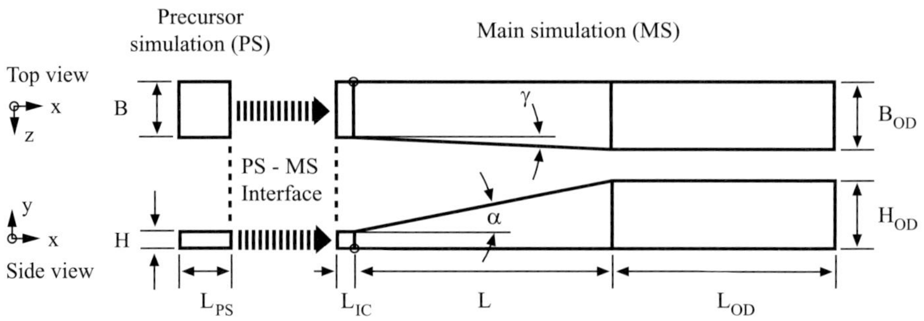

Fig. 2. Schematic of the computational domain including part of the experimental setup with flow in the positive x-direction (from left to right); circles indicate location of coordinate system

Table 1. Dimensions of diffusers D1 and D2 (lengths are made dimensionless with H)

	Dimension	D1	D2
Inlet channel	length (L_{PS})	3	
	length (L_{IC})	1	
	height (H)	1	
	width (B)	3.33	
Diffuser	length (L)	15	
	top angle of expansion (α)	11.3°	9°
	side angle of expansion (γ)	2.56°	4°
Outlet duct	length (L_{OD})	13	
	height (H_{OD})	4	3.37
	width (B_{OD})	4	4.51

coordinate system was placed at the intersection of the two non-expanding walls and the beginning of the expansion. The flow in the experiments and the simulations is incompressible with constant density $\rho = 1$. The Reynolds number based on U_{b} and H was $10,000$, and the Reynolds number based on the friction velocity u_τ in the inlet channel and H was 588.

In our computational setup, unsteady turbulent inflow data were generated by a precursor simulation (PS) running in parallel to the main simulation (MS). Both the PS and MS are connected via a common interface, see Fig. 2. The advantage of using a precursor technique is that only a relatively short part of the development channel with length $L_{\mathrm{PS}} = 3H$ needs to be simulated. This results in a significant cost reduction compared to simulating the whole development channel, while still providing time-dependant realistic flow structures. These time-dependant flow structures are generated by enforcing periodicity within the section of the inlet channel with length L_{PS} and by additionally using a controller to enforce the experimental mass flux. The data is fed at every time-step into the MS through the interface with the PS. The interface is placed one H upstream of the diffuser inlet. At the outlet, a convective BC was applied in conjunction with a buffer zone in which the viscosity is increased by a constant factor. At the walls, no-slip BC were applied.

2.2 Computational Grid and Averaging

All simulations were carried out on the same computational grid covering the solution domain $x/H \in [-4; 28]$ and containing 22×10^6 grid cells ($N_x \times N_y \times N_z = 896 \times 128 \times 192$). The grid was equidistantly spaced in the streamwise direction and stretched towards the no-slip walls, such that in the inlet channel $y^+_{\mathrm{wall}} = z^+_{\mathrm{wall}} = 1.0$, $\Delta x^+_{\mathrm{max}} = 21.0$, $\Delta y^+_{\mathrm{max}} = 6.0$, and $\Delta z^+_{\mathrm{max}} = 21.0$. The time step was $4 \times 10^{-3} H/U_{\mathrm{b}}$, resulting in a CFL limit of less than 0.65. In total, 600, 000 time-steps were computed. Averaging started after $10L/U_{\mathrm{b}}$, with a total averaging time T_{ave} of $150L/U_{\mathrm{b}}$.

2.3 Numerical Method and Turbulence Model

The simulations were performed with the Finite Volume Code LESOCC2 (Large Eddy Simulation On Curvilinear Coordinates) in Fortran 95, developed at KIT. LESOCC2 solves the incompressible, three-dimensional, time-dependent, filtered Navier–Stokes equations on body-fitted, collocated, curvilinear, block-structured grids using second-order accurate central differences for the discretization of the convective and viscous fluxes. Time advancement is accomplished by an explicit, low-storage Runge–Kutta method. Conservation of mass is achieved by the SIMPLE algorithm with the pressure-correction equation being solved using the strongly implicit procedure (SIP) of Stone. The momentum interpolation method of Rhie and Chow is employed to prevent pressure-velocity decoupling and associated oscillations. Parallelization is achieved via domain decomposition with the use of ghost cells and MPI for the data transfer. The standard Smagorinsky model with $C_{\mathrm{s}} = 0.065$ and van Driest wall-damping was employed as subgrid-scale model for the LES reported here.

LESOCC2 has been used in numerous studies and substantial experience has been gained with respect to its numerical properties and its performance on various hardware platforms, see [5] for more information on performance aspects. All simulations were run using 256 processor cores, i.e. 64 nodes, on the HP XC-4000 system at KIT. The computational cost per simulation was of the order of 100, 000 CPU hours.

2.4 Methodology and Simulation Overview

Before describing the methodology to manipulate the mean secondary vortices at the diffuser inlet, we briefly recall some characteristics of turbulent channel flow. In our approach, the fully developed turbulent channel flow is

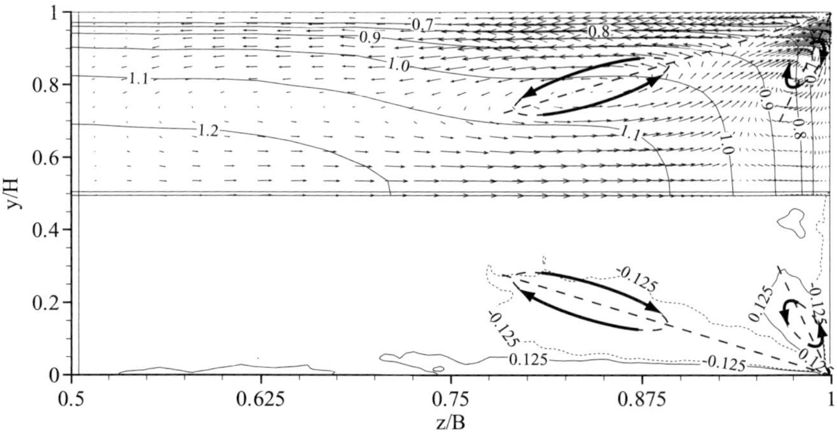

Fig. 3. Part of the cross section in the inlet channel $z/B \in [0.5; 1]$: mean streamwise velocity contours with superimposed mean tangential velocity vectors (upper half, $y/H \in [0.5; 1]$; every third and fifth vector is shown in the z- and y-direction, respectively); mean streamwise vorticity contours (lower half, $y/H \in [0; 0.5]$). Mean secondary vortices are schematically highlighted for clarity

generated by a PS of a short section with the dimensions given in Table 1. The instantaneous, 3D velocity field $\mathbf{u} = (u, v, w)^T$ is decomposed into the sum of its mean $\mathbf{U} = (U, V, W)^T$ and the fluctuating part $\mathbf{u'} = (u', v', w')^T$, i.e.

$$\mathbf{u} = \mathbf{U} + \mathbf{u'}. \tag{1}$$

The mean velocity component $\mathbf{U}$ can be rewritten as

$$\mathbf{U} = (U, 0, 0)^T + (0, V, W)^T, \tag{2}$$

where the first summand represents the streamwise velocity component and the second summand represents both the vertical and lateral velocity components. The mean streamwise velocity U is important for determining both the Reynolds number and the mass flux, whereas the mean vertical and lateral velocity components V and W, respectively, represent the secondary flow patterns within a cross-section. Figure 3 illustrates such secondary flow patterns in the inlet channel from the PS by showing mean tangential velocity vectors (upper half, $y/H \in [0.5; 1]$ and the mean streamwise vorticity ω_x contours (lower half, $y/H \in [0; 0.5]$), where

$$\omega_x = \frac{\partial W}{\partial y} - \frac{\partial V}{\partial z}. \tag{3}$$

Visual inspection of the ω_x contours and the velocity vectors reveals secondary vortices of the mean flow in the channel corners. These MSV are part of the equilibrium solution and are schematically highlighted for clarity in Fig. 3.

A pair of counter-rotating vortices can be identified in every corner of the channel. Both vortices are found to be elongated and inclined. However, the vortex nearer to the channel centerline is larger in size and extent than the outer vortex. Note that U/U_b is of the order of unity, whereas V/U_b or W/U_b are $\mathcal{O}(10^{-2})$. It follows that the kinetic energy of U/U_b is also $\mathcal{O}(1)$, while the kinetic energy of the MSV is only $\mathcal{O}(10^{-4})$.

The focus of our study was to investigate the importance of the MSV on the separation dynamics in the two diffuser geometries. Therefore, in order to test whether the MSV have an influence on the flow field, the structures were artificially altered in a controlled fashion. To this end, different inlet BC were applied at the interface between the PS and MS. All calculations involve fully-developed turbulence. Furthermore, the setup was chosen such that both the mass flux and the Reynolds number are kept constant. The only parameters which vary are the MSV at the PS–MS interface. This interface plays a crucial role in our computational campaign since it provides the means for manipulating the MSV. In order to comply with the above mentioned conditions, i.e. imposing time-dependant fluctuations, constant mass flux, constant Reynolds number, and variable MSV, the procedure for prescribing a consistent inlet BC for the MS reads as follows:

$$\mathbf{u}|_{\mathrm{MS}} = (U,0,0)^T|_{\mathrm{PS}} + \Phi \cdot (0,V,W)^T|_{\mathrm{PS}} + (u',v',w')^T|_{\mathrm{PS}}, \qquad (4)$$

where Φ is an adjustable parameter. For the results shown here, Φ was chosen as an integer constant.

In total, seven different simulations were conducted for the geometries D1 and D2 so far. Table 2 compiles more information such as denotation, purpose and value chosen for Φ. In the BSL case with $\Phi = 1$, no modifications to the MSV were applied. Hence, it corresponds to the reference case with equilibrium turbulent channel flow as inlet BC and a natural state of the secondary flow structures. In the SV0 case with $\Phi = 0$, the MSV were removed at the inlet boundary. This is what an eddy-viscosity RANS model would deliver. Contrary to a RANS calculation, however, the structures start to recover right after the inlet, but are still considerably weaker than in their natural state once they reach the diffuser. In the SVP2 case with $\Phi = 2$ the

Table 2. Overview of the simulation setup: D1, D2, BSL, SV, P and N denote diffuser geometry 1, diffuser geometry 2, baseline, secondary vortices, positive and negative respectively

Simulation	Purpose (modification of MSV)	Φ	Energy change of structures	D1	D2
BSL	baseline case (natural state)	1	0	√	√
SV0	weakening	0	$\mathcal{O}(10^{-4})$	√	√
SVP2	strengthening	2	$\mathcal{O}(10^{-4})$	√	√
SVN9	enhanced + opposite sense of rotation	−9	$\mathcal{O}(10^{-2})$	√	X

magnitude of the MSV was increased by a factor of two at the inlet boundary. Finally, in the SVN9 case with $\Phi = -9$, both the sense of rotation and the magnitude of the MSV were altered.

3 Results

In the following, computational results of mean flow quantities are shown for the diffuser D1 geometry. Results for diffuser D2, as well as instantaneous effects will be presented in [7]. Furthermore, comparisons with the experimental data are omitted since these are already reported for the natural case in our previous studies [4–6].

Figure 4 depicts mean streamwise velocity U/U_b contours with superimposed ω_x contours in the cross-section $x/H = 0$, i.e. where the diffuser starts. Recall that the inlet BC was placed one H upstream of this location and that the appropriate values for the (instantaneous and mean) velocities can be constructed with the information given in Table 2 and in (4). According to the inlet BC, there are considerable differences in the size, extent and sense of rotation of the MSV in the corners of the cross-section. The appearance of the MSV in the BSL case corresponds to their natural state, whereas all other MSV shapes result from artificially imposed BC. The MSV shown in Fig. 4 testify for the consistency of the inlet BC and demonstrate that the MSV are indeed altered. The MSV for case SV0 are very small in their size as they redevelop along L_IC. In case SVP2, the MSV are stronger but still bear a close resemblance in shape and rotation to those observed in the BSL case. This is consistent with the increase of their intensity by a factor of two. In contrast, the drastic change of the flow structures in case SVN9 results in considerably enlarged and intensified MSV with the opposite sense of rotation. In all simulations, the effect of the MSV on the U/U_b contour shape can be noted. Thereby, the most impact can be discerned for case SVN9. Furthermore, as expected, the flow field exhibits slight asymmetries due to the upstream effect of the one-sided expansions.

Figure 5 illustrates the consequences of the four different inlet BC farther downstream within diffuser D1 at selected streamwise locations. Mean streamwise velocity U/U_b contours are shown and the separation region is highlighted by the bold $U/U_\text{b} = 0$ contour. For all simulations, the separation starts to develop at the upper right corner of the cross-section. However, the separation bubble develops visibly differently as a function of the inlet BC. For the cases BSL, SV0 and SVP2, the separation zone is located at the upper corner of the diffuser developing its more or less horizontal alignment (as in the experiments). Although the change in the kinetic energy of the MSV is only $\mathcal{O}(10^{-4})$ for the SV0 and SVP2 cases, this change can cause a relocation of the core of the separation zone. This core is indicated by the dashed negative U/U_b contours, and shifts from the upper right in the BSL case to the upper left in case SV0 increasing the horizontal alignment. For case SVP2 the

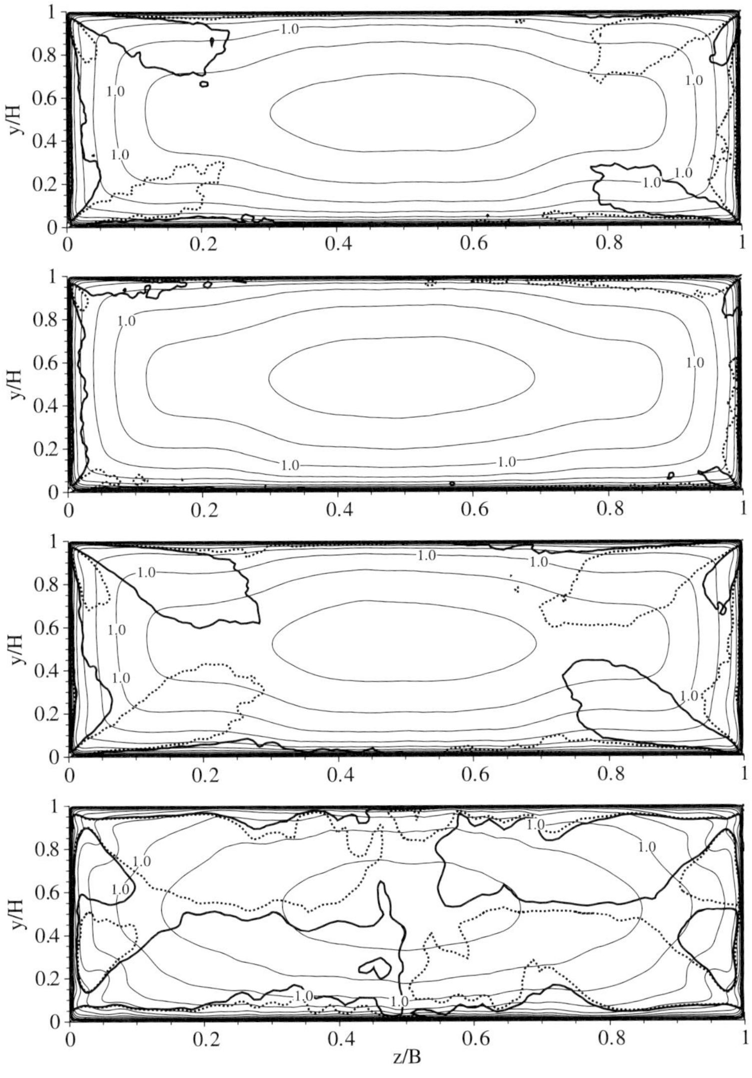

Fig. 4. Mean streamwise velocity U/U_b contours with superimposed mean streamwise vorticity ω_x contours in a cross-section at $x/H = 0$ for diffuser D1 and cases BSL, SV0, SVP2, SVN9 (from top to bottom); same U/U_b contours with interval 0.1 shown; thicker lines indicate $\omega_x \pm 0.125$ and dashed lines negative levels

opposite trend can be observed. For case SVN9, it is even possible to completely relocate the entire separation bubble. This is achieved by switching the sense of rotation of the MSV and by increasing their kinetic energy by two orders of magnitude to $\mathcal{O}(10^{-2})$. Still, the amount of energy in the MSV is rather small compared to the main flow. Nevertheless, Fig. 5 shows how the separation bubble is shifted to the right wall in case SVN9 (similar as in

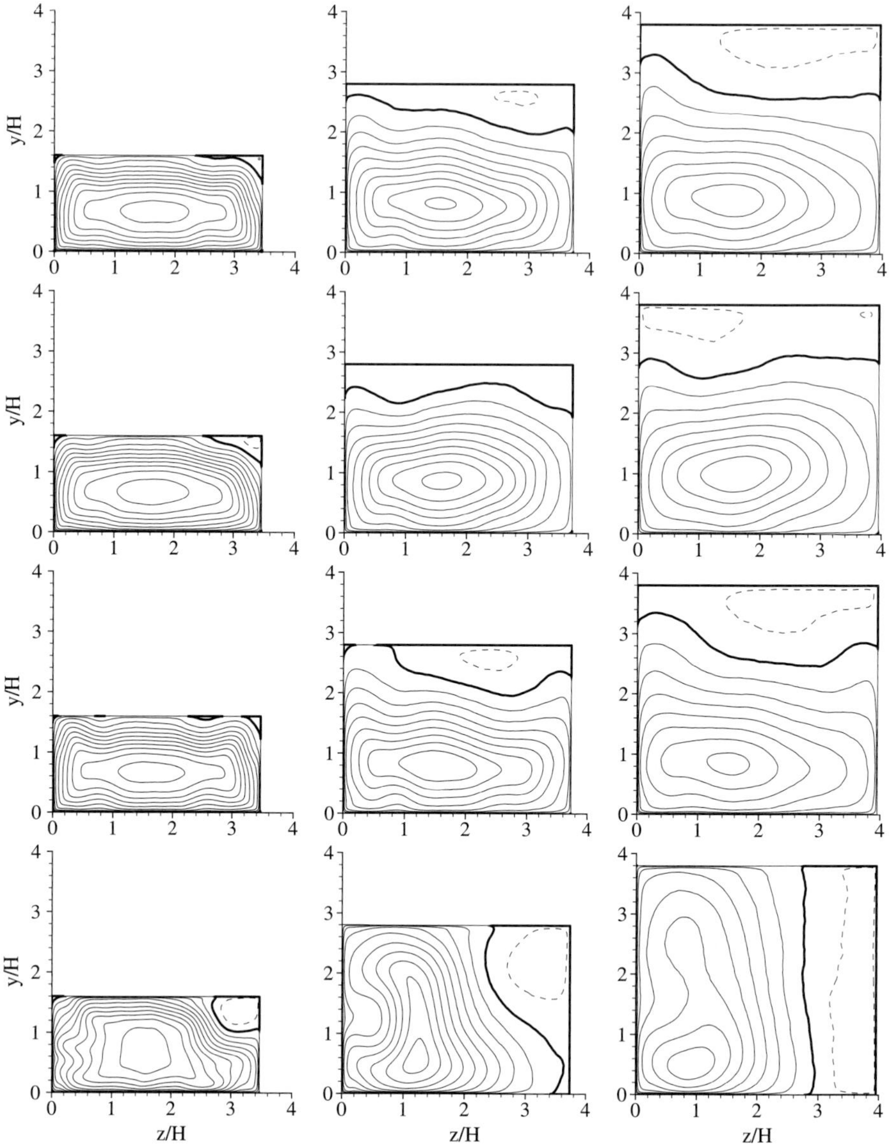

Fig. 5. Mean streamwise velocity U/U_b contours in cross-sections at $x/H = 3$, 9 and 14 (from left to right) for diffuser D1 and cases BSL, SV0, SVP2, and SVN9 (from top to bottom); same U/U_b contours with interval 0.1 shown; thicker lines indicate zero-velocity and dashed lines negative levels

the RANS prediction of Fig. 1). From the simulations so far, we cannot know whether it is the change in the sense of rotation or the increase in magnitude that exerts this drastic effect, but additional simulations with only one of these manipulations will be able to identify the cause.

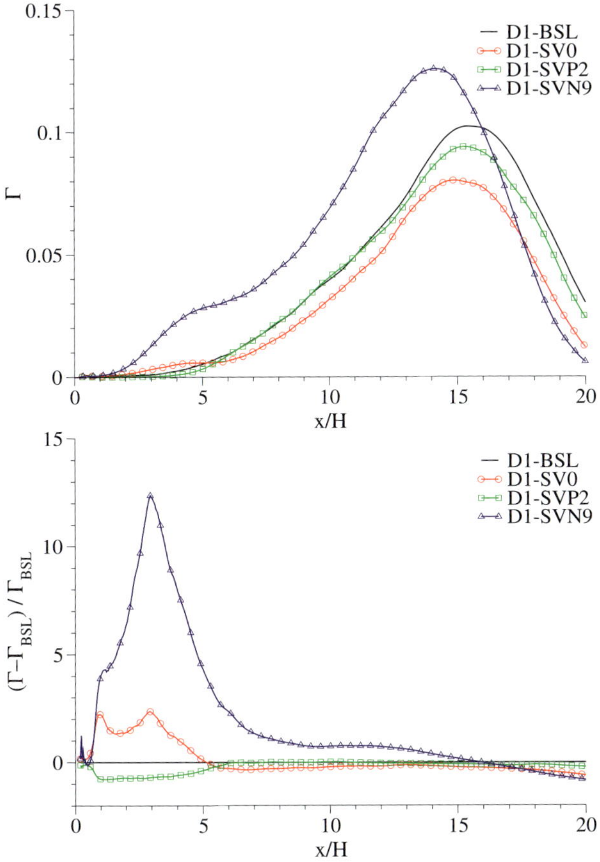

Fig. 6. Magnitude of negative mass flux Γ (top) and its relative deviation with respect to the baseline case $(\Gamma-\Gamma_{\mathrm{BSL}})/\Gamma_{\mathrm{BSL}}$ (bottom) along the streamwise coordinate x/H in diffuser D1

The U/U_{b} contour plots are valuable for determining the location and the approximate extent of the separation bubble at selected locations within the diffuser. However, it is difficult to exactly quantify the global performance of the device for a given setup. Will the manipulation only shift the separation zone within a cross-section or is their also a change in extent? Therefore, integral parameters that provide such information will be examined in the following. In order to quantify the amount of reverse flow, we introduce the magnitude of negative mass flux Γ per cross-section along the diffuser, and the relative deviation of Γ in the different cases SV0, SVP2 and SVN9 with respect to the BSL case. Both quantities allow for a global assessment of the performance variation due to the different BC and are shown in Fig. 6. As the cross-section of the diffuser increases with increasing streamwise coordinate x/H the integral amount of reverse flow in a cross-section, i.e. the magnitude

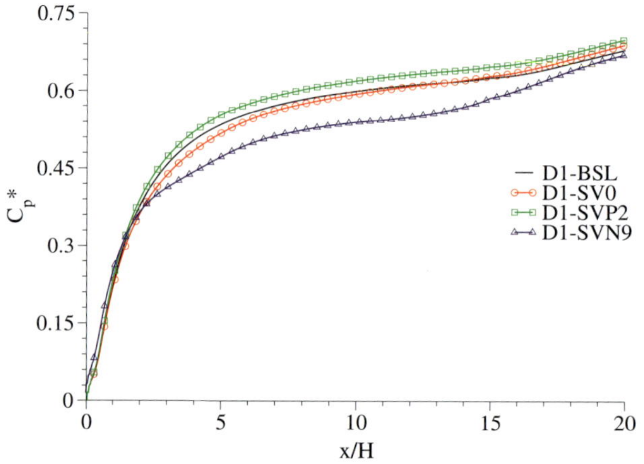

Fig. 7. Integral pressure coefficient C_P^* along streamwise coordinate x/H for diffuser D1

of negative mass flux Γ, increases due to the adverse pressure gradient. In all simulations, the location of the maximum reverse flow is found in the vicinity of the diffuser outlet at $x/H = 15$. While Γ of SVP2 follows the distribution of the BSL case, the amounts of reverse flow for cases SV0 and, in particular, SVN9 deviate more visibly from the reference case. The differences become even more evident by examining the relative deviation $(\Gamma - \Gamma_{\mathrm{BSL}})/\Gamma_{\mathrm{BSL}}$ between the simulations in Fig. 6.

A typical parameter for assessing the performance of a diffuser is the pressure coefficient C_P which allows for determining the pressure losses due to friction at walls. Commonly, C_P is taken along the streamwise coordinate for a single location in the cross-section, e.g. in the experiments in [8] or in the simulations in [4, 5]. However, in the 3D asymmetric diffuser, the pressure is not uniform across the cross-section. Therefore, for quantifying the losses, it is better to integrate the pressure along the circumferential direction s for each streamwise location weighted by the local circumference Λ. The integral pressure coefficient is then

$$C_P^* = \frac{1}{\Lambda} \oint \frac{p(s) - p_{\mathrm{ref}}}{\frac{1}{2}\rho U_{\mathrm{b}}^2} ds. \tag{5}$$

Note that both ρ and U_{b} are by definition unity. The results are shown in Fig. 7 and confirm the observations using the measure for the amount of reverse flow Γ: Manipulating the MSV can lead to a performance change of the diffuser. Reinforcing the natural state of the mean secondary vortices in the inlet duct increases the pressure recovery at the diffuser exit (at $x/H = 15$), whereas working against the natural MSV by switching their sense of rotation reduces the exit pressure.

4 Conclusions

In the present numerical experiments, it was demonstrated using LES that naturally occurring mean secondary flow patterns in the inlet duct of a 3D asymmetric diffuser impact strongly on separation inside the diffuser and on the performance of the device. By removing, enhancing or switching the sense of rotation of the MSV in the inlet duct the shape, location and extent of separation zones farther downstream were drastically altered. This in turn influenced the pressure recovery. The profound effect of the MSV can explain why experimentalists and engineers in industry have repeatedly reported a sensitivity of the separated flow in diffusers on the aspect ratio of the inlet. This is particularly true, since the mean vortices in the duct corners contain only very little kinetic energy, usually orders of magnitude lower than that of the mean flow in the streamwise direction. A relatively small amount of energy input can hence have an effect of the order of unity in the diffuser! These results also provide an explanation why eddy-viscosity based Reynolds-Averaged Navier–Stokes (RANS) models, that inherently cannot account for MSV, fail in predicting even the location of the separated flow in such diffusers.

Acknowledgements. The work reported here was carried out within the "Research Group Turbo-DNS" at the Institut für Thermische Strömungsmaschinen. Its financial support by means of the German Excellence Initiative and Rolls–Royce Deutschland is gratefully acknowledged. The authors are grateful for the provision of computer time by the Steinbuch Centre for Computing.

References

1. E.M. Cherry, C.J. Elkins, and J.K. Eaton. Geometric sensitivity of three-dimensional separated flows. *Int. J. Heat and Fluid Flow*, 29(3):803–811, 2008.
2. G. Brenn, S. Jakirlić, and H. Steiner, editors. 13th *SIG15 ERCOFTAC/IAHR Workshop on Refined Turbulence Modelling*, 2008.
3. S. Jakirlić, G. Kadavelil, E. Sirbubalo, and D. Borello, editors. 14th *SIG15 ERCOFTAC/IAHR Workshop on Refined Turbulence Modelling*, 2009.
4. H. Schneider, D.A. von Terzi, H.-J. Bauer, and W. Rodi. Reliable and accurate prediction of three-dimensional separation in asymmetric diffusers using Large-Eddy Simulation. *J. Fluids Eng.*, 132(3):031101–1–7, 2010.
5. D.A. von Terzi, H. Schneider, and J. Fröhlich. Diffusers with three-dimensional separation as test bed for hybrid LES/RANS methods. In W.E. Nagel, D.B. Kröner, and M.M. Resch, editors, *High Performance Computing in Science and Engineering '09*. Springer, Berlin 2010. ISBN 978-3-642-04664-3.
6. H. Schneider, D.A. von Terzi, H.-J. Bauer, and W. Rodi. Flow separation in a 3-d diffuser – 2. In S. Jakirlić, G. Kadavelil, E. Sirbubalo, and D. Borello, editors, 14th *SIG15 ERCOFTAC/IAHR Workshop on Refined Turbulence Modelling*, 2009.

7. H. Schneider, D.A. von Terzi, H.-J. Bauer, and W. Rodi. Impacts of secondary vortices on separation dynamics in three-dimensional asymmetric diffusers. In 8th *Workshop on Direct and Large Eddy Simulation*, Technical University of Eindhoven, The Netherlands, July 2010.
8. E.M. Cherry, C.J. Elkins, and J.K. Eaton. Pressure measurements in a three-dimensional separated diffuser. *Int. J. Heat and Fluid Flow*, 30(1):1–2, 2009.

Time-Dependent Three-Dimensional Simulation of the Turbulent Flow and Heat Transfer in Czochralski Crystal Growth Including the Three-Phase Boundary Movement

A. Raufeisen[1], M. Breuer[2], T. Botsch[1], and A. Delgado[3]

[1] Faculty of Process Engineering (VT), University of Applied Sciences Nuremberg, Wassertorstraße 10, 90489 Nuremberg, Germany
alexander.raufeisen@ohm-hochschule.de
[2] Department of Fluid Mechanics (PfS), Helmut Schmidt University Hamburg, Holstenhofweg 85, 22043 Hamburg, Germany, breuer@hsu-hh.de
[3] Institute of Fluid Mechanics (LSTM), University of Erlangen-Nuremberg, Cauerstraße 4, 91058 Erlangen, Germany

Summary. In the literature, numerical computations of the Czochralski process for crystal growth are conducted using a quasi-steady state assumption for the crystallization neglecting time-dependent effects. In the present work, an algorithm is developed, which allows to calculate the transient behavior of the crystallization interface including the movement of the three-phase boundary and the free surface of the melt. Thus, in conjunction with the computation of the turbulent melt flow and heat transfer, more realistic predictions of the crystal growth can be achieved. Test cases show that with the algorithm, realistic phenomena like crystal diameter increase and decrease during the growth process could be reproduced.

1 Introduction

For the industrial production of large silicon single crystals, the Czochralski (Cz) method is used most frequently. In this process, the liquid silicon is contained in an open crucible heated from the side (see Fig. 1). The crucible is rotating, while the counterrotating crystal is slowly pulled from the melt. Due to this setup, centrifugal and Coriolis forces, buoyancy, and Marangoni convection [9] occur in the fluid as well as thermal radiation from the surfaces. Furthermore, oxygen can be dissolved from the crucible and is transported in the melt [13]. This complex configuration generates a turbulent, three-dimensional and time-dependent flow [2]. The quality of the resulting crystal in terms of homogeneity of properties, perfect crystal lattice (no defects or

W.E. Nagel et al. (eds.), *High Performance Computing in Science and Engineering '10*, DOI 10.1007/978-3-642-15748-6_27,

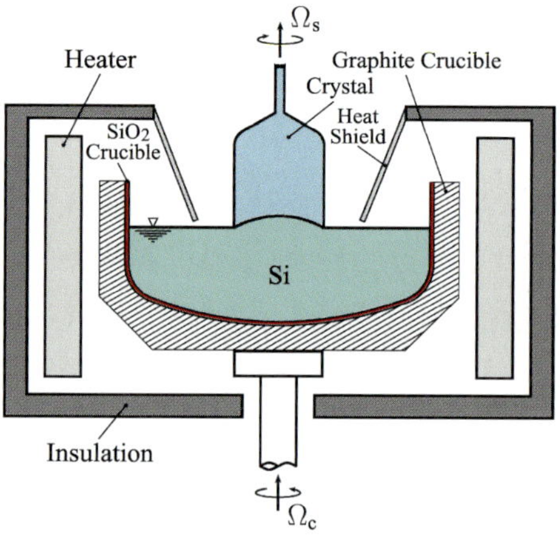

Fig. 1. Scheme of a Czochralski setup for crystal growth

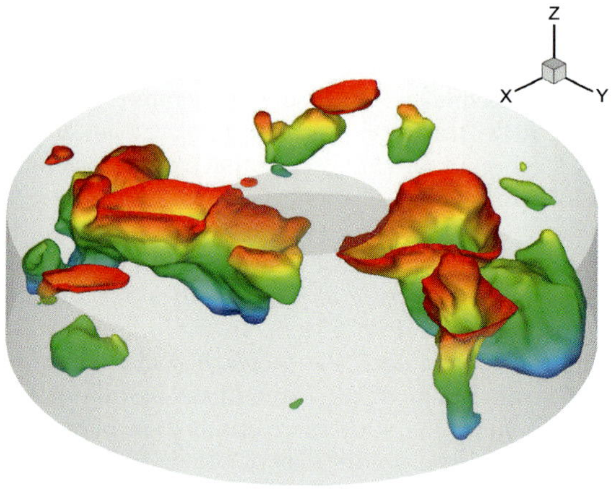

Fig. 2. DNS of the flow and heat transfer in a cylindrical Czochralski crucible [12]: Isosurfaces of dimensionless temperature difference $T_{instant} - T_{mean} = 0.1$, contour colored by z-coordinate. Turbulent thermal plumes can be recognized as the dominating flow structure

dislocations) and low thermal stresses, strongly depends on the shape of the phase interface between melt and crystal during the growth process. The interface in turn is essentially influenced by the complex melt flow.

The turbulent structure of the melt flow in an idealized configuration was analyzed in previous studies using Direct Numerical Simulations (DNS) and Large-Eddy Simulations (LES) [11, 12]. In Fig. 2, characteristic buoyant thermal plumes can be seen. In these investigations, however, the phase interface was fixed to an ideal flat shape.

In the present work, for a better understanding and a more accurate modeling of the process, the melt-crystal interface is computed explicitly. However,

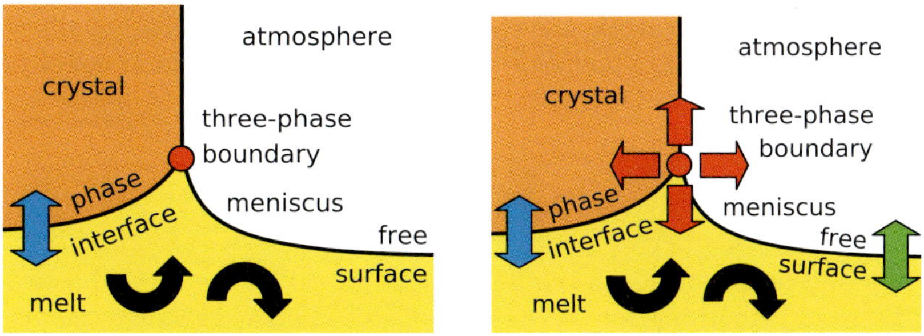

(a) Quasi-steady state (b) Fully dynamic

Fig. 3. 2D-sketch of the three-phase boundary in Czochralski crystal growth

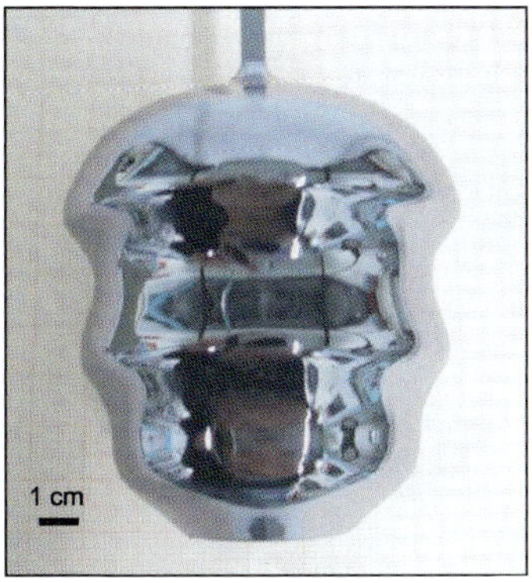

Fig. 4. GaAs crystal with diameter fluctuations due to unstable process conditions [14]

it is not enough to consider the movement of the melt-crystal interface alone. Also the dynamics of the three-phase boundary (TPB), where the melt, the crystal and the surrounding atmosphere meet, has to be taken into account (see Fig. 3), which furthermore necessitates the explicit computing of the free surface. This is a difficult task, and to the authors' knowledge, so far no-one has tried it in three dimensions. Instead, the three-phase boundary, which is a line in a general 3D geometry, is fixed in space in the simulations, assuming a "quasi-steady state" for the crystallization interface. This implies that the diameter of the crystal, the height of the melt, and the phase interface are remaining constant over time. This approach may be justified if the characteristic time of the growth process is much larger than the characteristic time for the heat transport. But especially in the starting and ending phase of the crystallization process (shouldering and tailing), this is not accurate enough, because the diameter changes significantly (see [3]). Also during the bulk growth of the crystal, variations can occur, see Fig. 4, e.g.

Thus, in the present work, the movement of the three-phase boundary will be computed within the simulation of the fluid flow and heat transfer in the crucible and crystal.

2 Mathematical Formulation

The flow and heat transfer in the melt are governed by the three-dimensional Navier-Stokes equations for an incompressible fluid expressing the conservation of mass, momentum and energy. Buoyancy is taken into account by the Boussinesq approximation. The simulations are conducted in a rotating frame of reference with the crucible as the reference system, thus additional source terms arise in the momentum equations due to Coriolis and centrifugal forces. The equations are non-dimensionalized with appropriate normalizations (see [11]) and read

$$\frac{\partial u_i}{\partial x_i} = 0, \tag{1}$$

$$\frac{\partial u_i}{\partial t} + \frac{\partial (u_i u_j)}{\partial x_j} = -\frac{\partial P}{\partial x_i} - \frac{1}{Re}\frac{\partial \tau_{ij}^{mol}}{\partial x_j} - \frac{Gr}{Re^2}T + \frac{Re_\omega}{Re}\epsilon_{ijk}\,\omega_j\,u_k - \frac{Re_\omega^2}{Re^2}\epsilon_{ijk}\,\omega_j\,(\epsilon_{klm}\,\omega_l\,x_m), \tag{2}$$

$$\frac{\partial T}{\partial t} + \frac{\partial (u_i T)}{\partial x_i} = \frac{1}{Re\,Pr}\frac{\partial^2 T}{\partial x_i \partial x_i}, \tag{3}$$

For the LES technique applied in the present study, the Navier-Stokes equations (1)–(3) are filtered in space, i.e., the flow quantities are divided into a “grid-scale” and a “subgrid-scale” (SGS) part. Here this is done implicitly, i.e., the filter width is the cell width of the computational grid. After filtering and reformulation one obtains the Navier-Stokes equations for the large scales and additionally a subgrid-scale stress tensor τ_{ij}^{SGS} and heat flux q_i^{SGS}, which have to be approximated. Using Smagorinsky's model [15], this is done by determining a turbulent eddy viscosity μ_t using the so-called Smagorinsky constant C_s and calculating a turbulent eddy conductivity k_t using a turbulent Prandtl number Pr_t. The idea of the dynamic procedure invented by Germano et al. [8], which is used in the present simulations, is to dynamically determine the SGS model parameters (here C_s, because the Smagorinsky model is chosen as the base model) from the smallest resolved scales of the filtered domain. Therefore, another (coarser) filter function, a so-called test-filter, is applied to the basic equations. Thus, another subgrid-scale stress tensor arises, which is used to determine C_s dynamically during the simulation based on the Germano identity. The turbulent Prandtl number is presently fixed to $Pr_t = 0.9$ [12].

For the tracking of the interface between melt and crystal, the free surface and the three-phase boundary, the boundaries of the computational domain are varying in time and thus the numerical grid has to be adjusted. Besides other numerical techniques to account for these variations, the most popular one is the so-called Arbitrary Lagrangian-Eulerian (ALE) formulation. Here the conservation equations for mass, momentum and energy are re-formulated for a temporally varying domain. The ALE formulation ensures that the space conservation law is fulfilled. Otherwise, artificial mass sources or sinks due to the grid movement would be generated. The space conservation law balances the volume change with the grid movement.

The variation of the boundaries necessitates the reconstruction of the numerical grid in the inner domain. This is achieved by a transfinite interpolation method [17] which redistributes the mesh nodes smoothly.

Physically, the movement of the melt-crystal interface can be determined by the Stefan condition calculating the difference of the heat flux densities in the liquid (l) and solid (s) part,

$$\rho \Delta H \, (u^I) n_i = \left(k \frac{\partial T}{\partial x_i} \right)_s n_i - \left(k \frac{\partial T}{\partial x_i} \right)_l n_i, \tag{4}$$

where ΔH denotes the latent heat of fusion, u^I the interface velocity, and k the thermal conductivity. Density changes across the interface (from liquid to solid) are neglected ($\rho_l \approx \rho_s$). Additionally, from the movement of the interface by crystal growth, the pulling velocity, with which the crystal is pulled in vertical direction, is subtracted. The pulling velocity is either fixed to a certain value or computed from the average of the crystal growth rate, such that an optimum control of the process is achieved.

Thermal radiation from the free surface of the melt and the crystal surface is considered by the Stefan-Boltzmann equation using an emissivity of $\varepsilon = 0.3$ for the melt and $\varepsilon = 0.6$ for the solid crystal, with an environmental temperature of $T_{env} = 1600$ K.

The movement of the free surface is computed by stating a force balance across the free surface in normal (n) and tangential (t) directions (Laplace-Young equations)

$$(n_i \, T_{ij})_l \, n_j + \sigma K = -(n_i \, T_{ij})_g \, n_j \tag{5}$$

$$(n_i \, T_{ij})_l \, t_j - \frac{\partial \sigma}{\partial x_t} = (n_i \, T_{ij})_g \, t_j, \tag{6}$$

where σ denotes the surface tension, K the curvature of the surface, T_{ij} the complete stress tensor including the pressure, and n_j and t_j the normal and tangential unit vectors, respectively. Due to the fact that the gas (g) density is much smaller than the density of the liquid (l), the viscous part of T_{ij} can be neglected, such that the equations are reduced to

$$- p_l + 2\mu \frac{\partial u_n}{\partial x_n} + \sigma K = -p_g \tag{7}$$

$$\mu \left(\frac{\partial u_t}{\partial x_n} + \frac{\partial u_n}{\partial x_t} \right) - \frac{\partial \sigma}{\partial x_t} = 0. \tag{8}$$

This is also called the "dynamic" condition. For the completeness of the description of the free surface, also the kinematic condition

$$u_{n,l} = u_{n,g} = u_{n,s}, \tag{9}$$

which means that the velocity is continuous across the surface (s), has to be fulfilled.

The algorithm for the movement of the free surface and the three-phase boundary together with the phase interface is explained in the following section.

3 Numerical Method

3.1 General Solver Features

For the numerical simulations the general-purpose CFD package FASTEST-3D developed by LSTM Erlangen [4, 5] is used.

With FASTEST-3D laminar as well as turbulent steady and unsteady flows including heat and mass transfer can be simulated numerically. The three-dimensional incompressible Navier-Stokes equations are solved based on a fully conservative finite-volume discretization on non-orthogonal curvilinear grids with a collocated arrangement of the variables. In order to resolve complex geometries, block-structured grids are used, i.e., the blocks are globally unstructured, but each block consists of a curvilinear structured grid [2, 6]. Second-order accurate central differences are applied for all terms. Flux blending (first-order upwind/second-order central) for the convective fluxes and deferred correction approaches are implemented but only the latter is used. For the temporal integration of the flow field, two second-order accurate schemes are available: a fully-implicit three-point backward formulation (presently used) and a Crank-Nicolson scheme. Due to the very sensitive free surface and three-phase boundary movement algorithm, which is described below, small time step sizes between $\Delta t = 0.01$ and 0.001 s must be chosen for the Cz predictions. The data are sampled for time averaging after every tenth time step. The statistics of the mean and fluctuating quantities are computed from these sampled data. Under-relaxation factors of 0.7, 0.3 and 0.9 were applied for velocity components, pressure and temperature, respectively. The SIMPLE algorithm is used to couple the velocity and pressure fields. The discretized conservation equations are solved in an iterative manner adopting the Strongly Implicit Procedure (SIP) of Stone [16]. In order to achieve convergence, the residual of each variable is brought down 5–6 orders of magnitude.

3.2 High-Performance Computing Approach

The algorithm is highly vectorized and additionally parallelized by domain decomposition with explicit message–passing based on MPI (Message Passing Interface) allowing efficient computations especially on vector–parallel machines and SMP (Symmetric Multi–Processing) clusters. Due to its recursive data structure, the SIP solver for the algebraic system of equations in FASTEST-3D is not vectorizable in a straightforward manner. However, Leister and Perić [10] showed that vectorization of the SIP solver can be achieved by avoiding data dependencies through indirect addressing and sweeping through the computational domain along diagonal planes, so called hyper-planes. Thus one sweep through the entire domain consists of hyper-planes having different vector length. Due to this variable vector length and the indirect addressing used, the performance of the vectorized SIP solver is slightly lower than the other parts of the code. Nevertheless, in a preceding project [1] it was shown that FASTEST-3D works very efficiently on NEC SX-machines with sustained performances of up to 40–50% of the peak performance. However, as the new algorithm for the three-phase boundary movement (see next section) is not fully parallelized yet, the performance of the code with the extensions is not optimal at the moment. Yet, it is expected to deliver similarly high performance when it is fully integrated in the parallel structure of the code. For the computation of the Direct Numerical Simulations (DNS) of the Czochralski case [11] containing more than 8.4 million CVs, an average performance of 8.2 GFlop/s (more than 50 % of the peak performance) and a vectorization ratio of 99.6 % was measured on the SX-8 using one node with 8 processors. The DNS was computed employing $\mathcal{LESOCC}$, which is a code with the same solution algorithm as FASTEST-3D, but with explicit time marching.

3.3 Problem-Specific Extensions

For the computation of the three-phase boundary movement, the following algorithm is employed, combining the movement of the phase interface and the free surface, see Fig. 3b.

After the SIMPLE algorithm, where the conservation equations are solved and thus a velocity and temperature field for the computational domain is obtained, the Stefan condition is applied at the phase boundary between melt and crystal. From that, the movement of the interface including the three-phase boundary is calculated. Because a collocated variable arrangement is used, the velocities are computed at the cell centers. However, the grid is defined at the cell vertices, which makes interpolation of the movement necessary. Presently this is done by trilinear interpolation.

The new position of the three-phase boundary is then transferred to the free surface algorithm, where it serves as the boundary condition for the computation of the free surface shape. At the crucible wall, a contact angle of 90° is assumed (no wetting) as the boundary condition.

For the movement of the free surface, the dynamic as well as the kinematic boundary conditions from the Laplace-Young equations are needed. The dynamic condition (7) is already applied in the SIMPLE algorithm via a static pressure boundary condition, adding the contributions from the normal

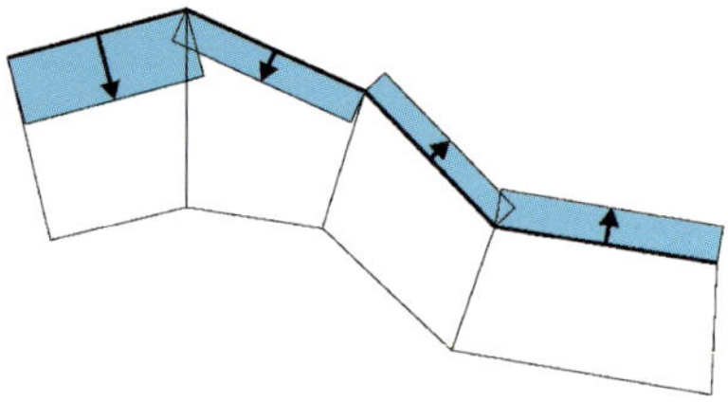

(a) Mass fluxes (blue) remaining at the free surface.

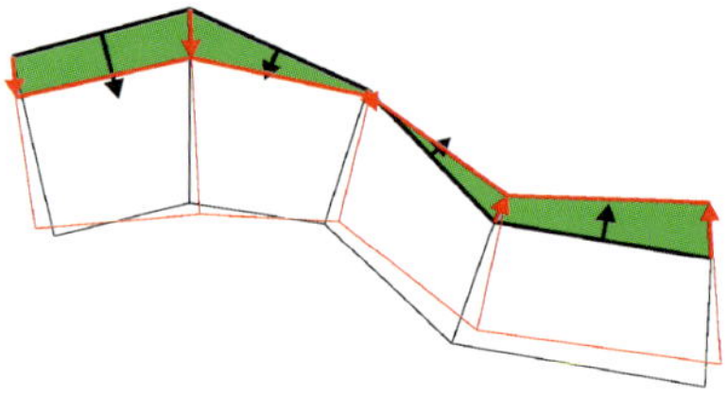

(b) Swept volumes (green) due to movement of free surface (red: new position).

Fig. 5. Schematic 2D-view of the free surface movement

stress, the curvature, the surface tension, and the hydrostatic pressure of the gas. In contrast to the classical boundary condition within the SIMPLE algorithm, which sets the gradient of the pressure correction and thus the velocity correction to zero, now the pressure correction itself has to be set to zero. That leads to a velocity correction at the boundary. Thus, after the solution of the momentum and continuity equation in the SIMPLE procedure, mass fluxes remain at the free surface (Fig. 5a). To achieve mass conservation, these fluxes have to be compensated by the movement of the free surface, i.e., by the volume swept by the boundary cells (Fig. 5b). In the ideal case, the swept volumes are equal to the mass fluxes at the free surface, i.e. the blue and green areas in Fig. 5 are equal:

$$\dot{m}_s + \rho \dot{V}_s = 0 \tag{10}$$

However, as the mass fluxes are located at the cell centers, yet the grid is defined by the vertices, the movement has to be obtained by interpolation, as with the Stefan condition. This leads to the problem of underdetermination, because there are more cell vertices than centers. Therefore, the free surface movement cannot be calculated directly, but needs to be approximated iteratively like the phase interface. Thus, the SIMPLE procedure has to be carried out several times within a single time step, until convergence is reached, i.e., no artificial mass fluxes occur at the free surface. A flow diagram of the coupled algorithm is depicted in Fig. 6.

4 Problem Details

As a test case, a cylindrical crucible geometry from Enger et al. [7] is used. The crucible has a diameter of 340 mm, the crystal of 100 mm, and the melt depth is 100 mm.

At the sidewall and the bottom of the crucible, a temperature profile taken from experiments is applied to account for the heating, ranging from approximately 1702 K to 1720 K. The crystal top is fixed to a temperature of 900 K.

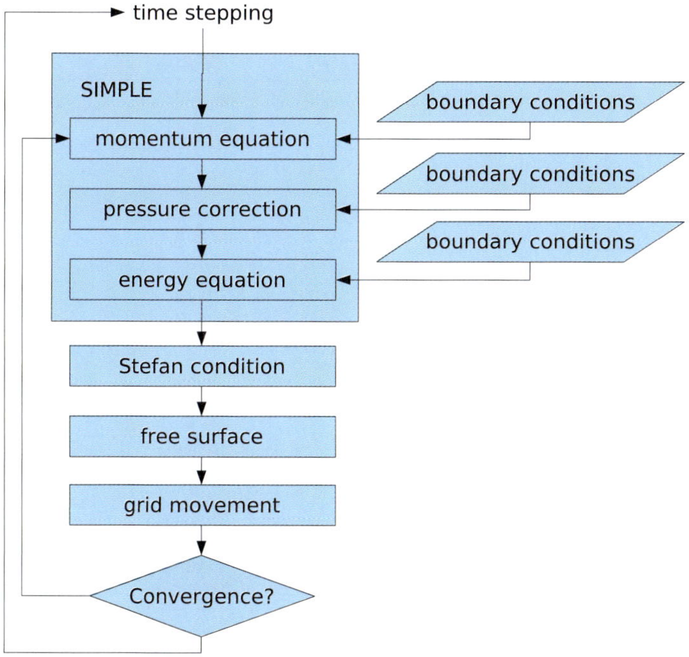

Fig. 6. Flow diagram of the coupled algorithm for the computation of the movement of the phase interface, the free surface and the three-phase boundary

The crucible and the crystal are rotating in opposite directions. Angular velocities of 5 rpm for the crucible and −20 rpm for the crystal are applied.

The relevant material properties of Si can be found in Table 1. From these values and the boundary conditions, the dimensionless numbers compiled in Table 2 are derived.

The geometry of the crucible and the crystal was spatially discretized by a block-structured mesh of O-grid type (see Fig. 7) to avoid the singularities of a polar or degenerated Cartesian grid [2]. The mesh is divided into 32 blocks for parallelization and to fit the system architecture of the NEC SX-8, whose computational nodes contain 8 processors each.

To achieve sufficiently accurate results using LES, the numerical grid was designed to contain about 600,000 control volumes (CVs). It was shown in a previous paper [12] that this resolution is sufficient to obtain results close to DNS predictions. The grid was also refined towards the walls to be able to resolve the thin boundary layers. Thus no-slip boundary conditions were applied at the solid walls.

As mentioned above, due to the lack of complete integration of the new algorithm in the parallel structure of the code, the performance of the code is not at an optimum level now. However, when the code is fully parallelized, it will show similar performance as in the DNS computations [11]. Using more

Table 1. Properties of liquid and solid silicon

Property	Symbol	Value	Unit
Liquid			
Density	ρ	2530	kg/m^3
Dynamic viscosity	μ	8.6×10^{-4}	kg/(m s)
Kinematic viscosity	ν	3.4×10^{-7}	m^2/s
Thermal expansion coefficient	β	1.4×10^{-4}	1/K
Thermal conductivity	k	67.0	J/(s m K)
Thermal diffusivity	α	2.65×10^{-5}	m^2/s
Heat capacity	c_p	1000	J/(kg K)
Temperature coefficient of surface tension	$d\sigma/dT$	-1.0×10^{-4}	N/(m K)
Melting temperature	T_{melt}	1685	K
Emissivity	ε	0.3	
Surrounding temperature	T_{env}	1600	K
Solid			
Thermal conductivity	k	22.0	J/(s m K)
Heat capacity	c_p	1060	J/(kg K)
Emissivity	ε	0.6	

Table 2. Dimensionless numbers

Number	Symbol	Formula	Value
Prandtl	Pr	$\frac{\nu}{\alpha}$	0.0128
Reynolds	Re	$\frac{R_c u_b}{\nu}$	3.24×10^4
Grashof	Gr	$\frac{\beta g R_c^3 \Delta T}{\nu^2}$	1.05×10^9
Marangoni	Ma	$-\frac{d\sigma}{dT}\frac{R_c \Delta T}{\mu\alpha}$	1.34×10^4
Rayleigh	Ra	$GrPr$	1.34×10^7

than 50 % of the peak performance of the NEC SX-8, the DNS took more than 8000 hours of CPU time to compute about 4 million time steps (ca. 1000 dimensionless units of simulation time) on a grid of ca. 8.4 million CVs. Thus, it is expected that even the calculations of the transient crystal growth process can be performed with moderate computational effort in the future.

5 Results and Discussion

Several simulations including the movement of the melt-crystal interface and the TPB were conducted. Different pulling velocities were applied to the crystal. In the first case, the pulling velocity was zero, i.e. the crystal was not pulled out of the melt. In Fig. 8, a snapshot of the local crystal growth rate V_g at the liquid-solid phase interface is depicted. It is quite high, which means

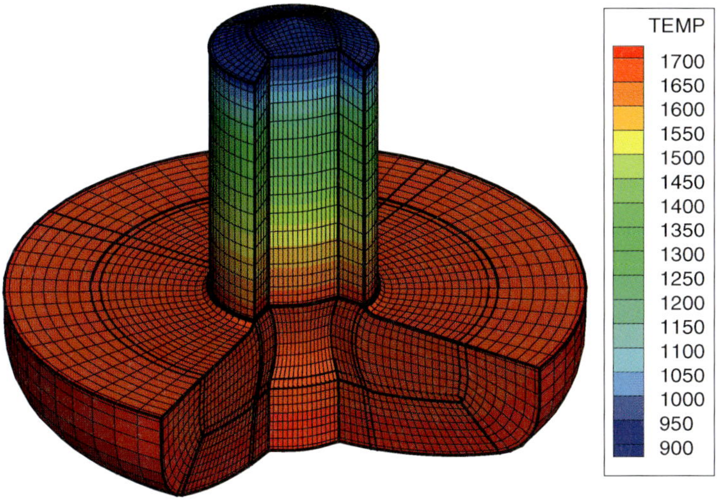

Fig. 7. Numerical grid of the crucible and crystal used for the simulations (coarsened by factor 2 in each coordinate direction). The color contours depict the temperature distribution after applying the boundary conditions and computing only heat conduction

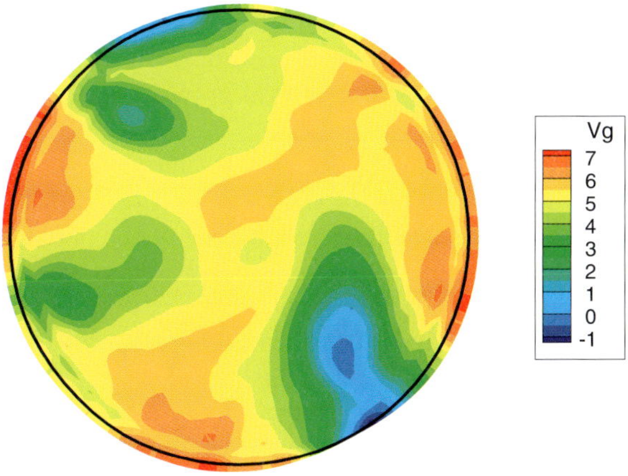

Fig. 8. Instantaneous local crystal growth rate V_g (mm/min.) at the melt-crystal interface. The black circle marks the crystal diameter at the beginning of the simulation

the crystal grows fast towards the melt, which meets the expectation for zero pulling velocity. Furthermore, the diameter increases, which can be seen by the black circle marking the diameter at the beginning of the simulation.

A more detailed view of the diameter evolution is offered in Fig. 9, where the displacement of the three-phase boundary at different time instants is depicted. It can clearly be seen that the diameter change strongly varies depend-

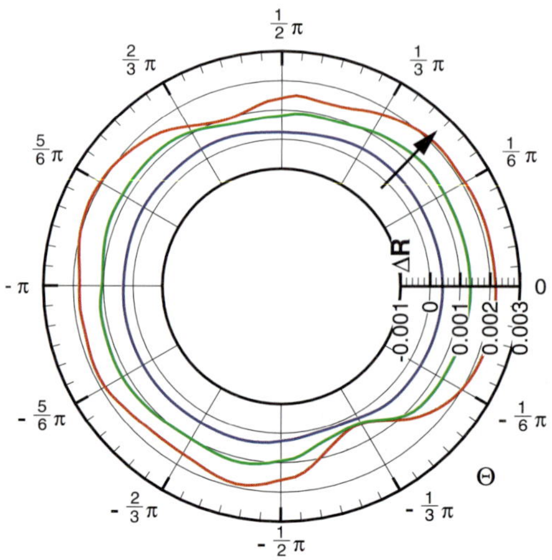

Fig. 9. Temporal evolution of the crystal diameter. Depicted is the movement ΔR (in m) of the TPB relative to the initial crystal diameter (at $\Delta R = 0$) in direction of the arrow at three different points in time of the simulated growth process

ing on the local crystal growth rate, which itself correlates with the thermal field. This can be observed in the direct comparison of Figs. 9 and 8, e.g. at the lower right corner, where the low crystal growth rate leads to the "bump" in the crystal diameter.

In Fig. 10, different stages of the simulated time-dependent crystal growth process are shown. The movement of the TPB and the resulting silhouette of the crystal can be observed as well as the history of the phase interface shape. Without pulling the crystal out of the melt, it would grow further downward to the bottom and towards the sides of the crucible, until in the end the whole melt in the crucible is solidified, if the crucible temperatures allow. Usually the heater power has to be lowered for this.

In a second simulation, a high pulling velocity of 5 mm/min. was applied. The results are shown in Fig. 11. With the higher pulling rate, the crystal comes into a higher position above the melt level, because the crystallization is not fast enough. Therefore, the meniscus shape is altered and the contact angle at the TPB is decreasing, such that the crystal diameter is also decreasing.

If the process is continued with these parameters, it is likely to happen that the crystal narrows down until it reaches a very small diameter and finally loses contact with the melt. This is the desired procedure in the last stage of the Cz process (tailing), when the crystal has to be pulled out of the melt and/or the melt is almost used up.

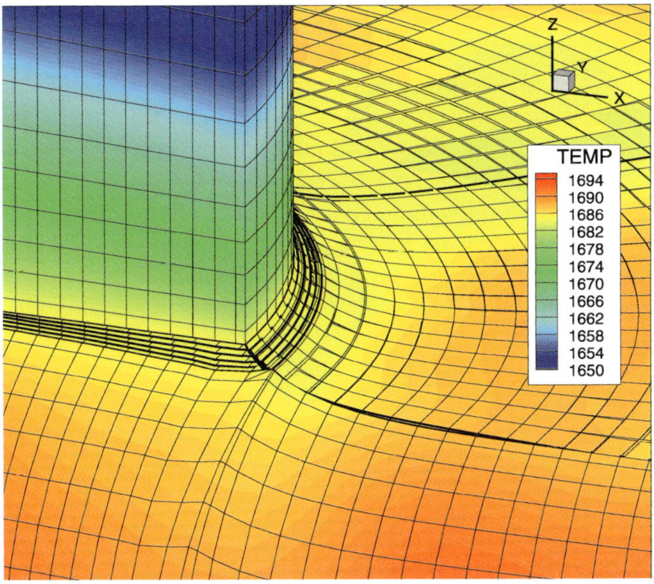

Fig. 10. Simulation of growing crystal with increasing diameter (no pulling). Different stages of the growth process with the shape of the phase interface moving downwards are shown in a vertical cut together with the temperature distribution and the computational grid

Fig. 11. Simulation of growing crystal with decreasing diameter (high pulling rate). Different stages of the growth process with the shape of the phase interface moving downwards are shown in a vertical cut

6 Conclusions

In the present work, a possibility is shown to overcome the restrictions due to the quasi-steady state assumption in Czochralski crystal growth. For that purpose, an iterative algorithm was designed, which integrates the computation of the Stefan condition for the crystallization and the simulation of the melt free surface, which both determine the movement of the three-phase junction. These new features completing realistic predictions of the CZ crystals growth process were combined with the numerical simulation of the turbulent flow and heat transfer in the melt using LES. In a test case, this algorithm was successfully applied. It could be shown that under certain conditions, diameter variations occur during the crystal growth process.

In future research, the algorithm will be used for the optimization of the process parameters in Czochralski crystal growth, especially for the development of new diameter control strategies.

Acknowledgements. The project is financially supported by the German Research Foundation (DFG) under contract numbers DE 634/15-1 and BO 3107/1-1. The simulations were performed on the national super computer NEC SX-8 at the High Performance Computing Center Stuttgart (HLRS) under the grant number CZ/12761. All kinds of support are gratefully acknowledged.

References

1. Bartels, C., Breuer, M., Wechsler, K., and Durst, F. (2001) CFD-Applications on Parallel-Vector Computers: Computations of Stirred Vessel Flows, *Computers and Fluids*, vol. **31**, pp. 69–97
2. Basu, B., Enger, S., Breuer, M., and Durst, F. (2000) Three-Dimensional Simulation of Flow and Thermal Field in a Czochralski Melt Using a Block–Structured Finite–Volume Method, *Journal of Crystal Growth*, vol. **219**, pp. 123–143
3. Derby, J.J., Brown, R.A. (1988) On the Quasi-Steady State Assumption in Modeling Czochralski Crystal Growth, *Journal of Crystal Growth*, vol. **87**, pp. 251–260
4. Durst, F., Schäfer, M. and Wechsler, K. (1996) Efficient Simulation of Incompressible Viscous Flows on Parallel Computers, In: Flow Simulation with High–Performance Computers II, ed. E.H. Hirschel, *Notes on Numer. Fluid Mech.*, vol. **52**, pp. 87–101, Vieweg Verlag, Braunschweig
5. Durst, F. and Schäfer, M. (1996) A Parallel Block–Structured Multigrid Method for the Prediction of Incompressible Flows, *Int. Journal Num. Methods in Fluids*, vol. **22**, pp. 549–565
6. Enger, S., Basu, B., Breuer, M., and Durst, F. (2000) Numerical Study of Three-Dimensional Mixed Convection due to Buoyancy and Centrifugal Force in an Oxide Melt for Czochralski Growth, *Journal of Crystal Growth*, vol. **219**, pp. 123–143

7. Enger, S., Gräbner, O., Müller, G., Breuer, M., and Durst, F. (2001) Comparison of Measurements and Numerical Simulations of Melt Convection in Czochralski Crystal Growth of Silicon, *Journal of Crystal Growth*, vol. **230**, pp. 135–142
8. Germano, M., Piomelli, U., Moin, P., and Cabot, W.H. (1991) A Dynamic Subgrid-Scale Eddy Viscosity Model, *Phys. Fluids A*, vol. **3**, pp. 1760–1765
9. Kumar, V., Basu, B., Enger, S., Brenner, G., and Durst, F. (2003) Role of Marangoni Convection in Si-Czochralski Melts, Part II: 3D Predictions with Crystal Rotation, *Journal of Crystal Growth*, vol. **255**, pp. 27–39
10. Leister, H.J. and Perić, M. (1993) Vectorized Strongly Implicit Solving Procedure for a Seven-Diagonal Coefficient Matrix, *Int. Journal of Heat and Fluid Flow*, vol. **4**, pp. 159–172
11. Raufeisen, A., Breuer, M., Botsch, T., and Delgado, A. (2008) DNS of Rotating Buoyancy- and Surface Tension-Driven Flow, *Int. Journal of Heat and Mass Transfer*, vol. **51**, pp. 6219–6234
12. Raufeisen, A., Breuer, M., Botsch, T., and Delgado, A. (2009) LES Validation of Turbulent Rotating Buoyancy- and Surface Tension-Driven Flow Against DNS, *Computers and Fluids*, vol. **38**, pp. 1549–1565
13. Raufeisen, A., Jana, S., Breuer, M., Botsch, T., and Durst, F. (2007) 3D Computation of Oxygen Transport in Czochralski Crystal Growth of Silicon Considering Evaporation, *Journal of Crystal Growth*, vol. **303**, pp. 146–149
14. Rudolph, P. and Kiessling, F.-M. (2006) Growth and Characterization of GaAs Crystals Produced by the VCz Method Without Boric Oxide Encapsulation, *Journal of Crystal Growth*, vol. **292**, pp. 532–537
15. Smagorinsky, J. (1963) General Circulation Experiments with the Primitive Equations, I, The Basic Experiment, *Mon. Weather Rev.*, vol. **91**, pp. 99–165
16. Stone, H.L. (1968) Iterative Solution of Implicit Approximations of Multidimensional Partial Differential Equations, *SIAM Journal of Numerical Analyses*, vol. **5**, pp. 530–558
17. Thompson, J.F., Warsi, Z.U.A., Mastin, C.W. (1985) Numerical Grid Generation. North Holland, New York
18. Togawa, S., Izunome, K., Kawanishi, S., Chung, S., Terashima, K., Kimura, S. (1996) Oxygen Transport From a Silica Crucible in Czochralski Silicon Growth, *Journal of Crystal Growth*, vol. **165**, pp. 362–371

Numerical Investigation of Shock Wave Boundary-Layer Interaction Using a Zonal RANS-LES Ansatz

Benedikt Roidl, Matthias Meinke, and Wolfgang Schröder

Institute of Aerodynamics, RWTH Aachen University, Wüllnerstr. 5a, 52062 Aachen, Germany, b.roidl@aia.rwth-aachen.de

Summary. In this paper a zonal RANS/LES approach is presented in which the regions with attached boundary-layers are computed via RANS and the regions with separated flows by using LES. The transition from RANS to LES takes place in an overlapping region between the RANS and LES zone. Two different turbulent inflow generation methods coupled with a controlled forcing ansatz are applied which enable a fast and smooth transition from two-dimensional RANS- to the three-dimensional unsteady LES solutions. Both approaches require local Reynolds shear stresses of a RANS solution which is located upstream of the LES. The inflow generation methods are validated for a boundary-layer flow and the fully coupled zonal approach is applied to a transonic flow over an airfoil both including a shock boundary-layer interaction.

1 Introduction

Most of all industrial CFD applications at high Reynolds numbers are nowadays based on solutions of the Reynolds averaged Navier-Stokes (RANS). The reasons for their application are obvious: they are simple to apply and computationally efficient. Therefore, they are used for the flow analysis at design and off-design conditions, for optimization and for cases where experimental data may not be easily obtained. However, it is known that solutions provided by one- or two equation eddy viscosity models are not reliably predicting complex flow cases [19]. Although various modifications and new concepts of turbulence modelling, i.e. algebraic Reynolds stress models, non-linear eddy-viscosity closures, etc., were proposed over the last decades, the optimum universal model has not been found yet.

The alternatives to RANS modelling are Direct- and Large-Eddy Simulation (DNS and LES). Todays available computer resources, however, prevent these methods to become standard simulation tools for high Reynolds number flows. Since in many applications complex flow regions are limited to a smaller part of the domain, hybrid methods, combining the computational efficiency

W.E. Nagel et al. (eds.), *High Performance Computing in Science and Engineering '10*, DOI 10.1007/978-3-642-15748-6_28,

of RANS with the higher accuracy of LES or DNS should in principle be able to provide better results at minimized additional costs.

An overview of hybrid and zonal RANS/LES approaches is given in [4]. There are at least two major techniques to couple RANS with LES in hybrid computations. The first approach uses a continuous turbulence model, which switches from RANS to LES mode in a unified domain, such as the detached eddy simulation (DES) proposed in [16, 18]. The other technique uses two separate computational domains which are linked via an overlapping zone where the transition from RANS to LES and vice versa occurs. This technique is also applied and further investigated with respect to the interface conditions in this paper. Within this method the turbulent flow properties at the LES inlet have to be prescribed by using information from the RANS solution. There are three distinct types of such inflow boundary conditions [9]: recycling a plane of flow characteristics downstream of the inflow plane; instantaneous flow fields from precursor simulations or fluctuations (random or physical) which are superimposed on a mean velocity profile. The latter method is well suited for the interface problem because the fluctuations, that contain the moments of the RANS simulation, could be either generated by local forcing, or computed separately and superimposed on a mean velocity profile that is also provided by the upstream RANS solution.

Recent applications that apply the synthetic eddy method (SEM) by Jarrin et al. [6] or that are based on the synthetic, homogeneous turbulence method by Kraichnan [11], which was extended to inhomogeneous flows by Smirnov et al. [14] and Batten et al. [1] suffer from long transition lengths for realistic turbulence to be generated. Following the idea of Keating et al. [8] this development region can be significantly shortened by combining STGM with controlled forcing [17] that is applied downstream of the LES inlet, which will be demonstrated in this paper.

The article is organized as follows: In Sect. 2, the numerical methods of the flow solver and the synthetic turbulence generation methods (STGM) are briefly described. Subsequently, in Sect. 3, the results are presented. First, the two STGM are compared in a zero-pressure gradient boundary-layer flow case. Then,the fully coupled zonal RANS-LES approach is compared with corresponding full domain LES and available experimental data for a transonic flow over an airfoil where shock wave boundary-layer interaction (SWBLI) occurs.

2 Mathematical Formulation

2.1 Numerical Methods

The Navier-Stokes equations for three-dimensional compressible flows are solved by a block-structured finite-volume flow solver. A modified AUSM method that was introduced by Liou et al. [12] is used for the Euler terms which are discretized to second-order accuracy by an upwind-biased approximation. For the non-Euler terms a centered approximation of second-order accuracy is used. The temporal integration from time level n to $n+1$ is done by a second-order accurate explicit 5-stage Runge-Kutta method, the coefficients of which are optimized for maximum stability. For a detailed description of the flow solver the reader is referred to Meinke et al. [13].

The subgrid scale modelling for the large-eddy simulations is based on an implicit ansatz, i.e., the MILES (monotone integrated LES) approach of Boris et al. [2] is used.

For the RANS computations the Spalart-Allmaras turbulence model (Spalart and Allmaras (1992)) is chosen to close the Reynolds-averaged Navier-Stokes equations.

2.2 Synthetic Turbulence Generation Methods (STGM)

A proper coupling of RANS and LES flow domains is considered to constitute the key feature of a zonal RANS/LES approach. In order to keep overlapping regions of both computational domains as small as possible, effective mechanisms for turbulence generation have to be applied in LES inflow regions. The turbulent characteristics coming from the RANS domain are introduced first at the LES inflow plane via synthetic turbulent eddies (Jarrin et al. [6], Batten et al. [1]) and controlled further downstream by employing control planes according to Spille and Kaltenbach [17]. The synthetic turbulence generation methods of Jarrin et al. and Batten et al. were implemented and tested for flat plate boundary-layer flows. In subsequent sections the STGM of Jarrin et al. combined with the control plane approach is referred to as 'zonal I' and the STGM of Batten et al. combined with the control plane approach is referred to as 'zonal II'.

2.3 Reconstruction of Eddy Viscosity for Transition from LES to RANS

For the correct and smooth transition from LES back to RANS the required eddy viscosity was reconstructed following the work of König et al. [10]. The eddy viscosity is determined by the ratio of the turbulent kinetic energy k and the turbulent frequency ω. The quantity ω is approximated by a generalized form of Bradshaw's hypothesis using the norm of the mean strain tensor. It is obvious that this method just needs the normal components of the Reynolds stress tensor to compute the turbulent kinetic energy k and the computation of the turbulent frequency ω requires only the derivatives of mean velocities and not of their fluctuations. A detailed discussion of the method is given in König et al. [10].

3 Results and Discussion

3.1 Validation of STGM

A zero-pressure gradient boundary layer was investigated and compared to reference solutions using two synthetic turbulence methods, based on the controlled forcing approach downstream of the inlet. Four simulations were carried out: a full domain LES (referred to as 'full LES'), a full domain RANS, and two synthetic turbulence-LES simulations with controlled forcing. All four cases were computed with the same flow and numerical configuration, $M_\infty = 0.4$, $Re_\delta = 10000$. The numerical details are given in Table 1. The reference full LES was computed using the rescaling method according to El-Askary et al. [3]. The full domain RANS calculation based on the Spalart-Allmaras model (referred to as 'RANS S-A') was performed for comparison purposes and to provide the target data for the synthetic turbulence inlet and for the control planes. In the zonal simulations four control planes are distributed over a length of one boundary layer thickness δ_0 downstream of the domain inlet.

Figure 1 (left) compares the evolution of the skin friction coefficient c_f for all four flow cases. The solutions of full LES and full RANS do not differ much regarding the wall shear stress for this simple zero-pressure gradient boundary layer. Thus, the applied rescaling method at this numerical configuration is valid for LES and RANS simulations and poses no difficulty to match them at

Table 1. Numerical details for the simulation of turbulent boundary layer at $Ma = 0.4$ and $Re_\delta = 10000$ for LES and RANS solutions

	Streamwise x	Wall normal y	Spanwise z
Domain size in δ_0	15	3	0.7
Grid points	484	65	49
Resolution, wall units	$\Delta x^+ \approx 21$	$\Delta y^+_{min} \approx 1.0$	$\Delta z^+ \approx 11$

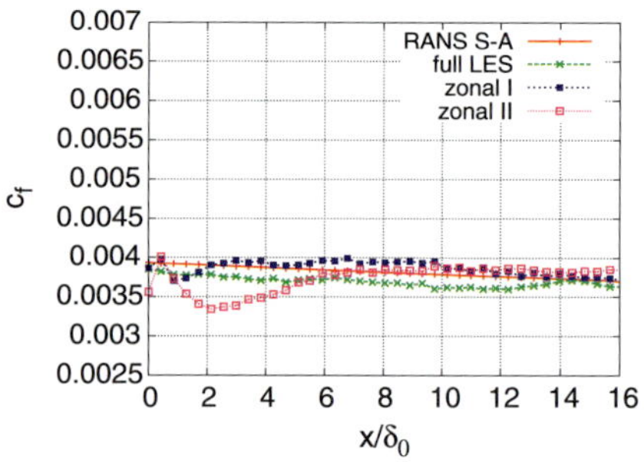

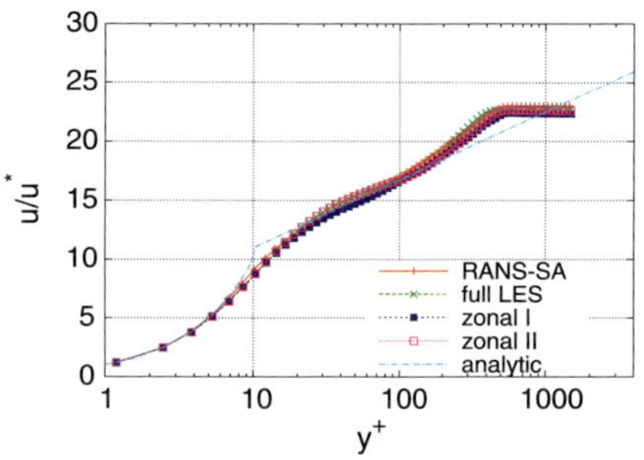

Fig. 1. Evolution of skin friction coefficient c_f (left) and van Driest velocity profile (right) for boundary layer flow at $Ma = 0.4$ and $Re_{\delta_0} = 10000$ for different computational configurations

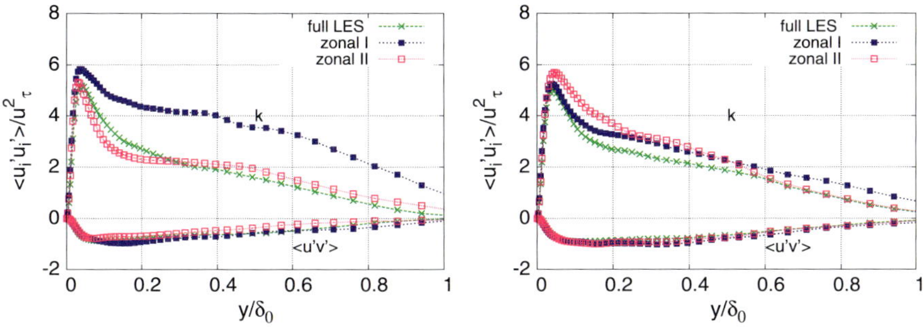

Fig. 2. Turbulent kinetic energy k and Reynolds shear stress $\langle u'v' \rangle$ at $x/\delta_0 = 1$ (left) and at $x/\delta_0 = 5$ (right)

the beginning of the computational domain. Both, the zonal I and the zonal II approach, show, despite their fundamental differences in their formulations, a comparable required length until they converge to the full LES solution. The van Driest velocity profiles obtained at $x/\delta_o = 5$ which are presented in Fig. 1 (right) show that both STGM produce the expected asymptotic near-wall behavior of a turbulent flow, but the results differ somewhat at the edge of the boundary layer.

Figure 2 depicts the turbulent kinetic energy k and Reynolds shear stress component $\langle u'v' \rangle$ at two different locations downstream of the inlet. It is shown that the flow generated by the zonal II approach undergoes a slight laminarization process downstream of the interface but the control planes increase the turbulent shear stress budget to the full LES level. The turbulent structures generated by the zonal I approach do not dissipate downstream of the inlet but the control planes introduce a local overshoot of the turbulence level which decreases to the full LES turbulence level at around 5 x/δ_0. By improving the response of the control planes to the local flow events, the turbulent shear stress level of the full LES could be reached within one boundary layer thickness δ_0.

λ_2-contours (Jeong et al. [7]) of both zonal cases and the full LES are visualized in Fig. 3. It is shown that the structures which are introduced into the domain by the zonal I approach are not dissipating. The structures at the inlet of the case computed with the zonal II approach fade away and the control planes downstream of the inlet have to enhance the locally rare events like turbulent bursts and sweeps to reach the turbulence level of the full LES computation.

Due to the low Reynolds number for this case it was expected that the 'artificial' turbulence would dissipate at the beginning of the domain to develop 'physical' turbulence further downstream after the transition process at about $x/\delta_0 \approx 10$. However, Figs. 1 (left) and 2 (left) show that when the zonal I approach is used the Reynolds shear stress $\langle u'v' \rangle$ does not decrease below the level of the full domain LES but generates an overshoot of turbulent

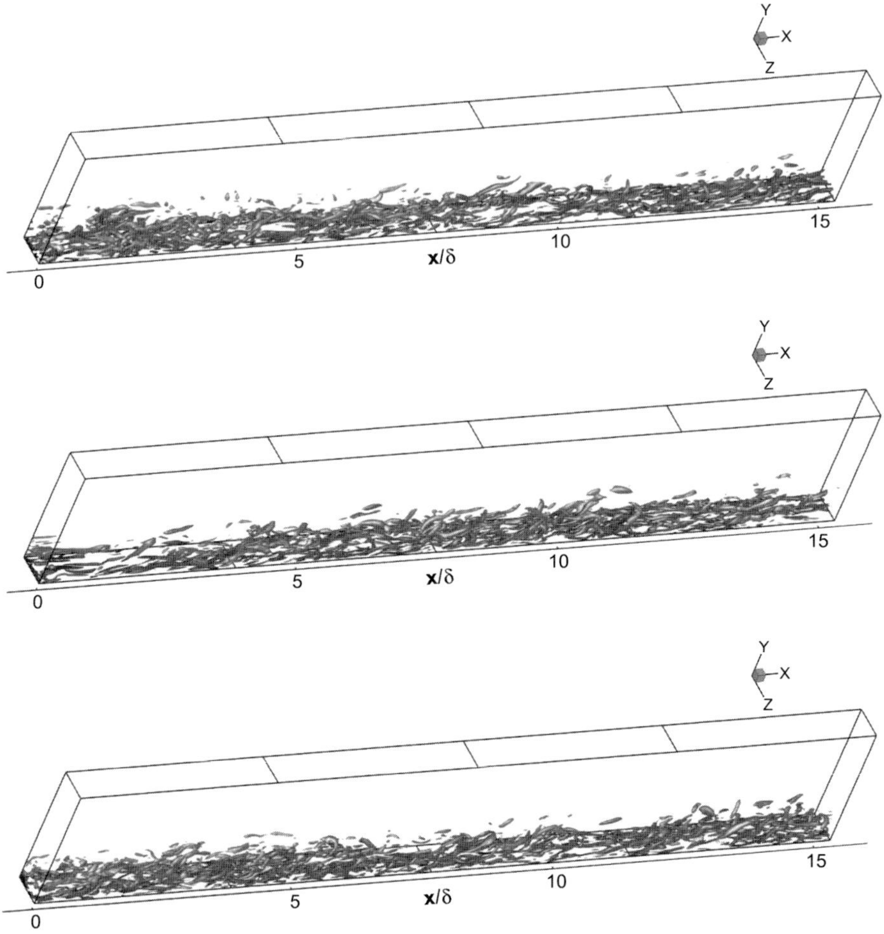

Fig. 3. λ_2 structures of a mildly compressible flat plate boundary layer flow computed by (top) a full LES, (middle) using zonal Batten approach and (bottom) applying the zonal SEM ansatz

kinetic energy. The zonal II approach tends to provide a lower Reynolds shear stress level which is to be increased by the control planes which are located downstream of the inlet. For the following computations the zonal I approach is used since the quality of the results is acceptable and it is computationally less expensive than the zonal II approach.

3.2 DRA2303 Transonic Profile

The DRA2303 transonic airfoil [5] was chosen as the aerodynamic reference case for the buffet phenomenon. Associated with buffet are self-sustained shock wave oscillations on airfoils at transonic flow. The flow configuration, which was to lie well within the buffet boundaries, was chosen with $Ma = 0.72$, $Re = 2.6 \cdot 10^6$ and $\alpha = 3°$. The flow configuration was chosen accordingly to available experiments that were recently conducted at the AIA Trisonic wind tunnel.

In this work the buffet is computationally targeted with three different simulations: a full domain LES which constitutes the reference solution and a fully coupled zonal RANS-LES solution. First, the RANS solution is compared with the full domain LES results of the DRA2303 airfoil and then, a fully coupled zonal RANS-LES solution yielding preliminary results is presented.

A grid resolution was chosen for the full LES and the LES domain of the zonal RANS-LES simulation according to the experience of Zhang et al. [20]. The numerical details of the full LES solution and the zonal RANS-LES are given in Table 2 and Table 3, respectively. The number of required grid points for the zonal RANS-LES simulation contains less than 50 % of the grid points used in the full domain LES simulation.

A full LES simulation was set up to examine the physical aspects of buffet phenomenon without using standard turbulence models. The simulation time for this configuration was about $40c/U_\infty$ which can be considered as a long term simulation where effects of initial perturbations or flow developing effects possess no influence anymore on the solution and a periodic flow behavior determines the result.

In Fig. 4 the instantaneous pressure coefficient c_p and the average pressure coefficient fluctuations at the upper side of the profile are presented. Note that the extension of the horizontal shock oscillation is $a \approx 0.07c$. The peak in the average pressure fluctuations is pronounced at about $x/c = 0.55$ and near the trailing edge the intensity of the fluctuations increases but they never exceed the strength of the shock.

The reduced frequency ω^* of the lift coefficient oscillation of the full LES solution presented in Fig. 5 is about 0.74. Oscillation of the lift coefficient is

Table 2. Numerical details for the of transonic flow over the DRA2303 airfoil for a full domain LES simulation

	Streamwise x	Wall normal y	Spanwise z
Domain size in c	20	20	0.021
Grid points ($30.4{\cdot}10^6$)	2364	130	99
Resolution, wall units	$\Delta x^+ \approx 100$	$\Delta y^+_{min} \approx 1.0$	$\Delta z^+ \approx 20$

Table 3. Numerical details of the LES domain for a zonal RANS-LES simulation of transonic flow over the DRA2303 airfoil

	Streamwise x	Wall normal y	Spanwise z
Grid points ($13.7{\cdot}10^6$)	1430	97	99
Resolution, wall units	$\Delta x^+ \approx 35$	$\Delta y^+_{min} \approx 1.0$	$\Delta z^+ \approx 20$

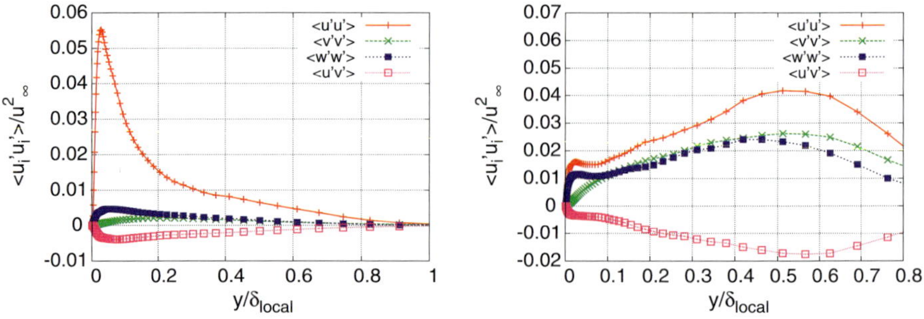

Fig. 4. Reynolds shear stresses at two different positions at the DRA2303-profile: at $x/c = 0.4$ (left) and $x/c = 0.6$ (right)

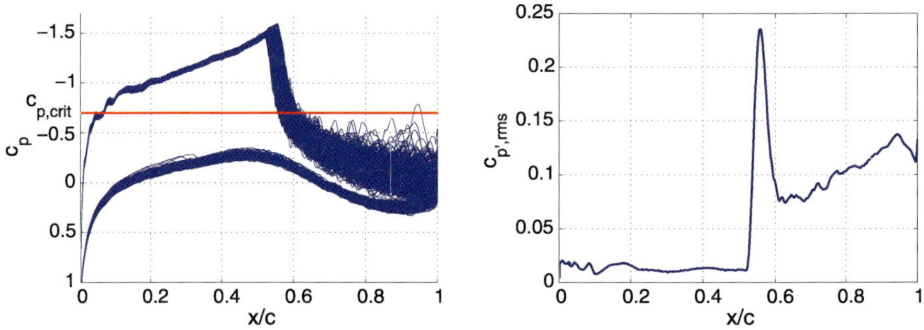

Fig. 5. Fluctuating wall pressure coefficient c_p (left) and corresponding *rms*-values at the upper side of the profile (right) for an LES simulation

highly periodic which is indicated by the peak in the frequency spectrum. The amplitude of the lift coefficient is $\Delta C_L \approx 0.03$.

Figure 6 compares the fluctuating pressure intensity at different locations. At the upper side at $0.25c$ the amplitude of the fluctuation is very low, however, a small distinctive bump is evident at $\omega^* \approx 0.7$. Near the shock at $0.55c$ a peak occurs at the reduced frequency $\omega^* = 0.74$ and the distribution of the values is very similar to that in Fig. 5. Close to the trailing edge at $0.9c$ the major peak is still at $\omega^* \approx 0.73$ but the pressure fluctuations at higher reduced frequencies have grown stronger compared to the position near the shock. This is due to the turbulent shear layer which is at this flow configuration maximum in size at the trailing edge. Although the pressure fluctuations are averaged in the spanwise direction the pressure fluctuations at very high reduced frequencies can be related to the turbulent shear layer. At the lower side of the profile at $0.9c$ the intensity level of the fluctuations is one order of magnitude smaller compared with the corresponding position at the upper side. However, a distinct peak at $\omega^* \approx 0.73$ is visible which is caused by the direct influence of the oscillating shock at the upper side.

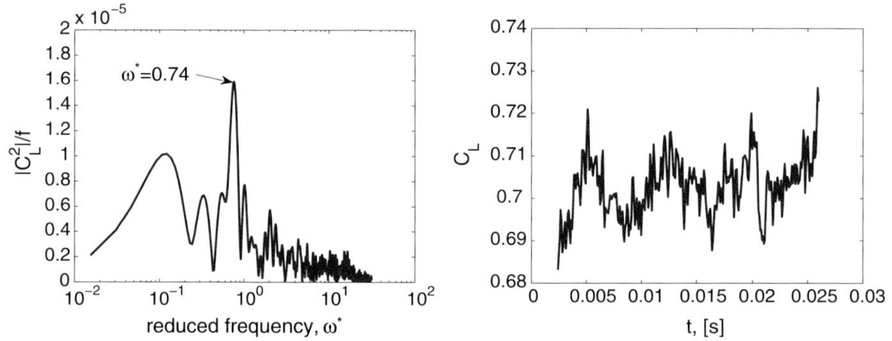

Fig. 6. Fourier-transformed fluctuations of lift coefficient C_L and corresponding fluctuations over time for an LES simulation

The Reynolds shear stresses of the averaged full LES solution at two different locations is exhibited in Fig. 7. At 0.4 c the distribution resembles that of a flat plate boundary-layer flow and its turbulent features. Behind the shock, however, the maxima of all components moves to the center of the boundary-layer and the intensity level of $\langle v'v' \rangle$, $\langle w'w' \rangle$, and $\langle u'v' \rangle$ is much higher compared with the position ahead of the shock. Downstream of the shock the intensities of $\langle u'u' \rangle$, $\langle v'v' \rangle$, and $\langle w'w' \rangle$ are at the same level which indicates that the turbulent structures show an isotropic behavior compared with the high level of anisotropy of the near-wall turbulence upstream of the shock.

The λ_2-contours [7] are shown in Fig. 8. After the interaction with the shock, the turbulent boundary-layer separates and a shear flow is formed that develops large coherent structures that convect near the trailing edge. These large structures are responsible for the high level pressure oscillations at high frequencies near the trailing edge.

The experimental findings showed a horizontal shock amplitude at the upper side of the airfoil of $a_{exp} \approx 0.05c$ with a corresponding reduced frequency of $\omega^*_{exp} = 0.68$. Hence, the full LES results regarding buffet dynamics agree well with the available experimental data ($a_{num} \approx 0.07c$ and $\omega^*_{num} = 0.74$).

In order to analyze the buffet phenomenon with a higher order turbulence model, i. e. a local LES, and to use a global efficient Ansatz the zonal RANS-LES approach was applied. The computational set up is shown in Fig. 9. The Spalart-Allmaras turbulence model [15] was used to close the RANS equations of the zonal RANS-LES approach. Due to the complexity of this particular case the results can be considered just preliminary.

RANS and LES domains used periodic boundary conditions in the spanwise direction and a no-slip, adiabatic condition was set at the wall. Non-reflective boundary conditions were applied at the far field boundaries. At the inlet of the LES domain on the upper and lower side of the airfoil, the zonal I approach according to Jarrin et al. [6] was used to generate synthetic

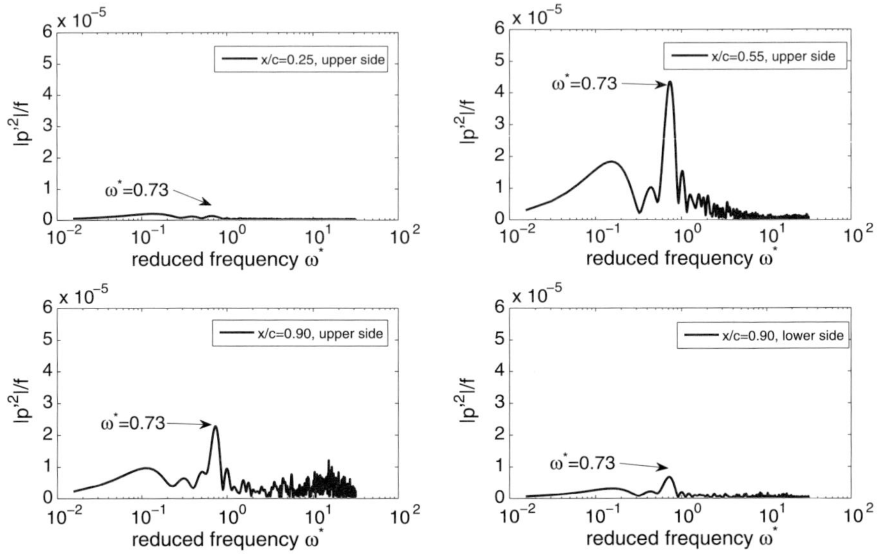

Fig. 7. Fourier transformed pressure fluctuations at different positions of the DRA2303-airfoil

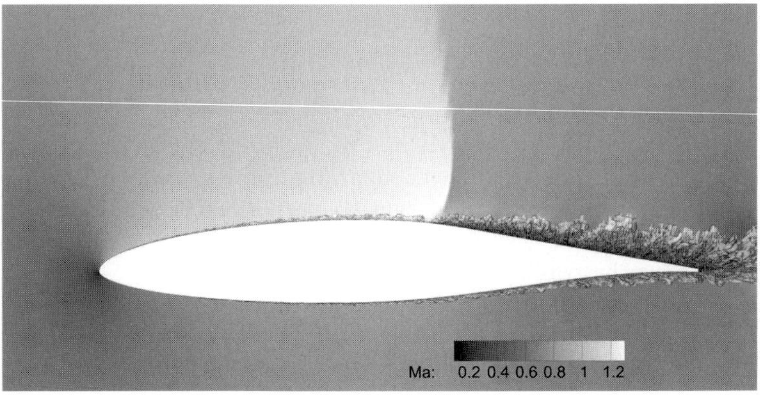

Fig. 8. λ_2 structures of transonic flow around a DRA2303 profile for a full domain LES simulation

turbulent structures. Downstream of the inlet four control planes are located between $0.37c$ and $0.4c$ at the upper side and between $0.7c$ and $0.73c$ at the lower side of the airfoil. The turbulent flow properties of the upstream RANS solution were used as target conditions for the zonal I approach and the control planes that were located downstream of the LES inlet. At the LES outlet the method of König et al. [10] was employed to reconstruct the required turbulent viscosity ν_t for the RANS domain that is located downstream of the LES region. At the RANS outflow a time averaged pressure from the

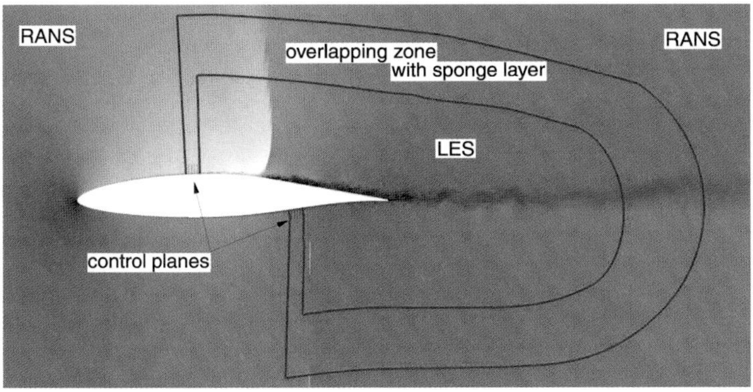

Fig. 9. Computational setup of a fully coupled zonal RANS-LES simulation

LES domain located downstream is prescribed whereas density and velocity profiles are extrapolated. At the LES inflow the density and velocity profiles are transferred from the RANS domain located upstream and the pressure is extrapolated from the interior of the LES-domain. The term 'fully coupled' refers to the above mentioned treatment of the in- and outflow boundaries of the RANS and LES domains. The LES domain is surrounded by a sponge layer to damp pressure fluctuations that otherwise might be reflected at the boundaries. Further details of such a sponge layer are given in [20].

Since pressure waves, caused by the transient shock behavior , travel from the LES domain to the RANS domain and vice versa the time window where the solutions are averaged has to be carefully defined. On the one hand, the turbulent flow properties of the LES solution have to be averaged properly over a sufficiently large time window before being transferred to the RANS domain. On the other hand, the amplitude and frequency of the travelling pressure waves caused by the shock must be captured in a time window which is as small as possible to prevent a significant alteration of the pressure wave signal. A time window of the size of 1 c/U_∞ was found to satisfy these requirements.

In Fig. 10 the instantaneous pressure coefficients c_p are given. The zonal RANS-LES solution shows a good agreement with the full LES solution concerning shock position and strength. Downstream of the shock the c_p evolution shows minor discrepancies near the trailing edge. However, upstream of the shock the zonal RANS-LES results agree pretty well with the findings of the full LES. Note the smooth transition from RANS to LES of the pressure coefficient near the overlapping zones at approximately $0.37c$ (upper side) and $0.7c$ (lower side).

The λ_2-contours of the instantaneous zonal RANS-LES solution are depicted in Fig. 11. They show the same features compared to the contours of the full LES simulation such as large coherent structures downstream of the shock interaction zone convecting towards the trailing edge.

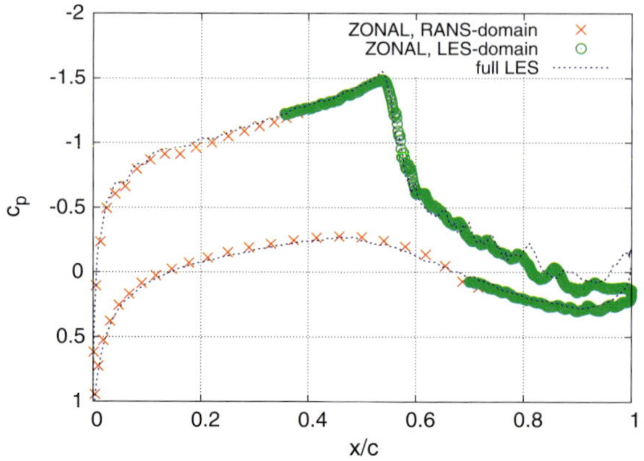

Fig. 10. Comparison of instantaneous wall pressure coefficient c_p between full domain LES and zonal RANS-LES computation

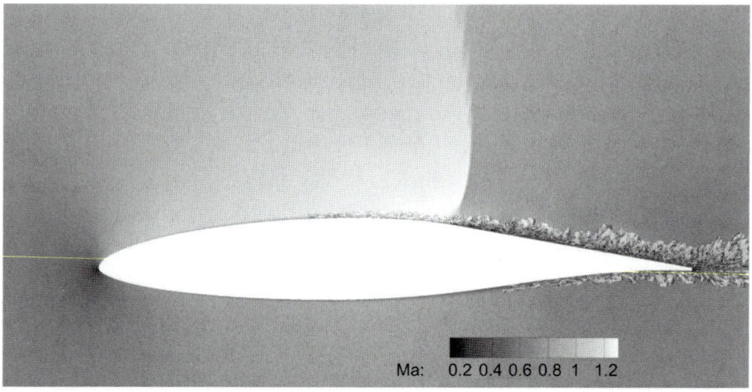

Fig. 11. λ_2 structures of transonic flow around a DRA2303 airfoil for a fully coupled zonal RANS-LES solution

4 Computational Resources

The simulations were carried out on the NEC SX-9 installed at the HLR Stuttgart. The full LES results for the SWBLI case and the DRA2303 airfoil presented in Chapt. 3 were computed on a domain of integration that is divided into 16 and 24 blocks, respectively, while each block resides on a single CPU. Data between the blocks is exchanged via MPI (message passing interface). The cluster is arranged in nodes with 16 CPUs each. The workload for each CPU, that is mesh points per block, is rather low in order to minimize the overall user time for the simulations and to achieve a maximized parallelized simulation, respectively. It was taken into account though that a rather high number of nodes would increase the turnaround time as the scheduling system prefers jobs with fewer nodes. In principle other distributions of the mesh points are possible. The DRA2303 case contains 31 million mesh points, respectively, that are distributed nearly equally to 24 blocks on two nodes. Due to the reduction of grid points the computational efficiency of the zonal RANS-LES case that is presented in the result Sect. 3 was about 50 % higher compared to the corresponding DRA2303 full LES case. Additionally the full LES of the DRA2303 profile with a high grid resolution has been investigated in a long term simulation (referred to as DRA2303-fine), however, the results are not shown here. The computational details are given in Table 4.

5 Conclusion

In this article, different inflow conditions for zonal RANS-LES simulations at the LES inlet where evaluated and compared. To couple the RANS with the LES domain, synthetic turbulence generating approaches and control planes were used to prescribe the turbulent intensities of the RANS solution for the LES domain. Two different STGM were tested and validated for a subsonic zero-pressure gradient boundary-layer flow. The zonal I (Jarrin et al. [6]) approach was found to be superior compared to the zonal II (Batten et al. [1]) approach and therefore used in subsequent computations.

The zonal RANS-LES was applied to simulate the transonic flow around a DRA2303 airfoil. The results of a pure RANS and a pure LES were compared. It was shown that the full LES findings showed a good agreement with preliminary experimental data and hence provided superior results concerning reduced frequency and horizontal amplitude of the shock oscillation compared with the full RANS results. The coupled zonal RANS-LES approach provided preliminary results which indicate that the quality of the solution is comparable to that of the full domain LES solution. For the fully coupled zonal

Table 4. Sample performance on NEC SX-9

Case	DRA2303	DRA2303-fine
Number of CPUs	24	32
Number of Nodes	2	2
grid points/CPU	$1.29 \cdot 10^6$	$3.16 \cdot 10^6$
total grid points	$31 \cdot 10^6$	$101 \cdot 10^6$
Avg. User Time [s]	39782	41095
Avg. Vector Time [s]	37142	39604
Vector Operations Ratio [%]	99.687	99.765
Avg. Vector length	249.445	254.4
Memory/CPU [MB]	1642	2125
total Memory [GB]	25.38	99.69
Avg. MFLOPS/CPU	13795.94	15100
Max. MFLOPS/CPU	14774.04	16460
total GFLOPS	331.080	483.185

RANS-LES simulation the computational costs in terms of grid size could be reduced by a factor of 2.

References

1. P. Batten, U. Goldberg, and S. Chakravarthy. Interfacing statistical turbulence closures with large-eddy simulation. *AIAA Journal*, 42(3):485–492, 2004.
2. J.P. Boris, F.F. Grinstein, E.S. Oran, and R.L. Kolbe. New insights into large eddy simulation. *Fluid Dynamics Research*, 10:199–228, 1992.
3. W.A. El-Askary, M. Meinke, and W. Schröder. Large Eddy Simulation of a Compressible Boundary Layer. Pap., Aerodyn. Inst. RWTH Aachen, 2002. Presented at GAMM conference 2002, March 25-28, Augsburg.
4. J. Fröhlich and D.V. Terzi. Hybrid les/rans methods for the simulation of turbulent flows. *Prog. Aerospace Sci.*, 44:349–377, 2008.
5. J.L. Fulker and M.J. Simmons. An Experimental Investigation of Passive Shock/Boundary Layer Interaction Control on an Aerofoil. *DRAIASIHWAICR 9521611 EUROSHOCK TR AER 2*, 4913(2), 1992.
6. N. Jarrin, S. Benhamadouche, D. Laurence, and R. Prosser. A synthetic-eddy-method for generating inflow conditions for large-eddy simulations. *International Journal of Heat and Fluid Flow*, 27:585–593, 2006.
7. J. Jeong and F. Hussain. On the identification of a vortex. *J. Fluid Mech.*, 285:69–94, 1995.
8. A. Keating, G. de Prisco, and U. Piomelli. Interface conditions for hybrid rans/les calculation. *International Journal of Heat and Fluid Flow*, 27:777–788, 2006.
9. A. Keating, U. Piomelli, E. Balaras, and H.-J. Kaltenbach. A priori and a posteriori tests of inflow conditions for large-eddy simulation. *Physics of Fluids*, 16(12):4696–4712, December 2004.
10. D. König, W. Schröder, and M. Meinke. Embedded les-to-rans boundary in zonal simulations. *Journal of Turbulence*, 11:1–25, 2010.
11. R.H. Kraichnan. Inertial Ranges in Two-Dimensional Turbulence. *Physics of Fluids*, 10(7):1417–1423, July 1967.
12. M.-S. Liou and C.J. Steffen. A new flux splitting scheme. *Journal of Computational Physics*, 107:23–39, 1993.
13. M. Meinke, W. Schröder, E. Krause, and T. Rister. A comparison of second- and sixth-order methods for large-eddy simulations. *Computers and Fluids*, 31:695–718, 2002.
14. A. Smirnov, S. Shi, and I. Celik. Random flow generation technique for large eddy simulations and particle dynamics modeling. *Journal of Fluids Engineering*, 123:359–371, 2001.
15. P.R. Spalart and S.R. Allmaras. A One-Equation Turbulence Model for Aerodynamic Flows. Paper 92-0439, AIAA, 1992. 30th Aerospace Sciences Meeting & Exhibit, Jan 6-9, Reno.
16. P.R. Spalart, W.H. Jou, M. Strelets, and S.R. Allmaras. Comments on the feasibility of les for wings, and on a hybrid rans/les approach. Advances on DNS/LES. Greyden Press, Columbus, OH, 1997.

17. A. Spille and H.-J. Kaltenbach. Generation of turbulent inflow data with a prescribed shear-stress profile. *Third AFSOR Conference on DNS and LES*, August 2001.
18. K.D. Squires. Detached-eddy simulation: current status and perspectives. Direct and Large-Eddy Simulation, pages 465–480. Kluwer, Dordrecht, 2004.
19. D.C. Wilcox. *Turbulence Modeling for CFD.* DCW Industries, 1993.
20. Q. Zhang, W. Schröder, and M. Meinke. A zonal rans/les method to determine the flow over a high-lift configuration. *Accepted for publication in Computers and Fluids*, 2010.

Large Eddy Simulation of the Cyclic Variations in an Internal Combustion Engine

F. Magagnato, A. Walcker, and M. Gabi

KIT, Kaiserstr. 12, Karlsruhe, Germany

Summary. A Large-Eddy Simulation of several cycles of a spark-ignition engine with port fuel injection has been performed. The investigated engine is a modified Rotax-engine with one cylinder and four valves, built and measured (Heidenreich, Private communication) by the Institute of Reciprocating Engines of the University of Karlsruhe (IFKM). A recently modified (Magagnato et al., 2008) progress variable combustion model of Schmid et al., 1998 has been used in this computation. A subgrid-scale model for the LES recently proposed by Stolz (Stolz et al., 2003) called High-pass filtered Eddy-Viscosity model has been employed, which appears to be ideal for moving meshes. In order to predict cyclic variations of the combustion process, an additional transport equation for the mixing of the fuel to air ratio inside the combustion chamber is proposed and investigated in this paper. The predicted cylinder pressure vs. crank angle is then compared to measured results (Heidenreich, Private communication).

Key words: CFD, LES, ICE, cyclic variation, combustion

1 Introduction

The combustion process in spark-ignition engines strongly depends on the mixing and turbulent fluctuations of the flow field during the ignition and combustion of the premixed fuel. The prediction and understanding of cyclic variability have been in the focus for many years in order to improve the efficiency of ICE. LES in principal offers the possibility to compute cyclic variations in ICE, provided that proper measures will be taken. In this investigation a number of this measures are realized, which could predict cyclic variations of ICE. In addition to a former investigation of the authors for one full cycle of the same engine, a model for the transport of the fuel-to-air mixing ratio into the cylinder and the mixing of it with the burnt gas has been proposed and used. In combination with the Large Eddy Simulation this allows a more realistic description of the inhomogeneous distribution of the fuel-to-air ratio ϕ during the combustion process.

W.E. Nagel et al. (eds.), *High Performance Computing in Science and Engineering '10*, DOI 10.1007/978-3-642-15748-6_29,

The investigated engine is a modified Rotax one-cylinder four-valve engine. The mesh generation was done very carefully in order to have a suitable mesh for a high quality computation. The finest mesh consists of about 7×10^6 points while during the startup process a mesh with about 9×10^5 points was used in order to save computer resources. A typical calculation of a full cycle took about 200h in our supercomputing center of KIT using 120 processors in parallel.

The fully compressible Navier-Stokes equations are solved on deforming meshes using an arbitrary Lagrangian-Eulerian (ALE) method with a 2nd order cell centered scheme in space and a dual time stepping approach for the integration in time. In order to prescribe the grid velocities accurately the method proposed by Lai and Przekwas [5] is applied. Details about it can be found in Magagnato et al. [6].

2 Numerical Setup

The code used in this investigation is our in-house Finite Volume code SPARC [7]. It is a block-structured and parallelized approach using second or fourth order central difference scheme in space and a second order implicit dual time stepping scheme in time. The classical explicit four stage Runge-Kutta time stepping scheme is also implemented but not used for this application since it is three times slower compared to the implicit scheme.

The mesh was generated with about 7×10^6 points in the finest mesh covering the combustion chamber and a fraction of the inlet valves and outlet valves (see Fig. 1).

The modified Rotax four-valve engine has a stroke and bore of 83×100 mm^2 and the compression ratio is 11.5. At an engine speed 2000 rpm the spark timing was at 12° (BTDC). At fully opened throttle, at the inlet a total pressure of $p_t = 99000$ N/m^2 and at the outlet a static pressure of $p = 101000$ N/m^2 are measured and set in the computation. The equivalence ratio was 1.123 of air/gasoline mixture.

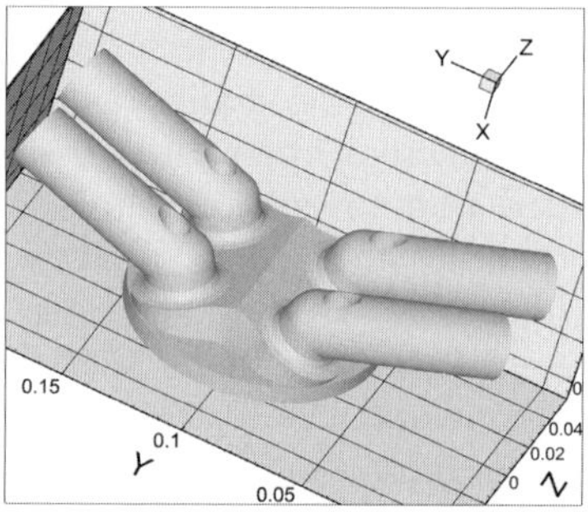

Fig. 1. Inlet and outlet valves of the engine

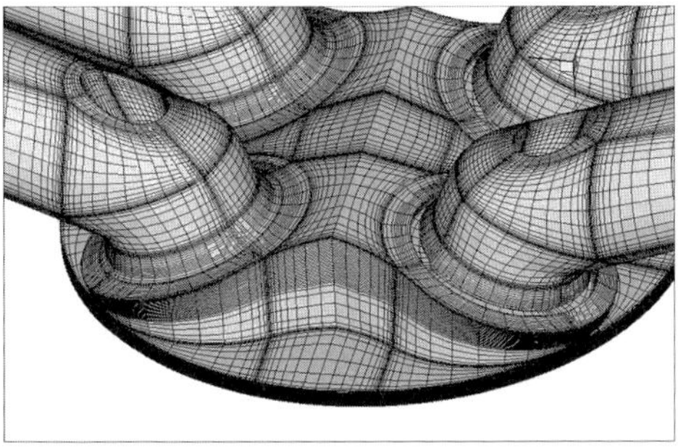

Fig. 2. Mesh on the walls of the four-valve engine (9×10^5)

The edge in the geometry complicates the generation of a sufficiently smooth block-structured grid suitable for a LES considerably. Especially in the area indicated in Fig. 2 the mesh points are unevenly distributed. There we had to apply a mesh stretching in the order of $q = 1.1$ instead of $q < 1.05$ to accommodate a reasonable number of mesh points. The block structured mesh consists of about 1008 blocks. This made the parallelization task relatively easy to accomplish.

We have used a time step of $\Delta t = 1 \times 10^{-6}$ s. This is smaller then physically necessary, but was found to be necessary due to stability problems arising at the highly deformed meshes near the edge (see Fig. 2).

3 Combustion Model

A simple but accurate combustion model has been proposed by Schmid et al. [3]. It is a reaction model solving an equation for the progress variable Θ as a function of the local turbulence and kinetic parameters like Da_t.

$$\frac{\partial \rho \Theta}{\partial t} + \frac{\partial (\rho u_i \Theta)}{\partial x_i} - \frac{\partial}{\partial x_i}\left(D_t \frac{\partial \Theta}{\partial x_i} \right) = S_\Theta \tag{1}$$

$$S_\Theta = 4.96 \cdot \frac{\epsilon}{k} \left(\frac{S_l}{\sqrt{\frac{2}{3}k}} + (1 + Da_t^{-2})^{-\frac{1}{4}} \right)^2 \cdot \Theta \cdot (1 - \Theta) \cdot \rho_u \tag{2}$$

Where S_l is the laminar burning velocity, k is the turbulent kinetic energy, ϵ is the dissipation rate of the k, D_t is the turbulent diffusion coefficient and Da_t is the turbulent Damköhler number. It is clear that the model has been proposed to be used in combination with a two equation k-ϵ turbulence model. In our case we want to extend it to be used with LES. Therefore k, ϵ and D_t has to be replaced by subgrid-scale quantities. We have modeled them as:

$$k = (c_{comb} \cdot \Delta \cdot |S_{ij}(\overline{u})|)^2 = u' \tag{3}$$

$$\epsilon = \frac{k^{\frac{3}{2}}}{\Delta} \tag{4}$$

Where Δ is the filter length and $S_{ij}(\overline{u})$ is the strain rate tensor. The constant $c_{comb} = 0.08$ has been calibrated using the bomb experiments of Abdel-Gayed et al. [8] for propane/air mixtures.

The laminar velocity is computed by using the correlations of Metghalchi and Keck [9] which depends on the static pressure and the temperature of the unburned mixture.

The original Schmid et al. model and the majority of progress variable models or flame surface density models have severe difficulties when applied to flames close to walls. There the turbulent kinetic energy becomes zero and therefore is not applicable there. The simple cure to limit the characteristic turbulent time $1/\tau_{turb} = \epsilon/k$ or $1/\tau_{turb} = u'/\Delta$ in the LES context by the chemical time $1/\tau_{chem}$ avoids effectively that the solution becomes singular but leads to unrealistic flame propagation velocities close to the wall. This well known problem has been known since years and a few attempts to solve this problem have been addressed in the past. In particular the Intermittent Turbulent Net Flame Stretch (ITNFS) of Meneveau and Poinsot [10] in which the flame/vortex interaction has been investigated via DNS and a so called efficiency function has been proposed which decreases the mean reaction rate in highly strained regions. A significant improvement when used in Eddy Break Up models or flame surface density models has been reported. We have used it also in the Schmid et al. model. The efficiency function Γ_k modifies the turbulent time according to:

$$\frac{1}{\tau_{turb}} = \Gamma_k \left(\frac{u'}{S_l}, \frac{l_t}{\delta_l} \right) \cdot \frac{u'}{\Delta} \tag{5}$$

The formulation of Γ_k can be found in [5]. Additionally to the inclusion of ITNFS formalism a Van Driest wall damping function has been used in the calculation of u'. It reduces the turbulent kinetic energy k or u' close to the wall additionally. In Fig. 3 the difference in the distribution of the flame front is shown between Schmid et al. model with and without ITNFS modification. The propagation of the flame close to the wall is clearly reduced with the ITNFS model in the left picture.

Another modification of the Schmid model was found to be necessary. In order to prevent the first term inside the parenthesis to become singular a lower limit of the turbulent velocity u' must be introduced. Here we limit it by the laminar flame speed S_l. The new source term S_Θ reads now:

$$S_\Theta = 4.96 \cdot \frac{\sqrt{u'}}{\Delta} \left(\frac{S_l}{\sqrt{\frac{2}{3} u'}} + \left(1 + Da_t^{-2}\right)^{-\frac{1}{4}} \right)^2 \cdot \Theta \cdot (1 - \Theta) \cdot \rho_u \tag{6}$$

$$u' = \left(1 - exp^{\frac{-y^+}{25}}\right) \cdot c_{comb} \cdot \Delta \cdot |S_{ij}(\overline{u})| \tag{7}$$

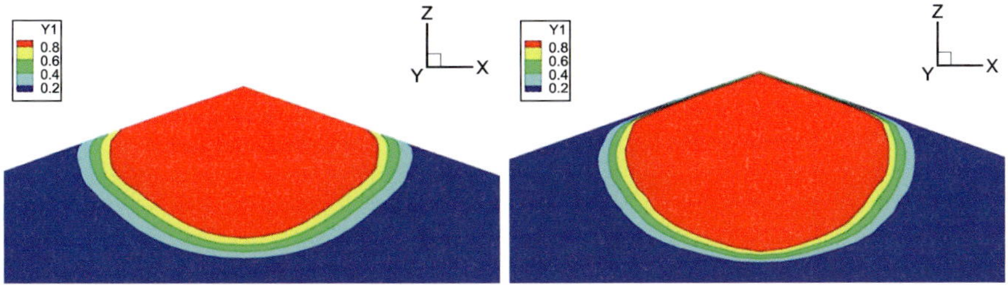

Fig. 3. Prediction of the flame front without ITNFS (left) and with ITNFS (right) model

4 Mixing Model

One of the reasons why cyclic variations in the combustion process arise is the variable mean fuel-to-air ratio ϕ due to turbulent variability between two consecutive cycles of an ICE. In addition to that also an inhomogeneous distribution of this ratio during the combustion process is possible. Since the computational time for an LES is very high compared to RANS the use of rather simple combustion models is appropriate for the simulation. For a big number of these simple models like Progress variable models, Flame surface density models etc. there is no influence on a variable fuel-to-air ratio and must therefore be modified in order to be used for premixed combustion processes.

We model this mixing by an additional transport equation of the fuel-to-air ratio ϕ and modifying the energy equations by multiplying the heat-release rate term of the combustion process by ω_T.

$$\frac{\partial \rho\phi}{\partial t} + \frac{\partial(\rho u_i \phi)}{\partial x_i} - \frac{\partial}{\partial x_i}\left(D_t \frac{\partial \phi}{\partial x_i}\right) = S_\phi \tag{8}$$

$$S_\Theta = \begin{Bmatrix} -\frac{10\rho_u}{\tau_t} & \rightarrow \phi > 0.99 \\ 0 & \rightarrow \phi \leq 0.99 \end{Bmatrix} \tag{9}$$

$$\omega_T = S_\Theta \cdot \phi \cdot \Delta h_{T_0} \tag{10}$$

As boundary condition for the ϕ-equation at the inlet we set the inflow value of the equivalence ratio $\phi = 1.123$ according to the measurements. The destruction term in the ϕ-equation is necessary in order to bring the fuel-to air ratio toward zero behind the flame front. This is necessary for the next cycles to enable the new prediction of the mixing process.

5 Results

We observe then in the simulation a mixture of the unburnt gas and the burnt residual gas inside the cylinder leading to a distribution of ϕ like in Fig. 4 during the intake stroke.

Fig. 4. Mixing of the fuel-to-air ratio ϕ during the intake stroke

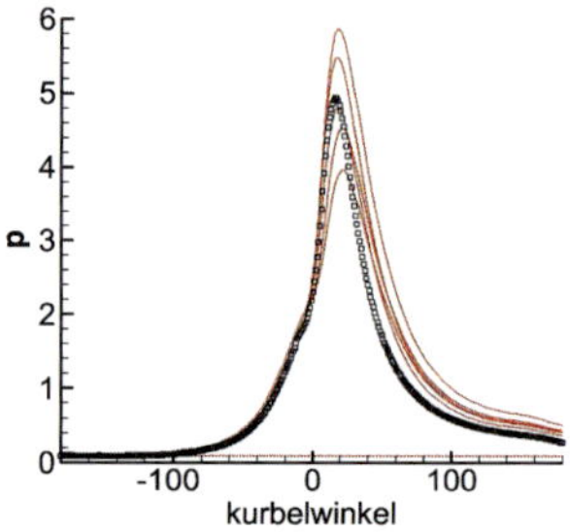

Fig. 5. Cylinder pressure over crank angle for different cycles (squares: measured; line: predicted)

After closing of the intake valve the fuel mixture will no longer flow into the cylinder and starts to homogenize until the ignition takes place at 12° BTDC. At that time the effective equivalence ratio reaches a value between 75 and 85 % depending on the different cycles.

In our former investigation with a single cycle we had to prescribe the effective equivalence ratio so that we could predict the cycle averaged cylinder pressure over crank angle. That means that using the additional equation for the equivalence ratio we are in the position to calculate instead of guessing the right equivalence ratio inside the cylinder. This is a big advantage.

In Fig. 5 we show the predicted cylinder pressure over a number of cycles. We observe that during the startup of the computation the peak pressure varies strongly but unfortunately after 6 to 7 cycles the pressure variations disappear and we predict always the same pressure rise over crank angle (in Fig. 5 this is difficult to see since a few lines are on top of each other).

This is disappointing since we wanted to simulate the cyclic variations. It is well known in the computing community that additionally to what we do now, we have to compute the unsteady inlet and outlet boundary conditions (mainly pressure) which develop due to the opening and closing of the valves in combination with the compressible gas inside the intake and exhaust channels.

6 Computational Efficiency

In the computation of these results we have been using up to 120 Opteron processors of the HP XC4000 in Karlsruhe. The in-house developed code Sparc is parallelized with the MPI 1.2.7 software. The computational time for one full cycle was about 700h. Since we were using 1008 blocks of the finite volume scheme we could efficiently distribute the blocks on the 120 processors with the domain decomposition technique. The load balancing was at about 96%. The parallel efficiency was also very good. Since in Karlsruhe the communication is done with the InfiniBand 4X DDR Interconnect the parallel efficiency was close to 98%. From our recent investigations we know that a higher resolution of the computational mesh is required. We think that using about 50 million points in the next phase will be adequate for a well resolved Large Eddy Simulation

7 Conclusions

In this investigation we have extended a former Large Eddy Simulation of a single full cycle of a spark-ignited engine by proposing and using an additional equation for the transport of the fuel-to-air mixture (equivalence ratio) in order to be able to predict the cycle to cycle variation observed in all internal combustion engines. In the former investigation we could predict a very good agreement of the measured cylinder mean pressure with our computations provided that we know or guess the averaged equivalence ratio inside the cylinder. With the new transport equation for the equivalence ratio we are now in the position to compute rather then guessing the exact value of it.

We were hoping that the computation of the variable equivalence ratio from cycle to cycle is large enough to predict the full range of the cylinder pressure observed in the experiment. But this didn't work because an important aspect of the reasons causing this variability namely the unsteady pressure and temperature distribution at the inlet and outlet valves has not been modeled yet.

As a result the observed variable cylinder pressures in Fig. 5 are due to the initialization process and die out completely after a few cycles.

We therefore will implement a one-dimensional extension of the flow field in front of the inlet valve as well as after the exhaust valve such that the oscillating pressure waves could be approximated and used as boundary condition in the inlet and outlet planes.

Acknowledgements. The present work is a part of the subproject B2 of the Collaborative Research Centre (CRC) 606—"Unsteady Combustion: Transport phenomena, Chemical Reactions, Technical Systems" at the Karlsruhe Institute of Technology (KIT). The project is supported by the German Research Foundation (DFG).The authors acknowledge the contribution of their colleagues to this work.

References

1. Heidenreich, T., Private communication
2. Magagnato, F., Yong, Y., Gabi, M.: Large Eddy Simulation of a Spark-Ignition Engine using a High-Pass filtered Eddy-Viscosity Model. Proceedings of the 7th Comodia Int. Conf., Sapporo, Japan (2008)
3. Schmid, H.P., Habisreuther, P., Leuckel, W.: A model for calculating heat release in premixed turbulent flames. Comb. Flame, **113**, 79–91 (1998)
4. Stolz, S., Schlatter, P., Meyer, D., Kleiser, L.: High-Pass filtered Eddy-Viscosity Models for LES. In: Friedrich, V.R., Geurts, B.J., Metais, O. (ed) Direct and Large-Eddy Simulation. Kluwer, Dordrecht (2003)
5. Lai, Y.G., Przekwas, A.J.: A Finite-Volume Method for fluid flow Simulations with Moving Boundaries. Comp. Fluid Dynamics, **2**, 19–40 (1994)
6. Magagnato, F., Gabi, M., Heidenreich, T., Velji, A., Spicher, U.: Large Eddy Simulation (LES) with moving meshes on a Rapid Compression Machine: Part 2: Numerical Investigations using Euler-Lagrange-Technique. In: Nagel, W., Kröner, D., Resch, M. (ed) High-Performance Computing in Science and Engineering '07. Springer, Berlin Heidelberg (2007)
7. Magagnato, F.: KAPPA—Karlsruhe Parallel Program for Aerodynamics. TASK Quarterly, **2** no. 2, Gdansk (1998)
8. Abdel-Gayed, R., Bradley, D., Hamid, M., Lawes, M.: Twentieth Symposium (International) on Combustion. The Combustion Institute, Pittsburgh, 505–512 (1984)
9. Metghalchi, M., Keck, J.C.: Burning velocities of mixtures of air with methanol, isooctane, and Indolene at high pressure and temperature. Comb. Flame, **48**, 191–210 (1982)
10. Meneveau, C., Poinsot, T.: Stretching and quenching of flamelets in premixed turbulent combustion. Comb. Flame, **86**, 311–332 (1991)

CFD-CSD-Coupled Simulations of Helicopter Rotors Using an Unstructured Flow Solver

Felix Bensing, Manuel Keßler, and Ewald Krämer

Institut für Aerodynamik und Gasdynamik, Universität Stuttgart, Pfaffenwaldring 21, D-70569 Stuttgart, Germany, bensing@iag.uni-stuttgart.de

Summary. Demands concerning geometrical detail arising in the simulation of certain flow phenomena in helicopter interactional aerodynamics can only be dealt with in an unstructured simulation environment. This work deals with the extension of the simulation capabilities at IAG to an unstructured flow solver. The new approach is validated against and compared to the standard structured flow chain in the context of isolated weakly CFD-CSD coupled simulations in wind tunnel conditions corresponding to the low-speed pitch up case of the GOAHEAD experiment. Generally, good agreement between the two respective toolchains is shown. All computations performed herein were conducted using the massively-parallel NEC Nehalem platform at HLRS.

1 Introduction

Helicopter aeromechanics, a research field involving strongly multi disciplinary aspects of both aerodynamics and aeroelasticity, has been a field of extensive research at the Institut für Aerodynamik und Gasdynamik (IAG) for several years now. Within this context, the topic of fluid-structure coupling at isolated main rotors has turned out essential for a realistic representation of the flow phenomena, especially in forward flight. Furthermore, a procedure for trimming the rotor towards a specified flight dynamic state has proven mandatory. In this sense, a rotor trim is defined by the action of numerically reproducing certain rotor parameters of a corresponding experiment. Only in fulfillment of these presuppositions can reasonable comparability of simulation and experiment be guaranteed. From simulations of isolated rotors, research at IAG steered towards the simulation of complete helicopter configurations [4, 7]. Up to now, all simulations were conducted following a structured grid approach. As quantitative investigations of determinate flight conditions in interactional helicopter aerodynamics, such as the well-known tail shake phenomenon, call for a substantial increase in geometrical detail, the structured grid generation process suffers from excessive manual time consumption and eventually becomes impossible. Therefore, IAG is currently building up a new

W.E. Nagel et al. (eds.), *High Performance Computing in Science and Engineering '10*, DOI 10.1007/978-3-642-15748-6_30,

toolchain based on the DLR unstructured flow solver TAU. As part of setting up this new simulation environment, the weak coupling methodology has been extended to unstructured grids of arbitrarily mixed element types. Results for an isolated rotor setup in a low speed forward flight condition are compared between two toolchains, where FLOWer is used as a structured and TAU as its unstructured counterpart. In both cases, structural dynamics and the trim of the rotor is done by the flight mechanics code HOST by Eurocopter [3].

2 Mathematical Formulation and Numerical Scheme

2.1 Flow and Structural Modeling

Aerodynamics—CFD

In the present work, two computational fluid dynamics codes for the solution of the three-dimensional, unsteady Reynolds-averaged Navier-Stokes equations (URANS) were used. Our standard toolchain uses the FLOWer code [8, 11] and has been used as a reference for validation of our new simulation environment built around the TAU code. FLOWer is based on a block-structured Finite-Volume approach, whereas TAU features an unstructured discretization, designed for massively parallel computations. Both schemes employ dual time-stepping according to Jameson [6], converging each timestep to quasi steady-state by means of Runge-Kutta time integration. Spatial discretization is performed using central differencing, stabilizing the convective terms with the help of second and fourth order difference artificial dissipation operators. Additional fluxes due to grid movement can be taken into account using an Arbitrary Lagrangian Eulerian (ALE) formulation, where accuracy and stability are enhanced by the satisfaction of a discrete Geometric Conservation Law (GCL). Large grid movements such as the rotation of rotor blades are rendered possible via the Chimera technique of overlapping grids.

Structural Dynamics—CSD

The CFD codes mentioned above were coupled to the Eurocopter flight mechanics code HOST, a general purpose computational environment for the simulation and stability analysis of complete helicopters with all their substructures, as well as isolated rotors. The software is capable of trimming the rotor towards prescribed objectives, based on lifting line methodology using 2D airfoil tables. HOST's internal representation of the elastic blade comprises a quasi one-dimensional Euler-Bernoulli beam where deflections in flap and lag directions as well as elastic torsion along the blade axis are permitted. Simplifying assumptions involve a linear material law and neglection of shear deformation as well as tension elongation. Mismatches of the local cross-sectional centers of gravity, tension and shear are however accounted for, resulting in couplings between bending and torsional degrees of freedom. The geometrically non-linear blade is modeled as a sequence of rigid elements which are connected by virtual joints. At each joint, rotations about the lag, flap and torsional axes are allowed. The large number of degrees of freedom resulting from this discretization is reduced by means of a modal Rayleigh-Ritz approach such that the deformation is finally described by a weighted sum of a limited set of mode-like deformation shapes.

2.2 Weak Coupling Methodology and Trim Procedure

In contrast to a strong coupling methodology [1, 2] where data exchange between CFD and CSD is done at each timestep, the weak coupling approach involves intercommunication on a per-period basis. For a n-bladed rotor, this would mean that n/rev periodicity of the flow solution has first to be established before passing these periodic parameters to the CSD solver. When communicating data from CFD to CSD, these parameters are blade-sectional forces and moments, while the corresponding deformations are transferred back from CSD to CFD. Concurrently, rotor control angles are updated in order to reach prescribed trim objectives. Usually, simulations of wind tunnel experiments require three control parameters be set free, namely main rotor collective θ_0 and the two cyclic pitches θ_c and θ_s. An equal number of trim objectives has to be specified. Most commonly, these trim objectives are global time-averaged rotor thrust, pitching and rolling moments. However, for this study a pure force trim was employed, i.e. the trim objectives consisted of the three components of the integral averaged rotor forces.

The coupling procedure can be broken down into the following steps:

1. HOST computes an initial rotor trim based on its internal 2D aerodynamics derived from airfoil tables. The complete blade dynamic response is fully described by the modal base and the related generalized coordinates.
2. A CFD computation is carried out taking into account the blade dynamic response by the reconstruction of the azimuth dependent blade deformation from the modal base and the respective grid deformation of the blade grids.
3. From the CFD calculation the radial 3D blade load distributions in the rotating hub rotor system (F_x, F_y, F_z in [N/m], M_x, M_y, M_z in [Nm/m]) are derived for each azimuth angle and radial station of the blade.
4. In the next trim HOST uses a load given by

$$F^n_{HOST} = F^n_{2D} + F^{n-1}_{3D} - F^{n-1}_{2D},$$

   where F^n_{2D} represents the free parameter for the actual HOST trim. A new dynamic blade response is obtained which is expressed by an update of the generalized coordinates.

5. Steps (2) to (4) are repeated until convergence has been reached, i.e. when the difference
$$\Delta F^n = F^n_{2D} - F^{n-1}_{2D} \rightarrow 0 \, .$$
Then trim loads depend solely on the three-dimensional CFD aerodynamics and no longer on HOST-internal two-dimensional airfoil data.

The available weak coupling algorithm for structured meshes has in the course of this work been augmented to accommodate unstructured meshes of arbitrary cell types. Radial stations, for which output of the coupling loads are requested, are specified via input file. In a preprocessing step, grid point data for each of these radial stations are extracted from the undeformed mesh. These data are then used in a postprocessing step to reconstruct relevant grid points in deformed state such as moment reference points or plane vectors. During the integration of the loads for HOST, the algorithm then slices the surface mesh cells according to the possibly flapped and lagged planes corresponding to the respective radii. Then, all cells emerging from this slicing process are subtriangulated and force integration is performed over all these subtriangles. Consequently, triangular as well as quadrilateral blade surface cells can be treated in a generalized manner. Furthermore, only certain so-called aerodynamic parts of the blade surface can be selected to count towards loads integration by simple flagging with the help of surface marker information contained in the mesh. This will provide the flexibility to simulate entire and detailed rotors, where the rotor hub region is not likely to be of interest for the pure blade force coupling. Also, fast-prototyped blade grids for testing purposes containing triangular cells on the surface are rendered possible following this approach.

3 Results

3.1 Experiment and Test Case Setup

A test case from the GOAHEAD test campaign (Generation of Advanced Helicopter Experimental Aerodynamic Database for CFD code validation) [10] was selected as a reference in the present study. Experiments were carried out on a fully equipped 1 : 3.9—scaled helicopter configuration mounted in the 6 × 8 m open test section of the German-Dutch wind tunnel (DNW). In the simulations an isolated four-bladed rotor was considered in low-speed pitch-up conditions featuring a flight Mach number of $Ma = 0.059$, a blade tip Mach number of $Ma_{tip} = 0.617$ and a corresponding rotatory speed of $\Omega = 954 \text{ min}^{-1}$, resulting in an advance ratio of $\mu = 0.0956$. Geometrically, the blade consists of a rectangular planform of chord length $c = 0.14$ m up to the radial station $r/R = 0.946$, followed by a parabolic tip with a reduction of chord length to $c = 0.046$ m at $r/R = 1.0$, leading to a rotor solidity of $\sigma = 0.085$. An airfoil of OA213-type was used up to $r/R = 0.75$ and OA209 above

$r/R = 0.9$ with a transition of airfoil geometry in between. The blade features a $-8.3°$ linear twist and was meshed using a CH-topology for the aerodynamic part of the blade and a HH-like topology in the root and tip regions. The rotor shaft angle was set to zero in this case and mesh sizes were 0.81 million for each blade and 2.85 million for the background grid. Simulations were carried out using both flow solvers FLOWer and TAU on identical meshes using a timestep corresponding to a $2°$ increment in rotor azimuth angle. For both flow solvers the RANS equations were closed employing the standard $k - \omega$ turbulence model according to Wilcox [12].

3.2 Trim Convergence

As stated above in Sect. 2.2, a pure force trim was employed for this investigation. Trim objectives are shown in Table 1. Convergence of the control angles obtained from both solutions using TAU and FLOWer are presented in Fig. 1a. The iterative trim process was stopped once the variations of all three control angles had fallen below $0.03°$. The corresponding evolution of the instationary rotor loads for the TAU computation is shown in Fig. 2 where thick vertical lines mark the individual trim iterations. From this it can be seen that differences in the force distributions between successive trim iterations are significant during the first two or three cycles, whereas in the last two

Table 1. Trim objectives and actual simulation results (Trim 4)

	Trim Objective	TAU Code	FLOWer Code
F_X [N]	−44.4	−44.5	−46.2
F_Y [N]	0.0	0.9	−1.4
F_Z [N]	4501.2	4541.3	4535.9

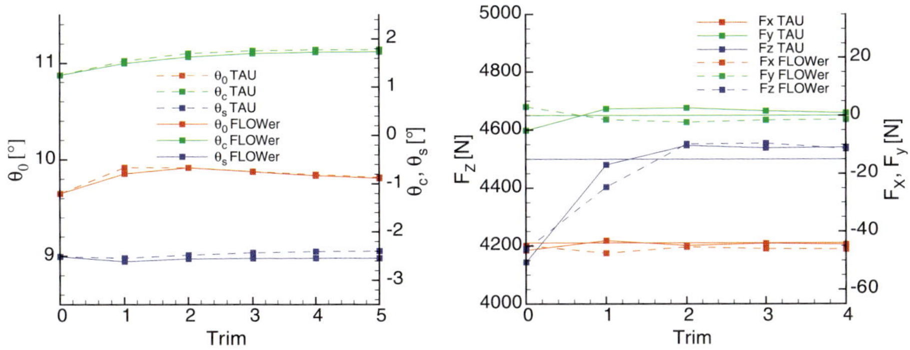

(a) Control angles (b) Average rotor loads

Fig. 1. Convergence of control angles and rotor loads: comparison FLOWer–TAU

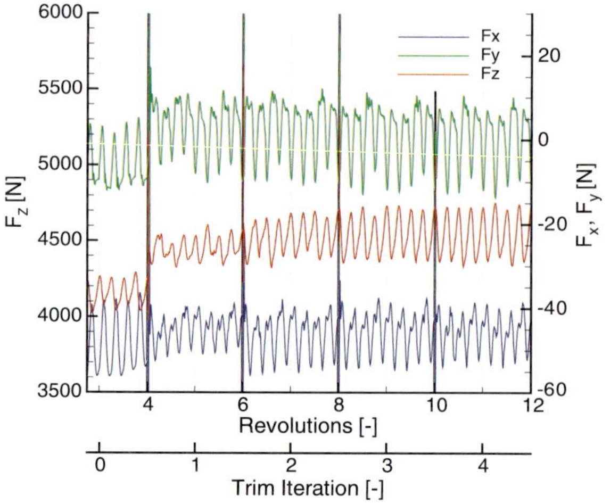

Fig. 2. Development of (instationary) rotor loads (TAU)

retrims (revolution 8–12) no further changes are apparent. When comparing the control angles obtained in the FLOWer and TAU simulations, differences are damping out during convergence of θ_0 and θ_c and only the lateral angle θ_s finally produces a slight offset of 0.15°. In order to confirm that the prescribed trim objectives were met, instationary forces of the last quarter revolution of each trim cycle were averaged. This is shown in Fig. 1b. Trim objectives are displayed as straight lines without symbols, the TAU computation as solid and the FLOWer simulation as dashed lines. The actual values of the rotor loads in converged trim conditions are shown in Table 1. While FLOWer and TAU produce a comparable 0.8% and 0.9% deviation from the trim objective in z-direction, differences in the lateral and longitudinal forces appear to be notably smaller when employing the new TAU toolchain.

3.3 Blade Dynamics and Rotor aerodynamics

Blade Dynamics

In Figs. 3a and 3b the development of the blade dynamics is shown. Again the differences between the last two trim iterations are neglegible for both motions, thus trim 3 and 4 are not shown here for clarity. The 4/rev-character of the tip torsion cannot be captured by the pure HOST calculation of trim 0 in contrast to the subsequent computations including corrections from three-dimensional CFD-data. While differences between both toolchains in the flapping motion are not very significant, somewhat larger deviations are observed in blade tip torsion during the trim process: Only slight differences in the vicinity of $\Psi = 150°$ can be observed in the flap motion. The greater differences that

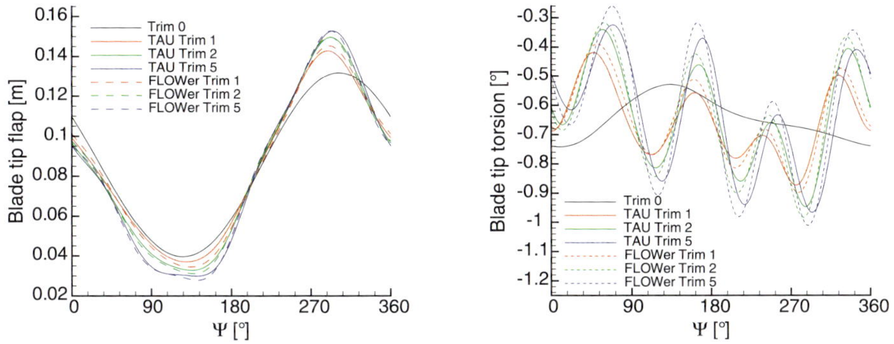

(a) Blade tip flap deflection (b) Elastic tip torsion

Fig. 3. Development of blade tip flap and torsion: comparison FLOWer–TAU

occur in the tip torsion are already suggested by the deviation in the lateral control angle of approximately 0.1° (see Fig. 1a).

Aerodynamics

Aerodynamics are first assessed by studying chordwise distributions of the inviscid forces. This is done by comparing the pressure coefficient at specified radii and azimuth angles for computations using FLOWer and TAU. In Fig. 4, sample results from this analysis are shown for radial stations $r/R = 0.47$ and $r/R = 0.89$. Both solutions appear to be in excellent agreement for the inner stations and only small differences occur on the outer radial position. TAU shows slight overpredictions of $c_{p,min}$ compared to FLOWer for most azimuthal positions. Only for $\Psi = 330°$ FLOWer shows a smaller leading-edge suction peak. Generally, good correlation between FLOWer and TAU has been achieved.

Azimuth- as well as radius-dependent distributions of the total vertical force on the rotor plane are plotted in Figs. 5a and 5b. It can be seen that TAU predicts slightly higher loads on the retreating side in the range $\Psi = 210 \ldots 270°$ and on the advancing side around $\Psi = 110°$. This is in accordance with the lower magnitude of the θ_s control angle in Fig. 1a in the TAU-computation since HOST tries to react to such higher loads by generating smaller amplitudes in blade torsion (Fig. 3b). The overall agreement between the two flow solutions however is good.

For a more in-depth analysis, force coefficients corresponding to the sectional tangential (drag-directed) and normal (lift-directed) forces are plotted in Figs. 6 and 7. Azimuthal variations of the Mach-normalized coefficients $c_n Ma^2$ and $c_t Ma^2$ are shown for radial stations $r/R = 0.31, 0.77$ and 0.90. Again, it can be observed that generally differences between the two codes are smaller at the inner sections of the blade. The maximum difference in the

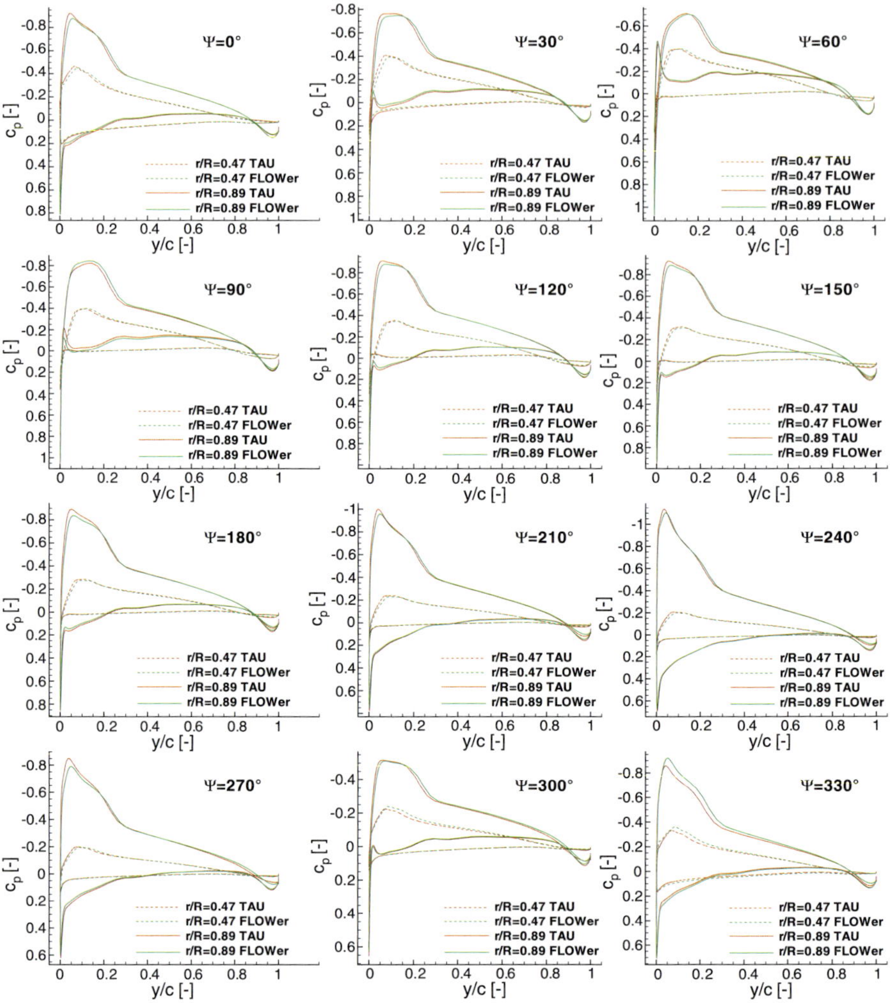

Fig. 4. Chordwise c_p-distributions at two radial stations ($r/R = 0.47$ dashed, $r/R = 0.89$ solid) along azimuth: comparison FLOWer–TAU (Trim 4)

normal force coefficient at the outer radius amounts to about 6 %. Although the close-to zero values of the tangential force coefficient of the FLOWer computation in the range $\Psi = 120° \ldots 180°$ are not exactly reproduced using TAU, trends are very similar. In Fig. 8, a vortex visualization of the flow field is shown using the λ_2-criterion. Contour plots of λ_2 are shown for two distinct vertical slice planes and the pressure coefficient is plotted on the blade surfaces.

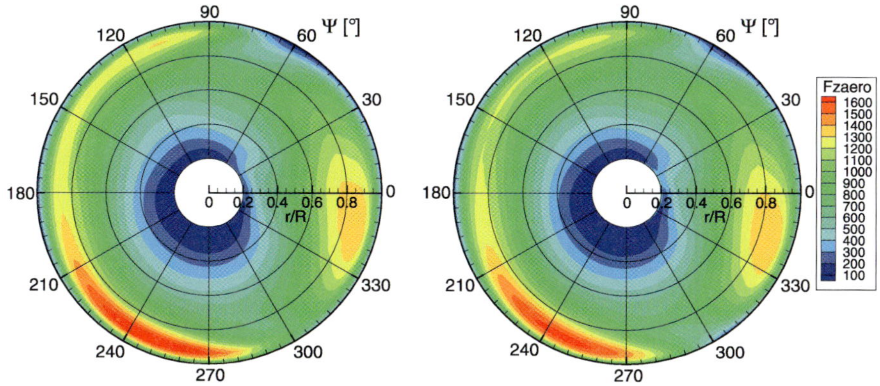

(a) Vertical force (TAU) (b) Vertical force (FLOWer)

Fig. 5. Distribution of the vertical force on the rotor plane F_z: comparison FLOWer–TAU

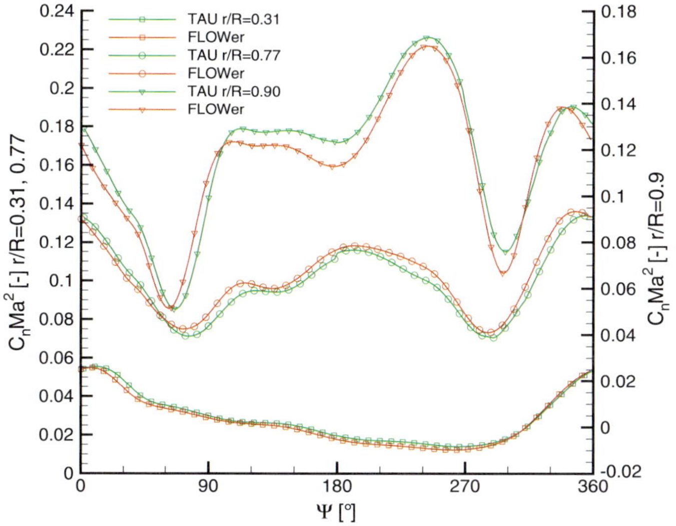

Fig. 6. Azimuthal variation of the sectional normal force coefficient $C_n Ma^2$ at radial stations $r/R = 0.31, 0.77$ and 0.90

3.4 Computational Performance

Code Architecture

Flow computations around helicopter configurations rank amongst the most elaborate simulations concerning code infrastructure. Solvers for these problems must incorporate certain algorithms that, if implemented inefficiently, will lead to prohibitively low performance. Examples for such algorithms are

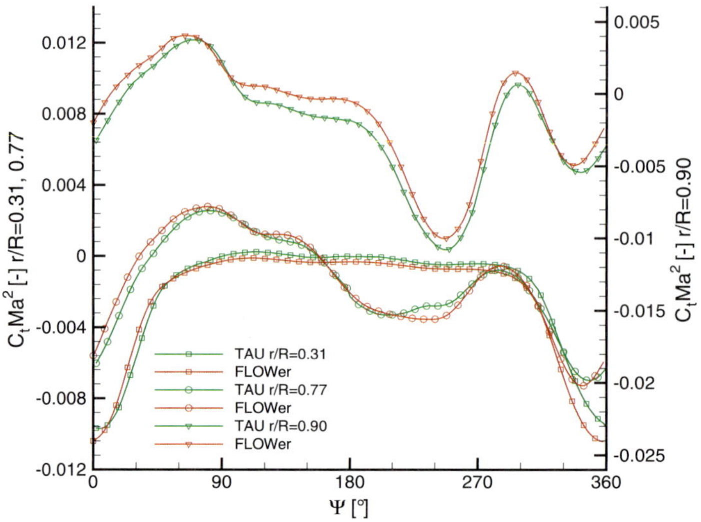

Fig. 7. Azimuthal variation of the sectional tangential force coefficient $C_t Ma^2$ at radial stations $r/R = 0.31, 0.77$ and 0.90

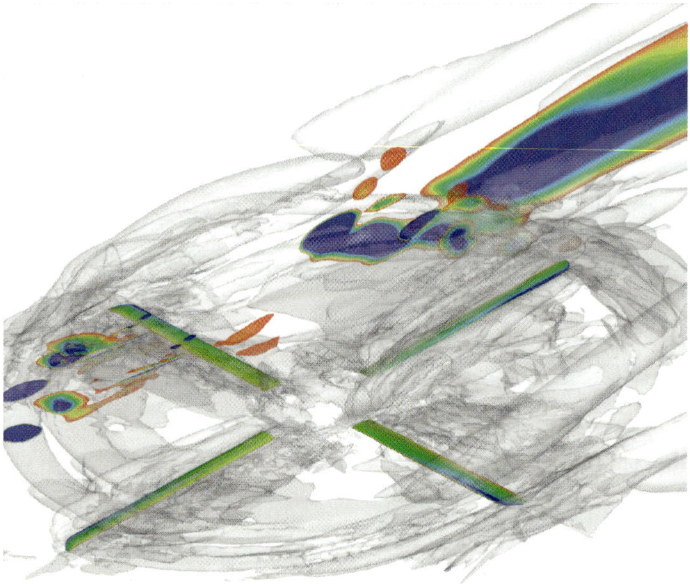

Fig. 8. Vortex visualization using the λ_2-criterion, blades: c_p-distribution

the Chimera search in case of meshes with large relative motion, or the mesh deformation. Especially in the context of Chimera search, TAU features an asynchronously parallelized search based on an Alternating Digital Tree algorithm [9], which in our simulations leads to a time consumption of less than

1% per typical time step cycle. Deformation in the TAU code is based on a purely algebraic approach. This methodology is less expensive than other algorithms commonly applied to unstructured grids such as spring analogy or linear elasticity methods which require the solution of discrete equations using iterative methods. It is also fully parallelized and uses a volume-spline formulation based on radial basis functions [5]. The number of base points for the computation of the radial basis functions was optimized for performance such that a minimum number was chosen while still maintaining high accuracy in the deformed geometry. This way deformation could be configured to only account for 2% of single time step cycle's CPU-time. While FLOWer is a FORTRAN-based code designed as a standalone software, where all parts of the solution process are incorporated into one package, the C-based TAU code features a highly modular architecture: different parts of the cycle, namely determination of blade grid deformations, grid deformation, solver preprocessing and the actual flow solution, are combined in a modern Python-based environment. This way very flexible arrangements become possible without compromising performance since large data sets such as mesh point information or the flow field solution are kept in memory. Furthermore, since FLOWer pursues a block-structured approach, arbitrary repartitioning is not possible without remeshing the whole domain. TAU on the other hand offers an easily resizable and rescalable alternative. Finally, the TAU code with its unstructured approach is perfectly suited for massively parallel platforms such as the NEC Nehalem cluster at HLRS.

Scalability

Figure 9 shows the strong scaling performance of the DLR TAU code applied to the above mentioned four bladed configuration. The test was carried out on the NEC Nehalem cluster at HLRS and featured computations on 2^n ($n = 3, \ldots, 7$) processes. The setup used herein consisted of 6 090 240 cells. As stated above, this simulation was performed using a weakly fluid-structure-coupled toolchain, i.e. computation of deformations, mesh deformation itself, solver preprocessing and flow solver had to be done each time step. Results plotted in red (square symbols) mark the scaling for pure flux computation which accounts for the largest fraction in CPU-time. Close-to linear scaling was obtained for the above mentioned number of MPI-processes. A similar trend is observed for the complete cycle (blue line, circles), which displays the time needed for one entire time step including deformation, solver preprocessing and flow solver. Further parallelization than 128 processes has not been considered here due to the comparably small setup. As a rule of thumb, each parallel subdomain should contain not less than about 10^5 cells which for the given mesh would yield a partitioning to roughly 60 domains.

Targeted simulation cases in the near future will consist of about 100 million, in certain cases of 500 million cells. While this case showed good

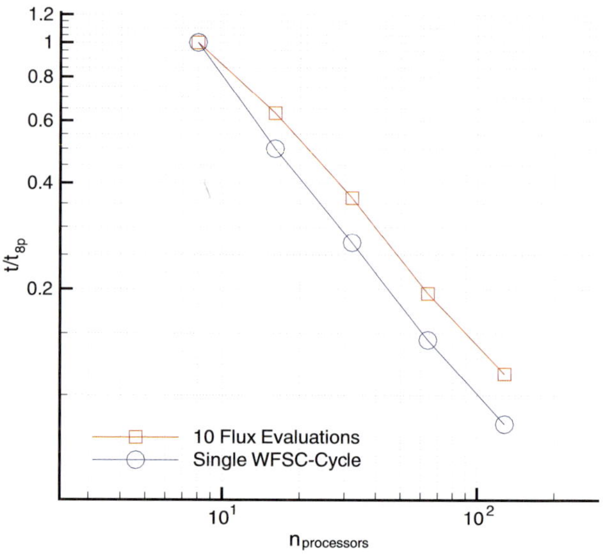

Fig. 9. Scaling of the TAU solver and the complete Python-based weak coupling toolchain

scalability up to about 128 processes, these configurations are expected to scale up to several thousand processors.

Performance Measures

During a 16-process computation, performance was measured for the flux evaluation process. In single-threading mode performance reached 8.5 GFLOPS and in hyperthreading mode improvements to 11 GFLOPS could be obtained. This translates into fractions of 9% and 12% of the node-wide peak performance, respectively. All numbers are subject to the base clock rate of 2.8 MHz of the Nehalem's Intel Xeon X5560 processor.

4 Conclusions and Outlook

Further enhancements concerning the accuracy of flow field predictions in the context of interactional helicopter aerodynamics call for a substantial increase in geometric complexity, making the use of an unstructured approach more and more attractive. In this work the extension of a weak fluid-structure coupling interface to the computational environment and potential advantages of this approach are discussed. The new toolchain has been implemented into the flexible Python-based environment of the TAU-code. Comparisons of the structured-solver based toolchain using FLOWer and the new solution technique are drawn and show good agreement in trim convergence, blade dynamics, and aerodynamic measures. The setup consisting of six million cells showed a good scalability on the NEC Nehalem cluster at HLRS up to 128 MPI-processes and a maximum performance of 12% peak.

Additional modifications within the Python-environment will be needed in order to further decrease the amount of file I/O in each cycle of the toolchain: so far, memory I/O has only been established between the preprocessing and flow solution steps of the solution cycle, while TAU's deformation module still re-reads the undeformed grid coordinates in each time step.

Once comparability of the two respective toolchains has been established, future work will include a more detailed modeling of the rotor hub regions as well as flow computations past an entire helicopter. Accurate predictions of interactional flow phenomena such as tail-shake are finally targeted.

Acknowledgements. This work has been supported by Deutsche Forschungsgemeinschaft (DFG) under grant KR 2959-1. We greatly acknowledge the provision of supercomputing time and technical support by the High Performance Computing Center Stuttgart (HLRS) for our project HELISIM.

References

1. Altmikus, A., Wagner, S., Beaumier, P., Servera, G.: A Comparison: Weak versus Strong Modular Coupling for Trimmed Aeroelastic Rotor Simulations, American Helicopter Society, 58th Annual Forum, Montreal, Canada (2004)
2. Altmikus, A.: Nichtlineare Simulation der Strömungs-Struktur-Wechselwirkung am Hubschrauberrotor, Institut für Aerodynamik und Gasdynamik, Universität Stuttgart, PhD thesis (2004)
3. Benoit, B., Dequin, Kampa, K. Grunhagen, W., Basset P.-M., Gimonet, B.: HOST, A General Helicopter Simulation Tool for Germany and France, American Helicopter Society, 56th Annual Forum, Virginia Beach, Virginia (2000)
4. Dietz, M.: Simulation der Umströmung von Hubschrauberkonfigurationen unter Berücksichtigung von Strömungs-Struktur-Kopplung und Trimmung, Institut für Aerodynamik und Gasdynamik, Universität Stuttgart, PhD thesis (2009)
5. Gerhold, T., Neumann, J.: The Parallel Mesh Deformation of the DLR TAU-Code, In: Notes on Numerical Fluid Mechanics and Multidisciplinary Design, Springer Berlin/Heidelberg (2008), Vol. 96, p. 162–169
6. Jameson, A.: Time dependent calculations using multigrid, with applications to unsteady flows past airfoils and wings, AIAA-Journal 91-1596, 1991
7. Khier, W., Dietz, M., Schwarz, Th., Wagner, S.: Trimmed CFD Simulation of a Complete Helicopter Configuration, 33rd European Rotorcraft Forum, Kazan, Russia (2007)
8. Kroll, N., Eisfeld, B., Bleecke, H.M.: FLOWer, In: Notes on Numerical Fluid Mechanics Vieweg Braunschweig (1999), Vol. 71, p. 58–68
9. Madrane, A., Raichle, A., Stuermer, A.: Parallel Implementation of a Dynamic Overset Unstructured Grid Approach, European Congress on Computational Methods in Applied Sciences and Engineering (ECCOMAS) 2004
10. Pahlke, K.: The GOAHEAD Project, 33rd European Rotorcraft Forum, Kazan, Russia (2007)

11. Schwarz, Th.O.: Ein blockstrukturiertes Verfahren zur Simulation der Umströmung komplexer Konfigurationen, Institut für Aerodynamik und Strömungstechnik, Universität Braunschweig, PhD thesis (2005)
12. Wilcox, D.C.: Turbulence Modeling for CFD, Second Edition, DCW Industries, Inc., La Canada, CA

Wake Signature of Finite-Span Flapping Rigid Wings

Joel E. Guerrero

University of Genoa, Via Montallegro 1, Genoa, 16145, Italy, joel.guerrero@unige.it

Summary. A numerical study is conducted in order to investigate the wake signature and aerodynamics forces of finite-span flapping wings. The unsteady laminar incompressible Navier-Stokes equations are solved on moving overlapping structured grids using a second-order accurate in space and time finite-difference scheme. Specifically, finite-span rigid wings undergoing pure heaving and root-flapping motions are studied. From the results presented, it is found that root-flapping wings produce wake structures similar to those of heaving wings, but with the difference that the latter wing kinematics generates larger vortices and forces than root-flapping wings; aside from this, similar wake regimes occurs at comparable values of the Strouhal numbers. The numerical simulations are performed at a Reynolds number of $Re = 250$ and at different values of Strouhal number and reduced frequency.

1 Introduction

Biologists, zoologists and engineers are actively investigating birds, bats, insects, fishes and cetaceans, as they represent illuminating examples of unsteady aerodynamics, high manoeuvrability, endurance, and large aero/hydrodynamics efficiency. The main reason of studying natural flyers and swimmers relies on the use of Mother Nature as the inspiration model for improving existing applications or developing new technologies by just mimicking nature evolutionary-optimization process (biomimetics). Such applications may include drag and noise reduction by using feather-like structures [1] or the development of new propulsion/lift generation systems for man-made technologies on land, air and sea [2–5].

When studying flapping flight, the term *"flapping"* is often used to refer to the complex motion of birds, bats and insects' wings. Flapping flight involves up and down movement of the wings about the body joint (rolling). During the downstroke (or power-stroke), the wings move downward and forward. During the upstroke (or recovery-stroke), the wings move upward and are drawn in towards the body to reduce drag. The wings also change their angle of attack depending on the stroke and also systematically change shape. Flapping flight

W.E. Nagel et al. (eds.), *High Performance Computing in Science and Engineering '10*, DOI 10.1007/978-3-642-15748-6_31,

is basically rowing in the air with the added complication that lift must be generated. In many experimental and numerical flapping wings studies [6–10] this very complex flapping motion has been simplified to pure heaving or coupled heaving-and-pitching motions, wherein the rolling motion of the wings is replaced by a heaving motion. These studies have given us useful but limited insight into the unsteady aerodynamics of flapping flight, as they do not take into account the root-flapping motion characteristic of birds, bats and insects.

Another important aspect of flying using flapping wings or swimming using oscillating fins is the ability of generating thrust with relatively high propulsive efficiency (defined as the ratio of aerodynamic power output to mechanical power input); early attempts at building fish-inspired mechanisms achieved disappointingly low propulsive efficiencies [11]. It was only through a deeper understanding of the vorticity and wake produced by swimming animals that significant progress was achieved [12].

At this point, an immediate question that comes to our mind when studying natural flyers and swimmers is the range of flapping parameters that may be chosen to optimize the aero/hydro-dynamic performance. Darwinian evolution is not guaranteed to find a solution that is globally optimized among the range of available parameters. However, a thorough examination of Mother Nature's technique is a logical starting point in defining guiding principles. In this field of study, several researchers [13–15], have found that flying and swimming animals cruise at Strouhal numbers St corresponding to a regime of vortex growth and shedding in which the propulsion efficiency peaks. St is a dimensionless parameter defined as

$$St = \frac{f \times \mathcal{A}}{U} \tag{1}$$

where f is the flapping frequency, $\mathcal{A}$ the peak to peak amplitude of the flapping stroke and U is the forward velocity. Biologists, zoologists and engineers, have observed that most flying and swimming animals oscillate their wings or tails in a narrow range corresponding to $0.2 < St < 0.4$ [13–15].

In the current numerical study, we aim to perform a comprehensive analysis of the wake signature of finite-span rigid wings undergoing root-flapping motion, a motion mode which seems not to have been sufficiently studied. In addition, we also check whether this motion mode generates a similar vortex growth and shedding regimes to those of pure heaving motion at similar St values. The unsteady laminar incompressible *3D* Navier-Stokes equations are numerically approximated and all unsteady, viscous and spanwise effects are solved. The simulations are conducted for St regimes between 0.15 to 0.5 and for two different reduced frequency values, one corresponding to high frequency and the other one to low frequency. This is done in order to study leading edge vortex (LEV) shedding dependency. The reduced frequency k, which is another dimensionless parameter that characterizes the unsteady aerodynamics of flapping wings or oscillating fins, is a measure of the residence time of a particle (or a vortex) convecting over the wing (or fin) chord, compared to the period of motion and is defined as

$$k = \frac{\pi \times f \times c}{U} \tag{2}$$

with c the wing cross-section chord.

The remainder of this paper is organized as follows. We begin by briefly describing the numerical method and gridding methodology. Then, we present a description of the computational domain and a short discussion of the results obtained from a grid dependence study. This is followed by a detailed discussion of the results. Finally, we present the conclusions and perspectives.

2 Solution Method

In this section, we briefly outline the solution methodology used to solve the governing equations on moving overlapping structured grids. The complete description of the numerical method and gridding methodology can be found in references [16] and [17], respectively. For the sake of simplicity, we consider the governing equations to be expressed in Cartesian coordinates $\mathbf{x} = (x, y, z)$. The transformation of the governing equations from physical space $\mathcal{P}$ to computational space $\mathcal{C}$ is simple done by replacing the Cartesian derivatives in the governing equations by their equivalent form in the transformed computational space $\mathcal{C}$ (refer to [10, 16–19] for a detailed derivation). In primitive variables *(u,v,w,p)*, the initial-boundary-value problem (IBVP) for the laminar incompressible Navier-Stokes equations is

$$\frac{\partial \mathbf{u}}{\partial t} + \mathbf{u} \cdot \nabla \mathbf{u} = \frac{-\nabla p}{\rho} + \nu \nabla^2 \mathbf{u} \qquad \mathit{for} \quad \mathbf{x} \in \mathcal{D}, \qquad t > 0, \tag{3}$$

$$\nabla \cdot \mathbf{u} = 0 \qquad \mathit{for} \quad \mathbf{x} \in \mathcal{D}, \qquad t > 0, \tag{4}$$

with the following boundary conditions and initial conditions

$$\begin{aligned} B(\mathbf{u}, p) &= \mathbf{g} && \mathit{for} \quad \mathbf{x} \in \partial\mathcal{D}, \qquad t > 0, \\ \mathcal{D}\mathring{\mathbf{Q}}(\mathbf{x}, 0) &= \mathbf{q}_0(\mathbf{x}) && \mathit{for} \quad \mathbf{x} \in \mathcal{D}, \qquad t = 0. \end{aligned} \tag{5}$$

In this IBVP, $\mathbf{x} = (x, y, z)$ (for $\mathbb{N} = 3$ where $\mathbb{N}$ is the number of space dimensions) is the vector containing the Cartesian coordinates in physical space $\mathcal{P}$, $\mathcal{D}$ is a bounded domain in $\mathcal{P} \in \Re^{\mathbb{N}}$ ($\mathbb{N} = 1, 2, 3$), $\partial\mathcal{D}$ is the boundary of the domain $\mathcal{D}$, t is the physical time, $\mathbf{u} = (u, v, w)$ is the vector containing the velocity field in $\mathcal{P}$, p is the pressure, ν is the kinematic viscosity, ρ is the density, B is a boundary operator, $\mathbf{g}$ is the boundary data and $\mathbf{q}_0$ is the initial solution. An alternative formulation to the system of (3)–(5), called the velocity-pressure formulation is

$$\frac{\partial \mathbf{u}}{\partial t} + \mathbf{u} \cdot \nabla \mathbf{u} = \frac{-\nabla p}{\rho} + \nu \nabla^2 \mathbf{u} \quad \textit{for} \quad \mathbf{x} \in \mathcal{D}, \quad t > 0, \tag{6}$$

$$\frac{\nabla^2 p}{\rho} + \nabla u \cdot \mathbf{u}_x + \nabla v \cdot \mathbf{u}_y + \nabla w \cdot \mathbf{u}_z = 0 \quad \textit{for} \quad \mathbf{x} \in \mathcal{D}, \quad t > 0, \tag{7}$$

with the following boundary and initial conditions

$$\begin{aligned} B(\mathbf{u}, p) &= \mathbf{g} && \textit{for} \quad \mathbf{x} \in \partial\mathcal{D}, && t > 0, \\ \nabla \cdot \mathbf{u} &= 0 && \textit{for} \quad \mathbf{x} \in \partial\mathcal{D}, && t > 0, \\ \mathcal{D}\mathring{\mathbf{Q}}(\mathbf{x}, 0) &= \mathbf{q}_0(\mathbf{x}) && \textit{for} \quad \mathbf{x} \in \mathcal{D}, && t = 0. \end{aligned} \tag{8}$$

The system (6)–(8) is the form of the equations that will be discretized. Hence, we look for an approximate numerical solution of (6) and (7) in a given domain $\mathcal{D}$ with prescribed boundary conditions and given initial conditions (8). For the system of (6)–(8) an extra boundary condition is required in order to make the problem well-posed. The boundary condition $\nabla \cdot \mathbf{u} = 0$ for $\mathbf{x} \in \mathcal{D}$ is added as the extra boundary condition. This latter condition is an essential boundary condition for this formulation and ensures that the system (6)–(8) is equivalent to the original formulation ((3)–(5)) [16, 20, 21]. As a boundary condition for the pressure-Poison equation (PPE) (7), we use the normal component $\hat{\mathbf{n}}$ of the momentum equation (6), as discussed in [16, 20–22]; in this way we obtain

$$\left.\frac{\partial p}{\partial \hat{\mathbf{n}}}\right|_{\partial\mathcal{D}} = \left.\hat{\mathbf{n}}\right|_{\partial\mathcal{D}} \cdot \left(-\mathbf{g}_t - (\mathbf{g} \cdot \nabla)\,\mathbf{u} + \nu \nabla^2 \mathbf{u}\right) \rho\,, \tag{9}$$

Equations (6)–(9) are solved in logically rectangular grids in the transformed computational space $\mathcal{C}$, using second-order centred finite-difference approximations on structured overlapping grids. Then, the discretized equations are integrated in time using a semi-implicit multi-step method, which uses a Crank-Nicolson scheme for the viscous terms and a second-order Adams-Bashforth/Adams-Moulton predictor-corrector approach for the convective terms. By using this time-stepping scheme, the velocity is advance in time by using a second-order Adams-Bahforth predictor step, followed by a second order Adams-Moulton corrector step, where we use two corrector steps. This yields a second-order accurate in space and time numerical scheme. We choose to implicitly treat the viscous terms because if they were treated explicitly we could have a severe time step restriction, proportional to the spatial discretization squared.

The overlapping structured grids method consists in generating a set of body-fitted conforming structured components grids $\mathcal{G}_g$ that completely cover the domain $\mathcal{D}$ that is being modelled in physical space $\mathcal{P} = \mathcal{P}(x, y, z, t)$ and overlap where they meet (see Fig. 1). The governing equations are then solved separately in computational space $\mathcal{C} = \mathcal{C}(\xi, \eta, \zeta, \tau)$ on each component grid $\mathcal{G}_g$ and domain connectivity is obtained through proper interpolation in the

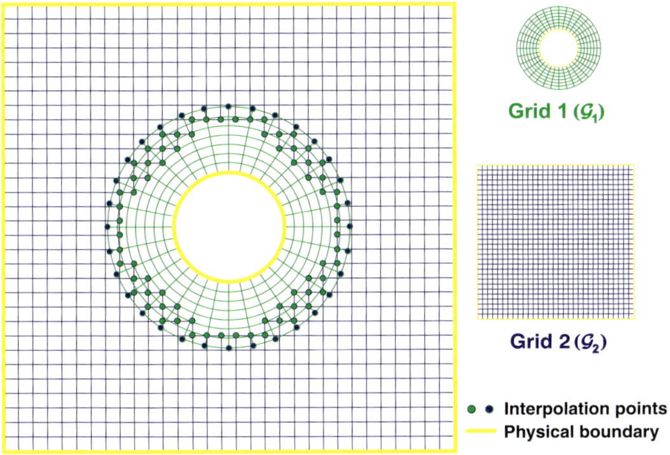

Fig. 1. Simple overlapping grid system $\mathbb{G}$ in physical space $\mathcal{P}$

overlapping areas. In this study, solution values at interpolation points are determined by using a non-conservative Lagrange interpolation scheme [18].

The presence of moving bodies changes the relative position of the overlapping grids continuously during the flow simulation. As the component grid (around a moving body) traverses through the computational domain, overlapping connectivity information, such as interpolation stencils and unused points regions, is recomputed. The automation of the hole cutting and interpolation stencils computation makes the present methodology a powerful tool for the simulation of flows with one or multiple moving bodies, since the grids do not have to be regenerated as the solution evolves. In general, the motion of the component grids $\mathcal{G}_g$ and/or boundaries may be an user-defined time dependent function, may obey the Newton-Euler equations for the case of rigid body motion or may correspond to the boundary nodes displacement in response to the stresses exerted by the fluid pressure for the case of fluid-structure interaction problems. When a component grid changes position during a moving grid computation, the overlapping grid generator is called at each time step in order to update the interpolation stencils and unused points regions. The component grids themselves do not have to be recomputed unless they deform in shape.

In moving overlapping grids, the governing equations are solved in a frame that moves with the grid. For moving overlapping grids, (6)–(7) are expressed in a reference frame moving with the component grid as follows,

$$\frac{\partial \mathbf{u}}{\partial t} + \left[\left(\mathbf{u} - \dot{\mathbf{G}}\right) \cdot \nabla\right] \mathbf{u} = \frac{-\nabla p}{\rho} + \nu \nabla^2 \mathbf{u} \quad \textit{for} \quad \mathbf{x} \in \mathcal{D}, \quad t > 0, \tag{10}$$

$$\frac{\nabla^2 p}{\rho} + \nabla u \cdot \mathbf{u}_x + \nabla v \cdot \mathbf{u}_y + \nabla w \cdot \mathbf{u}_z = 0 \quad \textit{for} \quad \mathbf{x} \in \mathcal{D}, \quad t > 0, \tag{11}$$

where $\dot{\mathbf{G}}$ is the rate of change of the position of a given set of grid points $\mathbf{x}_P^g$ in the physical space $\mathcal{P}$ (grid velocity). It is important to mention that the new governing equations expressed in the moving reference frame must be accompanied by proper boundary conditions. For a moving body with a corresponding moving no-slip wall, only one constraint may be applied and this corresponds to the velocity on the wall, such as

$$\mathbf{u}\left(\mathbf{x}_P^g|_{wall}, t\right) = \dot{\mathbf{G}}\left(\mathbf{x}_P^g|_{wall}, t\right), \quad where \quad \mathbf{x}_P^g|_{wall} \in \partial\mathcal{D}_{wall}\left(t\right). \tag{12}$$

On a moving no-slip wall the boundary condition for the pressure equation is obtained by dotting the normal $\hat{\mathbf{n}}$ into the momentum equation

$$\left.\frac{\partial p}{\partial \hat{\mathbf{n}}}\right|_{\partial\mathcal{D}_{wall}} = \left.\hat{\mathbf{n}}\right|_{\partial\mathcal{D}_{wall}} \cdot \left(-\ddot{\mathbf{G}} + \nu\nabla^2\mathbf{u}\right)\rho, \tag{13}$$

note that the acceleration $\ddot{\mathbf{G}}$ of the wall appears on the right hand side of (13).

To assemble the overlapping grid system $\mathbb{G}$ and solve the laminar incompressible Navier-Stokes equations in their velocity-pressure formulation the Overture[1] framework is used. Overture is a collection of C++ libraries that enable the use of finite-difference approximations to solve partial differential equations (PDEs) in overlapping structured grids. The large sparse non-linear system of equations arising from the discretization of the laminar incompressible Navier-Stokes equations is solved using the PETSc[2] library, which was interfaced with Overture. The system of non-linear equations is then solved using a Newton-Krylov iterative method, in combination with a suitable preconditioner.

3 Computational Domain and Grid Dependence Study

The overlapping grid system $\mathbb{G}$ layout used to conduct this parametric study is shown in Fig. 2. In this figure, the background grid (BG) extends $4.0 \times c$ away from the wing leading edge (LE), $10.0 \times c$ away from the wing trailing edge (TE), $2.0 \times c$ away from the left and right wing-tips (LH WT and RH WT respectively) and $4.0 \times c + h$ away from the point of maximum thickness of the upper and lower surfaces (where h is the flapping amplitude). These dimensions are the dimensions for the overlapping grid system $\mathbb{G}$ layout when the wing is in the mean position of the flapping cycle.

The wing aspect ratio is defined as $AR = S^2/A$, where S is the wing span (wingtip-to-wingtip distance) and A is the wing area and which for a rectangular wing is simplified to $AR = S/c$. In all the cases to be presented, a rectangular wing with aspect ratio equal to $AR = 2$ is used. Finally, the rectangular wing has an elliptical cross-section with a corresponding major axis $a = 0.25$ and minor axis $b = 0.025$, hence, the wing chord c is equal to $c = 0.5$.

[1] https://computation.llnl.gov/casc/Overture/
[2] http://www-unix.mcs.anl.gov/petsc/petsc-as/

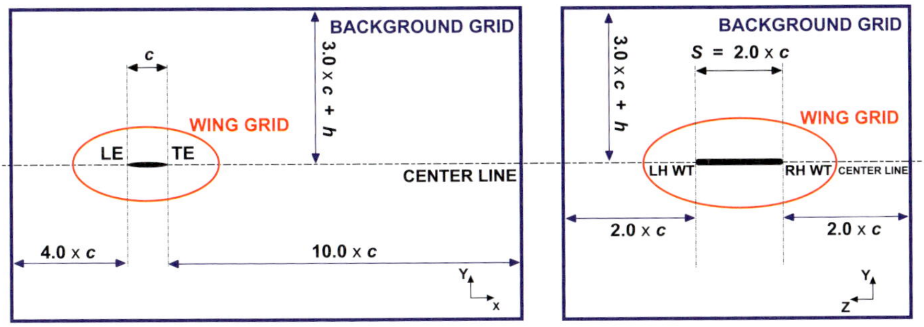

Fig. 2. Left figure: computational domain layout in the xy plane. Right figure: computational domain layout in the zy plane

The initial conditions used for all the flapping wing simulations are those of a fully converged solution of the corresponding fixed wing. In Fig. 2, the left boundary of the BG of the computational domain layout in the xy plane corresponds to an inflow boundary condition $[(\mathbf{u}) = (1.0, 0.0, 0.0),\ \partial_{\hat{\mathbf{n}}} p = 0]$ and the top, bottom and right boundaries of the BG are outflow boundaries (basically vanishing pressure gradient and velocity extrapolated from the interior points). All the boundaries of the BG of the computational domain layout in the zy plane correspond to outflow condition. On the wing surface we impose a no-slip boundary condition ($\mathbf{u} = \dot{\mathbf{G}}_x, \dot{\mathbf{G}}_y, \dot{\mathbf{G}}_z$). The rest of the boundaries are interpolation boundaries. The Reynolds number $Re = U \times c/\nu$ is equal to 250 for all the numerical experiments.

It is a very well known fact that the effect of mesh size is an important factor to consider when assessing the quality of a numerical solution. Here, we conduct a grid dependence study in order to determine the most suitable grid in terms of computing time and solution accuracy from a quantitative point of view. To conduct this study, we use the grid convergence index (GCI) method as described by Roache [23]. Here, a pure heaving wing is considered with different grid sizes, layouts and clustering. Several simulations were run at a Strouhal number $St = 0.3$ ($k = 1.570795$). In our calculations, unsteadiness is observed to disappear typically after 4 cycles of wing flapping motion and further calculations show negligible non-periodicity. In Table 1, the parameters for the three overlapping grids system used for the GCI study are presented. In order to better resolve the wing wake, we used inverse hyperbolic tangent grid stretching in the x, y and z directions of the BG, we also used inverse hyperbolic tangent grid stretching towards the LE, TE, LH WT and RH WT, since we expect vortices to be generated in these areas. Table 1 show the grid dimensions, grid spacing ratio (from the finest grid to the coarser grid) and position of the first node normal to the wing surface. The wing grid (WG) was used as the reference grid to conduct the GCI study and a grid spacing refinement ratio $r = 2$ was used. Each simulation was checked for acceptable iterative convergence.

Table 1. Grid dimensions used for the grid dependence study. Here, BG stands for background grid, WG-CS for wing grid center-section, WG-TS for wing grid tip-section, GSR for grid refinement spacing ratio and 1st NW for position of the first node normal to the wall

Grid $\mathbb{G}_g$	BG	WG-CS	WG-TS	GSR	1st NW
$\mathbb{G}_1$	$161 \times 121 \times 101$	$221 \times 121 \times 41$	$81 \times 61 \times 41$	1	$0.001 \times 2c$
$\mathbb{G}_2$	$161 \times 121 \times 101$	$201 \times 101 \times 31$	$61 \times 51 \times 31$	2	$0.002 \times 2c$
$\mathbb{G}_3$	$161 \times 121 \times 101$	$161 \times 81 \times 31$	$51 \times 41 \times 31$	4	$0.004 \times 2c$

Table 2. Observed values of the average drag coefficient $\overline{c_d}$ and rms lift coefficient c_l^{rms} for each grid used in the grid dependence study

Grid $\mathbb{G}_g$	GSR	$\overline{c_d}$	c_l^{rms}
$\mathbb{G}_1$	1	0.057892	0.905822
$\mathbb{G}_2$	2	0.058476	0.897486
$\mathbb{G}_3$	4	0.060914	0.865326

In unsteady aerodynamics, in order to quantify the flow characteristic, we compute the lift and drag coefficients (c_l and c_d respectively), defined as

$$c_l = \frac{L}{\frac{1}{2}\rho U^2 A}, \quad c_d = \frac{D}{\frac{1}{2}\rho U^2 A}. \tag{14}$$

Since in the study of flapping wings propulsion the main task is thrust production, it is more convenient to think in terms of thrust force T instead of drag force D. Thrust is equal in magnitude but opposite in direction to the drag force ($T = -D$), therefore one has

$$c_t = -c_d = \frac{T}{\frac{1}{2}\rho U^2 A}, \tag{15}$$

in (14) and (15), the lift force L and the thrust force T are computed by integrating the viscous and pressure forces over the wing surface. The lift, drag and thrust coefficients can be averaged in time as follows

$$\overline{c_l} = \frac{1}{\mathbb{T}} \int_t^{t+\mathbb{T}} C_l(t)dt, \quad \overline{c_d} = \frac{1}{\mathbb{T}} \int_t^{t+\mathbb{T}} C_d(t)dt, \quad \overline{c_t} = \frac{1}{\mathbb{T}} \int_t^{t+\mathbb{T}} C_t(t)dt, \tag{16}$$

where $\mathbb{T}$ is the period of flapping motion and is equal to $\mathbb{T} = 1/f$.

In Table 2, the observed values of the average drag coefficient $\overline{c_d}$ and rms lift coefficient c_l^{rms} are presented; these values are used to compute the observed order of convergence p, the value of the observed quantity at zero grid spacing $\chi_{spacing=0}$, the fine grid convergence index GCI_{12} and GCI_{23}, and the constancy of $\mathrm{GCI}_{23} = r^p \times \mathrm{GCI}_{12}$. Based on the GCI index values found (refer to Table 3), we can say that $\overline{c_d}$ is estimated to be 0.057708 with an error band of 0.397203% for grid $\mathbb{G}_1$ and within an error band of 1.641617% for grid $\mathbb{G}_2$; whereas c_l^{rms} is estimated to be 0.908738 with an error band of

Table 3. GCI study results

Outcome	$\overline{c_d}$	c_l^{rms}
p_{obs}	2.061650	1.947838
$\chi_{spacing=0}$	0.057708	0.908738
$\text{GCI}_{12}(\%)$	0.397203	0.402503
$\text{GCI}_{23}(\%)$	1.641617	1.567203
$(\text{GCI}_{23}/\text{GCI}_{12}) \times (1/r^p)$	0.990012	1.009249

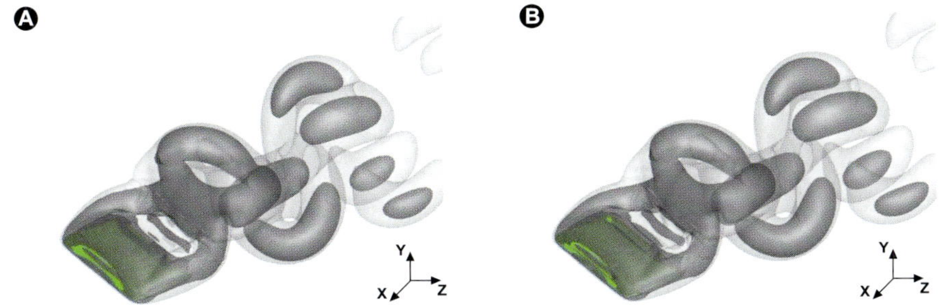

Fig. 3. Vortex topology at the beginning of the upstroke ($t = 5.0$). In light gray we show the iso-surfaces of $\mid \omega \mid$-*criterion* and in dark gray the iso-surfaces of *Q-criterion*. Heaving parameters: $Re = 250$, $St = 0.3$, $k = 1.570795$. Figure **A** corresponds to $\mathbb{G}_1$ and Fig. **B** to $\mathbb{G}_2$

0.402503% for grid $\mathbb{G}_1$ and within an error band of 1.567203% for grid $\mathbb{G}_2$. In Table 3, it can be also evidenced that the value of $(\text{GCI}_{23}/\text{GCI}_{12}) \times (1/r^p)$ for both observed quantities is approximately 1, which indicates that the results for grids $\mathbb{G}_1$ and $\mathbb{G}_2$ are well within the asymptotic range of convergence [23].

Additionally to the quantitative GCI study, we also conducted an extra grid dependence study but from the qualitative point of view (wake structures resolution). The results for grids $\mathbb{G}_1$ and $\mathbb{G}_2$ presented in Table 1 are illustrated in Fig. 3. In this figure, we used the iso-surfaces of $\mid \omega \mid$-*criterion* (vorticity magnitude) and the iso-surfaces of *Q-criterion* to capture the vortices and their corresponding cores. It can be seen from this figure that the $\mid \omega \mid$-criterion, although capable of capturing the general vortical structures, has the disadvantage of also showing the shear layers near the wing surface and between the vortices. The *Q-criterion*, show the vortical structure details more clearly and provide nearly identical structures, hence we choose this as the main criterion for wake topology characterization. In this criterion, Q is defined as the second invariant of the velocity gradient tensor $\nabla \mathbf{u}$. The *Q-criterion* defines a region to be a vortex if every point in this region has $Q > 0$ [24, 25]. As it can be seen in Fig. 3, there is no discernible difference between the solutions, indicating that the grids are adequate for wake structures resolution.

Summarizing the quantitative and qualitative results previously presented, it can be said that the solutions obtained by using grid $\mathbb{G}_1$ and $\mathbb{G}_2$ are grid

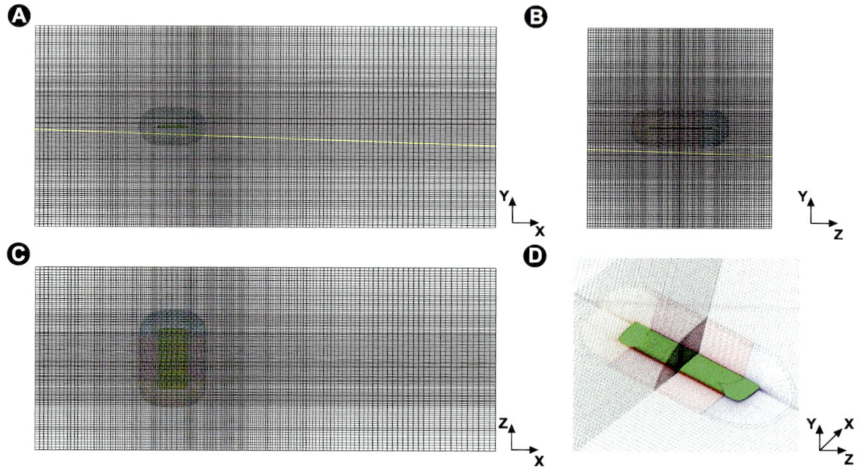

Fig. 4. Typical overlapping grid system $\mathbb{G}$ used in the wake topology study. **A** Side view. **B** Front view. **C** Bottom view. **D** Detailed perspective view. Notice that the wing grid is made up of three component grids, namely, wing grid center-section (in red), left wing grid tip-section (in blue) and right wing grid tip-section (in orange)

independent. Taking into account the computational resources available, CPU time restrictions and solution accuracy, $\mathbb{G}_2$ with a corresponding first node normal to the wall located at a distance equal to $0.001 \times 2c$ from the wing surface will be used as the base grid to perform all further computations. In the case of a smaller or bigger computational domain (refer to Fig. 2), the grid dimensions are scaled in order to keep the same grid spacing as for this domain. In Fig. 4, we illustrate a typical overlapping grid system $\mathbb{G}$ used in the wake topology study discussed in the next section.

4 Simulation Results

Hereafter we carry out a comprehensive parametric study in order to assess the wake signature of heaving and root-flapping rigid finite-span wings. We begin by studying a pure heaving wing case and then we compare the results obtained to those of root-flapping wings in order to study whether this motion generates similar wake structures to those of pure heaving wings at similar St and k values. For the pure heaving case, the wing heaves in the *y-direction* according to

$$y(t) = h \times sin(2 \times \pi \times f \times t). \quad (17)$$

In (17), $y(t)$ is the heaving motion, h the heaving amplitude, f the heaving frequency and t is the time. In Table 4, we present the kinematics parameters governing this numerical experiment, we also present the quantitative results, where $\overline{c_t}$ is the average thrust coefficient and $\widehat{c_l}$ is the maximum lift coefficient

Table 4. Simulation results for the pure heaving parametric study (positive $\overline{c_t}$ indicate thrust production whereas negative $\overline{c_t}$ indicates drag production)

Case number	St	k	$\overline{c_t}$	$\widehat{c_l}$
3DH-1	0.15	1.570795	−0.122675	0.560613
3DH-2	0.15	0.785397	−0.111949	0.422429
3DH-3	0.20	1.570795	−0.106101	0.784628
3DH-4	0.20	0.785397	−0.087877	0.597156
3DH-5	0.25	1.570795	−0.086136	1.040502
3DH-6	0.25	0.785397	−0.068118	0.798169
3DH-7	0.30	1.570795	−0.058294	1.238780
3DH-8	0.30	0.785397	−0.052475	1.024220
3DH-9	0.35	1.570795	0.022962	1.651670
3DH-10	0.35	0.785397	−0.036191	1.257870
3DH-11	0.40	1.570795	0.061370	2.030980
3DH-12	0.40	0.785397	0.008296	1.539740
3DH-13	0.45	1.570795	0.143680	2.417032
3DH-14	0.45	0.785397	0.010296	1.891345
3DH-15	0.50	1.570795	0.199366	2.944020
3DH-16	0.50	0.785397	0.024617	2.150710

(measured during the downstroke). By inspecting these results, we can observe that, as we increase St and k, the values of $\overline{c_t}$ and $\widehat{c_l}$ also increase. We can also notice that for values of St less than $St < 0.30$ we are in the drag producing regime, for values of St between $0.30 < St < 0.35$ we produce little or no drag (or thrust), whereas for values of St higher than $St > 0.35$ we are in the thrust producing regime. We also observe two different behaviours of the aerodynamic forces for high and low reduced frequencies k. Hence, it seems that for flapping wings, the flapping frequency also plays an important role in the vortex generation and shedding and, henceforth, on the aerodynamic forces. These observations of the frequency dependence are similar to those presented by Wang [26], but here we extend them to three-dimensional cases.

In Figs. 5 and 6, the vortex topology for a heaving case in the thrust producing regime is illustrated. These figures show that the downstream wake of this heaving wing consists of two sets of complex shaped vortex rings (VR) which convect at oblique angles about the centerline of the motion. Thus, the flow induced by each vortex ring along its axis is expected to have a net streamwise component. This net streamwise momentum excess in the wake is connected with the thrust production of flapping wings. The process by which the vortex rings are formed can be explained by examining the vortex formation and shedding in the vicinity of the wing surface. During the heaving motion and close to the wing, four vortices are formed; namely, leading edge vortex (LEV), trailing edge vortex (TEV) and two wing-tip vortices (WTV), that is to say, left and right WTV (LH-WTV and RH-WTV respectively). These four vortices are all connected and form a vortex loop (VL); as this vortex loop is convected downstream, it fully disconnects from the wing,

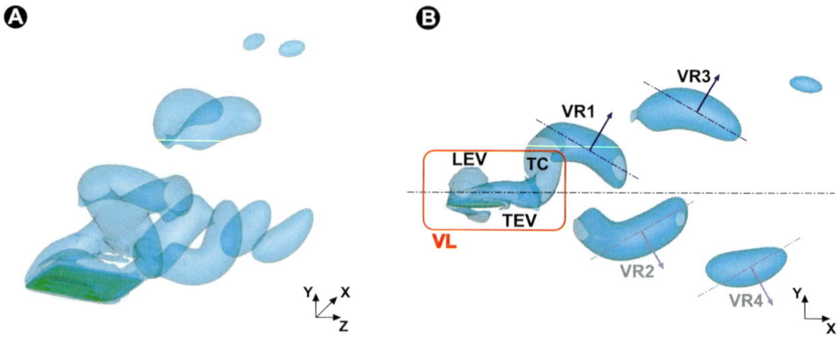

Fig. 5. Vortex wake topology at the beginning of the upstroke for case 3DH-11 ($t = 5.0$). Heaving parameters: $Re = 250$, $St = 0.4$, $k = 1.570795$ (thrust producing wake). An animation is available at the author's website, http://www.dicat.unige.it/guerrero/flappingsimulationsnew.html

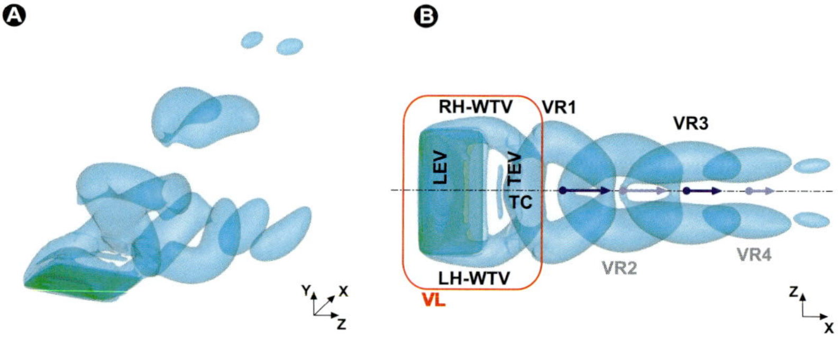

Fig. 6. Vortex wake topology at the beginning of the upstroke for case 3DH-11 ($t = 5.0$). Heaving parameters: $Re = 250$, $St = 0.4$, $k = 1.570795$ (thrust producing wake). An animation is available at the author's website, http://www.dicat.unige.it/guerrero/flappingsimulationsnew.html

forming a ring. It is also of interest the fact that each vortex loop has two sets of thin contrails (TC), that connect the VL to the VR generated in the previous stroke, these structures are segments of the wing-tip vortices and as the vortex loops are convected downstream; these contrails become weaker and ultimately disappear, as for vortex ring VR2 (see Fig. 5). During a complete cycle, two VRs are formed, one at the end of the upstroke and the other one at the end of the downstroke.

In Fig. 7, we present the wake topology for a drag producing case (3DH-3 in Table 4). it is clear from this figure that the wake topology is very different from the one of the thrust producing case. In this case, as the vortex loops are convected downstream, they do not convert into vortex rings; instead, they keep their original shape and are diffused. It can also be observed that the wake height is very compact, opposite to that of the thrust production case.

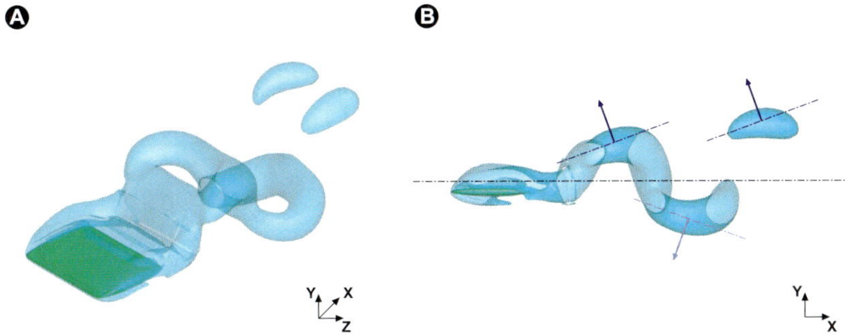

Fig. 7. Vortex wake topology at the beginning of the upstroke for case 3DH-3 ($t = 5.0$). Heaving parameters: $Re = 250$, $St = 0.2$, $k = 1.570795$ (drag producing wake). An animation is available at the author's website, http://www.dicat.unige.it/guerrero/flappingsimulationsnew.html

Also, the vortex loops are inclined in the direction opposite of the wing travel direction, resulting in a momentum surfeit linked to drag production.

We now proceed to simulate a wing rolling about its travelling axis (root-flapping motion). In this case, the wing is hinged at the wing-tip and is rolling about the travelling axis (where the travelling axis and the hinged point are collinear) as per equation

$$\phi(t) = \phi \times sin(2 \times \pi \times f \times t), \tag{18}$$

where $\phi(t)$ is the rolling motion, ϕ the dorsoventral stroke angle, f the rolling frequency and t is the time. The Strouhal number for this case is based on ϕ as proposed by Taylor [15] and is computed as follows

$$St = \frac{S \times sin(\phi/2) \times f}{U}. \tag{19}$$

In Table 5 we present the kinematics parameters governing this case, together with results for lift and thrust. As it can be seen, for values of St less than 0.30 we are in the drag production regime, for values approximately between to $0.30 < St < 0.35$ little or no drag (or thrust) is produced, whereas for values higher that $St = 0.35$ we are in the thrust production regime. From these results, it is clear that this behaviour is similar to that of heaving wings. Comparing these results with the results for pure heaving wings, we found that the latter motions generate larger vortices and forces than root-flapping motion, presumably because the average velocity is higher across the wing span, but otherwise the same wake regimes occurs at similar St.

In Figs. 8 and 9, the three-dimensional wake structures for a thrust producing case (3DR-11) and a drag producing case (3DR-1) are presented. The salient feature that needs to be pointed out from these figures is the absence of any link at all or the presence of a very weak link between the root-tip vortex (hinged extreme) and the LEV and TEV vortices. We also observe that for

Table 5. Simulation results for the rolling wing parametric study (positive $\overline{c_t}$ indicate thrust production whereas negative $\overline{c_t}$ indicates drag production)

Case number	St	k	$\overline{c_t}$	$\widehat{c_l}$
3DR-1	0.15	1.570795	−0.097292	0.549903
3DR-2	0.15	0.785397	−0.127856	0.518200
3DR-3	0.20	1.570795	−0.080279	0.770973
3DR-4	0.20	0.785397	−0.106010	0.708133
3DR-5	0.25	1.570795	−0.073560	1.197320
3DR-6	0.25	0.785397	−0.101956	1.011560
3DR-7	0.30	1.570795	−0.009014	1.577630
3DR-8	0.30	0.785397	−0.057741	1.495240
3DR-9	0.35	1.570795	0.008048	1.865820
3DR-10	0.35	0.785397	−0.009846	1.785210
3DR-11	0.40	1.570795	0.010457	2.584290
3DR-12	0.40	0.785397	0.009964	2.528980
3DR-13	0.45	1.570795	0.018115	3.115787
3DR-14	0.45	0.785397	–	–
3DR-15	0.50	1.570795	0.034010	3.433700
3DR-16	0.50	0.785397	–	–

the case 3DR-1, there is no connection between both wing-tip vortices and LEV and TEV vortices; this is due to the fact that the vortices generated at this low St have very low strength. Aside from these differences, we can observe that the downstream wake of the rolling wing have some common features with the heaving wing wake, such as the formation of sets of well shaped vortex rings and the convection of these vortex rings at oblique angles.

The vortex rings formation process for rolling wings is a little bit different from the one for heaving wings; here, the VL close to the rolling wing does not form a fully closed loop; instead, an open VL or horseshoe shaped vortex (HSV) is observed (Fig. 8), which wraps the wing and comprises a LEV, a wing-tip vortex associated to the free wing-tip (WTV-F) and a TEV. In Fig. 10, a sequence corresponding to case 3DR-9 during upstroke motion is illustrated. In this sequence and as the wing accelerates during the upstroke motion, the LEV and TEV grow in size and in strength, stretching from the free wing-tip towards the wing base (hinged wing-tip), presumably as a result of the onset of the WTV-F during the beginning of the upstroke. Subsequently and as the wing approaches to its horizontal position, the TEV begins to detach from the trailing edge, forming the starting vortex (STV), which remains connected to the WTV-F, forming a very complex *3D* structure around the free wing-tip. After the wing goes beyond its horizontal position and decelerates, the HSV fully detaches from the wing surface, while being convected in the streamwise direction. As the horseshoe shaped vortex is being shed, the open extremes of the HSV connect due to its induced rotation, forming in this way a well shaped VR far from the wing surface. As the wing reaches its topmost position and initiates its downstroke motion, a similar

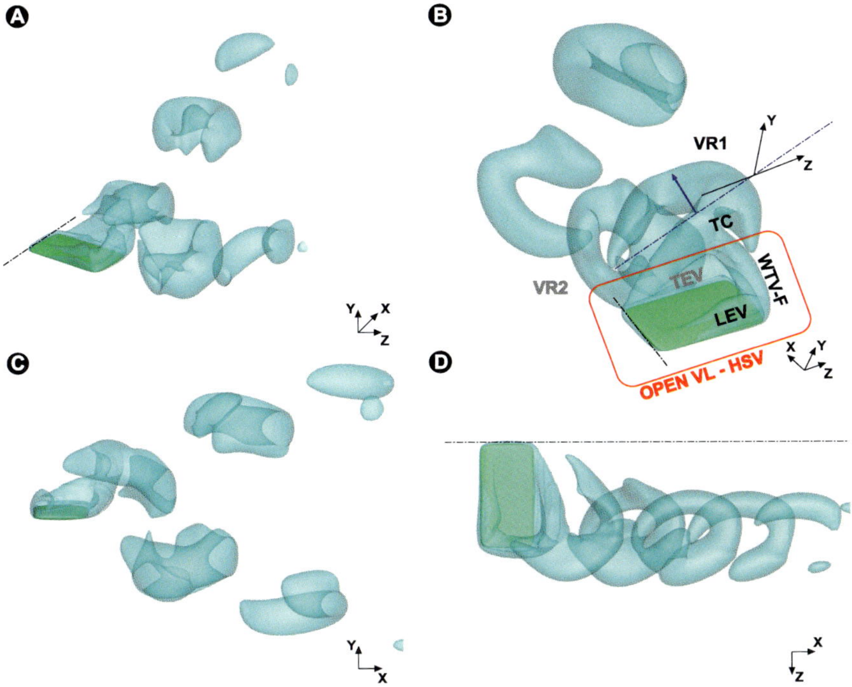

Fig. 8. Vortex wake topology at the middle of the upstroke for the rolling wing case 3DR-11 (t = 5.0). Flapping parameters: Re = 250, St = 0.4, k = 1.570795. The black dotted line represents the rolling axis (thrust producing wake). An animation is available at the author's website, http://www.dicat.unige.it/guerrero/flappingsimulationsnew.html

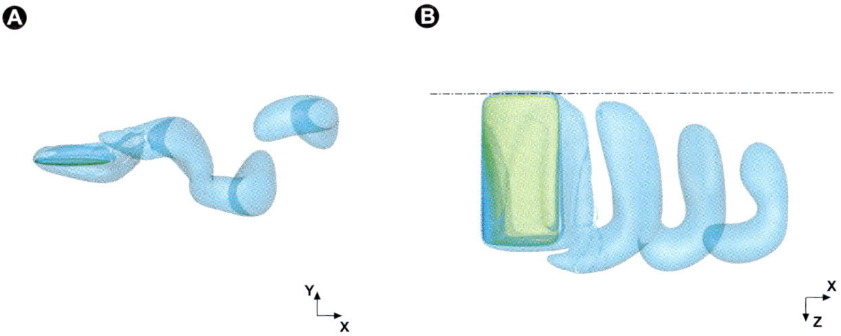

Fig. 9. Vortex wake topology at the middle of the upstroke for the rolling case 3DR-1 (t = 5.0). Flapping parameters: Re = 250, St = 0.15, k = 1.570795. The black dotted line represents the rolling axis (drag producing wake). An animation is available at the author's website, http://www.dicat.unige.it/guerrero/flappingsimulationsnew.html

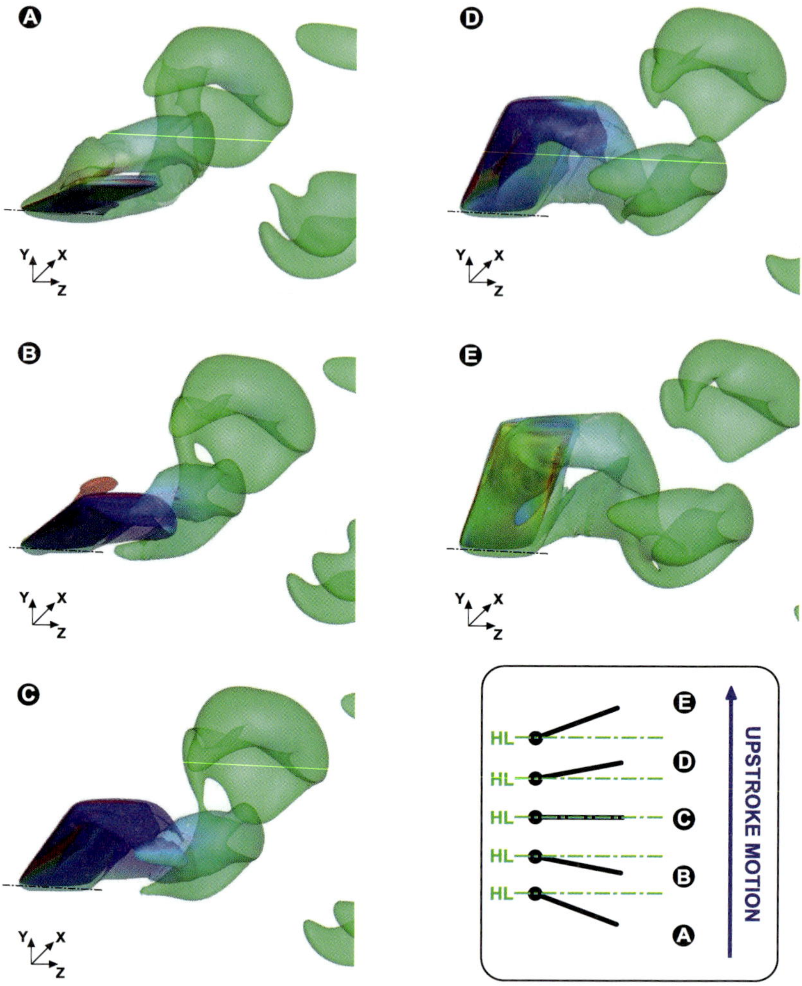

Fig. 10. Vortex shedding and formation sequence during upstroke motion. The *Q-criterion* iso-surfaces are coloured by pressure levels, where red corresponds to high pressure values and blue to low pressure values. The wing surface is coloured by pressure gradient magnitude levels. The sequence is from **A** to **E**, where **A** $0.75 \times \mathbb{T}$, **B** $0.875 \times \mathbb{T}$, **C** $1.0 \times \mathbb{T}$, **D** $1.125 \times \mathbb{T}$, **E** $1.25 \times \mathbb{T}$. Flapping parameters: $St = 0.35$, $k = 1.570\,795$. In Figs. **A** to **E**, the black dotted line represents the rolling axis. HL is the reference horizontal line (horizontal position)

vortex shedding and formation process is observed; hence, two sets of VRs are formed during a full rolling cycle. As for the heaving cases, the flow induced by each vortex ring along its core is expected to have a net streamwise component which, for this case (3DR-9), is connected to thrust production (momentum excess).

It is also observed that during the first part of the upstroke motion, the LEV initially remains attached to the wing surface, with no evidence of breaking down (separation) or shedding. As the wing accelerates to its horizontal position, the LEV becomes a large, conical spiral vortex with a strong axial flow at the core (Fig. 10). This leads to a steadily increasing negative pressure region on the lower surface and hence augmentation of the forces acting on the wing. The occurrence of this axial flow is thought to be closely related to the spanwise pressure gradient resulting from the different velocities along the wing due to the rolling motion. The existence of this axial flow at the core of the LEV delays the vortex shedding and likely stabilizes the LEV, independently of the chosen reduced frequency (Fig. 11). Shortly after the wing reaches its horizontal position, the LEV breaks down at approximately 80-90% of the wing span (measured from the wing base), as seen in Fig. 10. This break down is due to the reverse pressure gradient which destabilizes the LEV at the position where the spanwise flow disappears. A similar phenomenology is observed during the downstroke.

Additionally, in Fig. 11 we compare the wake structure and LEV shedding for cases 3DR-7 and 3DR-8 (same St and different values of k). As it can be seen in the sequences shown in this figure, the LEV vortex shedding process for both cases is very similar. In both cases, the LEVs are generated and shed approximately at the same rate for similar flapping periods. All these qualitative observations are also confirmed by the force measurement results shown in Table 5. From these results, we can observe that the computed forces are almost the same, independently of the value of k used. From this qualitative and quantitative results it is clear that the frequency selection does not introduce a frequency dependence into the results, opposite to heaving wings, where the aerodynamic performance is highly influenced by the LEV separation and convection, which provides a mechanism of optimal selection of heaving frequency (in the sense of maximum propulsive efficiency), as discussed by Wang [26] and Young and Lai [27].

5 Conclusions and Future Perspectives

In this manuscript, we have studied the unsteady aerodynamics of finite-span flapping rigid wings. The incompressible laminar Navier-Stokes equations were solved in their velocity-pressure formulation using a second-order accurate in space and time finite-difference scheme on overlapping structured moving grids. Both pure heaving and root-flapping motion were studied. All the simulations were conducted at different values of Strouhal number ($0.15 < St < 0.5$) and two reduced frequency values ($k = 0.785397$ and $k = 1.570795$).

The simulations show that the wake of thrust producing, rigid, finite-span heaving wings is formed by two sets of interconnected vortex loops that slowly convert into vortex rings as they are convected downstream. It was also

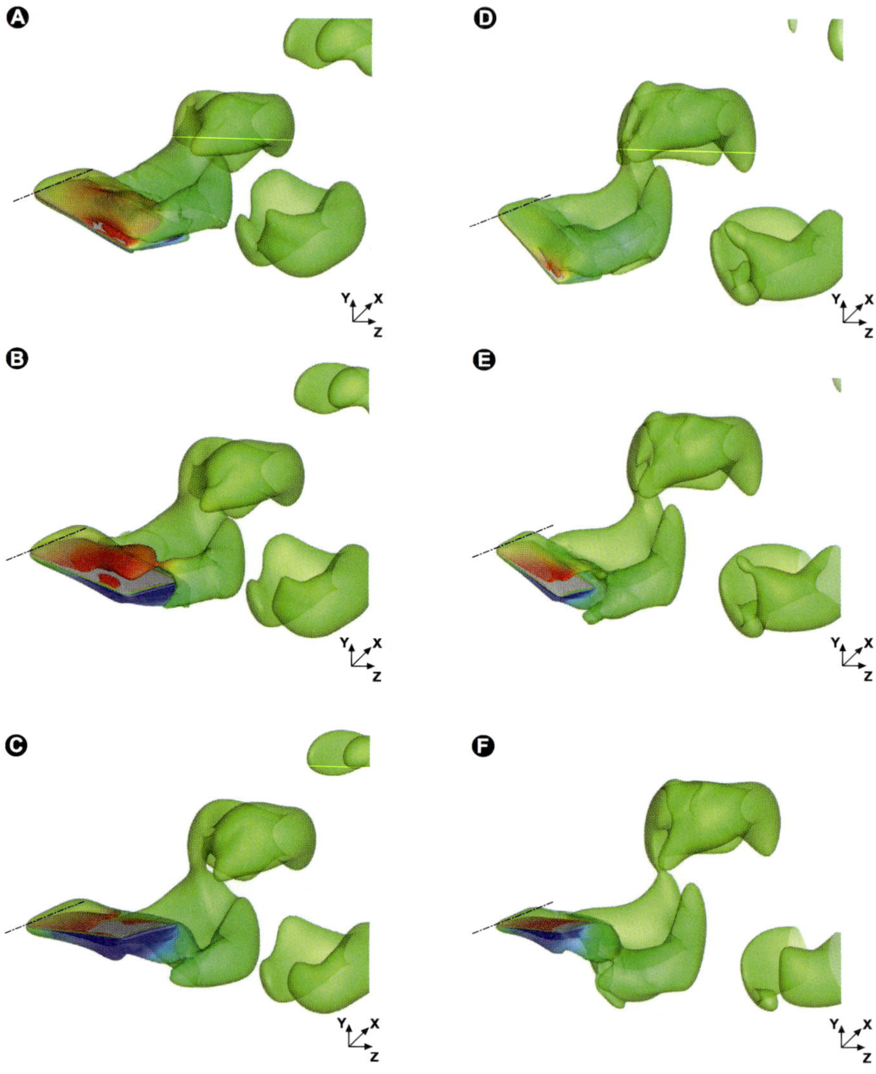

Fig. 11. Vortex shedding comparison for two cases rolling at different reduced frequency values. The *Q-criterion* iso-surfaces are coloured by pressure levels, where red corresponds to high pressure values and blue to low pressure values. Left column, flapping parameters: $St = 0.30$, $k = 1.570795$. The sequence is from **A** to **C**, where **A** $0.75 \times \mathbb{T}$, **B** $0.875 \times \mathbb{T}$, **C** $1.0 \times \mathbb{T}$ (upstroke motion). Right column, flapping parameters: $St = 0.30$, $k = 0.785397$. The sequence is from **D** to **F**, where **D** $0.75 \times \mathbb{T}$, **E** $0.875 \times \mathbb{T}$, **F** $1.0 \times \mathbb{T}$ (upstroke motion). The black dotted line represents the rolling axis

observed that the vortex rings are inclined with respect to the free-stream; the angle of inclination of the vortex rings is found to be in the direction of their travel and in the streamwise direction for thrust producing configurations;

whereas for drag producing configurations the angle of inclination is opposite to the direction of the streamwise flow. It was also noted the presence of thin contrails that link the vortex loops; these structures are segments of the wing-tip vortices and as the vortex loops are convected downstream, these contrails become weaker and ultimately disappear. In general, the observed structures are qualitatively similar to those observed in the experiments by Parker et al. [28] and the numerical simulations of Dong et al. [6] and Blondeaux et al. [7].

From the force measurement study for pure heaving wings, two different behaviours were observed for the average thrust coefficient $\overline{c_t}$ and maximum lift coefficient $\widehat{c}_l$, for high and low reduced frequencies k. Hence, it seems that for heaving wings, the oscillating frequency also plays an important role in the vortex generation and shedding and, henceforth, on the aerodynamic forces.

From the qualitative and quantitative study of the root-flapping motion characteristic of flying animals (which as far as the author is aware still remains virtually unexplored) it is found that, indeed, root-flapping motion produces wake structures similar to those of finite-span heaving wings, but with the difference that the latter motion generates larger vortices and forces than root-flapping motion; apart from this, similar wake regimes (in terms of thrust producing or drag producing wake) occurs at similar Strouhal numbers. It is also observed that the vortex rings are formed from an open vortex loop or horseshoe shaped vortex, which as it is convected downstream connects to form a well shaped vortex ring. As for the heaving wing case, two sets of complex shaped vortex rings are formed during each flapping cycle. The vortex rings are also convected at oblique angles about the centerline of the motion and with a lateral inclination angle. As for heaving wings, the oblique inclination of the vortex rings is a direct indication of momentum excess or surfeit (thrust or drag production respectively). The presence of only one thin contrail that links the horseshoe shaped vortex with the vortex rings, is also worthy of note; this thin contrail is generated in the free wing-tip.

Conversely to the heaving wings, it is observed that for rolling wings the frequency selection does not introduce a dependence onto the results. In fact, it is found that for different reduced frequency values the LEV shows a similar shedding process. The stabilization of the leading edge vortex is thought to be related to the axial flow that traverses the core of the LEV and delays the vortex break down and shedding. The LEV break down is found shortly after the wing reach its horizontal position and approximately at 80-90% of the wingspan (measured from the wing base). This LEV break down is due to the reverse pressure gradient, which destabilizes the LEV and cause rapid separation. These observations clearly suggest that an additional parameter, such as pitching angle or wing geometry, should be taken into account in order to avoid LEV break down. Nevertheless, prolonging or delaying the LEV break down does not necessarily translate into increased global aerodynamic efficiency.

Finally, for the limited range of St and k values and simplified wing geometry and flapping kinematics covered in this study, all the qualitative and quantitative results presented are in close agreement with the experimental observations of Nudds et al. [14], Parker et al. [28], Rohr and Fish [12], Taylor et al. [15] and Triantafyllou et al. [13]; this supports the hypothesis that *"flying and swimming animals cruise at a Strouhal number tuned for high power efficiency"* [15].

The results presented in this paper are limited to laminar flow; nevertheless, they provide an excellent insight into the wake signature of the unsteady aerodynamics of flapping wings. The extension of the current study to the turbulent case and the dynamics of separation bubbles in the turbulent regime are envisaged. Also, the use of more realistic flapping kinematics and wing geometries, and the use of flexible wings and feathered surfaces will be the subject of future investigations. Lastly, the selection of an *"optimal"* wing shape and kinematics in terms of the maximum global aerodynamic efficiency is an open issue that we hope to deal with in the future.

Acknowledgements. The simulations were performed on the national supercomputers NEC SX-9 and NEC Nehalem Cluster at the High Performance Computing Center Stuttgart (HLRS) under the grant number WTAP3DW/12832.

References

1. J. Favier, A. Dauptain, D. Basso, and A. Bottaro. Passive separation control using a self-adaptive hairy coating. *Journal of Fluid Mechanics*, 627:451–483, 2009.
2. K.D. Jones and M.F. Platzer. Experimental investigation of the aerodynamic characteristics of flapping-wing micro air vehicles. *AIAA Paper 2003-0418-CP*, 2003.
3. S.N. Singh, A. Simha, and T. Mittal. Auv maneuvering by pectoral fins: Inverse control design based on cfd parametrization. *IEEE Journal of Oceanic Engineering*, 29:777–785, 2004.
4. J.M. McMichael and M.S. Francis. Micro air vehicles—toward a new dimension in flight. Technical report, Defense Advanced Research Projects Agency (DARPA), 1997.
5. R. Michelson and M. Naqvi. Extraterrestrail flight (entompter-based mars surveyor). *Von Karman Institute for Fluid Dynamics, Lecture Series Nov 24-28, 2003.*
6. H. Dong, R. Mittal, and F.M. Najjar. Wake topology and hydrodynamic performance of low-aspect-ratio flapping foils. *Journal of F*, 566:309–343, 2006.
7. P. Blondeaux, F. Fornarelli, L. Guglielmini, M.S. Triantafyllou, and R. Verzicco. Numerical experiments on flapping foils mimicking fish-like locomotion. *Physics of Fluids*, 17:113601–113601–12, 2005.
8. R. Ramamurti and W. Sandberg. Computational study of three-dimensional flapping foil flows. *Journal of Experimental Biology*, 205:1507–1518, 2002.
9. K. Parker, K. von Ellenrieder and J. Soria. Flow structures behind a heaving and pitching finite-span wing. *Journal of Fluid Mechanics*, 490:129–138, 2003.

10. J. Guerrero. *Numerical Simulation of the Unsteady Aerodynamics of Flapping Flight.* PhD thesis, University of Genoa. Department of Civil, Environmental and Architectural Engineering, Italy, 2009.
11. G.S. Triantafyllou and M.S. Triantafyllou. An efficient swimming machine. *Scientific American*, 272:40–48, 1995.
12. J. Rohr and F. Fish. Strouhal number and optimization of swimming by odontocete cetaceans. *The Journal of Experimental Biology*, 207:1633–1642, 2004.
13. M.S. Triantafyllou, G.S. Triantafyllou, and R. Gopalkrishnan. Wake mechanics for thrust generation in oscillating foils. *Physics of Fluids*, 3:2835–2837, 1991.
14. R.L. Nudds, G.K. Taylor, and A.R. Thomas. Tuning of strouhal number for high propulsive efficiency accurately predicts how wingbeat frequency and stroke amplitude relate and scale with size and flight speed in birds. *Proc. Biol. Sci.*, 7:2071–2076, 2004.
15. G.K. Taylor, R.L. Nudds, and A.R. Thomas. Flying and swimming animals cruise at a strouhal number tuned for high power efficiency. *Letters to Nature*, 425:707–711, 2003.
16. W.D. Henshaw. A fourth-order accurate method for the incompressible Navier-Stokes equations on overlapping grids. *Journal of Computational Physics*, 113:13–25, 1994.
17. G. Chesshire and W. Henshaw. Composite overlapping meshes for the solution of partial differential equations. *Journal of Computational Physics*, 90:1–64, 1990.
18. W. Henshaw. Ogen: An overlapping grid generator for overture. Technical report, Lawrence Livermore National Laboratory, Research Report UCRL-MA-132237, 1998.
19. M. Vinokur. Conservation equations of gas-dynamics in curvilinear coordinate systems. *Journal of Computational Physics*, 14:105–125, 1974.
20. N. Petersson. Stability of pressure boundary conditions for Stokes and Navier-Stokes equations. *Journal of Computational Physics*, 172:40–70, 2001.
21. R.L. Sani, J. Shen, O. Pironneau, and P.M. Gresho. Pressure boundary condition for the time-dependent incompressible Navier-Stokes equations. *International Journal for Numerical Methods in Fluids*, 50:673–682, 2006.
22. H. Johnston and J. Liu. Finite difference schemes for incompressible flow based on local pressure boundary conditions. *Journal of Computational Physics*, 180:120–154, 2002.
23. P. Roache. *Verification and Validation in Computational Science and Engineering.* Hermosa Publishers, Socorro, USA, 1998.
24. J. Jeong and F. Hussain. On the identification of a vortex. *Journal of Fluids Mechanics*, 285:69–94, 1995.
25. R. Haimes and D. Kenwright. On the velocity gradient tensor and fluid feature extraction. *AIAA Paper 1999-3288-CP*, 1999.
26. Z.J. Wang. Vortex shedding and frequency selection in flapping flight. *Journal of Fluid Mechanics*, 410:323–341, 2000.
27. J. Young and J. Lai. Oscillation frequency and amplitude effects on the wake of a plunging airfoil. *AIAA Journal*, 42:2042–2052, 2004.
28. K. Parker, J. Soria, and K. von Ellenrieder. Thrust measurements from a finite-span flapping wing. *AIAA Journal*, 45:58–70, 2007.

Computational Design Study of a 3D Hypersonic Intake for Scramjet Demonstrator Testing

Birgit Reinartz and Marek Behr

Chair for Computational Analysis of Technical Systems (CATS), Center for Computational Engineering Science (CCES) RWTH Aachen University, Schinkelstr. 2, 52062 Aachen, Germany, reinartz@cats.rwth-aachen.de, behr@cats.rwth-aachen.de

Summary. Scramjets are hypersonic airbreathing engines that utilize the unique technology of supersonic combustion. The ultimate goal of developing such engines is to propel future space transportation systems safely through the atmosphere. Scramjets are dominated by highly complex aerothermodynamic processes that are difficult, if not impossible, to fully simulate with ground-based test facilities. Flight-testing is a means to overcome some of these difficulties, although current test programs can only provide useful data for a limited range of the flight envelope. Computational simulations therefore form a crucial element within the development and design of scramjet engines, combining all of the physical aspects necessary to extend small-scale ground-based testing to full-scale flight conditions.

Under DFG grant GA 1332-1, funding for an experimental test campaign of a complete scramjet demonstrator model has been secured. The testing will take place in a hypersonic test facility under real flight conditions at Mach 8. Prior to those tests, a numerical design study for various intake configurations was initiated, investigating compression ramp angles, side wall contraction, intake height and boundary layer bleed with respect to generated inflow for the supersonic combustion chamber. Because establishing stable combustion in a supersonic flow is a key technology to a scramjet engine, it is of extreme importance to guarantee certain inflow parameter for the combustor.

1 Introduction

Supersonic combustion ramjets (scramjets) are airbreathing propulsion systems that currently provide the greatest potential for achieving sustained economic hypersonic flight, adding safety and flexibility to space transportation. Like a conventional ramjet, a scramjet decelerates the incoming high-speed air by a series of shock waves, which provides the necessary compression to the flow for the combustion process. In contrast to a ramjet the scramjet compression process is not terminated by a normal shock, therefore the core flow

W.E. Nagel et al. (eds.), *High Performance Computing in Science and Engineering '10*, DOI 10.1007/978-3-642-15748-6_32,

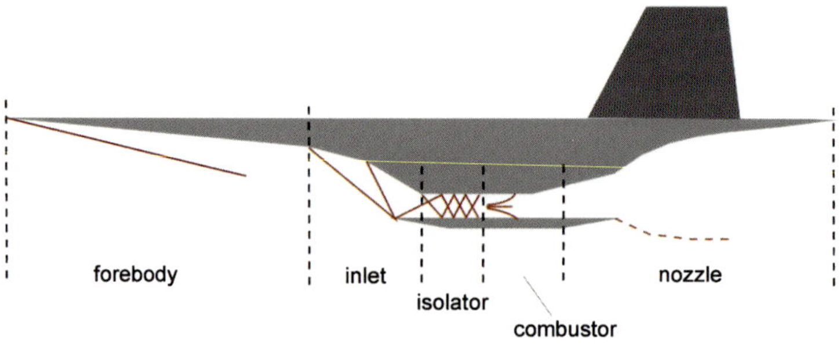

Fig. 1. Schematic of a hypersonic plane featuring a scramjet propulsion system

remains supersonic throughout the engine. It is the supersonic flow inside the engine that characterizes a scramjet engine. At hypersonic speeds a normal shock would cause extremely high total pressure losses and a static temperature so intense that any further heat addition due to the combustion process would result in thermal choking of the engine. Scramjets do not suffer from these deleterious effects, as a normal shock wave is not generated within the engine.

A scramjet consists of four highly interdependent components: inlet, isolator, combustion chamber and nozzle. As shown in Fig. 1, the scramjet is geometrically very simple. No rotating parts such as a compressor or turbine, as found in conventional aircraft engines, are required. Rather than the complexity of the system residing in the mechanical design, it is the necessity to consider the complex aerothermodynamic environment that provides the greatest design challenge. Aerodynamic difficulties arise because slight changes in the angle-of-attack or the flight velocity will change the position of the subsequent shock waves, which have a significant impact upon the performance of the engine (e.g., captured mass flow). Aerothermodynamic considerations are involved because the engine operates at hypersonic velocities (flight Mach number above 7), where extremely high heat loads and wall temperature effects become important.

The primary purpose of the inlet is to provide constant high-pressure air mass-flow to the engine with a minimum of aerodynamic losses (e.g., spillage drag). Compression is performed through a series of oblique shock waves and internal contraction that lead to a shock wave/expansion wave interaction pattern inside the inlet. Within the current analysis, the main research focus for intake design is the quality of the combustor entrance conditions. The experimental test campaign uses a central strut injector to injection gaseous hydrogen in streamwise direction into the combustion chamber. The ignition relies on thermal self-ignition, requiring an approximate temperature of 1000 K and a static pressure of around 1 bar. The tested demonstrator model is not investigated with respect to thrust, therefore the overall mass flow through the engine, which is normally considered a performance parameter of

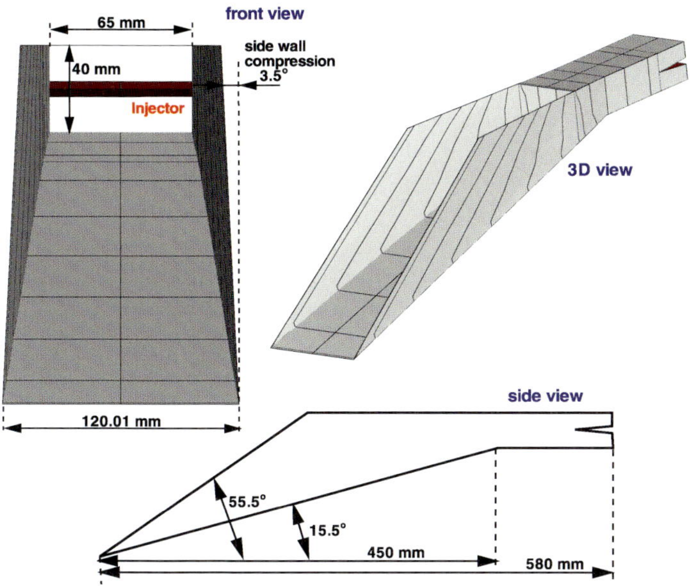

Fig. 2. Design studies of the current 3D inlet configuration (red: front half of injector)

the intake, is circumstantial in the current investigation. However, generating inflow conditions for the combustor that allow a stable combustion process to be initiated is the main research focus.

Different two-dimensional (2D) and three-dimensional (3D) intake designs have been investigated in preparation for the test campaign. Here, the so-called 2D intakes refer to an intake design with two outer compression ramps and no side wall compression whereas a 3D intake has only one outer compression ramp but also utilizes side wall contraction to increase the flow compression. The numerical analysis started out investigating 2D intakes, however it turned out that even though the performance of such intake geometries was quite good, they are of limited usage for the scheduled test campaign, because the intake models are very long and slim and, thus, difficult to be tested in hypersonic facilities. Therefore, an analysis of 3D intakes was initiated, focusing on understanding the effect of side wall compression and investigating the effect of laminar-turbulent transition on the computed flow field. The most recent test geometry is shown in Fig. 2 with a 15.5 degrees compression ramp angle and 3.5 degrees side wall compression. All leading edges (ramp and cowl) are assumed to be ideally sharp. Otherwise, a leading edge radius of approximately 0.5 mm needed to be resolved, increasing the grid size even furher. The interior area of the intake is 40 mm (height) times 65 mm (width). At the end of the intake, the front half of the central strut injector is also simulated, represented by a wedge with 43 mm in length and 7 mm in height as well as a sharp leading edge.

2 Physical Model

2.1 Conservation Equations

The governing equations for high-speed turbulent flow are the unsteady, compressible, Favre-averaged Navier-Stokes equations in integral form

$$\frac{\partial}{\partial t}\int_V \mathbf{U}\ \mathrm{d}V + \oint_{\partial V} \left(\mathbf{F}^{\mathrm{c}} - \mathbf{F}^{\mathrm{d}}\right)\mathbf{n}\ \mathrm{d}S = \mathbf{0} \tag{1}$$

where

$$\mathbf{U} = [\ \bar{\rho}\ ,\ \bar{\rho}\tilde{\mathbf{v}}\ ,\ \bar{\rho}\tilde{E}\ ]^{\mathrm{T}} \tag{2}$$

is the array of the mean conserved quantities: density, momentum density, and total energy density. The tilde and the bar over the variables denote the mean value of Favre-averaged and Reynolds-averaged variables, respectively. Furthermore, the double prime on a variable refers to its fluctuating component. The quantity V denotes an arbitrary control volume with the boundary ∂V and the outer normal $\mathbf{n}$. The flux array is divided into its inviscid part and its diffusive part

$$\mathbf{F}^{\mathrm{c}} = \begin{pmatrix} \bar{\rho}\tilde{\mathbf{v}} \\ \bar{\rho}\tilde{\mathbf{v}}\circ\tilde{\mathbf{v}} + \bar{p}\,\mathbf{1} \\ \tilde{\mathbf{v}}(\bar{\rho}\tilde{E} + \bar{p}) \end{pmatrix} \quad \text{and} \quad \mathbf{F}^{\mathrm{d}} = \begin{pmatrix} \mathbf{0} \\ \bar{\boldsymbol{\sigma}} - \overline{\rho\mathbf{v}''\circ\mathbf{v}''} \\ \tilde{\mathbf{v}}\bar{\boldsymbol{\sigma}} + \overline{\mathbf{v}''\boldsymbol{\sigma}} - \overline{\mathbf{q}} - c_p\,\overline{\rho\mathbf{v}''T''} \\ -\frac{1}{2}\,\overline{\rho\mathbf{v}''\,\mathbf{v}''\circ\mathbf{v}''} - \tilde{\mathbf{v}}\,\overline{\rho\mathbf{v}''\circ\mathbf{v}''} \end{pmatrix}$$

where $\mathbf{1}$ is the unit tensor and $\circ$ denotes the dyadic product[1]. Currently, the airflow is still considered to be a calorically perfect gas with constant ratio of specific heats, $\gamma = 1.4$, and a specific gas constant of $R = 287\,\mathrm{J/(kgK)}$, which yields the following expression for the specific total energy:

$$\tilde{E} = c_v\bar{T} + \frac{1}{2}\tilde{\mathbf{v}}\tilde{\mathbf{v}} + k\ . \tag{3}$$

The last term represents the turbulent kinetic energy

$$k := \frac{1}{2}\frac{\overline{\rho\mathbf{v}''\mathbf{v}''}}{\bar{\rho}}\ . \tag{4}$$

For isotropic Newtonian fluids, the mean molecular shear stress tensor

$$\bar{\boldsymbol{\sigma}} = 2\bar{\mu}\,\bar{\mathbf{S}} - \frac{2}{3}\bar{\mu}\,\mathrm{tr}\big(\bar{\mathbf{S}}\big)\ \mathbf{1} \tag{5}$$

[1] Scalar Products of dyadics formed by two vectors $\mathbf{a}$ and $\mathbf{b}$ with a vector $\mathbf{c}$ are defined as usual, i.e., $\mathbf{a}\circ\mathbf{b}\,\mathbf{c} = \mathbf{a}(\mathbf{b}\mathbf{c})$, $\mathbf{c}\,\mathbf{a}\circ\mathbf{b} = (\mathbf{c}\mathbf{a})\mathbf{b}$.

is a function of the mean strain rate tensor

$$\bar{\mathbf{S}} := \frac{1}{2}\left[\operatorname{grad}(\bar{\mathbf{v}}) + (\operatorname{grad}(\bar{\mathbf{v}}))^{\mathrm{T}}\right] \tag{6}$$

where the molecular viscosity $\bar{\mu} = \bar{\mu}(\bar{T})$ is determined by Sutherland's law. Similarly, the molecular heat flux is considered a linear isotropic function of the temperature gradient

$$\bar{\mathbf{q}} = -\frac{c_{\mathrm{p}}\bar{\mu}}{Pr}\operatorname{grad}(\bar{T}) \tag{7}$$

with the Prandtl number $Pr = 0.72$.

3 Numerical Method

3.1 Spatial Discretization

The DLR FLOWer code (version 116.22) [1] is applied, which solves the unsteady compressible Navier–Stokes equations using a cell–centered finite volume method on block–structured grids. Applying finite-volume discretization to (1) results in a consistent approximation to the conservation laws. The computational domain is divided into non-overlapping hexahedra in general curvilinear coordinates ξ, η, ζ, and the integral formulation (1) is then applied to each cell (i, j, k) separately. Semidiscretization of (1) results in a set of equations for the time rates of change of the volume-averaged conserved quantities $\mathbf{W}_{i,j,k}$ which are in balance with the sum of the corresponding area-averaged fluxes, $\mathbf{R}^{\mathrm{c}}_{i,j,k}$ and $\mathbf{R}^{\mathrm{d}}_{i,j,k}$, across the cell faces and the artificial dissipation $\mathbf{D}_{i,j,k}$:

$$\frac{\mathrm{d}\mathbf{W}_{i,j,k}}{\mathrm{d}t} = -\frac{1}{V_{i,j,k}}\left(\mathbf{R}^{\mathrm{c}}_{i,j,k} - \mathbf{R}^{\mathrm{d}}_{i,j,k} + \mathbf{D}_{i,j,k}\right) = \mathbf{Res}_{i,j,k} \ . \tag{8}$$

Standard central discretization schemes are used for the convective and diffusive terms in the presented supersonic computations. However, to account for the directed propagation of information in the inviscid part of the equations, the AUSM (Advection Upstream Splitting Method) scheme will be used for the approximation of the convective flux functions of hypersonic flows [2]. Higher-order accuracy and consistency with the central differences used for the diffusive terms is achieved by MUSCL (Monotonic Upstream Scheme for Conservation Laws) Extrapolation, and TVD (Total Variation Diminishing) property of the scheme is ensured by a modified van Albada limiter function.

3.2 Time-Stepping Scheme

The system of ordinary differential equations (8) is solved by an explicit five-stage Runge-Kutta time-stepping scheme in combination with different convergence acceleration techniques like multigrid and local time stepping for asymptotically steady-state solutions [3]. Additionally, for inviscid flows, the total enthalpy is a constant throughout the flow field and its numerical deviation is applied as forcing function to accelerate convergence. For turbulent flow, the time integration of the turbulence equations is decoupled from the mean equations and the turbulence equations are solved using a Diagonal Dominant Alternating Direction Implicit (DDADI) scheme. The implicit scheme increases the numerical stability of turbulent flow simulations which is especially important since the low Reynolds number damping terms as well as the high grid cell aspect ratios near the wall make the system of turbulent conservation equations stiff. Due to the CFL-condition for explicit schemes, the CFL number of the Runge-Kutta scheme has an upper limit of 4. Implicit residual smoothing allows to increase the explicit stability limit by a factor of 2 to 3 [3]. For the implicitly solved turbulence equations, the CFL number can be ten times higher.

3.3 Reynolds Stress Model

For wall dominated flows with thick boundary–layers, strong shock / boundary–layer interaction and with separation, as they are of interest here, the assumption of a linear dependence between the Reynolds stress tensor (9) and the strain rate tensor is not always valid. Therefore, a differential Reynolds stress models (RSM) [4, 5] is used in the simulations. This model solves transport equations for each component of the Reynolds stress tensor as well as for an additional length scale. Thus, it is computationally expensive. Furthermore, it decrease the stability of the numerical scheme. However, RSM computations show promising results, especially for separated flows [6]. The model uses a simplified version of the Launder-Reece-Rodi (LRR) model by Wilcox close to the wall, the Speziale-Sarkar-Gatski (SSG) in the farfield and Menter's ω-equation for closure. The initial intention behind this combination was to combine the well-performing SSG model with a model using the well-behaving omega-equation close to the wall. Because Menter's blending function is used to switch between the omega-equation close to the wall and the epsilon-equation in the farfield, to use also Menter's omega-equation was only natural. Accordingly, that model is called SSG/LRR-ω model.

$$\bar{\rho}\tilde{\mathbf{R}} = \begin{pmatrix} \overline{\rho u_1'' u_1''} & \overline{\rho u_1'' u_2''} & \overline{\rho u_1'' u_3''} \\ \overline{\rho u_2'' u_1''} & \overline{\rho u_2'' u_2''} & \overline{\rho u_2'' u_3''} \\ \overline{\rho u_3'' u_1''} & \overline{\rho u_3'' u_2''} & \overline{\rho u_3'' u_3''} \end{pmatrix} \tag{9}$$

3.4 Boundary Conditions

At the inflow boundary, the freestream conditions of the experimental investigation listed in Table 1 are prescribed. The turbulent values are determined by the specified freestream turbulence intensity Tu_∞: $k_\infty = 1.5(Tu_\infty u_\infty)^2$. The Reynolds stress matrix is initialized by placing $2/3\ k_\infty$ on the diagonal and

Table 1. Test conditions

Conditions	M_∞	Re_∞ [10^6/m]	p_0 [bar]	T_0 [K]	T_∞ [K]
flight	8.0	2.94	114	3130	227
I	8.0	2.66	110	3280	237.7
II	8.0	8.42	130	1650	120

the specific dissipation rate of the freestream is $\omega_\infty = k_\infty/(RLTU \cdot \mu_{lam})$ with RLTU being a measure for the ratio of turbulent to laminar viscosity in the freestream (here: RLTU=0.001). For the supersonic outflow, the variables are extrapolated from the interior. At solid walls, the no-slip condition is enforced by setting the velocity components and the normal pressure gradient to zero. Due to the short measurement times in a high-enthalpy facility it is assumed the model remains at a constant ambient temperature of $T_{wall} = 293K$. Additionally, the Reynolds stresses are set to zero at the wall and the respective length scale is prescribed based on the first grid spacing according to Menter.

3.5 Numerical Accuracy

A complete validation of the FLOWer code has been performed by the DLR prior to its release [3, 7] and continued validation is achieved by the analyses documented in subsequent publications [8–10]. Furthermore, FLOWer has already been successfully used in the analysis of 3D hypersonic intake flows [6, 11–13].

Due to the complexity of the problem at hand and because the parameter study did not allow for even larger grids, no grid convergence study has been performed. The discretization relies on experience, knowing that not all features of the flow have been sufficiently resolved. Figures 3 and 4 show the node distribution in the near wall region (symmetry plane) and the intake exit

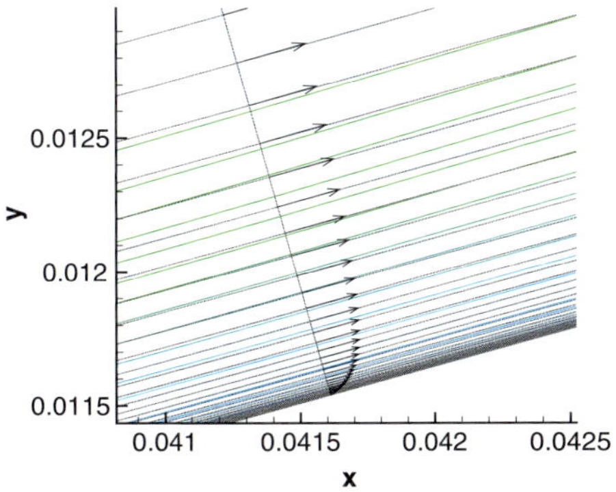

Fig. 3. Boundary layer resolution of the compression ramp in the symmetry plane

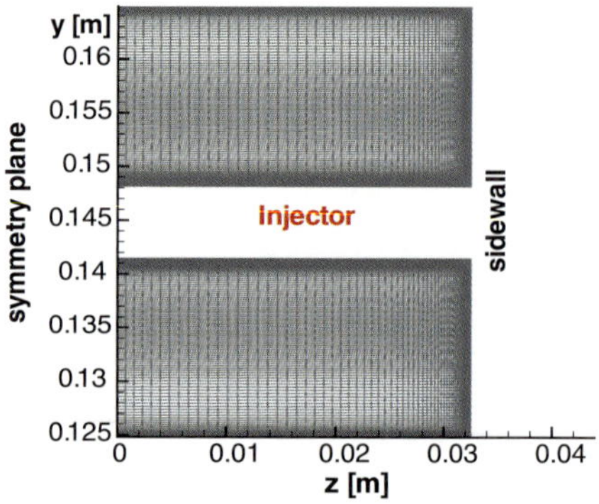

Fig. 4. Grid cell distribution at exit plane for 3D intake with 15.5 degrees compression angle

plane, respectively. In Fig. 3 the vectors represent the local velocity vectors and quite impressively show the strong gradients of hypersonic boundary layer flow. For the numerical analysis of the half model of the current intake approximately 10.7 million grid points are used: 488 in the streamwise direction and 85×265 in the cross plane. The grid spacing stretches away from the walls and is clustered at the leading edges of ramp and cowl. Otherwise, great care is taken to obtain an overall homogeneous grid spacing (see Fig. 4). Earlier analysis showed a certain grid sensitivity of hypersonic shock wave boundary layer interaction when local grid refinement is applied [11]. A minimum wall spacing of $\Delta = 1.\text{e-}06$ is used in all directions yielding a y^+ of 1.

4 Results

The design of the tested scramjet demonstrator is centered around the supersonic combustion process employing a central strut injector with hydrogen injection. The primary purpose of the intake is to provide high-pressure flow to the engine with a minimum of aerodynamic losses and a static temperature high enough to allow for stable combustion. Additionally, the combustion process itself causes a variation of the inlet back pressure which needs to be considered when analyzing the performance of the whole system. However, earlier analyses of the combustion chamber have shown the back pressure to be negligible due to the lean mixture of fuel and air currently used. Thus, only the area contraction due to the injector is considered in the current analysis. Another critical factor is the static temperature at the end of the compression process which has to be high enough to ensure self ignition of the hydrogen. Thus, the current focus is mainly on the combustor inflow as well as to obtain some inside concerning placement and measuring range of pressure transducers in the intake model.

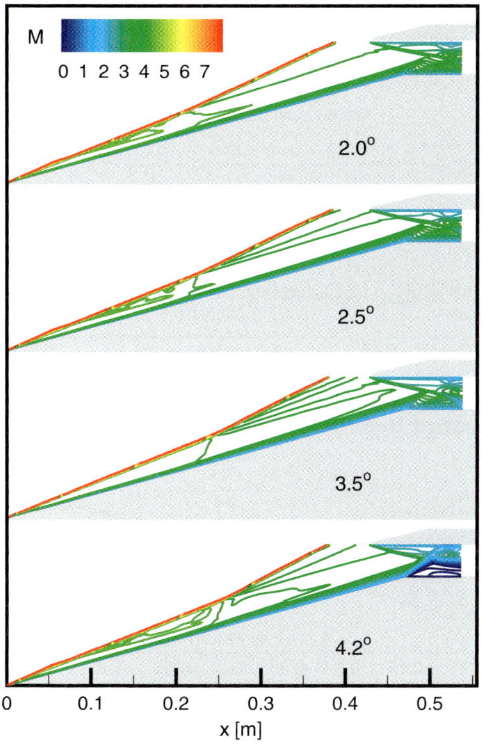

Fig. 5. Mach number contours at center plane for various sidewall compression angles

Investigation of the effect of varying the sidewall compression angle reveals a strong 3D effect, visible in Fig. 5 by the increasing flow blockage in the intake duct. The CFD analysis for 4.2 degrees sidewall compression is not converged but shows an intermediate result. Continuing the computation yielded an inlet unstart shortly after the shown solution [14]. Therefore a compression side angle using a straight side wall is currently being pursued.

Figure 6 shows the overall Mach number distribution in the center plane. The compression ramp shock is slightly bend due to the influence of the side wall compression, resulting in an increased spillage drag. This drag is caused by a mass flow loss between the ramp shock and the engine cowl. The shock generated at the leading edge of the cowl is interacting with the expanding ramp boundary layer and causes a large separation right before the onset of the injector. Currently, the installment of a bleed slot in this area is investigated in more detail to reduce if not avoid the flow separation. The reattachment shock then hits the underside of the strut injector right at the outflow boundary. This can be seen in Fig. 7 by the high compression of isolines at the exit area.

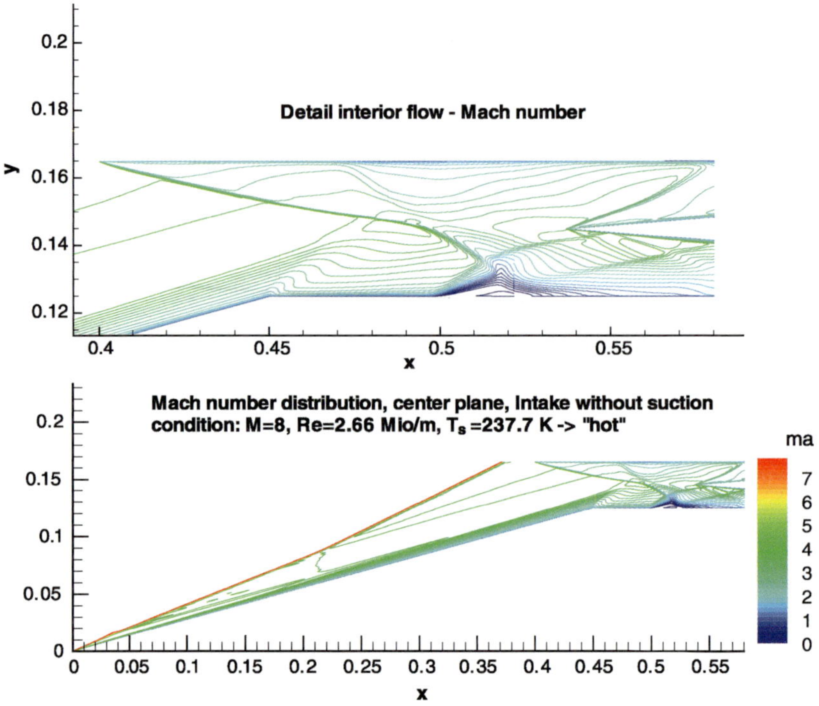

Fig. 6. Distribution of the Mach number at the symmetry plane, top: detail of interior flow, bottom: overall flow field

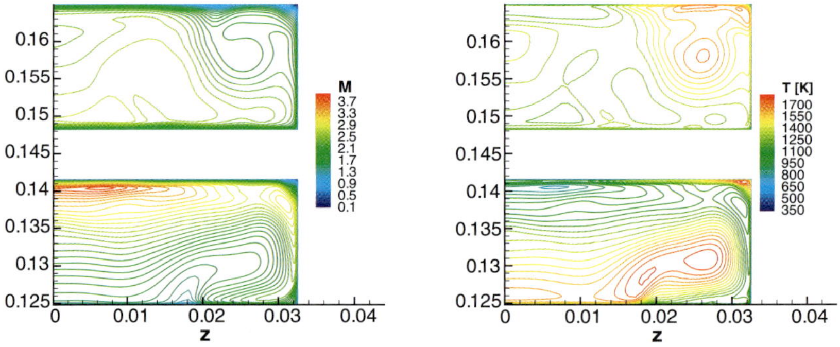

Fig. 7. Mach number and temperature isolines on exit plane of 3D intake (condition I)

The flow distribution established at the exit plane (see Fig. 7) is relatively homogeneous and the temperature levels are sufficiently high to guarantee self ignition and subsequently stable combustion. Thus, the current design is quite promising. However, the issue of when and if the boundary layer transitions from laminar to turbulent flow has not yet been completely investigated. All current flow computations assume fully turbulent inflow. Furthermore, the large separation might unstart the intake, therefore, the flexible installation of a bleed slot in this area is currently considered for the test model to be build.

5 Computational Considerations

The FLOWer computations are performed on a NEC SX-9 cluster using 16 processors. Due to the physical complexity of the underlying problem, the resolution had to be greatly increased from before. Thus, the grid which generally used to have a number of cells between three and four millions now has over 10 million gris cells. The memory requirement is around 80 gigabyte for a simulation of the 3D intake. For a typical problem to converge approximately 100000 iterations have to be performed. A single batch job performs approximately 12000 iterations and requires 10 hours of CPU-time per node, after which it is resubmitted into the batch queue. Scalability was tested by performing 100 iterations of a typical problem on 16, on 8 and on 4 processors. Accordingly the required user time increase from 100% for the 16 processors to 186% for 8 processors on to 363% for only 4 processors. This behavior can partially be explained by the comparatively large amount of time spend on the set-up of the problem, i.e., to perform one iteration on 16 processors already requires 79% of the user time required for 100 iterations.

FLOWer uses block based MPI parallelization where the communication between different grid blocks is performed using MPI. Therefore, the achievable load balancing depends on the discretization of the physical problem at hand. For a typical intake simulation, 4200 and 5100 MFLOPS are achieved as minimum and maximum value, respectively, resulting in a load difference of approx. 18 %. The vectorization level of the FLOWer FORTRAN program package is 98.6%. However, the average vector length lies between 116 and 163 whereas the hardware would allow for 256. Nevertheless the vector length is determined by the necessary physical resolution of the flow in the cross plane and, thus, could only be increased by a higher resolution. So far, more then 35 different cases have been investigated, varying either geometry, test conditions or turbulence modeling.

6 Conclusions

In this paper the performance analysis of the intake at a flight Mach number of 8 is performed. The role of the present numerical computations is to complement an experimental test campaign and to determine the intake geometry best applied for the hypersonic wind tunnel testing. Additionally, the numerical simulation completes the knowledge of the flow field in areas which are not accessible to measurements and allows for an overall performance analysis of the inlet geometry. The current analysis showed sidewall compression to be of great influence to the overall performance of a 3D intake. Thus, an easier to control version of straight side walls with moderate compression angle is currently the favorite design. Issues still investigated are the effect of flow transition and boundary layer bleed on the intake performance.

Acknowledgements. The authors would like to thank Dr. Uwe Gaisbauer, Stuttgart University, for the collaboration and support. This work was supported by the German Research Foundation under GRK 1095 and GSC 111. Furthermore, greatly appreciated is the support and computational time provided by the *High Performance Computing Center Stuttgart* (HLRS) under the project "Shykos".

References

1. Kroll, N., Rossow, C.-C., Becker, K., and Thiele, F., The MEGAFLOW Project, *Aerospace Science and Technology* , Vol. 4, No. 4, pp. 2000, 223–237.
2. Kroll, N. and Radespiel, R., An Improved Flux Vector Split Discretization Scheme for Viscous Flows, DLR-Forschungsbericht 93–53, 1993.
3. Radespiel, R., Rossow, C., and Swanson, R., Efficient Cell-Vertex Multigrid Scheme for the Three-Dimensional Navier-Stokes Equations, *AIAA Journal*, Vol. 28, No. 8, 1990, pp. 1464–1472.
4. Eisfeld, B. and Brodersen, O., Advanced Turbulence Modelling and Stress Analysis for the DLR-F6 Configuration, AIAA Paper 2005-4727, 2005.
5. Eisfeld, B., Implementation of Reynolds stress models into the DLR-FLOWer code, IB 124-2004/31, DLR, Institute of Aerodynamics and Flow Technology, 2004.
6. Reinartz, B. and Ballmann, J., Computation of Hypersonic Double Wedge Shock / Boundary Layer Interaction, *26th International Symposium on Shock Waves (ISSW 26), Göttingen, Germany 16-20 July 2007, 2007*, pp. 1099–1104.
7. Becker, N., Kroll, N., Rossow, C.C., and Thiele, F., Numerical Flow Calculations for Complete Aircraft - the Megaflow Project, *DGLR Jahrbuch 1998*, Vol. 1, Deutsche Gesellschaft für Luft- und Raumfahrt (DGLR), Bonn, Germany, 1998, pp. 355–364.
8. Reinartz, B.U., Ballmann, J., Herrmann, C., and Koschel, W., Aerodynamic Performance Analysis of a Hypersonic Inlet Isolator using Computation and Experiment *AIAA Journal of Propulsion and Power*, Vol. 19, No. 5, 2003, pp. 868–875.
9. van Keuk, J., Ballmann, J., Sanderson, S.R., and Hornung, H.G., Numerical Simulation of Experiments on Shock Wave Interactions in Hypervelocity Flows with Chemical Reactions, AIAA Paper 03-0960, January 2003.
10. Coratekin, T.A., van Keuk, J., and Ballmann, J., On the Performance of Upwind Schemes and Turbulence Models in Hypersonic Flows, *AIAA Journal*, Vol. 42, No. 5, May 2004, pp. 945–957.

11. Reinartz, B., Ballmann, J., Brown, L., Fischer, C., and Boyce, R., Shock Wave / Boundary Layer Interaction in Hypersonic Intake Flows, *2nd European Conference on Aero-Space Sciences (EUCASS), Brussels, Belgium 1–6 July 2007*, 2007.
12. Krause, M. and Ballmann, J., Numerical Simulations and Design of a Scramjet Intake Using Two Different RANS solver, AIAA Paper 2007-5423, July 2007.
13. Krause, M., Reinartz, B., and Ballmann, J., Numerical Computations for Designing a Scramjet Intake, *25th Congress of International Council of the Aeronautical Sciences (ICAS), Hamburg, Germany, 3–8 September 2006*, 2006.
14. Reinartz, B. and Behr, M., The influence of sidewall compression on hypersonic intakes, AIAA Paper 2009-7351, October 2009.

Characterization of Mixing in Food Extrusion and Emulsification Processes by Using CFD

M. Azad Emin, Karsten Köhler, Marc Schlender, and Heike P. Schuchmann

Institute of Food Process Engineering, Karlsruhe Institute of Technology, Karlsruhe, Germany, azad.emin@kit.edu, karsten.koehler@kit.edu

Summary. In this contribution, local flow of complex food matrices in an extrusion (i.e. twin screw extruder) and high-pressure homogenization processes (i.e. in micro-structured nozzles) were simulated by ANSYS POLYFLOW® and ANSYS FLUENT® respectively. In extrusion process, the efficiency of dispersive mixing of food-grade triglycerides of Newtonian behaviour in rheologically complex plasticized starch matrices was evaluated. Simulation results were validated quantitatively by experimental data. Furthermore, in order to understand the local mixing behaviour and its influence on the local temperature distribution within the homogenization nozzle, the flow conditions in the simultaneous Emulsification and Mixing (SEM)-nozzle were simulated by the average Navier-Stokes equations using the RNG-k-e turbulence model. Based on the results, the kinetics of the cooling effect, which can significantly reduce the coalescence of droplets, could be estimated.

1 Introduction

Optimizing food processes demands controlling flow of complex food matrices. CFD simulations are an important instrument in understanding it on a microscopic level, especially when measurement techniques do not really exist or cannot be applied as in processes in which high pressures, temperatures and flow velocities act on multi-phase systems. In this chapter results will be presented for homogenization and mixing processes. On the one hand, food-grade triglycerides of Newtonian behaviour are mixed into plasticized starch matrices of strongly non-Newtonian behaviour in an extruder, a process applying high shear stresses, pressures and temperatures in a rotating, 3-dimensional complex flow channel. On the other hand, sub-micron-sized droplets are produced by high-pressure homogenization, i.e. in micro-structured nozzles. Here a zone of laminar-turbulent transition flow is found, in which several flows are mixed in order to influence local shear, temperature and reactant concentration. In order to simulate local flow, ANSYS POLYFLOW® and ANSYS FLUENT® respectively were applied. Influence of process and material prop-

W.E. Nagel et al. (eds.), *High Performance Computing in Science and Engineering '10*, DOI 10.1007/978-3-642-15748-6_33,

erties on local flow velocities, resulting shear stresses and thus resulting droplet sizes is investigated.

2 Evaluation of Dispersive Mixing in the Fully Filled Zone of a Twin-Screw Extruder

M. Azad Emin, Heike P. Schuchmann

2.1 Motivation

Health and nutrition is the most demanding and challenging field in this era and will certainly be in the future as well. Beyond maintaining and increasing the nutritional quality, adding health-promoting and/or disease-preventing properties by adding bioactive substances to food during processing is a promising and potentially important area of research. Extrusion cooking is a high-temperature, short-time process in which starchy and/or proteinaceous food materials are plasticized, cooked and in some cases expanded by a combination of moisture, pressure, heating and mechanical shear, resulting in molecular transformation and chemical reactions. Compared to other food processing techniques it is of high economic interest as it is a continuous process with higher productivity and usually lower energy costs [1]. The process is used in numerous applications, including increasing numbers of ready-to-eat cereals; salty and sweet snacks; co-extruded snacks; indirect expanded products; croutons for soups and salads; an expanding array of dry feed for pets and fish; textured meat-like materials from defatted high-protein flours; nutritious precooked food mixtures for infant feeding; and confectionery products [2]. Parallel to the increased applications, interest has grown in the physicochemical, functional and nutritionally relevant effects of extrusion processing. Especially encapsulation of functional components, such as bioactive lipids (e.g. carotenoids and phytosterols), into edible food matrix promises potential health benefits [3].

For the encapsulation of lipophilic bioactive components a lipophilic carrier is required. In foods triglycerides serve as non-toxic, food-grade carrier. Especially when being composed of long-chained poly-unsaturated fatty acids (LCPUFAs) they themselves serve as bioactive compound, often enhancing the effect of the molecules dissolved in them. Triglyceride droplets, especially such of small sizes (e.g. $<1\ \mu$m) are known to encapsulate high concentrations of bioactive substances and increase their bioavailability [4, 5]. Encapsulating in oil/fat droplets should also result in an additional stabilizing effect [6] as e.g. diffusion is retarded. To control important properties of the systems, as the stability and release characteristics of the bioactive substances, the dispersed phase's (oil droplets') morphology (droplet concentration, size distribution, charge, interfacial characteristics and physical state) is crucial and a key in designing delivery systems with specific functional performances suitable for different types of lipophilic components and food matrices.

In laminar flow, as mainly found in extruders, shear and elongational stresses are responsible for droplet deformation and break-up [7] as well as the molecular status of starch bio-polymers presenting the encapsulating matrix material [8]. Resulting droplet sizes in addition depend on the viscosity ratio between the dispersed phase (lipophilic components) and the continuous phase (thermoplasticized starch matrix under extrusion conditions) being both temperature and shear rate dependant. Hence the knowledge of temperature, elongational and shear profiles inside the extruder channels is the key to control resulting matrix and emulsion characteristics, and thus bioavailability and stability of lipophilic bioactive ingredients encapsulated within the oil droplets and starch matrix. However, the complex geometry of extruder channels leads to a wide and inhomogeneous distribution of the stresses in the flow field. The measurement of the exposed stresses in extruders is a very difficult task, if not impossible. Computational fluid dynamics simulation (CFD) of flow characteristics and mixing ability inside the extruder thus is a promising tool improving the understanding of underlying mechanisms and depicting the influence of process parameters.

The scope of this study is to characterize the dispersive mixing of oil in plasticized starch matrix within a twin-screw extruder by computational simulation and—when possible—experimental procedures, which promises innovative solutions and new concepts for the encapsulation of lipophilic functional components into extruded foods.

2.2 Fundamentals

Droplet Deformation and Break-Up Within an Extruder Channel

When a droplet is suspended in a polymer melt (e.g. oil in starch melt) and subjected to laminar flow in a co-rotating twin-screw extruder, it will deform, orient, and possibly break-up into smaller droplets [9–11]. The final dispersed phase morphology, here the droplet size, can be predicted by means of two dimensionless parameters: The capillary (or Taylor) number:

$$Ca_{crit} = \frac{\dot{\gamma} \cdot \eta \cdot R}{\sigma} \tag{1}$$

and the viscosity ratio:

$$\lambda = \frac{\eta_d}{\eta_m} \tag{2}$$

where η_d and η_m are the dispersed phase and matrix viscosity respectively, $\dot{\gamma}$ is the shear rate, R is the initial droplet radius, and σ is the interfacial tension. The capillary number Ca may be regarded as the ratio of the deforming viscous stresses to the restoring stresses the latter resulting from the interfacial tensions. It indicates that the larger the local shear stress τ $(= \dot{\gamma} \cdot \eta)$

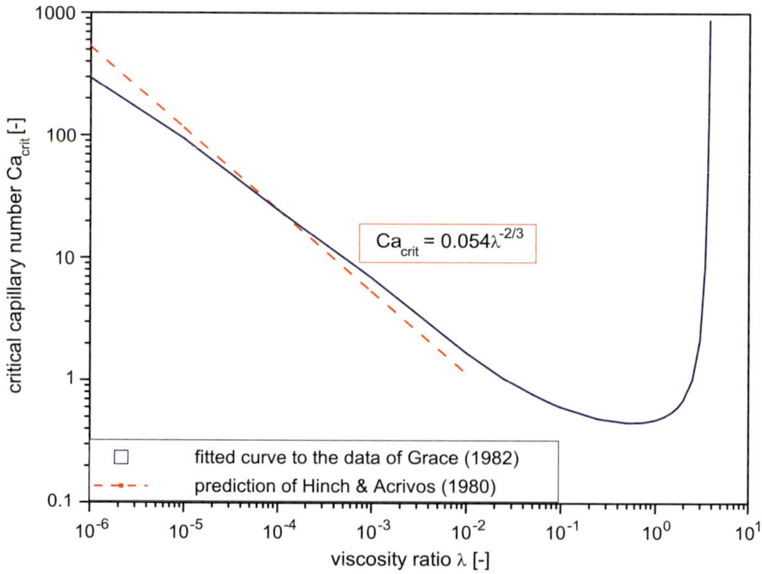

Fig. 1. Critical capillary number for droplet break-up in simple shear flow. The blue curve is fitted to experimental data of Grace [9] by Bruijn [13] and the red curve and box is the prediction of Hinch & Acrivos [14] for $Ca \gg 1$ and $\lambda < 1$

relative to the interfacial stress ($=\sigma/R$) is, the more the droplet will deform until it breaks-up. Correspondingly, for a given droplet and matrix fluid pair, there exists a critical capillary number, Ca_{crit}, at which the droplet breaks-up in a flow field of given shear stress. Figure 1 depicts the dependency of the critical capillary number on the viscosity ratio, as it was experimentally determined by Grace [9] for Newtonian fluids in a simple shear flow. Hereby, it is worth mentioning that these results were obtained via a stepwise increase in shear stress breaking droplets repeatedly under quasi-equilibrium conditions (at Ca_{crit}). However, the stepwise break-up mechanism is quite unlikely to occur in real mixing processes, most particularly in an extrusion process where a transient break-up mechanism is more realistic to be appeared [12].

For large capillary numbers ($Ca \gg Ca_{crit}$) and for viscosity ratios $\lambda < 1$, droplets are stretched into long, slender filaments. Local radii are decreased such that the interfacial tension starts playing a role, since it tends to minimize the interfacial area between the two phases. As a consequence, small disturbances present at the interface of the drop filaments grow and finally result in the disintegration of the thread into a line of drops. For viscous systems, sinusoidal disturbances grow exponentially in time; hence the deformation time plays also an important role on the break-up of the droplets [11]. Therefore, not only the capillary number but also the ratio of the capillary number to the critical capillary number ($Ca\ /Ca_{crit}$) and break-up time should be taken into account in order to evaluate the dispersive mixing in an extrusion process.

The nature of the flow should also be considered, since elongational flow is known to be more effective than simple shear flow for dispersive mixing, especially at high viscosity and low interfacial tension [10, 11]. The mixing index, λ_{MZ} [18], which can be considered to be a mapping of the mechanism of dispersive mixing within the extruder, is defined as:

$$\lambda_{MZ} = \frac{\dot{\gamma}}{\dot{\gamma} + \omega} \tag{3}$$

where is the rate of strain tensor and is the vorticity tensor, which are the symmetric and asymmetric components of the velocity gradient tensor. For a mixing index of 0, the system is undergoing purely rotational flow and no effective mixing can occur. A mixing index of 0.5 denotes simple shear flow, while a value of 1.0 denotes pure elongational flow.

Furthermore, recent studies [15–17] showed that the elasticity of the matrix fluid has a significant contribution on the deformation of dispersed droplets, most especially if the applied flow field is transient as in extruders. However, the problems regarding the measurement of elasticity of complex food matrices (e.g. thermomechanical plasticized starch) under extrusion conditions (e.g. high temperature and high pressure) and the constitutive modelling and computational simulation of such a flow in complex geometry of twin-screw extruder still remains very challenging. Therefore the influence of matrix elasticity on droplet break-up was not considered in this study.

Nevertheless, measurements that classify the elongational nature or strength of the flow, when combined with the shear stress and deformation time, allow that aspect of the mixing to be evaluated in future.

Simulation of Transitional Intermeshing Screw Elements

In a twin-screw extruder, the flow is 3-D and unsteady due to the screws' rotation. In order to model the flow in an extruder channel containing two rotating intermeshing screws, some type of remeshing or moving-mesh technique is required to take into account the time-dependent volume occupied by the fluid. Hence new meshes for each time-step can be created and time-dependent fields are then obtained by interpolation between the meshes. However this meshing task is cumbersome and many meshes or very complex meshing tools are generally required in case of a time-dependent geometry. Nevertheless, with this quasi-steady state method, good agreement between numerical and experimental results in terms of velocity and pressure profiles has been obtained at low screw speeds [19].

In order to simplify the set up of such a simulation and to avoid the use of a remeshing algorithm, flow domains and moving elements can be meshed separately and then superimposed as they would be positioned at any given time. Different authors presented similar techniques based on this principle, such as the mesh superposition technique introduced by Avalosse et al. [20], the fictitious domain method proposed by Bertrand et al. [21] and the mesh immersion technique recently proposed by Valette et al. [22]. The mesh superposition technique is based on a static mesh corresponding to the flow domain and a dynamic mesh describing the moving screws. At each time step, a procedure identifies the elements of the static mesh that contain elements of the dynamic mesh. The velocity of the moving part is imposed on the nodes of these static elements by means of a penalty technique. For this purpose, a penalty force term, $H(v - v_p)$, has been introduced where H is either zero outside the moving part or 1 within the moving part and v_p is the velocity of the moving part. The term is used by modifying the equation of motion as follows [23]:

$$H(v - v_p) + (1 - H)\langle -\nabla p + \nabla \cdot \tau + \rho g - \rho a\rangle \tag{4}$$

where p is pressure, τ is the extra-stress tensor, v is the velocity, ρ is the density, g is the volume force, and a is the acceleration term. When $H = 0$ the normal Navier-Stokes equations are left, but when $H = 1$ the equation degenerates into $v = v_p$. The value of H is determined by the generation of an "inside" field that depends on the position of the moving part. When an element or node is greater than 60 within the moving part, H is given the value of 1. Alsteen et al. [24] simulated the flow of polyamide in a twin-screw extruder by using the mesh superposition technique and showed that the design of the mesh and the selection of mixed interpolation have a significant influence on the reliability of the obtained results and should be treated carefully for such a simulation.

The fictitious domain method proposed by Bertrand et al. [21] offers similarities with mesh superposition technique. The main difference lies in the fact that, with the fictitious domain method, the kinematics of a moving part is enforced on the control points that describe it by means of Langrange multipliers and not by a penalty technique.

2.3 Material and Methods

Materials

Native maize starch was obtained from Cargill, Germany. Medium chain triglyceride oil was purchased from Schumann Sohn, Karlsruhe, Germany. Fluorescent dye was used to make the dispersed oil droplets visible under confocal laser scanning microscopy. For this purpose, nile red was purchased from Sigma-Aldrich, Germany.

Extrusion Trials

Experiments using a co-rotating high speed twin-screw extruder (Coperion ZSK 26 Mc, screw speed range 180–1800 rpm) have been performed varying process parameters such as the screw speed, feed rate and water content. In

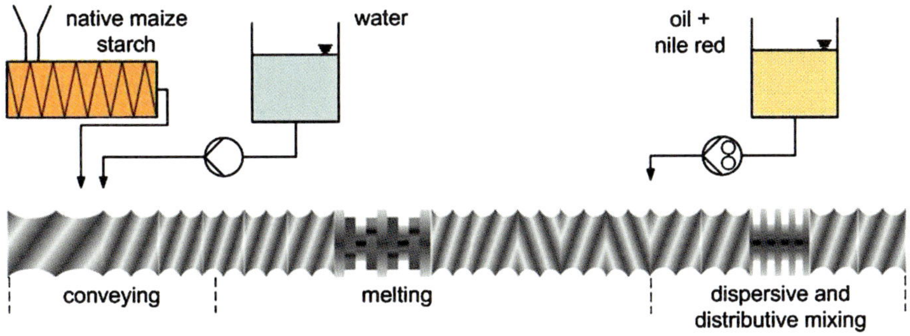

Fig. 2. Schematic illustration of the experimental set-up, screw geometry and mixing zone

the first part of the extruder, starch was mixed with water which was then plasticized under thermal and mechanical stresses generated by the rotation of screws. The oil phase mixed previously with nile red, was incorporated into the plasticized starch matrix at the end of the extruder. In this zone the extruder barrel was fully filled due to the pressure generated at extrusion die which was verified by inline mounted pressure sensors. Illustration of the experimental set-up, screw geometry and mixing zone can be seen on Fig. 2.

Confocal Laser Scanning Microscopy

In order to obtain the size distribution of the dispersed oil droplets in the extruded starch samples, a confocal laser scanning microscope (CLSM) was used. The sections cut out from every sample were placed on a glass slide and covered with a cover slide sealed with bi-distilled water. The oil was already stained with nile red before extrusion processing. An Ar-ion laser (488 nm) was used to excite the samples, and the images were digitally recorded with a 63x oil immersion objective (Carl-Zeiss, Plan-Apochromat®). The sections were analyzed using a Confocal Microscope (Zeiss LSM 510 Meta, Oberkochen, Germany). The images obtained from CLSM were further analyzed using an image processing software (Image pro plus 6.0, Media Cybernetics, Silver Spring, USA), in order to determine the droplet size distribution. Area weighted cumulative size distribution was chosen for plotting.

Rheological Measurements

Rheological data of plasticized starch under extrusion process conditions were determined online by a multiple step online slit die rheometer [B. Hochstein, Pers. Comm.]. Within the extruder channel thermo-mechanical stresses transform the biopolymer molecules of starch (especially amylopectin). The resulting rheological properties can be calculated from the pressure drop of the matrix within the slit-die being measured using flush-mounted transducers. Various shear rates necessary to obtain a viscosity curve were generated by increasing the cross section of the slit-die without any disturbance on the extrusion process parameters, thus the thermomechanical history of the product. According to Van den Einde (2004) [8] the transformation of starch molecules under mechanical stresses only depends on the applied maximum stress. For the screw geometry selected in this study, it is expected that the maximum stresses are generated at the inverse screw elements being located before the zone of interest in the measurement and simulation. Therefore it is assumed that the molecular structure, and thus, the rheological behaviour of the starch did not change further at the end of extruder (mixing zone), which could be measured by online slit die rheometer.

CFD Simulation of Mixing in Extrusion Process

The governing equations regarding a time-dependent, isothermal and incompressible flow of a generalized Newtonian fluid is

$$\rho \frac{Dv}{Dt} = -\nabla p + \nabla \cdot \tau + \rho g \qquad (5)$$

where D/Dt is the material derivative, v is the velocity, p is the pressure, ρ is the density, g is the gravity and τ is the extra stress tensor defined by:

$$\tau = 2\eta(\dot{\gamma})\dot{\epsilon} \qquad (6)$$

with the strain rate tensor $\dot{\epsilon}$ and the kinematic viscosity η. The shear rate is defined as the square root of the second invariant of the strain rate tensor,

$$\dot{\gamma} = \sqrt{2\dot{\epsilon} : \dot{\epsilon}} \qquad (7)$$

These governing equations and—based on them—mixing characteristics (shear rate, shear stress, mixing index, etc.) of thermo-plasticized starch matrix at the mixing region of the extruder were calculated by the CFD code ANSYS POLYFLOW® software package, which is a finite element code dedicated to highly viscous and viscoelastic flows. As emphasized in Sect. 2.2, the flow in a co-rotating twin-screw extruder is 3-D and unsteady due to moving and intermeshing parts (two rotating intermeshing screws in a non-moving barrel). In order to simplify the set up of such a simulation and to avoid the use of a remeshing algorithm, the mesh superimposition technique implemented by Avalosse et al. [20] into ANSYS POLYFLOW® was used. Since the dispersive and distributive mixing of the oil phase into the plasticized starch matrix takes place at the very end of the screws (see Fig. 2), only this part was chosen for simulation. Flow through the mixing elements (45° kneading blocks) is assumed to be isothermal in accordance with negligible temperature change of the material at this region. The rheological data of plasticized starch required for simulation were measured with the multiple step online slit die rheometer (see Sect. 2.3) and fitted to the Carreau viscosity model:

$$\eta(\dot{\gamma}) = \eta_o\left(1 + (\alpha\dot{\gamma})^2\right)^{(n-1)/2} \tag{8}$$

where η is the shear viscosity, η_o is the zero-shear-viscosity, α is the time constant, $\dot{\gamma}$ is the shear rate and n is the Power lax index.

Since the flow rate of the starch is relatively low (11 to 26 kg/h), creeping flow is assumed and gravity effects are neglected.

After investigating the influence of the mesh design and of the interpolation method on the pressure profile along the extruder, 503,000 bricks for mesh size, enriched mini-element velocity and linear pressure for mixed interpolations were chosen (results are not shown). In order to obtain a good resolution, the time dependent simulation was conducted with 40 time steps for half a turn of the screws.

In order to calculate the maximum shear stresses experienced by the droplets in the extruder channel, virtual particles were launched at the same time in the flow domain. It is assumed that the marker particles are massless, volumeless, and non-interacting with each other. Under these assumptions, the particles can be located by integrating the velocity vectors with respect to time. Initially, these particles are randomly distributed in an inlet section of the flow domain where the triglyceride oil is incorporated in experiments and their trajectories between the inlet and outlet are calculated from the velocity profiles. Along the trajectory, stresses experienced by each particle are computed. In order to obtain statistically good results, 1,500 particles were tracked.

2.4 Results and Discussion

Experimental Validation of the Computational Simulation of Flow of Plasticized Starch in the Mixing Zone of an Extruder

In order to check the physical relevance of the results obtained from numerical simulation and validate simulated data, pressure sensors were mounted on the inlet and outlet region of the mixing zone in the extruder being simulated. In order to monitor the rheological behaviour of plasticized starch matrix at different process conditions, the multiple step online slit-die rheometer was mounted on the extruder during the validation experiments. Obtained rheological data and applied process parameters (e.g. screw speed, flow rate and screw geometry) were used to conduct computational simulation of the flow in extruder parallel to the experiments. Pressure differences between inlet and outlet of the flow domain of the simulation were compared to the experimental values obtained by the inline pressure sensors. Pressure differences obtained from computation vary depending on the time step (screw position) chosen.

Table 1. Pressure difference in the simulated extruder zone: comparison of experimental and computational results for different process conditions

	Process condition 1	Process condition 2	Process condition 3
Flow rate (kg/h)	13	11	26
Screw speed (rpm)	200	200	400
Water content (%)	30	18	30
Viscosity (η $(Pa \cdot s)$ $= \mathrm{k} \cdot \dot{\gamma}^{(n-1)}$)	$n = 0.15$, $k = 21{,}583$	$n = 0.11$, $k = 41{,}300$	$n = 0.15$, $k = 21{,}583$
Δp experimental (bar)	14.5 ($\pm$ 0.2)	30 ($\pm$ 0.3)	16.2 ($\pm$ 0.2)
Δp computational (bar)	14.3 ($\pm$ 1.5)	30.5 ($\pm$ 3.2)	16 ($\pm$ 1.7)

Similarly, the values measured during the experiments showed deviations according to technical limitations of the pressure sensors. Therefore, the mean values were calculated for both cases, and used for the validation. Results and relevant deviations are given in Table 1, which show good agreement between experimental and computational values for different process conditions.

Evaluation of the Dispersive Mixing in the Extruder Channel by Using Computational Simulation: Influence of Screw Speed as an Exemplary Process Parameter

For evaluating the final droplet size distribution it is crucial to have knowledge on initial droplet size distribution and the stresses experienced by the droplets with their exposure time. Few data were published on this question (e.g. [25–27]). It still remains a very challenging task regarding the rheological complexity of the starch-based matrix and the wide distribution of stresses owing to the complex geometry of a twin-screw extruder. Based on the physically well grounded assumption, that the maximum stresses within the extruder channel are mainly generated at the tip of the screws, where the flow is dominated by simple shear flow, these values were taken from the simulation and used to evaluate the efficiency of dispersive mixing within the extruder at different process conditions.

Figure 3 depicts the maximum capillary numbers calculated from these maximum shear stresses which droplet will experience during their flow in the extruder channel for two different screw speeds (200 and 400 rpm). The initial droplet radius was assumed to be 5 μm. It is naturally expected that increasing screw speed should lead to an increase in maximum capillary number since it increases the shear rate, and thus the shear stress. However, there is a negative correlation between shear stress and the viscosity of the plasticized starch matrix since higher shear stresses lead to a higher degree of starch degradation [28], which further results in a decreased shear viscosity, and thus in lower maximum stresses at the mixing zone of extruder. Hereby, it is again worth to mention that the starch was exposed to maximum shear stresses

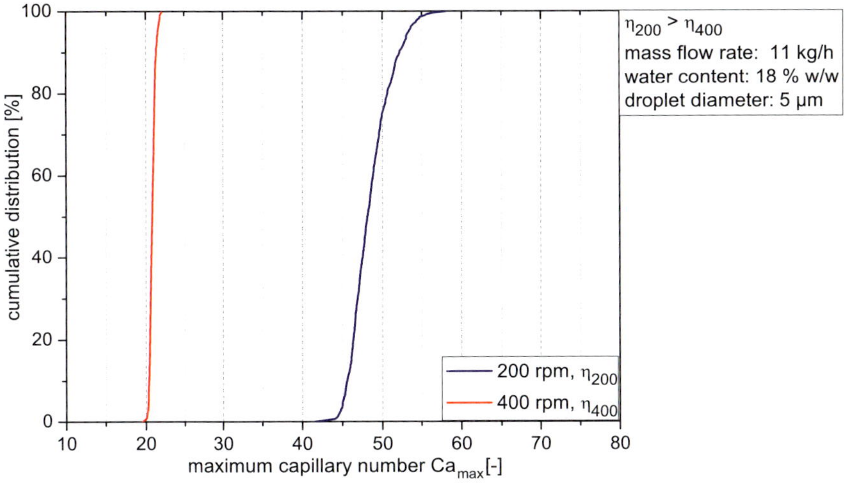

Fig. 3. Maximum capillary number experienced by droplets during their flow in the mixing zone of the extruder at two different screw speeds (200 rpm and 400 rpm)

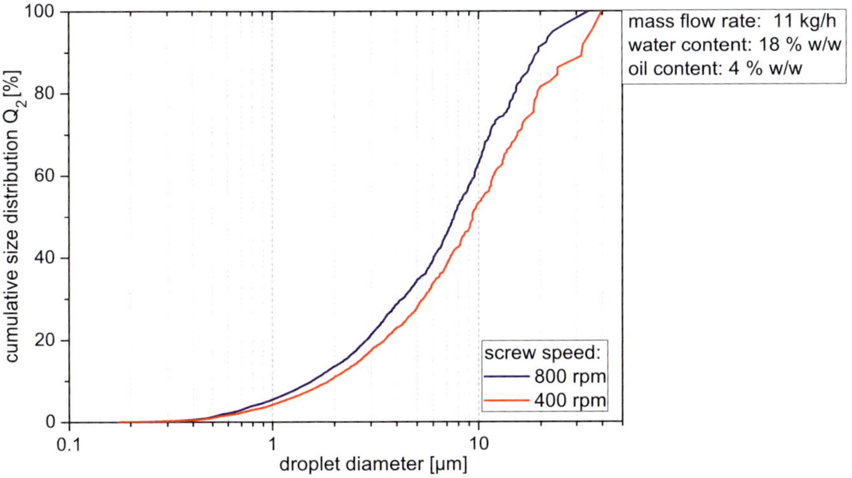

Fig. 4. Cumulative droplet size distribution obtained from extrusion experiments at two different screw speeds (400 rpm and 800 rpm)

at the inverse screw element region which is located before the simulated mixing region. Increasing screw speeds decrease the maximum shear stresses and maximum capillary number experienced by the particles tracked in the flow domain simulated owing to a dramatic decrease in starch viscosity at higher screw speeds.

As explained before (Sect. 2.2), neither the capillary number nor the critical capillary number can be used on their own to evaluate the dispersive mix-

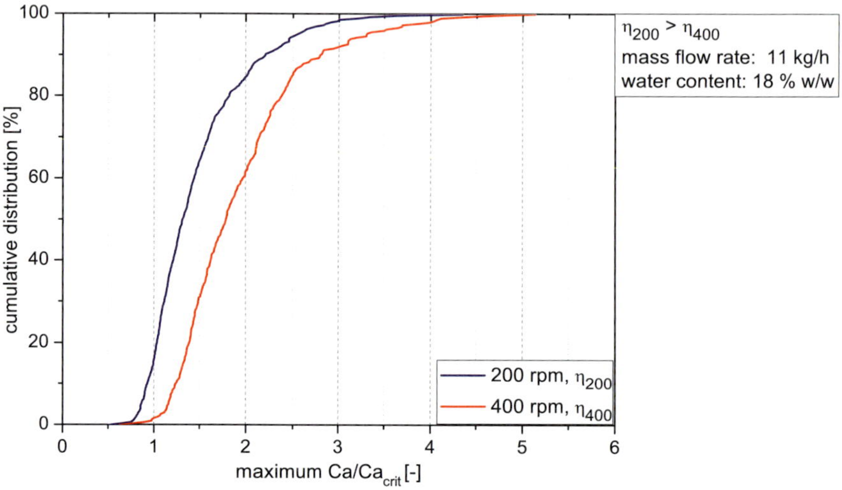

Fig. 5. Distribution of maximum capillary ratios experienced by the droplets within the extruder channel in the mixing zone at two different screw speeds (200 rpm and 400 rpm)

ing in the extruder channel. Of relevance for the dispersive mixing (droplet break-up) is the ratio of these two parameters, the capillary ratio (Ca/Ca_{crit}). Therefore, the critical capillary number was implemented in order to calculate the capillary ratio with the numerical simulation (see Fig 4). The viscosity of the plasticized starch, showing shear thinning flow behaviour, has a strong influence on the obtained shear stress profile in extruder. In Fig. 1 it can be seen that the matrix viscosity has also a strong influence on the viscosity ratio and thus the critical capillary number. Therefore, the viscosity of the plasticized starch has controversial effects on the droplet break-up which makes the evaluation of the dispersive mixing in plasticized starch very complicated. The computation of the capillary ratio which includes both controversial effects of the starch viscosity however enables us to evaluate the effect of process parameters on the droplet break-up in dispersive mixing at least quantitatively.

According to the results depicted in Fig. 4, it is expected to obtain smaller droplet sizes in plasticized starch matrix extruded at higher screw speeds, regardless of the fact that the droplets are exposed to lower maximum shear stresses (see Fig. 3). In order to validate this simulation result, experiments were carried out at different screw speeds. Oil was added into the mixing zone of the extruder and resulting droplet size distributions were measured according to Sect. 2.3. Results are depicted in Fig. 5. They confirm the computational results: Increasing the screw speed (here from 400 rpm to 800 rpm) led to smaller droplet sizes. Results for screw speeds of 200 rpm were not plotted, since the oil was dispersed that bad, that it dripped out of the starch matrix when leaving the extruder also confirming the simulation results.

2.5 Conclusions

Evaluation of the droplet break-up through dispersive mixing of oil in rheologically complex plasticized starch matrices within an extruder is a very challenging task, since several parameters such as the complex matrix rheological behaviour, initial droplet size distribution, nature and strength of the flow with the time of exposure, are necessary to be known. An experimental access is practically impossible. Flow simulation, however, offers an access to local data on flow velocities and resulting shear stresses, capillary numbers and capillary ratio, the latter being parameters of relevance for droplet break-up in laminar flow. Thus, the influence of process parameters on the efficiency of dispersive mixing in such a system can be reasonably investigated. Simulation results could be validated quantitatively by experimental data. Thus they can be used for process parameter adaptation and optimization in terms of dispersive mixing efficiency even for processes governed by complex flows as the homogenization of oil in thermoplastized starch matrices in an extruder.

3 Evaluation of Local Mixing in SEM (Simultaneous Emulsification and Mixing) Nozzles for Melt Emulsification Purposes

Karsten Köhler, Marc Schlender, Heike P. Schuchmann

3.1 Motivation

SEM (Simultaneous Emulsification and Mixing) nozzles were developed to intensify the milk homogenization process by enabling a rapid stabilization of fat droplets being disrupted in a homogenization nozzle against aggregation [29]. A SEM nozzle is based on a simple homogenization orifice (as published by e.g. Stang [30] or Aguilar [31]). This orifice is combined with a micro-mixer inducing a mixing zone. Here a side stream (SS) is mixed at low pressure to the main stream (MS) coming from the high pressure homogenization orifice. This technology was investigated in order to produce sub-micron-sized wax particles via a melt-emulsification step. In this process, patented [32], the main challenge to date is the choice of an appropriate stabilizer system. As depicted in the literature, emulsifiers and dispersing additives are used that are able to stabilize the wax particles in an aqueous phase both as droplets and solid particles. They also have to survive the high temperatures required for melting the waxes ([33], p. 553). Most often, emulsifiers do not fulfil both tasks, and/or are extremely high in price and/or interfere negatively with the products in which the wax particles are applied. In this work it was investigated whether it is possible to stabilized liquid wax droplets by cooling them rapidly with poor cold water below the melting temperature. This would allow us to work without any emulsifier [32]. To proof this concept and to design the SEM

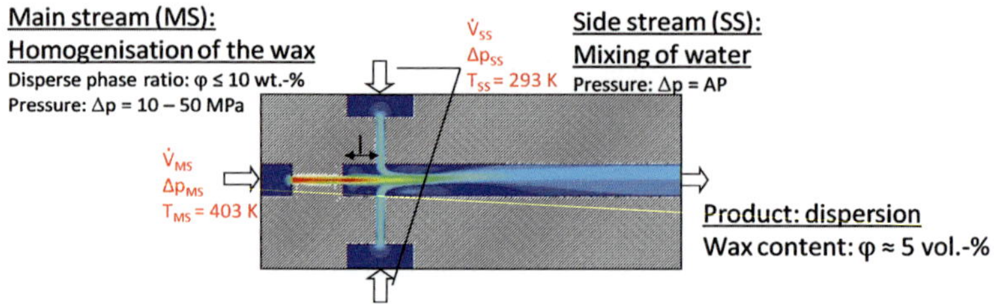

Fig. 6. Geometry of the studied SEM-nozzle [36]

valves, the cooling kinetics in the mixing zone of a SEM-nozzle has to be known. This was simulated in this work. First experimental results are also presented.

3.2 Materials and Methods

In Fig. 6 the geometry of the SEM-nozzle used in this study is depicted. The wax-in-hot water emulsion (MS) is homogenized through the orifice at elevated pressure (10–50 MPa) and diluted in the following mixing zone by a cold water side stream (SS) in order to quench the homogenized wax-in-water emulsions and thus ensure thermal stabilization by solidification. The side stream is mixed directly after the orifice outlet into the jet developing there. As shown ([31, 34, 35]) droplets are disrupted within this zone. Thus stabilization has to start exactly here. For further geometrical details see Fig. 1 in Köhler et al. [29].

For the experimental work a alkyl-ketene-dimer wax (Basoplast 90®, BASF SE, Ludwigshafen) was used as the disperse phase in the main stream. The solidification temperature according to the supplier is 335 K. The dispersed phase fraction in the final suspension was up to approximately 10 vol.-%, being at approx. 20 vol.-% in the main stream. As continuous phase and side stream de-mineralized water was used. An emulsifier was excluded.

In the experiments and the numerical simulation the main stream temperature T_{MS} was set to 403 K, being about 70 K above the melting point. The side-stream temperature T_{SS} was set to 293 K being well below the melting point. The homogenization inlet pressure was varied in three steps from 10 MPa to 50 MPa. The mixing ratio between main and side-stream was set to around 1. Light scattering with PIDS-Technology (Coulter LS 13.320, Beckmann Coulter, Krefeld) was used to measure the size distribution of the wax particles in aqueous suspension.

To elucidate the cooling kinetics inside the mixing zone of the SEM-nozzle the flow was numerical simulated using the CFD code Fluent® 6.23. To proof the concept, first trials were performed at wax contents below 1 %. Here the disperse phase can be neglected in the simulation. Thus, the simulation was conducted as 1-phase only, using the material data of water. The solid-

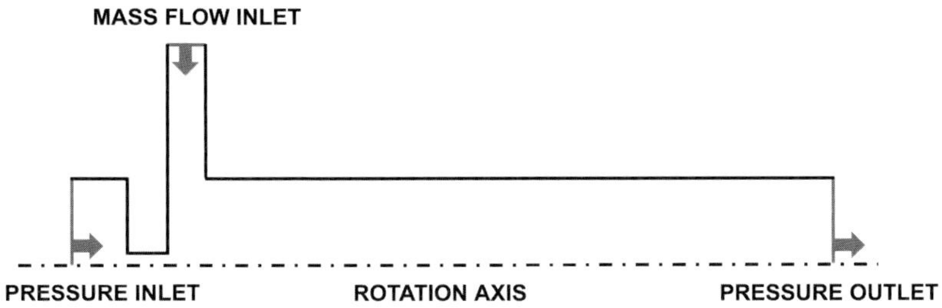

Fig. 7. Schematic of the axis-symmetrical boundary conditions

Table 2. Parameters of the turbulent flow model

Software package	Ansys Fluent®, double precision, bi-dimensional
Flow type	steady, incompressible
Viscosity model	RNG-k-ϵ turbulence model, with Standart values
Medium	water, constant density $\rho = 998{,}2\ \mathrm{kg\,m^{-1}}$ constant viscosity $\eta = 0{,}001003$ Pa s
Inlet	pressure inlet 10 MPa, 30 MPa, 50MPa; mass flow inlet; temperature T_{MS} 403 K, T_{SS} 298 K; turbulent intensity 5 %, hydraulic diameter 2,0 mm
Outlet	pressure outlet; turbulent intensity 5 %, hydraulic diameter 2,0 mm
Discretization scheme	Pressure: standard; Momentum: First order upwind; Turbulence Kinetic Energy: First order upwind; Turbulence Dissipation Rate: First order upwind
Geometry and grid	Axis-symmetric, quad; appr. 500000 cells of the grid

ification enthalpy of the wax could be neglected, too. Due to those simplifications the average Navier-Stokes equations were solved using the RNG-k-ϵ turbulence model. Because of the rotational symmetry of the SEM-nozzle axis-symmetrical boundary conditions were used as shown in Fig. 7. Detailed parameters of the computational models used for the forced steady turbulent flow model can be seen in Table 2.

3.3 Results

In order to get a better understanding of the SEM process while producing small wax particle the results were analyzed by simulations and experiments.

Simulation Results: Turbulence Intensity and Temperature Distribution in the Mixing Zone

Based on the hypothesis that the broken up droplets have to be cooled down below a critical temperature before colliding and coalescing with other droplets, the cooling of the main stream have to takes place in or directly after

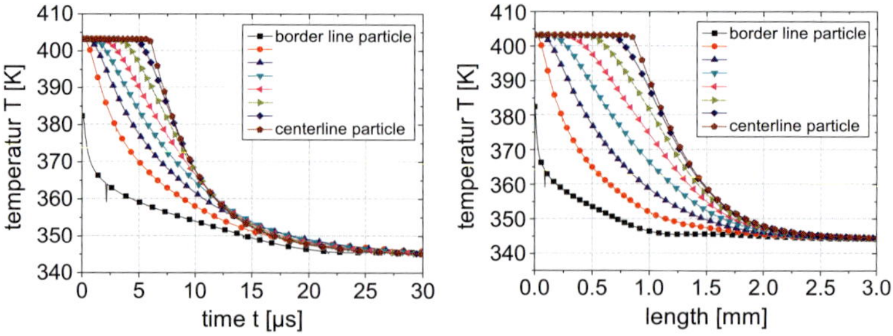

Fig. 8. Results of the particle tracking: temperature distribution over mixing zone length (left) and the time (right) for a temperature combination of T_{MS} = 403 K and T_{SS} = 293 K by a homogenisation pressure of 30 MPa

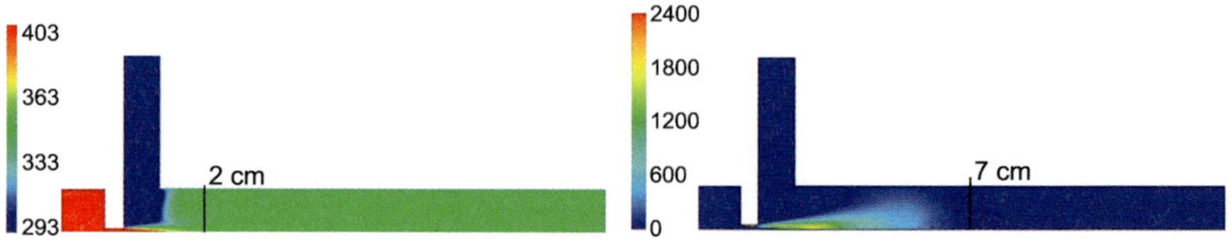

(a) (b)

Fig. 9. Numerical simulated temperature distribution [T in K] (**a**) and numerical simulated turbulence intensity [k in m^2/s^2] (**b**) inside a SEM-nozzle for a temperature combination of T_{MS} = 403 K and T_{SS} = 293 K at 30 MPa homogenization pressure

the laminar-turbulent transient jet in the mixing chamber. Particle tracking results are shown in Fig. 8. In each case the '0' on the ordinate stands for the nozzle exit point (see Fig. 7), the most left line describes the particle on the outer radius of the orifice exit and the most right line describes the particle following the center line. It clearly points out that the particle on the centre line needs the longest way and time in the mixing zone to cool down to equal temperature. For all homogenization pressures the mixing and thus the temperature equalization are completed after 10–25 μs. Compared to the time typical emulsifiers require typical emulsifiers require to adsorb at new interfaces and thus stabilize newly formed droplets this time is in the same range [37], indication that stabilization by cooling should work.

Figure 9a depicts the temperature distribution within a SEM-nozzle. The temperature profile is homogeneous at a distance of 2 mm downstream from the orifice exit. At the same processing condition the turbulent kinetic energy k drops to the zero at about 7 mm downstream the orifice exit only (see Fig. 9b). From this point on, the jet re-laminarises and droplet collision time increases allowing for droplet-droplet coalescence [38]. Based on these results it can be stated that the cooling and therewith the stabilization of wax droplets

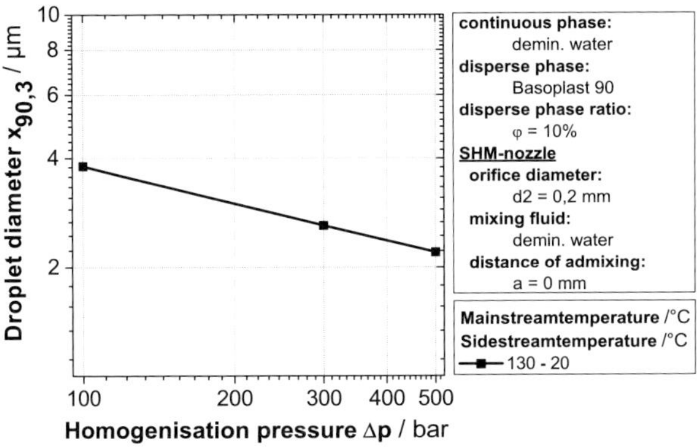

Fig. 10. Results of wax melt-emulsification at a disperse phase ratio of 10 vol.-% in the final product using the SEM-nozzle (temperature combination: T_{MS} = 403 K and T_{SS} = 293 K): Maxmila droplet diameter $x_{90,3}$ of the volume size distribution as a function of homogenization pressure

is territorial-wise completed before the turbulence is finished. This supports the hypothesis that it is possible to cool droplets well below the melting point before they may coalesce.

Experimental Validation of the Stabilization of Wax Droplets in the Novel SEM-Nozzle

Experiments were carried out at a low disperse phase ratio. They did confirm the numerical results: Droplets could be successfully stabilized by rapid cooling. In Fig. 10 the maximal particle diameter of the volume distribution $x_{90,3}$ is depicted over the homogenization pressure. The disperse phase ratio of the final product was set to 10 vol.-%, the main stream temperature to 403 K and the side stream temperature to 293 K. Also at these higher disperse phase ratios small wax particles, most of them well below 2 μm in size, could be produced. Increasing homogenization pressure results in decreasing wax particle.

3.4 Conclusion

The Simultaneous Emulsification and Mixing (SEM)-nozzle combining a disruption system, e.g. a simple homogenizing orifice, with a micro-mixing unit, e.g. a T-shaped micro-mixer, enables feeding of additional streams directly into the zone of droplet disruption. Thus droplet-droplet collisions followed by aggregation and coalescence as typically found in emulsions is significantly reduced by instant dilution and/or cooling and solidification of the droplets. The latter was examined in this study.

In order to understand local mixing behaviour and its influence on the local temperature distribution within the nozzle the flow conditions in the SEM-nozzle were simulated by the average Navier-Stokes equations using the RNG-k-ϵ turbulence model. Based on the simulation results the kinetics of the cooling effect could be estimated. The temperature in the mixing stream is homogenized within 10–25 μs which is well before turbulence is reduced and droplet collisions may result in coalescence. This supports the hypothesis that droplets may be stabilized by rapid cooling only in such a nozzle. As several assumptions (1-phase, static flow and no pulsation) are used in these simulations LES or DNS simulation should be accomplished to resolve the time dependency of the flow and the influence of the second phase (droplets).

In summary, the simulation could help in understanding the technique of molten wax droplets' stabilization by rapid cooling. Spherical submicron sized wax particle dispersions can thus be produced via a melt emulsification process without the need of any dispersing additives such as emulsifiers. For this, instant cooling of the dispersed droplets under the melting point and instant dilution by a second stream is a pre-requisite. The SEM-nozzle described here provides a possibility of technical implementation.

References

1. Guy, R. (ed.) (2001). Extrusion cooking, CRC Press, Boca Raton, and Woodhead Publishing Ltd., Cambridge.
2. Mercier, C., Linko, P., Harper, J.M., (eds) (1989). Extrusion cooking. *American Association of Cereal Chemists Inc.*, St. Paul, Minn.
3. Mcclements, D.J., Decker, Y., Weiss, J. (2007). Emulsion based delivery systems for lipophilic bioactive components. *Concise reviews in food science*, 72(8), 109–123.
4. Horn, D., Rieger, J. (2001). Organic Nanoparticles in the Aqueous Phase - Theory, Experiment, and Use. *Angew. Chem. Int. Ed.*, 40, 4330–4361.
5. Ribeiro, H.S., Schuchmann, H.P., Engel, R., Briviba, K., Walz, E. (2009). Encapsulation of Carotenoids and Vitamins. In: *Encapsulation Technologies for Food Active Ingredients and Food Processing*. Eds.: N.J. Zuidam, V.A. Nedović. Springer, New York, Chapter 8.
6. McClements, D.J. (2005). Food Emulsions: principles, practice and techniques. 2nd ed. Boca Raton, Fla: CRC Press.
7. Walstra, P. (1983). *Encyclopedia of emulsion technology*, Becher, P., Marcel Dekker Inc., New York.
8. Van den Einde, R.M. (2004). *Molecular modification of starch during thermomechanical treatment.* Dissertation, Wageningen University, The Netherlands.
9. Grace, H.P. (1982). Dispersion phenomena in high viscosity immiscible fluid systems and applications of static mixers as dispersion device in such systems. *Chemical Engineering Communications*, 14, 225–277.
10. Elmendorp, J.J. (1986). A study on polymer blending microrheology. *Polym. Eng. Sci.*, 26, 418.

11. Elemans, P.H.M., Bos, H.L., Janssen, J.M.H., Meijer, H.E.H. (1993). Transient phenomena in dispersive mixing. *Chemical Engineering Sc.*, 48, 267.
12. Janssen, J.M.H. (1993). *Dynamics of liquid-liquid mixing*, Dissertation, Eindhoven University of Technology, The Netherlands.
13. De Bruijn, R.A. (1989). *Deformation and break-up of drops in simple shear flow.* PhD Thesis. Eindhoven University of Technology, The Netherlands.
14. Hinch, E.J., Acrivos, A. (1980). Long slender drops in a simple shear flow. *J. Fluid Mech.*, 98, 305–328.
15. Sibillo, V., Simeno, M., Guido, S. (2004). Break-up of a Newtonian drop in a viscoelastic matrix under simple shear flow. *Rheol. Acta*, 43, 449–456.
16. Yue, P., Feng, J.J., Liu, C., Shen, J. (2005). Transient drop deformation upon startup of shear in viscoelastic fluids. *Phys. Fluids*, 17.
17. Verhulst, K., Cardinales, R., Moldenaers, P., Afkhami, S., Renardy, Y. (2009). Influence of viscoelasticity on drop deformation and orientation in shear flow. Part 2: Dynamics. *Non-Newtonian Fluid Mech.*, 156, 44–57.
18. Manas-Zloczower, I., Cheng, J.J. (1990). Flow field characterization in a banburry mixer. *Intern. Polymer Processing*, 5(3), 178–183.
19. Ishikawa, T., Amano, T., Kihara, S., Funatsu, K. (2001). 3-D non-isothermal flow field analysis and mixing performance evaluation of kneading blocks in a co-rotating twin screw extruder. *Polym. Eng. Sci.*, 41, 840.
20. Avalosse, T. (1996). Macromol. Symp. 112, p. 91.
21. Bertrand, F., Tanguy, P.A., Thibault, F. (1997). A three-dimensional fictitious domain method for incompressible fluid flow problems. *Intern. J. for numerical methods in fluids.* 25(6), 719–736.
22. Valette, R., Coupez, T., David, C., Vergnes, B. (2009). A Direct 3D Numerical Simulation Code for Extrusion and Mixing Processes. *Intern. Polymer Processing*, 24(2), 141–147.
23. Avalosse, T., Rubin, Y. (2000). Analysis of mixing in corotating twin-screw extruders through numerical simulation. *Intern. Polymer Processing*, 15, 117–123.
24. Alsteens, B., Legat, V., Avalosse, T. (2004). Parametric study of the mixing efficiency in a kneading block section of a twin-screw extruder. *Intern. Polymer Processing*, 19(3), 207–217.
25. Huneault, M.A., Shi, Z.H., Utracki, L.A. (1995). Development of polymer blend morphology during compounding in a twin-screw extruder. Part IV: A new computational model with coalescence. *Poly. Eng. Sci.*, 35(1), 115–127.
26. Potente, H., Bastian, M., Flecke, J. (1999). Design of a compounding extruder by means of the SIGMA simulation software. *Adv. Polymer Tech.*, 18(2), 147–170.
27. DeRoussel, P., Khakhar, D.V., Ottino, J.M. (2001). Mixing of viscous immiscible liquids. Part 1: Computational models for strong-weak and continuous flow systems. *Chemical Engineering Sc.*, 56, 5511–5529.
28. Schuchmann, H.P. (2008). Extrusion zur Gestaltung von Lebensmittelstrukturen. *Chemie Ingenieur Technik*, 80 (8), 1097–1106. (DOI: 10.1002/cite.200800065).
29. Köhler, K., Aguilar, F.A., Hensel, A., Schubert, K., Schubert, H., Schuchmann, H.P. (2007). Design of a Microstructured System for Homogenization of Dairy Products with High Fat Content. *Chemical Engineering & Technology.* 30 (11) 1590–159. (DOI: 10.1002/ceat.200700266).

30. Stang, M. (1998). *Zerkleinern und Stabilisieren von Tropfen beim mechanischen Emulgieren.* Dissertation, Universität Karlsruhe (TH).
31. Aguilar, F.A., Köhler, K., Schubert, H., Schuchmann, H.P. (2008). Herstellen von Emulsionen in einfachen und modifizierten Lochblenden: Einfluss der Geometrie auf dic Effizienz der Zerkleinerung und Folgen für die Maßtabsvergrößerung. *Chemie Ingenieur Technik.* 80 (5) 607–613.
32. Patent: Schuchmann, H.P., Köhler, Karsten (2009). Verfahren zur Herstellung einer Dispersion und Vorrichtung hierzu. Pat. nr. DE102009009060.
33. Schubert, H. (2005). Emulgiertechnik, Behr's Verlag Hamburg.
34. Tesch, S., Freudig, B., Schubert, H. (2003). Production of emulsions in high-pressure homogenizers - Part I: Disruption and stabilization of droplets. *Chemical Engineering & Technology.* 26 (5) 569–573.
35. Freudig, B., Tesch, S., Schubert, H. (2003). Production of emulsions in high-pressure homogenizers - Part II: Influence of Cavitation on Droplet Breakup. *Eng Life Sci.* 6 (3) 266–270.
36. Köhler, K., Aguilar, F.A., Hensel, A., Schubert, H., Schuchmann, H.P. (2009). Design of a Micro-Structured System for the Homogenization of Dairy Products at High Fat Content - Part III: Influence of Geometric Parameters. *Chem.-Ing.-Tech.* 32 (7) 1120–1126.
37. Schröder, V. (1999). *Herstellen von Öl-in-Wasser-Emulsionen mit mikroporösen Membranen.* Dissertation, Universität Karlsruhe (TH).
38. Danner, T. (2001). *Tropfenkoaleszenz in Emulsionen.* Dissertation, Universität Karlsruhe (TH).

Transport and Climate

Prof. Dr. Christoph Kottmeier

Institut für Meteorologie und Klimaforschung, Karlsruher Institut für Technologie, Wolfgang-Gaede-Str. 1, 76131 Karlsruhe, Germany

There is a gradually increasing number of high-performance computing projects making use of the HLRS in Stuttgart and of the SSC in Karlsruhe in the field of "Transport and Climate". All of these deal with models and data issues related to processes in the climate system. The topics cover a broad range of scales, ranging from downscaling global climate to subnational regions to processes driving large ocean circulation systems. The CPU time requirements of such models are continuously increasing, since in oceanic and atmospheric models computer power allows now for explicit resolution of energy-containing processes down to the mesoscale ($<$ 10–100 km). Processes on such scales interact considerably and progress in grid resolution means that the fields can be resolved more realistically than before. Approximations such as parametrizations can be avoided and net effects of small scales are calculated at grid resolution.

The HPC requirements for simulations of natural systems in general are still increasing. This is reflected by, e.g., the large storage amount and CPU times of regional climate simulations in the project "Modelling Regional Climate in Southwest Germany" (IMK-TRO at KIT), that provided for the first time an ensemble of high-resolution (7 km) simulations of climate change in the next decades. The study provided an assessment of the capabilities of regional climate models in simulating the observed climate of the last decades and the future in a highly structured mountaineous region. The new projects WRFClim (Univ. Hohenheim) and CEDIM-IFU (KIT) go along the same direction.

As part of the DFG-Forschergruppe PANDOWAE, operational COSMO-Forecasts were performed for realtime research flight planning within T-PARC. T-PARC is a multinational field campaign which aimed to investigate the full life-cycle of tropical cyclones in the West Pacific region. In support of T-PARC we computed numerical weather forecasts on the high performance computer of the federal state Baden-Württemberg HP XC4000 at the Steinbuch Centre for Computing (SCC), Karlsruhe Institute of Technology (KIT), and in collaboration with the German National Weather Service (DWD) using the COSMO model. The forecast operation started on 1 August 2008 and lasted until 20 October 2008. The model was run on 2 domains, which were chosen every day according to the current weather conditions. Within that domain a one-way nesting with 0.0625° (7 km) horizontal resolution on the coarser and 0.025° (2.8 km) resolution on the finer grid was performed. The HPC thus contributed to safe and successful hurricane research.

The project on Global Long-Term MIPAS Processing (IMK-ASF at KIT) provides a wealth of topics which can benefit from MIPAS data, as e.g. analyses of age of air and Brewer-Dobson circulation, validation of chemistry-climate models, influence of solar proton events on the upper atmosphere, stratospheric water vapour budget and its variability, the transport of water vapour and other species through the tropical tropopause layer, biomass burning, upper tropospheric pollution, and its impact on the upper tropospheric ozone budget. This has partly been demonstrated in the reports of the last years.

Other projects also reflect very well the high importance of the HLRS and SSC computing facilities for highly visible research programmes in actual research in meteorology and oceanography. They were not chosen for oral presentation, since are partly in an early stage, were presented in HLRS reports before, or principal investigators asked for skipping them.

The project GICS (Effects of Intentional and Inadvertent Hygroscopic Cloud Seeding; IMK-TRO at KIT) considered an innovative approach of cloud seeding in Israel, namely to shift the rain from the ocean closer to the land by reducing the speed of cloud droplet growth. This requires a very sophisticated treatment of microphysical processes in clouds as well as a full threedimensional atmospheric model (COSMO).

The project AGULHAS (The Agulhas system as a key region of the global oceanic circulation, IFM-Geomar Kiel) applies a model hierarchy that is based on the "Nucleus for European Modelling of the Ocean" (NEMO) set up in a global version as part of European model collaboration. For a full representation of the Agulhas dynamics a high spatial resolution with grid scales less than 10 km are needed. This is achieved here by the nesting capability AGRIF (Adaptive Grid Refinement in Fortran, directly embedded in the NEMO code. Details of this HLRS specific setup have been described in detail in the last year's report. During the year 2009 further experiments with the high-resolution model have been performed to allow systematic analysis; firstly, to unravel the role of the atmospheric forcing in respect to the internally generated variability and model drift; secondly, to explore the role of changes in the wind fields by applying IPCC-motivated southward wind shifts. Results from this project made it to the title page of NATURE on Nov. 26, 2009.

The project AMMA (African Monsoon Multidisciplinary Analysis; IMK-TRO at KIT) focusses on the sensitivity of the rare precipitation events in West Africa on soil moisture. In this region convective precipitation from only a few events contributes predominantly to the annual rainfall. The initiation and development of convective systems are still a great challenge. In the model runs mechanisms for the initiation of moist convection are analysed. It is found that inhomogeneous land surface conditions cause areas with a higher near-surface temperature, horizontal convergence and upward motion, which trigger convection.

Developing coupling tools for models of different compartments (e.g. ocean and atmosphere) requires substantial efforts in adaption and testing with many model test runs. Related projects therefore need relatively small HPC in the beginning of projects, but have much large CPU and storage requirements in a later stage.

Modelling Regional Climate Change in Germany

Peter Berg, Hans-Jürgen Panitz, Gerd Schädler, Hendrik Feldmann, and Christoph Kottmeier

Institut für Meteorologie und Klimaforschung, Karlsruhe Institut für Technologie (KIT), Karlsruhe, Germany, berg@kit.edu

1 Introduction

A series of regional climate simulations using the regional climate model COSMO-CLM (CCLM) have been carried out at the Institute for Meteorology and Climate Research (IMK) of Karlsruhe Institute of Technology (KIT) [1]. These simulations span the years 1971–2000 to represent the climate of the recent past and the years 2011–2040 to analyse the climate change during the next few decades. The focus of these investigations was on Southwest Germany and they have been conducted within the research program "Herausforderung Klimawandel" initiated by the Ministry of Environment of Baden-Württemberg. The aim of the research program was to asses the impacts and risks of climate change for the area of Baden-Württemberg [1].

In the meantime the spatial area of interest has been extended to the whole of Germany for the assessment of changes in flood risk for medium and small mountainous river catchments in the CEDIM-project "Flood hazard in a changing climate" (Hochwassergefahr durch Klimawandel). The river catchments in this project are the Ammer in southern Germany, the Mulde in the east and the Ruhr in the west. The (upper) Ammer catchment, e.g., has a small spatial extent of less than 600 km^2 and the need for high spatial resolution in the forcing data from the climate models is obviously necessary. The typical resolution of existing global climate model simulations is about 150 km, which is much too coarse since features like valleys and mountains that can have considerable influence on regional climate are not properly resolved by the model at this resolution. Therefore, the CCLM simulations have been performed with a spatial resolution of 0.0625 deg (about 7 km). To reach the goals of all projects and to assess the uncertainties of the climate projections, it is necessary to create multi-model, multi-member ensembles of climate simulations. In the CEDIM-project, downscaling simulations of three realisations, (global simulations with different initial conditions) of the ECHAM5 model for a control period (1971–2000) and a future scenario period (2021–2050) are

W.E. Nagel et al. (eds.), *High Performance Computing in Science and Engineering '10*, DOI 10.1007/978-3-642-15748-6_34,

carried out using the A1B SRES scenario [2]. Further downscaling simulations will be performed with the Canadian global circulation model (CGCM) in order to sample the uncertainty due to the global model used. It is desirable that an ensemble includes as many members as possible; the number which can be realised depends very much on the computer power available. It is obvious that such simulations require high computational power and large storage capacity for the results.

2 The CCLM Model

The regional climate model CCLM (COSMO-CLM) is the climate version of the operational weather forecast model COSMO (Consortium for Small-scale Modeling) of the German Weather Service (DWD). It is a three-dimensional non-hydrostatic model which means that spatial resolutions far finer than 10 km (which is considered the limit for hydrostatic models) are possible. The model solves prognostic equations for wind, pressure, air temperature, different phases of atmospheric water, soil temperature and soil water content.

Further details on COSMO can be found in [3], on the web-page of the COSMO consortium (http://www.cosmo-model.org), and in [4] and [5].

The model is coded in Fortran 90, making extensive use of the modular structures provided in this language. Code parallelisation is done via MPI (message passing interface) on distributed memory machines using horizontal domain decomposition with a 2-grid halo. The model runs on various platforms, including LINUX clusters, vector computers like NEC SX and Fujitsu VPP series as well as MPP machines like IBM-SP.

3 Regional Climate Simulations Using the HLRS Facilities

3.1 Simulation Setup and Downscaling Chain

As described in [4] and [5], dynamical downscaling is used to transfer large scale information to the regional scale. Basically, this method is a nesting of the regional climate model (RCM) into large-scale global circulation model (GCM) projections or reanalyses. This means that the model is initialized once with a state derived from the large scale information and that this information is updated at the lateral boundaries of the regional model domain at regular time intervals.

To validate the model, we perform downscaling simulations with the ERA40 reanalysis data [6]. This data set comes from a global simulation where also observations are assimilated, and it is therefore close to the observations while also filling in the gaps between the observations. The product covers the period 1957–2002 and is a good representation of climate for that period. The downscaling performed with the CCLM can therefore be directly compared to the observational record.

To assess climate change, GCM simulations without assimilation of observations are required. These are provided here by the ECHAM5 and the HadCM3 GCM models. For the Baden-Württemberg projects two realisations of the ECHAM5 model have been used which in the control simulation (1968–2000, the first three years are model spin-up time and not considered in further analyses) take into account both anthropogenic emissions and natural radiative forcing such as volcanoes and solar activity changes [7, 8]. The future projection [9, 11] is based on the SRES A1B scenario [2], which is considered as a middle of the road scenario. Further control and future scenario simulations were performed using the results of the standard version 3.0 and the Q3 realisation of the HadCM3 GCM [12]. These data have been provided by UK's Met Office Hadley Centre.

For the CEDIM-project, a slightly different data set is used. First of all, a more recent CCLM model version is used. The forcing GCM is again ECHAM5, which is forced by anthropogenic emissions [13–15]. The future scenario time-frame is furthermore increased to the period 2008–2050 [9–11].

The horizontal grid sizes of the global data sets are larger than 100 km (about 125 km for ERA40 and 150 km for ECHAM5). To avoid a too large step in the grid sizes, a double nesting technique is used for the dynamical downscaling. In a first step the large scale data are used to drive a CCLM simulation with a horizontal grid size of 0.44 deg (about 50 km at our latitudes). The chosen model domain for this first nest (Nest 1) is shown in Fig. 1 (left). In the West-East direction the domain consists of 118 grid-points, in

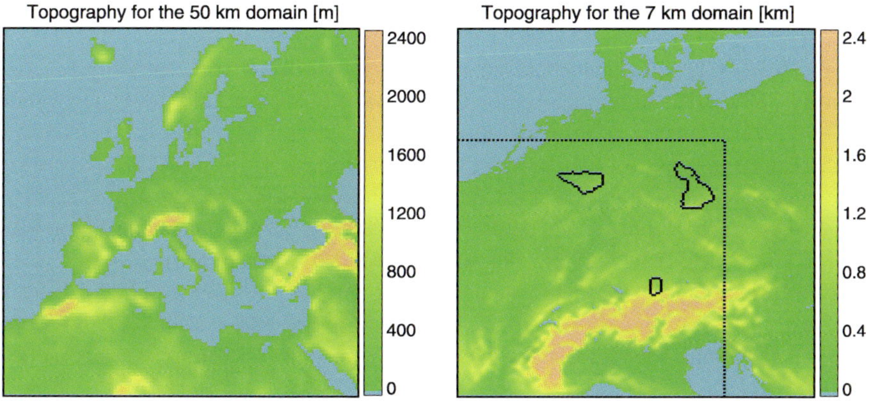

Fig. 1. Illustration of model domain of the Nest 1 (left) with a horizontal resolution of 50 km, and Nest 2* for the CEDIM-simulations (right) with a 7 km resolution. The northern and eastern boundaries of Nest 2 used for the earlier simulations are marked by dashed lines. The river catchments of the CEDIM-project are also outlined in the right figure

Table 1. Wall-clock times needed for CCLM simulations on NEC SX-8, NEC SX-9 and the BWGRID cluster at HLRS as well as performance indicators and numbers of CPUs used. The wall-clock times do not include waiting times in the queue and system maintenance times. Nest 1 is the 118*112 grid 50 *km* simulation, Nest 2 is the 124*140 grid 7 *km* simulation and Nest 2* is the 180*200 grid 7 *km* simulation

Platform	Nest No	Wall-Clock Time (d)	Average vector length	Average vector ratio (%)	Average MFLOPS	CPUs used
SX-8	1	9	112	86	1400	16
SX-8	2	22	121	93	1700	24
SX-8	2*	28	159	98	3600	16
SX-9	1	2.3	117	97	3680	24
SX-9	2	8	127	97	3275	32
BWGRID	1	9	–	–	–	80
BWGRID	2	13	–	–	–	168
BWGRID	2*	42	–	–	–	168

the South-North direction of 110 grid-points. In the second step the results of this 50 *km* simulation are used to drive the CCLM simulations with a grid spacing of 0.0625 *deg* (about 7 *km*). The corresponding model domains are shown in Fig. 1 (right). The domain for the recent CEDIM simulations with 180*200 horizontal grid points is denoted as Nest 2*. The Nest 2 domain of the earlier simulations ([1]) has 124*140 grid points, about a factor of two less than the Nest 2* domain. Its northern and eastern boundaries are also indicated in Fig. 1 by the dashed lines.

3.2 HPC Aspects

So far 16 CCLM simulations have been carried out within the research programme “Herausforderung Klimawandel” with focus on Baden-Württemberg ([1, 16]). The overall computational needs were rather large. The required wall-clock time was about 198 days, the total CPU-time amounted to about 20 years. Within the CEDIM project, eight additional simulations have already been performed at the High Performance Computing Centre Stuttgart (HLRS) facilities, and another six simulations are currently running or are in the pipeline for the next few months. Due to the larger domain size of Nest 2* the computational needs are even larger than for the Baden-Württemberg project. Performance details for the simulations are summarized in Table 1 for each platform.

The bulk of the simulations have been carried out using the NEC SX-8 high performance computer, and some are carried out on the SX-9 and on the BWGRID cluster. The Nest 2 simulations are more computationally demanding. This is due to a shorter numerical time-step and a larger grid-domain, which has a larger contribution for the newer simulations with a domain almost a factor of two larger than the previous ones. There is a strong increase in performance for the SX-9 compared to the SX-8, but the BWGRID cluster is also performing well although requiring several times more CPUs.

In order to optimize the numerical code of CCLM a cooperation between HLRS and IMK has been initiated, which is very much appreciated.

4 Results

4.1 Change of Number of Summer and Frost Days in Baden-Württemberg

Up to now analyses of the simulations with focus on Baden-Württemberg concentrated on the results for the present climate, namely the period from 1971 to 2000: comparison of the results with observational climatologies, studies of the added value of high resolution and ensemble modelling [1].

These analyses have now been extended to the future period. As an example of the impact of climate change in Baden-Württemberg, the changes of the occurrences of summer days and frost days between the periods 1971–2000 and 2011–2040 have been investigated. Summer days are days with maximum temperature above 25 °C, and frost days have a minimum temperature below 0 °C. The results, which are based on one representative simulation, are shown in Figs. 2, 3, 4 and 5. In the period 1971–2000 more than 50 summer days per year occurred in the Rhine valley and about 10 in the higher areas of the Black Forest and the Swabian Alps. During the future period 2011–2040 a general increase of the number of summer days is projected (Fig. 2). The strongest increase between 15 and 20 days occurs in the Rhine and Neckar Valleys and along the Lake Constance. In the mountains the increase is below 10 days. However, the changes relative to the present situation give quite an opposite picture: the relative changes are higher in the upper mountainous areas where the number of summer days is low (Fig. 3).

For the present climate the number of frost days lies below 80 in the Rhine valley and above 120 in the mountainous regions of the Black Forest and the Swabian Alps. During the period 2011–2040 these numbers generally decrease (Fig. 4). In the Rhine and Neckar valleys and around the Lake Constance the decrease is weakest (about 10 days). It is significantly stronger in mountainous regions, where a decrease between 15 and 20 days occurs. In the case of frost days, the relative changes (Fig. 5) are lower in the non-mountainous areas (with presently low numbers of frost days) and stronger in the higher regions where the number of frost days is larger for the present climate .

4.2 Future Changes of Strong Summer Precipitation Events

In mountainous regions like Southwest Germany strong precipitation events can produce dangerous flash floods. Strong precipitation is here defined as the daily sum of precipitation that, on the average, can be expected each 10

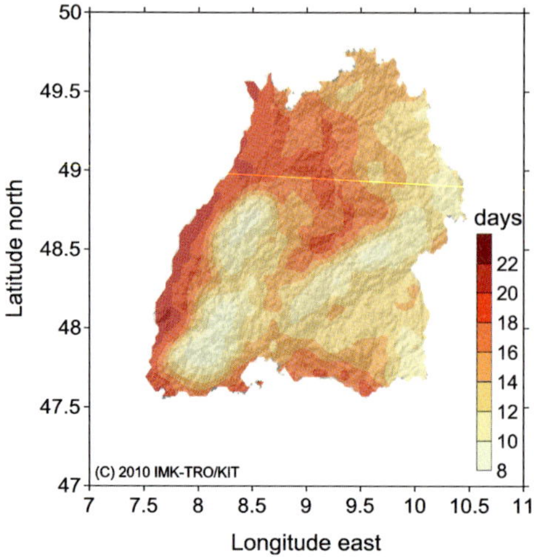

Fig. 2. Change of number of summer days between 1971–2000 and 2011–2040

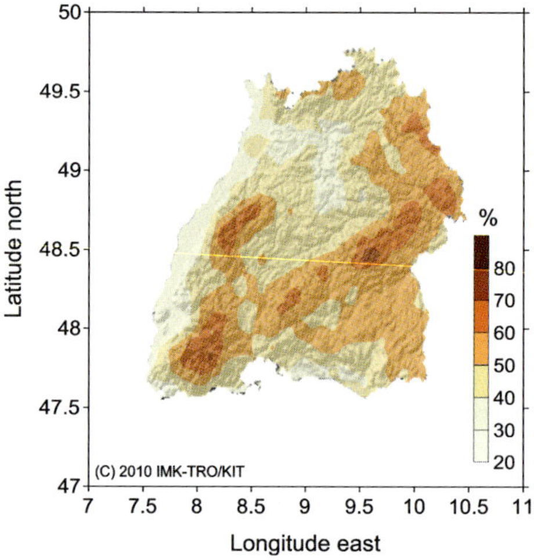

Fig. 3. Relative change of number of summer days (%) between 1971–2000 and 2011–2040

years. For Southwest Germany the values lie between 50 and 100 mm/d. The flood that occurred in June 2008 in the eastern part of the Black Forest and which caused human casualties and high damages was caused by precipitation amounts of between 50 and 80 mm/d. In order to investigate whether the frequency and intensity of such strong events change during the next decades under the influence of climate change, the ensemble of high resolution CCLM simulations for Baden-Württemberg has been used. The result is shown in Fig. 6 as a probability map.

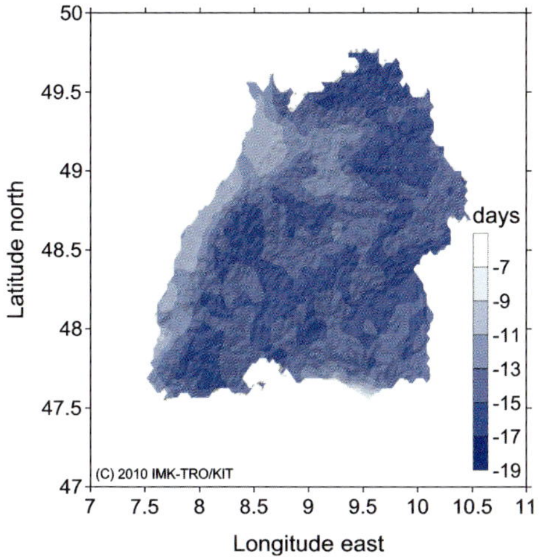

Fig. 4. Change of number of frost days between 1971–2000 and 2011–2040

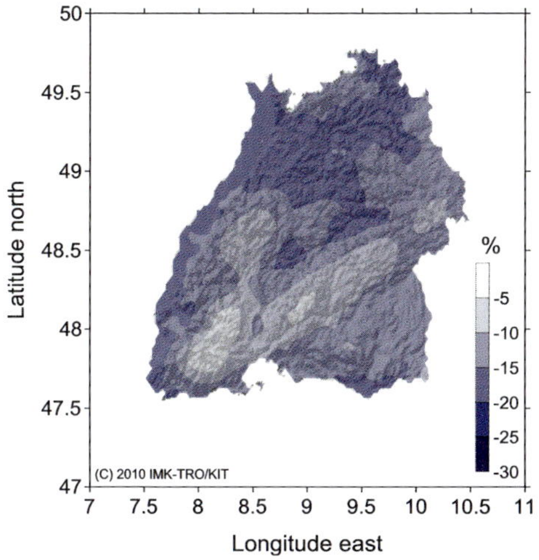

Fig. 5. Relative change of number of frost days (%) between 1971–2000 and 2011–2040

A probability map does not represent the change of a parameter (here the intensity of strong precipitation events). It represents the probability that the parameter changes by a certain, significant (here 10%) amount. Such a presentation does not only provide information about the change itself, but also about its relevance. This additional information can only be achieved by the analysis of ensemble simulations.The probability is derived from the ensemble of simulation results and is calculated here as the "ensemble consistency",

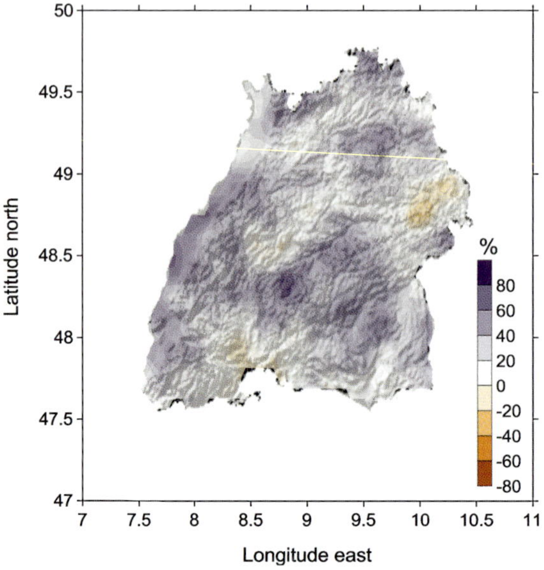

Fig. 6. Probability of change of intensity of strong summer precipitation events between 1971–2000 and 2011–2040

i.e. the percentage of the simulations indicating a change of more than 10%, respectively less than −10%.

For those regions depicted in dark blue a high probability exists that the strong precipitation events during summer increase, whereas the dark brown areas indicate a high probability for a decrease. For light coloured regions the probabilities are correspondingly weaker. Although a general increase of strong summer precipitation events can be expected, the map shows a large spatial variability of changes. However, trend analyses of observations during the last decades confirm this spatial variability, and they also show that the simulated tendencies agree well with the observations in many of the regions

4.3 Validation Results for the CEDIM-Project Simulations

The recent simulations for the CEDIM-project show an improved performance when it comes to precipitation. Figure 7 presents some first results of the comparison of the ERA40 reanalysis driven simulations (CE40_7k) at a 7 km resolution for the three river catchments under study. The figures point out one difficult problem when validating an RCM, namely that the uncertainties in the observational data is fairly large. Here we have used three different sources of observations; E-OBS [18] which is originally at a 25 km grid, REGNIE [19] at a 1 km grid, and a 7 km gridded version of a homogenized PIK/DWD station data set [20], which we refer to as "PIK". All three data sets are interpolated to the CCLM 7 km grid for easier comparison. Besides the resolution,

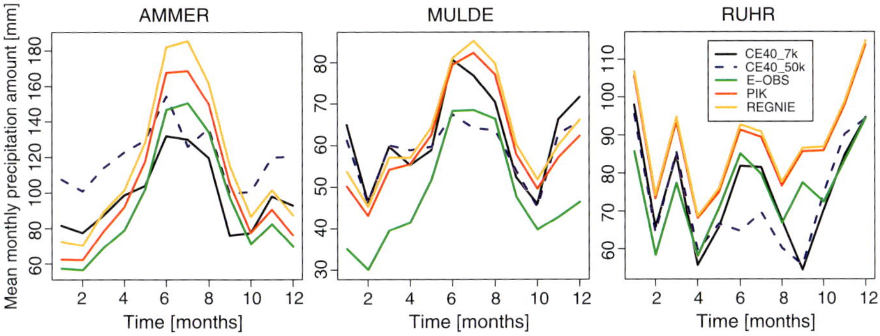

Fig. 7. Mean monthly precipitation sum for 30 years simulation with the CCLM model (black), compared to different observational data sets as explained in the text, for the Ammer (left), Mulde (middle) and Ruhr (right) river catchments. The dashed dark blue lines depict the results of the 50 km simulations being interpolated to the 7 km grid

the data sets differ due to two main reasons: (1) the number of station observations underlying the data set; REGNIE and PIK have high densities of stations while E-OBS has an order of magnitude less, (2) under-catch correction; REGNIE is corrected for the problems that the precipitation gauges have to catch all precipitation for windy and snowy conditions [21], while the other two data sets are based on direct measurements. ([17]).

The uncertainty in the observed value from these data sets is evident from Fig. 7. E-OBS shows generally lower monthly precipitation by 10–30% compared to the other data sets. PIK and REGNIE are relatively close with REGNIE having slightly higher values, possibly due to the under-catch correction. The seasonal cycle of precipitation in the catchments can be classified into "summer type" for the Ammer and the Mulde, while the Ruhr has more of a "low-mountain type" precipitation characteristic [22].

For the semi-alpine Ammer catchment, Fig. 7 (left) shows that the 7 *km* simulation is under-estimating the precipitation in summer, while over-estimating in winter. A large part of this is due to the weak performance in the mountains, which is a general problem for most RCMs even at this high resolution. The performance is better for the Mulde, but for the Ruhr there is a general underestimation of precipitation throughout the year. The figure also shows the results of the simulations at a 50 *km* resolution, which have been interpolated to the 7 *km* grid in the same way as the observations (Fig. 7, dashed dark blue lines). It can be seen that,in general, the performance for the 7 *km* simulation is improving on the seasonal cycle. The improvement is especially large for the summer-time precipitation, where the convection is better reproduced by the high resolution simulation (Mulde and Ruhr catchments). For the semi-alpine Ammer region the improvement occurs during the winter, spring, and autumn seasons.

5 Future Work

The ensemble modelling aspect of climate modelling is central to assessing probabilities for future changes in e.g. precipitation. It is important to include several GCM-simulations in any regional assessment of future climate, and the more models used the better the sampling of the uncertainty range, assuming that the models cluster around the "truth". One additional member to the IMK CCLM ensemble will be simulations based on the results of the third GCM generation of the Canadian Centre for Climate Modelling and Analysis (CGCM3, [23, 24]) which are presently being prepared.

IMK is also contributing to the Fifth Assessment Report (AR5) of IPCC with simulations with the CCLM over Africa, which are currently prepared.

Further large demands for High Performance Computing will also be necessary within the MIKLIP (Mittelfristige Klimaprognosen) program initiated by the Federal Ministry for Science and Education (BMBF). The IMK will participate in joint projects which aim in the assessments of the regional decadal predictability of climate for Central Europe, and the West African monsoon region and the Atlantic region of tropical cyclogenesis.

References

1. Panitz, H.-J., G. Schädler, and H. Feldmann (2010): Modelling Regional Climate Change in Southwest Germany, In: High Performance Computing in Science and Engineering '09 [W.E. Nagel, D. Kröner, M. Resch (Eds.)]. ISBN 978-3-642-04664-3, Springer Berlin Heidelberg New York.
2. Nakicenovic, N. et al. (2000). Special Report on Emissions Scenarios: A Special Report of Working Group III of the Intergovernmental Panel on Climate Change, Cambridge University Press, Cambridge, U.K., 599 pp. Available online at: http://www.grida.no/climate/ipcc/emission/index.htm.
3. Doms, G. and U. Schättler (2002): A description of the nonhydrostatic regional model LM, Part I: Dynamics and Numerics. COSMO Newsletter, 2, 225–235.
4. Meissner, C. and G. Schädler (2007): Modelling the Regional Climate of Southwest Germany: Sensitivity to Simulation Setup. In: High Performance Computing in Science and Engineering '07 [W.E. Nagel, D. Kröner, M. Resch (Eds.)]. ISBN 978-3-540-74738-3, Springer Berlin Heidelberg New York.
5. Meissner, C., G. Schädler, H.-J. Panitz, H. Feldmann, and Ch. Kottmeier (2009): High resolution sensitivity studies with the regional climate model COSMO-CLM. Meteoroligsche Zeitschrift, 18, 543–557, doi: 10.1127/0941-2948/20090400.
6. Uppala, M., P.W. Kallberg, A.J. Simmons, U. Andrae, V. Da Costa Bechtold, M. Fiorino, J.K. Gibson, J. Haseler, A. Hernandez, G.A. Kelly, X. Li, K. Onogi, S. Saarinen, N. Sokka, R.P. Allan, E. Andersson, K. Arpe, M.A. Balmaseda, A.C.M. Beljaars, L. Van de Berg, J. Bildlot, N. Bormann, S. Caires, F. Chevallier, A. Dethof, M. Dragosavac, M. Fischer, M. Fuentes, S. Hagemann, E. Holm, B.J. Hoskins, L. Isaksen, P.A.E.M. Janssen, R. Jenne, A.P. McNally, J.-F. Mahfouf, J.-J. Morcrette, N.A. Rayner, R.W. Saunders, P. Simon, A. Sterl,

K.E. Trenberth, A. Untch, D. Vasiljevic, P. Viterbo, and J. Woolen (2005): The ERA40 re-analysis, Q.J.R Meteorol. Soc., 131, 2961–3012.
7. Röckner, E. (2005a): IPCC MPI-ECHAM5_T63L31 MPI-OM_GR1.5L40 20C3M_all run no.1: atmosphere 6 HOUR values MPImet/MaD Germany. World Data Center for Climate. CERA-DB "EH5-T63L31_OM_20C3M_1_6H" http://cera-www.dkrz.de/WDCC/ui/Compact.jsp?acronym=EH5-T63L31_OM_20C3M_1_6H.
8. Röckner, E. (2005b): IPCC MPI-ECHAM5_T63L31 MPI-OM_GR1.5L40 20C3M_all run no.3: atmosphere 6 HOUR values MPImet/MaD Germany. World Data Center for Climate. CERA-DB "EH5-T63L31_OM_20C3M_3_6H" http://cera-www.dkrz.de/WDCC/ui/Compact.jsp?acronym=EH5-T63L31_OM_20C3M_1_6H.
9. Röckner, E., M. Lautenschlager, and H. Schneider (2006a): IPCC-AR4 MPI-ECHAM5_T63L31 MPI-OM_GR1.5L40 SRESA1B run no.1: atmosphere 6 HOUR values MPImet/MaD Germany. World Data Center for Climate. doi: 10.1594/WDCC/EH5-T63L31_OM-GR1.5L40_A1B_1_6H.
10. Röckner, E., M. Lautenschlager, and H. Schneider (2006c): IPCC-AR4 MPI-ECHAM5_T63L31 MPI-OM_GR1.5L40 SRESA1B run no.2: atmosphere 6 HOUR values MPImet/MaD Germany. World Data Center for Climate. doi: 10.1594/WDCC/EH5-T63L31_OM-GR1.5L40_A1B_2_6H.
11. Röckner, E., M. Lautenschlager, and H. Schneider (2006b): IPCC-AR4 MPI-ECHAM5_T63L31 MPI-OM_GR1.5L40 SRESA1B run no.3: atmosphere 6 HOUR values MPImet/MaD Germany. World Data Center for Climate. doi: 10.1594/WDCC/EH5-T63L31_OM-GR1.5L40_A1B_3_6H.
12. Collins M., B.B.B. Booth, G.R. Harris, J.M. Murphy, D.M.H. Sexton, and M.J. Webb (2006): Towards quantifying uncertainty in transient climate change. Climate Dynamics, 27, 127–147, doi: 10.1007/s00382-006-0121-0.
13. Röckner, E. (2006a): IPCC-AR4 MPI-ECHAM5_T63L31 MPI-OM_GR1.5L40 20C3M run no.1: atmosphere 6 HOUR values MPImet/MaD Germany. World Data Center for Climate. doi: 10.1594/WDCC/EH5-T63L31_OM-GR1.5L40_20C_1_6H.
14. Röckner, E. (2006b): IPCC-AR4 MPI-ECHAM5_T63L31 MPI-OM_GR1.5L40 20C3M run no.2: atmosphere 6 HOUR values MPImet/MaD Germany. World Data Center for Climate. doi: 10.1594/WDCC/EH5-T63L31_OM-GR1.5L40_20C_2_6H.
15. Röckner, E. (2006c): IPCC-AR4 MPI-ECHAM5_T63L31 MPI-OM_GR1.5L40 20C3M run no.3: atmosphere 6 HOUR values MPImet/MaD Germany. World Data Center for Climate. doi: 10.1594/WDCC/EH5-T63L31_OM-GR1.5L40_20C_3_6H.
16. Panitz, H.-J., G. Schädler, and H. Feldmann (2009): Modelling Regional Climate Change in Southwest Germany. Oral presentation, The 12th Results and Review Workshop of the HLRS, October 8–9, 2009, at the High Performance Computing Center Stuttgart.
17. KLIWA (2006): Langzeitverhalten der Starkniederschläge in Baden-Württemberg und Bayern. KLIWA-Projekt A 1.1.3 "Trenduntersuchungen extremer Niederschlagsereignisse in Baden-Württemberg und Bayern". KLIWA-Berichte, Heft 8, Hrsg.: Arbeitskreis KLIWA, Landesanstalt für Umwelt, Messungen und Naturschutz Baden-Württemberg, Bayerisches Landesamt für Umwelt, Deutscher Wetterdienst, Juni 2006, 94 pp. ISBN 3-

88148-412-4. Bezug ber: Deutscher Wetterdienst, Abteilung Hydrometeorologie, Postfach 100465, 64004 Offenbach.
18. Haylock, M.R., N. Hofstra, A.M.G. Klein Tank, E.J. Klok, P.D. Jones, and M. New (2008): A European daily high-resolution gridded data set of surface temperature and precipitation for 1950-2006. J. Geophys. Res., 113, D20119, doi: 10.1029/2008JD010201.
19. Deutscher Wetterdienst (DWD), Germany (2008).
20. Deutscher Wetterdienst (DWD)/Potsdam-Institut für Klimafolgenforschung (PIK), Germany (2008): Meteorologischer Datensatz für Deutschland 1951–2006.
21. Yang, D., E. Elomaa, A. Tuominen, A. Aaltonen, B. Goodison, T. Gunther, V. Golubev, B. Sevruk, H. Madsen, and J. Milkovic (1999): Wind-induced precipitation undercatch of the Hellmann gauges. Nordic Hydrology, 30, 57–80.
22. BMU (2003): Hydrologischer Atlas von Deutschland, Freiburg, ISBN:3-00-005624-6.
23. McFarlane, N.A., J.F. Scinocca, M. Lazare, R. Harvey, D. Verseghy, and J. Li (2005): The CCCma third generation atmospheric general circulation model. CCCma Internal Rep., 25 pp.
24. Scinocca, J.F., N.A. McFarlane, M. Lazare, J. Li, and D. Plummer 2008: The CCCma third generation AGCM and its extension into the middle atmosphere. Atmos. Chem. and Phys., 8, 7055–7074.

Modelling the Extratropical Transition of Tropical Cyclones and Its Downstream Impact

Christian M. Grams and Sarah C. Jones

Institut für Meteorologie und Klimaforschung, Karlsruher Institut für Technologie, Wolfgang-Gaede-Str. 1, 76131 Karlsruhe, Germany, sarah.jones@kit.edu

1 Introduction

Accurate numerical weather forecasts are essential for the mitigation of the threat to life and property presented by high-impact weather systems (HIW). In Europe a variety of different systems present significant challenges for numerical weather prediction. The focus of this HLRS-SCC project is on tropical cyclones and their transformation into extratropical systems. Tropical cyclones are responsible for some of the most severe weather impacts in the Atlantic. Their direct impact on land as a tropical system is confined almost entirely to the western Atlantic, although in an active year such as 2005 southern Europe can be affected by weak tropical cyclones (Vince made landfall in Spain, Delta in the Canary Islands). A more significant threat for Europe is presented by those systems that interact with the midlatitude flow and undergo extratropical transition. When a tropical cyclone moves polewards it interacts with the midlatitude flow such that it undergoes significant structural changes and may transform into an extratropical weather system (Jones et al., 2003 [12] and references therein). Transformed tropical cyclones that reach Europe can still result in severe weather (e.g. ex-Hurricane Lili in 1996 Anna Agustí-Panareda et al. (2005) [1]), but these events are relatively infrequent. Arguably a more significant threat for Europe is posed by the excitation of a Rossby wave train as a tropical cyclone interacts with the midlatitude flow (hereafter downstream impact). Hoskins and Berrisford (1988) [11] attributed the development of a tropopause step that was crucial for the development of the October Storm in 1987 to the outflow from Hurricane Floyd (1987).

PANDOWAE (Predictability ANd Dynamics Of Weather systems in the Atlantic-European sector)[1] is a Research Unit of the German Research Foundation (DFG). It aims to advance the knowledge and understanding of the dynamical processes responsible for the development of high impact weather

[1] http://www.pandowae.de

W.E. Nagel et al. (eds.), *High Performance Computing in Science and Engineering '10*, DOI 10.1007/978-3-642-15748-6_35,

systems, to elucidate the factors that limit the predictability of such systems, and to contribute the basic research that will lead to the development of innovative interactive weather forecasting systems. PANDOWAE represents an important German contribution to the ten year international research programme of the World Meteorological Organisation THORPEX: a World Weather Research Programme. It promotes cooperation between academic institutions and operational centres through close collaboration with operational weather services, in particular with the Deutscher Wetterdienst (DWD) and the European Centre of Medium Range Weather Forecasts (ECMWF). PANDOWAE was applied for six years and will be organised in two phases for each three years. Research is divided into three areas:

- Upper-level Rossby wave trains: generation, propagation and wave-breaking
- Moist processes and diabatic Rossby waves
- Ensembles and adaptivity

Amongst the different scientific institutions research is organised in seven individual projects and three associated projects. In the project PANDOWAE-ET which is located at the Institute for Meteorology and Climate Research (IMK), Karlsruhe Institute of Technology (KIT) we investigate the mechanisms responsible for the downstream impact of the extratropical transition (ET) of tropical cyclones. The main tool for this investigation is the numerical weather forecast model COSMO[2] and potential vorticity (PV) diagnostics. We run COSMO on the HP XC4000 super-computer of the SCC. In support of the THORPEX Pacific Asian Regional Campaign (T-PARC) we performed operational COSMO forecasts at SCC in summer 2008.

2 The COSMO Model

COSMO is a non-hydrostatic regional numerical weather prediction model designed for the meso-β (5–50 km) and meso-γ (500 m–5 km) scale [18, 19]. COSMO is the operational weather forecast model of the German Weather Service (DWD) and is therefore optimised for usage on parallel computers. The non-hydrostatic thermo-hydrodynamical equations for compressible flow are formulated on a rotated geographical coordinate system, using an Arakawa-C grid. In the vertical a hybrid vertical coordinate is used, terrain following in the lower half of the troposphere. In our simulations the height and number of model levels was increased compared to the standard setup to allow for deep tropical convection and the vertical layers were defined to be as close as possible to the vertical layers of the input fields. The basic equations are solved using the time-splitting technique of Klemp and Wilhemson (1978) [14]. The various physics package of COSMO include a turbulence and surface layer scheme using a prognostic turbulent kinetic energy equation with a 2.5 order

[2] http://www.cosmo-model.org

closure by Mellor and Yamada (1982) [15] with extensions by Raschendorfer (2001) [16] and a two-category bulk model cloud microphysics scheme [6, 8]. Precipitation formation is treated by a Kessler-type bulk microphysics parametrisation [13] including water vapour, cloud water, rain and cloud ice with column equilibrium for the precipitating phase. The subgrid-scale clouds, on the other hand, are parametrised by an empirical function depending on relative humidity and height. A δ two-stream radiation scheme (Ritter and Geleyn, 1992 [17]) is used for the short- and longwave fluxes and the full cloud radiation feedback. The surface layer is parametrised by a stability-dependent drag-law formulation of momentum, heat and moisture fluxes according to similarity theory. The initial and boundary conditions are taken from the ECMWF operational global model IFS and from the DWD global model GME.

3 The Downstream Impact of Tropical Cyclones Undergoing Extratropical Transition

In this sub-project the impact of tropical cyclones on the midlatitude weather patterns is investigated using COSMO simulations along with potential-vorticity (PV) diagnostics. In summer 2008 a huge effort was made to provide mesoscale numerical weather forecasts for the THORPEX Pacific Asian Regional Campaign (T-PARC), running COSMO operationally on HP XC4000. Here we summarise this forecast operation (Sect. 3.1). The T-PARC forecasts enabled a more detailed case study on the interaction of the outflow of Typhoon Jangmi and the midlatitude jet on which we focus at the end of this section (Sect. 3.3). Many of the figures in Sect. 3.3 are taken from Grams and Jones (2009) [9]. The physical processes involved in ET in different ocean basins are similar. Thus studying a Typhoon in the Pacific helps to get generally more insight in the dynamics of the ET process and this knowledge can be applied to Atlantic cases.

3.1 Operational COSMO-T-PARC Forecasts

T-PARC is a multi-national field campaign which aimed to investigate the full life-cycle of tropical cyclones in the West Pacific region. In support of T-PARC we computed numerical weather forecasts on the high performance computer of the federal state Baden-Württemberg HP XC4000 at the Steinbuch Centre for Computing (SCC), Karlsruhe Institute of Technology (KIT), and in collaboration with the German National Weather Service (DWD) using the COSMO model (see also Grams et al., 2008 [10]). The forecast operation started on 1 August 2008 and lasted until 20 October 2008. The model was run on 2 domains, which were chosen every day according to the current weather conditions. Within that domain a one-way nesting with 0.0625° (7 km) horizontal resolution on the coarser and 0.025° (2.8 km) resolution on the finer grid was

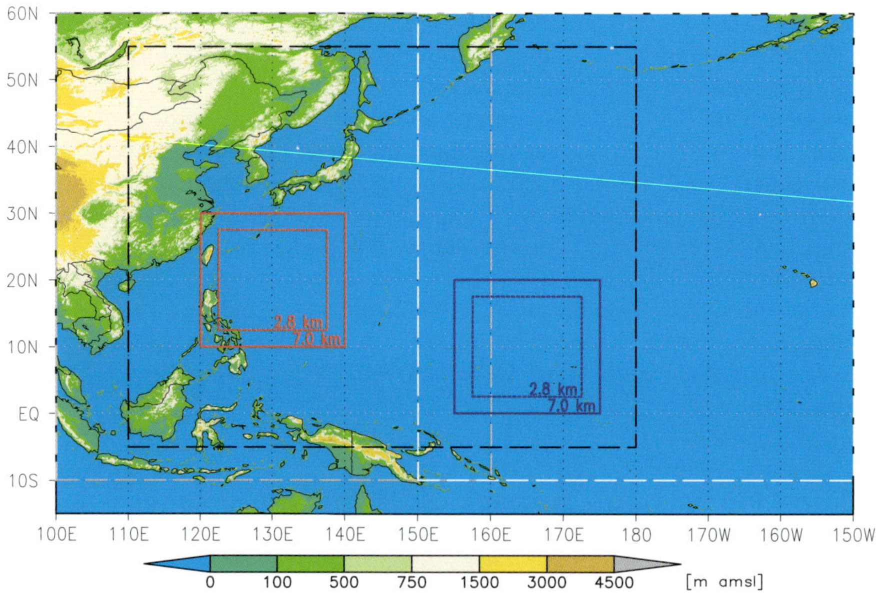

Fig. 1. Domain of T-PARC operations. COSMO-T-PARC forecasts at 0.0625° resolution were possible in the whole domain. Dashed lines indicate the subsets of external data at 0.025° resolution. The red and blue boxes show examples of the one-way nesting domains. The shading indicates model orography (in metres above mean sea level) at 0.0625° resolution

performed (Fig. 1). A typical model domain with reasonable resource usage in terms of computing time and data amount covers 20° × 20° in longitude and latitude at 0.0625° horizontal resolution and 15° × 15° at 0.025°. The 0.025° high resolution forecasts were necessary to explicitly resolve moist convective processes. 51 levels were used in the vertical. The model was initialised at 00 UTC and 12 UTC using the global forecasts from the GME (DWD) and IFS (ECMWF) models. The two forecast basetimes enabled the most recent forecasts to be available at the different stages of the flight planning. Together with the two driving global models a mesoscale "mini-ensemble" with 4 differently initialised COSMO forecasts was available. This helped to assess uncertainties in the deterministic forecasts. Thus, up to 16 COSMO forecasts were computed every day. The forecasts were computed for 72 hours and hourly model output made available. Selected model products for the first domain were sent to the online T-PARC field catalogue. As numerical weather prediction is always a time-critical task, the different steps of the COSMO runs from the global data retrieval to the publication of COSMO products had to be coordinated effectively. Seven different computer systems were involved in the forecast process: A schematic sketch of the forecast suite in Fig. 2 shows the data flow.

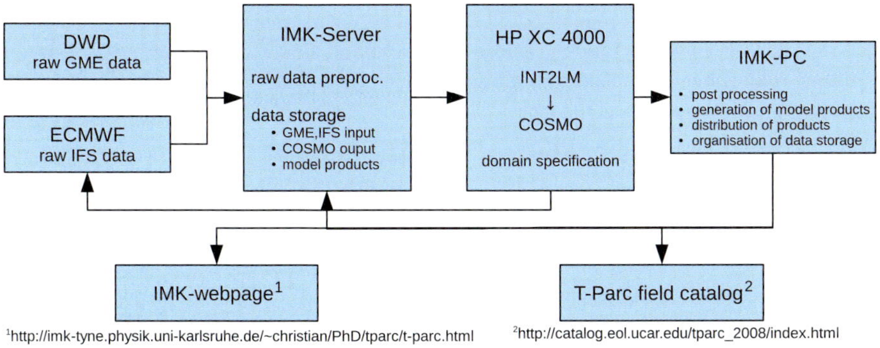

Fig. 2. Schematic sketch of COSMO-T-PARC forecast organisation

Table 1. COSMO-T-PARC model configuration and computing resources

dx, dy	dt	nx, ny	nz	♯ CPU	CPU-time	duration	data
0.0625°	20–40 s	321	51	32–64	72 h	1.5–2.5 h	5 GB
0.025°	8–16 s	601	51	96–144	720 h	4.5–6.0 h	16 GB

3.2 Performance Tests and System Data

The choice of the model domain extent and the number of processors was a compromise between the request for covering a large area, computing time, and the availability of computing and data storage resources (Table 1). Prior to the campaign we made tests with different model domain configurations. Soon it was obvious that a horizontal extent of the model domain of 20° (321 gridpoints) in longitude and latitude was reasonable at 0.0625° horizontal resolution and 15° (601 grid points) at 0.025° horizontal resolution. Rough performance tests showed that the distribution of the COSMO job on a number of tasks which is a multiple of 4 (4 is the number of processors per node on HP XC4000) was advantageous. The maximum benefit from speed-up due to parallelisation was most obtained when dividing the COSMO job in 8×4 (32) tasks for the 0.0625° resolution runs and 12×12 (144) tasks for the 0.025° resolution runs. Nevertheless we occasionally computed the 0.0625° resolution runs with 64 CPUs to decrease the total forecast duration while moderately increasing the total CPU-time. Also, we had to reduce the number of tasks for the 0.025° resolution runs on 12×8 (96) CPUs when computing resources were limited. The maximum physical memory used by any process was less than 900 MB. The maximum virtual memory used by any process was less than 2800 MB. Table 2 shows a selection of the performance tests effected. Following the experiences from the performance tests not more but more powerful CPUs would help to reduce total CPU time on future HPC systems.

Table 2. Performance for selected COSMO configurations: typical 0.0625° (0.025°) run on the left (right). The duration of computing one forecast hour is given (in minutes real time (2nd row), in minutes CPU time (3rd row)) depending on the number of CPUs selected (1st row)

0.0625° (7 km), 321 × 321 × 51 GPs, dt = 40 s			0.025° (2.8 km), 601 × 601 × 51 GPs, dt = 16s		
♯CPU	min/1 h fcst	CPU time/1 h fcst	♯CPU	min/1 h fcst	CPU time/1 h fcst
4 × 4 = 16	4	64	8 × 8 = 64	14	896
8 × 4 = 32	2	64	12 × 8 = 96	6	576
8 × 8 = 64	1.5	96	12 × 12 = 144	4	576

3.3 The Outflow-Jet Interaction During the Extratropical Transition of Typhoon Jangmi

Super Typhoon Jangmi was a major tropical cyclone reaching category 4 on the Saffir-Simpson scale. It formed in the Philippine Sea on 24 September 2008 and while intensifying headed northwestwards towards Taiwan. The minimum central pressure was 905 hPa and the maximum surface wind speed reached more than 210 km h^{-1}. On 28 September 2008 Jangmi made landfall at the northeastern tip of Taiwan at the time of maximum intensity. After landfall the typhoon soon weakened and the structural changes indicated its transformation in an extratropical system. Jangmi recurved in the East China Sea and passed south of Kyushu before it decayed.

As Jangmi approached the midlatitudes a maximum of the jet stream which was located across the northern part of Honshu became rather stationary from around 18 UTC 28/09/2008 to 18 UTC 30/09/2008. The jet streak extended southwards and a strong increase in maximum wind speed occurred. This is illustrated with the wind field at 200 hPa for two time steps[3] (Fig. 3). At 00 UTC 29/09/2008 westerly flow north of 36° N indicates a zonally-oriented jet (Fig. 3a) with maximum wind speed of more than 65 ms^{-1} over northern Honshu. The recurving tropical cyclone is located north of Taiwan. North of the typhoon the anticyclonic outflow of Jangmi is evident as southwesterly flow towards the jet. 36 hours later the typhoon outflow merged with a southwestward extension of the jet core and it appears that the southwesterly outflow extends into the jet core (Fig. 3b). At this time the wind speed in a wide region of the jet core increased to more than 75 ms^{-1}. Backward trajectories calculated with the LAGRANTO trajectory model (Wernli and Davies, 1997; Wernli, 1997 [20, 21]), for a 36h time interval and starting in a box in the jet core show that Jangmi advects tropical air from mid and lower levels into the jet core (Fig. 4). Trajectory calculations also indicate that the zonal flow over the Yellow Sea is deflected into the jet (not shown). The temporal evolution of the trajectories shows, that most of the lifting occurs in

[3] For details on the model runs used in this case study refer to Sect. 5.

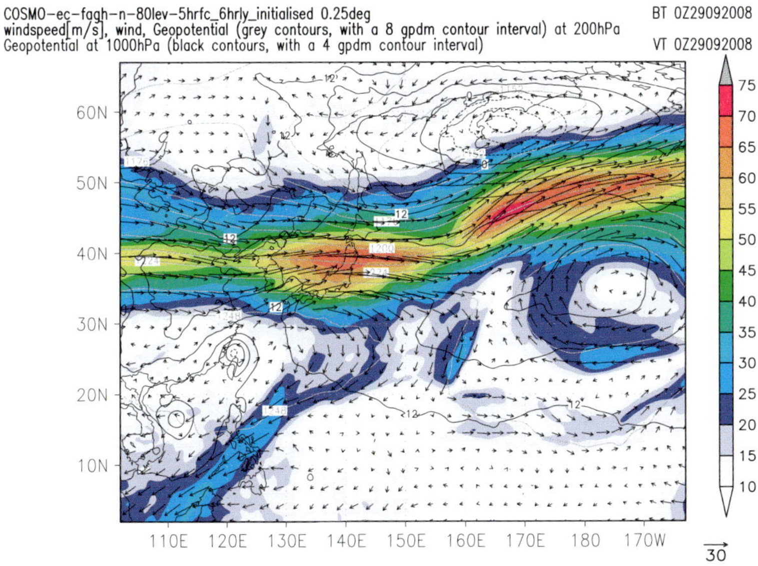

(a)

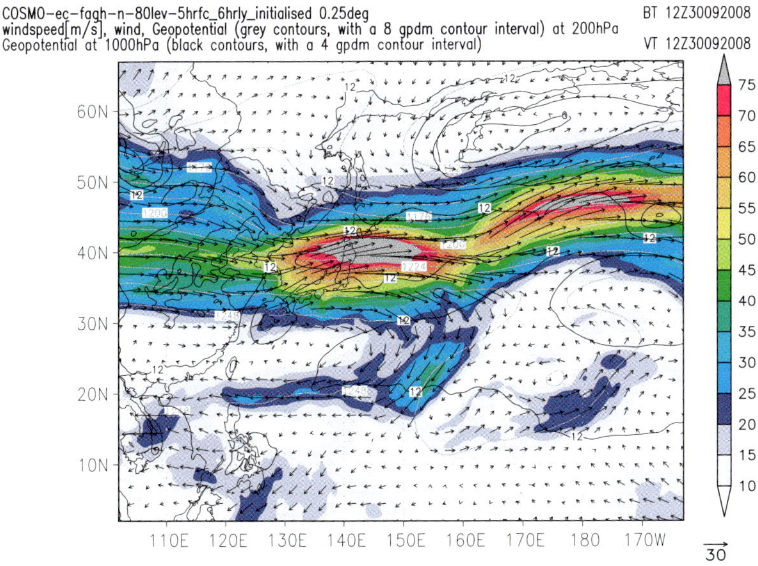

(b)

Fig. 3. Wind vectors (black), wind speed (shaded, with a 5 ms^{-1} contour interval starting at 10 ms^{-1}) at 200 hPa, and geopotential at 1000 hPa (black contours, with a 4 gpdm contour interval), at 00 UTC 29/09/2008 (**a**) and 12 UTC 30/09/2008 (**b**). Data from 0.25° COSMO-IFS pseudo-analysis

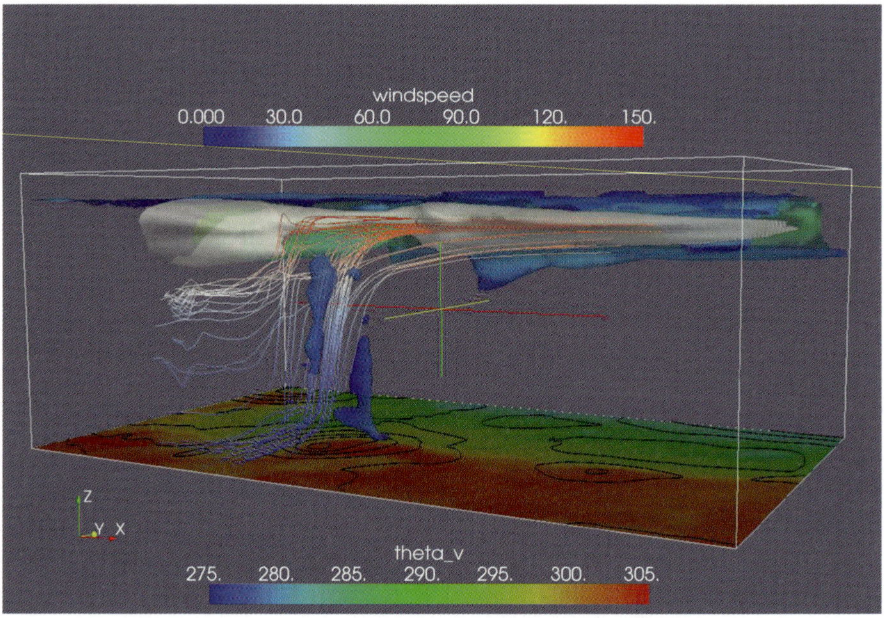

Fig. 4. Contour of potential vorticity at 2 PVU shaded with horizontal windspeed (magUV in ms^{-1}). The transparent grey contour indicates the midlatitude jet (mag(u,v) > 60 ms^{-1}). Virtual potential temperature in K and contours of geopotential (with a 250 gpm contour interval) are shown at 1000 hPa. The lines indicate 36 h backward trajectories calculated from a box (138–140° E, 38–40° N, 9000–13000 m amsl) in the jet core region at 12 UTC 30/09/2008. The criterion for trajectory calculations is "ascent $>$ 5000 m". Blue-red contour shading indicates the height. Every 10th trajectory is plotted. View is from the Southeast. The domain covers 110–160° E, 20–50° N and 1000–50 hPa. Data are taken from 0.25° COSMO-GME forecast initialised at 00 UTC 29/09/2008 and shown at 12 UTC 30/09/2008

a small region at the baroclinic zone which is indicated by a strong gradient of virtual potential temperature at 1000 hPa (at the sharp transition from red to green). At an early stage trajectories at all levels show that parcels are advected northward cyclonically around the PV tower associated with the core of Jangmi (seen as the leftmost trajectories in Fig. 4). At this time vertical motion is weak. Later when the air parcels reached the baroclinic zone strong lifting occurs especially for the lowest parcels (almost vertical trajectories in Fig. 4). This lifting is almost entirely confined to a small region at the baroclinic zone where a new PV anomaly evolves. In the following hours this PV anomaly evolved to a second PV tower in the low and mid troposphere, whose remnants are seen as the rightmost PV tower in Fig. 4. The air parcels continue their ascent and head towards the jet core. At 12 UTC 30/09/2008 (Fig. 4) the new PV tower erodes at mid levels but becomes wider at low levels. All trajectories reached the jet core. Note that there are three

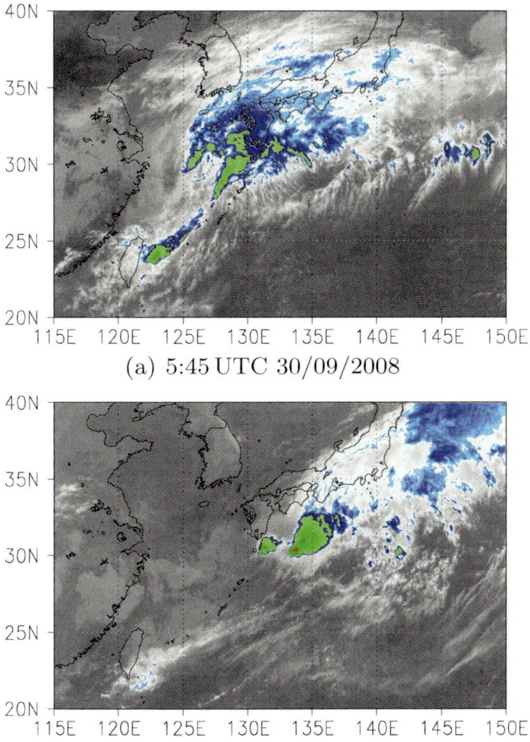

Fig. 5. Sections of MTSAT infrared satellite imagery in the region 20–40° N, 115–150° E at 5:45 UTC (**a**) and at 23:45 UTC (**b**) 30/09/2008. The colours indicate brightness temperature at the cloud tops: White/grey: $T < -30°$ C, blue: $-50°$ C $< T < -30°$ C, green: $-70° < T < -50°$ C, red: $-80°$ C $< T < -70°$ C, yellow: $T < -80°$ C. The remnants of Jangmi are located in the centre of the sections. Satellite data courtesy of Japan Meteorological Agency

distinct paths into the jet core: firstly, the early air reaching jet level is from the mid levels (above 500 hPa) and is lifted in the vicinity of the PV tower associated with the TC core, secondly, neutral ascent in the convection associated with the new PV tower at the baroclinic zone enables lifting of air from mid levels (below 500 hPa) and low levels to the jet level. Thirdly, further air reaches the lower jet level through gradual upgliding along (moist) isentropes. The new PV tower at the baroclinic zone is most likely diabatically forced from strong latent heat release. Satellite imagery indicates new deep clouds at the baroclinic zone from around 12 UTC 30/09/2008 whereas convection in the typhoon region had almost collapsed earlier. This is shown exemplary for 5:30 UTC and 23:30 UTC 30/09/2008 (Fig. 5).

4 Details of the Outflow-Jet Interaction of Typhoon Jangmi

4.1 PV Perspective

Potential vorticity (PV) (Ertel, 1942 [7]) is a key quantity for the investigation of ET. Ertel's potential vorticity q for fully baroclinic, compressible flow is defined as the product of the absolute vorticity vector $\boldsymbol{\eta} = f\mathbf{k} + \nabla \times \mathbf{v}$ and the gradient of potential temperature Θ scaled by the density of the fluid ρ: $q = 1/\rho\ \boldsymbol{\eta} \cdot \nabla\Theta$. PV is conserved following three-dimensional, adiabatic and inviscid motion. Different features involved in the ET process can easily be determined by their characteristic PV structure (Fig. 6) and the conservation principle allows them to be tracked. The core of a tropical cyclone constitutes a strong positive PV anomaly. Above the storm, a region of low or negative PV characterises the outflow region. During ET, latent heat release due to condensation in the warm sector leads to a secondary positive PV anomaly at lower levels. At upper levels a sharp PV gradient with high (stratospheric) values at the poleward side marks the transition to the midlatitudes. This is where the jet stream occurs and deflections result in upper tropospheric troughs (positive PV anomaly) and ridges (negative PV anomaly)—a so-called Rossby wave train (RWT). In this study PV is calculated from the COSMO model fields.

In the following we make use of the model PV fields to get more insight in the advection of tropical air in the jet core. The isentropic PV at $\Theta = 315$ K shows the centre of Jangmi located at around 29° N, 125° E (Fig. 7).

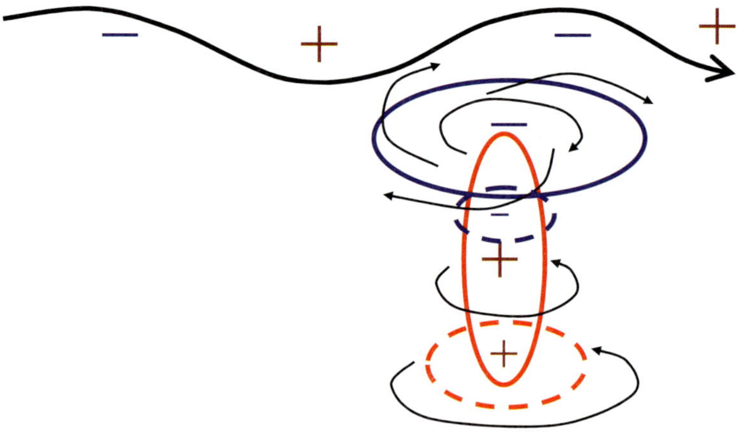

Fig. 6. Schematic sketch illustrating the different PV features involved in the ET process: positive anomalies are shown in red, negative anomalies in blue. The red solid line symbolises the positive PV anomaly associated with the TC core, the blue solid line shows the negative outflow anomaly. Dashed lines indicate anomalies associated with latent heat release. The upper level RWT is shown by a black line

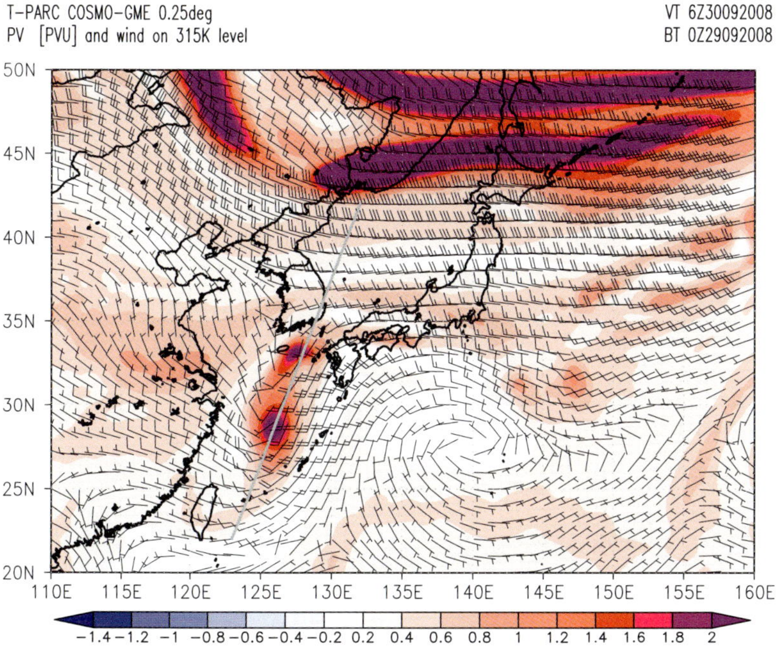

Fig. 7. Potential Vorticity (shaded) [PVU] and wind barbs at $\Theta = 315$ K at 06 UTC 30/09/2008. The grey line indicates the location of the cross sections in Fig. 8. Data from 0.25° COSMO-GME forecast initialised at 00 UTC 29/09/2008. Note the different scaling for positive and negative PV

However, other PV features in the vicinity of Jangmi and the baroclinic zone are striking, namely the PV centre at around 33° N, 127° E which reflects the new PV tower seen in Fig. 4 and to the East a band of high PV at 34° N, 127–136° E where Jangmi advects warm moist air from the South towards the baroclinic zone. This is the region where warm frontogenesis occurs, also seen in the high cloud tops of satellite imagery (Fig. 5b). The new PV tower and the ascent at the baroclinic zone become evident in a cross section through Jangmi's core towards the jet (Fig. 8). The remnants of the TC core can be identified as a broad region of PV > 2 PVU at around 800 km (Fig. 8a). The dense contour lines of equivalent potential temperature Θ_e at 1400 km indicate the baroclinic zone which has only a small poleward tilt up to 500 hPa. The new PV tower evolved directly at the equatorward side of the baroclinic zone at 1350 km from the surface up to 500 hPa. This is also a region where CAPE has a maximum (not shown). In the new PV tower Θ_e is neutrally stratified and thus enables fast pseudo-adiabatic ascent into the jet level as indicated by the wind vectors in the plane of section (see also the trajectories in Fig. 4). Above 500 hPa and north of 1500 km the contour lines of Θ_e (which here almost coincide with dry isentropes, not shown) are directed into the jet core and thus enable (moist) isentropic upgliding directly into the jet core. The

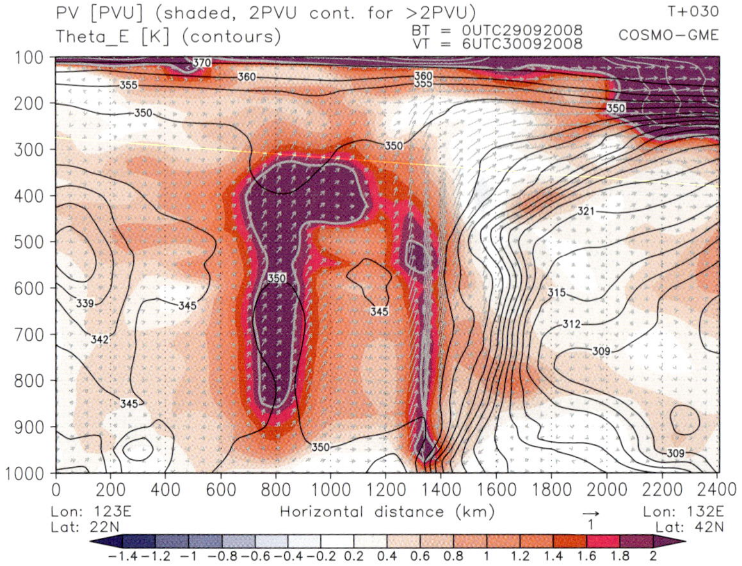

(a)

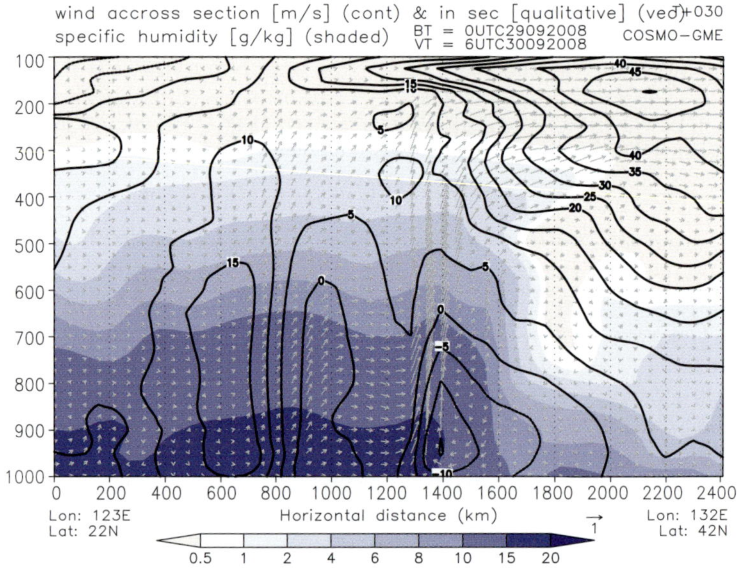

(b)

Fig. 8. Vertical cross sections from 22° N, 123° E to 42° N, 132° E at 06 UTC 30/06/2008. **a** potential vorticity (shaded, with a 0.2 PVU contour interval), equivalent potential temperature (black contours, with a 3 K contour interval), and wind in the plane of section (black arrows). **b** specific humidity (shaded in gkg^{-1}), wind across the plane of section (black contours, with a 5 ms^{-1} contour interval), and wind in the plane of section (black arrows). Data from 0.25° COSMO-GME forecast initialised at 00 UTC 29/09/2008. Note the different scaling for positive and negative PV

latter is located approximately at the step of the dynamic tropopause (PV = 2 PVU) at 2000 km (Fig. 8a, see also the contours of wind speed across the section in Fig. 8b). In the jet level the flow is almost along the isentropes. As isentropes become denser towards the jet core the flow has to accelerate. This may explain the increase in jet core wind speed. The low PV values from 400 to 200 hPa and 1400–1700 km may reflect the negative PV anomaly associated with the outflow. The tongue of moist air in Fig. 8b between 400–300 hPa and 1600–2000 km indicates the advection of tropical air towards the jet core. In summary this more detailed look in the model fields confirms the three distinct paths of lifting into jet level, already seen with the help of trajectory calculations. More details on the lifting into jet level is given in Grams and Jones (2009) [9].

4.2 Quantification of the Impact of Jangmi on the Midlatitude Jet Using PV Surgery

Making use of an independent relation between the wind and temperature field and using the invertibility principle of PV it is possible to calculate the flow and temperature field—the "balanced fields"—associated with the PV field. We use an algorithm which assumes non-linear balance originally implemented by Davis and Emanuel (1991) [4, 5]. The non-linear balance equation was derived by Charney (1955) [3] by taking the horizontal divergence of the horizontal momentum equation, effecting a Helmholtz-Partitioning of the horizontal wind field $\mathbf{v} = \mathbf{v}_\Psi + \mathbf{v}_\Phi$ in a nondivergent part $\mathbf{v}_\Psi = \mathbf{k} \times \nabla\Psi$ associated with the streamfunction Ψ and an irrotational part $\mathbf{v}_\Phi = \nabla\Phi$ associated with the geopotential Φ. Scaling assumptions enable to neglect the irrotational part, yielding to the non-linear balance equation in spherical coordinates:

$$\nabla^2\Phi = \nabla \cdot (f\nabla\Psi) + \frac{2}{a^4 cos^2\varphi}\frac{\partial(\partial\Psi/\partial\lambda, \partial\Psi/\partial\varphi)}{\partial(\lambda,\varphi)}, \tag{1}$$

where a is the Earth radius, and λ, φ the longitude,latitude in geographical coordinates. Replacing the vertical derivative of the total wind by the vertical derivative of the nondivergent wind in the definition of PV in spherical coordinates and using the Exner function

$$\pi = \left[c_p\,(p/p_o)^{\frac{R_d}{c_p}}\right], \Theta = \frac{d\Phi}{d\pi} \tag{2}$$

as vertical coordinates gives a second diagnostic relation of Ψ and Φ:

$$q = \frac{g\kappa\pi}{p}\left[\left(f + \nabla^2\Psi\right)\frac{\partial^2\Phi}{\partial\pi^2} - \frac{1}{a^2 cos^2\varphi}\frac{\partial^2\Psi}{\partial\lambda\partial\pi}\frac{\partial^2\Phi}{\partial\lambda\partial\pi} + \frac{1}{a^2}\frac{\partial^2\Psi}{\partial\varphi\partial\pi}\frac{\partial^2\Phi}{\partial\varphi\partial\pi}\right]. \tag{3}$$

This definition of PV (3) along with the non-linear balance equation (1) and suitable boundary conditions form a complete system for the streamfunction Ψ and geopotential Φ which is solved using a successive over-relaxation algorithm (SOR). At the horizontal boundaries we use von Neumann boundary conditions, at the lateral boundaries Ψ and Φ are prescribed. The PV inversion code can be applied on model data with approximately 20 km and coarser horizontal resolution. The code runs on a workstation computer. For more detail on the PV inversion technique see [5].

The balanced fields are used as initial fields for a new COSMO run. In a first step PV inversion is effected on the original PV field and a new model run—the control run (CNTRL)—is initialised from the balanced fields. In a second step the tropical cyclone is removed from the PV field of the same domain, a PV inversion is effected, and the new balanced fields without the tropical cyclone are used for a new model run, which we call the NOTC run. Extracting the tropical storm at a time before interaction with the midlatitude flow occurs enables to study the impact of the tropical storm on the midlatitude flow. The PV inversion code can be applied on model data with approximately 20 km and coarser horizontal resolution. This method of using the balanced fields of PV inversion as initial data for a new model run is very similar to the method used in the studies on the ET of Hurricane Irene (1999) and Tropical Storm Lili (1996) by Anna Agustí-Panareda et al. (2004, 2005) [1, 2].

Three different methods for removal of a PV anomaly were implemented. Firstly a box in geographical and pressure coordinates is defined around the PV anomaly to be extracted. Secondly at each grid point within the box the PV is set to a distinct critical value, if the PV exceeds/under-cuts that critical value. There are three options to define the critical PV values:

1. the critical PV is given as a constant value for all levels,
2. the critical PV equals the spatial average PV at each level,
3. the critical PV equals the zonal average at each level.

To assure a kind of PV conservation (we are aware that PV is not conserved on pressure levels) the sum of removed PV at each level is distributed equally to all grid points of this level. The modified field serves as input for PVI which results in the balanced fields of PV, Ψ, Θ and thus U, V, T, Θ without the anomaly. Here we present results for a PV inversion at 00 UTC 27/09/2008 and the data is from a COSMO-IFS forecast[4] initialised at 00 UTC 27/09/2008. (Thus the PV inversion is effected on the IFS analyses interpolated on the COSMO grid). The balanced fields from a PV inversion without modification of the model fields gives the initial fields for the CNTRL run. For the NOTC run we only show results using the PV average method. The critical PV value was chosen to be 0.3 PVU and the box for PV modification in the NOTC run covers 122–130° E, 16–23° N, and 900–100 hPa. Further the mois-

[4] For details on the model runs used in this case study refer to Sect. 5.

ture anomaly can be modified before rerunning COSMO. For removal of the moisture anomaly (in the fields of specific humidity q_v, specific cloud ice content q_i, and specific cloud water content q_c) a rectangular box in geographical and model level coordinates is defined around the storm and within that box $q_x; x = i, v, c$ is set to the spatial average of the level multiplied by a factor. Typically the box for removal of the moisture anomaly is larger than the one for removal of the PV anomaly. For the NOTC run shown here the factor was chosen to be 1.4 an the box for moisture modification covers 119–133° E, 14–26° N, and all model levels.

An overview of the PV modification is given in Fig. 9 which shows the low-level PV along with the pressure at mean sea level (pmsl) after 3 hours forecast time. In the CNTRL run (Fig. 9a) Jangmi becomes evident as a strong PV anomaly centred at around 126° E, 20° N and a minimum pressure of less than 978 hPa. Contrary the PV and pressure anomaly of Jangmi is extinct in the NOTC run (Fig. 9b). Note that at this time the large scale low-level PV and pmsl structure in the midlatitudes is hardly modified by the PV surgery (see also Fig. 10, top for the upper tropospheric fields). The impact of the outflow of Jangmi on the midlatitude jet becomes evident when looking at the temporal sequence of the wind field at 200 hPa of the CNTRL and NOTC run (Fig. 10). At 12 UTC 27/09/2008 there are hardly differences in the upper level midlatitude flow of both runs (Fig. 10, top). A trough with its axis over Japan is evident and a jet streak with wind speed up to 65ms^{-1} is located over Korea. 36 hours later in the control run as well as in the COSMO pseudo-analysis (not shown) this jet streak became stationary over Japan (Fig. 10, middle). Maximum wind speed slightly weakened. A first interaction of Jangmi and this jet streak lead to a broader region of high wind speed in the control run compared with the NOTC run. The impact of Jangmi on the midlatitude jet becomes evident at 12 UTC 30/09/2008 when the outflow of Jangmi was directed straight into the jet core (Fig. 10, bottom; see also Figs. 3b, 4). In the CNTRL run the jet core wind speed downstream of Jangmi significantly increased to more than 75 ms^{-1} in a vast region (Fig. 10, bottom left). Further the jet streak is located rather North at around 41° N. A weak ridge evolved downstream of Jangmi. Against in the NOTC run the jet streak is located over Japan at around 38° N and much weaker with wind speed exceeding 60 ms^{-1} only in a small region (Fig. 10, bottom right). Also a rather weak trough is evident over Japan than a ridge West of Japan like in CNTRL. In summary the PV surgery turned out to be very useful to quantify the impact of Jangmi on the midlatitude jet. The refinement of the PV surgery technique is a major topic of our ongoing research.

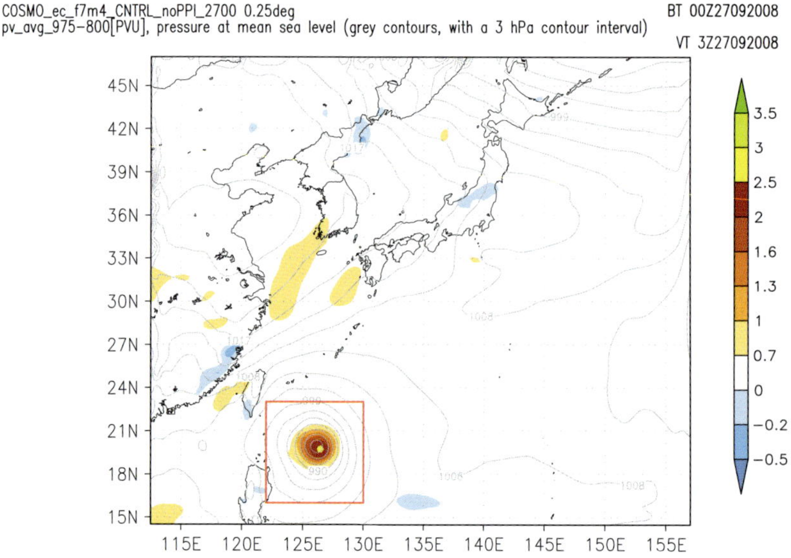

(a) Control run

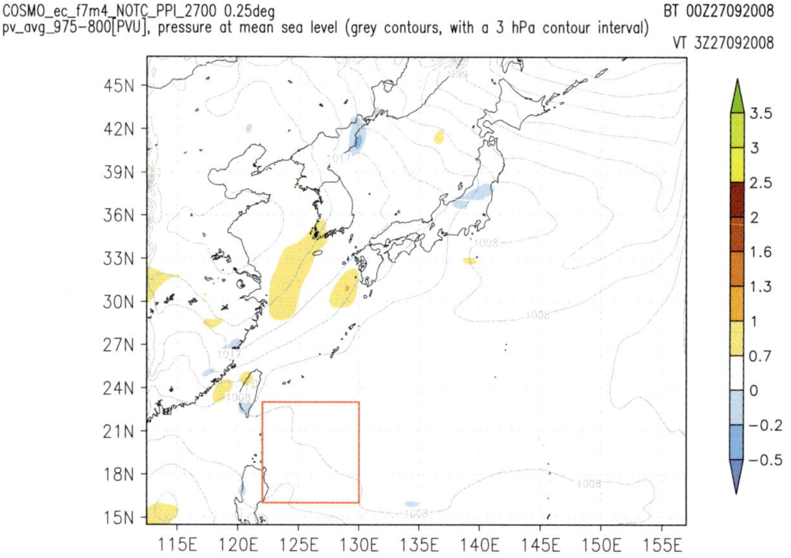

(b) NOTC run

Fig. 9. Potential vorticity (shaded, in PVU) vertically averaged from 975–800 hPa, pressure at mean sea level (grey contours with a 3 hPa contour interval). **a** CNTRL run with non-modified PV field. **b** NOTC run with PV associated with Jangmi extracted in the box, indicated by the red contour. Time shown is 03 UTC 27/09/2008

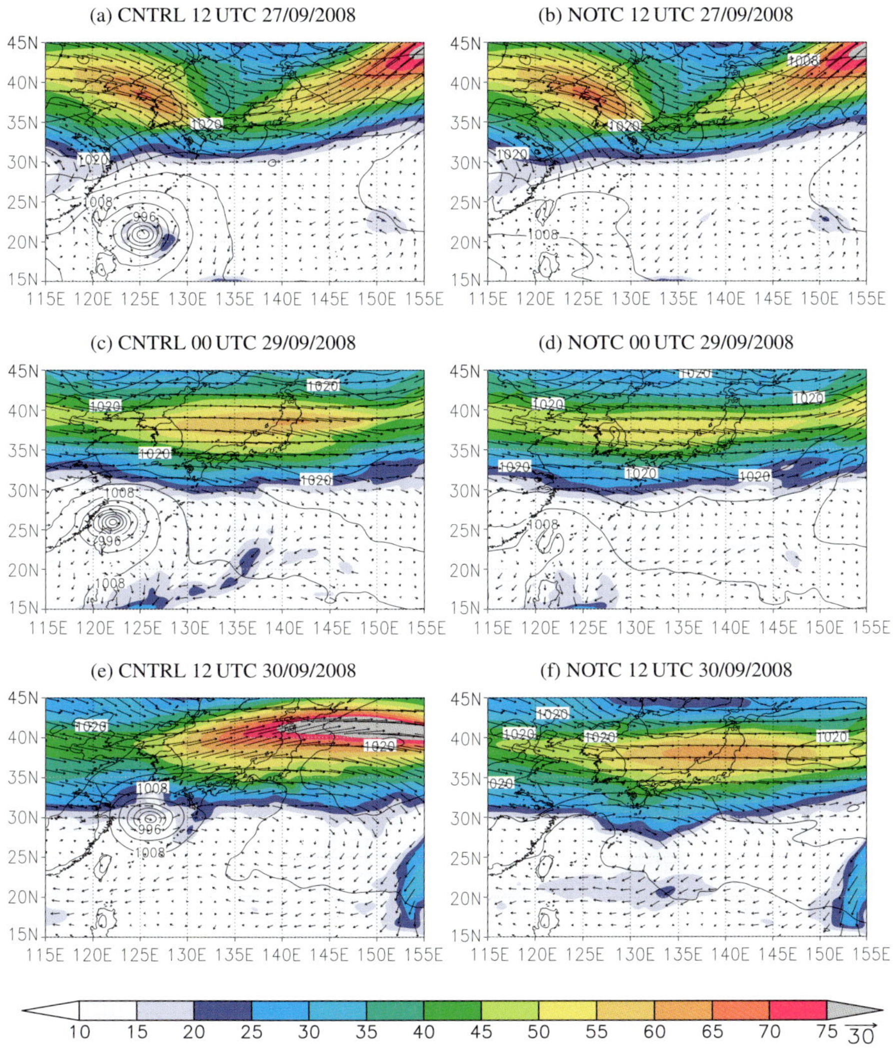

Fig. 10. Wind vectors (black), wind speed (shaded, with a 5 ms^{-1} contour interval starting at 10 ms^{-1}) at 200 hPa, and pressure at mean sea level (black contours, with a 4 hPa contour interval), at 12 UTC 27/09/2008 (top), 00 UTC 29/09/2008 (middle), and 12 UTC 30/09/2008 (bottom). Data in the left column is from CNTRL run, data in the right column is from NOTC run

5 Details of the COSMO Runs Used in the Case Study of Typhoon Jangmi

In this section the different model runs used in this case study are briefly described and information on the performance of XC2 for the different model configurations are summarised in Tables 3 and 4. The use of different model

Table 3. Configuration of COSMO runs used in the case study of Jangmi: name of the model run as in text, initialisation time, forecast range, domain in geographical coordinates, number of grid points in longitudinal, latitudinal, and vertical direction

name	BT	fcst.hrs	dt	domain	nx, ny, nz
COSMO-GME	29/00Z	72	40	110–160° E, 20–50° N	201, 121, 50
COSMO-IFS	27/00Z	90	40	110–160° E, 12–50° N	201, 153, 50
CNTRL	27/00Z	90	80	112.5–157° E, 14.5–47° N	179, 131, 50
NOTC	27/00Z	90	80	112.5–157° E, 14.5–47° N	179, 131, 50
pseudo-an.	27/00Z	6	60	102–197° E, 2–67° N	381, 261, 79

Table 4. Details on computational cost of COSMO runs: name for the model run, number of nodes x number of CPUs per node, CPU time, real time, maximum virtual memory by any process

name	♯ CPU	CPU time	real time	virt. mem
COSMO-GME	$8 \times 4 = 32$	17 h 24 min	33 min	250 MB
COSMO-IFS	$8 \times 4 = 32$	27 h 18 min	51 min	265 MB
CNTRL	$8 \times 4 = 32$	14 h 27 min	27 min	255 MB
NOTC	$8 \times 4 = 32$	12 h 14 min	23 min	255 MB
pseudo-an.	$8 \times 4 = 32$	6 h 25 min	12 min	460 MB

runs represents the historical evolution of the investigation of Jangmi which ends in the creation of a common reference data set on which in a final step the whole case study will be based. The horizontal resolution for all runs used in this study was 0.25°, due to constraints of the non-linear balance condition to be valid. The first model run for the investigation of Typhoon Jangmi was a 72 hour COSMO-GME forecast at 0.25° horizontal resolution initialised at 00 UTC 29/09/2008 with the GME analysis and 3 hourly GME forecasts as lateral boundary conditions. In comparison with IFS analyses the forecast was found to represent the structure of the troposphere in the first 48 forecast hours accurately. This forecast was initially used to look at the advection of tropical moist air into the jet core (Figs. 4, 7, 8). The need of a COSMO forecast for Jangmi before interaction with the midlatitudes took place motivated the calculation of a 90 hour COSMO-IFS forecast at 0.25° horizontal resolution initialised at 00 UTC 27/09/2008 and initial and boundary data was taken from 6 hourly IFS analyses. In comparison with IFS analyses this forecast also showed to be useful to analyse the outflow jet interaction of Jangmi. This forecast was used to provide the initial fields for the PV surgery as well as the lateral boundaries for the CNTRL and NOTC runs (Figs. 9, 10). The CNTRL run was initialised from the output of the PV inversion at 00 UTC 27/09/2008 and 6 hourly lateral boundaries were provided by the COSMO-IFS forecast. The input fields of the PV inversion was the 0 forecast hours fields of the COSMO-IFS run, in other words this is the IFS analyses at 00 UTC 27/09/2008 interpolated on the COSMO grid. The model configuration for the NOTC run was the same as for CNTRL except that the initial fields were modified as described in Sect. 4.2. Finally a reference data set—the COSMO pseudo analyses—were produced which is thought to be the best available model representation of the study period. The COSMO pseudo-analyses are a data set consisting of 6 hourly COSMO runs initialised every 6 hours with 6 hourly IFS operational analyses as initial and boundary data. One hourly model output of the individual 6 hour forecast are combined to cover the period 00 UTC 27/09/2008 to 23 UTC 03/10/2008. For each forecast hour 0 is the interpolation of the IFS analysis on the COSMO grid. Thus every 6 hours we have the real IFS analysis and the hours in between are bridged by 1–5 hour COSMO forecasts. The vertical levels were adjusted to be as close to the IFS vertical levels as possible. Currently the whole case study of Typhoon Jangmi is set on a common model data basis using the COSMO pseudo-analyses. Details on the model configuration and the computational costs of the different COSMO runs are listed in Tables 3 and 4.

6 Summary and Outlook

In this study a series of COSMO runs were performed on the HP XC4000 super computer of the SCC in order to investigate the downstream impact of extratropical transition. The case study of Typhoon Jangmi showed that a tropical cyclone undergoing ET can have a major impact on the structure of the midlatitude jet. Currently the results of this case study are set on a common model data basis. For this purpose new COSMO runs with similar configurations will be performed on HP XC4000. The diagnostics which were developed in the course of this first case study will be applied on a future case study which focuses on an Atlantic system and involves even higher resolved numerical simulations with the COSMO model.

Acknowledgements. This project is supported by the German Research Foundation (DFG) as part of the research unit PANDOWAE (FOR 896). We would like to thank the Steinbuch Centre for Computing in particular Vincent Heuveline, Horst Gernert and Hartmut Häfner for enabling the COSMO forecasts in support of T-PARC. We would also like to thank the Deutscher Wetterdienst (DWD) in particular Ulrich Schättler, Detlev Majewski and Michael Gertz, for support with the COSMO model, providing GME data, and enabling access to ECMWF data. Thanks to Maxi Böttcher and Heini Wernli, who helped with the trajectory calculations. Special thanks to Chris Davis and Micheal Riemer who provided and helped with the implementation of the PV inversion code.

References

1. A. Agustí-Panareda, S.L. Gray, G.C. Craig, and C. Thorncroft. The Extratropical Transition of Tropical Cyclone Lili (1996) and Its Crucial Contribution to a Moderate Extratropical Development. *Monthly Weather Review*, 133:1562–+, 2005.

2. A. Agustí-Panareda, C.D. Thorncroft, G.C. Craig, and S.L. Gray. The extratropical transition of hurricane Irene (1999): A potential-vorticity perspective. *Quarterly Journal of the Royal Meteorological Society*, 130:1047–1074, April 2004.
3. J.G. Charney. The use of primitive equations of motion in numerical prediction. *Tellus*, 7:22–26, 1955.
4. C.A. Davis. Piecewise Potential Vorticity Inversion. *Journal of Atmospheric Sciences*, 49:1397–1411, August 1992.
5. C.A. Davis and K.A. Emanuel. Potential Vorticity Diagnostics of Cyclogenesis. *Monthly Weather Review*, 119:1929–+, 1991.
6. G. Doms and U. Schättler. A description of the nonhydrostatic Regional Model LM. Part I: Dynamics and Numerics. November 2002. available at http://www.cosmo-model.org.
7. H. Ertel. Ein neuer hydrodynamischer Erhaltungssatz. *Meteorologische Zeitschrift*, 59:277–281, 1942.
8. A. Gassmann. Numerische Verfahren in der nichthydrostatischen Modellierung und ihr Einfluss auf die Güte der Niederschlagsvorhersage. *Berichte des Deutschen Wetterdienstes*, 221:1–96, 2002.
9. C.M. Grams and S.C. Jones. The interaction of Typhoon Jangmi (2008) and the midlatitude jet (extended abstract). 9 2009.
10. C.M. Grams, S.C. Jones, and J. Keller. Im Auge des Taifuns—Numerische Wettervorhersage für Forschungsexperiment. *SCC News*, 3:14–17, 2008.
11. B.J. Hoskins and P. Berrisford. A potential vorticity perspective of the storm of 15-16 October 1987. *Weather*, 43:122–129, 1988.
12. S.C. Jones, P.A. Harr, J. Abraham, L.F. Bosart, P.J. Bowyer, J.L. Evans, D.E. Hanley, B.N. Hanstrum, R.E. Hart, F. Lalaurette, M.R. Sinclair, R.K. Smith, and C. Thorncroft. The Extratropical Transition of Tropical Cyclones: Forecast Challenges, Current Understanding, and Future Directions. *Weather and Forecasting*, 18:1052–+, 2003.
13. E. Kessler. On the distribution and continuity of water substance in atmospheric circulation models. *Meteorological Monographs*, 10, 1969.
14. J.B. Klemp and R.B. Wilhelmson. The Simulation of Three-Dimensional Convective Storm Dynamics. *Journal of Atmospheric Sciences*, 35:1070–1096, June 1978.
15. G.L. Mellor and T. Yamada. Development of a turbulence closure model for geophysical fluid problems. *Reviews of Geophysics and Space Physics*, 20:851–875, November 1982.
16. M. Raschendorfer. The new turbulence parametrization of LM. *COSMO Newsletter*, 1:89–97, 2001.
17. B. Ritter and J.F. Geleyn. A comprehensive radiation scheme for numerical weather prediction models with potential applications in climate simulations. *Monthly Weather Review*, 120:303–325, February 1992.
18. J. Steppeler, G. Doms, U. Schättler, H.W. Bitzer, A. Gassmann, U. Damrath, and G. Gregoric. Meso-gamma scale forecasts using the nonhydrostatic model LM. *Meteorology and Atmospheric Physics*, 82:75–96, 2003.
19. U. Schättler, G. Doms, and C. Schraff. A description of the nonhydrostatic regional COSMO-model. Part VII: User's Guide. December 2009. available at http://www.cosmo-model.org.

20. B.H. Wernli and H.C. Davies. A Lagrangian-based analysis of extratropical cyclones. I: The method and some applications. *Quarterly Journal of the Royal Meteorological Society*, 123:467–489, January 1997.
21. H. Wernli. A Lagrangian-based analysis of extratropical cyclones. II: A detailed case-study. *Quarterly Journal of the Royal Meteorological Society*, 123:1677–1706, July 1997.

Global Long-Term MIPAS Data Processing: Some Aspects of the Dynamics of the Atmosphere from Lower Stratosphere to Lower Thermosphere

M. Kiefer[1], B. Funke[2], U. Grabowski[1], and A. Linden[1]

[1] Institut für Meteorologie und Klimaforschung, Karlsruhe Institute of Technology, Karlsruhe, Germany
[2] Instituto de Astrofísica de Andalucía, Granada, Spain

Summary. The Michelson Interferometer for Passive Atmospheric Sounding (MIPAS) is a Fourier transform mid-infrared limb scanning high resolution spectrometer which allows for simultaneous measurements of more than 30 atmospheric trace species related to atmospheric chemistry and global change. MIPAS is operated by ESA since mid of 2002 and the mission will be extended through 2013. At the Institute for Meteorology and Climate Research (IMK), MIPAS spectra are used for retrieval of altitude-resolved profiles of abundances of trace species of the atmosphere. These 4-D trace gas distributions are used for the assessment of e.g. stratospheric ozone chemistry, stratospheric cloud physics and heterogeneous chemistry, tropospheric stratospheric exchange, intercontinental transport of pollutants in the upper troposphere, mesospheric stratospheric exchange, effects of solar proton events on stratospheric chemistry, and climate-chemistry models. Over the last year the XC4000 supercomputer has become a major contributor to the total amount of MIPAS data processed at IMK, and hence has helped a lot in filling the gap which results from ESA's failure to produce altitude-resolved species profiles for the time since March of 2004. Due to the extremely low administration/communication overhead within the processing system, up to 800 processors could be used (and have been many times) in parallel, 50% of the projects used more than 200 processors in parallel. In the last year the processing of MIPAS data on the XC4000 became more focussed on species which are strongly influenced by NLTE (non-local thermodynamic equilibrium). Two corresponding examples of scientific exploitation of MIPAS data are given. 1. The distribution of CO, which essentially behaves like a tracer, from upper troposphere to mesosphere allows to gain insight into several aspects of middle atmosphere dynamics. 2. The evolution of the temperature field from stratosphere to lower thermosphere hints at a dynamic coupling of these altitude regions via planetary wave activity during a major warming event.

W.E. Nagel et al. (eds.), *High Performance Computing in Science and Engineering '10*, DOI 10.1007/978-3-642-15748-6_36,

1 The MIPAS/Envisat Mission

The Michelson Interferometer for Passive Atmospheric Sounding (MIPAS) is a space-borne cooled Fourier transform spectrometer which operates in the mid-infrared spectral region 4.15 to 14.6 μm. The design spectral resolution is 0.035 cm^{-1}. It measures thermal emission spectra of the Earth's limb, whereby variation of the tangent altitude provides altitude-resolved information (Fischer et al. [1, 2]).

MIPAS is part of the core-payload of the Envisat research satellite. Envisat is a sun-synchronous polar orbiter which passes the equator at 10:00 am local time in southward direction 14.4 times a day. The MIPAS mission intends to improve the understanding of the composition of the Earth's atmosphere by measurement of 4D distributions of more than 30 trace species relevant to atmospheric chemistry and climate change. Operation of satellite and instrument and level-1b (L1b) data processing (Fourier transformation, phase correction, calibration) is done by the European Space Agency (ESA). MIPAS was operational in its original, high spectral resolution specification from June 2002 to 26 March 2004 and provided about 1000 limb scans a day, each consisting of 17 times 60000 spectral radiance measurements. In summary 4 TB of data are available for scientific analysis from this period. Due to an instrument problem the measurements could not be resumed before the beginning of 2005. The spectral resolution had to be reduced to about 40% of the initial value since then. After recovery the duty cycle was initially lowered to approximately 50%, however, due to the very good instrument conditions, meanwhile the full 100% duty cycle is reached again.

Up to March 2004 level-2 (L2) data processing (distributions of temperature, H_2O, O_3, CH_4, N_2O, HNO_3, and NO_2) was done operationally under responsibility of the European Space Agency. Retrieval of further species ($ClONO_2$, ClO, HOCl, (H)CFCs, NO, N_2O_5, CO, C_2H_6 and many others) was left to scientific institutions. L2 data for dates after March 2004 was up to now performed exclusively by several scientific institutions. This is due to severe difficulties on ESA's side to adapt their processing software to the reduced spectral resolution of MIPAS.

2 Data Analysis

Retrieval of atmospheric constituents abundances requires the inverse solution of the radiative transfer equation. Measured radiances are compared to radiative transfer calculations and residuals are minimized in a least squares sense by adjustment of the constituents abundances which are fed into a forward model. The Retrieval Control Program (RCP), together with the forward model, is the core of the processing chain.

The forward model, the Karlsruhe Optimized and Precise Radiative Transfer Algorithm (KOPRA) (Stiller [15]) is a computationally optimized line-by-line model which simulates radiative transfer through the Earth's atmosphere under consideration of all relevant physics: the spectral transitions of all involved molecules, atmospheric refraction, line-coupling and non-local thermodynamic equilibrium (NLTE). Along with the spectral radiances, KOPRA also provides the Jacobian matrices, i.e. the sensitivities of the spectral radiances to changes in the atmospheric state parameters.

Atmospheric state parameters are retrieved by constrained multi-parameter non-linear least-squares fitting of simulated and measured spectral radiances. The inversion is regularized with a Tikhonov-type constraint which minimizes the first order finite differences of adjacent profile values (von Clarmann et al. [16]).

Instead of simultaneous retrieval of all target parameters from a limb sequence, the retrieval is decomposed in a sequence of analysis steps. First, spectral shift is corrected, then the instrument pointing is corrected along with a retrieval of temperature, using CO_2 transitions. Then the dominant emitters in the infrared spectrum are analyzed one after the other (H_2O, O_3, HNO_3...), each in a dedicated spectral region where the spectrum contains maximum information on the target species but least interference by non-target species. Finally, the minor contributors are analyzed, whereby pre-determined information on the major contributors is used. The analysis is done limb sequence by limb sequence, i.e. limb scans are processed independently from each other by separate calls of the RCP. Typically some ten thousands to hundred thousand of limb sequences are processed for several species in one project.

3 Computational Considerations

3.1 Processing System Overview

The retrieval of atmospheric trace constituents is performed for each geolocation separately. A geolocation is determined by the mean values of the geographic coordinates and of the times of the spectral radiance measurements of the corresponding limb scans. ESA delivers L1b data as binary files with one file containing spectra plus additional information for all geolocations of one Envisat orbit.

The processing steps consist of:

- preprocessing with 8 single steps performed by 8 different programs, always performed at IMK;
- transfer of preprocessed data from IMK to the XC4000
- retrieval (core processing step) with the Retrieval Control Program (RCP) on the XC4000 supercomputer;
- postprocessing on the XC4000 to generate some elementary result diagnostics and corresponding plots.
- transfer of result and diagnostics data from XC4000 to IMK.

A detailed description of the above steps, as well as of the adaptation of the retrieval software to the XC4000 supercomputers, already has been given in Kiefer et al. [9]. A new concept for the MIPAS data processing was developed together with W. Augustin of the scientific supercomputing group of the Karlsruhe University's computing center, and subsequently implemented in the first half of 2008. Details of this have been given in the last year's report.

3.2 Throughput

The increased performance of the processing system since mid 2008 has made the XC4000 a major contributor to the total amount of data processed at IMK during the year 2008 and the beginning of 2009. However, the computing capabilities at IMK have considerably increased since then and as a consequence the focus of data processing on the XC4000 shifted towards the species, which require the most computing time per geolocation, namely those which require costly non-local thermodynamic equilibrium (NLTE) calculations during the retrieval.

In the last year's report there was given a throughput of 6–7 species per full data day results. This means that, during that reporting period, processing on the XC4000 would have allowed to keep pace with the data delivery of MIPAS if only 6–7 species were to be processed.

For the current reporting period this number has decreased to 4–5 species per full data day results. This decrease is due to the shift of focus towards NLTE-species, which are computationally expensive.

3.3 Processor Usage

As already said, the retrieval of one species at one geolocation is a task completely independent from retrievals at other geolocations. Hence there is no overhead like e.g. message or data passing between the parallel threads of a retrieval job. This allows to use as many XC4000 processors as available for MIPAS data processing. Figure 1 shows the corresponding data: The abscissa is the number of simultaneously used processors in steps of 16, because currently 16 processors are required per job. The ordinates give the number of projects which achieved this processor usage (upper plot) and the accumulated number of projects (lower plot).

A total of 270 projects were processed during 22 months. Peak numbers of 800 simultaneously used processors could be achieved for 24 projects. 70 projects used 400 processors or more. Half of the projects used more than 160 processors simultaneously. Projects with just 16 processors (the minimum number to be required) mainly consists of small test jobs, which were performed to check the setup, e.g. in case an atmospheric constituent has not been processed on the XC4000 before.

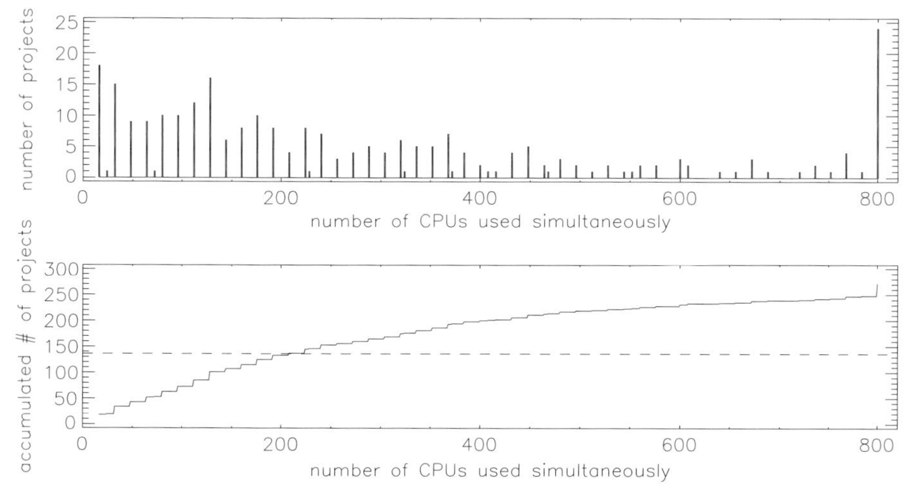

Fig. 1. Frequency (number of projects, top) and accumulated frequency (bottom) of peak numbers of simultaneously used processors. The dashed line marks the value of half the number of projects

4 Examples of Scientific Work with MIPAS Data

The European Space Agency (ESA), which runs the MIPAS instrument, has never done operational level 2 processing for reduced resolution data, due to enduring problems with the implementation of the necessary changes in their retrieval processor. Meanwhile for the greater part of the MIPAS mission, namely the time from March 2004 until now, therefore there is no ESA level 2 product. To compensate at least partly for this, IMK has concentrated mainly on processing of data of this time range.

There is a wealth of topics which can benefit from MIPAS data, as e.g. analyses of age of air and Brewer-Dobson circulation; validation of chemistry-climate models; influence of solar proton events on the upper atmosphere; stratospheric water vapour budget and its variability; the transport of water vapour and other species through the tropical tropopause layer; biomass burning, upper tropospheric pollution, and its impact on the upper tropospheric ozone budget. This has partly been demonstrated in the reports of the last years. In this report we shall focus on the scientific exploitation of the new NLTE data set which has been performed in close collaboration with colleagues of the Instituto de Astrofísica de Andalucía (Granada, Spain).

4.1 Carbon Monoxide Distributions from the Upper Troposphere to the Mesosphere

We have derived vertically resolved global CO distributions from the upper troposphere to the mesosphere which have been derived from MIPAS limb emission measurements at 4.7 μm taken during September 2003 to March 2004. These measurements, while offering the advantage of dense spatial sampling independent on illumination, are strongly affected by spatial inhomogeneities related to the high variability of CO abundances, NLTE-effects, and spectral saturation, making the retrieval of CO mixing ratios challenging. Our retrieval approach, included in the IMK/IAA retrieval processor, addresses these issues successfully by means of several adjustments with respect to the standard algorithm. The derived data (version V3O_CO_9+10) provides thus accurate CO vmr distributions for any atmospheric conditions from approximately 8 to 70 km with a typical vertical resolution of 4–7 km. The estimated total retrieval error for a single limb scan is 5–30 ppbv (15–40% for altitudes greater than 40 km and lower than 15 km and 30–90% within 15–40 km).

The temporal evolution of zonally averaged CO abundances (see Fig. 2) during the observation period provides a detailed picture of the meridional circulation in the middle atmosphere, particularly of the polar winter descent. We have observed very efficient descent in the mesospheric and upper stratospheric NH polar vortex in January 2004 with vertical velocities of about 1.2 km per day. This finding will help to shed some light on the ongoing discussion about the origin of the extra-ordinary NO_x enhancements observed in the same region and period (Randall et al. [13], Rinsland et al. [14], Hauchecorne et al. [8], Funke et al. [4]).

The scientific value of MIPAS CO observations for dynamical studies in the stratosphere and mesosphere, as well as in the upper troposphere and lower stratosphere (UTLS) has been further demonstrated by dedicated case studies. Quasi-synoptic stratospheric and mesospheric CO distributions observed during the NH major warming episode during December 2003 to January 2004 have provided new information on the polar vortex dynamics under perturbed conditions, particularly with respect to vertical transport processes. An example of isentropic CO distributions on several days during this warming event is given in Fig. 3. We have observed accelerated descent inside the middle stratospheric vortex in the first phase of the warming event, when the vortex distortion was most pronounced in the USM. In the second phase, very efficient descent took place in the recovered vortex in the USM, while the vortex disruption had propagated down to the middle stratosphere. Our findings support that dynamical perturbations of the polar winter stratosphere reduce the efficiency of mesospheric tracer descent in a first instance, but often provoke accelerated descent as an indirect effect which may even overcompensate the direct implications.

The analysis of MIPAS CO distributions in the UTLS benefit from the combination of dense spatial sampling and high accuracy, enabling the representation of global distributions on a close-to-daily basis. Our case study, focussed on two 3-day periods (see Fig. 4) during the SH biomass burning season in September and October 2003 confirms the general picture of UTLS dynamics obtained from previous studies with respect to transport patterns of polluted air masses and uplift mechanisms. In particular, we have observed

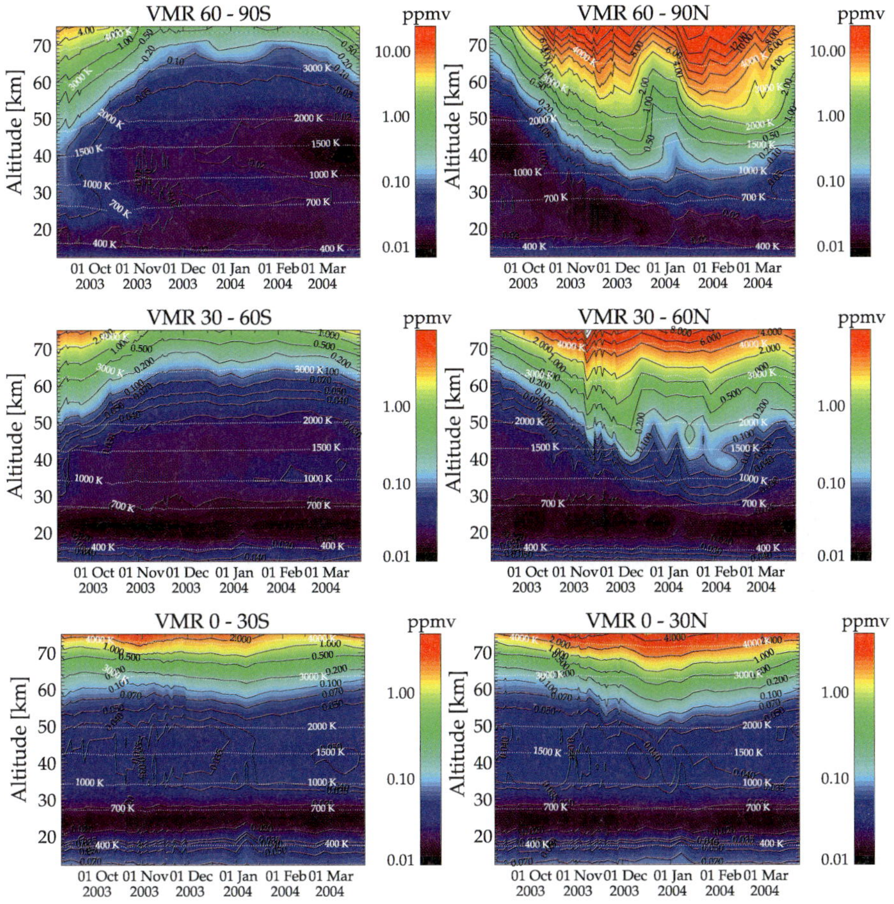

Fig. 2. Temporal evolution of retrieved CO vmrs averaged over equivalent latitudes 60–90° S, 60–90° N, 30°–60° S, 30°–60° N, 0°–30° S, and 0°–30° N (upper left to lower right panels) from September 2003 until March 2004 (note the logarithmic color scale). An area-weighting factor (cosine of equivalent latitude) has been applied. Mean potential temperatures are indicated by dotted white lines. Note that the apparent higher variability during October/November 2003 is related to a finer temporal sampling in this period. Reproduced from Funke et al. [5]

the trapping of polluted air masses in the Asian monsoon anticyclone during the Asian monsoon season which extends well up to the tropical tropopause and possibly above. We have demonstrated by means of backward trajectory calculations that the major fraction of upper tropospheric CO plumes observed by MIPAS during 9–11 September, predominantly located in the NH, was related to Southeast Asian pollution. Transpacific transport of polluted air masses was responsible for enhanced CO abundances observed over the west coast of North America. During 20–22 October, in contrast, SH biomass

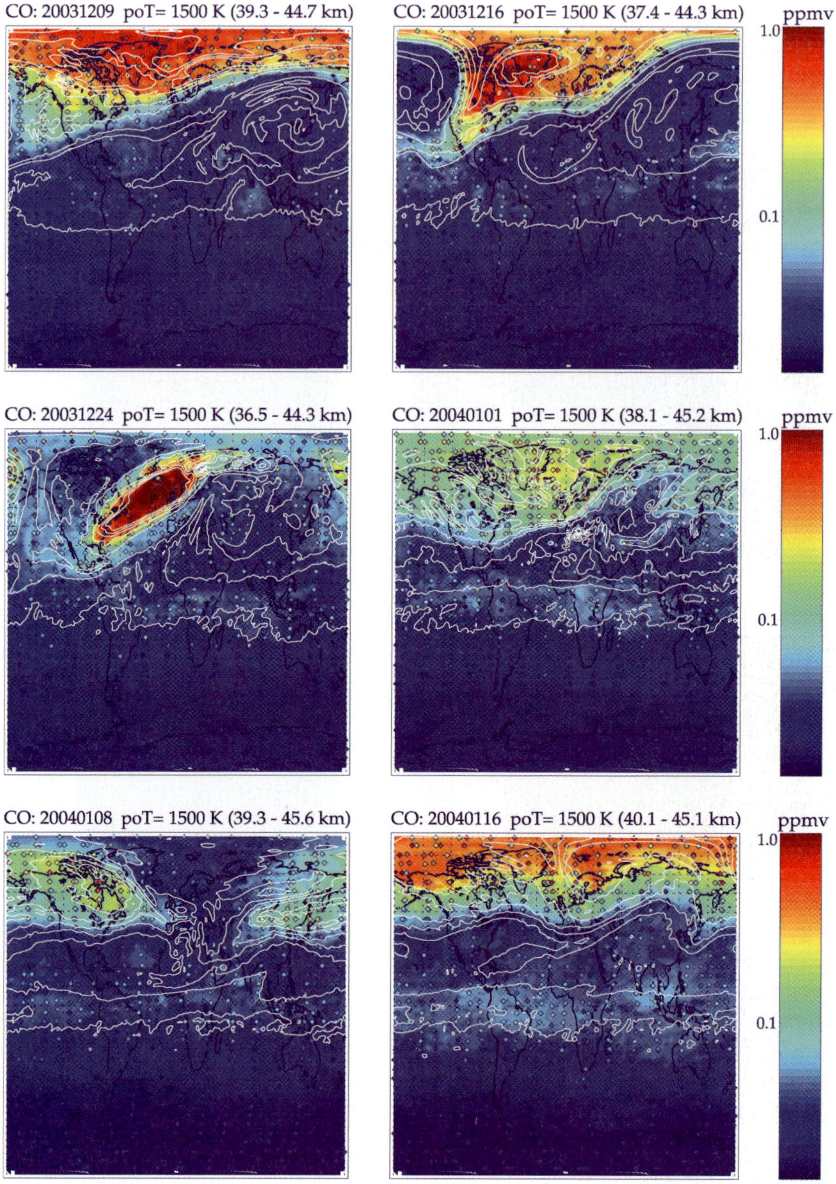

Fig. 3. Observed CO abundances at the 2500 K isentropic surface on several days in December 2003 and January 2004, covering the NH major warming (from top left to bottom right: 9 December, 16 December, 24 December, 1 January, 8 January, and 16 January). Individual MIPAS observations are shown by colored diamonds. The underlaid CO fields are constructed by averaging all measurements within $\pm 10^{\circ}$ latitude/$\pm 20^{\circ}$ longitude around each grid point, weighted by the PV-difference of the measurement's location and the actual grid point. Note the logarithmic color scale. Representative PV contours are shown as white lines. Reproduced from Funke et al. [5]

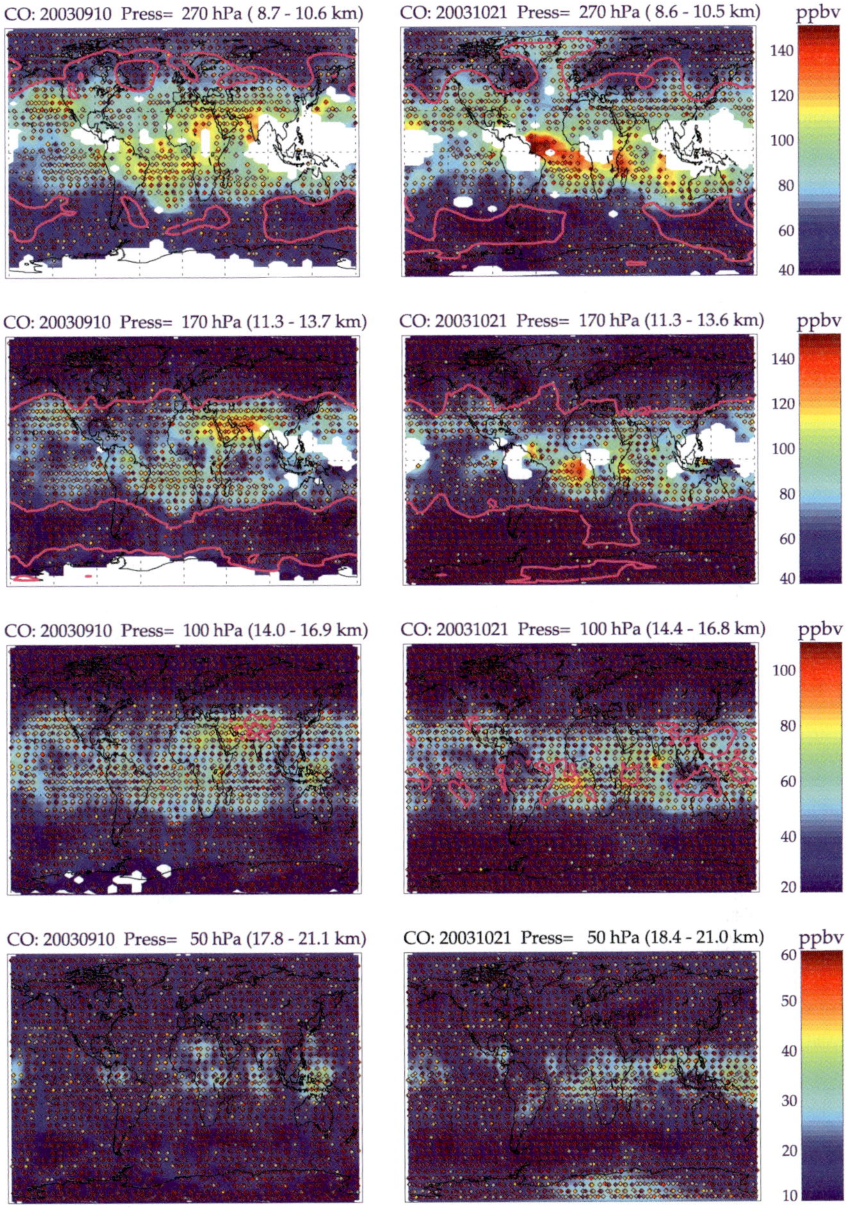

Fig. 4. Averaged CO fields constructed from observations by distance-weighted averaging of all measurements within ±10° latitude/±20° longitude around each grid point at pressure levels 100, 170, and 270 hPa (top to bottom) on 9–11 September 2003 between 45° S and 45° N. Areas with OLR <220 W m^{-2} indicating deep convection are enclosed by purple contour lines. The vector wind fields at the corresponding pressure levels are also shown. OLR and wind field data are taken from the NCEP reanalysis

burning was the most important source of the major CO plumes observed over the Southern Atlantic and Indian Ocean. Polluted air masses were lofted up to the tropical tropopause layer most efficiently over Central African convective areas.

The high-lightened results obtained from approximately half a year of available V3O_CO_9+10 data open a promising perspective for future scientific studies. Data availability is currently being increasing, though progress is constrained by the high demands on computational time required by the inversion of IR limb emission measurements under NLTE conditions. Despite of this restriction, the coverage of the "high resolution" period 2002–2004, as well as the extension to "reduced resolution" measurement obtained since 2005 is envisaged in the near future, being particularly important for the analysis of the inter-annual dynamical variability in the lower and middle atmosphere.

4.2 Temperature Distributions from the Stratosphere to the Lower Thermosphere During a Major Warming Event

We used temperature data from MIPAS to analyze the temperature responses in the mesosphere and thermosphere up to 170 km to a major stratospheric sudden warming (SSW) which occurred in January 2009 (Funke et al. [3]). Temperature profiles are inferred from MIPAS calibrated and geolocated limb emission spectra (level 1-B version 4.67 data) with the IMK/IAA (Institut für Meteorologie und Klimaforschung/Instituto de Astrofísica de Andalucía) scientific data processor in a two-step approach. First, temperature and pointing information is retrieved from the 12–15 μm spectral region (von Clarmann et al. [17]) up to 100 km under consideration of NLTE effects by inclusion of the Generic RAdiative traNsfer AnD non-LTE population Algorithm (GRANADA) model (Funke et al. [7]). In a second step, spectrally resolved thermospheric 5.3 μm NO emissions are used to retrieve simultaneously NO abundances and kinetic temperature from 100 to 170 km. Independent temperature information is obtained from the rotational and spin-orbit structure of the NO fundamental band (Funke et al. [6]) taking into account vibrational, rotational, and spin-orbit NLTE populations which are calculated by GRANADA. Single measurement precision and vertical resolution of the temperatures retrieved at 12–15 μm (40–100 km) range from 0.5–7 K and 3–10 km, respectively, increasing with altitude. Systematic errors, mainly due to NLTE-related uncertainties are below 5 K. In the thermosphere, precision and vertical resolution of the temperatures retrieved at 5.3 μm (100–170 km) range from 5–60 K and 6–20 km, respectively. Systematic errors there, particularly due to uncertain oxygen concentrations, have been assessed by means of retrieval simulations and were found to be small compared to the noise error.

The temperature observations (see Fig. 5) show clear signatures of a mesospheric cooling, providing a further confirmation of the classic mesospheric cooling response to SSWs (Labitzke [11]). We have also observed a thermospheric warming peaking at 120–140 km in concert with the stratospheric

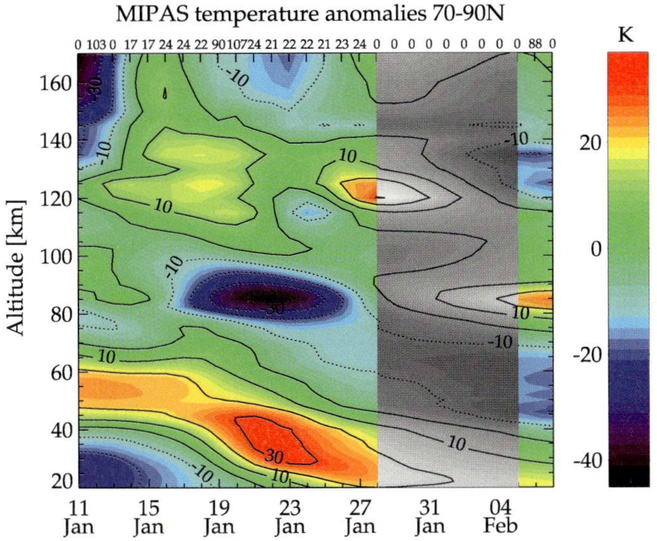

Fig. 5. Temporal evolution of zonally averaged MIPAS temperature anomalies (with respect to the temporal average) within 70°–90° N. During the MIPAS data gap between 28 January and 5 February (grey-shaded) temperature anomalies are interpolated. Note that data below approximately 37 km reflect ECMWF reanalysis temperature anomalies. Numbers at the top x-axis reflect the number of single observations included in the daily averages

warming pulse. The thermospheric temperature increases observed by MIPAS are also consistent with the nighttime ion temperatures observed by the EISCAT UHF radar during the same period (Kurihara et al. [10]). Our observations provides an experimental confirmation of a thermospheric response to SSWs predicted by TIME-GCM simulations (Liu and Roble [12]) From the analysis of the zonal temperature structure during the SSW a pronounced wave 1 pattern was found in the entire middle and upper atmosphere with maximum amplitudes around 50 and 140 km. In the mesosphere, the wave amplitude is significantly damped. The wave amplification above is most likely produced by in situ forced planetary waves in the mesosphere and lower thermosphere (MLT) region. Our observations represent the first experimental evidence of a dynamical coupling from the lower atmosphere up to the thermosphere by means of satellite data.

5 Conclusions and Outlook

For the reporting period the XC4000 supercomputer again has proven to be a highly valuable tool for MIPAS data processing as we have demonstrated in this report for two exemplary applications of data of CO and temperature, which were retrieved using computationally expensive NLTE calculations.

For the future there a two aspects to be taken into consideration: First, operation of MIPAS will be continued until 2013, and therefore there is a need to do further processing. Second, the whole mission L1b data from 2002 through early 2010 is currently reprocessed by ESA with the most recent L1b-processor software. This data version, besides removing known errors, contains several significant improvements which makes it highly desirable to have the species retrieval done with that L1b versions. Since it is our aim to have the best atmospheric data available, IMK will submit a follow-on proposal to be able to benefit from the computing power of the XC4000 to tackle the task of MIPAS L2 data re-processing.

Acknowledgements. We want to express our sincere thanks to W. Augustin of the scientific supercomputing group of the Karlsruhe University's computing center, who has provided the core batch management program and supported the development of the new processing system at IMK.

References

1. Fischer, H., Birk, M., Blom, C., Carli, B., Carlotti, M., von Clarmann, T., Delbouille, L., Dudhia, A., Ehhalt, D., Endemann, M., Flaud, J.M., Gessner, R., Kleinert, A., Koopmann, R., Langen, J., López-Puertas, M., Mosner, P., Nett, H., Oelhaf, H., Perron, G., Remedios, J., Ridolfi, M., Stiller, G., and Zander, R. (2008). MIPAS: an instrument for atmospheric and climate research. *Atmos. Chem. Phys.*, 8:2151–2188.
2. Fischer, H., Blom, C., Oelhaf, H., Carli, B., Carlotti, M., Delbouille, L., Ehhalt, D., Flaud, J.-M., Isaksen, I., López-Puertas, M., McElroy, C.T., and Zander, R. (2000). *Envisat-MIPAS, an instrument for atmospheric chemistry and climate research.* European Space Agency-Report SP-1229, C. Readings and R. A. Harris (eds.), ESA Publications Division, ESTEC, P. O. Box 299, 2200 AG Noordwijk, The Netherlands.
3. Funke, B., López-Puertas, M., Bermejo-Pantaleón, D., García-Comas, M., Stiller, G.P., von Clarmann, T., Kiefer, M., and Linden, A. (2010). Experimental evidence for dynamical coupling from the lower atmosphere to the thermosphere during a major stratospheric warming. *Geophys. Res. Lett.* submitted.
4. Funke, B., López-Puertas, M., Fischer, H., Stiller, G.P., von Clarmann, T., Wetzel, G., Carli, B., and Belotti, C. (2007). Comment on 'Origin of the January-April 2004 increase in stratospheric NO_2 observed in northern polar latitudes' by J.-B. Renard et al. *Geophys. Res. Lett.*, 34(7).
5. Funke, B., López-Puertas, M., García-Comas, M., Stiller, G.P., von Clarmann, T., Höpfner, M., Glatthor, N., Grabowski, U., Kellmann, S., and Linden, A. (2009). Carbon monoxide distributions from the upper troposphere to the mesosphere inferred from 4.7 μm non-local thermal equilibrium emissions measured by MIPAS on Envisat. *Atmos. Chem. Phys.*, 9(7):2387–2411.
6. Funke, B., López-Puertas, M., Stiller, G.P., von Clarmann, T., and Höpfner, M. (2001). A new non–LTE retrieval method for atmospheric parameters from MIPAS–ENVISAT emission spectra. *Adv. Space Res.*, 27(6–7):1099–1104.

7. Funke, B., Martín-Torres, F.J., López-Puertas, M., Höpfner, M., Hase, F., López-Valverde, M.Á., and García-Comas, M. (2002). A generic non–LTE population model for MIPAS–ENVISAT data analysis. In *Geophys. Res. Abstracts*, volume 4. Abstracts of the Contributions of the European Geophysical Society, Nice, France, 21–26 April 2002, CD-ROM, ISSN:1029–7006.
8. Hauchecorne, A., Bertaux, J.-L., Dalaudier, F., Russell III, J.M., Mlynczak, M.G., Kyrölä, E., and Fussen, D. (2007). Large increase of NO_2 in the north polar mesosphere in January-February 2004: Evidence of a dynamical origin from GOMOS/ENVISAT and SABER/TIMED data. *Geophys. Res. Lett.*, 34.
9. Kiefer, M., Grabowski, U., and Fischer, H. (2008). Global Long-Term MIPAS Processing. In Nagel, W.E., Kröner, D., and Resch, M., editors, *High Performance Computing in Science and Engineering '07, Transactions of the High Performance Computing Center, Stuttgart (HLRS) 2007*, pages 519–532. Höchstleistungsrechenzentrum Stuttgart, Springer, Berlin Heidelberg.
10. Kurihara, J., Oyama, S., Nozawa, S., Fujii, R., Tsutsumi, M., Ogawa, Y., and Tomikawa, Y. (2010). Impacts of a stratospheric sudden warming on thermal structures in the high-latitude mesosphere, lower thermosphere, and ionosphere: 2. ground-based observations. *Geophys. Res. Lett.* to be submitted.
11. Labitzke, K. (1981). Stratospheric-mesospheric midwinter disturbances: A summary of characteristics. *J. Geopys. Res.*, 86:9665–9678.
12. Liu, H.-L. and Roble, R.G. (2002). A study of a self-generated stratospheric sudden warming and its mesospheric–lower thermospheric impacts using the coupled TIME-GCM/CCM3. *J. Geophys. Res.*, 107(D23):4695.
13. Randall, C.E., Harvey, V.L., Manney, G.L., Orsolini, Y.J., Codrescu, M., Sioris, C., Brohede, S., Haley, C.S., Gordley, L.L., Zawodny, J.M., and Russell III, J.M. (2005). Stratospheric effects of energetic particle precipitation in 2003–2004. *Geophys. Res. Lett.*, 32.
14. Rinsland, C.P., Boone, C., Nassar, R., Walker, K., Bernath, P., McConnell, J.C., and Chiou, L. (2005). Atmospheric Chemistry Experiment (ACE) Arctic stratospheric measurements of NO_x during February and March 2004: Impact of intense solar flares. *Geophys. Res. Lett.*, 32.
15. Stiller, G.P., editor (2000). *The Karlsruhe Optimized and Precise Radiative Transfer Algorithm (KOPRA)*, volume FZKA 6487 of *Wissenschaftliche Berichte*. Forschungszentrum Karlsruhe.
16. von Clarmann, T., Glatthor, N., Grabowski, U., Höpfner, M., Kellmann, S., Kiefer, M., Linden, A., Mengistu Tsidu, G., Milz, M., Steck, T., Stiller, G.P., Wang, D.Y., Fischer, H., Funke, B., Gil-López, S., and López-Puertas, M. (2003a). Retrieval of temperature and tangent altitude pointing from limb emission spectra recorded from space by the Michelson Interferometer for Passive Atmospheric Sounding (MIPAS). *J. Geophys. Res.*, 108(D23).
17. von Clarmann, T., Glatthor, N., Grabowski, U., Höpfner, M., Kellmann, S., Kiefer, M., Linden, A., Mengistu Tsidu, G., Milz, M., Steck, T., Stiller, G.P., Wang, D.Y., Fischer, H., Funke, B., Gil-López, S., and López-Puertas, M. (2003b). Retrieval of temperature and tangent altitude pointing from limb emission spectra recorded from space by the Michelson Interferometer for Passive Atmospheric Sounding (MIPAS). *J. Geophys. Res.*, 108(D23).

Miscellaneous Topics

Univ.-Prof. Dr.-Ing. Wolfgang Schröder

Institute of Aerodynamics, RWTH Aachen University, Wüllnerstr. 5a,
52062 Aachen, Germany, office@aia.rwth-aachen.de

In this chapter, contributions on structure mechanics, computational engineering, molecular modeling, and geophysics complete the research fields such as physics, solid state physics, reacting flows, computational fluid dynamics, and transport and climate which have been tackled before. The articles widen the field where numerical simulations represent useful tools to gain novel results and to improve the scientific knowledge. The findings evidence the close link between natural science and computer science. Furthermore, these results provide a sound basis to develop novel scientific models. Compact mathematical descriptions will be solved by highly sophisticated and efficient algorithms using high-performance computers. This interdisciplinary collaboration between several scientific fields defines the extremely intricate numerical challenges and as such drives the progress in fundamental and applied research. The subsequent articles are a survey over the projects being linked with HLRS. The computations are used to obtain some quantitative results and to corroborate physical models and to even derive new theoretical approaches. However, to substantiate numerical simulations experimental investigations and analytical solutions are a must. That is, there has to be a close link between analytical, numerical, and experimental findings.

The Institut für Mechanik of the Karlsruhe Institut für Technologie analyzes in the first article the implosion of buildings which is particularly efficient for very high and large buildings. Controlled demolitions have been investigated by the DFG Forschergruppe 500 "Computergestützte Destruktion komplexer Tragwerke durch Sprengung". The contribution describes the simulation of building implosions with finite elements. These simulations serve in principle as verification for simplified models.

The Institut für Anthropomatik of the Karlsruhe Institut für Technologie participated in the second Quaero evaluation for automatic speech recognition (ASR) and Machine Translation (MT). State-of-the-art ASR research systems usually employ techniques which require the parallel execution of several recognition systems for the purpose of system combination. The use of unsupervised adaptation techniques further requires the execution of several stages or passes of ASR systems. This leads to the fact that modern research systems process speech only in a multitude of real-time, under certain circumstances up to 100 times real-time. The process of speech recognition in this form can be easily parallelized at a speaker level in independent processes without the need for interprocess communication. Therefore, the scheduling system of the HP XC4000 in combination with its global, high performing file space is an ideal environment for executing such an evaluation.

In the project on molecular modeling and simulation of thermodynamic properties of hydrogen bonding fluids the Lehrstuhl für Thermodynamik und Energietechnik of the Universität Paderborn and the Lehrstuhl für Thermodynamik of the Technische Universität Kaiserslautern have developed numerous molecular models. Starting from quantum mechanical calculations of a single molecule in the vacuum, the geometry is determined via the Hartree-Fock approach. On top of this geometry optimization, also the electrostatics of the molecule can be reasonably determined by quantum mechanics. The most attractive capability of molecular models is their transferability. This allows obtaining information on other types of properties or how the molecules behave in other environments like in mixtures. In the third contribution, the two hydrogen bonding molecules ammonia and cyclohexanol will be investigated in this sense. A fully predictive application of the ammonia model to obtain the transport properties self-diffusion coefficient and shear viscosity over a range of temperature and up to very high pressures will be discussed. Furthermore, an application of three molecular models for carbon dioxide, cyclohexane, and cyclohexanol is given to describe and predict the phase behavior of the mixtures.

Thermohaline-driven flow, i.e., the combination of temperature- and density-driven groundwater flow, was investigated by the Zentrum für Wissenschaftliches Rechnen of the Goethe-Universität Frankfurt in the fourth article. A porous medium is macroscopically described by a biphasic mixture consisting of a fluid and a solid phase. The fluid phase filtrating the pores is itself a mix of pure water and hypersaline water called brine. The dependence of the density of the fluid-phase on its brine mass fraction and its temperature generates the flow phenomena to be analyzed. Depending on salinity and temperature of a small brine placed in the flow field in one case positive and in the other case negative buoyancy occurs which induces in both cases a characteristic and interesting fluid flow.

The implementation of tailored analysis algorithms has gained considerable significance in the context of spaceborne gravity field recovery. This problem is addressed by the Institut für Geodäsie of the Universität Stuttgart in the fifth contribution. The optimization of individual software modules and the efficient parallelization of the procedures as a whole are key to access state-of-the-art science results within reasonable time frames. Thus, high performance computing technologies play an important role in modern satellite geodesy. Typically, space-borne geodesy requires the estimate of tens of thousands of unknown gravity field parameters from tens of millions of observations. To assemble and solve the systems of equations in terms of a least-squares adjustment the algorithms are adjusted to the NEC SX-8 and SX-9 systems. Implementation details and runtime results using the NEC systems as single-node architectures are presented.

In the last article, the Institut für Geowissenschaften of the Friedrich-Schiller-Universität in Jena presents a dynamic 3-D spherical-shell model for the chemical evolution of the Earth's mantle. The model is based on the solution of the balance equations of mass, momentum, energy, angular momentum, and four sums of the number of atoms of the pairs ^{238}U-^{206}Pb, ^{235}U-^{207}Pb, ^{232}Th-^{208}Pb, and ^{40}K-^{40}Ar. The details of the continental distribution of the model are largely stochastic, but the spectral properties are quite similar to the present real Earth. The comparison of the observational results of the present-day proportion of depleted MORB mantle with the model shows a similar order of magnitude.

Computer Simulation for Building Implosion Using LS-DYNA

Georgios Michaloudis, Steffen Mattern, and Karl Schweizerhof

Institute of Mechanics, Otto-Ammann-Platz 9, Geb. 10.30, 76131 Karlsruhe, Germany, Georgios.Michaloudis@kit.edu, Steffen.Mattern@kit.edu, Karl.Schweizerhof@kit.edu

1 Introduction

The destruction of buildings by placing explosive material at strategic positions is particularly efficient for very high and large buildings. Controlled demolitions have been subject of the research done by the research group DFG Forschergruppe 500 "Computergestuetzte Destruktion komplexer Tragwerke durch Sprengung" [1] which is supported by the DFG. This contribution describes the continuation of simulations of building implosions with finite elements. These simulations are only part of the general project and serve in principle as verification for simplified models [2].

2 Aspects of Numerical Simulation

The finite element program LS-DYNA [5, 6] is used for the generation of the models and the simulations which must consider all the significant phenomena and uncertainties [7–9], though for the latter some series of analysis are needed. In order to apply the so-called Node Split algorithm for modelling local failure a modification of the finite element mesh is necessary. To avoid performing these modifications all by "hand" since no preprocessor supports such a feature, an automatic algorithm has been programmed which at first creates coinciding nodes, so that every element consists of eight unique nodes. Within the algorithm then the element connectivity can be given up depending on some conditions. In the following simulations contact is then checked between the new created surfaces. Some additional modifications are necessary, e.g. applying the proper boundary conditions for the splitted model. The resulting meshes need to be fairly fine to achieve a reasonably good approximation of the geometry of the failure surfaces. All computations were made in parallel on a HP X6000-Cluster of the Karlsruhe Institute of Technology.

W.E. Nagel et al. (eds.), *High Performance Computing in Science and Engineering '10*, DOI 10.1007/978-3-642-15748-6_37,

3 Simulation Models

3.1 Previous Models

Within the research unit FOR500 five reference models are studied with varying parameters. The first model is a simple academic fictional model, which was created in order to test the basic numerical algorithms before applying them to the large, time-consuming models. Both algorithms for the modelling of failure, element-erosion and node-split, were applied in this model [2]. The second model is a storehouse in Thueringen, demolished in 1998, which has been computed previously with the element-erosion-algorithm [10] and is now computed in this contribution with the node-split-algorithm. The third model is a silo-building in Borna, for which up to now only the element-erosion-algorithm has been used [11] with satisfactory results for the entire process. These three first models represent typical cases of blasting strategies which lead to inclination and finally to the collapse of the structure.

The fourth reference model is a fictional model which was created in order to test the simulation of the demolition of the fifth large reference model, the Sparkasse building in Hagen. This model incorporated some basic characteristics of the Sparkasse building but was much smaller in size, not having more than 150.000 elements. A similar demolition strategy was applied and the basic algorithms were first tested in this analysis, since applying them directly to the large model would be too time consuming. After certain modifications and when the fictional building was computed [12] with satisfactory results all algorithms were used in the simulation of the large model. Both models (fourth reference model and the building in Hagen) represent vertical collapse scenarios, in which structural failure results mainly from strong compressive forces and shear.

The experience gained from these examples showed firstly that the element-erosion-algorithm in the case of non vertical collapse scenarios, in which failure occurs mostly due to strong bending of elements and not due to high pressures, can provide reliable results, at least for the initial collapse kinematics. On the other hand, the erosion scheme should not be applied in the case of a vertical collapse, since as a result of removing elements it does not deliver reliable results. The node-split-algorithm, however, is more time consuming and imposes several changes in the discretization, but is a better alternative which delivers realistic results even in the case of vertical collapse with high pressures and material failure due to strong compression.

3.2 Sparkasse Building in Hagen

The demolition of the Sparkasse building in Hagen was one of the most complex blasting processes which ever took place in Germany [13]. The size and especially the height of the building in combination with the limited space available for the final debris demand a sophisticated blasting strategy which

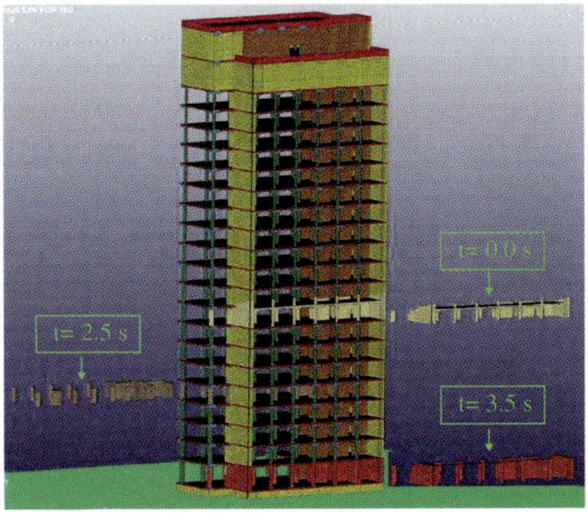

Fig. 1. Sparkasse building in Hagen—Computational model and blasting strategy

results in triaxial collapse kinematics. The structural system of the building consists of frames and shear walls; it is 93 m high (20 floors) and its surface has dimensions 37,20 m × 18,60 m. The blasting strategy, which consists of three phases, is shown in Fig. 1.

The finite element model consists of 392481 8-node hexahedral solid elements. At first, the CAD geometry was created from the building drawings. Then, using the preprocessor Hypermesh [14], the finite element mesh was constructed and imported into LS-DYNA. At this final stage, all the numerical algorithms were applied, such as material model, element formulation etc. In the computations to be presented a rather simple, piecewise linear, plasticity model is used (LS-DYNA MAT24) for efficiency reasons. However, some limitations have to be mentioned resulting from the fairly simple failure criterion. In addition to that, it must be noted that also in the case when Node-Split is applied, a rather high resolution of the model is necessary in order to capture sufficiently the entire cracking process. Furthermore, by using continua for the modelling of reinforced concrete the effect of the steel bars is "smeared" in the continuum. E.g. the presence of steel bars, after the mechanical failure of the concrete, would in reality prevent structural parts from flying away from the structure. However, in the simulations when the continua fail, a complete dissolution is obtained. This results in general in an overly brittle behaviour and some elements are flying away from the structure, In the simulations we try to damp this motion which is not a perfect solution. The ground plate is simulated as a rigid body and the building collapses under the effect of gravity.

The results shown in the Fig. 2 are from a first analysis in which the element erosion algorithm was applied. As expected modelling material failure with erosion of elements leads to the removal of almost the entire structure and, due to that, the final results do not-by far-have the required accuracy and reliability.

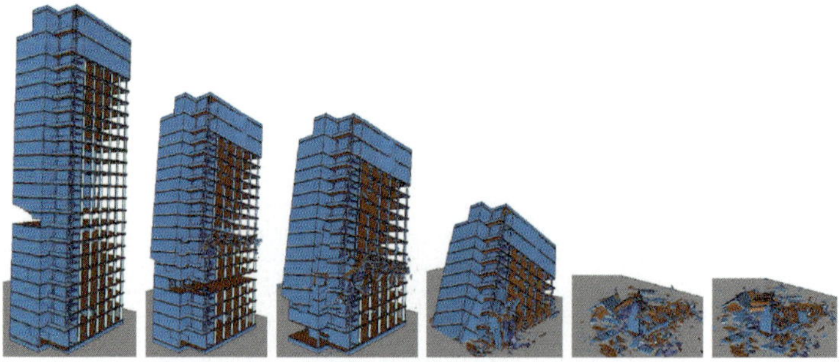

Fig. 2. Sparkasse in Hagen—Evolution of collapse kinematics, element erosion as failure mechanism

Fig. 3. Sparkasse in Hagen—Evolution of collapse kinematics, Node-Split as failure mechanism vs Real demolition, t = 0.0, 2.5, 3,5, 5.0, 7.0, 7.9 seconds

In Figs. 3 and 4 the results of the simulation when applying the Node-Split algorithm are presented and compared with photos taken during the real demolition. When the Node Split algorithm is used, each element consists of eight nodes which do not belong to any other element. Nodes with the same positions get merged by certain constraints to enable separating them easily if the failure criteria is reached. The chosen failure criteria checks for the volume weighted plastic strain in adjacent elements. If the given limit for

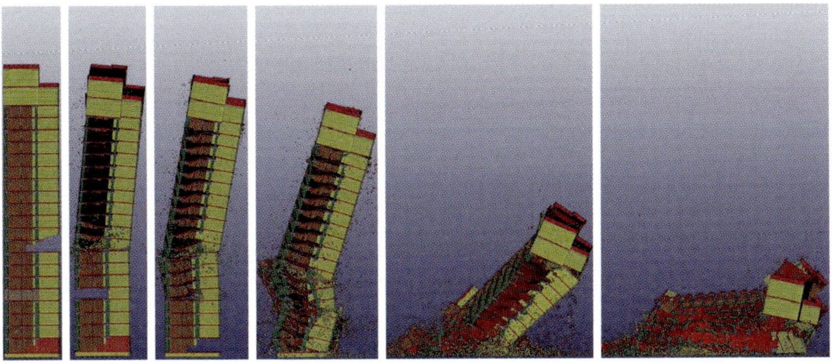

Fig. 4. Sparkasse in Hagen—Evolution of collapse kinematics, side view, t = 0.0, 2.5, 3.5, 5.0, 7.0, 7.9 seconds

the plastic strain is reached not all connections of the element are broken, but only the nodes of neighboring elements, which carry this plastic strain, are disconnected, while the other tied connections of the element remain unchanged. The first three figures show the beginning of each of the three phases of the demolition. As expected, this approach provides a much more realistic simulation of the entire process, compared to the results previously shown. All contact conditions are captured and well described since no structural parts are removed due to failure. This allows a good prediction of the collapse kinematics throughout the demolition as well as a reliable prediction of the final rest position at the end of the collapse. It is shown that for a vertical collapse scenario results are achieved which are by far closer to reality than with the erosion algorithm.

The computation was performed on the HP XC6000-Cluster of the KIT using the MPP-Parallel version of LS-DYNA on 4 processors. The computational time for this problem of approximately 9.500.000 unknowns, 663807 extra constraint conditions (for the modelling of failure), computed time of 7.9 seconds with timestep size of approximately $6{\cdot}10^{-6}$ seconds (1.310.000 time steps) reaches 43000 minutes. Although the time step size does not change dramatically throughout the analysis the necessary numerical effort is increasing as the problem evolves. This is due to the large deformations and to the fact that more and more elements are splitted due to the failure algorithm. Then e.g. in the last phase of the analysis (computed time 7.0 seconds until 7.9 seconds) the computation for 0.1 seconds of the problem takes almost 15 hours. For the first five seconds of the analysis in order to compute 0.1 seconds the process did not need more than 8 hours. This occurs mainly because of the extreme increase in the number of the contact surfaces due to the splitting of elements. New contact interfaces are created and included in the search contact algorithm which is very time consuming and as a consequence slows down the entire computation. In order to remove the major part of the artificial energy which was added to the model during the application of the initial

Fig. 5. Sparkasse in Hagen—Validation of results by video overlapping, t = 1.0, 3.0, 5.0, 6.0, 7.0, 7.5 seconds

gravitation with explicit time integration, the system had to be damped for 0.1 second, otherwise large vibrations were leading to error terminations.

It proved to be particularly important to avoid the "inversion" of solid elements leading to negative volumes and finally to error terminations, as a result of large element deformations. In order to handle this problem additional contact checks were added which consider contacts between interior surfaces of the hexahedral elements to avoid an inversion. Nevertheless, in the results the limitation in the applicability of the Node Split algorithm is observed from the very strong dissolution of element connections due to material failure, which creates for some parts unrealistic phenomena (structural parts flying e.t.c). The latter would not be possible with reinforcing steel which mostly prevents a complete dissolution of parts.

The validation of the results of the finite element simulations is an essential issue because it defines the accuracy of the performed computations. In Fig. 5 the outcome of the validation procedure is presented, in which the video of the simulation is overlapped with the video of the real demolition.

It is shown that the results are not only realistic but they also describe sufficiently the entire collapse kinematics. Although there is a small delay at the beginning of the simulation, soon this disappears. Furthermore, even though some elements after the dissolution with their neighboring elements, seem to fly in a distance from the main structure, the final rest position is predicted with an acceptable accuracy and the resulting debris on the ground appears in the simulation as well.

3.3 Storehouse in Thueringen

The second example is a seven storey storehouse in Thueringen with a height of 22 m and surface dimensions 22 m × 12 m. It is the second reference model of the research unit FOR500. Two explosions initiate the collapse, as shown

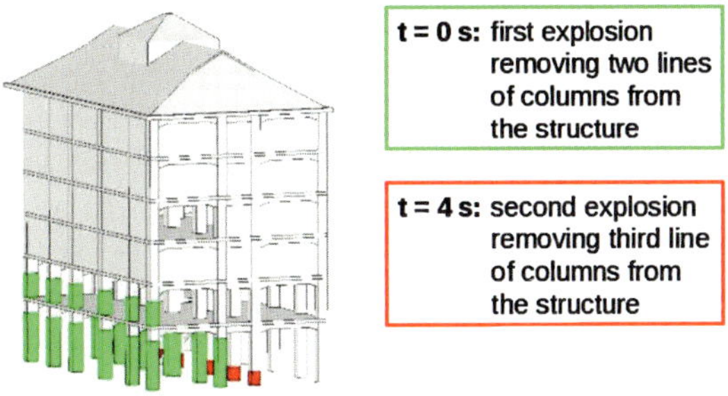

Fig. 6. Storehouse in Weida—Blasting strategy

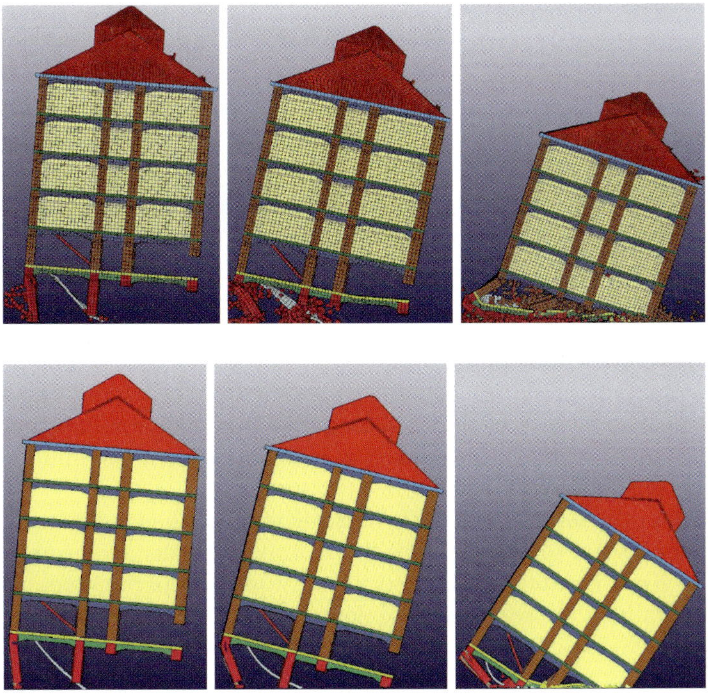

Fig. 7. Storehouse in Weida—Initial collapse kinematics, Node Split strategy (top) vs Element Erosion (bottom), at t = 4.6, 5.0, 5.6 seconds

in Fig. 6. The numerical model consists of 82867 hexahedral elements and the node-split-algorithm is chosen in order to model material failure.

In Fig. 7 the initial collapse kinematics are presented, which have good correlation with the results of a first analysis in which the element-erosion-algorithm was applied. The node split strategy seems to have a physically better result. 18 hours on 16 processors were necessary for this computation. Simulations with parameter variations are still running for this model. Computing several buildings and different collapse scenarios aims at gaining important informations about the whole process and providing these informations to the engineering offices. For each model a large number of simulations is performed in order to define the influence of some algorithmic values as well as process parameters (value of failure criterion, hourglass stabilization, blasting strategy etc) in the final result. This is necessary for a reliable judgement of the results. Testing all these aspects provides the possibility to define general limitations of the algorithms as well as to find the suitable tools for each different case, e.g. that the element-erosion-algorithm has strong limitations and is not suitable for vertical collapse scenarios. Then robust computations can be performed that solve problems which the engineering consultants face very often, e.g. testing different blasting strategies, rearranging the time intervals between explosions etc. Particularly, in the case of vertical collapse scenarios aspects such as the location of placing the explosives or the time difference between the explosions become very significant. In the case of the fourth and fifth reference models the blasting strategy imposes kinematics which combine first inclination of the upper part of the building and then vertical collapse. Delaying the second explosion even for one second can lead to complete different dynamical behaviour and even to a failed demolition.

This entire process involves several problems which must be solved. The finite element meshes must be fairly fine, mostly for the modelling of structural failure. Using structural elements (e.g. beams) instead of continuum elements would be insufficient. Due to that, the computations become very time consuming and another difficulty is imposed from the fact that all the preprocessors do not support mesh refinement with meshes for the node-split-algorithm. Hence, it was important to find a reasonable balance between accuracy and computational effort by creating meshes which consist of the minimum number of necessary elements.

Even when the mesh is ready, the computation of the model is not straight forward, since a lot of details must be defined in order to ensure a robust simulation. Appropriate contact algorithms, hourglass parameters [15], values for material failure, are some of these details which influence strongly the final result. In the Node Split scheme the value of failure denotes the plastic strain for which elements are disconnected, but this does not necessarily mean that the elements have mechanically failed. In order to define a realistic value for this failure criterion several computations have been performed which were then validated with the video of the real demolition.

Experience is gained through the large number of computations and several numerical algorithms have been improved or substituted by others, more efficient ones. For example, the first general contact algorithm is substituted from three more specialized algorithms. One for describing the contact between structural parts and the ground plate (CONTACT AUTOMATIC SURFACE TO SURFACE), one for the contact just between structural parts (CONTACT ERODING SINGLE SURFACE) and an additional check for preventing solid elements to get inverted (CONTACT INTERIOR), which otherwise would be avoided with finer meshes. Similarly, the parameters which control the very efficient hourglass stabilization method have been updated since the previous values proved appropriate for the general contact algorithm and the element-erosion-technique, but caused instabilities when the new contact schemes were used and mostly in combination with the huge number of constraints imposed from the application of the node-split-algorithm. This is because of the fact that the elements are not removed and undertake very large deformations due to compression. Thus, in order to apply the Belytschko-Bindemann hourglass control form, the hourglass coefficient was reduced by a factor of 10, followed by appropriate changes in other coefficients as well. Other hourglass control types were tested as well (e.g. Flanagan-Belytschko), which in order to deliver reliable results imposed the reduction of the timestep size by a factor of 2.5. As a conclusion, the Belytschko-Bindemann hourglass control was taken, as it proved more efficient and more effective.

References

1. DFG Forschergruppe FOR500 www.sprengen.net 2006.
2. Hartmann D, Breidt M, Nguyen vV, Stangenberg F, Höhler S, Schweizerhof K, Mattern S, Blankenhorn G, Michaloudis G, Möller B, Liebscher M (2008) On a fundamental concept of structural collapse simulation taking into account uncertainty phenomena. NATO Advanced Research Workshop 983112, Sarajevo, Bosnia and Herzegovina.
3. Mattern S, Blankenhorn G, Breidt M, Nguyen vV, Höhler S, Schweizerhof K, Hartmann D (2006) Comparison of building collapse analysis simulation from finite element and rigid body models. In: Eberhard P (Ed.), IUTAM Symposium on multiscale problems in multibody system contacts, IUTAM Bookseries, Springer, Berlin.
4. Mattern S, Blankenhorn G, Schweizerhof K (2006) Numerical analysis of building collapse—Case scenarios and validation. Proceedings of the NATO Advanced Research Workshop, PST.ARW981641, ed.: A. Ibrahimbegovic and I. Kozar.
5. Hallquist JO (2006) LS-DYNA Theory Manual. Livermore Software Technology Corporation.
6. Hallquist JO (2006) LS-DYNA Keyword User's Manual. Livermore Software Technology Corporation.
7. Möller B, Beer M (2004) Fuzzy Randomness—Uncertainty in civil engineering and computational mechanics, Springer, Berlin.
8. Hartmann D, Breidt M, Nguyen vV, Stangenberg F, Höhler S, Schweizerhof K, Mattern S, Blankenhorn G, Möller B, Liebscher M (2008) Structural collapse simulation under consideration of uncertainty—Fundamental concept and results. Computers and Structures 86, 2064–2078.
9. Möller B, Liebscher M, Schweizerhof K, Mattern S, Blankenhorn G (2008) Structural Collapse under Consideration of Uncertainty—Improvement of Numerical Efficiency. Computer and Structures 86, 1875–1884.
10. Mattern S, Blankenhorn G, Schweizerhof K (2007) Numerical simulation of controlled building collapse with finite elements and rigid bodies—Case studies and validation. ECCOMAS Thematic Conference in Structural Dynamics and Earthquake Engineering, Rethymno, Crete, Greece.

11. Blankenhorn G, Mattern S, Schweizerhof K (2007) Controlled building collapse-Analysis and validation. Proceedings LS-DYNA Anwenderforum, Frankenthal.
12. Michaloudis G, Blankenhorn G, Mattern S, Schweizerhof K (2009) Modelling structural failure with finite element analysis of controlled demolition of buildings by explosives using LS-DYNA. High Performance Computing in Science and Engineering '09, In: W.E. Nagel, D.B. Kröner, M.M. Resch (Eds.), pp. 539–551, Springer, Berlin.
13. Hartmann D, et al. (2010) Computersimulation für Bauwerkssprengungen—ein Einblick in laufende Arbeiten. Sprenginfo 32, Nr 01, 27–35.
14. HyperMesh Users Manual, Version 8.0.
15. Belytschko T, Bindemann LP (1993) Assumed strain stabilization of the eight node hexahedral element. Computer Methods in Applied Mechanical Engineering 105, 225–260.

Quaero Speech-to-Text and Text Translation Evaluation Systems

Sebastian Stüker[1], Kevin Kilgour[1], and Jan Niehues[2]

[1] Research Group 3-01 'Multilingual Speech Recognition', Karlsruhe Institute of Technology, Karlsruhe, Germany, sebastian.stueker@kit.edu, kevin.kilgour@kit.edu

[2] Interactive Systems Laboratories, Karlsruhe Institute of Technology, Karlsruhe, Germany, jan.niehues@kit.edu

1 Introduction

Our laboratory has used the HP XC4000, the high performance computer of the federal state Baden-Württemberg, in order to participate in the second Quaero evaluation for automatic speech recognition (ASR) and Machine Translation (MT).

State-of-the-art ASR research systems usually employ techniques which require the parallel execution of several recognition systems for the purpose of system combination. The use of unsupervised adaptation techniques further requires the execution of several stages or passes of ASR systems. This leads to the fact that modern research systems process speech only with a run-time of many times realtime, under certain circumstances up to 100 times realtime. The process of speech recognition in this form can be easily parallelized at a speaker level in independent processes without the need for inter-process communication. Therefore the scheduling system of the XC4000 in combination with its global, high performing file space, is an ideal environment for executing such an evaluation.

The training of machine translation systems also requires the processing of large amounts of data in order to train statistical models.

2 Quaero

Quaero (http://www.quaero.org) is a French research and development program with German participation. It targets to develop multimedia and multilingual indexing and management tools for professional and general public applications such as the automatic analysis, classification, extraction, and exploitation of information. The projects within Quaero address five main application areas:

W.E. Nagel et al. (eds.), *High Performance Computing in Science and Engineering '10*, DOI 10.1007/978-3-642-15748-6_38,

- Multimedia Internet search
- Enhanced access services to audiovisual content on portals
- Personalized video selection and distribution
- Professional audiovisual asset management
- Digitalization and enrichment of library content, audiovisual cultural heritage, and scientific information.

Also included in Quaero is basic research in the technologies underlying these application areas, including automatic speech recognition, machine translation, and speech-to-speech translation. The vision of Quaero is to give the general public as well as professional user the technical means to access various information types and sources in digital form, that are available to everyone via personal computers, television, and handheld terminals, across languages.

Quaero is organized as a program consisting of seven projects. Five projects are concerned with applications. In addition, one project, the *Core Technology Cluster* (CTC), conducts basic research in the technologies underlying the application projects, and one project is concerned with providing the data resources necessary for the research within CTC.

Our laboratory is mainly involved in the CTC project. Two of the technologies under investigation are *Automatic Speech Recognition* (ASR), i.e. the automatic transcription of human speech into written records and *Machine Translation* (MT). Within Quaero research is driven by competitive evaluation and sharing of results and technologies employed. This process is called *coopetition.* Evaluations are conducted once a year on a predefined domain and a set of languages. As the project continues the number of languages to address will grow. Also the performance of the recognition systems developed within the project is expected to improve.

The second evaluation conducted in August 2009 was intended as a first, real evaluation after the baseline evaluation in 2008. Five languages were addressed: English, French, German, Russian, and Spanish. We participated in the languages English, German, Russian, and Spanish. The test data for the evaluation consisted of various audio files collected from the World Wide Web, including broadcast news, lectures, and video blogs.

3 English Evaluation Recognition Systems

For the English evaluation within Quaero we participated with a recognition system that is a further development of the ISL 2007 English Speech Transcription System for European Parliament Speeches from the European Integrated Project Technology and Corpora for Speech-to-Speech Translation (TC-STAR) [21]. The system has been trained and tested with the help of the Janus Recognition Toolkit that features the IBIS single pass decoder [22]. In general all recognition systems employ left-to-right Hidden Markov Models (HMMs), modeling phoneme sequences with 3 HMM states per phoneme.

3.1 Front-End

We trained systems for two different kinds of acoustic front-ends. One is based on the traditional Mel-frequency Cepstral Coefficients (MFCC) obtained from a fast Fourier Transform and the other on the warped minimum variance distortionless response (MVDR). The second front-end replaces the Fourier transformation by a warped MVDR spectral envelope [25], which is a time domain technique to estimate an all-pole model using a warped short time frequency axis such as the Mel scale. The use of the MVDR eliminates the overemphasis of harmonic peaks typically seen in medium and high pitched voiced speech when spectral estimation is based on linear prediction.

For training, both front-ends have provided features every 10 ms. During decoding this was changed to 8ms after the first stage. In training and decoding, the features were obtained either by the Fourier transformation followed by a Mel-filterbank or the warped MVDR spectral envelope.

For the MVDR front-end we used a model order of 22 without any filterbank since the warped MVDR already provides the properties of the Mel-filterbank, namely warping to the Mel-frequency and smoothing. The advantage of this approach over the use of a higher model order and a linear-filterbank for dimensionality reduction is an increase in resolution in low frequency regions which cannot be attained with traditionally used Mel-filterbanks. Furthermore, with the MVDR we apply an unequal modeling of spectral peaks and valleys that improves noise robustness, due to the fact that noise is mainly present in low energy regions.

Both frond ends apply vocal tract length normalization (VTLN) [26]. For MFCC this is done in the linear domain, for MVDR in the warped frequency domain. The MFCC front-end uses 13 cepstral coefficients, the MVDR front-end uses 15. The mean and variance of the cepstral coefficients were normalized on a per-utterance basis. For both front-ends seven adjacent frames were combined into one single feature vector. The resulting feature vectors were then reduced to 42 dimensions using linear discriminant analysis (LDA).

3.2 Acoustic Model Training

We trained acoustic models for two different kinds of phoneme sets *P1* and *P2*. P1 is a version of the Pronlex phoneme set which consists of 44 phonemes and allophones while P2 is a version of the phoneme set used by the CMU dictionary that consists of 45 phonemes and allophones. We trained models for all four combinations of the two phoneme sets and the two acoustic front-ends described above.

We used two different kinds of set of training data two train different acoustic models.

The first training set (*train1*) contains approximately 80h of English EPPS data provided by RWTH Aachen within the TC-STAR project [7], 9.8h of TED data [13], and 167h of unsupervised EPPS training material that had been collected within TC-STAR by RWTH Aachen but had not been manually transcribed. Transcriptions for the unsupervised training material were obtained by adapting an acoustic model of last year's system on automatic transcriptions provided by RWTH Aachen on that data. We then decoded the data, using the segmentation provided by RWTH Aachen.

The second set of training data (*train2*) contains in addition to the data from the first training data set 140h of BroadCast News data from the HUB-4 corpus. Exploratory experiments showed that the use of broadcast news data leads to improvements on the web data in the Quaero development data, while the use of meeting data from the CHIL project did not lead to improvements.

All models are semi-continuous quinphone systems that use 16000 distributions over 4000 codebooks. They were trained using incremental splitting of Gaussians training, followed by 2 iterations of Viterbi training. For all models we used one global semi-tied covariance (STC) matrix after LDA [6] as well as Vocal Tract Length Normalization. In addition to that feature space constraint MLLR (cMLLR) speaker adaptive training [5] was applied on top.

We improved the acoustic models further with the help of Maximum Mutual Information Estimation (MMIE) training [19]. We applied MMIE training firstly to the models after the 2 viterbi iterations, and secondly to the models after the FSA-SAT training, taking the adaptation matrices from the last iteration of the maximum likelihood FSA-training and keeping them unchanged during the MMIE training.

This all resulted in fourteen different acoustic models: for each combination of front-end, MVDR and MFCC, and phoneme set, P1 and P2, one set of models trained with VTLN plus MMIE, and one with FSA-SAT plus MMIE. From now on we refer to these models as *P1-train1-MFCC-VTLN*, *P2-train1-MFCC-VTLN*, *P1-train1-MVDR-VTLN*, *P2-train1-MVDR-VTLN*, *P2-train2-MFCC-VTLN*, *P2-train2-MVDR-VTLN*, *P1-train1-MFCC-SAT*, *P2-train-MFCC-SAT*, *P1-train1-MVDR-SAT*, *P2-train1-MVDR-SAT*, *P1-train2-MFCC-SAT*, *P2-train2-MFCC-SAT*, *P1-train2-MVDR-SAT*, and *P2-train2-MVDR-SAT*.

Segmentation and Clustering

Segmenting the input data into smaller, sentence-like chunks used for recognition was performed with the help of a fast decoding pass on the unsegmented input data in order to determine speech and non-speech regions. Segmentation was then done by consecutively splitting segments at the longest non-speech region that was at least 0.3 seconds long. The resulting segments had to contain at least eight speech words and had to have a minimum duration of six seconds.

In order to group the resulting segments into several clusters, with each cluster, in the ideal case, corresponding to one individual speaker we used the same hierarchical, agglomerative clustering technique as last year which

Table 1. English text sources

Corpus	**Wordcount**
UK parliament debares (Hansard)	49,681,456
EPPS acoustic training data	749,668
EPPS text data	33043959
UN Parallel Text (English)	40,991,279
Gigaword 4rth Edition without 2008 texts	3,618,495,574
Hub4 Broadcast News data	832,068
Web dump (pre 2008)	144,062,839
total	3,887,856,843

is based on TGMM-GLR distance measurement and the Bayesian Information Criterion (BIC) stopping criterion [11]. The resulting speaker labels were used to perform acoustic model adaptation in the multipass decoding strategy described below.

Language Model and Test Dictionary

To select the vocabulary the development data text was split into a tuning set and a test set with each containing approximately half the text of every *show*. For each of our English text sources (see Table 1) we built a Witten-Bell smoothed unigram language model using the union of the text sources' vocabulary as the language models' vocabulary (global vocabulary). With the help of the maximum likelihood count estimation method described in [24] we found the best mixture weights for representing the tuning set's vocabulary as a weighted mixture of the sources' word counts thereby giving us a ranking of all the words in global vocabulary by their relevance to the tuning set. While the baseline 64k vocabulary had an OOV rate of 3.9% when measured on the validation set, the OOV rate of the vocabulary containing only the top ranked 64k words was 2.9%. This vocabulary was slowly increased until the OOV rate was under 1%. The final 130k vocabulary had a case sensitive OOV rate of 0.73%.

Pronunciations missing from the initial dictionary were created either manually or automatically with the help of Bill Fisher's tool [3] for P1 and Festival [2] for P2 respectively.

For each of the text sources in Table 1 we built, using the SRI Language Modeling Toolkit [23], a modified Kneser-Ney smoothed 4 gram language model. These were then interpolated using interpolation weights estimated on the aforementioned tuning set which was extendend with some general data to avoid overfitting problems. Because the resulting language model contained over 3×10^8 3 grams and 4 grams we decided to prune it to reduce its memory footprint and speed up decoding time. The final language model contained only slightly more than 6×10^7 3 grams and 4 grams.

Decoding Strategy and Results

Decoding within our recognition system was performed in two stages. The acoustic models of the second stage were adapted on the output(s) from the previous stage using Maximum Likelihood Linear Regression (MLLR) [14], Vocal Tract Length Normalization (VTLN) [26], and feature-space constrained MLLR (cMLLR) [5]. For the second and third stage the frame shift during recognition was changed to 8ms.

In the first stage we used the acoustic models P2-train1-MFCC-VTLN, P2-train-1-MVDR-VTLN, P2-train2-MFCC-VTLN, P2-train-2-MVDR-VTLN. The models trained on train1 were run on the EPPS portion of the test data only, the other models on the rest of the test data. The resulting word lattices were then combined via confusion network combination to the output *o1*. In this first stage we adapted the acoustic models using incremental VTLN and incremental fMLLR on a per speaker basis.

For the second stage all SAT systems were adapted on o1. The result of the different models were combined via confusion network combination to the final output.

On the official 2009 development set the system achieved a word error rate of 31.7%.

4 Translation System

In the Quaero text translation evaluation 2009 we participated in the following tasks: German-English, English-German, French-English, English-French, German-French, French-German. The submitted translations were generated by the following systems.

4.1 Data

Our translation systems were trained on the parallel corpora from the following sources: Europarl, Project Syndicate, Voltaire, and Who corpora. Table 2 shows few numbers about the training data.

The development sets for our systems were created by joining the nc-dev and nc-devtest sets from the WMT. The News Commentary 2007 test sets form the WMT were utilized as test sets for our systems.

Multiple monolingual corpora were used to train the language models. For every language at least two language models were trained: the target part of the parallel data and the News corpus. In addition to that, for English and French a language model was also trained on the Gigaword corpus.

For Discriminative Word Alignment training, hand-aligned data was used. For instance, DE-EN and EN-DE systems used a set of around 500 sentences from the Europarl domain annotated with “sure” and “possible” links. Whereas EN-FR and FR-EN used the data from the 2003 NAACL shared task and which consists of 443 sentences.

Table 2. Brief parallel corpora statistics

		Sydicate		Voltaire	
		Sentences	Words	Sentences	Words
EN-DE,	DE	41.2K	966.8K	189	4.4K
DE-EN	EN	41.2K	919.9K	189	4.4K
FR-DE,	FR	24.8K	659.4K	94	2.0K
DE-FR	DE	24.8K	607.5K	94	1.6K
EN-FR,	FR	28.5K	753.3K	17.9K	335.4k
FR-EN	EN	28.5K	666.7K	17.9K	340.1k
		Who		**Europarl**	
		Sentences	Words	Sentences	Words
EN-DE,	DE	–	–	1.42M	33.10M
DE-EN	EN	–	–	1.42M	35.69M
FR-DE,	FR	–	–	1.28M	33.73M
DE-FR	DE	–	–	1.28M	30.01M
EN-FR,	FR	122.3K	2.5M	1.43M	37.65M
FR-EN	EN	122.3K	2.1M	1.43M	36.20M

4.2 Preprocessing

The training data was preprocessed before used for training. In this step different normalizations like mapping different types of quotes were done. In the end the first word of every sentence was smartcased.

For the German text additional preprocessing steps were applied. First, the old German data uses the "Alte Deutsche Rechtschreibung" whereas the newer parts of the corpus use the "Neue Deutsche Rechtschreibung". We tried to normalize the text by converting the whole text to the "Neue Deutsche Rechtschreibung". In a first step, we search for words that are only correct according to the old writing rules. Therefore, we selected all words in the corpus, that are correct according to the hunspell lexicon using the old rules, but not correct according to the hunspell lexicon using the new rules. In a second step we tried to find the correct spelling according to the new rules. Therefore, we first applied some rules describing how words changed from one spelling system to the other like replacing 'ß' by 'ss'. If the new word is a correct word according to the hunspell lexicon using the new spelling rules, we map the words.

If translating from German to French or English, we apply compound splitting as described in [12] to the German corpus.

As a last preprocessing step we remove sentences that are too long and empty lines to get the final corpus.

4.3 Language Model

For all tasks and languages we use 4-gram language models generated using the SRILM Toolkit [23]. Different types and amounts of monolingual data were used to train the language models. We experimented with the EPPS data, the News corpus, the News-Commentary corpus as in-domain data and the Gigaword corpus for those languages for which it is available (i.e. English and French).

Three different language model setups were applied: (1) Using a language model trained on one of the above mentioned data types. The factor for the language model is optimized during Minimum Error Rate Training. (2) Accessing two different language models at the same time from within the decoder. Here, two factors are optimized, one for each language model. (3) Combining two language models into one through linear interpolation. We choose the interpolation weights such that the perplexity of the resulting language model on the development set is minimized. In this case one language model factor was optimized with MERT.

4.4 Translation Model

In order to generate the translation model, i.e. the SMT phrase table, we first used the GIZA++ Toolkit [17] to calculate a word alignment for the training corpus. This was done in both directions and then the alignments were combined using the "grow-diag-final-and" heuristic. Afterwards the Moses Toolkit was used to build the phrase table.

The relative frequencies of the phrase pairs are a very important feature of the translation model, but they often overestimate rare phrase pairs. Therefore, the raw relative frequency estimates found in the phrase translation tables are smoothed by applying modified Kneser-Ney discounting as described in [4].

In addition to this baseline approach, some of the systems used a discriminative word alignment approach as described in [16] instead of using the word alignment generated by the "grow-diag-final-and" heuristic. For German-English and English-German we trained the discriminative word alignment on hand-aligned data and used the lexical probabilities as well as the fertilities generated by the GIZA++ Toolkit and POS information generated by the TreeTagger. We used all local features, the GIZA and indicator fertility features as well as the first order features for 6 directions. The model was trained in three steps first using the maximum likelihood optimization and afterwards it was optimized towards the alignment error rate. For more details see [10].

For French-German and German-French no hand-aligned data was available. To be able to use the discriminative word alignment for those language pairs as well, we reused a model trained on the English-German or German-English data. In this case we used only the lexical probabilities as well as the fertilities generated by the GIZA++ Toolkit as knowledge sources. To generate the alignment for French-German we then used the model trained on the English-German hand-aligned data and replace the English-German knowledge sources by the French-German ones. We did the same for German-French using the German-English model.

Since the test data was from the same source as part of the training data, we adapted the translation model towards this domain. Therefore, we trained a second phrase table only on the data from the project syndicate. Afterwards we combined both phrase tables in a log-linear way.

In addition we tried to trigger the relative frequencies in the phrase table by other words in the source sentence. Therefore, we calculated for every source word that cooccurs with a phrase pair, the relative probabilities for that phrase pair given that this source word occurs. In a next step, we then selected the trigger words, that led to a significant change in the relative frequencies. We then sorted the triggers by the information gain from this trigger and used the top N triggers in the phrase table. This was defined as

$$log(\frac{cooc(Trigger, PP)}{\#Trigger}) * \frac{cooc(Trigger, PP)}{\#Trigger} \quad (1)$$

where $cooc(Trigger, PP)$ is the number of cooccurrences of the trigger word and the phrase pair and $\#Trigger$ is the number of times the trigger occurs in the corpus. In the experiments the parameter N was set to 100.

During decoding we then checked if a trigger word of the phrase pair occurs in the source sentence. If this is the case, we used the triggered relative frequencies instead of the original ones.

4.5 Reordering Model

In addition to the built-in reordering in the decoder with a reordering window of two words, we also applied the POS-based reordering model presented in [20] to account for the different word orders in the languages. This model learns rules from a parallel text on how to reorder the source side such that it matches the word order of the target side. The aim is to generate a reordered source side that can be translated in a more monotone way.

In this framework, first, POS information is added to the source side and reordering rules are extracted from an aligned parallel corpus. These rules are of the form *VVIMP VMFIN PPER* → *PPER VMFIN VVIMP* and describe how the source side has to be reordered to match the target side. Then the rules are assigned scores according to their relative frequencies.

In a preprocessing step to the actual decoding different reorderings of the source sentences are encoded in a word lattice. Therefore, for all reordering rules that can be applied to a sentence the resulting reorderings are added to the lattice if the score is better than a given threshold. The decoding is then performed on the resulting word lattice.

This approach does model the reordering well if only short-range reorderings occur. But especially when translating from and to German, there are also long-range reorderings that require the verb to be shifted nearly across the whole sentence. During this shift of the verb, the rest of the sentence remains mainly unchanged. It does not matter which words are in between, since they are moved as a whole. Furthermore, rules including an explicit sequence of POS-tags spanning the whole sentence would be too specific. A lot more rules would be needed to cover long-range reorderings with each rule being applicable only very sparsely. Therefore, we model long-range reordering by generalizing over the unaffected sequences and introduce rules with gaps. (For more details see [15]). These are learned in a way similar to the other type of reordering rules described above, but contain a gap representing one or several arbitrary words. It is, for example, possible to have the following rule *VAFIN * VVPP* → *VAFIN VVPP **, which puts both parts of the German verb next to each other.

4.6 Decoder

All our systems are currently based on the STTK toolkit for statistical machine translation, which is being developed at our institute (Interactive Systems Laboratories). The current UKA translation system is a phrase-based statistical MT system that combines several models in the log-linear domain. The standard models combine multiple lexical scores, word and phrase translation probabilities, relative phrase frequencies, word and phrase count features, and one or several n-gram language models.

The scaling factors for all system configurations are determined automatically and optimized with respect to BLEU score via Minimum Error Rate Training [1].

In some configurations, a lattice is used as input for the decoder. This lattice encodes different possible reorderings of the source sentence according to the POS-based reordering model described above.

4.7 Re-Ranking

For n-best list re-ranking we combined two unique 500-best lists, which may have less than 500 entries for some sentences. The two lists were generated by the same system using two different word re-ordering strategies.

We used several features computed from different information sources such as additional language models trained on the full word form, lemmatized text and POS tags, IBM-4 word lexica and the n-best list itself.

- Language models
- POS and Lemma LMs
- Statistical Word Lexica
- Position Dependent Word Agreement
- N-best List N-gram Agreement
- N-best List N-gram Probability
- Sentence Length Features

In each case we trained two language models, one from the target side of the bilingual data and on from all available target language data. For the translation into German we also used a language model from POS tags and one from lemmatized text. We calculated 4 features from the word lexica: the word probability sum as well as the maximum word probability in both language directions. The A position-dependent word agreement, the position-independent n-gram agreement, the n-best list n-gram probability and the sentence length features are calculated solely from information in the n-best list itself. The features are described in more detail in [9].

We used MER training [18] to optimize the feature weights on a n-best list for a tuning set.

In the evaluation we only used the n-best list re-ranking for the translation into German, because preliminary experiments had not shown stable improvements for other language directions. This might be due to the additional information from the POS and lemma LMs for German.

5 Parallelization Utilized

For the two different tasks described above, Automatic Speech Recognition and Machine Translation, the XC4000 was utilized at different stages in the development and application of the system. For ASR the XC4000 was used during decoding, while for MT the XC4000 was used during the training stage of the systems.

5.1 Automatic Speech Recognition

During the period described in this paper, only the actual decoding and adaptation algorithms for the ASR systems have been run on the cluster, but not the training algorithms. Decoding can run in parallel on a per-speaker basis without any need for inter-process communication with other parallel jobs.

The SLURM scheduler was thus used to simply start several processes in parallel that in theory can run independently of each other. The English test set contains 303 speakers, so that up to 303 parallel processes can be started for the decoding of the test set. In reality usually an amount of ca. 100 jobs was run in parallel. The runtime of the different jobs per speaker can vary greatly, depending on how much speech is associated with one speaker. Due to the accounting system of the queue on the cluster—one process still running on one CPU on one node, while all the other processes belonging to one scheduled job have already finished, will lead to being charged the same amount of CPU time as if all processes were still running—as little processes per scheduled job as possible were sought. However, due to the fact that only 10 jobs in the production environment per user are allowed—no matter how many nodes or CPUs are actually employed—this would mean that only 40 parallel jobs could be started, if 4 CPUs per node in the cluster are assumed.

In order to make use of more parallelization, therefore up to 8 or 16 processes per scheduled job were committed to the queue, even though that potentially means that one is being charged for runtime of nodes on which all processes have already finished.

5.2 Machine Translation

For the Machine Translation systems we used the XC4000 during the training phase of the systems. As described above, one of the main components of our training is the use of the GIZA++ training toolkit. The original training toolkit is not programmed to be executed in parallel but needs to run in a single process on all the training data.

For our experiments we used an adaptation of the GIZA++ training toolkit that has added the possibility of parallelization to the toolkit [8] and that can run a computing cluster such as the XC4000. Using the slurm scheduler deployed on the XC4000 we were then able to execute the training in parallel. Data sharing and result combination is done at file level using the shared file space of the XC4000 so that no inter-process combination is necessary.

Acknowledgements. This work was realized as part of the Quaero Programme, funded by OSEO, French State agency for innovation. The Research group '3-01 Multilingual Automatic Speech Recognition' received financial support by the 'Concept of the Future' of Karlsruhe Institute of Technology within the framework of the German Excellence Initiative.

References

1. Andreas Zollman, Ashish Venugopal, and Alex Waibel. Training and Evaluation Error Minimization Rules for Statistical Machine Translation. In *Proc. of ACL 2005, Workshop on Data-drive Machine Translation and Beyond (WPT-05)*, Ann Arbor, MI, 2005.
2. A.W. Black and P.A. Taylor. The festival speech synthesis system: System documentation. Technical report, Human Communication Research Centre, University of Edinburgh, Edinburgh, Scotland, United Kingdom, 1997.
3. W.M. Fisher. A statistical text-to-phone function using ngrams and rules. In *Proceedings the 1999 IEEE International Conference on Acoustics, Speech, and Signal Processing*, Phoenix, AZ, USA, December 1999. IEEE.
4. George Foster, Roland Kuhn, and Howard Johnson. Phrasetable Smoothing for Statistical Machine Translation. In *Proc. of Empirical Methods in Natural Language Processing*, Sydney, Australia, 2006.
5. M.J.F. Gales. Maximum likelihood linear transformations for hmm-based speech recognition. Technical report, Cambridge University, Engineering Department, May 1997.
6. M.J.F. Gales. Semi-tied covariance matrices for hidden Markov models. Technical report, Cambridge University, Engineering Department, February 1998.

7. Christian Gollan, Maximilian Bisani, Stephan Kanthak, Ralf Schlüter, and Hermann Ney. Cross domain automatic transcription on the tc-star epps corpus. In *Proceedings of the 2005 IEEE International Conference on Acoustics, Speech, and Signal Processing (ICASSP'05)*, Philadelphia, PA, USA, March 2005.
8. Qin Gao and Stephan Vogel. Parallel implementation of word alignment tool. In *Proceedings of the ACL Workshops Software Engineering, Testing, and Quality Assurance for Natural Language Processing*, pages 49–57, Columbus, Ohio, June 2008. ACL.
9. Almut Silja Hildebrand and Stephan Vogel. Combination of machine translation systems via hypothesis selection from combined n-best lists. In *MT at work: Proceedings of the 8th Conference of the AMTA*, pages 254–261, Waikiki, Hawaii, October 2008.
10. Muntsin Kolss, Jan Niehues, Teresa Herrmann, and Alex Waibel. The Universität Karlsruhe Translation System for the EACL-WMT 2009. In *Fourth Workshop on Statistical Machine Translation (WMT 2009)*, Athens, Greece, 2009.
11. Qin Jin and Tanja Schultz. Speaker segmentation and clustering in meetings. In *Proceedings of the 8th International Conference on Spoken Language Processing (Interspeech 2004 — ICSLP)*, Jeju Island, Korea, October 2004. ISCA.
12. Philipp Koehn, Franz Josef Och, and Daniel Marcu. Statistical Phrase-Based Translation. In *HLT/NAACL 2003*, 2003.
13. E. Leeuwis, M. Federico, and M. Cettolo. Language modeling and transcription of the ted corpus lectures. In *International Conference on Acoustics, Speech, and Signal Processing, Hong Kong*, China, March 2003.
14. C.J. Leggetter and P.C. Woodland. Maximum likelihood linear regression for speaker adaptation of continuous density hidden Markov models. *Computer Speech and Language*, 9:171–185, 1995.
15. Jan Niehues and Muntsin Kolss. A POS-Based Model for Long-Range Reorderings in SMT. In *Fourth Workshop on Statistical Machine Translation (WMT 2009)*, Athens, Greece, 2009.
16. Jan Niehues and Stephan Vogel. Discriminative Word Alignment via Alignment Matrix Modeling. In *Proc. of Third ACL Workshop on Statistical Machine Translation*, Columbus, USA, 2008.
17. Franz J. Och. GIZA++: Training of statistical translation models. http://www.fjoch.com/GIZA++.html, 2000.
18. Franz Josef Och. Minimum error rate training in statistical machine translation. In *Proceedings of the 41st Annual Meeting of the Association for Computational Linguistics*, pages 160–167, Sapporo, Japan, July 2003. Association for Computational Linguistics.
19. D. Povey and P.C. Woodland. Improved discriminative training techniques for large vocabulary continuous speech recognition. In *International Conference on Acoustics, Speech, and Signal Processing*, Salt Lake City, UT, USA, May 2001.
20. Kay Rottman and Stephan Vogel. Word reordering in statistical machine translation with a pos-based distortion model. In *TMI '07*, 2007.
21. Sebastian Stüker, Christian Fügen, Florian Kraft, and Matthias Wölfel. The isl 2007 English speech transcription system for European parliament speeches. In *Proceedings of the 10th European Conference on Speech Communication and Technology (INTERSPEECH 2007)*, pages 2609–2612, Antwerp, Belgium, August 2007.

22. H. Soltau, F. Metze, C. Fügen, and A. Waibel. A one pass-decoder based on polymorphic linguistic context assignment. In *Proceedings of the IEEE Workshop on Automatic Speech Recognition and Understanding (ASRU '01)*, pages 214–217, Madonna di Campiglio Trento, Italy, December 2001.
23. A. Stolcke. SRILM – An Extensible Language Modeling Toolkit. In *Proceedings of the 7th International Conference on Spoken Language Processing (ICSLP 2002)*, pages 901–904, Denver, CO, USA, 2002. ISCA.
24. A. Venkataraman and W. Wang. Techniques for effective vocabulary selection. *Arxiv preprint cs/0306022*, 2003.
25. M.C. Wölfel and J.W. McDonough. Minimum variance distortionless response spectralestimation, review and refinements. *IEEE Signal Processing Magazine*, 22(5):117–126, September 2005.
26. Puming Zhan and Martin Westphal. Speaker normalization based on frequency warping. In *Proceedings of the 1997 IEEE International Conference on Acoustics, Speech, and Signal Processing*, Munich, Germany, April 1997.

Molecular Modeling of Hydrogen Bonding Fluids: Transport Properties and Vapor-Liquid Coexistence

Jadran Vrabec[1], Gabriela Guevara-Carrion[2], Thorsten Merker[2], and Hans Hasse[2]

[1] Lehrstuhl für Thermodynamik und Energietechnik (ThEt), Universität Paderborn, Warburger Str. 100, 33098 Paderborn, Germany*, jadran.vrabec@upb.de
[2] Lehrstuhl für Thermodynamik (LTD), Technische Universität Kaiserslautern, Erwin-Schrödinger-Str. 44, 67663 Kaiserslautern, Germany

Summary. Predictions of the transport properties self-diffusion coefficient and shear viscosity are presented for a recently developed molecular ammonia model. These data show mean unsigned deviations to the experiment over a temperature range from 200 to 500 K of 8 % for the self-diffusion coefficient and 12% for the shear viscosity. Furthermore, the vapor-liquid equilibria of the ternary system carbon dioxide+cyclohexanol+cyclohexane and its binary subsystems are investigated. The modified Lorentz-Berthelot combination rule with one state-independent binary interaction parameter was used for the pairwise unlike dispersive interactions. Per binary subsystem, the parameter was adjusted to a single experimental vapor pressure. The binary subsystems are in good agreement with experimental data throughout the entire composition range. For the ternary system, the vapor pressure is underpredicted by about 12 %.

1 Introduction

In this project on molecular modeling and simulation of thermodynamic properties of hydrogen bonding fluids, numerous molecular models have been developed in recent years. In order to set up a new molecular model for a given substance, a methodology is now at hand. Starting from quantum mechanical calculations of a single molecule in the vacuum, the geometry is determined via the Hartree-Fock approach. On top of this geometry optimization, also the electrostatics of the molecule can reasonably be determined by quantum mechanics. By following the COSMO idea to a single molecule is embedded into a dielectric cavity that mimics the electrostatics of a liquid, a Moller-Plesset 2 calculation yields the electron density distribution of the molecule.

* Author to whom correspondence should be addressed: Prof. Dr.-Ing. habil. J. Vrabec.

W.E. Nagel et al. (eds.), *High Performance Computing in Science and Engineering '10*, DOI 10.1007/978-3-642-15748-6_39,

On the basis of this data, polar moments are derived. To finalize a molecular model, the Lennard-Jones parameters have to be determined which can be done by optimization to vapor-liquid equilibrium (VLE) data. Particularly experimental data on the vapor pressure and the saturated liquid density of the pure fluid from the triple point to the critical point should be considered.

It was shown for numerous substances that hydrogen bonding itself may also reasonably be modeled by electrostatic interaction sites, i.e. partial charges that are located eccentrically within Lennard-Jones sites. Examples for fluids which were described in this way are water, methanol, methylamine, ammonia and cyclohexanol. The achieved quality of optimization to experimental VLE data gives valuable information whether the model structure, e.g. the chosen number and type of interaction sites, was reasonable. If a chosen structure is oversimplified, an accurate adjustment to the experimental VLE data is not feasible.

Nonetheless, the most attractive capability of molecular models is their transferability. This allows obtaining information on other types of properties or how the molecules behave in other environments like in mixtures. In the present report, the two hydrogen bonding molecules ammonia and cyclohexanol, which were modeled in preceding periods, are investigated in this sense. The following section, discusses a fully predictive application of the ammonia model to obtain the transport properties self-diffusion coefficient and shear viscosity over a range of temperature and up to very high pressures, being 2000fold of the ambient pressure. Thereafter, an application of three molecular models for carbon dioxide, cyclohexane and cyclohexanol is given, to describe and predict the phase behavior of the mixtures that can be formed out of these components.

2 Transport Properties of Ammonia

Transport properties play an important role in many fields of science and engineering. They are generally obtained from experiments, however, at high temperatures and pressures, usually sophisticated and expensive equipment is required. Moreover, traditional phenomenological equations of state are often unable to even reproduce the static thermodynamic behavior of hydrogen bonding fluids under these conditions. In cases when direct measurements are difficult, molecular simulation offers an attractive alternative route to predict these properties [1].

Different methods for calculating transport properties using molecular dynamics simulations are available. The equilibrium methods (EMD) using the using either the Green-Kubo formalism or the Einstein relations, analyze the time dependent response of a fluid system to spontaneous fluctuations. In this work, EMD and the Green-Kubo formalism were used to predict the self-diffusion coefficient and the shear viscosity of liquid ammonia. For details on these approaches, see [2].

Ammonia is a well-known chemical intermediate, mostly used in fertilizer industries but also as refrigerant. It is a strongly hydrogen-bonding molecule that possesses a simple symmetric structure, which makes it attractive for experimental and theoretical studies.

Numerous potential models for the intermolecular interactions of ammonia have been proposed in the literature, e.g. [3–5]. These models have mostly been used to study hydrogen bonding in liquid ammonia. Just a few of them have been tested with respect to their ability to predict transport properties. E.g., Impey and Klein [3] predicted the self-diffusion coefficient of liquid ammonia at 260 K, underestimating it by around 45%. Sagarik et al. [4] also tested their model with respect to the self-diffusion coefficient obtaining slightly better results than [3]. Only Mansour and Murad [5] predicted the self-diffusion coefficient and the shear viscosity along the coexistence curve achieving a good agreement to experimental data.

The present work assesses the predictive power of the rigid, non-polarizable molecular model for ammonia developed in preceding work of our group [6] with respect to transport properties over a wide range of thermodynamic conditions. It was also reported in the present book series in 2005 [7] on the development of this ammonia model. The ammonia model was optimized on the basis of quantum mechanical data as well as experimental data on vapor pressure and saturated liquid density. No experimental transport property data was thus taken into account.

Two types of transport properties, i.e. self-diffusion coefficient and shear viscosity, were studied here using EMD and the Green-Kubo formalism. Both properties were predicted for ammonia at 10, 50, 75, 100, 150 and 200 MPa in the temperature range from 203 to 473 K. Exemplarily, Fig. 1 shows the temperature dependence of the predicted self-diffusion coefficient at a pressure of 50 MPa in comparison with experimental data from the literature. The predicted temperature dependence of the shear viscosity is shown in Fig. 2 at 50 MPa together with data taken from the literature.

As can be seen, the predicted transport properties show a good agreement with the data from the literature. This is the case throughout the regarded wide range of temperature and pressure with a similar accuracy. Molecular modeling and simulation correctly reproduces the temperature dependence for both properties. The predicted self-diffusion coefficient deviates from the literature data by 8% on average and the shear viscosity by 12%. Higher deviations are observed for the shear viscosity at low temperatures, where the molecular approach tends to overestimations. This can be explained as an effect of the long-time behavior of the shear viscosity autocorrelation function, cf. [2]. In case of the self-diffusion coefficient, the molecular model exhibits at high temperatures a slightly lower molecular mobility than indicated by the experiment. Nonetheless, it was confirmed that molecular modeling is a robust approach to predict transport properties also for strongly hydrogen bonding fluids like ammonia.

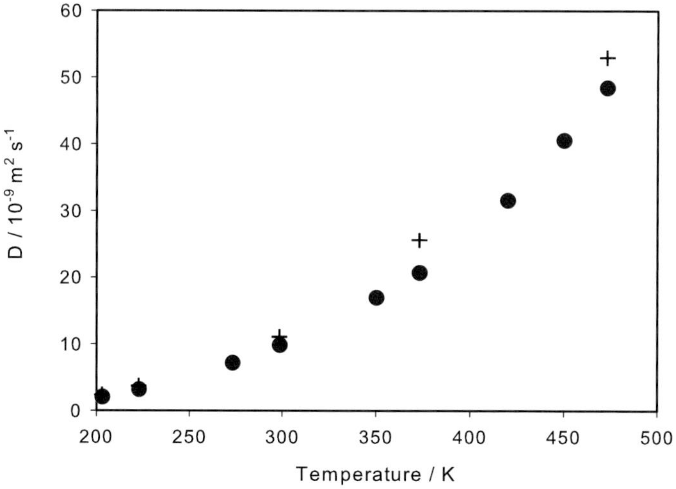

Fig. 1. Temperature dependence of the self-diffusion coefficient of ammonia at 50 MPa. Present simulation results (•) are compared to literature data (+) from [8]

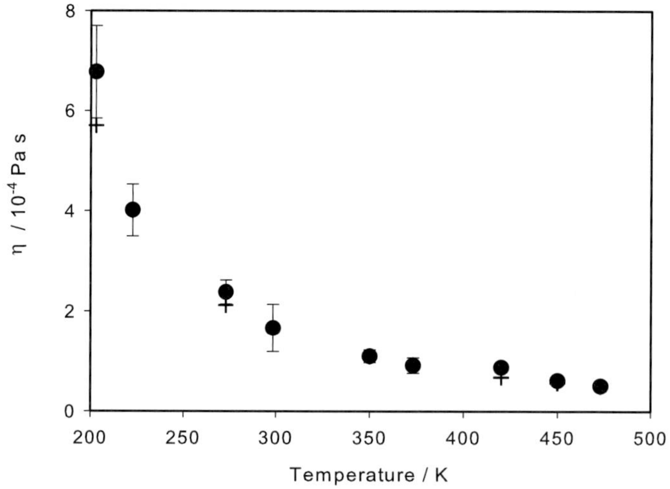

Fig. 2. Temperature dependence of the shear viscosity of pure ammonia at 50MPa. Present simulation results (•) are compared to literature data (+) from [9]

3 Vapor-Liquid Equilibria of Carbon Dioxide + Cyclohexane + Cyclohexanol and Its Subsystems

In the Collaborative Research Centre 706 (SFB 706) at the University of Stuttgart, novel octahedral molecular sieves for the heterogeneously catalyzed selective oxidation of cyclohexane are investigated. There, supercritical fluids and carbon dioxide-expanded liquids are used as innovative reaction media.

For a rational planning of catalytic experiments and process design, especially at higher pressures, reliable thermodynamic data are needed. Due to the fact that systems containing both oxygen and cyclohexane are highly reactive, they can hardly be investigated with respect to the phase behavior. Most groups used nitrogen as a substitute for oxygen to predict the phase behavior of the reacting system. Alternatively, for predictive applications, e.g. the Peng-Robinson equation of state has been used. Molecular modeling and simulation is an approach to generate data on a physically more sound basis.

Using the molecular model for cyclohexanol that was reported in the present book series in 2008 [10], predictions of VLE of the ternary mixture carbon dioxide + cyclohexane + cyclohexanol were carried out. The molecular models for carbon dioxide [11] and cyclohexane [12] were taken from prior work of our group as well. For an accurate modeling of mixtures, the pairwise unlike interactions have to be determined. While the unlike polar interactions are fully specified by the laws of electrostatics, the unlike dispersive interaction is not straightforward. In order to specify this often weak cross-interaction, combining rules are applied. Here, the modified Lorentz-Berthelot combining rule with one state-independent binary interaction parameter ξ was used

$$\sigma_{ab} = \frac{\sigma_a + \sigma_b}{2}, \tag{1}$$

$$\epsilon_{ab} = \xi\sqrt{\epsilon_a \epsilon_b}. \tag{2}$$

For VLE, ξ should be adjusted to a single experimental binary vapor pressure, as shown in [13]. In the following, the adjustment and predictions for the three binary subsystems and the ternary system are presented.

3.1 Carbon Dioxide + Cyclohexane

The binary interaction parameter ξ was adjusted for this binary mixture at 410.9 K to the vapor pressure, yielding a value of $\xi = 0.978$. Figure 3 presents the simulation results together with experimental data [14–16] and the Peng-Robinson equation of state [17] over a range of temperature from 344.4 to 473.15 K. For 410.9 K the agreement between simulation and experiment is excellent, however, at 473.15 K simulation overpredicts the vapor pressure and at 344.4 K simulation somewhat underpredicts the vapor pressure. Interestingly, the Peng-Robinson equation of state shows the same trend.

3.2 Carbon Dioxide + Cyclohexanol

For the binary mixture carbon dioxide + cyclohexanol the binary interaction parameter was adjusted to the vapor pressure at 393 K, yielding $\xi = 0.938$. Of these components, cyclohexanol is hydrogen bonding. For mixtures with hydrogen bonds, classical equations of state, like the one by Peng and Robinson

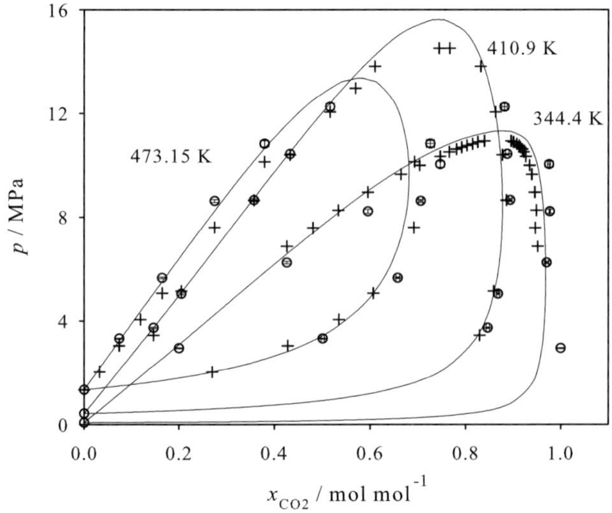

Fig. 3. Vapor-liquid equilibria of carbon dioxide + cyclohexane at 344.4, 410.9 and 473.15 K: (•) present simulation data; (+) experimental data [14–16]; (—) Peng-Robinson equation of state with $k_{ij} = 0.123$.

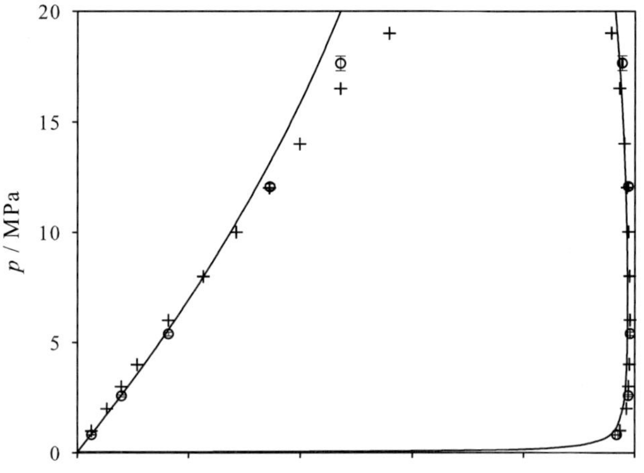

Fig. 4. Vapor-liquid equilibria of carbon dioxide + dyclohexanol at 393 K: (•) present simulation data; (+) experimental data [18]; (—) Peng-Robinson equation of state with $k_{ij} = 0.15$.

[17], are often inaccurate in predicting the VLE. In Fig. 4 the simulation results are compared to experimental data [18] and the Peng-Robinson equation of state. The simulation results are in a very good agreement with the exper-

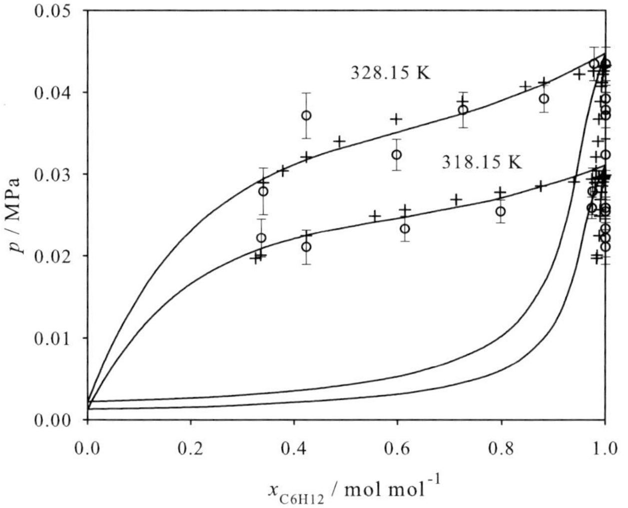

Fig. 5. Vapor-liquid equilibria of cyclohexane + cyclohexanol at 318.15 and 328.15 K: (•) present simulation data; (+) experimental data [19]; (—) Peng-Robinson equation of state with $k_{ij} = 0.0718$.

imental data, whereas the equation of state overpredicts the vapor pressure at high carbon dioxide contents in the liquid.

3.3 Cyclohexane + Cyclohexanol

The last binary mixture presented in this section is cyclohexane + cyclohexanol. Due to the fact that both molecules are large, which results in numerous interaction sites, and that cyclohexanol forms hydrogen bonds, simulating this mixture is computationally very demanding. It was not necessary to adjust ξ, thus the results presented in Fig. 5 are strictly predictive, i.e. $\xi = 1$. Figure 5 shows the simulation results together with experimental data [19] and the Peng-Robinson equation of state for 318.15 and 328.15 K. Note that experimental data is available for these temperatures only. Because these temperatures are near the triple point temperature of both components and thus the vapor pressure is low, the simulation results have high relative statistical uncertainties. Nevertheless, a good agreement with the experimental data was achieved. Also the predicted vapor phase composition from simulation is in a significantly better agreement with the experiment than the equation of state.

3.4 Carbon Dioxide + Cyclohexane + Cyclohexanol

Experimental data for ternary mixtures are very rare. For carbon dioxide + cyclohexane + cyclohexanol only a single VLE data point at 423 K and

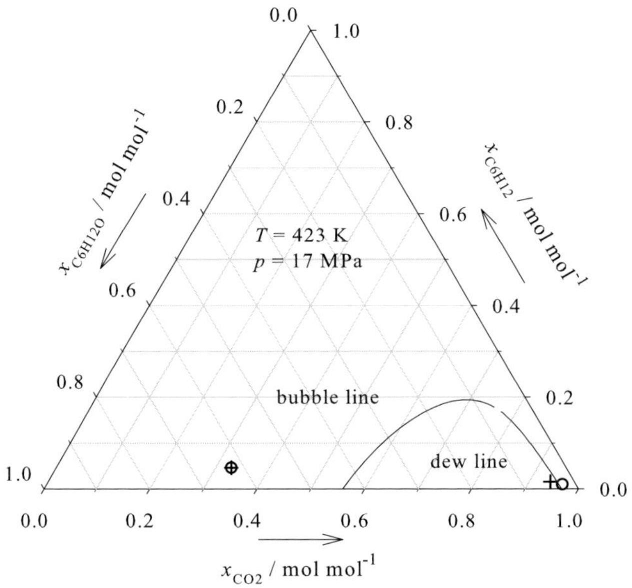

Fig. 6. Vapor-liquid equilibria of carbon dioxide + cyclohexane + cyclohexanol at 423 K and 17 MPa: (•) present simulation data; (+) experimental data [20]; (—) Peng-Robinson equation of state

17 MPa can be found [20]. The simulation result for this data point is predicted with the adjusted binary interaction parameters from the binary subsystems. In Fig. 6 the simulation result, the experimental data and the Peng-Robinson equation of state are compared. A good agreement for the vapor phase concentration between simulation and experiment was found. Nevertheless, the simulative vapor pressure is underpredicted by around 12 %. The Peng-Robinson equation of state is not capable to predict the ternary VLE of this mixture.

4 Conclusion

It was shown that molecular models which are based on quantum mechanical information on geometry and electrostatics and are optimized to experimental pure substance VLE data can be transferred to different predictive applications. Thereby, it was confirmed that hydrogen bonding can be described by eccentric point charges.

For liquid ammonia, transport data that describe the response of a system to external gradients were predicted over a wide range of temperature and pressure. Such data are generally rare, particularly for mixtures. Thus, it was confirmed that molecular modeling and simulation opens up a reliable route to such properties. Further work will investigate multi-component mixtures where hardly any experimental data is available.

The phase behavior of multi-component mixtures is essential for many process engineering applications, like reactions or separations. Particularly reactive, toxic or explosive mixtures are little attractive for experimental approaches so that *in silico* experiments are preferred. For the highly non-ideal mixture carbon dioxide+cyclohexan+cyclohexanol it was shown here that molecular modeling and simulation allow gaining access to necessary thermodynamic data. Subsequent work will include the component oxygen which is reactive and thus problematic *in vivo*, but can well be handled in computer experiments.

References

1. Allen, M.P. *Introduction to Molecular Dynamics Simulation*, Computational Soft Matter, Vol. 23, John von Neumann-Institute for Computing, Julich, 2004
2. Guevara-Carrion, G., Vrabec, J., Hasse, H. *J. Phys. Chem. B* **2008**, *112*, 16664.
3. Impey, R.W., Klein, M.L. *Chem. Phys. Lett.* **1984**, *104*, 579.
4. Sagarik, K.P., Ahlrichs, R., Brode, S. *Mol. Phys.* **1986**, *57*, 1247.
5. Mansour, K.A., Murad, S. *Fluid Phase Equil.* **1987**, *37*, 305.
6. Eckl, B., Vrabec, J., Hasse, H. *Mol. Phys.* **2008**, *106*, 1039.
7. Schnabel, T., Vrabec, J., Hasse, H. in Nagel, W. E.; Jäger, W.; Resch, M. (Eds.). *High Performance Computing in Science and Engineering '05* **2006**, 319.
8. Gross, T., Buchhauser, J., Price, W., Tarasov, I.N., Lüdemann, H. *J. Mol. Liq.* **1997**, *73*, 433.
9. Lemmon, E.W., Huber, M.L., McLinden, M.O. *Standard Reference Database 23: Reference Fluid Thermodynamic and Transport Properties -REFPROP;* NIST: 2007.
10. Merker T., Guevara-Carrion G., Vrabec J., Hasse H. in Nagel, W. E.; Kröner, D. B.; Resch, M. (Eds.). *High Performance Computing in Science and Engineering '08* **2009**, 529.
11. Vrabec, J., Stoll, J., Hasse, H. *J. Phys. Chem. B* **2001** *105*, 12126.
12. Eckl, B., Vrabec, J., Hasse H. *J. Phys. Chem. B* **2008** *112*, 12710.
13. Schnabel, T., Vrabec, J., Hasse H. *J. Mol. Liq.* **2007**, *135*, 170.
14. Nagarajan, N., Robinson, R.L. Jr. *J. Chem. Eng. Data* **1987**, *32*, 369.
15. Shibata, S.K., Sandler, S.I. *J. Chem. Eng. Data* **1989**, *34*, 419.
16. Krichevski, I.R., Sorina, G.A. *Russ. J. Phys. Chem.* **1960**, *34*, 679.
17. Smith, J.M., Van Ness, H.C., Abbott, M.M. *Introduction to chemical engineering thermodynamics*, McGraw-Hill, New York, 1996.
18. Chen, J.-T., Lee, M.-J. *J. Chem. Eng. Data* **1996**, *41*, 339.
19. Chevalley, J. *Bull. Soc. Chim. Fr.* **1961**, *52*, 510.
20. Mukhapadhay, M., Srinivas, P. *Ind. Eng. Chem. Res.* **1996**, *35*, 4713.

Software Framework UG: Parallel Simulation of a Three-Dimensional Benchmark Problem for Thermohaline-Driven Flow

Michael Lampe, Alfio Grillo, and Gabriel Wittum

G-CSC, Goethe-Universität Frankfurt, Kettenhofweg 139, 60325 Frankfurt (M.), Germany, lampe@gcsc.uni-frankfurt.de, grillo@gcsc.uni-frankfurt.de, wittum@gcsc.uni-frankfurt.de

$\mathcal{UG}$ [2] is a software framework for the numerical solution of problems that are described by systems of partial differential equations. It provides functionality for *u*nstructured grids, adaptive local grid refinement, robust multigrid methods for the solution of systems of linear equations, and parallelization of all these algorithms on MIMD-style computers. In this contribution, we present results of an application built upon $\mathcal{UG}$, a finite-volume simulator for thermohaline-driven flow called d^3f [3], on HLRS' Nehalem cluster.

1 Introduction

An often encountered problem in environmental science is the modelling and simulation of temperature- and density-driven groundwater flow, which combination we call *t*hermohaline-driven flow. To this end, we consider a porous medium macroscopically described by a biphasic mixture consisting of a fluid- and a solid-phase. The former is a two-constituent fluid experiencing single-phase flow, and the latter is a porous solid, which may be identified with the rocky matrix of a soil. The fluid-phase filtrating the pores is itself a mix of pure water and hypersaline water (referred to as *b*rine). As the density of the fluid-phase depends on both its brine mass fraction and its temperature, this gives rise to the flow phenomena we study here.

Our two numerical experiments consist of placing a relatively small brine "parcel" in the middle of a cuboid porous medium saturated by a fluid and assumed to be homogeneous and in thermo-mechanic equilibrium prior to the insertion of the parcel. Depending on salinity and temperature of this parcel, we observe in one case positive and in the other case negative buoyancy which induces in both cases a characteristic and interesting fluid flow.

W.E. Nagel et al. (eds.), *High Performance Computing in Science and Engineering '10*, DOI 10.1007/978-3-642-15748-6_40,

2 Mathematical Model

Our mathematical model consists of three partial differential equations, that describe (in this order) conservation of fluid mass as a whole, conservation of salt mass, and conservation of energy. [1, 5]

$$\frac{\partial}{\partial t}(\phi\rho) + \nabla \cdot (\rho \boldsymbol{q}) = 0 \tag{1}$$

$$\frac{\partial}{\partial t}(\phi\rho\omega) + \nabla \cdot (\rho\omega\boldsymbol{q} - \rho\boldsymbol{D}\nabla\omega) = 0 \tag{2}$$

$$\frac{\partial}{\partial t}\Big((\phi\rho C + (1-\phi)\rho_r C_r)\Theta\Big) + \nabla \cdot \big(\rho C\Theta\boldsymbol{q} - \nabla(\Lambda\Theta)\big) = 0 \tag{3}$$

Here, ϕ denotes the porosity of the medium, ρ the fluid density, ω the brine mass fraction, $\boldsymbol{D}$ the diffusion tensor, C and C_r the thermal capacity of the fluid-phase and the rock, respectively, Θ the temperature, ρ_r the density of the rock, and Λ the thermal conductivity of the mixture as a whole. Finally, the filtration velocity $\boldsymbol{q}$ is given by Darcy's law,

$$\boldsymbol{q} = -\frac{K}{\mu}(\nabla p - \rho\boldsymbol{g}) \tag{4}$$

with K being the permeability of the porous medium, μ the viscosity of the fluid-phase, $\boldsymbol{g}$ the gravity vector, and pressure p. If we add a constitutive law for the fluid-phase density, we end up with a system of three equations in the three unknowns p (pressure), ω (brine mass fraction), and Θ (temperature). Following [7], we consider the constitutive law for the fluid-phase density to be as follows:

$$\frac{1}{\rho} = \frac{1-\omega}{\rho_w} + \frac{\omega}{\rho_b} \tag{5}$$

This equation assumes volume additivity for the fluid components pure water and pure brine with densities ρ_w and ρ_b, respectively. Temperature dependency is expressed by

$$\rho_{w/b} = \rho_{0;w/b} \exp\big(-\alpha_{w/b}(\Theta - \Theta_{0;w/b})\big) \tag{6}$$

with $\rho_{0;w/b}$ being reference densities for pure water and brine at temperatures $\Theta_{0;w/b}$, and $\alpha_{w/b}$ thermal expansion coefficients.

The values we used for all these parameters in our simulations can be found in Table 1.

3 Numerical Solution

For the numerical solution of the governing equations (1)–(3), we are using a version of d^3f [3] that we are modifying—an ongoing effort—to include temperature in addition to pressure and brine mass fraction. The implementation

Table 1. Parameters for our simulations

Quantity	Symbol	Units	Value
Porosity	ϕ	–	0.1
Permeability	K	m^2	$5.0 \cdot 10^{-12}$
Viscosity	μ	$kg \cdot m^{-1}s^{-1}$	$1.0 \cdot 10^{-4}$
Molecular diffusivity	D	$m^2\,s^{-1}$	$1.0 \cdot 10^{-8}$
Thermal conductivity	Λ	$W \cdot m^{-1}\,K^{-1}$	1.8
Heat capacity of the solid-phase (rock)	C_r	$J\,kg^{-1}\,K^{-1}$	$1.0 \cdot 10^3$
Heat capacity of the fluid-phase	C	$J\,kg^{-1}\,K^{-1}$	$4.184 \cdot 10^3$
Mass density of the solid-phase (rock)	ρ_r	$kg\,m^{-3}$	$2.650 \cdot 10^3$
Gravity acceleration	g	$m\,s^{-2}$	9.81
Reference density of pure water	$\rho_{0;w}$	$kg\,m^{-3}$	800
Reference density of pure brine	$\rho_{0;b}$	$kg\,m^{-3}$	1000
Thermal expansion coefficient of water	α_w	K^{-1}	$1.45 \cdot 10^{-3}$
Thermal expansion coefficient of brine	α_b	K^{-1}	$8.13 \cdot 10^{-4}$
Reference temperature for water	$\Theta_{0;w}$	K	523.15
Reference temperature for brine	$\Theta_{0;b}$	K	563.15

basically follows the known "vertex centred" Finite Volume Method, which has the advantage of carrying over the conservation properties of the continuous equations to their discretized form. For a detailed description of the steps involved, the interested reader is referred to the paper by Frolkovič [4].

4 Benchmark Description

As in [8], we consider the following benchmark problem: Let $\mathcal{C}$ be a cubic region made of a porous medium, saturated by a fluid, and $L = 2500\,\text{m}$ the length of the edges. We then imagine to place in the centre of this cube a brine "parcel" $\mathcal{P}$ with volume ℓ^3 ($\ell = 500\,\text{m}$), and its faces parallel to those of the outer cube $\mathcal{C}$. Initially, the fluid in $\mathcal{R} := \mathcal{C} \setminus \mathcal{P}$ is pure water of constant temperature, while the fluid in $\mathcal{P}$ is a prescribed mixture of water and brine at another temperature. This leads to the following initial conditions:

$$\omega = \begin{cases} \omega_{\mathcal{P}} \text{ in } \mathcal{P} \\ 0 \quad \text{in } \mathcal{R} \end{cases} \quad \text{and} \quad \Theta = \begin{cases} \Theta_{\mathcal{P}} \text{ in } \mathcal{P} \\ \Theta_{\mathcal{R}} \text{ in } \mathcal{R} \end{cases} \tag{7}$$

Two sets of initial conditions are used, one which leads to positive, and one which leads to negative buoyancy of the initial parcel, see Table 2. Dirichlet boundary conditions fix the values for ω and Θ at their initial values at the top and bottom faces of the cube, while no-flow conditions are prescribed at the lateral faces. For pressure, the value is fixed at $p_0 = 0$ (normalisation, the absolute values for pressure don't feature in our model) at the top four corners, otherwise no-flow conditions are applied as well.

Table 2. Initial conditions for our simulations

Quantity	Subdomain	Symbol	Units	Value
Initial mass fraction	$\mathcal{R}$	$\omega_{\mathcal{R}}$	–	0.00
Initial temperature	$\mathcal{R}$	$\Theta_{\mathcal{R}}$	K	473.15
Initial fluid-phase mass density	$\mathcal{R}$	$\rho_{\mathcal{R}}$	$\mathrm{kg\,m^{-3}}$	875.00
Initial positive buoyancy				
Initial mass fraction	$\mathcal{P}$	$\omega_{\mathcal{P}}$	–	0.47
Initial temperature	$\mathcal{P}$	$\Theta_{\mathcal{P}}$	K	573.15
Initial fluid-phase mass density	$\mathcal{P}$	$\rho_{\mathcal{P}}$	$\mathrm{kg\,m^{-3}}$	831.00
Initial negative buoyancy				
Initial mass fraction	$\mathcal{P}$	$\omega_{\mathcal{P}}$	–	0.55
Initial temperature	$\mathcal{P}$	$\Theta_{\mathcal{P}}$	K	523.15
Initial fluid-phase mass density	$\mathcal{P}$	$\rho_{\mathcal{P}}$	$\mathrm{kg\,m^{-3}}$	919.00

5 Results and Discussion

Our first results with these benchmarks are described in [6] and were done on an in-house parallel computer on 64 processor cores with a grid resolution of $8^8 \approx 16$ million elements. 200 time steps for a simulation time of 20 years took about 50 hours of real time. Data files produced in VTK file format resulted in 0.33 TB per run. The visualisations shown in Figs. 1 and 2 were done with ParaView on that same in-house parallel computer with software rendering on the nodes.

The text cited below from [6] sums up our physical interpretation of the results:

For the following discussion we introduce the retardation factor

$$R := \frac{\phi \rho C + (1 - \phi)\rho_r C_r}{\phi \rho C}, \tag{8}$$

as the quotient of the effective thermal capacity of the mixture and the thermal convection term [8]. *Physically, it describes the fact that the solid is able to store thermal energy but not brine. With the parameters we have chosen for our simulations, we get* $R \approx 6.7$, *which means that the thermal front moves at approximately* 1/7*th the velocity of the brine front.*

As described in Sect. 4, we consider two cases, one in which the initial parcel is subject to negative *buoyancy (cf. Fig.* 1*) and one in which it is subject to* positive *buoyancy (cf. Fig.* 2*). Each of these two figures shows (from left to right) brine mass fraction, temperature, and density, with the initial values depicted in the first row, the evolution after ten years in the second row, and the final result after 20 years in the third row. In these pictures, we show the values of these quantities as a colour plot on the surface of the computational domain with the front cuboid (from top to bottom, front quarter of top/bottom face) removed.*

In the case of negative buoyancy, brine is transported downward faster than heat because of the retardation factor. The distribution of the brine mass

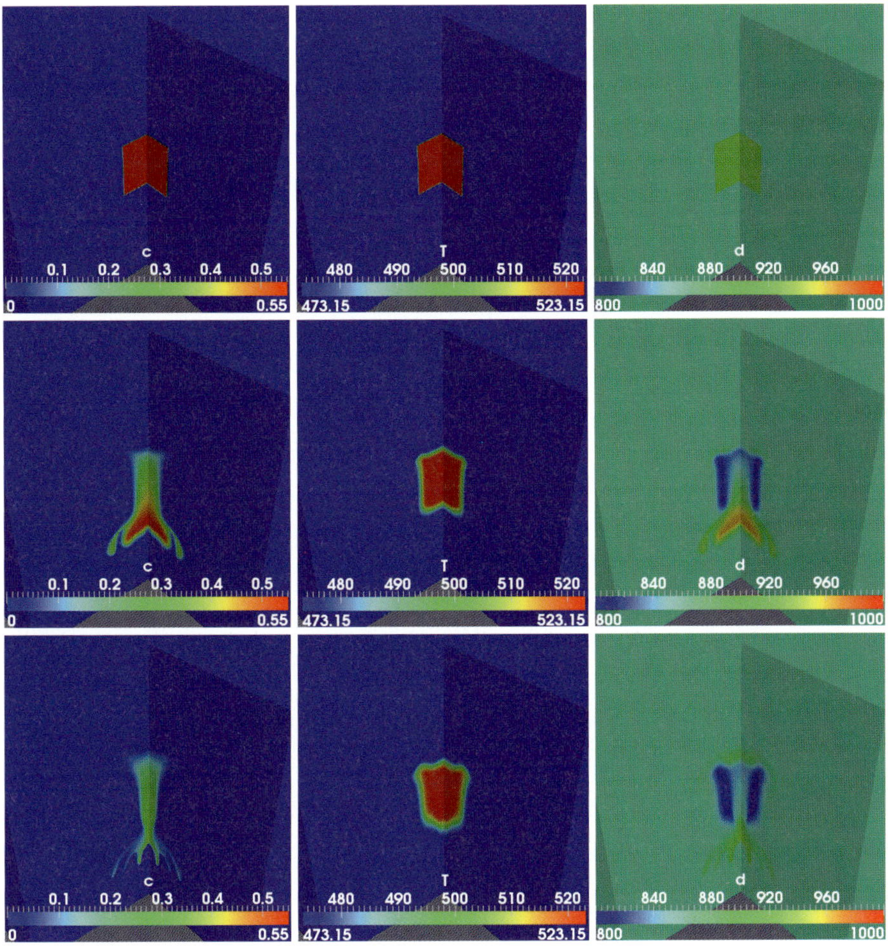

Fig. 1. Evolution of parcel: case of negative buoyancy

fraction features several "fingers", which are generated by the interplay of down- and up-swelling of fluid motions. Heat, on the other hand, is (as it is delayed) mainly confined to the initial parcel. When the brine starts to leave the initial parcel, fluid density in the parcel becomes smaller and eventually this now low-brine heat parcel begins to show even positive buoyancy. Thus we observe a separation of brine and heat from the initial parcel: brine moving downward and heat slowly upward. Since the formation of brine "fingers" occurs mainly at the sides of the initial parcel, there is a "cylindrical" zone in the middle of the parcel that contains less brine than at the beginning, but has still preserved (almost) its initial temperature. In this zone, we observe a small "fountain"-like motion: brine, that is moved upward by temperature, enters

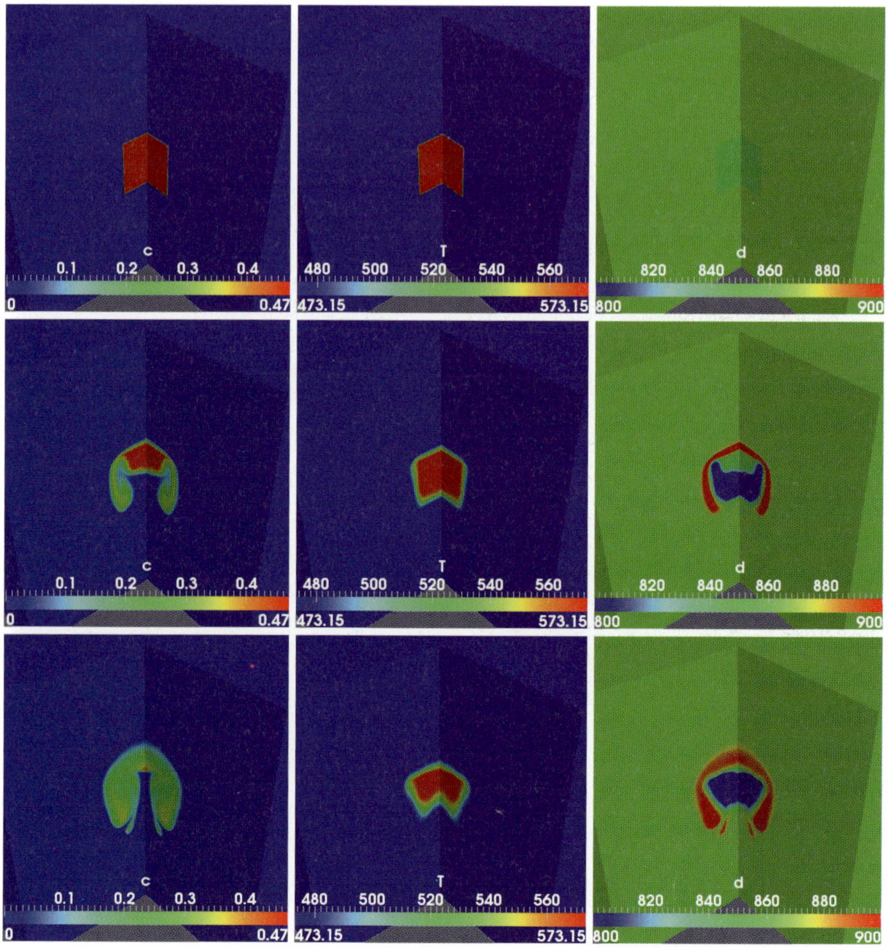

Fig. 2. Evolution of parcel: case of positive buoyancy

colder regions where it again leads to higher fluid density, and is consequently convected downward.

In the case of positive buoyancy, again due to the retardation factor, brine is transported upward faster than heat. Above the initial parcel, brine is cooled down, leading to higher density of the fluid. The induced flow moves brine downward along the sides of the parcel. This is comparable to what was described as the "fountain"-like motion above, but is now the main phenomenon. Despite positive buoyancy of the initial parcel, brine leaves this parcel and moves downward in this round-about way. Since the heat is again confined mostly to the initial parcel because of retardation, fluid density in this zone is decreasing, which even intensifies its positive buoyancy. However, the sep-

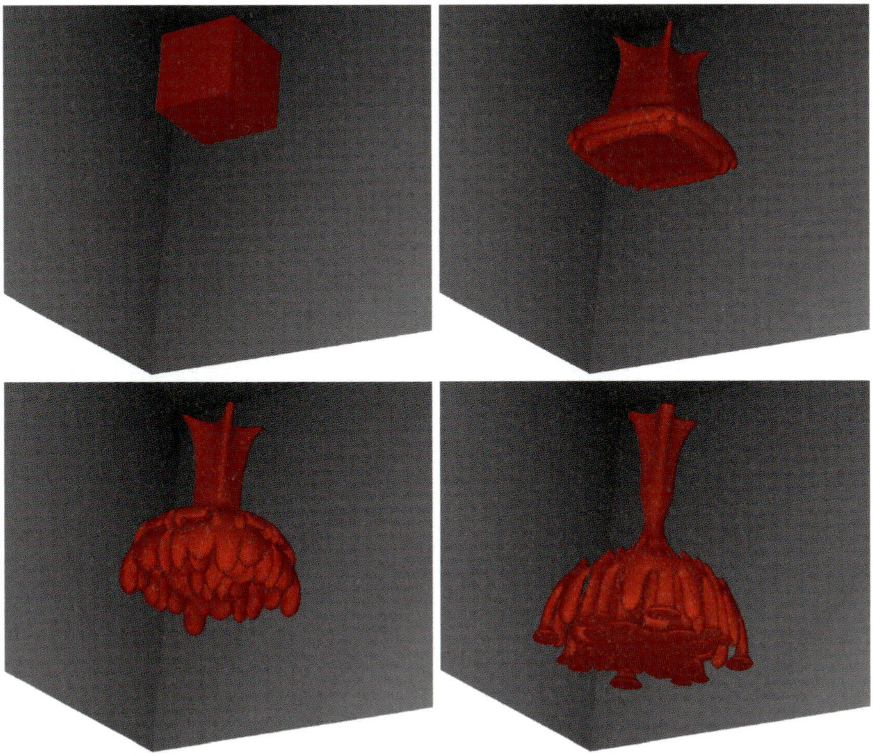

Fig. 3. Isosurfaces of brine mass fraction ($\omega = 0.15$) in case of negative buoyancy

aration of heat from the initial parcel is hindered by a density "lid", which is built up by the cold brine (highest density) flowing around the parcel (hot, low-brine, relatively low density).

It may be noted that such a density "lid" is also produced by the central "fountain"-like motion in the case of negative buoyancy, although it is less marked.

While 16 million elements sounds somehow impressive, it isn't in 3D. We therefore repeated the simulation for the case of negative buoyancy on HLRS' Nehalem cluster with a grid resolution of $8^9 \approx 134$ million elements. The 200 time steps (20 years) this time took about 18 hours on 2048 processor cores. The result is shown as an isosurface of brine mass fraction for $\omega = 0.15$, see Fig. 3. This value is supposed to be a good indicator for the location of the brine front.

How good is our parallel performance? For equal numbers of cores, our in-house cluster is almost exactly as fast as the HLRS Nehalem machine. So, if we make the problem eight times bigger and use 32 times as many processors, one might expect to quarter the running time, i.e. about $12\frac{1}{2}$ hours could be anticipated. But this is a little naive, because with the finer grid, we are also having a higher spacial resolution, and are able to detect new phenomena that were not present in the coarser version. In other words: we are not only making the problem bigger, we are also changing it. We therefore conclude that the actual 18 hours are acceptable and that programs built upon $\mathcal{UG}$ can be sensibly used on 2048 processor cores.

A final word with regard to visualisation: as the expected 2.64 TB of VTK data files seemed unmanageable, we have used $\mathcal{UG}$'s parallel built-in visualisation capabilities to produce the pictures shown without exporting any data files. While this avoids the size problem with output files altogether, it has the rather unsatisfactory side effect of leaving nothing but a few images from a simulation run.

Acknowledgements. This work has been supported by the Goethe-Universität Frankfurt and by the German Ministry of Economy and Technology (BMWi) via grant #02E10326. And of course, we thank HLRS for the opportunity to use their Nehalem cluster.

References

1. Bear, J. Hydraulics of groundwater. Dover Publications, INC., Mineola, New York (USA) (1979)
2. Bastian, P., Birken, K., Eckstein, K., Johannsen, K., Lang, S., Neuss, N., Rentz-Reichert, H. $\mathcal{UG}$—a flexible software toolbox for solving partial differential equations. Computing and Visualization in Science, **1** (1), 27–40 (1997)
3. Fein, E. d^3f – ein Programmpaket zur Modellierung von dichtegetriebenen Strömungen. GRS, Braunschweig, GRS-139, ISBN 3-923875-97-5 (1998)
4. Frolkovič, P. Finite volume discretization of density driven flow in porous media. In Benkhaldoun F., Vilsmeier, R. (ed): Finite Volumes for Complex Applications. Hermes, Paris, 433–440 (1996)
5. Grillo, A., Lampe, M., Wittum, G. Modelling and Simulation of temperature-density-driven flow and thermodiffusion in porous media. Journal of Porous Media. Accepted (2010)
6. Grillo, A., Lampe, M., Wittum, G. Three-dimensional simulation of the thermohaline-driven buoyancy of a brine parcel. Computing and Visualization in Science. Submitted (2010)
7. Oldenburg, C.M., Pruess, K. Layered Thermohaline Convection in Hypersaline Geothermal Systems. Transport in Porous Media, **33**, 29–63 (1998)
8. Oldenburg, C.M., Pruess, K. Plume separation by transient thermohaline convection in porous media. Geophysical Research Letters, **26** (19), 2997–3000 (1998)

Tailored Usage of the NEC SX-8 and SX-9 Systems in Satellite Geodesy

M. Roth, O. Baur, and W. Keller

University of Stuttgart, Institute of Geodesy, Geschwister-Scholl-Str. 24D, D-70174 Stuttgart, Germany, O.Baur@gis.uni-stuttgart.de

Summary. In the context of space-borne gravity field recovery, in recent years the implementation of tailored analysis algorithms has gained considerable significance. Both the optimization of individual software modules and the efficient parallelization of the procedures as a whole are key to access state-of-the-art science results within reasonable time frames. As such, HPC technologies play an important role in modern satellite geodesy. Typically, space-borne geodesy requires the estimation of tens of thousands of unknown gravity field parameters from tens of millions of observations. In order to assemble and solve the (linear) systems of equations in terms of a least-squares adjustment, we adapted our algorithms to the NEC SX-8 and SX-9 systems. We present implementation details and runtime results using the NEC systems as single-node architectures. As a result, compared to previous implementations, we accelerated design matrix assembly by a factor of ten. Moreover, we parallelized our algorithms successfully on both systems.

1 Introduction

Knowledge about the high-resolution gravity field of the earth is key to assess the complex interactions and dynamics of our planet. It considerably contributes to improve climate-forecasting models and to bring forward earthquake prognosis. The GOCE (Gravity field and steady-state Ocean Circulation Explorer) satellite, launched on March 17, 2009, has been specifically designed to meet the state-of-the-art geoscientific requirements. The gradiometer mission is supposed to recover the short-scale features of the terrestrial gravity field with a geoid accuracy of about 2 cm, and a spatial resolution around 100 km (ESA [7]). Therefore, GOCE will provide valuable contributions to the solution of up-to-date problems, such as arising in ocean circulation modeling, density variation mapping of the earth's interior, unification of height reference systems and high-precision satellite orbit determination.

Naturally, space-borne gravity field recovery involves the solution of large-scale systems of equations. GOCE data analysis, for instance, requires processing up to 100 000 000 observations in order to resolve more than 90 000

W.E. Nagel et al. (eds.), *High Performance Computing in Science and Engineering '10*, DOI 10.1007/978-3-642-15748-6_41,

unknown gravity field parameters. Solution quality issues such as decorrelation (satellite data are typically contaminated by correlated errors) and regularization (space-borne gravity field determination is an ill-posed problem) make repeated analysis inevitable. As a consequence, the computational burden can only be tackled by a tailored parallel processing scheme.

In former contributions (Austen et al. [2], Baur and Keller [5]), we presented HPC implementation and results operating (i) the iterative LSQR solver (representative for the group of Krylov-space methods) on distributed memory systems (e.g., the CRAY cluster "Strider" or the NEC cluster "Cacau") using MPI, and (ii) "brute-force" normal equations system inversion on shared memory and ccNUMA (cache-coherent Non-Uniform Memory Access) architectures (e.g., SGI system "Aster") using OpenMP.

Here, we present our first results adopting the NEC SX-8 and SX-9 systems for normal equations system inversion. To this end, we addressed a single SX node as shared memory system. Noteworthy, we payed special attention to the highly efficient implementation of the individual calculation steps involved in the overall process.

2 Methodology

Parameterized in spherical harmonics, the gravitational potential of the earth reads

$$V(\lambda, \varphi, r) = \frac{GM}{a} \sum_{l=0}^{L} \sum_{m=0}^{l} \left(\frac{a}{r}\right)^{l+1} \bar{P}_{lm}(\sin\varphi)[\bar{c}_{lm}\cos m\lambda + \bar{s}_{lm}\sin m\lambda] \quad (1)$$

(Heiskanen and Moritz [8]). Therein (λ, φ, r) denote spherical polar coordinates with λ east longitude, φ latitude and r distance from the geocenter. GM is the geocentric constant, a indicates the major semi-axis of a reference ellipsoid of revolution. Furthermore, the latitude-dependent functions $\bar{P}_{lm}(\sin\varphi)$ represent the 4π-normalized associated Legendre functions of the first kind (e.g., Hobson [10]). The series coefficients $\bar{c}_{lm}$ and $\bar{s}_{lm}$ constitute unknown model parameters. $L = l_{\max}$ is the maximum spherical harmonic degree (spectral resolution); in practice, the spectral resolution depends on data sensitivity. The GOCE mission allows resolving the geopotential coefficients up to $L = 250$.

Gravitational gradients, as observed by GOCE, are obtained by second-order (partial) derivation of (1). They are summarized in the 3×3 gravitational tensor (or Eötvös tensor) with main diagonal elements $V_{\lambda\lambda}$, $V_{\varphi\varphi}$ and V_{rr} (Rummel [14]). Exemplary for the radial component V_{rr}, the GOCE observation equation reads

$$V_{rr}(\lambda, \varphi, r) = \frac{GM}{a^3} \sum_{l=0}^{L} \sum_{m=0}^{l} \left(\frac{a}{r}\right)^{l+3} (l+1)(l+2) \times \bar{P}_{lm}(\sin\varphi)[\bar{c}_{lm}\cos m\lambda + \bar{s}_{lm}\sin m\lambda]. \quad (2)$$

Moreover, the partial derivatives with respect to the series coefficients become

$$\left\{ \begin{array}{c} \frac{\partial V_{rr}}{\partial \bar{c}_{lm}} \\ \frac{\partial V_{rr}}{\partial \bar{s}_{lm}} \end{array} \right\} = \frac{GM}{a^3} \sum_{l=0}^{L} \sum_{m=0}^{l} \left(\frac{a}{r}\right)^{l+3} (l+1)(l+2)\bar{P}_{lm}(\sin\varphi) \left\{ \begin{array}{l} \cos m\lambda \\ \sin m\lambda \end{array} \right. . \quad (3)$$

Reformulation of the (linear) functional model of gravitational gradients in matrix-vector notation yields

$$\mathbf{y} = \mathbf{A}\mathbf{x} + \mathbf{r} \quad (4)$$

with the vector of observations $\mathbf{y}(n \times 1)$, the vector of unknown parameters $\mathbf{x}(u \times 1)$ and the design matrix $\mathbf{A}(n \times u)$, i.e., the matrix of partial derivatives according to (3). In addition, (4) accounts for the case $n > u$, hence representing an overdetermined, inconsistent system of equations. The vector of residuals (or inconsistencies) is denoted as $\mathbf{r}(n \times 1)$. In order to obtain the "best" solution $\hat{\mathbf{x}}$ to (4), we minimized the square sum of residuals subject to $\min_{\mathbf{x}} \|\mathbf{r}\|^2 = \min_{\mathbf{x}} \|\mathbf{y} - \mathbf{A}\mathbf{x}\|^2$ (Koch [11]), resulting in the least-squares (LS) estimate

$$\hat{\mathbf{x}} = (\mathbf{A}^{\mathrm{T}}\mathbf{A})^{-1}\mathbf{A}^{\mathrm{T}}\mathbf{y} = \mathbf{N}^{-1}\mathbf{b}. \quad (5)$$

Moreover, the variance-covariance matrix of the parameter estimate, $D(\hat{\mathbf{x}})$, is obtained by inversion of the normal equations matrix $\mathbf{N} = \mathbf{A}^{\mathrm{T}}\mathbf{A}$ according to

$$D(\hat{\mathbf{x}}) = \hat{\sigma}^2(\mathbf{A}^{\mathrm{T}}\mathbf{A})^{-1} = \hat{\sigma}^2\mathbf{N}^{-1}, \quad (6)$$

$$\hat{\sigma}^2 = \frac{\hat{\mathbf{r}}^{\mathrm{T}}\hat{\mathbf{r}}}{n-u}, \quad \hat{\mathbf{r}} = [\mathbf{A}(\mathbf{A}^{\mathrm{T}}\mathbf{A})^{-1}\mathbf{A}^{\mathrm{T}} - \mathbf{I}]\mathbf{y}. \quad (7)$$

In order to estimate unknown parameters by means of LS adjustment, iterative solvers have gained considerable popularity. In satellite-based gravity field research, Krylov-space methods such as CGLS (Hestenes and Stiefel [9]) and LSQR (Paige and Saunders [13]) are widely adopted (e.g. Schuh [15], Ditmar et al. [6], Baur et al. [3]). They avoid explicit normal matrix assembly, hence are characterized by well manageable core memory requirements.

Unfortunately, however, iterative solvers are naturally unable to provide the accuracy information of the parameter estimate. As a matter of fact, the variance-covariance matrix is of crucial significance. It judges the reliability of the estimated coefficients as well as quantities derived from them (in terms of error propagation). For this reason, here we follow (5) and (6) to compute $\hat{\mathbf{x}}$ and $D(\hat{\mathbf{x}})$. A more detailed discussion on the pros and cons of iterative solvers opposed to "brute-force" normal equations system inversion is provided by Baur [4] and Baur and Keller [5].

3 Tailored Implementation on NEC SX Systems

In order to adapt our algorithms to the NEC SX systems, we performed initial optimization and testing on a standard personal computer (PC) with 1 GB RAM and a double core x64 CPU. In a second step, we achieved further optimization by analyzing medium- to large-scale problems within multi-processor environments. To this end, we conducted comprehensive test studies on the SX-8 front-end machine before running the tailored implementations on the SX-8 and SX-9 systems themselves. From the computational point of view, we split the coding in five main modules:

- data input,
- design matrix assembly,
- setup of the normal equations system
- solution of the normal equations system and
- data output.

For algebraic operations we made use of BLAS (Basic Linear Algebra Subroutines) and LAPACK (Linear Algebra PACKage) routines. Additional functions we coded by ourselves, paying special attention on optimization issues. In the following we give a detailed description of the tailored implementation. In its current development stage, our algorithms are optimized for shared memory and ccNUMA systems in combination with OpenMP. Hence, we addressed one node of the SX-8 (up to 8 CPUs) and SX-9 (up to 16 CPUs) systems. Forthcoming work will be concerned with the adaption of our implementations to multi-node HPC environments in combination with MPI. Amongst others, this will involve the replacement of LAPACK by ScaLAPACK (Scalable Linear Algebra PACKage) routines.

3.1 Data Input

Originally, the data is provided in ASCII format. In a first try, the data has been read directly to the memory, hence without any prior conversion done. Whereas on the local PC data input took a few seconds only, it turned out to be very time consuming on the SX systems. As an example, 100 000 data streams (each stream consisting of ten columns) kept a SX-8 CPU busy for $\approx 1/4$ h. The reason for slow data reading was that the scalar unit speed of the SX-8 CPUs is considerably low.

As mentioned before, the total number of observations, i.e., input data streams, is in the order of 10^8; therefore, data conversion while reading is out of question. We could overcome the obstacle by converting the data beforehand on a local PC from ASCII to binary format. Noteworthy, for the SX-8 and SX-9 systems, the binary format has to be big-endian (most significant byte first) opposed to standard x86-type computers which use the little-endian (most significant byte last) format. We chose to provide the data in one binary format only, namely big-endian for obvious reasons. For alternative HPC architectures, if necessary the program converts the data to little-endian while reading. In binary representation, the input of 100 000 data steams took less than one second on the SX systems.

3.2 Design Matrix Assembly

The computation of the design matrix **A** involves the evaluation of the fully normalized associated Legendre functions of the first kind, $\bar{P}_{lm}(\sin\varphi)$, for each observation. The $\bar{P}_{lm}(\sin\varphi)$ follow the recursive relations (all $\bar{P}_{lm}(\sin\varphi)$ with at least one negative index are set to zero)

$$
\begin{aligned}
\bar{P}_{0,0}(\sin\varphi) &= 1, \\
\bar{P}_{m,m}(\sin\varphi) &= W_{m,m}\cos\varphi\bar{P}_{m-1,m-1}(\sin\varphi), \\
\bar{P}_{l,m}(\sin\varphi) &= W_{l,m}\left[\sin\varphi\bar{P}_{l-1,m}(\sin\varphi) - W_{l-1,m}^{-1}\bar{P}_{l-2,m}(\sin\varphi)\right] \\
&= W_{l,m}\sin\varphi\bar{P}_{l-1,m}(\sin\varphi) - W_{l,m}^{*}\bar{P}_{l-2,m}(\sin\varphi), \\
W_{1,1} &= \sqrt{3}, \\
W_{m,m} &= \sqrt{\frac{2m+1}{2m}}, \\
W_{l,m} &= \sqrt{\frac{(2l+1)(2l-1)}{(l+m)(l-m)}} = \sqrt{\frac{4l^2-1}{l^2-m^2}}, \\
W_{l,m}^{*} &= W_{l,m}W_{l-1,m}^{-1} = \sqrt{\frac{(2l+1)(l+m-1)(l-m-1)}{(l^2-m^2)(2l-3)}}.
\end{aligned}
$$

(Heiskanen and Moritz [8]). The W-factors are only dependent on degree l and order m. In an initial step, we compute these factors and save them to a table for later use. The initial step accelerates the computation of the $\bar{P}_{lm}(\sin\varphi)$ considerably opposed to the repeated evaluation of the W-factors.

Further runtime optimization we achieved by rearranging (3) according to

$$
\left\{ \begin{array}{c} \frac{\partial V_{rr}}{\partial \bar{c}_{lm}} \\ \frac{\partial V_{rr}}{\partial \bar{s}_{lm}} \end{array} \right\} = \sum_{m=0}^{L} \left\{ \begin{array}{c} \cos m\lambda \\ \sin m\lambda \end{array} \right\} \sum_{l=m}^{L} \underbrace{\frac{GM}{a^3}\left(\frac{a}{r}\right)^{l+3}(l+1)(l+2)}_{=:B_l} \bar{P}_{lm}(\sin\varphi). \quad (8)
$$

In (8), the term B_l is computed once for each observation and saved to a table. Consequently, the summation over l ($l = m \ldots L$) only requires the multiplication of trigonometric functions with pre-evaluated values.

Altogether, we could speed up design matrix assembly up to 900% compared to previous implementations (cf. Table 1). As each observation, i.e., each row of the design matrix, can be handled independently, parallelization of the procedure is straightforward. The individual threads of a multi-processor architecture successively assemble the design matrix row by row.

3.3 Setup of the Normal Equations System

Recalling (5) and (6), memory requirements are dominated by the normal matrix dimension. Although only one triangle of the symmetric matrix has to be computed, the BLAS functions require the whole matrix to be kept in the memory. Thus, a spectral resolution of $L = 200$ (40 398 gravity field parameters) requires around 13 GB RAM, whereas for $L = 300$ (90 598 gravity field parameters) 65.7 GB RAM has to be available (the number of parameters increases approximately quadratically with increasing spectral resolution).

We assembled $\mathbf{N}$ by splitting the n observations in j blocks. With a suitable choice of j, design matrix storage requirements are negligible. $\mathbf{N}$ and $\mathbf{b}$ assembly becomes

$$\mathbf{N} = \mathbf{A}_1^{\mathrm{T}}\mathbf{A}_1 + \cdots + \mathbf{A}_j^{\mathrm{T}}\mathbf{A}_j = \sum_{i=1}^{j} \mathbf{A}_i^{\mathrm{T}}\mathbf{A}_i = \sum_{i=1}^{j} \mathbf{N}_i, \tag{9}$$

$$\mathbf{b} = \mathbf{A}_1^{\mathrm{T}}\mathbf{y}_1 + \cdots + \mathbf{A}_j^{\mathrm{T}}\mathbf{y}_j = \sum_{i=1}^{j} \mathbf{A}_i^{\mathrm{T}}\mathbf{y}_i = \sum_{i=1}^{j} \mathbf{b}_i. \tag{10}$$

In previous implementations, we first chose the dimension k of the individual blocks, and afterwards fixed the number of blocks j subject to $n = kj$. Hence, the number of observations had to be an integer multiple of both k and j. The improved implementation handles an arbitrary number of observations, allowing the dimension of the last block to be smaller than k. Moreover, k and j are now computed automatically from the memory size of the design matrix, which is the only information to be provided by the user.

In our initial coding we used the BLAS routine `DGEMM` for matrix-matrix multiplication, $\mathbf{N}_i = \mathbf{A}_i^{\mathrm{T}}\mathbf{A}_i$. We replaced it by the much more efficient `DSYRK` routine. The `DGEMV` routine was used to evaluate the vector $\mathbf{b}_i = \mathbf{A}_i^{\mathrm{T}}\mathbf{y}_i$. The setup of the normal equations system is the most time-consuming part of the algorithm. In order to achieve performance as best as possible, special implementations of the BLAS library exist that are tuned to match different parallel architectures. Noteworthy, the versions for the SX systems efficiently use their CPU vector units.

3.4 Solution of the Normal Equations System

For the solution of the normal equations system we used the LAPACK routine `DPOSV`.

4 Results

In order to evaluate both the quality and runtime performance of our implementations, we conducted a series of closed-loop simulation studies (GOCE real data is supposed to be released to the public mid 2010). The test data set consists of $n = 518\,400$ synthetic GOCE-like gravitational gradients, corresponding to an observation period of 30 days with a 0.2 Hz sampling rate. It was generated using the EGM96 (Lemoine et al. [12]) gravity field model. Exemplary for the spectral resolution of $L = 300$ for synthesis and $L = 200$ for analysis, Fig. 1 displays the accuracy of the closed-loop simulation in terms of degree-error RMS (Root Mean Square) values. In the following, we concentrate ourselves on the presentation of parallelization and runtime results; noteworthy, all our least-squares estimates meet the solution quality requirements in terms of degree-error RMS values.

4.1 Design Matrix Assembly

As stated before, we investigated initial code optimization and runtime testing on a local PC. Due to memory restrictions (1 GB RAM), we focused on small-scale systems up to a spectral resolution of $L = 70$ (5038 unknown gravity field parameters). Table 1 summarizes the speed-up we achieved due to the tailored implementation/optimization as outlined in Sect. 3. The speed-up increases continuously with increasing problem dimension. It is up to 900%, and expected to be even higher for large-scale problems.

4.2 Parallelization Performance

In this section, we present runtime results within parallel programming environments (using OpenMP for parallelization). In a first experiment, we performed a series of closed-loop computations on the front-end machine

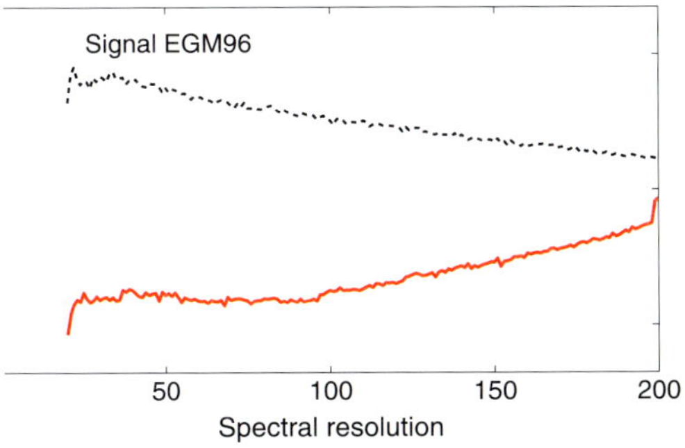

Fig. 1. Closed-loop simulation accuracy in terms of degree-error RMS values (red solid line) compared to simulation input signal (black dashed line). Besides spectral leakage effects, the geopotential parameter estimates are solely affected by numerical rounding errors. Synthesis up to spectral resolution $L = 300$. Analysis up to $L = 200$

Table 1. Runtime results for design matrix assembly on a local PC, dependent on the spectral resolution L; 100 000 observations (input data streams) considered

L	Initial implementation [s]	Optimized implementation [s]	Speed-up
30	8.43	1.20	702.5%
40	14.59	1.90	767.9%
50	22.94	2.75	834.2%
60	33.55	3.78	887.6%
70	44.55	4.93	903.7%

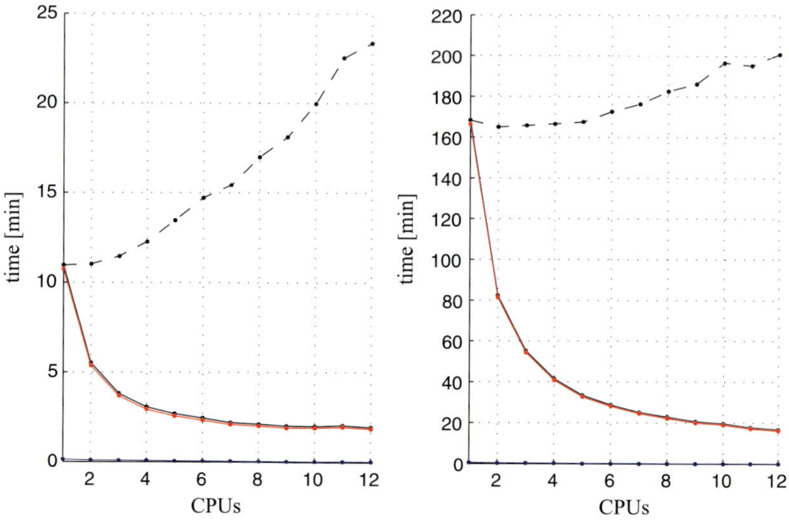

Fig. 2. Runtime results on front-end ASAMA (500 000 observations): design matrix assembly (blue), normal equations system assembly (red), overall time (black), user time (dashed black). Left panel $L = 50$; right panel $L = 100$

ASAMA. Based on 500 000 observations (input data streams) the gravity field parameters were solved up to degree and order $L = 50$, $L = 100$ respectively, using one to twelve CPUs. Figure 2 displays the corresponding runtime results. The real time (overall time) is clearly dominated by the computation of the normal equations matrix. It decreases with increasing number of CPUs, demonstrating the computations are split on different processors. On the other hand, the user time increases with increasing number of CPUs due to the more extensive communication between the individual threads.

How do our algorithms perform on the SX-8 and SX-9 systems? This is what we investigated in a second experiment. In particular, we were interested in runtime scaling dependent on the number of observations and number of CPUs considered. According to Fig. 3 (right panel), the SX-9 shows an expected behavior. Doubling the problem dimension results in doubling the runtime. Moreover, the algorithm scales well using different numbers of CPUs.

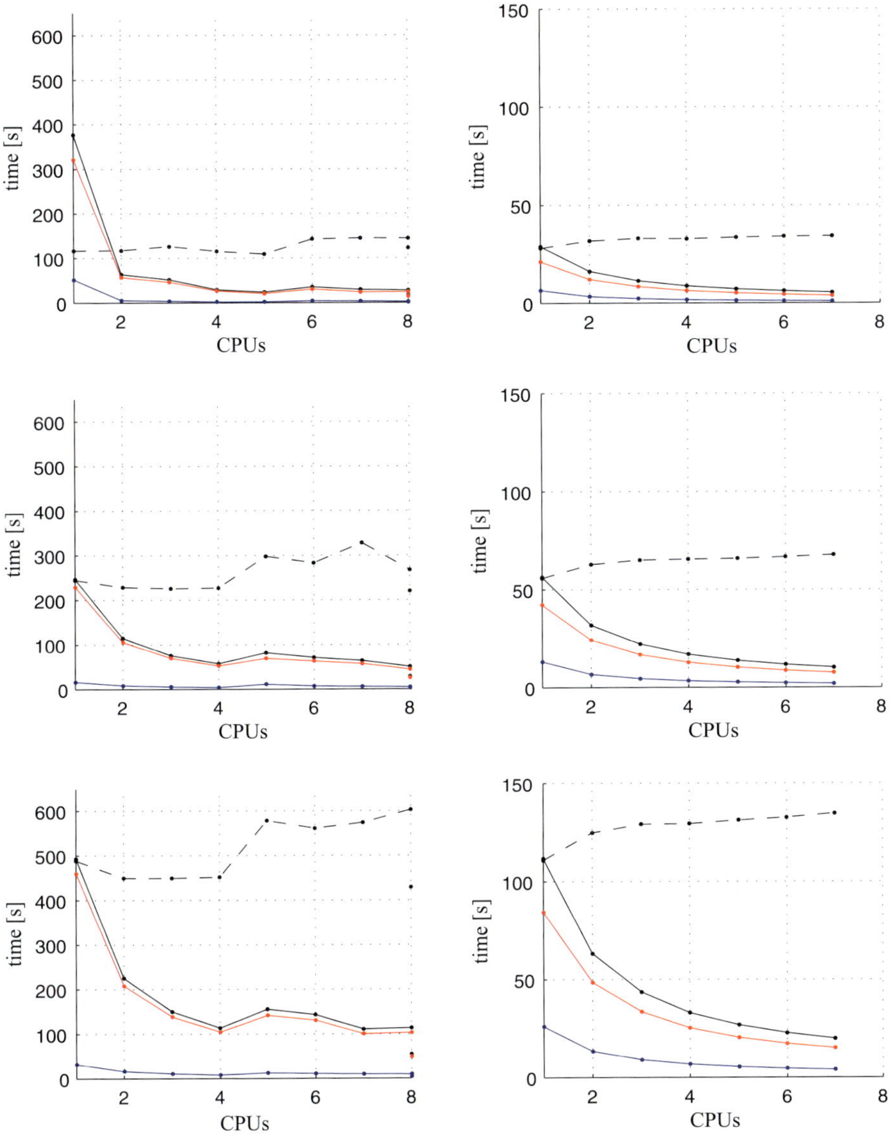

Fig. 3. Runtime results on SX-8 (left panel) and SX-9 (right panel) systems (number of observations: top 125 000, middle 250 000, bottom 500 000): design matrix assembly (blue), normal equations system assembly (red), overall time (black), user time (dashed black). Spectral resolution $L = 50$. The single dots in the left panel figures represent runtimes with exclusive use of one computing node (8 CPUs) on the SX-8 system

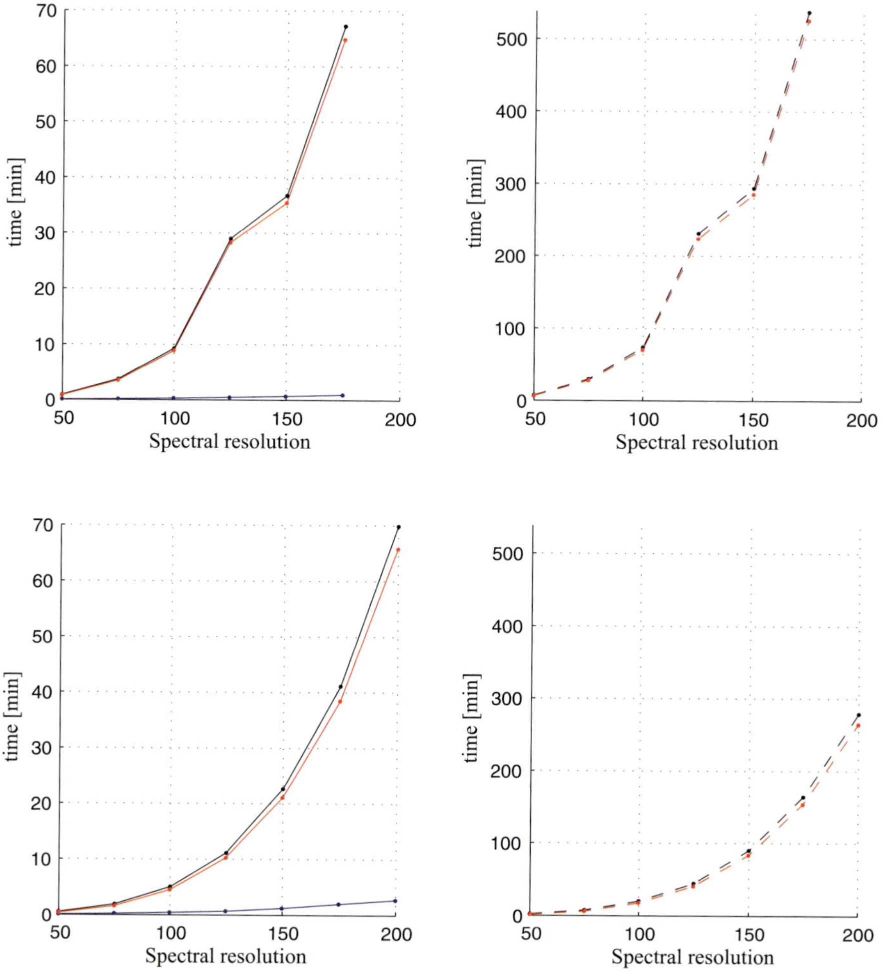

Fig. 4. Runtime results on SX-8 (upper panel, 8 CPUs) and SX-9 (lower panel, 4 CPUs) systems (500 000 observations). Left panel: design matrix assembly (blue), normal equations system assembly (red), overall time (black). Right panel: user time (dashed black), vector time (dashed red)

However, the graphs related to the SX-8 seem rather weird. As we found out, this is due to the `dq` queue we attached the (single-node) jobs to. `dq` is not designed for exclusive use, but various processes share the CPUs. The top left curves in Fig. 3, for instance, reveal the user time to be lower than the real time for one CPU. This indicates that several processes were running simultaneously. Therefore we did some additional tests using the `multi` queue for exclusive access to one node. The corresponding results for eight CPUs are highlighted by the single dots in Fig. 3. As expected, the runtimes are lower opposed to jobs attached to the `dq` queue.

Finally, in a third experiment, we focused on the parallelization performance with regard to medium- to large-scale problems. To this end, starting from $L = 50$, we increased the spectral resolution in steps of 25 to $L = 175$ (SX-8), $L = 200$ (SX-9) respectively. In order to avoid too long queuing times, we only addressed four CPUs on the SX-9 system. For SX-8, we chose the `multi` queue, requesting a full node (eight CPUs) for exclusive use. Figure 4 shows the runtimes on both systems. As a result, the algorithm makes good use of the vector units. The right panel in Fig. 4 reveals the user time and the vector time (time used by the vector unit) only to differ slightly. Moreover, we pinpoint the increasing computational costs with increasing spectral resolution, i.e., increasing number of unknown gravity field parameters.

Although we thought to have overcome the weird SX-8 behavior by choosing the `multi` queue, the runtime graphs again show features we presently can not explain. We expected an exponential behavior as provided by the SX-9 results.

5 Discussion and Conclusion

In order to adapt our algorithms to the NEC SX-8 and SX-9 systems, we achieved both code optimization and parallelization. Concerning optimization we accelerated design matrix assembly considerably. Compared to previous implementations, the efficiency increased by several hundred percent. This is due to the benefit of pre-evaluated tables and reordering of computations. Both result in the reduction of overall operations.

Using OpenMP for parallelization, runtime turned out to scale well up to eight CPUs; for a higher number of processors, communication costs balance out the benefit of additional computation power. Both the SX-8 and SX-9 system outperform the front-end machine ASAMA. The SX-8 is twice to three times faster than ASAMA. The SX-9 is roughly four times faster than the SX-8 (hence, twelve times faster than ASAMA). Altogether, we substantially improved the individual modules of the software.

With regard to large-scale problems, in near future we will extend our algorithms for multi-node distributed memory processing. We started first tests using MPI on the NEC SX systems. In this context, the ScaLAPACK parallel programming environment will be incorporated. Furthermore, the evaluation of the variance-covariance matrix will be added as an additional module.

Acknowledgements. The authors thank the High Performance Computing Center Stuttgart (HLRS) for the opportunity to use their computing facilities; furthermore, we gratefully acknowledge the helpful technical support. This work was supported by the German Ministry of Education and Research (BMBF) within the geoscientific R+D program GEOTECHNOLOGIEN under grant 03G0726G.

References

1. Alkhatib H., Schuh W.-D. (2007) Integration of the Monte Carlo covariance estimation strategy into tailored solution procedures for large-scale least squares problems, J. Geod. 81: 53–66, doi: 10.1007/s00190-006-0034-z
2. Austen G., Baur O., Keller W. (2005) Use of High Performance Computing in Gravity Field Research, in: Nagel W.E., Jäger W., Resch M. (Eds.) High Performance Computing in Science and Engineering '05, 305–318, Springer Berlin Heidelberg New York
3. Baur O., Austen G., Kusche J. (2008) Efficient GOCE satellite gravity field recovery based on least-squares using QR decomposition, J. Geod. 82: 207–221, doi: 10.1007/s00190-007-0171-z
4. Baur O. (2009) Tailored least-squares solvers implementation for high-performance gravity field research, Computers and Geosciences 35: 548–556, doi: 10.1016/j.cageo.2008.09.004
5. Baur O., Keller W. (2009) Computational considerations for satellite-based geopotential recovery, in: Nagel W.E., Kröner D.B., Dietmar B., Resch M. (Eds.) High Performance Computing in Science and Engineering '09, 511–522, Springer Berlin Heidelberg New York
6. Ditmar P., Klees R., Kostenko F. (2003) Fast and accurate computation of spherical harmonic coefficients from satellite gravity gradiometry data, J. Geod. 76: 690–705, doi: 10.1007/s00190-002-0298-x
7. ESA SP-1233 (1999) The four candidate Earth explorer core missions—gravity field and steady-state ocean circulation mission, European Space Agency Report SP-1233(1), Granada
8. Heiskanen W.A., Moritz H. (1967) Physical Geodesy, W.H. Freeman and Company San Francisco
9. Hestenes M.R., Stiefel E. (1952) Methods of conjugate gradients for solving linear systems, J. Res. Nat. Bur. Stand. 49: 409–436
10. Hobson E.W. (1931) The theory of spherical and ellipsoidal harmonics, University Press, Cambridge
11. Koch K.-R. (1999) Parameter estimation and hypothesis testing in linear models, Springer Berlin Heidelberg New York
12. Lemoine F.G., Kenyon S.C., Factor J.K., Trimmer R.G., Pavlis N.K., Chinn D.S., Cox C.M., Klosko S.M., Luthcke S.B., Torrence M.H., Wang Y.M., Williamson R.G., Pavlis E.C., Rapp R.H., Olson T.R. (1998) The Development of the Joint NASA GSFC and NIMA Geopotential Model EGM96, NASA Goddard Space Flight Center, Greenbelt
13. Paige C.C., Saunders M.A. (1982) LSQR: An algorithm for sparse linear equations and sparse least squares, ACM T. Math. Software 8: 43–71
14. Rummel R. (1986) Satellite gradiometry, in: Sünkel H. (Ed.) Mathematical and numerical techniques in physical geodesy, Lect. Notes Earth Sci., 7: 317–363, Springer, Berlin
15. Schuh W.-D. (1996) Tailored numerical solutions strategies for the global determination of the Earth's gravity field, Mitteilungen der Universität Graz 81

A Geodynamic Model of the Evolution of the Earth's Chemical Mantle Reservoirs

Uwe Walzer and Roland Hendel

Institut für Geowissenschaften, Friedrich-Schiller-Universität, Burgweg 11, 07749 Jena, Germany, u.walzer@uni-jena.de, roland.hendel@uni-jena.de

Summary. A dynamic 3-D spherical-shell model for the chemical evolution of the Earth's mantle is presented. Chemical differentiation, convection, stirring, and thermal evolution constitute an inseparable dynamic system. Our model is based on the solution of the balance equations of mass, momentum, energy, angular momentum, and four sums of the number of atoms of the pairs ^{238}U-^{206}Pb, ^{235}U-^{207}Pb, ^{232}Th-^{208}Pb, and ^{40}K-^{40}Ar. Similar to the present model, the continental crust of the real Earth was not produced entirely at the start of the evolution but developed episodically in batches. The details of the continental distribution of the model are largely stochastic, but the spectral properties are quite similar to the present real Earth. Fig. 6 reveals that the modelled present-day mantle has *no* chemical stratification but we find a marble-cake structure. If we compare the observational results of the present-day proportion of depleted MORB mantle with the model then we find a similar order of magnitude. The MORB source dominates under the lithosphere. In our model, there are nowhere pure unblended reservoirs in the mantle. It is, however, remarkable that, in spite of 4500 Ma of solid-state mantle convection, certain strong concentrations of distributed chemical reservoirs continue to persist in certain volumes, although without sharp abundance boundaries. Section 4 presents results regarding the numerical method, implementation, scalability and performance.

1 Introduction: Geochemical Mantle Reservoirs

Chemical differentiation in the uppermost parts of the silicate shell of the Earth alters the distribution of radioactive elements and volatiles and generates geochemical heterogeneity. On the other hand, convective stirring by solid-state creep diminishes and annihilates these chemical heterogeneities. For example, convection transports and mixes oceanic crust and oceanic mantle lithosphere into the deeper mantle. Where mantle viscosity is high, however, mixing is partly suppressed and chemical signatures tend to be partly preserved. Generation, preservation and destruction of mantle heterogeneities are obviously influenced by an interplay among chemical differentiation, convection, and secular cooling. It is evident that the chemical evolution of the

W.E. Nagel et al. (eds.), *High Performance Computing in Science and Engineering '10*, DOI 10.1007/978-3-642-15748-6_42,

mantle cannot be modeled without accounting for the convective process and the mixing it generates. The fact that chemical differentiation causes the distribution of heat-producing elements in the mantle to be non-uniform, in turn, also influences convection. This paper seeks to account for both differentiation and convection/mixing together. The *focus* is the distribution of the MORB source in the Earth's mantle. See also Hofmann [13, 14].

In this connection, there are four major questions.

1. How large is the mass ratio of the depleted MORB mantle (DMM) to that part of the mantle which is richer in incompatible elements, yet? MORB denotes the mid-oceanic ridge basalts.
2. Is there a chemical layering, e.g., an upper DMM spherical shell and a lower shell rich in incompatible elements?
3. Or are the chemical mantle reservoirs totally mixed by convection?
4. Is the distribution of depleted parts of the mantle like that of pancakes in a matrix or do we observe a marble-cake structure where DMM dominates only in the asthenosphere?

In the Ogawa [27] models, chemical stratification develops regardless of the strength of the olivine barrier and lithospheric plates move if the mass exchange through the 660-km discontinuity is not strongly impeded by the olivine barrier. For their case LIHST, the final solution of the compositional distribution is reminiscent of the model by Kellogg et al. [20] in regard to its geometry but not to how it arises. Despite large advances in numerical modeling, there is currently no consensus regarding the present-day spatial distribution of chemically distinct reservoirs in the Earth's mantle. The classical chemical two-layer model was a depleted mantle above the 660-km seismic discontinuity and a primordial mantle below it. But many modifications have since been suggested. Becker et al. [2] proposed a distribution of the primitive mantle in blobs plus an enriched recycled crust (ERC) at the bottom of the mantle. Tackley [37, 38] suggested that enriched piles sit on top of an ERC layer that is at the bottom. Because P-wave and S-wave velocity differences are not correlated deep in the lower mantle [22, 36], these variations cannot be accounted for by temperature differences. So, Kellogg et al. [20] proposed a primordial layer at the base of the mantle with an undulating upper surface.

The present picture of the chemistry and dynamics of the mantle is somewhat complex. Moreover, with respect to the provenance of MORB and Oceanic Island Basalts (OIB) there are two different fundamental interpretations.

(a) As illustrative for the first, Ito and Mahoney [15, 16] inferred that the MORB and OIB systematics arise from a heterogeneous mantle without large-scale reservoirs. Lower melt fractions and smaller sampling volumes generate the relatively heterogeneous OIB, whereas larger melt fractions and larger sampling volumes produce the relatively homogeneous MORB. The mean mantle of their model generates less than 30% of the mantle's heat budget so a large heat flow from the core is required. A similar interpretation is that, under slightly different P-T conditions, different degrees of partial melting, and different degrees of homogenization prior to eruption, MORB and OIB can be produced from the same statistical upper mantle assemblage [24–26, 32].

(b) As illustrative for the second major view, Kellogg et al. [19] adopt an extended conventional geochemical reservoir model and conclude that the isotopic differences between MORB and OIB cannot simply be due to differences in sampling volume but must also reflect differences in the source reservoirs and/or the melting processes. They claim to demonstrate that the argument that the absence of samples with a primitive-mantle isotopic signature means an absence of primitive material now in the mantle is not valid.

In the meantime, the theory of geochemical reservoirs has been further refined (Hofmann [14]; Bennett [3]) such that the reservoirs are no longer necessarily understood as separate large-scale volumes. For example, Stracke et al. [35] and Willbold and Stracke [42] propose a new FOZO similar to the traditional FOZO advanced by Hart et al. [12] except that this new FOZO is a small-scale component, ubiquitously dispersed throughout the entire mantle. However, in most *current* reservoir models, the percentages of the reservoirs can systematically vary across the mantle. The existence of sharp chemical boundaries is no longer deemed necessary. We remark that, according to our present dynamical model, the percentage of FOZO in the uppermost part of the mantle should be less than in the lower mantle. This result corresponds with Wilson's and Spencer's [43] conclusion that FOZO is the characteristic signal of mantle plumes. Using this conception, the mantle is everywhere heterogeneous and only the mixing ratios of the reservoirs vary. Whether or not such a distributed geochemical reservoir theory is viable is still an open issue. Sections 1 and 2 of [41] give lots of further information regarding the geochemical foundations of our numerical model.

2 Model

We use the numerical model by Walzer and Hendel [39] in which the differential equations of infinite Prandtl-number convection within a 3-D spherical-shell mantle are solved using a finite element method. We integrated the chemical differentiation of plateau basalts and continental growth calculating the whole thermal history of about 4500 Ma for those times when the mantle creep was mainly dominated by solid-state creep with only a very low percentage of magma chambers. Unlike other mantle-convection papers with continents, our continents are not artificially imposed but evolve by chemical differentiation of which the process has been represented by a tracer approach. A full derivation of the equations and a presentation of the model parameters is given by Walzer et al. [41]. Nevertheless, the present companion paper presents exclusively unpublished material.

The viscosity law is based on experimental results of Karato and Li [17], Karato and Wu [18], and Li et al. [23]. Therefore, we assume Newtonian solid-state creep for the Earth's mantle. The shear viscosity, η, is calculated by

$$\eta(r,\theta,\phi,t) = 10^{r_n} \cdot \frac{\exp(c\,\overline{T_m/T_{av}})}{\exp(c\,\overline{T_m/T_{st}})} \cdot \eta_3(r) \cdot \exp\left[c_t \cdot T_m \left(\frac{1}{T} - \frac{1}{T_{av}}\right)\right] \quad (1)$$

where r is the radius, θ the colatitude, ϕ the longitude, t the time, r_n the viscosity-level parameter, T_m the melting temperature, T_{av} the laterally averaged temperature, T_{st} the initial temperature profile, T the temperature as a function of r, θ, ϕ, t. The quantity r_n of the first factor of the right-hand side of (1) has been used for a stepwise shift of the viscosity profile to vary the averaged Rayleigh number from run to run. The second factor attempts to capture the increase in the viscosity profile resulting from the cooling of the Earth. According to Yamazaki and Karato [44], an appropriate value for $MgSiO_3$ perovskite is $c = 14$, and for MgO wüstite is $c = 10$. So, the lower-mantle c should be somewhere between these two values. For numerical reasons, we are restricted to a value of $c = 7$. The quantity $\eta_3(r)$ is the viscosity profile at the initial temperature and for $r_n = 0$. That is, $\eta_3(r)$ describes the dependence of viscosity on pressure and mineral phase. The derivation of $\eta_3(r)$ is provided in Walzer et al. [40]. In that paper, we start from a self-consistent theory using the Helmholtz free energy, the Birch-Murnaghan equation of state, the free-volume Grüneisen parameter and Gilvarry's [9] formulation of Lindemann's law. The viscosity is calculated as a function of melting temperature provided by Lindemann's law. We use pressure, P, bulk modulus, K, and $\partial K/\partial P$ from the seismic model PREM [8] to obtain the relative variation in radial viscosity. To set the absolute scale of the viscosity profile, we choose $\eta_3 = 3.45 \times 10^{20}$ Pa·s at 367 km depth. This assumption yields an asthenospheric average viscosity of 10^{21} Pa·s corresponding to the widely accepted value from postglacial uplift. Because we allow variation of r_n, the exact choice is not so crucial.

For the uppermost 285 km of the mantle (including crust), an effective viscosity, η_{eff}, was implemented where

$$\eta_{eff} = \min\left[\eta(P,T), \frac{\sigma_y}{2\dot{\varepsilon}}\right], \quad (2)$$

where P is pressure and $\dot{\varepsilon}$ is the second invariant of the strain-rate tensor. The quantity σ_y is a viscoplastic yield stress. This introduction of yield stress for the portion of the domain that includes the cold boundary layer allows our model to display plate-like behavior at the surface. Plate-like behavior would not occur without such a rheological treatment and without an asthenosphere.

3 Solution of Numerical Problems Regarding the Tracers Representing the Incompatible Elements in the Earth's Mantle and Results

3.1 General Remarks

Schmalzl [33] demonstrated more than a decade ago that the stirring behavior of the numerical simulation of a convective system depends on whether the simulation is 2-D or 3-D. See also [11]. For 2-D models, side-wall effects can raise additional difficulties since there are no such side walls in the real Earth. Even if we compute in 3-D, but in a Cartesian box, the balance of the atoms of a chemical element is problematic because the area of the core-mantle boundary (CMB) is considerably smaller than the Earth's surface. Therefore we decided to use the 3-D spherical-shell code Terra. As a resolution test, we performed each run twice with differing numbers of tracers. For our basic run (B-run) we used about 10.5 million tracers, and for our comparative run (C-run) we used about 84 million tracers.

The focus of this paper is the generation of continental material via oceanic plateaus by chemical differentiation and the *distribution of reservoirs in the mantle*. This differentiation creates complementary a depleted part of the mantle below the lithosphere. Observational evidence suggests that the contribution of OIBs to continental growth seems to have been considerably larger than that of MORB in the bulk of earlier geological epochs compared to the present [30]. In this paper, we do not *simultaneously* treat all the relevant problems but restrict our attention to that of the chemical differentiation coupled with thermal convection. In particular, we do not include ridge magmatism that produces basaltic oceanic crust and the complementary harzburgitic layer below of it. This has been done by Ogawa [28] who, on the other hand, did not include the continent generation. Ogawa's and our model complement each other to a certain degree. Therefore, our model certainly does not deal with all relevant aspects of chemical differentiation.

We ran each of our cases to a time of 4.5 billion years to compare our model results with observations of the present Earth. We track the time evolution of Rayleigh number, Ra, Urey number, Ur, and the laterally averaged heat flow, qob, in our runs and compare them with the results of other authors. For this purpose, we vary systematically not only the viscosity-level parameter, r_n, or the temporally averaged Rayleigh number but also the viscoplastic yield stress, called σ_y or ys. We do this not only in that region of $r_n - \sigma_y$ space that is geologically relevant but also outside it to study different solution regimes.

3.2 Thermal and Chemical Evolution Using a Pair of Reference Runs

We begin by presenting what we call our two reference runs 816B and 816C. They are representative of the results we obtain in a moderately extensive

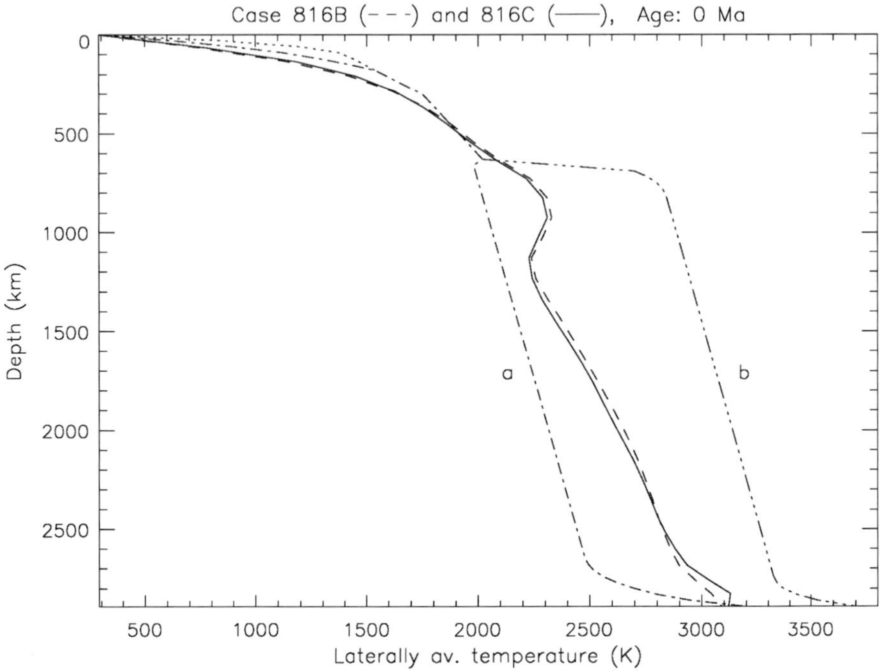

Fig. 1. The laterally averaged temperature of the geological present time as a function of depth for the reference B-run (dashed) and the reference C-run (solid curve) both with a viscoplastic yield stress, $\sigma_y = 120$ MPa, and a viscosity level parameter, $r_n = -0.5$. Cf. (1) and (2). A range of realistic mantle geotherms using parameterized models of the mantle's thermal history given by Schubert et al. [34] is depicted for comparison. Label a and b signify geotherms of whole-mantle and layered convection, respectively. The dotted line denotes a mid-oceanic ridge geotherm

region of Rayleigh number-yield stress parameter space. Our chosen reference runs are defined by a viscoplastic yield stress, $\sigma_y = 120$ MPa, and a viscosity-level parameter, $r_n = -0.5$. Now, we present the Figures, in each case immediately followed by the corresponding discussion. In Fig. 1, the solid line represents the laterally averaged present-day temperature profile of the C-run, whereas the dashed line displays the corresponding B-run profile. The differences between the two curves are a result of fluctuations produced by the active tracers over the course of the runs. In the beginning of the evolution of the model, the two curves are nearly identical. Both curves lie closer to the geotherm of the parameterized whole-mantle convection model than to the corresponding layered-convection temperature. This is understandable since the present model displays whole-mantle flow. However, the flow is somewhat impeded by a high-viscosity transition zone and by the endothermic 660-km phase boundary. Therefore, the temperature is slightly augmented, especially

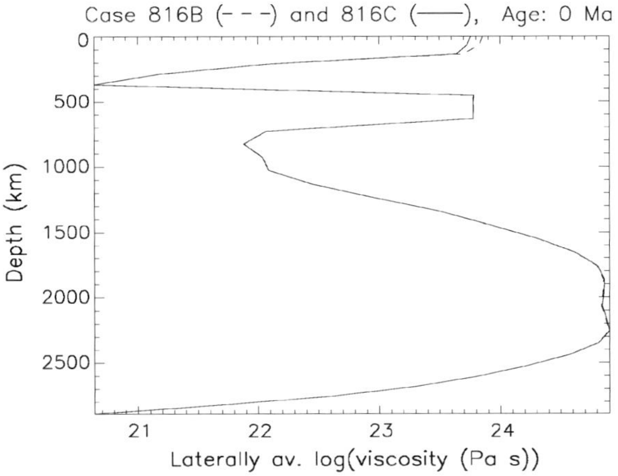

Fig. 2. The laterally averaged shear viscosity of the reference B-run (dashed) and the reference C-run (solid curve) as a function of depth for the present geological time

immediately beneath the 660-km discontinuity. Our model also includes the full effect of the phase-boundary deflections of the upper mantle.

Figure 2 shows the laterally averaged viscosity for the present epoch. Deviations between B-run and C-run are distinguishable only for the lithosphere. The definition of the viscosity profiles is given in Sect. 2. Fig. 3 reveals the production of juvenile continental mass by chemical differentiation of the mantle. The total growth rate within each 25-Ma interval is converted to 10^{18} kg/Ma and graphically represented by a point. The points have been linearly connected. At least in the first half of the evolution, the conversion curves are similar. For this reason and because of the large number of tracers, it is evident that the curves are *not* merely an expression of numerical discretization noise despite the episodic character of the curves. It is obvious that stochastic processes occur both in the real mantle and in the model at bifurcation points etc. That is why it is not to be expected that the two curves of Fig. 3 should exactly coincide and also why the distribution of continents is not predictable in detail *for the present geological epoch.* This applies also for a start with the same slight temperature deviation from the spherically symmetric temperature distribution. We can, however, show that the Fourier spectra of the continent-production curves correspond closely with each other. Notably, the curves of Fig. 3 qualitatively resemble those of observed episodic continental growth ([7, 21]). Parman [30] showed that not only the production peaks of OIB but also those of MORB can be correlated to the zircon-age peaks. This observation argues for *episodic* mantle melting and crustal growth. This suggests that, similar to our model, the continental crust was not generated entirely at the beginning but instead evolved episodically in batches, distributed in time over Earth's history.

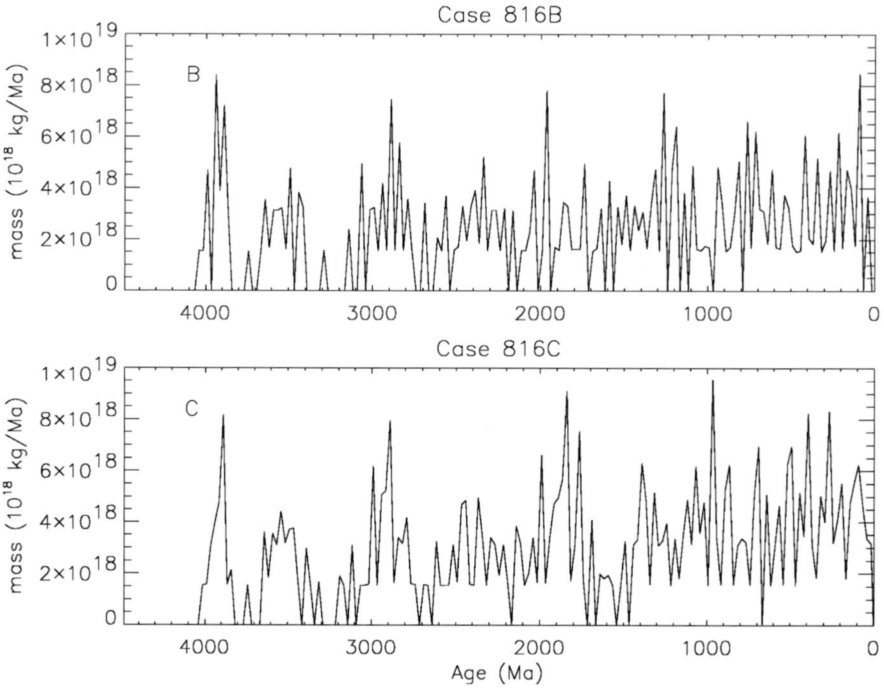

Fig. 3. Time evolution of the juvenile contributions to the total mass of the continents. Comparison of B-run (upper panel) and C-run (lower panel). The rate of the converted continental-tracer mass has been averaged for every 25 Ma and plotted in discretized form. We converted the mass into units of 10^{18} kg/Ma

Figure 4 reveals the present-time distribution of continents (red) of our reference B-run. The oceanic plateaus (black dots) are carried along by self-consistently generated, moving oceanic lithosphere (yellow). Whenever the plateaus touch a continent they join with it. This is the only additional implementation. At first, we did only visual comparisons between the reference B-run and the reference C-run continental distributions (Fig. 5b). Later we decided to represent both continental distributions in terms of spherical harmonics.

$$\{A_n^m \quad or \quad B_n^m\} = \frac{\sqrt{2n+1}}{\pi\sqrt{2}} \sqrt{\frac{(n-m)!}{(n+m)!}} \cdot$$

$$\cdot \int_0^{2\pi} \{\cos m\phi \quad or \quad \sin m\phi\} \cdot \left[\int_0^{\pi} f(\theta,\phi) \cdot P_{n,m}(\cos\theta) \cdot \sin\theta \cdot d\theta \right] d\phi, \quad (3)$$

respectively, where $f(\theta,\phi)$ is 1 for continents and 0 for oceanic lithosphere. While the individual coefficients A_n^m or B_n^m depend on the position of the pole of the grid (θ,ϕ), the quantity h_n^* given by

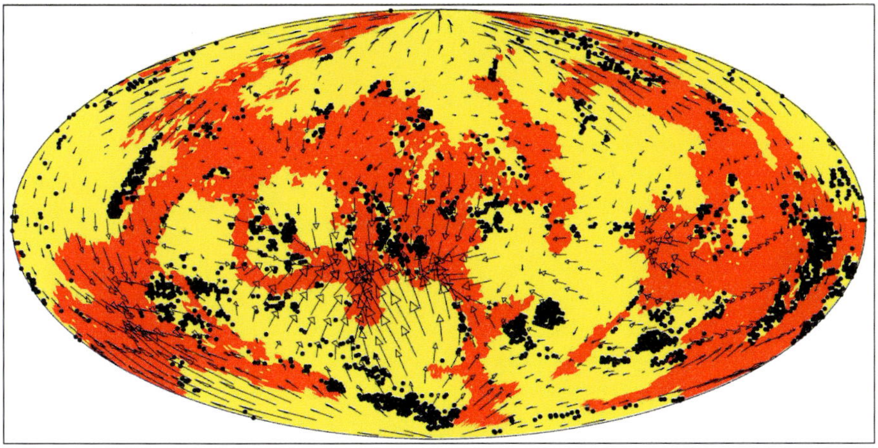

Fig. 4. The distribution of continents (red) and oceanic lithospheric plates (yellow) at the Earth's surface for the geological present time according to the reference B-run. Black dots represent oceanic plateaus. Arrows denote velocity

$$h_n^* = \frac{\sqrt{n \cdot (n+1)}}{2} \cdot \left\{ \sum_{m=0}^{n} \left[(A_n^m)^2 + (B_n^m)^2 \right] \right\}^{1/2} \tag{4}$$

is orientation-independent. We find the h_n^*-spectra of B-run and C-run to be similar. We conclude therefore that, whereas the details of the continental distribution are largely stochastic, the spectral properties are quite similar.

3.3 Continents and Mantle Inhomogeneities

Figure 5 shows that, not so much for the early Earth evolution but for the geological present time, the distribution of continents and velocity vectors of the plates vary for only a slight variation of input parameters, in this case the yield stress, σ_y. A detail in Fig. 5c is a convergent plate boundary displaying an aggregation of oceanic plateaus along a zone of an ocean-ocean collision. Such a phenomenon, of course, requires previous generation of oceanic plateaus on those plates that are moving toward the convergent zone. In these models, of course, it is only the h_n^*-spectrum and not such details that is predictable. The same is also likely to be applicable to the Earth's mantle. We are not aware of other papers on spherical-shell mantle convection in which continents evolve entirely as a consequence of physical laws rather than simply being applied to the surface.

Six different cases of the equatorial section of the distribution of ubiquitously dispersed geochemical reservoirs (cf. last paragraph of Sect. 1) for

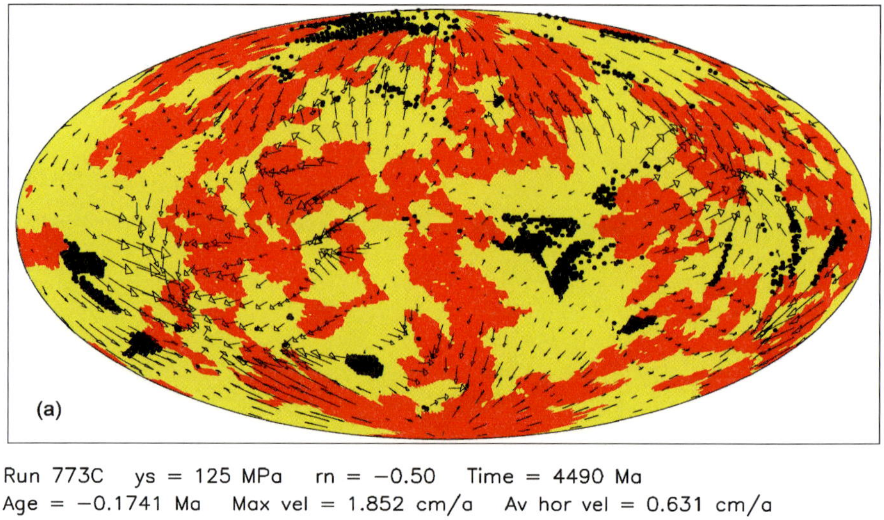

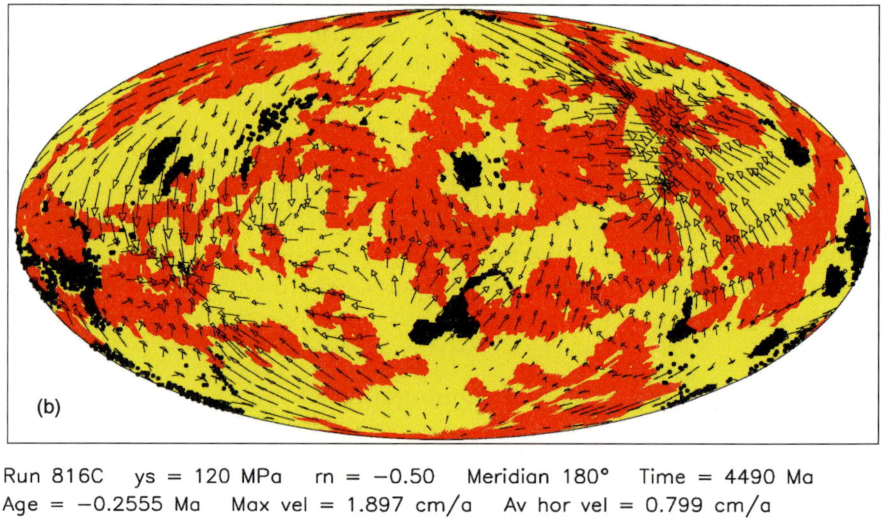

Fig. 5. Continents (red), oceanic lithosphere (yellow) and oceanic plateaus (black dots) for the reference C-run and its neighbors. These C-runs have the same $r_n = -0.50$. First panel, (**a**), second panel, (**b**), and third panel, (**c**), differ among each other only by the yield stress 125, 120 and 115 MPa, respectively

the present epoch are given by Fig. 6. The second panel corresponds to the reference C-run. In the first three panels of Fig. 6, the viscosity-level parameter is kept constant, namely, $r_n = -0.50$. The first panel corresponds to a yield stress of 125 MPa, the second one to 120 MPa, and the third panel to 115 MPa. So, these three panels correspond to the three continental distributions of Fig. 5. The panels (d) to (f) of Fig. 6 correspond to the same succession

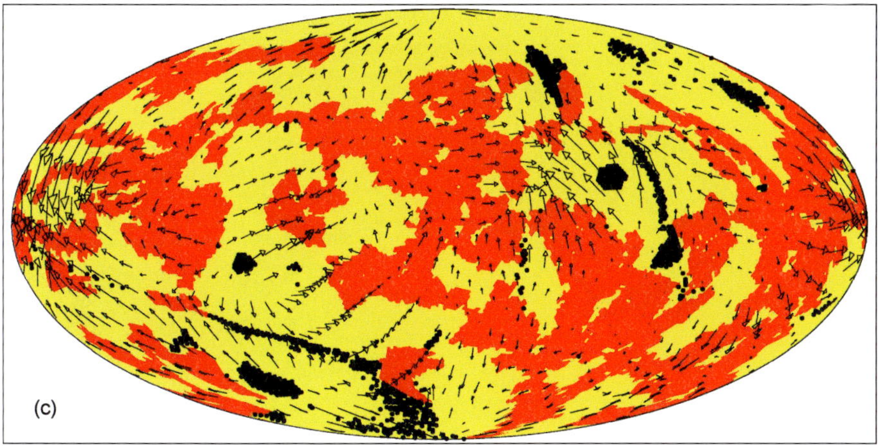

Fig. 5. (Continued)

of yield-stress values but for $r_n = -0.60$. As implied by (1), this means that the viscosity of this second triad is somewhat lower. Depleted parts of the mantle with more than 50% depleted MORB mantle are depicted in yellow. DMM dominates near the surface because DMM is generated near the surface. The depleted, "yellow" material is subducted jointly with the high-viscosity surface plates into the mantle. When cut crosswise by the equatorial section, a partially disintegrated slab is visible as a yellow stripe. It is remarkable that, in spite of 4500 Ma of solid-state mantle convection, chemical reservoirs continue to persist, although without sharp concentration boundaries. In our model there are nowhere pure unblended reservoirs, and this may also be true of the Earth's mantle. DMM predominates immediately below the continents (red) and beneath the oceanic lithosphere. This is a realistic feature of the model since where the real oceanic lithosphere is rifted, MORB magma is formed by decompression melting. The MORB source (DMM) is not only depleted in incompatible elements but also relatively homogenized. It is homogenized not only with respect to its major geochemical components (SiO_2, MgO, FeO, Al_2O_3, CaO) (Palme and O'Neill [29]) but also with respect to isotope ratios $^{87}Sr/^{86}Sr$, $^{143}Nd/^{144}Nd$, $^{206}Pb/^{204}Pb$, $^{207}Pb/^{204}Pb$ and $^{208}Pb/^{204}Pb$. As a consequence, the standard deviation of these isotope ratios and of the major element compositions is small for MORBs in comparison to OIBs (Allègre and Levin [1]) although Hofmann [14] has modified this conclusion somewhat. Figure 6 represents a marble-cake mantle, similar to what was suggested by Coltice and Ricard [6] and Becker et al. [2]. It is remarkable that the yellow-orange boundary does not produce simply connected volumes. There is no clear chemical stratification for the present. The present-day proportion of

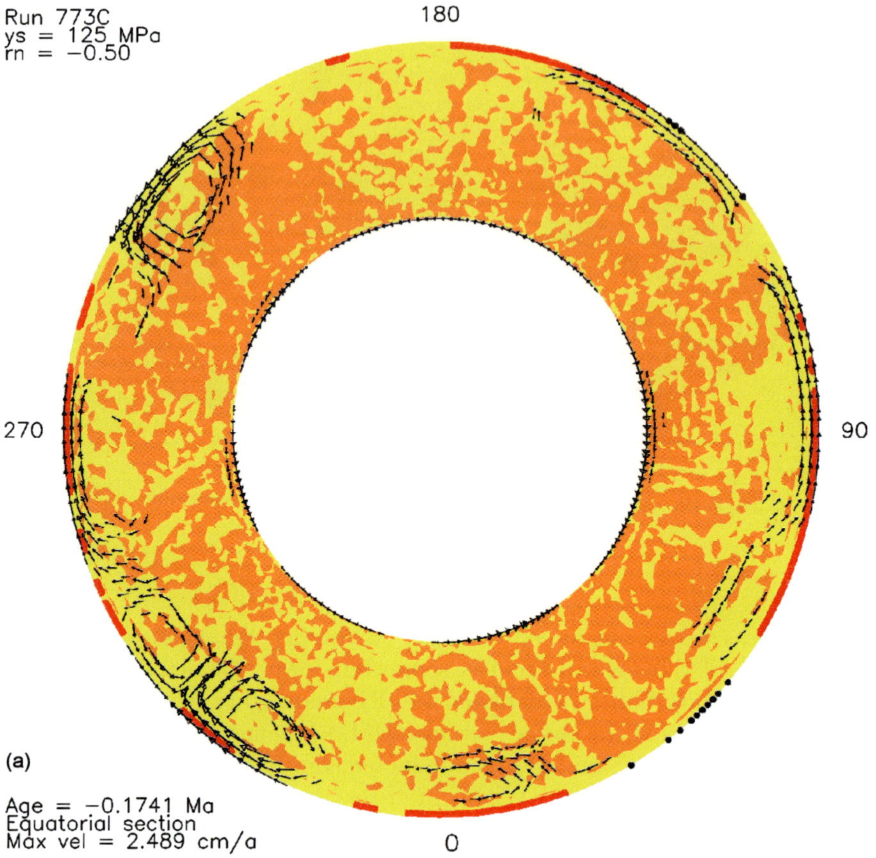

Fig. 6. A comparison of six equatorial sections showing the present-time state of the chemical evolution of incompatible elements of the Earth's mantle. We use a modernized reservoir theory. Cf. Sect. 1. The depleted MORB mantle (DMM) and a mantle which is rich in incompatible elements, yet, are strongly intermixed. Strongly depleted parts of the mantle which include more than 50% DMM are represented by yellow areas. Relatively rich mantle parts with less than 50% DMM are orange-colored. In general, the yellow-orange boundary does *not* correspond to a discontinuity of the abundances of U,Th,K, etc. The cross sections through the continents are red. Black dots represent the oceanic plateaus.
a The yield stress is 125 MPa, the viscosity-level parameter is −0.50

DMM in the real Earth's mantle is not exactly known. Estimates depend on which element is used and its assumed abundance in CC and DMM. If samarium and neodymium are not so extremely enriched in CC, then smaller volumes of DMM are possible. Calculations using Cs, Rb, Th and K suggest 50% of the mantle is DMM. Bennett [3] estimated that between 30% and 60% of the mantle is depleted if an intermediate degree of depletion of DMM is assumed. Hofmann [14] deduced a depleted reservoir of the mantle between

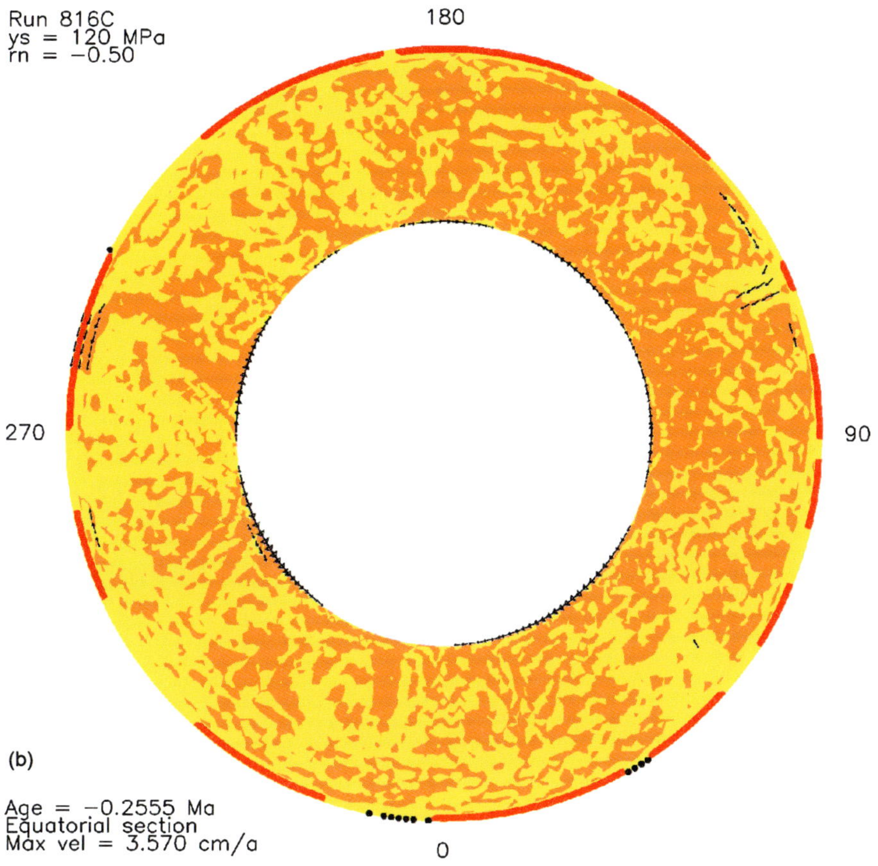

Fig. 6. b The yield stress is 120 MPa, the viscosity-level parameter is −0.50. Further Text see **a**

30% and 80%. If we compare these observational results with the first three panels of Fig. 6, i.e., with the reference C-run and its neighbors, we conclude that the yellow, depleted proportions are of the right order of magnitude. In the model, DMM prevails everywhere immediately under the oceanic or continental lithosphere. So, it is understandable why, as a rule, MORB is generated at sites of rifting of the oceanic lithosphere.

4 Numerical Method, Implementation, Scalability, and Performance

We treat the mantle as a thick spherical shell. The grid for this domain is constructed by projection of the edges of a regular icosahedron onto concentric spherical shell surfaces with different radial distances from the center. These

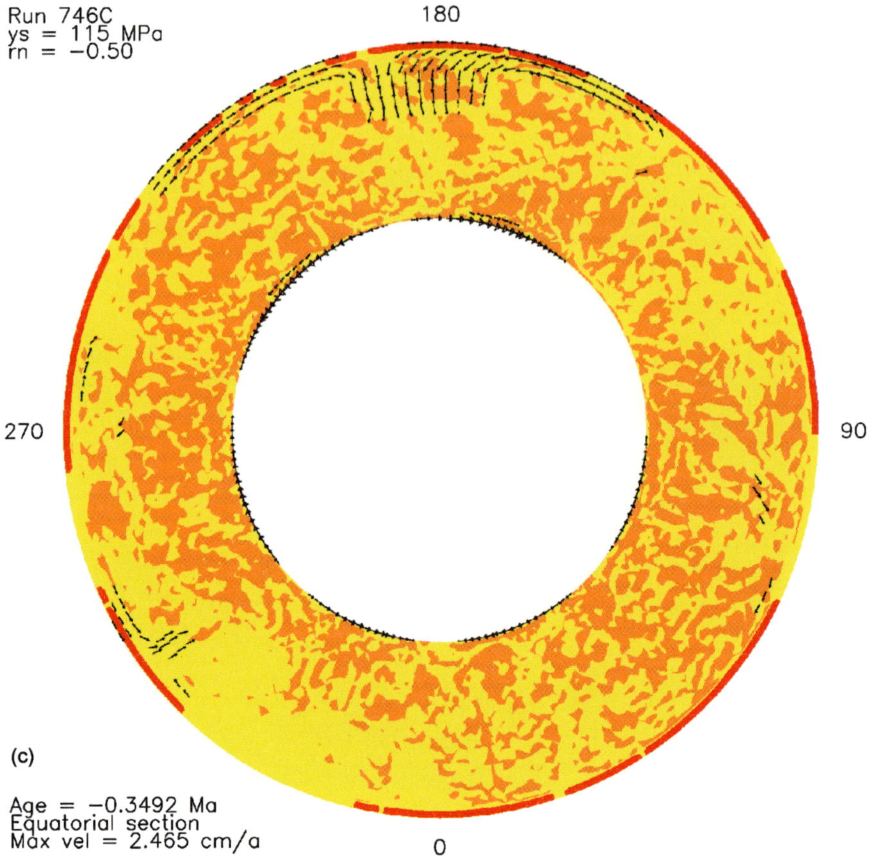

Fig. 6. c The yield stress is 115 MPa, the viscosity-level parameter is −0.50. Further Text see **a**

surfaces subdivide the mantle into thin shells. A first step of grid refinement consists of bisecting the edges of the resulting spherical triangles into equal parts. Connecting the new points with great circles, we obtain four smaller triangles from each starting triangle. The process can be repeated by successive steps to obtain a grid with the desired horizontal resolution. We replicate the resulting almost uniform triangular grid at different radii to generate the 3-D grid for a spherical shell. We can use different formulae for the distribution of the radial distances of the spherical grid surfaces. Here, we used exclusively a radially nearly equidistant grid with a superposed sinoidal half-wave length to refine the grid near the upper and lower boundaries of the spherical shell. The Navier-Stokes equations as well as pressure and creeping velocity are discretized using finite elements. We apply piecewise linear basis functions for the creeping velocity and either piecewise constant or piecewise linear basis functions for the pressure. We solve the equations for pressure and

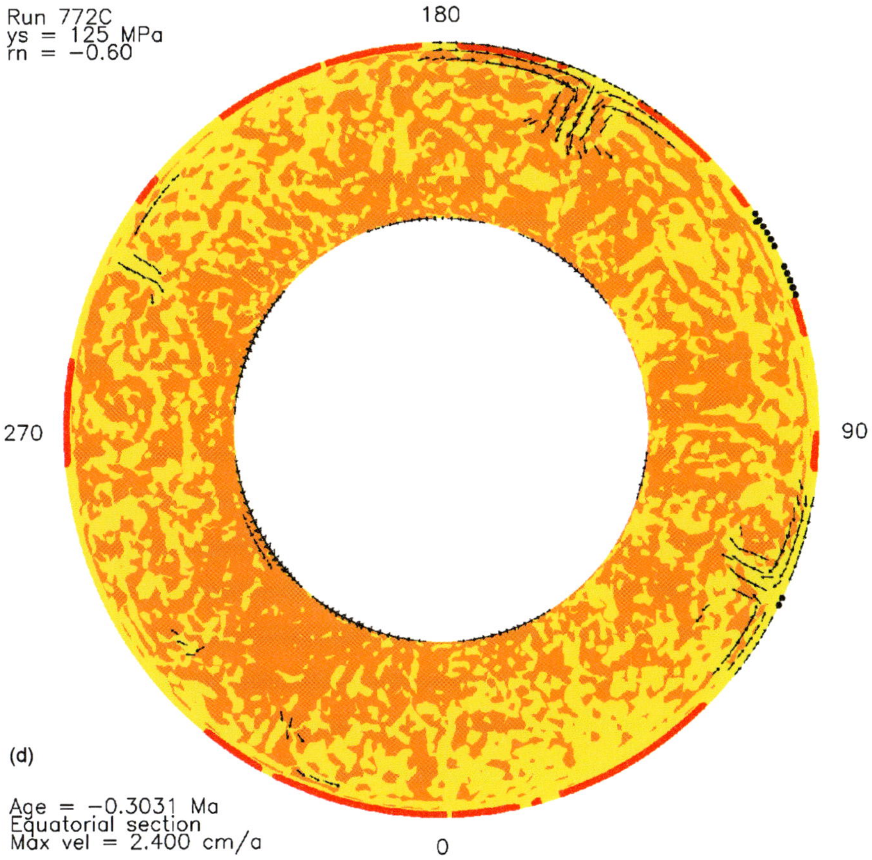

Fig. 6. d The yield stress is 125 MPa, the viscosity-level parameter is −0.60. Further Text see **a**

velocity simultaneously by a Schur-complement conjugate-gradient iteration [31]. This is a further development of an Uzawa algorithm. We solve the energy equation using an iterative multidimensional positive-definite advection-transport algorithm with explicit time steps (Bunge and Baumgardner [4]). Within the Ramage-Wathen procedure, the resulting equation systems are solved by a multigrid procedure that utilizes radial line Jacobi smoothing. In the multigrid procedure, prolongation and restriction are handled in a matrix-dependent manner. In this way, it is possible to handle the strong variations and jumps of the coefficients associated with the strong viscosity gradients (Yang [45]). For the formulation of chemical differentiation, we modified a tracer module developed by Dave Stegman. This module contains a second-order Runge-Kutta procedure to move the tracer particles in the velocity field. Each tracer carries the abundances of the radionuclides. In this sense, tracers are active attributes which determine the heat production rate per unit vol-

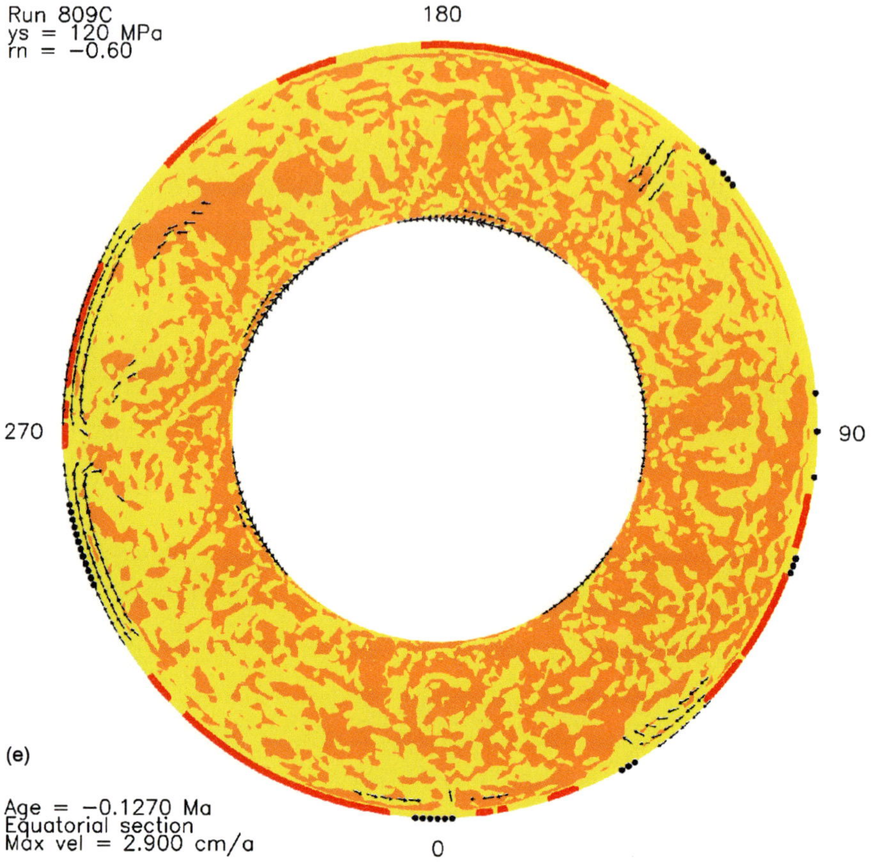

Fig. 6. e The yield stress is 120 MPa, the viscosity-level parameter is −0.60. Further Text see **a**

ume that varies with time and position. For convergence tests we compared the results of runs with 1 351 746 and 10 649 730 grid points. The deviations concerning Rayleigh number, Nusselt number, Urey number, and the laterally averaged surface heat flow, *qob*, were smaller than 0.5 %. Benchmark tests of the Terra code [5, 10] showed deviations of less than 1.5 %. Terra is parallelized by domain decomposition according to the dyadic grid refinement and using explicit message passing (MPI). In Table 1 we present measurements of scalability and performance. Using the performance measuring tool `jobperf`, we obtained an average of 1201 MFlop/s with 8 processors, 1116 MFlop/s with 32 processors, and 935 MFlop/s with 128 processors, respectively. In both resolutions the speedup was almost linear, in some cases slightly superlinear due to cache usage. With the high resolution, at least 4 processors are necessary to make efficient use of the cache memory.

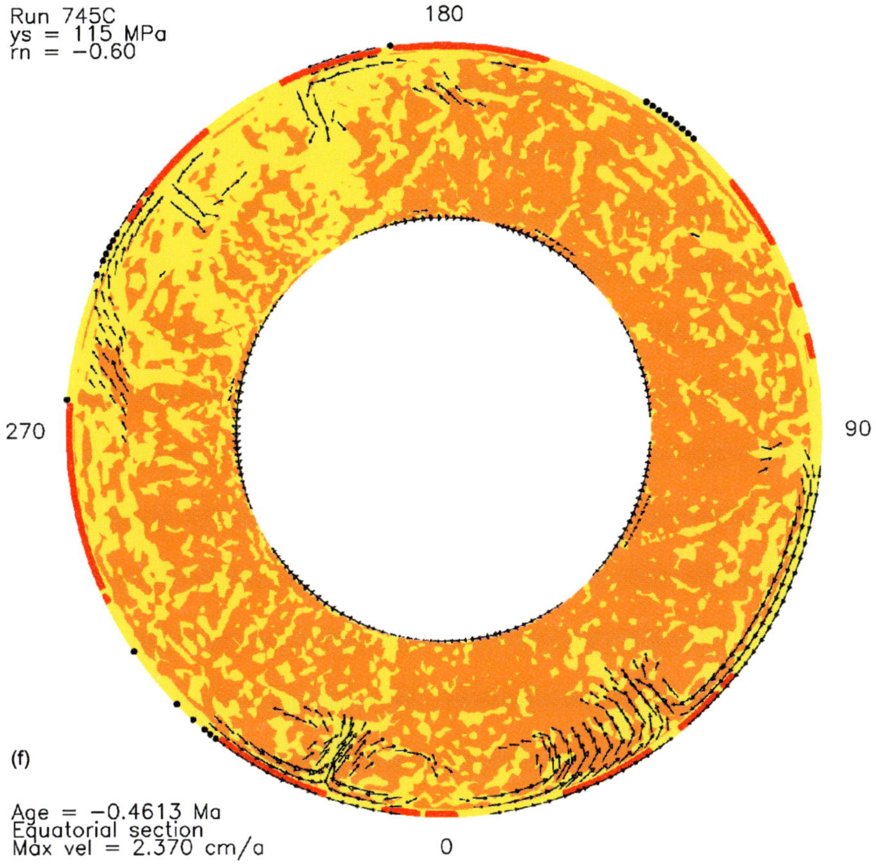

Fig. 6. f The yield stress is 115 MPa, the viscosity-level parameter is −0.60. Further Text see **a**

Table 1. CPU-time, walltime and speedup for runs with 100 time steps on 1 351 746 nodes (a) and on 10 649 730 nodes (b). For comparison, Speedup (b) for 4 processors has been deliberately set to 4.00

Procs	CPU-time (a)	Walltime (a)	Speedup (a)	CPU-time (b)	Walltime (b)	Speedup (b)
1	00:27:13	00:27:17	1.00			
4	00:29:12	00:07:33	3.61	05:01:23	01:16:09	4.00
8	00:26:08	00:03:31	7.76	04:53:26	00:37:32	8.12
16	00:25:36	00:02:02	13.42	05:08:51	00:19:35	15.69
32	00:21:52	00:01:02	26.40	04:34:40	00:09:11	33.17
64				05:43:25	00:05:36	54.39
128				05:34:24	00:03:04	100.42

For further tasks, we need an augmentation of grid resolution by a factor of ten in each of the three directions. To run such kind of jobs, a high number of processors, in the order of 10000, is necessary. Up to now, the particle module was the bottleneck of our code. Some parts of this module were not parallelized. But now this task has been solved.

Acknowledgements. We kindly acknowledge the confidential cooperation with John Baumgardner, Markus Müller, and Christoph Köstler. We are very thankful for Dave Stegman for providing his particle code. Furthermore, we thank for the use of supercomputing facilities at SCC Karlsruhe and LRZ München. This work was supported by the Deutsche Forschungsgemeinschaft under grants WA 1035/5-3 and KL 495/16-1.

References

1. C.J. Allègre and E. Levin. Isotopic systems and stirring times of the Earth's mantle. *Earth Planet. Sci. Lett.*, 136:629–646, 1995.
2. T.W. Becker, J.B. Kellogg, and R.J. O'Connell. Thermal constraints on the survival of primitive blobs in the lower mantle. *Earth Planet. Sci. Lett.*, 171:351–365, 1999.
3. V.C. Bennett. Compositional evolution of the mantle. In R.W. Carlson, editor, *Treatise on Geochemistry, Vol. 2: The Mantle and the Core*, pages 493–519. Elsevier, Amsterdam, 2003.
4. H.-P. Bunge and J.R. Baumgardner. Mantle convection modelling on parallel virtual machines. *Computers in Physics*, 9:207–215, 1995.
5. H.-P. Bunge, M.A. Richards, and J.R. Baumgardner. A sensitivity study of three-dimensional spherical mantle convection at 10^8 Rayleigh number: Effects of depth-dependent viscosity, heating mode and an endothermic phase change. *J. Geophys. Res.*, 102:11991–12007, 1997.
6. N. Coltice and Y. Ricard. On the origin of noble gases in mantle plumes. *Phil. Trans. Royal Soc. A: Math. Phys. Engng. Sci.*, 360:2633–2648, 2002.
7. K.C. Condie. Episodic continental growth and supercontinents: a mantle avalanche connection? *Earth Planet. Sci. Lett.*, 163:97–108, 1998.
8. A.M. Dziewonski and D.L. Anderson. Preliminary reference Earth model. *Phys. Earth Planet. Int.*, 25:297–356, 1981.
9. J.J. Gilvarry. The Lindemann and Grüneisen laws. *Phys. Rev.*, 102:307–316, 1956.
10. G.A. Glatzmaier. Numerical simulations of mantle convection: Time-dependent, three-dimensional, compressible, spherical shell. *Geophys. Astrophys. Fluid Dyn.*, 43:223–264, 1988.
11. K.-D. Gottschaldt, U. Walzer, R. Hendel, D.R. Stegman, J.R. Baumgardner, and H.-B. Mühlhaus. Stirring in 3-d spherical models of convection in the Earth's mantle. *Philosophical Magazine*, 86:3175–3204, 2006.
12. S.R. Hart, E.H. Hauri, L.A. Oschmann, and J.A. Whitehead. Mantle plumes and entrainment: Isotopic evidence. *Science*, 256:517–520, 1992.
13. A.W. Hofmann. Chemical differentiation of the Earth: The relationship between mantle, continental crust and oceanic crust. *Earth Planet. Sci. Lett.*, 90:297–314, 1988.

14. A.W. Hofmann. Sampling mantle heterogeneity through oceanic basalts: Isotopes and trace elements. In R.W. Carlson, editor, *Treatise on Geochemistry, Vol. 2: The Mantle and the Core*, pages 61–101. Elsevier, Amsterdam, 2003.
15. G. Ito and J.J. Mahoney. Flow and melting of a heterogeneous mantle: 1. Method and importance to the geochemistry of ocean island and mid-ocean ridge basalts. *Earth Planet. Sci. Lett.*, 230:29–46, 2005.
16. G. Ito and J.J. Mahoney. Flow and melting of a heterogeneous mantle: 2. Implications for a chemically nonlayered mantle. *Earth Planet. Sci. Lett.*, 230:47–63, 2005.
17. S.-I. Karato and P. Li. Diffusion creep in perovskite: implications for the rheology of the lower mantle. *Science*, 255:1238–1240, 1992.
18. S.-I. Karato and P. Wu. Rheology of the upper mantle: a synthesis. *Science*, 260:771–778, 1993.
19. J.B. Kellogg, S.B. Jacobsen, and R.J. O'Connell. Modeling the distribution of isotopic ratios in geochemical reservoirs. *Earth Planet. Sci. Lett.*, 204:183–202, 2002.
20. L.H. Kellogg, B.H. Hager, and R.D. van der Hilst. Compositional stratification in the deep mantle. *Science*, 283:1881–1884, 1999.
21. A.I.S. Kemp, C.J. Hawkesworth, B.A. Paterson, and P.D. Kinny. Episodic growth of the Gondwana supercontinent from hafnium and oxygen isotopes in zircon. *Nature*, 439:580–583, 2006.
22. B.L.N. Kennett, S. Widiyantoro, and R.D. van der Hilst. Joint seismic tomography for bulk sound and shear wave speed in the Earth's mantle. *J. Geophys. Res.*, 103:12469–12484, 1998.
23. P. Li, S.-I. Karato, and Z. Wang. High-temperature creep in fine-grained polycristalline $CaTiO_3$, an analogue material of $(Mg,Fe)SiO_3$ perovskite. *Phys. Earth Planet. Int.*, 95:19–36, 1996.
24. A. Meibom and D.L. Anderson. The statistical upper mantle assemblage. *Earth Planet. Sci. Lett.*, 217:123–139, 2003.
25. A. Meibom, N.H. Sleep, K. Zahnle, and D.L. Anderson. Models for noble gases in mantle geochemistry: Some observations and alternatives. In G.R. Foulger et al., editors, *Plumes, Plates and Paradigms*, volume 388, pages 347–363. Geological Society of America Special Paper, Boulder, Colorado, 2005.
26. J.P. Morgan and W.J. Morgan. Two-stage melting and the geochemical evolution of the mantle: a recipe for mantle plum-pudding. *Earth Planet. Sci. Lett.*, 170:215–239, 1999.
27. M. Ogawa. Chemical stratification in a two-dimensional convecting mantle with magmatism and moving plates. *J. Geophys. Res.*, 108(B12):2561, 2003.
28. M. Ogawa. Superplumes, plates, and mantle magmatism in two-dimensional numerical models. *J. Geophys. Res.*, 112:B06404, 2007.
29. H. Palme and H.S.C. O'Neill. Cosmochemical estimates of mantle composition. In R.W. Carlson, editor, *Treatise on Geochemistry, Vol. 2: The Mantle and the Core*, pages 1–38. Elsevier, Amsterdam, 2003.
30. S.W. Parman. Helium isotopic evidence for episodic mantle melting and crustal growth. *Nature*, 446:900–903, 2007.
31. A. Ramage and A.J. Wathen. Iterative solution techniques for finite element discretizations of fluid flow problems. In *Proceedings of the Copper Mountain Conference on Iterative Methods*, volume 1, Copper Mountain, Colorado, 1992.

32. J.F. Rudge, D. McKenzie, and P.H. Haynes. A theoretical approach to understanding the isotopic heterogeneity of mid-ocean ridge basalt. *Geochim. Cosmochim. Acta*, 69(15):3873–3887, 2005.
33. J. Schmalzl. Mixing properties of thermal convection in the earth's mantle. *Geologica Ultraiectina*, 140:104pp., 1996.
34. G. Schubert, D.L. Turcotte, and T.R. Olson. *Mantle Convection in the Earth and Planets*. Cambridge Univ. Press, Cambridge, UK, 2001.
35. A. Stracke, A.W. Hofmann, and S.R. Hart. FOZO, HIMU and the rest of the mantle zoo. *Geochem. Geophys. Geosys.*, 6:Q05007, 2005.
36. W.-J. Su and A.M. Dziewonski. Simultaneous inversion for 3-D variations in shear and bulk velocity in the mantle. *Phys. Earth Planet. Int.*, 100:135–156, 1997.
37. P.J. Tackley. Three-dimensional simulations of mantle convection with a thermochemical basal boundary layer: D"? In M. Gurnis et al., editors, *The Core-Mantle Boundary Region, Geodyn. Ser., vol. 28*, pages 231–253. AGU, Washington, D. C., 1998.
38. P.J. Tackley. Mantle convection and plate tectonics: Towards an integrated physical and chemical theory. *Science*, 288:2002–2007, 2000.
39. U. Walzer and R. Hendel. Mantle convection and evolution with growing continents. *J. Geophys. Res.*, 113:B09405, doi:10.1029/2007JB005459, 2008.
40. U. Walzer, R. Hendel, and J. Baumgardner. The effects of a variation of the radial viscosity profile on mantle evolution. *Tectonophysics*, 384:55–90, 2004.
41. U. Walzer, R. Hendel, and J. Baumgardner. Whole-mantle convection, continent generation, and preservation of geochemical heterogeneity. In W.E. Nagel, D.B. Kröner, and M.M. Resch, editors, *High Perf. Comp. Sci. Engng. '07*, pages 603–645. Springer, Berlin, 2008.
42. M. Willbold and A. Stracke. Trace element composition of mantle end-members: Implications for recycling of oceanic and upper and lower continental crust. *Geochem. Geophys. Geosys.*, 7:Q04004, 2006.
43. M. Wilson and E.A. Spencer. The origin and evolution of the fozo/prema and himu mantle components—the carbonatite perspective. http://www.geo.uw.edu.pl/ERASMUS/files/wilson3.ppt, 2003.
44. D. Yamazaki and S.-I. Karato. Some mineral physics constraints on the rheology and geothermal structure of the Earth's lower mantle. *Am. Min.*, 86:385–391, 2001.
45. W.-S. Yang. *Variable viscosity thermal convection at infinite Prandtl number in a thick spherical shell.* PhD thesis, University of Illinois, Urbana-Champaign, 1997.

Three-Dimensional Simulation of Rarefied Plasma Flows Using a High Order Particle in Cell Method

J. Neudorfer[4], T. Stindl[1], A. Stock[3], R. Schneider[5] D. Petkow[1], S. Roller[4], C.-D. Munz[3], and M. Auweter-Kurtz[2]

[1] Institut für Raumfahrtsysteme, Abt. Raumtransporttechnologie, Universität Stuttgart, Stuttgart, Germany, petkow@irs.uni-stuttgart.de
[2] Steinbeis Transferzentrum Plasma- und Raumfahrttechnologie, Stuttgart, Germany, info@plasma-raumfahrt.de
[3] Institut für Aerodynamik und Gasdynamik, Universität Stuttgart, Stuttgart, Germany, munz@iag.uni-stuttgart.de
[4] German Research School for Simulation Sciences GmbH, Aachen, Germany, s.roller@grs-sim.de
[5] Forschungszentrum Karlsruhe, Institut für Hochleistungsimpuls- und Mikrowellentechnik, Karlsruhe, Germany, rudolf.schneider@ihm.fzk.de

Summary. A three-dimensional Particle In Cell scheme for unstructured grids is presented. Since simulations of this kind require large computational resources, the solver was parallelized. The scalability of two parallel simulations is shown and an engineering application as well as two validation test cases for the scheme are presented.

1 Introduction

A hybrid particle code currently under development for the fully kinetic modelling of the complete Boltzmann equation has been described previously [4]. In the present report, the focus is on the advancements regarding the Particle In Cell (PIC) part of the code. The newly incorporated high order discontinuous Galerkin (DG) approach for solving Maxwell's equations is described in Sect. 2. Section 3 presents the charge and current deposition and interpolation methods. The performance of a newly implemented fourth-order low storage Runge-Kutta scheme to compute the particle movement is compared to that of the previously used second order Leap-Frog scheme in Sect. 4. In Sect. 5, the parallelization concept is described and results of scalability tests are shown. Finally, preliminary results of different simulation examples for the PIC solver are given in Sect. 6, followed by a summary and an outlook on future activities.

W.E. Nagel et al. (eds.), *High Performance Computing in Science and Engineering '10*, DOI 10.1007/978-3-642-15748-6_43,

2 Maxwell Solver

The PIC method models the plasma by tracking individual particles in a Lagrangian fashion. In general, these particles are electrons, ions and neutral particles with phase space coordinates $\mathbf{x}$ and $\mathbf{v}$. Due to the presence of charged particles in a plasma, electromagnetic fields ($\mathbf{E}$ and $\mathbf{B}$) are formed and propagate along with the particle flow. These fields again interact with the charges q of the plasma by the Lorentz force $\mathbf{F} = q\,(\mathbf{E} + \mathbf{v} \times \mathbf{B})$. Hence, a PIC cycle consists of four steps: First, the charge and current given by the particles charges and velocities are gathered and deposited onto a computational grid to obtain the charge and current density. With these source terms the Maxwell's equations are then solved on this grid. In a third step, the resulting electromagnetic fields $\mathbf{E}$ and $\mathbf{B}$ are evaluated at the positions of the particles. Finally, according to the Lorentz force the charged particles are redistributed in phase space, closing the self-consistent interplay of the charges and electromagnetic fields in a plasma.

Here, Maxwell's equations are solved in the purely hyperbolic form [10]

$$\frac{\partial \mathbf{U}}{\partial t} + \sum_{n=1}^{3} \frac{\partial}{\partial x_n} \left[\mathcal{K}_n\, \mathbf{U}\right] = \mathbf{g} \tag{1}$$

where $\mathbf{U}$ denotes the state vector, $\mathbf{g} = \mathbf{g}(\mathbf{j}, \rho)$ represents the source term which is established by the charge and current density, ρ and $\mathbf{g}$. The $\mathcal{K}_n$ are block-structured matrices with constant entries. To solve the full Maxwell's equations system, a discontinuous Galerkin scheme is used [11].

2.1 The Discontinuous Galerkin Scheme

The discontinuous Galerkin Maxwell solver that was implemented in the context of this project replaces the formerly used Finite Volume solver (as presented in [4]). The DG method has the advantage of maintaining a local scheme even for simulations of high convergence order. While a Finite Volume solver relies on the reconstruction of a high order solution using the cell mean values of neighboring elements, the discontinuous Galerkin method uses local basis functions, eliminating the need for information from neighboring elements. This local approach even for high convergence order simulations results in a scheme with an efficient parallelization (see below). The discontinuous Galerkin method has been investigated before in the two-dimensional PIC context by Jacobs and Hesthaven [7], who used a nodal DG method with Runge-Kutta time stepping. The presented scheme similarly uses a nodal formulation and a low storage explicit Runge-Kutta scheme.

In the three-dimensional nodal formulation, each element with the polynomial degree p needs $N = (p+1)(p+2)(p+3)/6$ degrees of freedom (DOF) for the definition of its basis function. N also equals the number of nodal points of such an element, as depicted exemplarily for a tetrahedron of polynomial

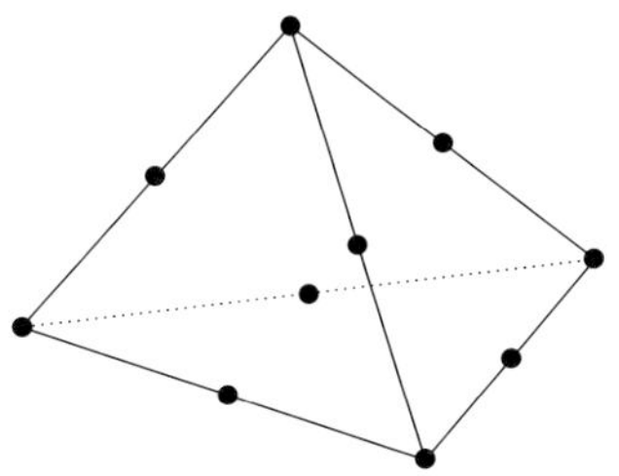

Fig. 1. Interpolation points for a tetrahedron with polynomial degree p = 2

degree $p = 2$ in Fig. 1. The element-local solution $\mathbf{U}(\mathbf{x}, t)$ for N DOF takes the form

$$\mathbf{U}(\mathbf{x}, t) = \sum_{j=1}^{N} \mathbf{U}(\mathbf{x}_j, t) L_j(\mathbf{x}) \tag{2}$$

with the multivariate Lagrange basis $\{L_j\}_{j=1,\ldots,N}$ and the solution $\mathbf{U}(\mathbf{x}_j, t)$ evaluated at N nodal points $\mathbf{x}_j$.

3 Particle-Grid Coupling

Since the electric field and magnetic induction are calculated on the grid while the particles move freely in space, an interface procedure is needed to couple the charge and current of the particles with the node-based charge and current density which are the sources for the Maxwell equations on the computational mesh. From these values, the DG source polynomials are constructed, thus allowing a high local resolution of the sources inside a high-order element. Similarly, the electromagnetic fields have to be determined at the position of the particles after being computed on the grid in order to determine the Lorentz force and the corresponding particle movement.

3.1 Deposition of Charges and Currents

The lowest order approach for the deposition of charges and currents consists of adding the contribution of charge and current of all particles within a grid cell and applying the corresponding densities uniformly to all grid nodes within a cell. While this method is fast, it leads to strong oscillations in the source terms due to particles moving between cells. In order to distribute the charge and current of a particle over a larger volume, thereby reducing the oscillations and allowing a higher order deposition, shape functions are used to approximate a particle cloud. Jacobs and Hesthaven [7] have used a two-dimensional polynomial shape function that proved to be optimal in terms of performance and accuracy of charge deposition. For the application in a three-dimensional solver, the shape function $S_{pol,3D}(r_p)$ has to be normalized by its volume integral yielding a polynomial shape function

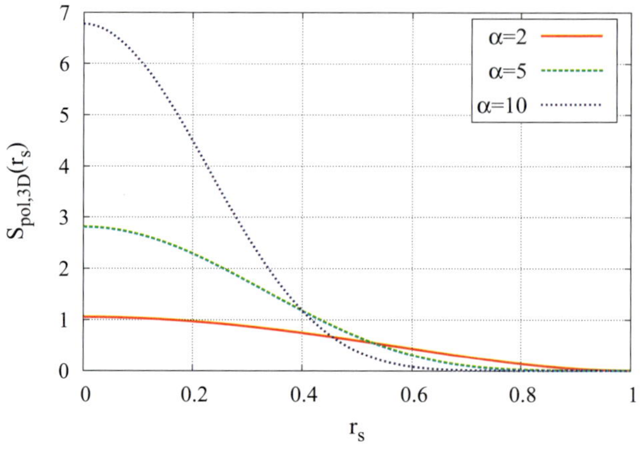

Fig. 2. Polynomial shape function (3) for shape radius $R = 1$, with different polynomial exponents α

$$S_{pol,3D}(r_p) = \frac{1}{2B(1.5, \alpha + 1)\pi R^3} \left[1 - \left(\frac{r_p}{R}\right)^2\right]^\alpha, r_p = [0, R], \tag{3}$$

for use in three-dimensional solvers. Here, B is the Beta function, $r_p = |\mathbf{x} - \mathbf{x}_p|$ is the distance from the center of the particle cloud $\mathbf{x}_p$ to position $\mathbf{x}$ and α is the polynomial exponent. For a low exponent α the distribution is very broad while for a high α the shape is getting more focused on the center. The shape function $S_{pol,3D}(r_p)$ is evaluated within $r_p \leq R$ with a cutoff radius R that can be chosen arbitrarily. Figure 2 shows the shape function for different exponents α. A different approach towards higher order particle deposition, introducing a background mesh, is currently being investigated. Here, the charges and currents are not deposited directly from the particles to the grid nodes but instead are first deposited on the superimposed Cartesian background mesh. From there, the charge and current densities are then interpolated to the grid nodes. Due to the Cartesian nature of the background mesh, the corresponding computations are a lot less costly. A volume weighting method and a higher order weighting method based on B-Splines are implemented and currently under investigation.

3.2 Evaluation of Electromagnetic Fields at Particle Positions

Computation Based on DG Polynomials

Since the DG polynomials can be evaluated at any point in space, the electromagnetic fields for a given particle can be computed by evaluating the polynomials at the position of the particle. This is the most accurate but also computationally most expensive method.

Cartesian Background Mesh

Instead of evaluating the DG polynomials at all particle positions, they can be evaluated at the nodes of the Cartesian background mesh already used for the deposition. This has the benefit of lower computational costs due to the simpler interpolation algorithms feasible for a Cartesian grid and due to the possibility to retain the fixed parameters for the evaluation of the polynomials. From the Cartesian mesh, the fields can then be interpolated to the particle positions using for example the volume or B-Splines weighting method.

4 Particle Pusher

The classic particle pusher is the Leap-Frog algorithm described in [2, 3]. Here, the particle velocities are shifted by half a time step compared to the particle positions. This can be compared to a midpoint rule which guarantees a second-order method. A higher order alternative method for the time integration of the particles is the fourth-order five stage low storage explicit Runge-Kutta method (LSERK4) from Kennedy et al. [9], which was also used for PIC by Jacobs and Hesthaven [7]. Compared to a classic fourth-order Runge-Kutta method the memory usage is halved. The higher order is definitely gained by decreased performance, which is investigated below.

4.1 Performance Analysis

In this section, the performance of the two different time steppers is examined. For this purpose, an equidistant particle distribution is initialized in a unit cube with $100,000$ particles. The time step for both particle pushers is equal as well as the final simulation time. The computations are performed on an AMD Opteron 8378 64-bit (4x Quad-core). Output of analyzed data has been disabled in order to gain the maximum performance. A shape function as described in Sect. 3.1 and the evaluation of the electromagnetic field DG polynomials directly at the position of the particles is used to achieve the currently highest accuracy possible for this code. Table 1 shows the computation time of both methods. It is found that the LSERK4 method is of a factor of ≈ 5 slower than the Leap-Frog method. This can be explained by the fact that LSERK4 has five evaluations of the right-hand side of the PIC system, whereas the Leap-Frog method only has one evaluation. It has to be determined from case to case whether the increase in order is worth the loss in performance.

Table 1. Computational time for a standard example with the Leap-Frog and LSERK4 method

LSERK4	661 [s]
Leap-Frog	135 [s]

5 Parallelization

The parallelization of the high order PIC solver is MPI based and uses the domain decomposition approach. Particles are assigned to the process that computes the domain containing the particle. Generally, the implementation of the parallelization can be grouped into three tasks:

1. Parallelization of the grid-based DG Maxwell solver
2. Particle parallelization
3. Parallelization of the particle-grid coupling

The parallelization of the discontinuous Galerkin Maxwell solver needs to provide the exchange of information on the domain surfaces only. This is done using persistent communication.

We first consider the parallel efficiency of the Maxwell solver itself. Investigations are performed on an Intel® Nehalem cluster with InfiniBand network. The Maxwell testcase that was used in this study has been described previously [11].

The parallelized Maxwell solver shows satisfying scalability both in strong and in weak scaling. Figure 3a shows the parallel efficiencies. For the weak scaling, a simulation with 28,672 DOF per MPI process was performed. Reference time is the elapsed time Δt_{weak}, obtained for a run on a single, but fully used node with 8 CPU cores hosting 8 MPI processes. The parallel efficiency $PE_{weak}(n_{procs})$ for n_{procs} cores was then obtained by scaling the problem size by the number of cores n_{procs} divided by 8, yielding the elapsed time $\Delta t_{weak}(n_{procs})$. The parallel efficiency is then calculated as

$$PE_{weak}(n_{procs}) = \frac{\Delta t_{weak}}{\Delta t_{weak}(n_{procs})}. \tag{4}$$

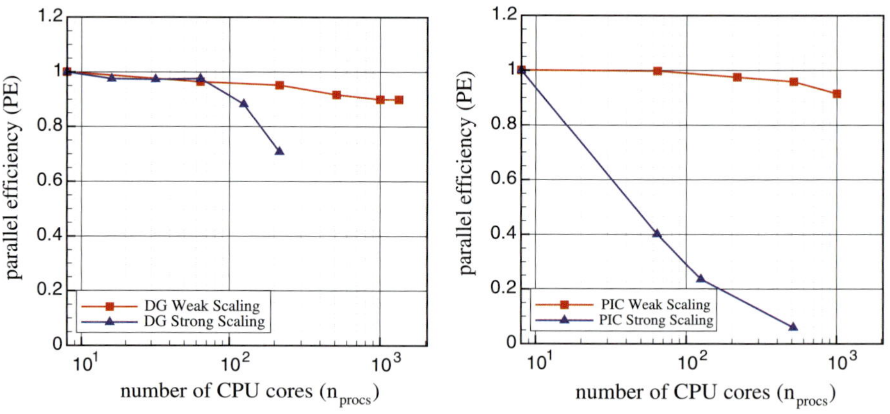

(a) Parallel efficiency of the DG Maxwell MPI parallelization (b) Parallel efficiency of the particle MPI parallelization

Fig. 3. Parallel efficiency of the MPI parallelization

For strong scaling, the problem size was kept constant at 229,376 DOF (i.e. $8 \times 28,672$ DOF, the number of the reference simulation for a single, full node). The resulting elapsed time on a single core of Δt_{strong} is the reference for the measurement of parallel efficiency $PE_{strong}(n_{procs})$ for the strong scaling

$$PE_{strong}(n_{procs}) = \frac{8}{n_{procs}} \frac{\Delta t_{strong}}{\Delta t_{strong}(n_{procs})}. \tag{5}$$

For computations on more than 100 CPU cores, the strong scaling for the considered problem is less efficient, due to the limited problem size of 229,376 DOF. For the large problem sizes targeted in the presented work, weak scaling is much more relevant, since the problem size we are targetting at is much larger than the current one. In weak scaling, computations on more than 1,000 CPU cores still show a satisfactory parallel efficiency of almost 90 percent.

The parallelization of the particles handles the exchange of particles over the domain boundaries. If, in one time step, a particle leaves one computational domain and enters another one, the properties of the particle are communicated. In this case, non-blocking peer-to-peer communication is currently used on a single buffer containing the properties of all particles that are transferred between the two communicating processes.

The challenge regarding the particle-grid coupling lies in correctly distributing the particles near domain boundaries. The Cartesian background mesh has been built in such a way that it seamlessly fits over domain boundaries with sufficiently overlapping mesh points. Initially, a mask for each domain is built that identifies for each mesh point whether information needs to be transmitted to neighboring domains. As for the particle parallelization, non-blocking peer-to-peer communication is used here as well. The same principle is used to provide the field information from neighboring domains needed for interpolation.

In order to describe the parallel performance of the coupled code, the elapsed times of several parallel computations conducted on the bwGRiD Intel® Xeon cluster at HLRS were recorded. For the node interconnect, the installed InfiniBand network was used. In these simulations, $1,000$ DG cells of polynomial degree $n = 3$ (i.e. $20,000$ DOF) per process were computed. Additionally, 100 particles per grid cell with a uniform spatial distribution and a Maxwellian velocity distribution were simulated. The particle-grid-coupling was realized using a third order B-spline interpolation on a Cartesian background mesh.

These simulations involve all three parallelizations described above. For a better characterization of the performances of the three tasks, the profiles of the computations were evaluated. In the profiles, about 50 percent of the simulation time was attributed to the DG solver. The remaining 50 percent are attributed to the particle treatment itself (30%) and the particle-grid-coupling (20%).

The parallel efficiency for weak and strong scaling based on these simulation times is shown in Fig. 3b.

The strong scaling in this case shows a poor efficiency. The total speedup remains below a factor of 4. This behaviour, though, is problem dependent. For a problem involving more grid cells and more particles, a better strong scaling has been observed.

As mentioned above, for the large problem sizes targeted in this project, strong scaling is of minor interest. The more interesting weak scaling of the particle simulation shows a very good efficiency. However, the scalability varies with the simulation parameters. Especially for simulations with a non-uniform distribution of the particles, an acceptable scaling can not yet be achieved. Therefore, we currently work on a dynamic load balancing, taking the particle density into account. It is also planned to improve the scalability of simulations with non-uniformly distributed particles by adding a shared memory parallelization.

The largest coupled problem that was simulated apart from this scaling analysis contained 8.192 million grid cells of polynomial degree $p = 1$, yielding a total of 32.768 million DOF plus 819.2 million simulated particles. This simulation was performed on 1024 cores of the bwGRiD Intel® Xeon cluster at HLRS.

6 Simulation Examples

6.1 Gyrotron Launcher

Before the fully coupled scheme is applied, the performance of the solver's basic building block shall be investigated using an engineering application. In the context of fusion engineering, the fusion plasma has to be heated, i.e. energy has to be coupled into the plasma. This can be done by directing high power microwaves into the plasma. Microwaves of high power can be provided by a gyrotron. The desired spatial distribution of the electromagnetic wave modes emitted by a gyrotron should be of a Gaussian form. An approximately Gaussian distribution is achieved by propagating wave modes generated by a resonator through a wave guide, the so-called launcher. The launcher is of a cylindrical form with small perturbations to shape the wave modes. The propagation of the electromagnetic waves in the launcher is described by Maxwell's equations and can be simulated with the presented Maxwell solver.

The necessary spatial discretization of the device for the simulation depends on the length of the waves being propagated through the launcher. For the conducted simulation, the wave length of the incoming waves was $\lambda = 1.7$ mm. With a minimal discretization of 5 points per wave length, the launcher with a length of $l = 350$ mm and a radius of $r = 32.5$ mm needs a minimum of 30 million degrees of freedom. The performed fifth order computation used about 6 points per wave length, yielding a total of 55 million degrees of freedom. The simulation was performed using 2048 MPI processes on an Intel® Nehalem cluster. Results of this simulation will be published soon.

6.2 Plasma Wave 3D

The plasma wave test case describes the plasma oscillation frequency, or Langmuir oscillation, named after Tonks and Langmuir [12], who first discovered these oscillations. The physical background for this experiment can be found in [6]. The plasma wave oscillation frequency ω_s for a particle species is given by

$$\omega_s = \sqrt{\frac{n_s q_s^2}{m_s \varepsilon_0}}, \tag{6}$$

where n_s is the particle density, q_s is the charge of the species, m_s is the particle mass and ε_0 is the permittivity of free space, also known as dielectric constant.

To demonstrate the capability of our code to show the plasma frequency as computed analytically, a setup has been used that goes back to Birdsall and Langdon [1] and was also used by Jacobs and Hesthaven [7]. A plasma wave can be created by a distribution function that deviates a set of equidistant particles $\mathbf{x}_{eq}$ by a sine function, yielding

$$\mathbf{x} = \mathbf{x}_{eq} + A \cdot \sin(k\mathbf{x}_{eq}), \tag{7}$$

where the amplitude of the deviation is A and k is the wave number. In this case, $A = 0.001[m]$ and $k = 2$ have been chosen. In y- and z-direction the one-dimensional distribution function is copied so that two-dimensional slices exist in the y-z-plane, which consist of a uniform distribution of 100 particles in x-, and 15 particles in y- and z-direction. The particles within these slices have to be equidistant in order to prevent a potential difference in the plane, so that the potential difference occurs only in the x-direction. The computational domain is $2\pi\,\mathrm{m}$ in x- and $1.5\,\mathrm{m}$ in y- and z-direction. A second order DG field solver on a Cartesian mesh with $60 \times 10 \times 10 = 6000$ elements is used. To achieve neutrality, two species are used—negatively charged electrons and positively charged protons—$45,000$ in total. The protons have no deviation but an equidistant distribution in all dimensions, with $A = 0$ and $k = 0$. The mass and charge of each particle is multiplied by a macro particle factor of 10^{10}, i.e. each simulated particle represents 10^{10} real particles. From the number density n_e of the negatively charged particles, a plasma frequency of $\omega_e = 3.38 \cdot 10^8 \, \frac{\mathrm{rad}}{\mathrm{s}}$ which gives a period of $T = 1.86 \cdot 10^{-8}\,\mathrm{s}$ is expected. Figure 4 shows the result for the given setup using the Leap-Frog method.

The simulated period is $\approx 1.86 \cdot 10^{-8} s$, which corresponds to the analytic calculation and indicates that the code is able to model basic plasma phenomena qualitatively accurate.

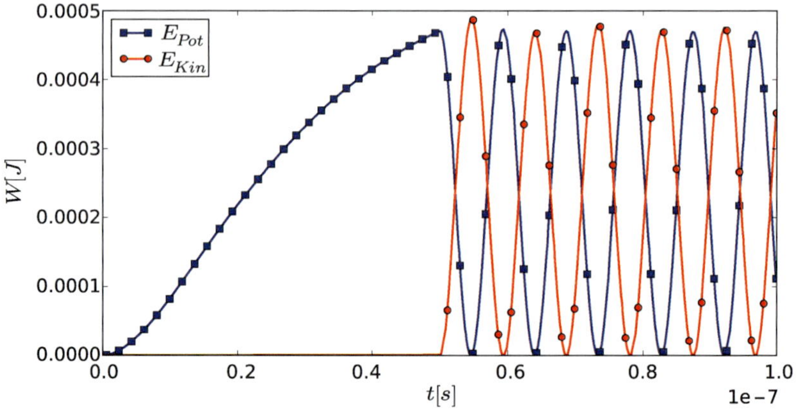

Fig. 4. Potential and kinetic energies for the plasma wave test

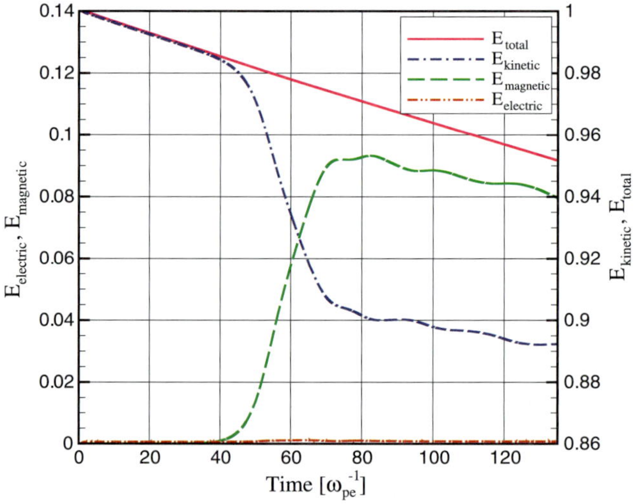

Fig. 5. Kinetic energy is converted into magnetic energy in the Weibel instability

6.3 Weibel Instability 3D

For the test of the three-dimensional, fully coupled scheme, a case from astrophysics is revisited. The Weibel instability has been numerically investigated previously by several researchers, among them Jacobs et al. [8] and Fonseca et al. [5]. This instability describes the time evolution of a plasma of bi-Maxwellian distribution. Kinetic energy is converted into magnetic energy as seen in Fig. 5.

180 million particles were simulated on a domain of size $15\frac{c}{\omega_{pe}}$ by $15\frac{c}{\omega_{pe}}$ by $\frac{c}{\omega_{pe}}$. This computational domain was discretized by $300 \times 300 \times 20$ grid cells of

polynomial degree $p = 1$. The simulation was performed on 225 cores of the LAKI cluster at the High Performance Computing Center Stuttgart (HLRS).

A preliminary result is the plot of the energy history of the simulation shown in Fig. 5, where the energy conversion from kinetic to magnetic energy is depicted. The plot also reflects the results obtained in the two-dimensional simulations described by Jacobs et al. [8].

7 Summary and Outlook

Maxwell's equations are solved using a newly incorporated discontinuous Galerkin method. The complete PIC solver has been tested successfully with first-order particle-grid-coupling. Higher-order coupling has been implemented and is currently being investigated. Several simulations with high numbers of particles and degrees of freedom in the grid solver have been conducted and the results show good agreement with the expected values. The code has been completely parallelized and shows an acceptable parallel efficiency for uniformly distributed particles. For non-uniformly distributed particles, the scaling depends on the distribution. In order to improve the scalability for non-uniform particle densities, the implementation of a dynamic load balancing weighted by the particle density is being pursued. Also planned is a shared memory parallelization of the particles to further amend any load imbalances due to inhomogeneous particle distributions.

Acknowledgements. We gratefully acknowledge the Landesstiftung Baden-Württemberg who funded the development within the program "Modeling and Simulation on High Performance Computers" from 2003–2005 and the Deutsche Forschungsgemeinschaft (DFG) for funding within the project "Numerische Modellierung und Simulation hochverdünnter Plasmaströmungen". T. Stindl wishes to thank the Landesgraduiertenförderung Baden-Württemberg and the Erich-Becker-Stiftung, Germany, for their financial support. Computational resources have been provided by the Bundes-Höchstleistungsrechenzentrum Stuttgart (HLRS).

References

1. C.K. Birdsall and A.B. Langdon. *Plasma Physics via Computer Simulation.* Adam Hilger, Bristol, Philadelphia, New York, 1991.
2. J.P. Boris. Relativistic plasma simulations—Optimization of a hybrid code. In *Proc. 4th Conf. on Num. Sim. of Plasmas*, pages 3–67, NRL Washington, Washington DC, 1970.
3. M. Fedoruk, C.-D. Munz, P. Omnes, and R. Schneider. A Maxwell-Lorentz solver for self-consistent particle-field simulations on unstructured grids. *Forschungszentrum Karlsruhe—Technik und Umwelt*, **FZKA 6115**, 1998.
4. M. Fertig, D. Petkow, T. Stindl, M. Quandt, C.-D. Munz, J. Neudorfer, S. Roller, D. D'Andrea, and R. Schneider. Hybrid code development for the numerical

simulation of instationary magnetoplasmadynamic thrusters. High Performance Computing in Science and Engineering '08, Springer, Berlin, Heidelberg, pp. 585–597, 2009.
5. R.A. Fonseca, L.O. Silva, J.W. Tonge, W.B. Mori, and J.M. Dawson. Three-dimensional Weibel instability in astrophysical scenarios. *Physics of Plasmas*, 10(5):1979–1984, 2003.
6. D.A. Gurnett and A. Bhattacharjee. *Introduction to Plasma Physics: With Space and Laboratory Applications*. Cambridge University Press, Cambridge, January 2005.
7. G.B. Jacobs and J.S. Hesthaven. High-order nodal discontinuous Galerkin particle-in-cell method on unstructured grids. *J. Comput. Phys.*, 214:96–121, 2006.
8. G.B. Jacobs, J.S. Hesthaven, and G. Lapenta. Simulations of the Weibel instability with a high-order discontinuous Galerkin Particle-In-Cell solver. 44th AIAA Aerospace Sciences Meeting and Exhibit, 2006.
9. Christopher A. Kennedy, Mark H. Carpenter, and R. Michael Lewis. Low-storage, explicit Runge-Kutta schemes for the compressible Navier-Stokes equations. *Applied Numerical Mathematics*, 35(3):177–219, November 2000.
10. C.-D. Munz, P. Omnes, and R. Schneider. A three-dimensional finite-volume solver for the Maxwell equations with divergence cleaning on unstructured meshes. *Computer Physics Communications*, 130:83–117, 2000.
11. J. Neudorfer, C.-D. Munz, T. Stindl, M. Fertig, S. Roller, R. Schneider, and M. Auweter-Kurtz. Particle In Cell Simulation Using a Discontinuous Galerkin Method. IEPC-2009-237, Proceedings of the 31th International Electric Propulsion Conference, Ann Arbor, MI, USA, 2009.
12. Lewi Tonks and Irving Langmuir. Oscillations in ionized gases. *Physical Review*, 33(2):195, February 1929.